Optical Network Design and Implementation

Vivek Alwayn, CCIE No. 2995

Cisco Press

Cisco Press
800 East 96th Street
Indianapolis, IN 46240 USA

Optical Network Design and Implementation

Vivek Alwayn, CCIE No. 2995

Copyright © 2004 Cisco Systems, Inc.

Published by:
Cisco Press
800 East 96th Street
Indianapolis, IN 46240 USA

All rights reserved. No part of this book may be reproduced or transmitted in any form or by any means, electronic or mechanical, including photocopying, recording, or by any information storage and retrieval system, without written permission from the publisher, except for the inclusion of brief quotations in a review.

Printed in the United States of America 10 9 8

Eight Printing August 2013

Library of Congress Cataloging-in-Publication Number: 2002104849

ISBN: 1-58714-150-7

Warning and Disclaimer

This book is designed to provide information about optical network design and implementation. Every effort has been made to make this book as complete and as accurate as possible, but no warranty or fitness is implied.

The information is provided on an "as is" basis. The authors, Cisco Press, and Cisco Systems, Inc. shall have neither liability nor responsibility to any person or entity with respect to any loss or damages arising from the information contained in this book or from the use of the discs or programs that may accompany it.

The opinions expressed in this book belong to the author and are not necessarily those of Cisco Systems, Inc.

Trademark Acknowledgments

All terms mentioned in this book that are known to be trademarks or service marks have been appropriately capitalized. Cisco Press or Cisco Systems, Inc. cannot attest to the accuracy of this information. Use of a term in this book should not be regarded as affecting the validity of any trademark or service mark.

Corporate and Government Sales

Cisco Press offers excellent discounts on this book when ordered in quantity for bulk purchases or special sales. For more information, please contact: **U.S. Corporate and Government Sales**   1-800-382-3419
corpsales@pearsontechgroup.com

For sales outside of the U.S. please contact: **International Sales**   1-317-581-3793   international@pearsontechgroup.com

Feedback Information

At Cisco Press, our goal is to create in-depth technical books of the highest quality and value. Each book is crafted with care and precision, undergoing rigorous development that involves the unique expertise of members from the professional technical community.

Readers' feedback is a natural continuation of this process. If you have any comments regarding how we could improve the quality of this book, or otherwise alter it to better suit your needs, you can contact us through e-mail at feedback@ciscopress.com. Please make sure to include the book title and ISBN in your message.

We greatly appreciate your assistance.

Publisher	John Wait
Editor-in-Chief	John Kane
Executive Editor	Jim Schachterle
Cisco Representative	Anthony Wolfenden
Cisco Press Program Manager	Nannette M. Noble
Production Manager	Patrick Kanouse
Development Editor	Andrew Cupp
Copy Editor	Keith Cline
Technical Editors	Russ Esmacher, Wayne Hickey, Doug Starr
Team Coordinator	Tammi Barnett
Book and Cover Designer	Louisa Adair
Composition	Interactive Composition Corporation
Indexer	Brad Herriman

Corporate Headquarters
Cisco Systems, Inc.
170 West Tasman Drive
San Jose, CA 95134-1706
USA
www.cisco.com
Tel: 408 526-4000
    800 553-NETS (6387)
Fax: 408 526-4100

European Headquarters
Cisco Systems International BV
Haarlerbergpark
Haarlerbergweg 13-19
1101 CH Amsterdam
The Netherlands
www-europe.cisco.com
Tel: 31 0 20 357 1000
Fax: 31 0 20 357 1100

Americas Headquarters
Cisco Systems, Inc.
170 West Tasman Drive
San Jose, CA 95134-1706
USA
www.cisco.com
Tel: 408 526-7660
Fax: 408 527-0883

Asia Pacific Headquarters
Cisco Systems, Inc.
Capital Tower
168 Robinson Road
#22-01 to #29-01
Singapore 068912
www.cisco.com
Tel: +65 6317 7777
Fax: +65 6317 7799

Cisco Systems has more than 200 offices in the following countries and regions. Addresses, phone numbers, and fax numbers are listed on the Cisco.com Web site at www.cisco.com/go/offices.

Argentina • Australia • Austria • Belgium • Brazil • Bulgaria • Canada • Chile • China PRC • Colombia • Costa Rica • Croatia • Czech Republic Denmark • Dubai, UAE • Finland • France • Germany • Greece • Hong Kong SAR • Hungary • India • Indonesia • Ireland • Israel • Italy Japan • Korea • Luxemburg • Malaysia • Mexico • The Netherlands • New Zealand • Norway • Peru • Philippines • Poland • Portugal Puerto Rico • Romania • Russia • Saudi Arabia • Scotland • Singapore • Slovakia • Slovenia • South Africa • Spain • Sweden Switzerland • Taiwan • Thailand • Turkey • Ukraine • United Kingdom • United States • Venezuela • Vietnam • Zimbabwe

Copyright © 2003 Cisco Systems, Inc. All rights reserved. CCIP, CCSP, the Cisco Arrow logo, the Cisco *Powered* Network mark, the Cisco Systems Verified logo, Cisco Unity, Follow Me Browsing, FormShare, iQ Net Readiness Scorecard, Networking Academy, and ScriptShare are trademarks of Cisco Systems, Inc.; Changing the Way We Work, Live, Play, and Learn, The Fastest Way to Increase Your Internet Quotient, and iQuick Study are service marks of Cisco Systems, Inc.; and Aironet, ASIST, BPX, Catalyst, CCDA, CCDP, CCIE, CCNA, CCNP, Cisco, the Cisco Certified Internetwork Expert logo, Cisco IOS, the Cisco IOS logo, Cisco Press, Cisco Systems, Cisco Systems Capital, the Cisco Systems logo, Empowering the Internet Generation, Enterprise/Solver, EtherChannel, EtherSwitch, Fast Step, GigaStack, Internet Quotient, IOS, IP/TV, iQ Expertise, the iQ logo, LightStream, MGX, MICA, the Networkers logo, Network Registrar, *Packet*, PIX, Post-Routing, Pre-Routing, RateMUX, Registrar, SlideCast, SMARTnet, StrataView Plus, Stratm, SwitchProbe, TeleRouter, TransPath, and VCO are registered trademarks of Cisco Systems, Inc. and/or its affiliates in the U.S. and certain other countries.

All other trademarks mentioned in this document or Web site are the property of their respective owners. The use of the word partner does not imply a partnership relationship between Cisco and any other company. (0303R)

Printed in the USA

About the Author

Vivek Alwayn, CCIE No. 2995, is a senior manager with Accenture. He has more than 15 years of experience working with core transport technologies, MPLS, and optical networking. He has designed and implemented several large-scale WAN-switched, MPLS, and optical networks for service provider and enterprise customers worldwide. He holds a Bachelor of Science degree in electronics engineering and is currently working on his Masters in telecommunications engineering. He has authored the *Advanced MPLS Design and Implementation* title for Cisco Press and has published several technical papers. Vivek Alwayn is also an active member of the IEEE and IETF. He can be reached at vivek@alwayn.com.

About the Technical Reviewers

Russ Esmacher is a product manager responsible for optical cross-connect and DWDM platforms for the Optical Technology Group of Cisco Systems. He holds Bachelor of Science and Master of Science degrees from Clemson University's Center for Optical Materials Science and Engineering Technology (COMSET).

Wayne Hickey has more than 20 years of telecommunications, computer, and data experience with product expertise in SONET, SDH, DWDM, IP, ATM, Frame, HFC, voice, video, and wireless. Wayne has held various positions with Cisco Systems as a CSE, SEM, SSEM, and now as a product manager for the Optical Technical Business Unit (where Wayne's focus is the ONS 15600 MSSP optical switch). Previously, Wayne spent 19 years working for Aliant Telecom (NBTel), the third largest telecommunications provider in Canada, where he was focused on transmission network design and the evaluation of emerging access and transmission technologies. He has co-authored and authored several papers on PMD, long-haul transmission systems, and has applied for several patents on primary and secondary protect for HFC.

Doug Starr is a systems engineer with Cisco Systems, supporting the service provider and channel markets. He has more than ten years of telecom experience in transport and L2/L3 VPN solutions. He assisted the development of the Cisco optical CCIP/CCIE/CQS exams. In prior experience with Ericsson, Doug was a staff engineer supporting mobile wireless and metro optical networks. He has a Bachelor of Science degree in electrical engineering from Texas A&M, College Station. He also has a lovely wife, Thuy, and two beautiful daughters, McKenzie and Kennedi.

Dedications

This book is dedicated in loving memory of Dr. Sr. Melanie FMM. We miss you.

To my wife, Sarita C. Alwayn, for her continuous support, without which this book would not have been possible.

To my mother, Belinda Alwayn, whose support and prayers have made this endeavor possible.

In fond memory of my father, Urban Alwayn, whose words of encouragement are still with me and continue to be my inspiration.

I thank you all.

—Vivek

"These fundamentals have got to be simple."

—Lord Ernest Rutherford, circa 1908

Acknowledgments

This book is a result of various inputs and is essentially a combined effort. There are several people I would like to acknowledge for their contribution to this book.

The Cisco Press team—John Kane, the Cisco Press editor-in-chief who was responsible for the planning and development of the book, from proposal stage to publication. Andrew Cupp, development editor at Cisco Press, for his meticulous editing, which has helped create a high quality error-free manuscript. I would also like to thank the editorial, production, and design team at Cisco Press.

The technical reviewers—Wayne Hickey, Russ Esmacher, and Doug Starr of Cisco Systems for their valuable technical feedback and comments that helped evolve the manuscript into a comprehensive document, covering all aspects of optical technology.

Doug Starr—Systems engineer with Cisco Systems for his ongoing help and support with all sections of the manuscript.

Adam Holloway—Systems engineer with Cisco Systems for his help with the voice technologies section of the book.

Sumeet Gohri—Principal engineer of Sumtertech for helping with the ML-Series section of the book.

Jim Wallace—Program manager with Unisys for helping with the MAN and ATM sections of the book.

Sohan Lal—For his support and encouragement, and most of all, for the meaningful philosophical discussions.

Contents at a Glance

Foreword xxvi

Introduction xxvii

Chapter 1 Introduction to Optical Networking 3
Chapter 2 Time-Division Multiplexing 19
Chapter 3 Fiber-Optic Technologies 49
Chapter 4 Wavelength-Division Multiplexing 93
Chapter 5 SONET Architectures 141
Chapter 6 SDH Architectures 209
Chapter 7 Packet Ring Technologies 275
Chapter 8 Multiservice SONET and SDH Platforms 307
Chapter 9 Provisioning the Multiservice SONET MSPP 357
Chapter 10 Provisioning the Multiservice SDH MSPP 447
Chapter 11 Ethernet, IP, and RPR over SONET and SDH 557
Chapter 12 Optical Network Case Studies 695
Appendix A ML-Series Command Reference 735
Appendix B References 759
Glossary 763
Index 783

Table of Contents

Foreword xxvi

Introduction xxvii

Chapter 1 Introduction to Optical Networking 3

    Introduction to SONET/SDH 3

    SONET/SDH 5
        Legacy SONET/SDH 6
        SONET/SDH Multiservice Provisioning Platforms 8
            Improving SONET/SDH Bandwidth Efficiency 10
            QoS 11
            SONET/SDH Encapsulation of Ethernet 11
            Packet Ring Technologies 12
            Provisioning 12
            Signaling 12

    Dense Wavelength-Division Multiplexing 13

    The Future of SONET/SDH and DWDM 16

    Summary 16

Chapter 2 Time-Division Multiplexing 19

    An Introduction to Time-Division Multiplexing 19

    Analog Signal Processing 20
        Analog Signal Generation and Reception 21
        Analog-to-Digital Conversion 21
            Filtering 22
            Sampling 22
            Quantization 22
            Encoding 23
            μ-law and A-Law Coding 24
            Echo Cancellation 24

    Circuit-Switched Networks 25
        TDM Signaling 26
            Channel-Associated Signaling (CAS) 27
            Common Channel Signaling (CCS) 27

The T-Carrier 27
    DS Framing 29
    DS Multiframing Formats 29
        D4 Superframe 30
        D5 Extended Superframe 30
    SF and ESF Alarms 31

The E-Carrier 32
    E1 Frame Alignment Signal (FAS) 34
    E1 MultiFrame Alignment Signal (MFAS) 35
    E1 CRC Error Checking 36
    E1 Errors and Alarms 36

ISDN 39
    ISDN BRI 39
    ISDN PRI 40
    ISDN Layer 1 41
    ISDN Layer 2 42
        ISDN Link-Layer Establishment 44
    ISDN Layer 3 44
        ISDN Call Setup 45

TDM Network Elements 46
    Repeaters 46
    CSU/DSU 46
    Digital Access and Cross-Connect Systems 47
    Channel Bank 47

Summary 47

Chapter 3 Fiber-Optic Technologies 49

A Brief History of Fiber-Optic Communications 49

Fiber-Optic Applications 50

The Physics Behind Fiber Optics 51
    Performance Considerations 53
    Optical-Power Measurement 53

Optical-Cable Construction 54
    Glass Fiber-Optic Cable 55
    Plastic Fiber-Optic Cable 55
    Plastic-Clad Silica (PCS) Fiber-Optic Cable 56
    Multifiber Cable Systems 56

Propagation Modes 58
    Multimode Step Index 59
    Single-Mode Step Index 60
        Single-Mode Dual-Step Index 61
    Multimode Graded Index 62

Fiber-Optic Characteristics 63
    Interference 64
    Linear Characteristics 64
        Attenuation 64
        Chromatic Dispersion 66
        Polarization Mode Dispersion 67
        Optical Signal-to-Noise Ratio 68
    Nonlinear Characteristics 69
        Self-Phase Modulation 69
        Cross-Phase Modulation 69
        Four-Wave Mixing 69
        Stimulated Raman Scattering 70
        Stimulated Brillouin Scattering 70

Fiber Types 70
    Multimode Fiber with a 50-Micron Core (ITU-T G.651) 71
    Nondispersion-Shifted Fiber (ITU-T G.652) 71
    Low Water Peak Nondispersion-Shifted Fiber (ITU-T G.652.C) 71
    Dispersion-Shifter Fiber (ITU-T G.653) 71
    1550-nm Loss-Minimized Fiber (ITU-T G.654) 72
    Nonzero Dispersion Shifted Fiber (ITU-T G.655) 72

Fiber-Optic Cable Termination 73
    FC Connectors 73
    SC Connectors 74
    ST Connectors 74
    LC Connectors 74
    MT-RJ Connectors 74
    MTP/MPO Connectors 75

Splicing 75

Physical-Design Considerations 76
    Tight Buffer Versus Loose Buffer Cable Plants 76
    Bend Radius and Tensile Loading 76
    Submarine Cable Systems 77

Fiber-Optic Communications System 78
    Transmitter 78
    Receiver 81

Fiber Span Analysis 83
    Transmitter Launch Power 83
    Receiver Sensitivity and Dynamic Range 84
    Power Budget and Margin Calculations 84
    Case 1: MMF Span Analysis 86
    Case 2: SMF Span Analysis 88

Summary 90

Chapter 4 Wavelength-Division Multiplexing 93

The Need for Wavelength-Division Multiplexing 93

Wavelength-Division Multiplexing 94
    Wavelength-Division Multiplexing Fundamentals 95
    Unidirectional WDM 96
    Bidirectional WDM 96
        Band-Separation Method 98
        Interleaving-Filter Method 99
        Circulator Method 99
    Channel Spacing 99

Coarse Wavelength-Division Multiplexing 100

Dense Wavelength-Division Multiplexing 101

The ITU Grid 102

Wavelength-Division Multiplexing Systems 104
    Transmitters 106
        Distributed Feedback Lasers 106
        Distributed Bragg Reflector Lasers 107
        Tunable Lasers 108
        Vertical Cavity Surface Emitting Lasers 110
        Chirp 111
        Modulators 111
    Optical Multiplexers and Multiplexers 113
        Thin Film Filter 114
        Fiber Bragg Grating 115
        Arrayed Waveguide 115
        Fabry Perot Cavity Filter 117
        Acousto Optical Tunable Filter 117
        Mach-Zehnder Interferometers 118
        Couplers, Circulators, and Isolators 119
        Periodic Filters, Frequency Slicers, and Interleavers 120

Amplifiers   121
    Erbium-Doped Fiber Amplifiers   123
    Raman Fiber Amplifiers   124
    Hybrid and Distributed Amplifiers   124
Optical-Fiber Media   125
Receivers   125
    PIN Photodiode   126
    Avalanche Photodiode   127

WDM Characteristics and Impairments to Transmission   127
    Forward Error Correction   128
        In-Band FEC   128
        Out-of-Band FEC   129
    Optical Signal-to-Noise Ratio   130
        OSNR Calculation   132

Dispersion and Compensation in WDM   133
    Chromatic Dispersion   134
    Chromatic Dispersion Compensation   134
    Polarization Mode Dispersion   136
    Polarization Mode Dispersion Compensation   138

Summary   138

Chapter 5   SONET Architectures   141

SONET Integration of TDM Signals   141
    Plesiochronous Digital Hierarchy   143

SONET Electrical and Optical Signals   144
    SONET Synchronization   144

SONET Layers   146
    Path Layer   146
        Path-Terminating Equipment (PTE)   147
    Line Layer   147
        Line-Terminating Equipment (LTE)   147
    Section Layer   147
        Section-Terminating Equipment (STE)   147
    Photonic Layer   148

SONET Framing   148
    SONET SPE   150
    STS-N Framing   151

SONET Transport Overhead   153
    SONET Section Overhead   154
        Framing Bytes (A1, A2)   154

Section Trace or Growth Byte (J0/Z0)   154
BIP-8 Byte (B1)   154
Orderwire Byte (E1)   155
User Byte (F1)   155
DCC Bytes (D1, D2, and D3)   155
Line Overhead   155
   Pointer Bytes (H1, H2)   156
   Pointer Action Byte (H3)   156
   BIP-8 Byte (B2)   157
   APS Bytes (K1, K2)   157
   DCC Bytes (D4 to D12)   158
   Synchronization Status Byte (S1)   158
   REI-L Byte (M0 or M1)   159
   Growth Bytes (Z1, Z2)   159
   Orderwire Byte (E2)   159
SONET Path Overhead   159
   STS Path Trace Byte (J1)   160
   BIP-8 Byte (B3)   160
   STS Path Signal Label Byte (C2)   160
   Path Status Byte (G1)   161
   Path User Channel Byte (F2)   162
   Indicator Byte (H4)   162
   Z3 to Z5 Bytes   162
   Tandem Connection Byte (Z5)   162

SONET Alarms   163
   Loss-of-Signal (LOS) Alarm   163
   Out-of-Frame (OOF) Alignment   163
   Loss-of-Frame (LOF) Alignment   163
   Loss-of-Pointer (LOP)   163
   Alarm Indication Signal (AIS)   164
   Remote Error Indication (REI)   164
   Remote Defect Indication (RDI)   164
   Remote Failure Indication (RFI)   164
   B1 Error   164
   B2 Error   165
   B3 Error   165
   BIP-2 Error   165
   Loss of Sequence Synchronization (LSS)   165

Virtual Tributaries   165
   VT Groups   166
   VT Superframe   171
   VT Path Overhead   173

SONET Multiplexing   174

SONET Network Elements   175
   Regenerator   175
   Terminal Multiplexer   176
   Add/Drop Multiplexer   176
   Broadband Digital Cross-Connect   178
   Wideband Digital Cross-Connect   178
   Digital Loop Carrier   179
   Multiservice Provisioning Platforms (MSPPs)   180

SONET Topologies   180
   Point-to-Point Topology   181
   Point-to-Multipoint Topology   181
   Hub Topology   182
   Ring Topology   183
   Mesh Topology   184
   Fiber Routing and Diversity   184

SONET Protection Architectures   187
   Automatic Protection Switching   187
      1+1 Protection   187
      1:1 and 1:N Protection   188

SONET Ring Architectures   189
   Unidirectional Versus Bidirectional Rings   190
   Two-Fiber Versus Four-Fiber Rings   191
   Path and Line Switching   193
   Dual-Ring Interconnect   194
   Unidirectional Path Switched Rings   196
   Asymmetrical Delay   198
   Bidirectional Line-Switched Rings   198
      Two-Fiber BLSR   198
      BLSR Node Failure   201
      Four-Fiber BLSR   202

SONET Network Management   205

Summary   206

Chapter 6   SDH Architectures   209

   SDH Integration of TDM Signals   209
      Plesiochronous Digital Hierarchy (PDH)   210
      SDH Synchronization   212
      SDH Electrical and Optical Signals   213

SDH Layers 214
   Path Layer 215
      Path-Terminating Equipment (PTE) 215
   Multiplex Section 215
      Multiplex Section-Terminating Equipment (MSTE) 215
   Regenerator Section Layer 216
      Regenerator Section-Terminating Equipment (RSTE) 216
   Photonic Layer 216

SDH Multiplexing 216
   SDH Multiplexing of E1 Signals 218
   SDH Multiplexing of DS1 Signals 218
   SDH Multiplexing of DS2 Signals 219
   SDH Multiplexing of E3 Signals 219
   SDH Multiplexing of DS3 Signals 220
   SDH Multiplexing of E4 Signals 220

SDH Framing 220
   STM-1 Frame Creation 221
   SDH Tributaries 222
      TU Multiframe 223
   SDH Pointers 223

SDH Transport Overhead 224
   SDH Regenerator Section Overhead 225
      Framing Bytes (A1, A2) 225
      Regenerator Section Trace Byte (J0) 225
      RS Bit-Interleaved Parity Code (BIP-8) Byte (B1) 225
      RS Orderwire Byte (E1) 226
      RS User Channel Byte (F1) 226
      RS Data Communications Channel (DCC) Bytes (D1, D2, D3) 226
      Reserved Bytes 226
   AU Pointers 226
   SDH Multiplex Section Overhead 226
      Bit-Interleaved Parity Code (BIP-24) Byte (B2) 226
      Automatic Protection Switching Bytes (K1, K2) 227
      Data Communications Channel (DCC) Bytes (D4 to D12) 228
      Synchronization Status Message Byte (S1) 228
      MS Remote Error Indication Byte (M1) 228
      MS Orderwire Byte (E2) 229
      Reserved Bytes 229
   SDH High-Order Path Overhead 229
      Higher-Order VC-N Path Trace Byte (J1) 230
      Path Bit-Interleaved Parity Code (Path BIP-8) Byte (B3) 230

Path Signal Label Byte (C2)   230
Path Status Byte (G1)   231
Path User Channel Byte (F2)   231
Position and Sequence Indicator Byte (H4)   231
Path User Channel Byte (F3)   232
APS Signaling Byte (K3)   232
Network Operator Byte (N1)   232
SDH Low-Order Path Overhead   232
VT Path Overhead Byte (V5)   232
Path Trace Byte (J2)   233
Network Operator Byte (N2)   233
APS Signaling Byte (K4)   233
SDH and SONET Interworking   233

SDH Alarms   235
Loss of Signal (LOS)   237
Out-Of-Frame (OOF) Alignment   237
Loss-Of-Frame (LOF) Alignment   237
Loss-Of-Pointer (LOP)   237
Alarm Indication Signal (AIS)   237
Remote Error Indication (REI)   237
Remote Defect Indication (RDI)   238
Remote Failure Indication (RFI)   238
B1 Error   238
B2 Error   238
B3 Error   238
BIP–2 Error   238
Loss of Sequence Synchronization (LSS)   238

SDH Higher-Level Framing   239

SDH Network Elements   241
Regenerator   241
Terminal Multiplexer   241
Add/Drop Multiplexer   242
Broadband Digital Cross-Connect   243
Wideband Digital Cross-Connect   244
Digital Loop Carrier   245
Multiservice Provisioning Platforms (MSPPs)   245

SDH Topologies   246
SDH Point-to-Point Topology   246
SDH Point-to-Multipoint Topology   247
SDH Hub Topology   247
Ring Topology   248

Mesh Topology   249
  Fiber Routing and Diversity   250

SDH Protection Architectures   253
  Automatic Protection Switching   253
    1+1 Protection   254
    1:1 and 1:N Protection   254

SDH Ring Architectures   255
  Unidirectional Versus Bidirectional Rings   256
  Two-Fiber Versus Four-Fiber Rings   257
  Path and Multiplex Section Switching   259
  Dual-Ring Interconnect   260
  Subnetwork Connection Protection Rings   262
  Asymmetrical Delay   264
  Multiplex Section-Shared Protection Rings   264
    Two-Fiber MS-SPRing   264
    MS-SPRing Node Failure   267
    Four-Fiber MS-SPRing   268

SDH Network Management   271

Summary   272

Chapter 7   Packet Ring Technologies   275

Ethernet Services   275

Ethernet over SONET/SDH   276
  Ethernet over SONET/SDH Encapsulation   277

Shared Packet Ring   277
  SPR Design Constraints   281

Resilient Packet Ring   287
  Dynamic Packet Transport   288
  RPR Operation   290
  RPR Topology Discovery   290
  RPR CoS   291
  RPR Fairness Algorithm   292
  RPR Bandwidth Management   292
    Spatial Reuse   292
  RPR Traffic Protection and Rerouting   293
    RPR Protection Hierarchy   294
  RPR Media Access Control   294
    RPR Data Frame   294

RPR Control Frame 296
RPR Fairness Control Frame 298
RPR MAC Operation 299
RPR OAM and Layer Management 303

Summary 303

Chapter 8 Multiservice SONET and SDH Platforms 307

Next-Generation SONET and SDH Platforms 307
ONS 15100 Series 308
ONS 15200 Series 308
ONS 15300 Series 309
ONS 15500 Series 309
ONS 15600 Series 310
ONS 15800 Series 310

ONS 15400 Series of Optical Platforms 311
ONS 15454 SONET MSPP 311
The ONS 15454 SONET Platform 312
ONS 15454 Card Protection 320
ONS 15454 SDH MSPP 321
The ONS 15454 SDH Platform 321
ONS 15454 SDH Card Protection 331
ONS 15454 MSTP 332
ONS 15454 MSTP Platform 334

Cisco Transport Controller (CTC) 336
CTC Node View 338
CTC Network View 340
Card View 342

Cisco Transport Manager (CTM) 343
CTM Client 345
CTM Domain Explorer 346
CTM Node View 348
CTM Network Map 350
CTM NE Explorer 351
NE Explorer for the ONS 15200 351
NE Explorer for the ONS 15327, ONS 15454, ONS and 15600 352
NE Explorer for the ONS 15540 352
NE Explorer for the ONS 1580x 353
CTM Control Panel 353

Summary 354

Chapter 9 Provisioning the Multiservice SONET MSPP 357

    Provisioning the SONET MSPP 357

    Initial Provisioning Tasks 357
        The ONS Craft Interface 357
        Set Up Basic Node Information 359
        Set Up Network Information 361
        User and Security Provisioning 362
            Create New Users 365
            Edit a User 366
            Delete a User 366

    Provisioning of Protection Groups 366
        Create a Protection Group 368
            Enable Ports 369
            Edit Protection Groups 369
            Delete Protection Groups 369

    ONS 15454 Timing 370
        Provisioning ONS 15454 Timing 371
        Setup of Internal Timing 373

    Node Inventory 374

    IP Networking of ONS Nodes for OAM&P 375
        CTC and ONS Nodes on the Same IP Subnet 375
        CTC and ONS Nodes on Separate IP Subnets 376
        Using Proxy ARP to Enable an ONS 15454 Gateway 377
        Default Gateway on the CTC 377
        Static Route on the ONS Node 378
        Static Routes for Multiple CTCs 380
        ONS OSPF Configuration 380

    UPSR Configuration 385
        Install the UPSR Trunk Cards 386
        Configure the UPSR DCC Terminations 387
        Enable the UPSR Ports 388
        Adding and Removing UPSR Nodes 388
            Switch UPSR Traffic 389
            Add a UPSR Node 389
            Remove a UPSR Node 391

    BLSR Configuration 392
        Install the BLSR Trunk Cards 394
        Create the BLSR DCC Terminations 395
            Enable the BLSR Ports 396

Remapping the K3 Byte   397
Provision the BLSR   397
Upgrading from Two-Fiber to Four-Fiber BLSR   400
Adding and Removing BLSR Nodes   403
Moving BLSR Trunk Cards   409

Subtending Ring Configuration   413
Subtend a UPSR from a BLSR   414
Subtend a BLSR from a UPSR   415
Subtend a BLSR from a BLSR   417

Linear ADM Configurations   418
Create a Linear ADM   419
Convert a Linear ADM to UPSR   420
Convert a Linear ADM to a BLSR   423

Pat-Protected Mesh Networking (PPMN)   427

Circuit Provisioning   428
Create an Automatically Routed Circuit   429
Create a Manually Routed Circuit   433
Creating Multiple Drops for Unidirectional Circuits   436
Creating Monitor Circuits   437
Searching for ONS 15454 Circuits   438
Editing UPSR Circuits   439
Creating a Path Trace   440
Provisioning SONET DCC Tunnels   442

Summary   444

Chapter 10   Provisioning the Multiservice SDH MSPP   447

Provisioning the SDH MSPP   447

Initial Provisioning Tasks   447
The ONS Craft Interface   447
Set Up Basic Node Information   449
Set Up Network Information   450
User and Security Provisioning   452
Create New Users   456
Edit a User   457
Delete a User   457

Provisioning of Protection Groups   457
Create a Protection Group   459
Enable Ports   460
Edit Protection Groups   460
Delete Protection Groups   460

ONS 15454 Timing for SDH   461
   Provisioning ONS 15454 SDH Timing   462
   Set Up of Internal Timing   464

Node Inventory   466

IP Networking of ONS 15454 SDH Nodes for OAM&P   467
   CTC and ONS Nodes on the Same IP Subnet   468
   CTC and ONS Nodes on Separate IP Subnets   468
   Using Proxy ARP to Enable an ONS 15454 SDH Gateway   468
   Default Gateway on the CTC   470
   Static Routes on the ONS Node   470
   Static Routes for Multiple CTCs   473
   ONS OSPF Configuration   474

SNCP Configuration   478
   Install the SNCP Ring Trunk Cards   479
   Configure the SNCP Ring DCC Terminations and Place Ports in Service   480
   Enable the SNCP Ports   481
   Adding and Removing Nodes from an SNCP Ring   482
      Switch SNCP Ring Traffic   482
      Add an SNCP Node   483
      Remove an SNCP Node   484

MS-SPRing Configuration   485
   Install the MS-SPRing Trunk Cards   486
   Create the MS-SPRing DCC Terminations and Place Ports in Service   488
   Remap the K3 Byte   490
   Provision the MS-SPRing   490
   Adding Nodes to an MS-SPRing   492
      Install Cards and Configure the New MS-SPRing Node   492
      Switch MS-SPRing Traffic Before Connecting a New Node   493
      Connect Fiber to the New Node   495
      Provision the Ring for the New Node   495
   Removing Nodes from an MS-SPRing   496
   Upgrading a Two-Fiber MS-SPRing to a Four-Fiber MS-SPRing   497
   Moving MS-SPRing Trunk Cards   500

Subtending Ring Configurations   502
   Subtend an SNCP Ring from an MS-SPRing   504
   Subtend an MS-SPRing from an SNCP Ring   505
   Subtend an MS-SPRing from an MS-SPRing   507

Linear ADM Configurations   508
   Create a Linear ADM   509
   Convert a Linear ADM to an SNCP Ring   510
   Convert a Linear ADM to an MS-SPRing   514

Extended SNCP Mesh Networks   517

SDH Circuit Provisioning   518
   Create an Automatically Routed Circuit   520
   Create a Manually Routed Circuit   526
   Creating VC Low-Order Path Tunnels for Port Grouping   530
   Creating Multiple Drops for Unidirectional Circuits   533
   Creating Monitor Circuits   535
   Searching for ONS 15454 SDH Circuits   536
   Editing SNCP Circuits   537
   Creating a Path Trace   538
   Monitor a Path Trace on STM-N Ports   541
   Create a Half Circuit Using an STM-N Card as a Destination in an MS-SPRing or 1+1 Topology   542
   Create a Half Circuit Using an STM-N as a Destination in an SNCP   544
   Filtering, Viewing, and Changing Circuit Options   546
      Filter the Display of Circuits   546
      View Circuits on a Span   547
      Change a Circuit State   548
      Edit a Circuit Name   549
      Change Active and Standby Span Color   549
   Creating SDH DCC Tunnels   550

Summary   553

Chapter 11   Ethernet, IP, and RPR over SONET and SDH   557

Ethernet and IP Services over SONET/SDH   557

G-Series Provisioning of Ethernet over SONET   557
   G1K-4 Port Provisioning   560
   G1K-4 Point-to-Point Circuit Provisioning   561
   G1K-4 Manual Cross-Connect Provisioning   563

E-Series Provisioning of Ethernet over SONET   565
   Provision E-Series Ethernet Ports   566
   E-Series EtherSwitch Point-to-Point Circuit Provisioning   567
   E-Series Shared Packet Ring Provisioning   571
   E-Series Hub-and-Spoke Ethernet Circuit Provisioning   574
   Provision an E-Series Single-Card EtherSwitch Manual Cross-Connect   577
   Provision an E-Series Multicard EtherSwitch Manual Cross-Connect   579
   Provision Ethernet Ports for VLAN Membership   581
   Enable E-Series Spanning Tree on Ethernet Ports   582
   Retrieve the MAC Table Information   584
   Creating Ethernet RMON Alarm Thresholds   584

G-Series Provisioning of Ethernet over SDH   585
   Provision G1K-4 Ethernet Ports   587

Provision a G1K-4 EtherSwitch Circuit   589
Provision a G1K-4 Manual Cross-Connect   592

E-Series Provisioning of Ethernet over SDH   595
Provision E-Series Ethernet Ports   596
Provision an E-Series EtherSwitch Point-to-Point Circuit (Multicard or Single Card)   597
Provision an E-Series Shared Packet Ring Circuit   603
Provision an E-Series Hub-and-Spoke Ethernet Circuit   607
Provision an E-Series Single-Card EtherSwitch Manual Cross-Connect   611
Provision an E-Series Multicard EtherSwitch Manual Cross-Connect   615
Provision Ethernet Ports for VLAN Membership   618
Enable E-Series Spanning Tree on Ethernet Ports   620
Retrieve the MAC Table Information   621
Creating Ethernet RMON Alarm Thresholds   622

ML-Series Provisioning of Ethernet, IP, and RPR over SONET/SDH   624
Accessing the ML-Series Card   628
ML-Series IOS Command Modes   632
ML-Series Fast Ethernet Interface Configuration   634
Gigabit Ethernet Interface Configuration   635
POS Interface Configuration   636
    ML-Series POS: Case 1   639
    ML-Series POS: Case 2   640
    ML-Series POS Configuration: Case 3   642
ML-Series Bridge Configuration   643
    ML-Series Bridging Example   644
ML-Series STP and RSTP Configuration   645
    Disabling and Enabling STP-RSTP   647
    Configuring the Port Priority of an Interface   647
    Configuring the Path Cost of an Interface   648
    Configuring the Switch Priority of a Bridge Group   649
    Configuring the Hello Time   649
Configuring the Forwarding-Delay Time for a Bridge Group   650
    Configuring the Maximum-Aging Time for a Bridge Group   650
    Monitoring STP and RSTP Status   651
ML-Series VLAN Configuration   651
    ML-Series VLAN Example   653
ML-Series IEEE 802.1Q and Layer 2 Tunneling   656
    IEEE 802.1Q Tunneling Design Constraints   658
    IEEE 802.1Q Tunneling Port Configuration   659
    ML-Series 802.1Q Tunneling Example   660
    Layer 2 Protocol Tunneling   661
    Layer 2 Protocol Tunneling Design Constraints   662
    ML-Series Layer 2 Protocol Tunneling Configuration   662

Link Aggregation on the ML-Series Cards 663
ML-Series POS Channel Configuration 664
ML-Series POS Channel Example 665
Configuring VRF Lite on the ML-Series 667
ML-Series VRF Lite Example 668
Configuring IP Protocols for the ML-Series Card 672
ML-Series Quality of Service 673
ML-Series Resilient Packet Ring 674
ML-Series RPR Operation 674
RPR Ring Wrapping 675
RPR VLAN Support 676
Dual RPR Interconnect 682
ML-Series Virtual Concatenation 691
ML-Series Switching Database Manager 691

Summary 692

Chapter 12 Optical Network Case Studies 695

Network Design Strategies 695
Customer Demographics 697
Customer Service Requirements 697
Customer Service Levels 697
Fiber Plant 698
Technology Selection 698
Vendor Selection 699

Case Study: Multiservice Metro Optical SONET/SDH Network 699
Case Study Solution 701
Implementing BLSR/MS-SPRing on the Network 723
Scaling Up the Network 723
Scalability Using the 15216 DWDM Filter Mux/Demux 724
Scalability Using the 15454 MSTP 729

Case Study: SAN Services 731
Case Study Solution 731

Summary 732

Appendix A ML-Series Command Reference 735

ML-Series Commands 735
[no] bridge *bridge-group-number* protocol {drpri-rstp | ieee | rstp} 735
[no] clock auto 736
interface spr 1 736
[no] pos flag c2 *value* 737
[no] pos report *alarm* 738

[no] pos trigger defects *condition*   739
[no] pos trigger delay *time*   740
[no] pos scramble-spe   741
show controllers pos *interface-number* [details]   742
show interface pos *interface-number*   742
show ons alarm   743
show ons alarm defect eqpt   744
show ons alarm defect port   745
show ons alarm defect pos *interface-number*   746
show ons alarm failure eqpt   746
show ons alarm failure port   747
show ons alarm failure pos *interface-number*   748
spr drpri-id {0 | 1}   748
spr-intf-id *shared-packet-ring-number*   749
spr station-id *station-id-number*   750
spr wrap [immediate | delayed]   751

Cisco IOS Commands Not Supported on the ML-Series Cards   752

Appendix B   References   759

Glossary   763

Index   783

Foreword

In 1999, Cisco took a major step forward in its entry into the optical market with the acquisition of Cerent Corporation, a pioneer and innovator of next-generation multiservice provisioning platforms (MSPPs). This began the optical revolution rendering legacy SONET and SDH systems a thing of the past by bringing advanced packet techniques to a traditional circuit- and voice-based network. The Cisco optical family has changed the way enterprises and service providers deliver services by combining familiar enterprise Ethernet networking technologies with optical transports and converging traditional voice, data, and video applications.

Since that eventful day in 1999, the optical industry in general and the original Optical Networking Systems (ONS) family at Cisco have expanded and evolved at Internet speed. At Cisco, the number of customers has grown from tens to in excess of a thousand customers worldwide. More than 45,000 ONS systems have been deployed in the past four years. A next-generation IP+Optical platform with built-in Cisco IOS services now delivers innovative VoIP, MPLS, and VPN capabilities at multigigabit line rates to achieve an industry first. The software code base has grown from several hundred thousand lines to more than five million. The engineering staff has grown from the original 100 engineers to more than 1000 engineers.

Many components comprise the Cisco Complete Optical Multiservice Edge and Transport (COMET) solution. The content of this book is the first public glimpse into the depth of Cisco COMET. The author offers insight into fundamental design, its evolution, and the motives behind the Cisco COMET family and architecture. Over the course of its five-year life, Cisco ONS products, and in particular the ONS 15454, have undergone several major transitions and now deliver Ethernet, SONET, SDH, 10GE/OC192, L2/L3/IP services, RPR, Fibre Channel, FICON, and wavelength multiplexing in a single system. This has been complemented with the ONS 15216 DWDM Flexlayer product family at both the metro edge and metro core, the ONS 15327 at the metro edge, the ONS 15600 at the metro core, the ONS 15530/540 for metro storage extensions, and OSS provisioning and management by Cisco Transport Manager (CTM). Despite the turbulent changes in the telecom industry, the key to success is the ability to adapt at optical speeds to provide more than just transport connectivity. Transcending Layers 1, 2, and 3 of the OSI model, Cisco COMET delivers the five Ps: photonics, protection, packets, protocols, and provisioning.

The author of this book, Vivek Alwayn, is not only a CCIE expert in routing, switching, and MPLS protocols, but should also receive my newly coined title of "CCOE" (Cisco Certified Optical Engineer). This represents the first book written that offers a comprehensive and technical guide to the unique IP+Optical innovations of Cisco COMET.

Jayshree V. Ullal
Senior VP, Optical Networking Group
Cisco Systems, Inc.

Introduction

The massive growth of the Internet has fueled technologies that can scale rapidly upwardly in terms of bandwidth. Optical networking and DWDM provide fiber relief without the high infrastructure costs associated with deploying new fiber plants. From a base technology perspective, many carriers and service providers have a huge SONET and SDH installed base that is currently used to deliver TDM voice. There is a strong tendency to run parallel data networks. The existing SONET/SDH fiber infrastructure of many carriers can be leveraged to offer new services, such as managed Ethernet services, SAN transport services, MPLS, DWDM wavelengths, and more. All of this is possible while retaining legacy TDM voice services offered by the service provider. Furthermore, multiservice SONET/SDH enables service providers to offer VoIP over the same infrastructure. This book discusses optical networking technologies as well as optical design and its implementation, from fiber-optic technologies to DWDM, SONET, SDH, and RPR implementations.

Who Should Read This Book?

This book is meant to cover the entire spectrum of optical networking technologies from the physical layer to the network layer. This book can be considered as a complete guide to optical networking. Network engineers, network architects, network design engineers, systems engineers, presales engineers, network consultants, network managers, and consultants who design, deploy, operate, and troubleshoot optical and DWDM networks should read this book. This book treats SONET and SDH with equal importance.

How This Book Is Organized?

The book is partitioned into 12 modular chapters. Each chapter deals with an important facet of optical networking. For example, an international reader could skip the SONET-related chapters and focus on SDH. The book is structured in such a way that the reader is initially exposed to physical layer (Layer 1) technologies, such as photonics and WDM. The reader is then eased into Layer 1.5 technologies, such as SONET and SDH. Then there is a move to Layer 2 technologies, such as RPR and Ethernet over SONET. Layer 3 technologies, such as VRF VPNs, are covered last. The book wraps up with a case study that consolidates the concepts presented throughout the book.

Chapter 1, "Introduction to Optical Networking"

Synchronous Optical Network/Synchronous Digital Hierarchy (SONET/SDH) is used as a Layer 1.5 time-division multiplexed (TDM) transport mechanism for voice as well as data. This chapter introduces SONET/SDH and dense wavelength-division multiplexing (DWDM) to the reader. Legacy SONET/SDH networks use add/drop multiplexers (ADMs) that add or drop OC-N/STM-N circuits between ADM nodes on the ring. Multiservice provisioning platforms integrate ADM, DACS, RPR, SAN transport, DWDM, and Ethernet switching functionality within the device.

Chapter 2, "Time-Division Multiplexing"

TDM is a common mechanism used for transmitting multiple signals or streams of information over a medium at the same time in the form of a single, complex signal, and then recovering the separate signals at the receiving end. This chapter discusses TDM concepts in detail and examines the T-carrier and E-carrier systems. This chapter also discusses ISDN.

Chapter 3, "Fiber-Optic Technologies"

This chapter discusses the physics behind fiber-optic cables. It examines various linear effects, such as attenuation and dispersion, as well as nonlinear effects at high bit rates with WDM signaling. Various fiber types are presented along with their refractive index profiles. Fiber loss budget analysis is also presented with examples.

Chapter 4, "Wavelength-Division Multiplexing"

This chapter discusses wavelength-division multiplexing principles, coarse wavelength-division multiplexing, dense wavelength-division multiplexing, the ITU grid, WDM systems, WDM characteristics, impairments to transmission, and dispersion and compensation in WDM systems. OSNR calculations for fiber amplifiers are also discussed.

Chapter 5, "SONET Architectures"

The SONET format allows different types of signal formats to be transmitted over the fiber-optic cable. This chapter discusses SONET framing, multiplexing, virtual tributaries, SONET network elements, SONET topologies, SONET protection mechanisms, APS, two-fiber UPSR, DRI, and two-fiber and four-fiber BLSR rings.

Chapter 6, "SDH Architectures"

The SDH format allows different types of PDH signal formats to be transmitted over the fiber-optic cable. This chapter discusses SDH framing, multiplexing, tributary units, SDH network elements, SDH topologies, SDH protection mechanisms, APS, two-fiber SNCP, DRI, and two-fiber and four-fiber MS-SPRings.

Chapter 7, "Packet Ring Technologies"

Shared packet ring (SPR) and resilient packet ring (RPR) technologies can be leveraged to carry Ethernet services over SONET/SDH. SPRs are built by concatenating SONET STS-1s or SDH VC-3s. This chapter discusses various Ethernet over SONET/SDH encapsulation schemes, such as Ethernet over SONET/SDH using ANSI T1X1.5 147R1 Generic Framing Procedure (GFP) headers, Ethernet over Packet over SONET/SDH using ITU-T x.86 LAPS, and IEEE 802.17 RPR.

Chapter 8, "Multiservice SONET and SDH Platforms"

This chapter discusses the various products in the Cisco ONS 15000 family. It focuses on the ONS 15454 MSPP, 15454 MSTP, and the 15454 SDH. The various electrical and optical cards associated with the ONS 15454 are discussed in detail. The E-Series Ethernet switch cards and the ML-Series Layer 2/3 cards are also discussed in detail.

Chapter 9, "Provisioning the Multiservice SONET MSPP"

This chapter discusses the various SONET provisioning aspects of the ONS 15454. It examines optical provisioning and the creation of TDM circuits over the optical layer. The Cisco Transport Controller (CTC) tool is presented as the GUI for provisioning. This chapter also discusses UPSR, DRI, and BLSR ring provisioning.

Chapter 10, "Provisioning the Multiservice SDH MSPP"

This chapter discusses the various SDH provisioning aspects of the ONS 15454 SDH. It examines optical provisioning and the creation of TDM circuits over the optical layer. The CTC tool is presented as the GUI for provisioning. This chapter also discusses SNCP, DRI, and MS-SPRing provisioning.

Chapter 11, "Ethernet, IP, and RPR over SONET and SDH"

This chapter discusses the various Ethernet, IP, and RPR provisioning aspects of the ONS 15454 for SONET and SDH. It examines SONET/SDH optical provisioning and the creation of Ethernet circuits over the optical layer. The Cisco Transport Controller (CTC) tool is presented as the GUI for provisioning the E-Series and ML-Series cards. This chapter also discusses VRF and RPR provisioning.

Chapter 12, "Optical Network Case Studies"

This chapter discusses the case study of a service provider involved in the analysis, design, and implementation of a large optical network that provides services to an excess of 350,000 users. The case study commences with an analysis of service requirements. This discussion is followed up with a capacity planning exercise, fiber plant analysis, delay analysis, technology analysis, logical design, and the complete physical design of the network. Scalability considerations are presented with the use of DWDM for scaling bandwidth.

Icons Used in This Book

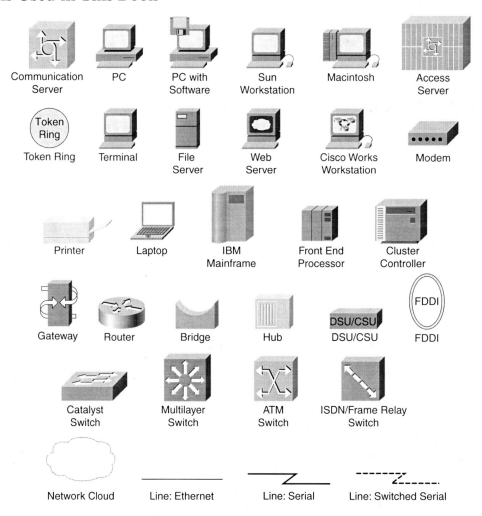

Command Syntax Conventions

The conventions used to present command syntax in this book are the same conventions used in the IOS Command Reference. The Command Reference describes these conventions as follows:

- **Boldface** indicates commands and keywords that are entered literally as shown. In actual configuration examples and output (not general command syntax), boldface indicates commands that are manually input by the user (such as a **show** command).
- *Italics* indicate arguments for which you supply actual values.
- Vertical bars (|) separate alternative, mutually exclusive elements.
- Square brackets [ ] indicate optional elements.
- Braces { } indicate a required choice.
- Braces within brackets [{ }] indicate a required choice within an optional element.

This chapter includes the following sections:

- **Introduction to SONET/SDH**—This section introduces Synchronous Optical Network (SONET) and the Synchronous Digital Hierarchy (SDH), which form the backbone of many service provider, carrier, and large enterprise networks. SONET/SDH is used as a Layer 1.5 time-division multiplexed (TDM) transport mechanism for voice and data. This section also offers an insight into and brief history of the evolution of SONET/SDH.

- **SONET/SDH**—This section discusses SONET/SDH topologies and compares legacy SONET/SDH equipment with next-generation multiservice provisioning platforms. Various SONET/SDH terminology used throughout this book is also discussed including packet ring technologies, SONET/SDH bandwidth efficiency, quality of service, SONET/SDH encapsulation of Ethernet, as well as provisioning and signaling.

- **Dense Wavelength-Division Multiplexing**—Dense wavelength-division multiplexing (DWDM) uses multiple channels or wavelengths to transmit signals over a single optical fiber. DWDM maximizes the use of installed fiber cable and allows new services to be quickly and easily provisioned over existing infrastructure.

- **The Future of SONET/SDH and DWDM**—This section briefly covers the future of SONET/SDH and DWDM.

CHAPTER 1

Introduction to Optical Networking

Introduction to SONET/SDH

The Synchronous Optical Network (SONET) forms the backbone of many service provider and carrier networks. The European, Asian, and Latin American implementation of SONET is known as Synchronous Digital Hierarchy (SDH). SONET/SDH is used as a Layer 1.5 time-division multiplexed (TDM) transport mechanism for voice and data. Frame- and cell-switched data networks, such as Frame Relay and ATM/SMDS, also depend on SONET/SDH TDM circuits for transport or back-haul to their respective point of presence (POP) nodes.

To be truly appreciated, SONET/SDH must be understood. Fear of the unknown results in avoidance of a product, service, or technology. It is precisely this mindset that calls for the overthrow of rugged technologies, such as SONET/SDH, which have withstood the test of time. The investment in SONET/SDH has been made over several decades and is in the order of billions of dollars. Carriers are not willing to trade the reliability and familiarity of proven SONET/SDH TDM-based backbone networks and replace them with technologies such as Gigabit Ethernet (GE). GE is being touted for its multiservice capability. However, the truth of the matter is that GE in itself is suitable only for data transport. It does not provide for the native transport of TDM voice or video. GE is not the answer or panacea for a truly multiservice network. By transporting GE, ATM, and packet as a service over SONET/SDH, carriers achieve the best of both worlds and retain their TDM capability for transport of voice. Multiservice SONET/SDH also carries SAN traffic, such as Enterprise Systems Connection (ESCON), Fibre Connectivity (FICON), and Fibre Channel deterministically, without distance limitations. Furthermore, Multiservice SONET/SDH can carry VoIP over Ethernet or TDM that future-proofs Multiservice SONET/SDH with respect to packet voice technology.

SONET/SDH's capability to transport TDM, ATM, packet, and Ethernet within Layer 1.5 gives it true multiservice capability. Legacy SONET/SDH systems carried TDM and ATM in a deterministic, nonflexible manner and did not provide any multiservice capability. Legacy SONET/SDH also used external digital access cross-connect systems (DACS) and DWDM components, such as digital access cross-connect switches for DS0 and N*64-kbps cross-connections between T1 and E1 ports and E1-to-T1 format-conversion functions. Dense wavelength-division multiplexers are used for combining multiple wavelengths over a single fiber pair, thereby providing bandwidth without the need for laying additional fiber. Next-generation SONET/SDH multiservice provisioning platforms (MSPPs) carry TDM, ATM, packet, and 10/100/1000 Ethernet services. MSPPs also integrate DACS and DWDM functionality within the chassis, enabling the provision of multiple services using just one platform. Integrated platforms also reduce the number of *single points of failure* on the network and increase overall network reliability. Furthermore, MSPPs offer graphical user interface (GUI)-based A to Z provisioning, thereby eliminating the need for provisioning technicians to master complicated Transaction Language 1 (TL1) commands, which are used to provision conventional SONET/SDH equipment.

Early copper-based electrical transmission systems used time-division multiplexing to resolve capacity issues. However, with the massive increase in voice traffic, and commercialization of optical technology in the early 1980s, fiber-optic cable technology turned out to be the medium of choice. Early fiber systems could carry 400 Mbps at 1300-nanometer (nm) wavelengths. Fiber-optic cables were also far less susceptible to electromagnetic interference (EMI) and radio frequency interference (RFI) from external sources.

Information that previously required hundreds of copper pairs could be transmitted down a glass fiber only slightly thicker than a human hair. Carriers adopted fiber-optic technology and started testing and deploying fiber networks. These first-generation fiber-optic systems in the Public Switched Telephone Network (PSTN) used proprietary architectures, equipment, line codes, multiplexing formats, and maintenance procedures.

Interconnection and interoperability between two carriers (each using different hardware vendors) was almost impossible to achieve.

The Regional Bell Operating Companies (RBOCs) and inter-exchange carriers (IXCs) needed an interoperability standard so that they could mix and match equipment from different vendors. In 1984, the Exchange Carriers Standards Association (ECSA) took up the task of creating such a standard—to establish a standard for interconnecting fiber systems to each other.

Bellcore extended the original ECSA idea in 1985 and proposed what we now know as SONET. In 1988, the initial SONET standards were approved as ANSI documents T1.105, which described optical rates and data format, and T1.106, which described the physical interface. The Consultative Committee for International Telegraph and Telephone (CCITT) had established a similar standard, Synchronous Digital Hierarchy (SDH), in Europe. The CCITT organization reorganized and formed the International Telecommunication Union Telecommunication Standardization Sector (ITU-T) in 1993.

The ANSI SONET standard has been widely accepted and implemented by service providers and equipment vendors in North America, whereas the ITU-T SDH standards and equipment are implemented in Europe, Middle East, and Africa (EMEA), Latin America, and in the Asia Pacific region. The SONET and SDH standards were designed to be compatible and carry complementary payloads. The differences between SONET and SDH are detailed in later chapters.

SONET/SDH

SONET/SDH networks are typically built in a hierarchical topology. The campus network could be GE or even an OC-3/STM-1 or OC-12/STM-4 SONET/SDH ring. Campus-to-central office (CO) traffic is normally carried over the metro access ring. CO-to-CO traffic is commonly carried over metro core rings and, finally, if the traffic is required to leave the metro core, long-haul traffic is typically carried over DWDM circuits.

As shown in Figure 1-1, customer rings are known as access rings and typically span a campus. The access rings converge and interconnect at major network traffic collection points. These collection points are referred to as *points of presence (POPs)* by carriers or as headends in the cable industry. The collector rings aggregate the access ring traffic and groom this traffic into the core rings, which are often referred to as interoffice facility (IOF) or metro core rings because they interconnect these collection points. Access rings reach further out to customer premises locations and are said to *subtend* off the larger collector rings. The collector rings subtend off the larger core rings.

Figure 1-1 *SONET/SDH Hierarchical Topology*

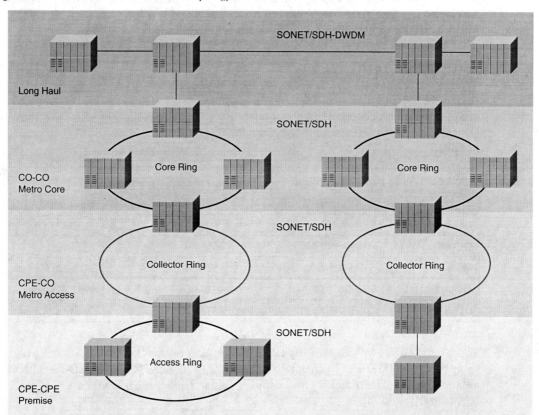

In legacy-based SONET/SDH time-division multiplexing (TDM), the sum of all subtending access ring bandwidth equals the total bandwidth required at the collector. Similarly, the sum of all subtending collector ring bandwidths equals the total bandwidth required at the core backbone.

Legacy SONET networks use automatic protection switching (APS), 1+1 protection, linear APS, two-fiber unidirectional path-switched ring (UPSR), two-fiber bidirectional line-switched ring (BLSR), or four-fiber BLSR protection mechanisms. Legacy SDH networks use multiplex section protection (MSP) 1+1, MSP 1:1, and MSP 1:N. They also implement two-fiber subnetwork connection protection (SNCP), two-fiber multiplexed section protection ring (MS-SPRing), or four-fiber MS-SPRing protection mechanisms. These protection mechanisms are also used by next-generation SONET/SDH, and are discussed in greater detail in later chapters.

Traffic flows in the access rings are typically of a hub-and-spoke nature, consolidating back at the local CO. UPSR/SNCP architectures are well suited for such multiple point-to-point or two-node traffic flows. The hub-and-spoke architecture also extends to 1+1 and linear access networks. Collector and core rings, however, support large amounts of traffic between access rings. As such, core ring traffic travels in a mesh, from any CO to any other CO. Because of their inherent potential for bandwidth reservation, BLSR/MS-SPRing architectures work well for such distributed "mesh" and node-to-node traffic applications.

Legacy SONET/SDH

Legacy SONET/SDH networks use add/drop multiplexers (ADMs) that add or drop OC-N or STM-N circuits between ADM nodes on the ring. The relationship between the SONET Optical Channel (OC-N) and SDH-Synchronous Transport Signal levels (STM-N) is presented in Table 1-1.

Table 1-1 *SONET OC-N and Its SDH Equivalent*

Signal Level	T-Carrier Equivalent	SDH Equivalent	Bandwidth
OC-3	84 * T1	STM-1	155.52 Mbps
OC-12	336 * T1	STM-4	622.08 Mbps
OC-48	1344 * T1	STM-16	2488.32 Gbps
OC-192	5376 * T1	STM-64	9953.28 Gbps
OC-768	21,504 * T1	STM-256	39,813.12 Gbps

SONET topologies typically use digital cross-connect systems (DCS) or DACS to groom lower-bandwidth DS-0 or DS-1 circuits to higher DS-3, OC-3, or STM-1 levels. SDH architectures use the term DXC for a digital cross-connect switch. Higher-order DXCs are used to cross-connect or switch traffic in 155-Mbps (STM-1) blocks, whereas lower-order DXCs are used to cross-connect traffic at 1.544 (DS-1) or 2.048 (E1) rates. Next-generation MSPPs integrate DCS/DXC functionality within the chassis.

Various CPE services, such as T1 or FT1 private line services, terminate on the DACS. Ethernet services could be provided using routers directly connected to the DACS, as shown in Figure 1-2. Voice services could be carried over TDM circuits by attaching the switches or private branch exchanges (PBXs) directly to the ADMs or via the DACS. Attaching ATM core switches directly to the ADMs provides ATM transport. In the case of ATM, the underlying SONET/SDH concatenated circuit would be completely transparent and the provider would need to provision permanent virtual circuits (PVCs) or switched virtual circuits (SVCs) as per customer requirements.

Figure 1-2 *Legacy SONET/SDH Applications*

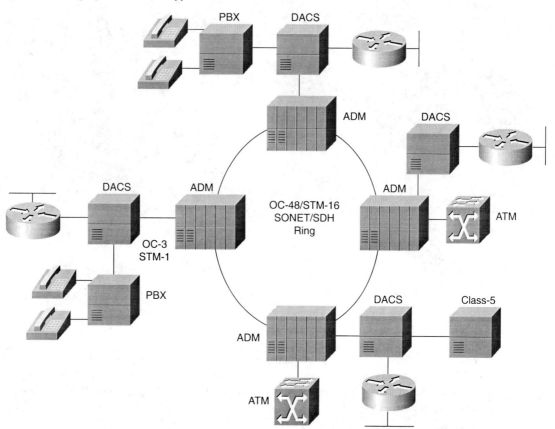

SONET/SDH Multiservice Provisioning Platforms

In recent times, since the late 1990s, the distinction between metro core and access rings has been blurred with the advent of next-generation SONET/SDH devices known as multiservice provisioning platforms or MSPPs. As illustrated in Figure 1-3, high-bandwidth core rings can aggregate customer traffic and perform a CO-to-CO function. The MSPP can perform the duties of an ADM and DCS/DXC on access rings and metro core rings.

Figure 1-3 *Next-Generation SONET/SDH MSPP Topology*

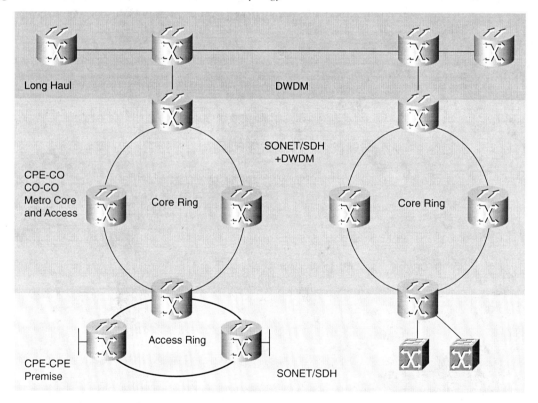

The current drivers for increasing optical bandwidth include unicast data (including voice over IP), TDM voice, videoconferencing, and multicast distance-learning applications. The optical infrastructure provides a true broadband medium for multiservice transport. Current optical technologies in use can be broadly classified, as shown in Table 1-2.

Table 1-2 *Classification of Optical Technology*

Technology	Application
Gigabit Ethernet	Metro access or metro core
Legacy SONET/SDH	Metro access, metro core, and long haul
Multiservice SONET/SDH	Metro access, metro core, and long haul
Packet over SONET/SDH	Metro access, metro core, and long haul
DWDM	Metro access, metro core and long haul

Legacy SONET/SDH TDM bandwidth summation no longer applies when packet- or frame-based traffic is statistically multiplexed onto a SONET/SDH ring. MSPPs can share SONET/SDH bandwidth among TDM, Ethernet, and other customer premises equipment (CPE) services. The inherent reliability of SONET/SDH is extended to Ethernet services when provisioned over MSPPs. These data services can be implemented across UPSR/SNCP, BLSR/MS-SPRing, linear, unprotected, and path-protected meshed network (PPMN) topologies. Furthermore, SONET/SDH 50-ms recovery is provided for these Ethernet services in the same manner as is done currently for TDM-based DS-N and OC-N circuits. The MSPP also includes support for resilient packet ring (IEEE 802.17) and has a roadmap for Generalized Multiprotocol Label Switching (GMPLS) with support for automatically switched optical networks (ASONs).

MSPPs enable carriers to provide packet-based services over SONET/SDH platforms. These services can be offered with varying service level agreements (SLAs) using Layer 1.5, 2, or 3 switching and quality of service (QoS) mechanisms. The optical network QoS includes the following parameters:

- Degree of transparency
- Level of protection
- Required bit error rate
- End-to-end delay
- Jitter requirements

As illustrated in Figure 1-4, multiservice provisioning platforms integrate DCS/DXC and Ethernet switching functionality within the device. However, ATM services would need an external core ATM switch to provision end-user PVCs or SVCs. The MSPP can provide private line TDM services, 10/100/1000-Mbps Ethernet services, and Multiprotocol Label Switched (MPLS) IP-routed services. This means that the service provider could build Layer 2 Ethernet virtual LAN (VLAN) virtual private networks (VPNs) or Layer 2.5 MPLS VPNs. Such versatility positions the MSPP as the solution of choice for metro access and core applications. Integration of DWDM capability also extends core and long-haul transport as an application for the MSPP.

Figure 1-4 *Next-Generation MSPP Applications*

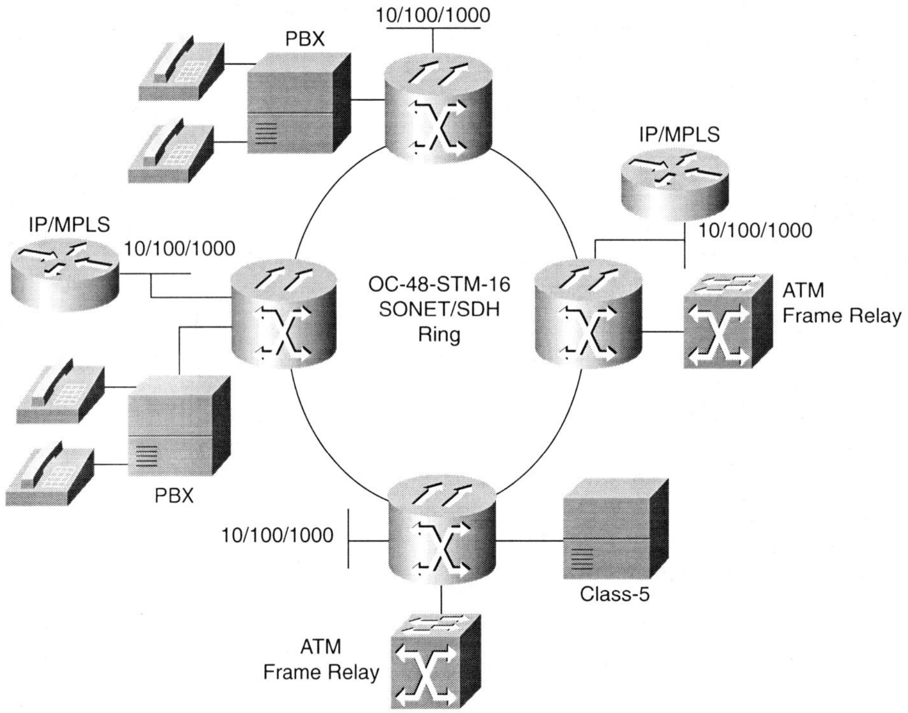

Improving SONET/SDH Bandwidth Efficiency

Legacy SONET/SDH networks were designed to transport TDM traffic in a highly predictable and reliable manner. Today's traffic patterns are shifting from TDM to an increasing percentage of bursty data traffic. Internet and data network growth in the past six years has highlighted legacy SONET/SDH's inefficiency in transporting data. Its rigid data hierarchy limits connections to fixed increments that have steep gaps between them. For example, an OC-3/STM-1 translates to 155 Mbps, but the next standard increment that is offered is OC-12/STM-4, which is 622 Mbps.

Inefficiency of bandwidth use in transporting Ethernet over SONET/SDH has been overcome by concatenation techniques. If one were to transport 100-Mbps Fast Ethernet over a SONET/SDH channel, for example, the legacy SONET/SDH channel selected would be an OC-3/STM-1. The OC-3/STM-1 channel consumes about 155 Mbps of bandwidth. This would result in a loss of an OC-1 or 51.84 Mbps worth of bandwidth. Concatenation supports nonstandard channels such as an STS-2. Transporting 100-Mbps Ethernet within an STS-2 (103.68 Mbps) optimizes bandwidth efficiency. Virtual concatenation (VCAT) and the link capacity adjustment scheme (LCAS) are techniques used to further enhance network efficiencies.

VCAT is an inverse multiplexing procedure whereby the contiguous bandwidth is broken into individual synchronous payload envelopes (SPEs) at the source transmitter that are logically represented in a virtual concatenation group (VCG). The VCG members are transported as individual SPEs across the SONET/SDH network and recombined at the far-end destination VCG receiver. VCAT is used to provision point-to-point connections over the SONET network using any available capacity to construct an (N * STS-1)-sized pipe for packet traffic.

LCAS is a protocol that ensures synchronization between the sender and receiver during the increase or decrease in size of a virtually concatenated circuit, in a hitless manner that doesn't interfere with the data signal.

QoS

The capability to classify packets, queue them based on that classification, and then schedule them efficiently into Synchronous Transport Signal (STS) channels is necessary to enable services that create and maintain sustainable service provider business cases. QoS is necessary in a service provider environment, to maintain customer SLAs. The various protection mechanisms used in optical networks such as APS, 1+1, two-fiber UPSR/SNCP, two-fiber BLSR/MS-SPRing, and four-fiber BLSR/MS-SPRing also determine the QoS and consequent SLA that a carrier can guarantee the customer. For example, circuits provisioned over a four-fiber BLSR/MS-SPRing ring can be offered with a higher QoS guarantee and SLA than a circuit provisioned over UPSR/SNCP, because four-fiber BLSR/MS-SPRing provides maximum redundancy.

SONET/SDH Encapsulation of Ethernet

Various methods for encapsulating Ethernet packets into SONET/SDH payloads have been discussed in the industry. The MSPP strategy focuses on delivering a single encapsulation scheme for both Ethernet and storage-area network (SAN) extension services while enabling interoperability between the transport components and the Layer 2 and 3 devices, which can exist within service provider networks. The vendor-accepted standard for encapsulation of Ethernet within SONET/SDH is the ANSI T1X1.5 Generic Framing Procedure (GFP). GFP provides a generic way to adapt various data traffic types from the client interface onto a synchronous optical transmission channel, such as SONET/SDH or WDM. GFP works in conjunction with VCAT and LCAS schemes, described earlier.

Packet Ring Technologies

Various technologies enable the transport of Ethernet services over SONET/SDH. Shared packet ring (SPR) and resilient packet ring (RPR) implementations vary by vendor. The only true standard is the IEEE 802.17 RPR specification. RPR technology uses a dual-counter rotating fiber ring topology to transport working traffic between nodes. RPR uses spatial reuse of bandwidth, which ensures that bandwidth is only consumed between the source and destination nodes. Packets are removed at their destination, leaving bandwidth available to downstream nodes on the ring.

Proactive span protection automatically avoids failed spans within 50 ms, thereby providing SONET/SDH-like resiliency in RPR architectures. RPR provides support for latency- and jitter-sensitive traffic, such as voice and video. RPR supports topologies of more than 100 nodes per ring with an automatic topology-discovery mechanism that works across multiple, interconnected rings.

SPR architectures are essentially Switched Ethernet over SONET/SDH optical transport topologies that follow the rules of bridging and Ethernet VLANs. SPR supports dual 802.1Q VLAN tagging and up to eight 802.1P classes of service. SPR and RPR are further discussed in later chapters.

Provisioning

MSPPs use GUI-based craft interfaces, management platforms, and the familiar IOS command-line interface (CLI) to simplify the provisioning task of SONET/SDH circuits, Ethernet circuits, IP routing, RPR, MPLS, and DWDM. Carriers and service providers that have experienced the complexities involved with Transaction Language 1 (TL-1) provisioning truly appreciate the ease of MSPP provisioning. Automated GUI-based provisioning is intuitive and reduces the learning curve associated with mastering TL-1. It also reduces the risk associated with incorrectly provisioning circuits that could result in breach of SLAs.

Signaling

MSPPs use signaling-based circuit provisioning using the user-network interface (UNI) signaling protocol, a standards-based unified control plane, and GMPLS signaling. GMPLS is also referred to as multiprotocol lambda switching. GMPLS supports packet switching devices as well as devices that perform switching in the time, wavelength, and space domains. GMPLS provides the framework for a unified control and management plane for IP and optical transport networks. The ITU G.ASON framework includes support for automated routing and signaling of optical connections at the UNI, network-network interface (NNI), and connection-control interface (CCI) level.

Dense Wavelength-Division Multiplexing

Dense wavelength-division multiplexing (DWDM) is a method to insert multiple channels or wavelengths over a single optical fiber. DWDM maximizes the use of the installed fiber base and allows new services to be quickly and easily provisioned over the existing fiber infrastructure. DWDM offers bandwidth multiplication for carriers over the same fiber pair. DWDM alleviates unnecessary fiber build-out in congested conduits and provides a scalable upgrade path for bandwidth needs.

As illustrated in Figure 1-5, various wavelengths are multiplexed over the fiber. End or intermediate DWDM devices perform amplification, reshaping, and timing (3R) functions. Individual wavelengths or channels can be dropped or inserted along a route. DWDM open architecture systems allow a variety of devices to be connected including SONET/SDH ADMs, ATM switches, and IP routers.

Figure 1-5 *DWDM Schematic*

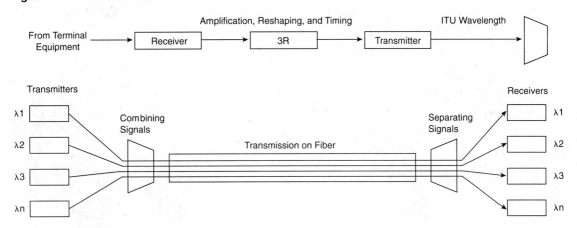

DWDM platforms provide the following:

- Optical multiplexing/demultiplexing to combine/separate ITU-T grid wavelengths launched by optical transmitters/transponders
- Optical filtering to combine ITU-T grid wavelengths launched by MSPPs
- Optical ADM functionality to exchange wavelengths on SONET/SDH spans between the MSPP and the DWDM device
- Optical performance monitoring (OPM)
- Fiber-optic signal amplification and 3R functionality

Long-haul DWDM is commonly divided into three categories with the main differentiator being unregenerated transmission distance. The three main long-haul DWDM classifications are long haul (LH), which ranges from 0 to 600 km; extended long haul (ELH), which ranges from 600 to 2000 km; and ultra long haul (ULH), which ranges from 3000+ km.

Storage networking is one of the key drivers for DWDM. The amount of data that enterprises store, including content or e-commerce databases, has increased exponentially. This has, in turn, driven up the demand for more storage connectivity. Information storage also includes backing up servers and providing updated, consistent mirror images of that data at remote sites for disaster recovery. Storage-area networking uses protocols such as ESCON, FICON, Fibre Channel, or Gigabit Ethernet.

The availability of fiber plants has become a key challenge for many companies that need multiple connections across a metropolitan-area network (MAN). Before DWDM technology was available, a company that wanted to connect data centers had to provide fiber for each individual connection. For small numbers of connections, this was not a problem. However, as shown in Figure 1-6, eight pairs of fiber-optic cable would be required if an organization were to connect two data centers via Gigabit Ethernet along with multiple ESCON channels, and FICON over Fibre Channel.

Figure 1-6 *Storage-Area Topology*

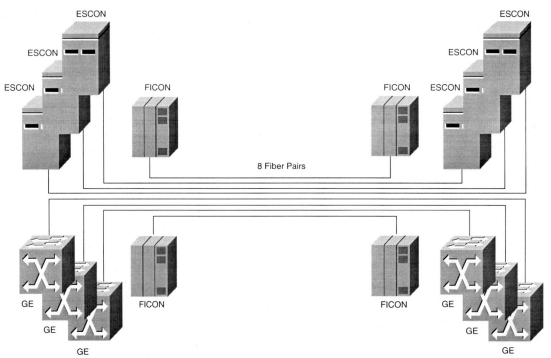

If the organization owned the fiber plant, they would be responsible for the underground installation of the fiber-optic cable and its maintenance. Many organizations outsource such work to dark-fiber providers. Fiber providers charge per strand per kilometer of fiber. Therefore, networks such as that in Figure 1-6 could be extremely expensive to build and maintain.

The metro DWDM platform enables service providers to deliver managed wavelength-based ESCON, FICON, Fibre Channel, and Ethernet services to customers offering outsourced storage or content services. This facilitates the convergence of data, storage, and SONET/SDH networking and provides an infrastructure capable of reliable, high-availability multiservice networking in the MAN at very economical levels.

Using DWDM technology, the service providers can strip off wavelengths and assign them to each connection as shown in Figure 1-7. Each connection is now assigned a wavelength, instead of being assigned to its own fiber pair. As illustrated in Figure 1-7, eight wavelengths are assigned to a single pair of fibers. This way, numerous data streams can be multiplexed at different speeds, across a single fiber pair. This saves the organization considerable expense. In addition, service providers can provision wavelengths to enterprise customers and charge for the number of wavelengths used.

Figure 1-7 *Storage-Area Topology Using DWDM*

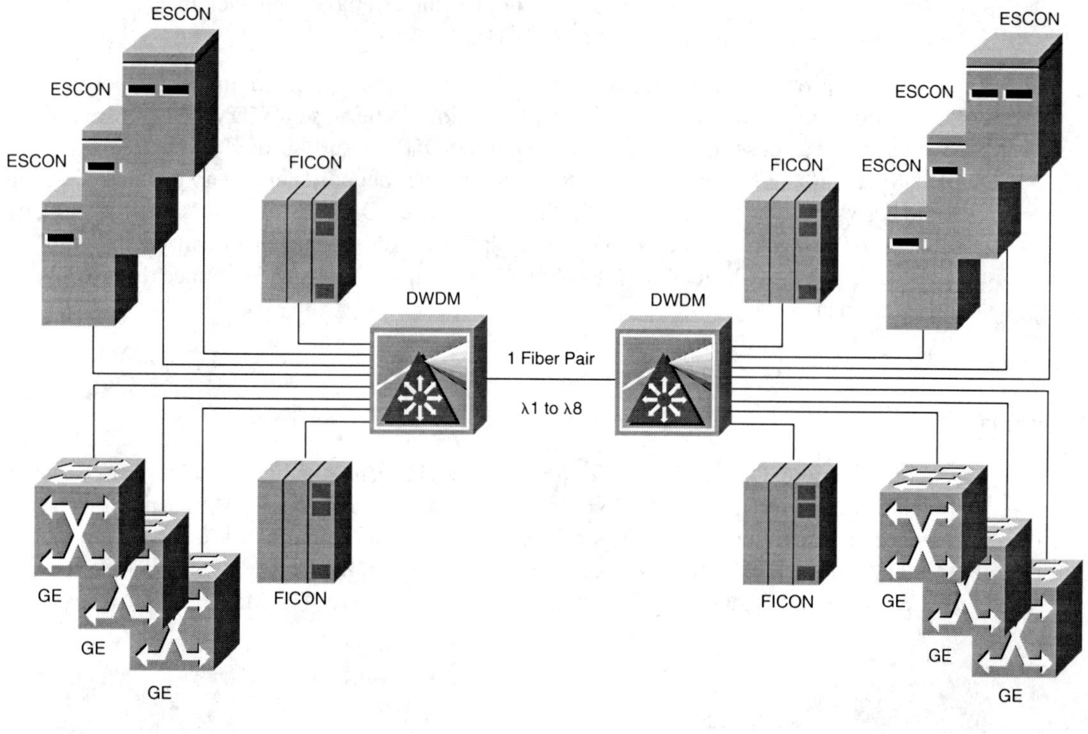

Consider a DWDM platform that provides 32 wavelengths multiplexed over a single fiber pair. By supporting speeds from 10 Mbps up to OC-192 (10 Gbps), the system could provide up to 320 Gbps of bandwidth. To increase the density of signals on the fiber-optic cable, most users would start by aggregating their existing traffic, such as Gigabit Ethernet, ESCON (136 Mbps/200 Mbps), FICON (1.062 Gbps), or Fibre Channel (640 Mbps/1.062 Gbps/ 2.125 Gbps) via DWDM.

Users also have the ability to increase the bandwidth on each of the channels (wavelengths)—for example, by moving from OC-3 to OC-48. Another key benefit is protocol transparency, which alleviates the need for protocol conversion, the associated complexity, and the transmission latencies that might result. Protocol transparency is accomplished with 2R networks and enables support for all traffic types, regardless of bandwidth and protocol.

The Future of SONET/SDH and DWDM

As telecommunication infrastructures continue to evolve, SONET/SDH and DWDM will continue to play a major role in the transport of high-bandwidth applications. The future bandwidth drivers include increased Internet and intranet usages, applications such as video on demand, online gaming, remote SANs, high-volume remote data backup, and integrated voice and video over fiber to the home (FTTH).

MSPPs will provide metro access, core, and long-haul transport. Many people believe that SONET/SDH and DWDM are competing technologies, because DWDM designs are trying to incorporate a subset of services that SONET/SDH currently provides. The reality is that DWDM and SONET are in some ways not competing, but complementary technologies. The future lies with standards-based MSPPs that consolidate SONET/SDH TDM, 100/1000 Ethernet services, ATM, packet, and DWDM on one platform. Additional functionality will continue to be added to MSPP devices. High-bandwidth SAN applications will continue to drive DWDM in the metro area.

Summary

The Synchronous Optical Network or SONET/SDH forms the backbone of many service provider and carrier networks. SONET/SDH is used as a Layer 1.5 TDM transport mechanism for voice and data. Legacy SONET/SDH networks use add/drop multiplexers (ADMs) that add or drop OC-N or STM-N circuits between ADM nodes on the ring. MSPPs integrate ADM, DACS, and Ethernet switching functionality within the device. The MSPP can provide private line TDM services, ATM services, 10/100/1000-Mbps Ethernet services, and Multiprotocol Label Switched (MPLS) IP-routed services. DWDM uses multiple channels or wavelengths to transmit signals over a single optical fiber. DWDM maximizes the use of installed fiber cable and allows new services to be quickly and easily provisioned over existing infrastructure. Long-haul transport, managed wavelength services, metro fiber exhaust, and storage-area networking (SAN) are key drivers for DWDM.

This chapter includes the following sections:

- **An Introduction to Time-Division Multiplexing**—TDM is a common way to send multiple signals or streams of information over a medium at the same time in the form of a single, complex signal, and then recover the separate signals at the receiving end. In a TDM scheme, several low-speed channels are multiplexed into a single high-speed channel for transmission over a network.

- **Analog Signal Processing**—This section presents the various stages involved with analog signal generation and reception, analog-to-digital conversion, signal filtering, sampling, quantization, encoding, and μ-law and A-law coding for pulse code modulation (PCM).

- **Circuit-Switched Networks**—Circuit-switched networks and a description of leased lines from a customer and carrier perspective are presented in this section. TDM signaling methods, such as channel-associated signaling (CAS) and common channel signaling (CCS), are also introduced.

- **The T-Carrier**—The North American T-carrier is discussed in detail with a description of digital signal (DS) single framing and multiframing formats. The D4 superframe and D5 extended superframe are presented along with a listing of various SF and ESF alarms.

- **The E-Carrier**—The Europe, Middle East, and Africa (EMEA), Asia Pacific, and Latin American E-carrier is discussed in detail with a description of the E1 frame alignment signal (FAS), E1 multiframe alignment signal (MFAS), and E1 cyclic redundancy check (CRC) error checking. Various E1 errors and alarms are also listed.

- **ISDN**—This section discusses the Integrated Services Digital Network (ISDN) basic rate and primary rate interfaces. It also discusses ISDN framing at Layer 2 and 3, Layer 2 session establishment, and Layer 3 call setup.

- **TDM Network Elements**—Various TDM network elements are used to build a TDM infrastructure. These include repeaters, channel service units/digital service units (CSU-DSUs), digital access and cross-connect systems (DACS), and channel banks. This section discusses the various TDM elements and their interaction.

CHAPTER 2

Time-Division Multiplexing

An Introduction to Time-Division Multiplexing

A medium can natively carry only one signal at any moment in time. For multiple signals to share a medium, the medium must somehow be divided, giving each signal a portion of the total bandwidth. The current techniques that can accomplish this include time-division multiplexing (TDM), frequency-division multiplexing (FDM), code-division multiple access (CDMA), and wavelength-division multiplexing (WDM). Variations of these four multiplexing techniques are used in data transmission systems.

TDM is a common way to send multiple signals or streams of information over a medium at the same time in the form of a single, complex signal and then recover the separate signals at the receiving end. In this scheme, several low-speed channels are multiplexed into a single high-speed channel for transmission over a network. TDM operates by dividing the network bandwidth into fixed-bandwidth segments. Each segment or channel is assigned to an end-user device and is given its own time slot for using the network. TDM also implies the sharing of a common facility. TDM is widely used in telephone communication systems along with digital encoding techniques such as pulse code modulation (PCM), which has been the mainstay of TDM communications systems.

As illustrated in Figure 2-1, information from channel A is sampled and transmitted first, information from channel B is then sampled and transmitted, and so on in a regular sequence, cycling back to channel A and continuing. Bandwidth allocation is static, in the sense that each channel has a fixed, predetermined bandwidth. Because TDM is protocol insensitive, it is capable of combining various higher-layer protocols onto a single high-speed transmission link.

Figure 2-1 *TDM Schematic*

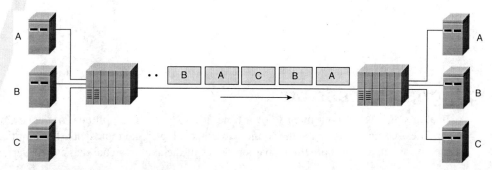

If an end-user device does not have data to transmit, empty time slots are transmitted. Such a TDM mechanism is also known as synchronous time-division multiplexing. You can see this in Figure 2-2, where device B is not transmitting data. Statistical multiplexing overcomes this inefficiency by only transmitting data from active end devices.

Figure 2-2 *Synchronous TDM*

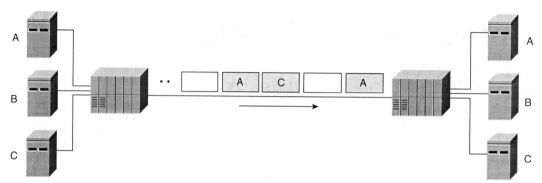

Figure 2-3 shows statistical TDM. If an end-user device is not active, no space is wasted on the multiplexed stream. A statistical multiplexer accepts the incoming data streams and creates a frame containing only the data to be transmitted. In synchronous TDM, each signal is given a unique but equal time slot for its information that is interleaved with the others. In statistical TDM (also known as asynchronous TDM), however, the amount of time per slot is variable. Conventional TDM systems usually use either bit-interleaved or byte-interleaved multiplexing schemes. Each time slot accommodates either a bit (1 or 0) or a byte (usually 8 bits long to represent a character, number, or symbol).

Figure 2-3 *Statistical TDM*

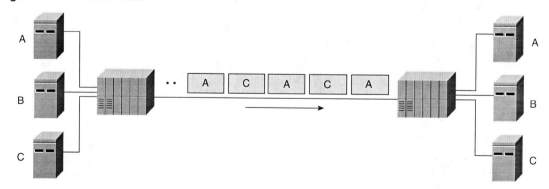

Analog Signal Processing

An analog signal varies continuously over time. It could vary in amplitude, frequency, or phase. These components define the sound wave an analog signal represents. The amplitude, frequency, and phase shift are three characteristics of the analog signal that can be varied to convey information. Analog signals are inherently susceptible to attenuation as they progress along the transmission medium. Analog signals are also susceptible to electromagnetic interference (EMI), radio frequency interference (RFI), and other noise sources. This results in signal distortion with changes in frequency characteristics.

Analog telephony signals span the 200-Hz to 3.4-kHz frequency band. Such analog signals are referred to as narrowband due to their narrow frequency response.

Analog video signals operate in a frequency band from flat response (0 Hz) up to 60 MHz. Such analog signals are referred to as broadband due to their wide frequency response. The National Television System Committee (NTSC) and PAL broadcast (radio frequency [RF] transmission) standards impose a limit on the bandwidth of the video signal of about 6 to 10 MHz. Video bandwidth is, effectively, the highest-frequency analog signal a monitor can handle without distortion. Amplification can be used to compensate for signal attenuation. However, narrowband repeaters cannot distinguish between the signal and distortion components of the analog signal. The repeater amplifies the entire input signal, thereby amplifying the noise along with the original signal. The effects of noise and distortion are cumulative along the analog transmission system.

Analog Signal Generation and Reception

The generation of an analog telephony signal takes place when a person speaks into the transmitter of a telephone set. Changes in the air pressure result in sound waves that are sensed by the diaphragm. The diaphragm responds to changes in air pressure and varies circuit resistance by compressing or decompressing carbon in the transmitter. The change in resistance causes a variation in the output voltage, thereby creating an electrical signal analogous to the sound wave. The phone connects to a central office (CO) in the caller's neighborhood through a subscriber line interface circuit (SLIC) that executes functions, such as powering the phone, detecting when the caller picks up or hangs up the receiver, and ringing the phone when required. A codec at the CO converts the analog voice signals to digital data for easy routing through the voice network and delivery to the CO located in the recipient's neighborhood. At the recipient's CO, the digital data stream is converted back into an electrical analog signal. During reception, a varying current flows through the coil and vibrates the receiver diaphragm that reproduces the sound wave. Digital transmission systems overcome the basic analog issue of the cumulative effects of noise and distortion by regenerating rather than amplifying the transmitted signal. The regenerative repeater detects the presence of a pulse (signal) and creates a new signal based on a sample of the existing signal. The regenerated signal duplicates the original signal and eliminates the cumulative effects of noise and distortion inherent in analog facilities.

Analog-To-Digital Conversion

Converting an analog telephony signal to a digital signal involves filtering, sampling, quantization, and encoding. The following example involves an audio frequency (AF) signal.

Filtering

Audio frequencies range from 20 Hz to 20,000 Hz. Telephone transmission systems are designed to transmit analog signals between 200 Hz and 3400 Hz. End frequencies below 200 Hz and above 3400 Hz are removed by a process called filtering.

As indicated in Figure 2-4, a band pass filter (BPF) is used to filter the audio telephony band for analog-to-digital (A/D) conversion. BPFs are constructed using analog electronic components, such as capacitors and inductors.

Figure 2-4 *Filtering of the Analog Telephony Waveform*

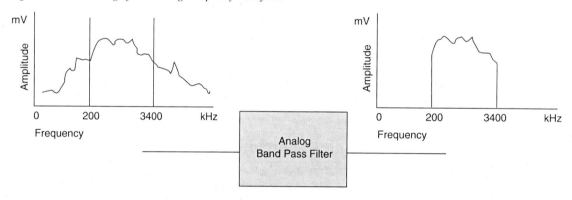

Sampling

In the sampling process, portions of a signal are used to represent the whole signal. Each time the signal is sampled, a pulse amplitude modulation (PAM) signal is generated. According to the Nyquist theorem, to accurately reproduce the analog signal (speech), a sampling rate of at least twice the highest frequency to be reproduced is required. Because the majority of telephony voice frequencies (200 to 3400 Hz) are less than 4 kHz, an 8-kHz sampling rate has been established as the standard. As illustrated in Figure 2-5, the PAM sampler measures the filtered analog signal 8000 times per second, or once every 125 microseconds. The value of each of these samples is directly proportional to the amplitude of the analog signal at the time of the sample (PAM, as mentioned previously).

Quantization

Quantization represents the original analog signal by a discrete and limited number of digital signals. When the original signal is in a quantized state, it can be safely relayed for any distance without further loss in quality. To obtain the digital signal, the PAM signal is measured and coded. As shown in Figure 2-5, the amplitude or height of the PAM is measured to derive a number that represents its amplitude level. Quantization essentially matches the PAM signals to one of 255 values on a segmented scale. The quantizer measures the amplitude or height of each PAM signal coming from the sampler and assigns it a value from –127 to +127. In telephony systems, each amplitude value (sample) is expressed as a 13-bit code word. Comparing the sample to a companding characteristic, which is a nonlinear formula, forms an 8-bit byte.

Figure 2-5 *Pulse Amplitude Modulation (PAM)*

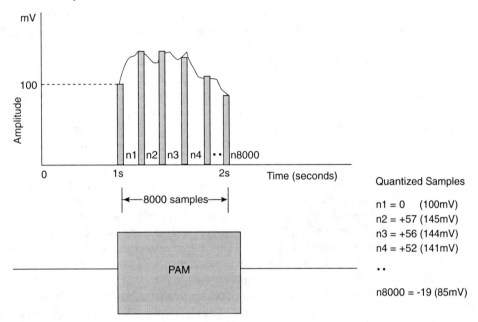

Encoding

The decimal (base 10) number derived via quantization is then converted to its equivalent 8-bit binary number. As illustrated in Figure 2-6, the output is an 8-bit "word" in which each bit can be either a 1 (pulse) or a 0 (no pulse). This process is repeated 8000 times a second for a telephony voice channel service. The output (8000 samples/second * 8 bits/sample) is a 64-kbps PCM signal. This 64-kbps channel is called a DS0, which forms the fundamental building block of the digital signal level (DS level) hierarchy.

Figure 2-6 *Pulse Code Modulation (PCM)*

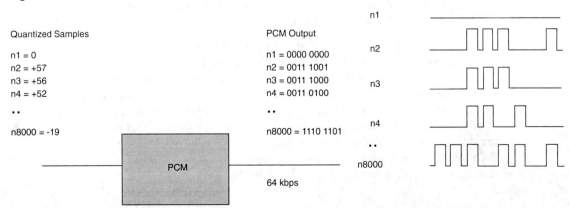

μ-law and A-Law Coding

Voice signals are not uniform, and some signals are weaker than others. The dynamic range is the difference in decibels (dB) between weaker (softer) and stronger (louder) signals. The dynamic range of speech can be as high as 60 dB. This does not lend itself well to efficient linear digital encoding. G.711 μ-law and A-law encoding effectively reduce the dynamic range of the signal, thereby increasing the coding efficiency and resulting in a signal-to-noise ratio (SNR) superior to that obtained by linear encoding for a given number of bits. The μ-law and A-law algorithms are standard compression algorithms used in digital communications systems to optimize and modify the dynamic range of an analog signal for digitizing. The μ-law is typically used on T1 facilities, whereas the A-law is used on E1 facilities.

Companding (compression and expansion) is a method commonly used in telephony applications to increase dynamic range while keeping the number of bits used for quantization constant. The compression is lossy, but provides lower quantization errors at smaller amplitude values than at larger values. Basically, the voice is sampled at 8000 samples per second and converted into a 14-bit word (μ-law) or 13-bit word (A-law) that goes into the compander. The samples are processed using a nonlinear formula to transform them into 8-bit words. The compander also inverts all even bits in the word. In A-law companding, for instance, the 13-bit word 1111111111111 is converted to 11111111 (+127) using compression, resulting in the PCM word 10101010 (AA hex). Telephony PCM words use a polarity, chord, and step makeup. Nonlinear coding uses more values to represent lower-volume levels and fewer values for higher-volume levels. This way, μ-law and A-law companding algorithms permit subtleties of a voice conversation to be captured.

Echo Cancellation

Line echo is created when a signal encounters an impedance mismatch in the telephone network, such as that typically caused by a two- to four-wire (hybrid) conversion in an analog system. The *hybrid* is a transformer located at the facility that connects the two-wire local loop coming from homes or businesses to the four-wire trunk at the CO for inter-exchange carrier (IXC) interconnectivity. The echo is intensified by distance and impedance-mismatched network equipment. In circuit-switched long-distance networks, echo cancellers reside in the metropolitan COs that connect to the long-distance network. These echo cancellers remove electrical echoes made noticeable by delay in the long-distance network. To eliminate echo, echo cancellation devices use adaptive digital filters, nonlinear processors, and tone detectors.

The adaptive filter is made up of an echo estimator and a subtractor. The echo estimator monitors the receive path and dynamically builds a mathematical model of the line that creates the returning echo. The echo estimate is then fed to the subtractor, which subtracts the linear part of the echo from the line in the send path. The nonlinear processor evaluates the residual echo, removes all signals below a certain threshold, and replaces them with simulated background noise that sounds like the original background noise without the echo. Echo cancellers also include tone detectors that disable echo cancellation by user equipment upon receipt of certain tones during data and fax transmission. As an example, the echo-cancellation function is turned off upon receipt of the high-frequency tone that precedes a modem connection.

Circuit-Switched Networks

Figure 2-7 shows an example of a circuit-switched network from a customer's perspective. Such a topology is also referred to as a point-to-point line or nailed-up circuit. Typically such lines are leased from a local exchange carrier (LEC) or IXC and are also referred to as leased lines. One leased line is required for each of the remote sites to connect to the headquarters at the central site.

Figure 2-7 *Leased Lines from a Customer Perspective*

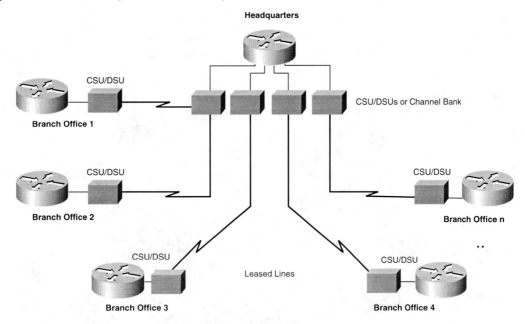

The private nature of leased line networks provides inherent privacy and control benefits. Leased lines are dedicated, so there are no statistical availability issues associated with oversubscription, as there are in public packet-switched networks. This is both a strength and weakness. The strength is that the circuit is available on a permanent basis and does not require that a connection be set up before traffic is passed. The weakness is that the bandwidth is paid for even if is not used, which is typically about 40 to 70 percent of the time. In addition to the inefficient use of bandwidth, a major disadvantage of leased lines is their mileage-sensitive nature, which makes it a very expensive alternative for networks spanning long distances or requiring extensive connectivity between sites.

Leased lines also lack flexibility in terms of changes to the network when compared to alternatives, such as Frame Relay. For example, adding a new site to the network requires a new circuit to be provisioned end to end for every site with which the new location must communicate. If there are a number of sites, the costs can mount quickly. Leased lines are priced on a mileage and bandwidth basis by a carrier, which results in customers incurring large monthly costs for long-haul leased circuits.

In comparison, public networks (such as Frame Relay) require only an access line to the nearest CO and the provisioning of virtual circuits (VCs) for each new site with which it needs to communicate. In many cases, existing sites will require only the addition of a new VC definition for the new site.

From the carrier perspective, the circuit assigned to the customer (also known as the local loop) is provisioned on the digital access and cross-connect system (DACS) or channel bank. The individual T1 circuits are multiplexed onto a T3 and trunked over terrestrial, microwave, or satellite links to its destination, where it is demultiplexed and fanned out into individual T1 lines. Figure 2-8 shows this scheme. The T-carrier hierarchy, DS1, and DS3 are covered later in this chapter.

Figure 2-8 *Leased Lines from a Carrier Perspective*

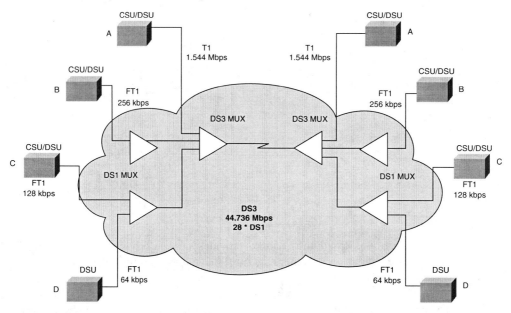

TDM Signaling

Signaling in the TDM telephony world provides functions such as supervising and advertising line status, alerting devices when a call is trying to connect, and routing and addressing information. Two different types of signaling information are within the T1/E1 system:

- Channel-associated signaling (CAS)
- Common channel signaling (CCS)

Channel-Associated Signaling (CAS)

CAS is the transmission of signaling information within the information band, or in-band signaling. This means that voice or data signals travel on the same circuits as line status, address, and alerting signals. Because there are 24 channels on a full T1 line, CAS interleaves signaling packets within voice packets; therefore, there are 24 channels to use for voice. Various types of CAS signaling are available in the T1 world. The most common forms of CAS signaling are loopstart, groundstart, and ear and mouth (E&M) signaling. CAS signaling is often referred to as *robbed-bit signaling* because signaling bits are robbed from every 6th and 12th frame in a D4 superframe (SF) or 6th, 12th, 18th, 24th frame, and extended superframe (ESF). This is explained in greater detail in a later section.

Common Channel Signaling (CCS)

CCS is the transmission of signaling information out of the information band. The most notable and widely used form of this signaling type is ISDN. One disadvantage to using an ISDN primary rate interface (PRI) is the removal of one DS0, or voice channel (in this case, for signaling use). Therefore, one T1 would have 23 DS0s, or bearer B channels for user data, and one DS0, or D channel for signaling. It is possible to control multiple PRIs with a single D channel, each using non-facility-associated signaling (NFAS). This enables you to configure the other PRIs in the NFAS group to use all 24 DS0s as B channels.

The T-Carrier

The North American DS1 consists of 24 DS0 channels that are multiplexed. The signal is referred to as DS1, whereas the transmission channel over the copper-based facility is called a T1 circuit. The T-carrier is used in the United States, Canada, Korea, Hong Kong, and Taiwan.

TDM circuits typically use multiplexers, such as channel service units/digital service units (CSUs/DSUs) or channel banks at the CPE (customer premises equipment) side, and they use larger programmable multiplexers, such as DACS and channel banks, at the carrier end. The T-carrier system is entirely digital, using PCM and TDM. The system uses four wires and provides duplex capability. The four-wire facility was originally a pair of twisted-pair copper wires, but can now also include coaxial cable, optical fiber, digital microwave, and other media. A number of variations on the number and use of channels is possible. The T-carrier hierarchy used in North America is shown in Table 2-1 and illustrated in Figure 2-9. The DS1C, DS2, and DS4 levels are not commercially used. The SONET Synchronous Transport Signal (STS) levels have largely replaced the DS levels above DS3.

Table 2-1 *T-Carrier Hierarchy*

Digital Signal Level	Number of 64 kbps Channels	Equivalent	Bandwidth
DS0	1	1 * DS0	64 kbps
DS1	24	24 * DS0	1.544 Mbps
DS1C	48	2 * DS1	3.152 Mbps
DS2	96	4 * DS1	6.312 Mbps
DS3	672	28 * DS1	44.736 Mbps
DS4	4032	6 * DS3	274.176 Mbps

Figure 2-9 *T-Carrier Multiplexed Hierarchy*

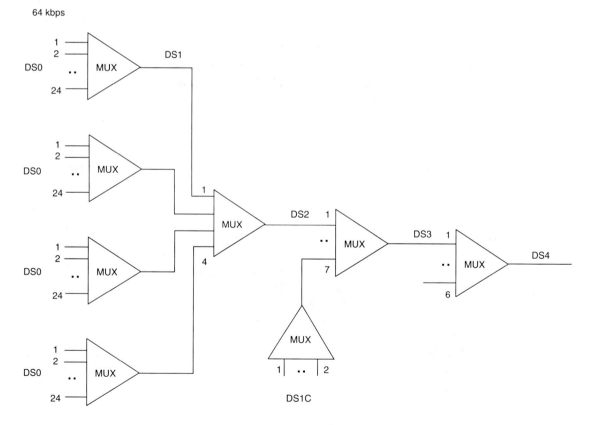

| NOTE | Some TDM systems use 8 kbps for in-band signaling. This results in a net bandwidth of only 56 kbps per channel. Japan uses the North American standards for DS0 through DS2, but the Japanese DS5 has roughly the circuit capacity of a U.S. DS4. |

DS Framing

The DS1 frame of Figure 2-10 is composed of 24 DS0 (8-bit) channels, plus 1 framing bit, which adds up to 193 bits. The DS1 signal transports 8000 frames per second, which results in 193 * 8000 bits per second or 1,544,000 bps (1.544 Mbps). The first bit (bit 1), or F bit, is used for frame alignment, performance-monitoring cyclic redundancy check (CRC), and data linkage. The remaining 192 bits provide 24 8-bit time slots numbered from 1 to 24.

Figure 2-10 *DS Frame*

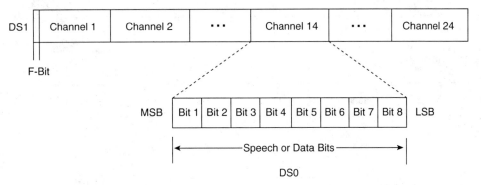

DS systems use alternate mark inversion (AMI) or binary 8 zero substitution (B8ZS) for line encoding. In AMI, every other 1 is a different polarity, and the encoding mechanism does not maintain a "1s density." In B8ZS, the encoding mechanism uses intentional bipolar violation to maintain a "1s density." Bipolar violations are two "1s" of the same polarity. T1 physical delivery is over two-pair copper wires—one pair for RX (1+2) and one pair for TX (4+5). For the CPE, RX means data from the network, whereas TX means data to the network.

DS Multiframing Formats

Two kinds of multiframing techniques are used for DS-level transmissions:

- D4 or superframe (SF)
- D5 or extended superframe (ESF)

D4 multiframing typically uses AMI encoding, whereas ESF uses B8ZS encoding. However, B8ZS line coding could be used with D4 framing as well as ESF. The multiplexer (mux) terminating the T1 usually determines the multiframing option.

D4 Superframe

In the original D4 (SF) standard, the framing bits continuously repeated the sequence 110111001000. In voice telephony, errors are acceptable, and early standards allow as much as one frame in six to be missing entirely. As shown in Figure 2-11, the SF (D4) frame has 12 frames and uses the least significant bit (LSB) in frames 6 and 12 for signaling (A, B bits). This method of in-band signaling is called *robbed-bit signaling*. Each frame has 24 channels of 64 kbps. Within an SF, F bits delineate the basic frames within the multiframe. In channel-associated signaling, bits are robbed from time slots to carry signaling messages. Figure 2-11 shows the D4 SF format.

Figure 2-11 *D4 SF Format*

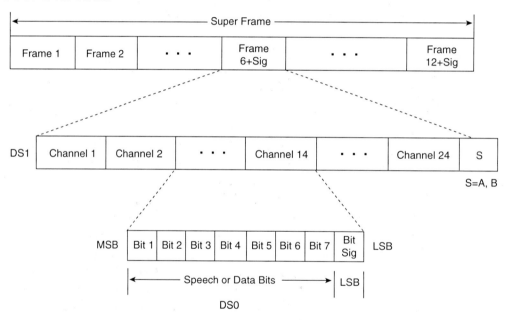

D5 Extended Superframe

To promote error-free transmission, an alternative called the D5 or extended superframe (ESF) of 24 frames was developed. As shown in Figure 2-12, the ESF frame has 24 frames and uses the LSB in frames 6, 12, 18, and 24 for signaling (A, B, C, D bits). Each frame has 24 channels of 64 kbps. In this standard, 6 of the 24 framing bits provide a 6-bit cyclic redundancy check (CRC-6), and 6 provide the actual framing. The other 12 form a VC of 4096 bps for use by the transmission equipment, for call progress signals such as busy, idle, and ringing. DS1 signals using ESF equipment are nearly error free, because the CRC detects errors and allows automatic rerouting of connections. Within an ESF, the F bits provide basic frame and multiframe delineation, performance monitoring through CRC-6-based error detection, a 4-kbps data link to transfer priority operations messages, and other maintenance or operations messages. The F bits also provide periodic terminal performance reports, or an idle sequence. In CAS, bits are robbed from time slots to carry signaling messages. Figure 2-12 shows the ESF format.

Figure 2-12 *ESF Format*

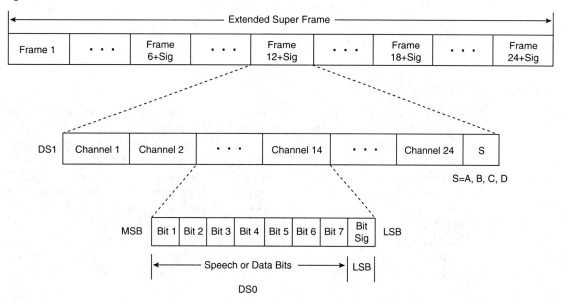

SF and ESF Alarms

It is important to understand D4 and ESF alarm conditions, in order to interpret the behavior of a TDM transmission system on the CPE as well as on the network side. The alarms listed here are commonly used with CPE equipment, such as CSUs/DSUs, T1 repeaters, DACS devices, and multiplexers.

- **AIS (alarm indication signal)**—The AIS is also known as a "Keep Alive" or "Blue Alarm" signal. This consists of an unframed, all-1s signal sent to maintain transmission continuity. The AIS carrier failure alarm (CFA) signal is declared when both the AIS state and red CFA persist simultaneously.

- **OOF (out-of-frame)**—The OOF condition occurs whenever network or DTE equipment senses errors in the incoming framing pattern. Depending upon the equipment, this can occur when 2 of 4, 2 of 5, or 3 of 5 framing bits are in error. A reframe clears the OOF condition.

- **Red CFA (carrier failure alarm)**—This CFA occurs after detection of a continuous OOF condition for 2.5 seconds. This alarm state is cleared when no OOF conditions occur for at least 1000 milliseconds. Some applications (certain DACS services) might not clear the CFA state for up to 15 seconds of no OOF occurrences.
- **Yellow CFA (carrier failure alarm)**—When a device enters the red CFA state, it transmits a "yellow alarm" in the opposite direction. A yellow alarm is transmitted by setting bit 2 of each time slot to a 0 (zero) space state for D4-framed facilities. For ESF facilities, a yellow alarm is transmitted by sending a repetitive 16-bit pattern consisting of 8 marks (1) followed by 8 spaces (0) in the data-link bits. This is transmitted for a minimum of 1 second.
- **LOS (loss of signal)**—A LOS condition is declared when no pulses have been detected in a 175 +/− 75 pulse window (100 to 250 bit times).

The E-Carrier

The basic unit of the E-carrier system is the 64-kbps DS0, which is multiplexed to form transmission formats with higher speeds. The E1 consists of 32 DS0 channels. The *E-carrier* is a European digital transmission format devised by the International Telecommunication Union Telecommunication Standardization Sector (ITU-T) and given the name by the Conference of European Postal and Telecommunication Administration (CEPT). E2 through E5 are carriers in increasing multiples of the E1 format. The E1 signal format carries data at a rate of 2.048 Mbps and can carry 32 channels of 64 kbps each. Unlike T1, it does not bit-rob and all 8 bits per channel are used to code the signal. E1 and T1 can be interconnected for international use. The E-carrier hierarchy used in EMEA, Latin America, South Asia, and the Asia Pacific region is shown in Table 2-2 and illustrated in Figure 2-13. The E2, E4, and E5 levels are not commercially used. The Synchronous Digital Hierarchy (SDH) levels have largely replaced the DS levels above E4.

Table 2-2 *E-Carrier Hierarchy*

Digital Signal Level	Number of 64 kbps Channels	Equivalent	Bandwidth
E1	32	32 * DS0	2.048 Mbps
E2	128	4 * E1	8.448 Mbps
E3	512	4 * E2	34.368 Mbps
E4	2048	4 * E3	139.264 Mbps
E5	8192	4 * E4	565.148 Mbps

As depicted in Figure 2-14, a 2.048-Mbps basic frame is comprised of 256 bits numbered from 1 to 256. These bits provide 32 8-bit time slots numbered from 0 to 31. The first time slot is a framing time slot used for frame alignment, performance monitoring (CRC), and data linkage. Time slot 0 carries framing information in a frame alignment signal as well as remote alarm notification, 5 national bits, and optional CRC bits. Time slot 16 is a signaling time slot and carries signaling information out of band. However, time slot 16 could carry data as well.

The E-Carrier 33

Figure 2-13 *E-Carrier Multiplexed Hierarchy*

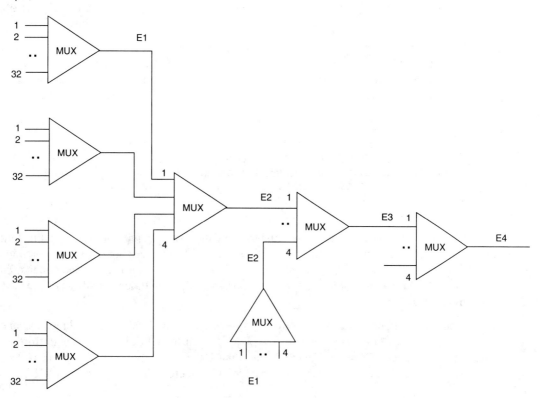

Figure 2-14 *E1 Frame Structure*

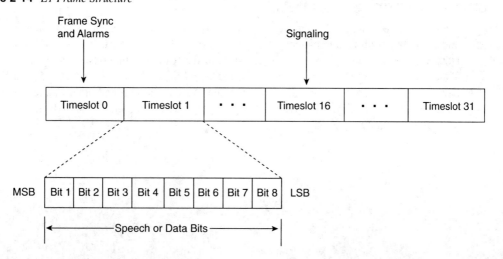

Like all basic frames used in telecommunications, the E1 basic frame lasts 125 microseconds. The full E1 bit rate is 2.048 Mbps. We calculate this bit rate by multiplying the 32-octet E1 frame by 8000 frames per second. Subtracting time slots 0 and 16, we see that E1 lines offer 30 time slots to carry user data or a payload-carrying capacity of 1.920 Mbps.

E1 uses AMI or high-density bipolar 3 (HDB3) for line encoding. In AMI, every other 1 is a different polarity, and the encoding mechanism does not maintain a "1s density." AMI is used to represent successive 1s' values in a bit stream with alternating positive and negative pulses to eliminate any direct current (DC) offset.

NOTE AMI is not used in most 2.048-Mbps transmission systems because synchronization loss can occur during long strings of 0s, because there are no pulses.

In HDB3, every other 1 is a different polarity and the encoding mechanism uses a bipolar violation to maintain a "1s density." The HDB3 coded signal does not have a DC component. Therefore, the signal can be transmitted through balanced transformer-coupled circuits. The clock recovery circuits of the receivers can operate well, even though the data contains long strings of 0s.

Unbalanced E1 physical delivery is over two-pair copper wires with 120-ohm line impedance—one pair for RX (1+2) and one pair for TX (4+5). For the CPE, RX means data from the network, whereas TX means data to the network. Balanced E1 physical delivery is over a pair of 75-ohm coaxial cables. One coax is used for TX, whereas the other one is for RX.

E1 Frame Alignment Signal (FAS)

Framing is necessary so that any equipment receiving the E1 signal can synchronize, identify, and extract the individual channels. The 2.048-Mbps E1 frame consists of 32 individual time slots (numbered 0 through 31). Each time slot consists of individual 64-kbps channels of data. Time slot 0 of every even frame is reserved for the FAS. As shown in Figure 2-15, odd frames have the NFAS word that contains the distant alarm indication bit and other bits reserved for national and International use. Thirty-one time slots remain for bearer channels, into which customer data can be placed.

Figure 2-15 *E1 Frame Alignment Signal*

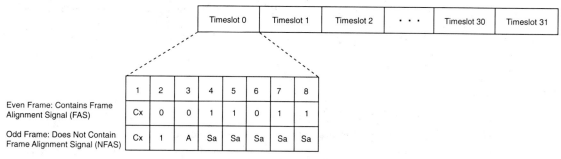

E1 MultiFrame Alignment Signal (MFAS)

Sixteen E1 consecutive frames form a new structure called an E1 multiframe. The frames in a multiframe are numbered 0 to 15. Multiframe structure is used for two purposes: CAS signaling and CRC. Each of these modes is independent from the use of the other. CAS is carried in time slot 16, and CRC is carried in time slot 0. The purpose of the multiframe is to have sufficient overhead bits to support two key functions in time slot 16, which carries signaling information when an E1 is transmitting digital voice streams. MFAS framing is used for CAS to transmit ABCD bit information for each of the 30 channels, as illustrated in Figure 2-16.

Figure 2-16 *E1 Multiframe Alignment Signal*

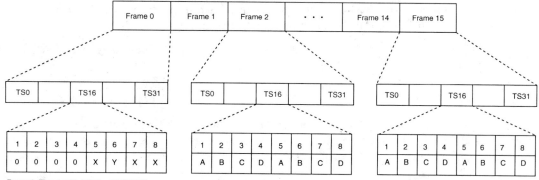

Frame 0, Timeslot 16: MFAS = 0000XYXX
X = Spare Bits (=1, If Not Used)
Y = MFAS Remote Alarm (=1, If MFAS Synchronization Is Lost)

Notes:
1) Frames are transmitting with 30 voice channels on timeslots 1-15 and 17-31...
2) Timeslot 16 (TS16) contains ABCD bits for signaling (CAS)...
3) MFAS framing still includes the orginal FAS frames and FAS framing information in TS0...

This method uses the 32 time slot frame format with time slot 0 for the FAS and time slot 16 for the MFAS and CAS. When a PCM-30 multiframe is transmitted, 16 FAS frames are assembled together. Time slot 16 of the first frame is dedicated to MFAS bits, and time slot 16 of the remaining 15 frames is dedicated to ABCD bits.

E1 CRC Error Checking

A cyclic redundancy check-4 (CRC-4) is often used in E1 transmission to identify possible bit errors during in-service error monitoring. *CRC-4* is a checksum calculation that allows for the detection of errors within the 2.048-Mbps signal while it is in service. A discrepancy indicates at least one bit error in the received signal. The equipment that originates the E1 data calculates the CRC-4 bits for one submultiframe. It inserts the CRC-4 bits in the CRC-4 positions in the next submultiframe.

The receiving equipment performs the reverse mathematical computation on the submultiframe. It examines the CRC-4 bits that were transmitted in the next submultiframe. It then compares the transmitted CRC-4 bits to the calculated value. If there is a discrepancy in the two values, a CRC-4 error is reported via E-bits indication. Each individual CRC-4 error does not necessarily correspond to a single bit error, which is a drawback. Multiple bit errors within the same submultiframe will lead to only one CRC-4 error for the block. Thirty-one time slots remain for bearer channels, into which customer data can be placed.

Errors could occur such that the new CRC-4 bits are calculated to be the same as the original CRC-4 bits. CRC-4 error checking provides a most convenient method of identifying bit errors within an in-service system, but only provides an approximate measure (93.75 percent accuracy) of the circuit's true performance. Consider the MFAS framing shown in Figure 2-17. Each MFAS frame can be divided into "submultiframes." These are labeled SMF1 and SMF2, and consist of eight frames apiece. We associate 4 bits of CRC information with each submultiframe. The CRC-4 bits are calculated for each submultiframe, buffered, and then inserted into the following submultiframe to be transmitted.

ITU-T specifications G.704 and G.706 define the CRC-4 cyclic redundancy check for enhanced error monitoring on the E1 line.

E1 Errors and Alarms

It is important to understand E1 error and alarm conditions, in order to interpret the behavior of a TDM transmission system on the CPE as well as on the network side. The alarms listed here are commonly used with CPE equipment such as CSUs/DSUs, E1 repeaters, DCS devices, and multiplexers:

- **Alarm indication signal (AIS)**—Alarm indication signal is an unframed, all-1s signal.
- **Background block error (BBE)**—A background block error is an error block (a *block* is a set of consecutive bits associated with a path) that does not occur as part of a severely errored second (SES).

Figure 2-17 *E1 CRC Error Checking*

Multiframe	Sub-Multiframe	Frame Number	Timeslot 0							
			Bit 1	Bit 2	Bit 3	Bit 4	Bit 5	Bit 6	Bit 7	Bit 8
	SMF 1	0	C1	0	0	1	1	0	1	1
		1	0	1	A	Sa4	Sa5	Sa6	Sa7	Sa8
		2	C2	0	0	1	1	0	1	1
		3	0	1	A	Sa4	Sa5	Sa6	Sa7	Sa8
		4	C3	0	0	1	1	0	1	1
		5	1	1	A	Sa4	Sa5	Sa6	Sa7	Sa8
		6	C4	0	0	1	1	0	1	1
		7	Sa5	Sa5	Sa5	Sa4	Sa5	Sa6	Sa7	Sa8
	SMF 2	8	C1	0	0	1	1	0	1	1
		9	1	1	A	Sa4	Sa5	Sa6	Sa7	Sa8
		10	C2	0	0	1	1	0	1	1
		11	1	1	A	Sa4	Sa5	Sa6	Sa7	Sa8
		12	C3	0	0	1	1	0	1	1
		13	E	1	A	Sa4	Sa5	Sa6	Sa7	Sa8
		14	C4	0	0	1	1	0	1	1
		15	E	1	A	Sa4	Sa5	Sa6	Sa7	Sa8

SMF1: Sub-Multiframe No.1
SMF2: Sub-Multiframe No. 2
Sa: Spare Bit Reserved for National Use
A: Remote Alarm (FAS Remote Alarm Indication)
Frame Alignment Signal Pattern: 0011011
CRC-4 Frame Alignment Signal: 001011
CRC Multiframe Is Not Aligned with MFAS Timeslot 16 Multiframe
E: E-Bit Indicator
C1, C2, C3, C4: CRC Bits

- **Bit errors**—Bit errors are bits that are in error. Bit errors are not counted during unavailable time.
- **Bit slip**—A bit slip occurs when the synchronized pattern either loses a bit or has an extra bit stuffed into it.
- **Clock slips**—Clock slips occur when the measured frequency deviates from the reference frequency by a one-unit interval.
- **Code errors**—A code error is a violation of the coding rules: two successive pulses with the same polarity. In HDB3 coding, a code error is a bipolar violation that is not part of a valid HDB3 substitution.

- **Cyclic redundancy check (CRC) errors**—CRC-4 block errors. This measurement applies to signals containing a CRC-4 check sequence.
- **Degraded minutes**—A degraded minute (DM) occurs when there is a 10 to 6 or worse bit error rate during 60 available, nonseverely bit-errored seconds.
- **Errored block**—A block in which one or more bits are in error.
- **E-bit indication**—An E-bit is transmitted by the receiving equipment after detecting a CRC-4 error.
- **Errored second (ES)**—An errored second is any second in which one or more bits are in error. An errored second is not counted during an unavailable second. For G.826, an errored second contains one or more blocks with at least one defect.
- **Frame alarm (FALM)**—Frame alarm seconds is a count of seconds that have had far-end frame alarm (FAS remote alarm indication [RAI]), which is when a 1 is transmitted in every third bit of each time slot 0 frame that does not contain the FAS.
- **Frame alignment signal (FAS)**—A count of the bit errors in the frame alignment signal words received. It applies to both PCM-30 and PCM-31 framing.
- **Frequency**—Any variance from 2.048 Mbps in the received frequency is recorded in hertz or parts per million.
- **Loss of frame seconds (LOFS)**—Loss of frame seconds is a count of seconds since the beginning of the test that have experienced a loss of frame.
- **Loss of signal seconds (LOSS)**—Loss of signal seconds is a count of the number of seconds during which the signal has been lost during the test.
- **Multiframe alarm (MFAL)**—Multiframe alarm seconds is a count of seconds that have had far-end multiframe alarm (MFAS RAI).
- **Multiframe alignment signal (MFAS) distant alarm**—In this alarm, a 1 is transmitted in every sixth bit of each time slot 16 in the 0 frame.
- **Severely errored second (SES)**—A severely errored second has an error rate of 10^{-3} or higher. Severely errored seconds are not counted during unavailable time. For G.826 block measurements, an SES is a 1-second period containing 30 percent or greater errored blocks.
- **Time slot 16 AIS**—In this alarm, all 1s are transmitted in time slot 16 of all frames.
- **Unavailable seconds (UAS)**—Unavailable time begins at the onset of 10 consecutive severely errored seconds. Unavailable seconds also begin at a loss of signal or loss of frame.
- **Wander**—This is the total positive or negative phase difference between the measured frequency and the reference frequency. The +wander value increases whenever the measured frequency is one unnumbered information (UI) frame larger than the reference frequency. The –wander increases whenever the measured frequency is one UI frame less than the reference frequency.

> **NOTE** The following ITU-T recommendations are commonly used with TDM systems: G.703, physical/electrical characteristics of hierarchical digital interfaces; G.704, synchronous frame structures used at 1544, 6312, 2048, 8488, and 44,736 kbps; G.706, frame alignment and CRC procedures relating to basic frame structures defined in Recommendation G.704; G.711, PCM of voice frequencies.

ISDN

Integrated Services Digital Network (ISDN) is a digital system that allows voice and data to be transmitted simultaneously using end-to-end digital connectivity. ISDN allows multiple digital channels to be transmitted simultaneously over the same wiring infrastructure used for analog lines. Two kinds of channels are defined in ISDN. The B channel or bearer channel carries user traffic, whereas the D channel or data channel carries CCS signaling data. The bandwidths of B channels are 64 kbps. Some switches limit B channels to a capacity of 56 kbps. The D channel handles signaling at 16 kbps or 64 kbps, depending on the service type. Original recommendations of ISDN were in Consultative Committee for International Telegraph and Telephone (CCITT) Recommendation I.120 (1984), which described some initial guidelines for implementing ISDN. As regards ISDN in North America, members of the industry agreed to create the National ISDN 1 (NI-1) standard as an interoperable ISDN standard. A more comprehensive standardization initiative, National ISDN 2 (NI-2), was later adopted. Two basic types of ISDN services are offered: basic rate interface (BRI) and primary rate interface (PRI).

ISDN BRI

ISDN BRI (2B+D) consists of two 64-kbps B channels and one 16-kbps D channel for a total of 144 kbps. BRI service is designed to meet the needs of most individual users. BRI ISDN also uses a channel-aggregation protocol, such as BONDING or Multilink PPP, that supports an uncompressed data transfer speed of 128 kbps, plus bandwidth for overhead and signaling.

As illustrated in Figure 2-18, the U interface is a two-wire (single-pair) interface from the ISDN switch, the same physical interface provided for plain old telephone service (POTS) lines. It supports full-duplex data transfer over a single pair. Echo cancellation is used to reduce noise, and data-encoding schemes, such as 2 binary 1 quaternary (2B1Q) in North America and 4B3T in Europe, permit a relatively high data rate of 160 kbps over ordinary single-pair local loops.

The U interface is terminated with a network termination 1 (NT-1) device at the CPE end. North American carriers provide customers with a choice of U or S/T interfaces. EMEA and Asia Pacific phone companies supply NT1s, thereby providing their customers with an S/T interface. The ISDN NT-1 converts the two-wire U interface into the four-wire S/T interface. The S/T interface supports up to seven devices on the full-duplex S/T bus. The BRI NT-1 provides timing, multiplexing of the B and D channels, and power conversion.

Figure 2-18 *ISDN Basic Rate Interface*

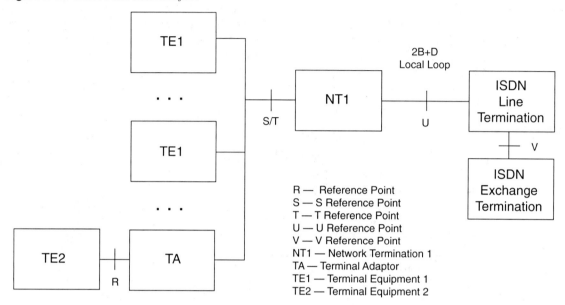

Devices that connect to the S/T interface include ISDN-capable telephones, videoconferencing equipment, routers, and terminal adapters. All devices that are designed for ISDN are designated terminal equipment 1 (TE-1). All other communication devices that are not ISDN capable, but have an asynchronous serial (EIA-232) or POTS telephone interface—including ordinary analog telephones, modems, and terminals—are designated terminal equipment 2 (TE-2). A terminal adapter (TA) connects a TE-2 to the ISDN S/T bus. ISDN services can be deployed as OPX services by carriers and service providers that operate carrier class 5 switches capable of ISDN PRI and BRI services. There are local loop distance limitations of 18,000 feet (5.5 km) of the CO point of presence (POP) for BRI service. Repeater devices are required for distances exceeding these guidelines.

ISDN PRI

ISDN PRI service is offered as T1/PRI or E1/PRI. T1/PRI (23B+D) has a channel structure that is 23 B channels plus one 64-kbps D channel for a total of 1536 kbps. In EMEA and the Asia Pacific, E1/PRI (30B+D) consists of 30 B channels plus one 64-kbps D channel for a total of 1984 kbps. It is also possible to support multiple PRI lines with one 64-kbps D channel using NFAS. H channels provide a way to aggregate B channels. They are implemented as follows:

- H0 = 384 kbps (6 B channels)
- H10 = 1472 kbps (23 B channels)

- H11 = 1536 kbps (24 B channels)
- H12 = 1920 kbps (30 B channels)

ISDN PRI services are offered over a two-pair T1/PRI or E1/PRI unbalanced facility. As shown in Figure 2-19, in the case of ISDN PRI, the NT-1 is a CSU/DSU-like device, whereas the NT-2 devices provide customer premises switching, multiplexing, or other forms of concentration. If a device performs NT-1 and NT-2 functions, it might be referred to as an NT-12. The NT-2 device converts the T interface into the S interface. The ISDN S and T interfaces are electrically equivalent. The NT-2 communicates with terminal equipment, and handles the Layer 2 and 3 ISDN protocols. The U interface local loop connects to ISDN line-termination equipment that provides the LT function. The connection between switches within the phone network is called exchange termination (ET). The LT and ET functions communicate via the V interface.

Figure 2-19 *ISDN Primary Rate Interface*

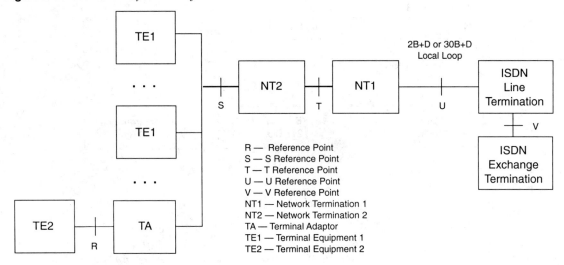

ISDN Layer 1

The ITU I-Series and G-Series documents specify the ISDN physical layer. Echo cancellation is used to reduce noise, and data encoding schemes, such as 2B1Q and 4B3T, are used to encode data.

As illustrated in Figure 2-20, 2B1Q is the most common signaling method on U interfaces. In this method, each pair of binary digits represents four discrete amplitude and polarity values. This protocol is defined in detail in ANSI spec T1.601. In summary, 2B1Q provides 2 bits per baud, which results in 80-kilo baud (baud = one modulation per second) or a transfer rate of 160 kbps. This means that the input voltage level can be one of four distinct levels. These levels are called quaternaries. Each quaternary represents 2 data bits, because there are 4 possible ways to represent 2 bits, as shown in Figure 2-18. Each U interface frame is 240 bits long. At the prescribed data rate of 160 kbps, each frame is therefore 1.5 ms long. Each frame consists of a 16-kbps frame overhead, 16-kbps D channel, and two B channels at 64-kbps each.

Figure 2-20 *ISDN Layer 1*

Bits	Quaternary Symbol	Voltage Level
00	-3	-2.5
01	-1	-0.833
02	+3	+2.5
11	+1	+0.833

Quaternary symbols

Sync 18 Bits	12 * (B_1 + B_2 + D) 216 Bits	Maintenance 6 Bits

ISDN Frame Format

The Sync field consists of 9 quaternaries (2 bits each) in the quaternary symbolic pattern +3 +3 −3 −3 −3 +3 −3 +3 −3. The (B1 + B2 + D) represent 18 bits of data consisting of 8 bits from the first B channel, 8 bits from the second B channel, and 2 bits of D-channel data. The Maintenance field contains CRC information, block error detection flags, and embedded operator commands used for loopback testing without disrupting user data. Data is transmitted in a superframe consisting of 8 * 240-bit frames for a total of 1920 bits (240 octets). The Sync field of the first frame in the superframe is inverted (−3 −3 +3 +3 +3 −3 +3 −3 +3).

ISDN Layer 2

The ISDN data link layer is specified by the ITU Q-Series documents Q.920 through Q.923. All the signaling on the D channel is defined in the Q.921 spec. ISDN uses the Link Access Protocol - D channel (LAP-D) as its Layer 2 protocol. LAP-D is almost identical to the X.25 LAP-B protocol. Figure 2-21 shows the LAP-D frame format.

Figure 2-21 *ISDN Layer 2*

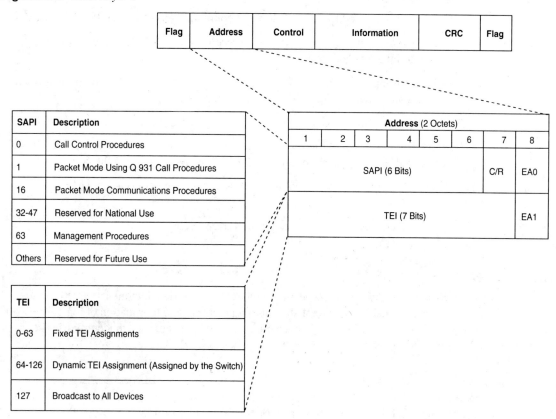

The Start Flag field is 1 octet long and its value is always 7E (hex) or 0111 1110 (binary). The Control field is 2 octets long and indicates the frame type (information, supervisory, or unnumbered) and sequence numbers (N(r) and N(s)). The Information field contains Layer 3 protocol information and user data. The CRC field is a 2-octet field that provides cyclic redundancy checks for bit errors on the user data. The End Flag field is also 1 octet long and its value is always set to 7E (hex) or 0111 1110 (binary).

The Address field contains the Service Access Point Identifier (SAPI) subfield, which is 6 bits wide; the C/R (command/response) bit, which indicates whether the frame is a command or a response; the EA0 (address extension) bit, which indicates whether this is the final octet of the address or not; the TEI (terminal endpoint identifier) 7-bit device identifier; and the EA1 (address extension) bit, which is similar to the EA0.

As detailed in Figure 2-21, the Service Access Point Identifier (SAPI) is a 6-bit field that identifies the point where Layer 2 provides a service to Layer 3. Terminal endpoint identifiers (TEIs) are unique IDs given to each device (TE) on an ISDN S/T bus. This identifier value could be dynamic, or the value can be assigned statically when the TE is installed.

ISDN Link-Layer Establishment

The following steps are used to establish Layer 2 communication between ISDN devices:

1. The TE and the network initially exchange receive ready (RR) frames, listening for someone to initiate a connection.
2. The TE sends an unnumbered information (UI) frame with a SAPI of 63 (management procedure, query network) and TEI of 127 (broadcast).
3. The network assigns an available TEI (in the range 64 to 126).
4. The TE sends a set asynchronous balanced mode (SABME) frame with a SAPI of 0 (call control, used to initiate a setup) and a TEI of the value assigned by the network.
5. The network responds with an unnumbered acknowledgement (UA), SAPI = 0, TEI = assigned.
6. The Layer 2 connection is now ready for a Layer 3 setup.

ISDN Layer 3

The ISDN network layer is also specified by the ITU Q-Series documents Q.930 through Q.939. Layer 3 is used for the establishment, maintenance, and termination of logical network connections between two devices. Service profile IDs (SPIDs) are used to identify what services and features the ISDN switch provides to the attached ISDN device.

NOTE The reader must not confuse the ISDN Layer 3 with Layer 3 of the OSI model. Protocols, such as ISDN and X.25, have their own Layer 3. Network layer protocols, such as IP, perceive such protocol stacks as the data link layer.

SPIDs are accessed at device initialization prior to call setup. The SPID is usually the 10-digit phone number of the ISDN line along with a prefix or suffix. The suffix is also known as a tag identifier (TID). SPIDs are used to identify features on the line, but in reality they can be whatever the carrier decides the value(s) should be. If an ISDN line requires a SPID, but it is not correctly supplied, Layer 2 initialization will take place, but Layer 3 will not, and the device will not be able to place or accept calls. ITU spec Q.932 provides greater details on SPIDs.

The Information field is a variable-length field that contains the Q.931 protocol data. Figure 2-22 describes the various subfields contained in the Information field. The following fields are contained in the Q.931 header:

- **Protocol Discriminator (1 octet)**—Identifies the Layer 3 protocol. If this is a Q.931 header, this value is always 08_{16}.
- **Length (1 octet)**—Indicates the length of the next field, the CRV.

- **Call Reference Value (CRV) (1 or 2 octets)**—Used to uniquely identify each call on the user-network interface. This value is assigned at call setup, and this value becomes available for another call when the call is cleared.
- **Message Type (1 octet)**—Identifies the message type (setup, connect, and so forth). This determines what additional information is required and allowed.
- **Mandatory and Optional Information Elements (variable length)**—Are options that are set depending on the message type.

Figure 2-22 *ISDN Layer 3*

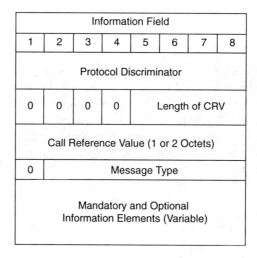

ISDN Call Setup

The following steps are used to establish ISDN calls from an ISDN Layer 3 perspective:

1 Caller sends a setup to the ISDN switch.

2 If the setup is okay, the switch sends a call proceeding to the caller, and a setup to the receiver.

3 The receiver gets the setup. If it is okay, it sends an alerting message to the switch.

4 The switch forwards the alerting message to the caller.

5 When the receiver answers the call, is sends a connect message to the switch.

6 The switch forwards the connect message to the caller.

7 The caller sends a connect acknowledge message to the switch.

8 The switch forwards the connect ack message to the receiver.

9 The call is now set up.

TDM Network Elements

A variety of TDM-based network elements are used to build TDM systems. Some of these elements are discussed in this section. Common handoff to optical systems takes place at the DS1/DS3 levels in the case of the T-carrier, and E1/E3 levels in the case of the E-carrier. Note that the various individual network elements presented in this section, such as repeaters, CSUs/DSUs, DACS, and channel banks, are commercially available as integrated units supporting a wide variety of low- and high-speed interfaces, encoding, signaling, and protocols. The TDM network elements integrate to form the digital loop carrier (DLC) supporting various TDM architectures and topologies.

Repeaters

Repeaters are four-wire T1/E1 unbalanced amplifiers and signal processors for use on T1 or E1 lines. Repeaters are used to extend in-house T1/E1 lines in campus and high-rise environments. Repeaters can also be used to extend the distance between any T1/E1 equipment, such as DSUs, channel banks, and routers with built-in CSU/DSUs. A pair of repeaters can be located up to 5000 feet apart. Solid copper 22 AWG two-twisted-pair is the preferred cable for connection between repeaters. Smaller wire sizes will reduce the functional distance between the repeater pairs. Connection is made through RJ-45 modular connectors or four-wire, screw-down barrier strips. Both types of connectors are commonly used standards.

CSU/DSU

Channel service units/digital service units are essentially CPE multiplexers that can assign channels or time slots to a circuit. For example, a 256-kbps circuit will have four time slots assigned to it (N1 to N4). Each of these time slots is 64 kbps. Most CSUs/DSUs also support 56-kbps time slots. Some CSUs/DSUs are equipped with multiple ports. This enables the user to allocate time slots to each physical port that might be attached to routers or other CPE equipment. For example, a CSU/DSU connected to a 256-kbps line could assign N1 to port 1 and N2+N3+N4 to port 2. This would allocate 64-kbps bandwidth to the CPE device attached to port 1, and 192 kbps to the CPE device attached to port 2. Another function supported by CSUs/DSUs is the drop and insert function. Drop and insert is used to terminate one or more DS0 channels of a T1 at the digital RS-530/V.35 interface of the FT. One or more of the remaining DS0 channels can be passed on to other equipment, typically a system using voice lines. For example, a single 112-kbps channel (56 kbps * 2) might be dropped off to support a router, and up to 22 of the remaining DS0 channels passed on to a private branch exchange (PBX) for voice lines. CSU/DSU devices are regarded as a demarcation point by some carriers. In such a case, the carrier would own and manage the CSU/DSU, permitting them to perform loopback tests in the event of local loop circuit outages.

Digital Access and Cross-Connect Systems

The modern DACS is truly an integrated access device (IAD) that integrates channel bank cross-connect and multiplexer functionality in one device. DACS cross-connect functionality enables carriers to physically wire user circuits and electrically groom these 64-kbps voice or data circuits to higher T1 or E1 levels. The higher-level T1 or E1s can be groomed into DS3s or E3s for back-haul to a carrier class 5 switch. The time-slot interchange (digital cross-connect) functionality of a DACS enables you to assign DS0s to higher-level T1/E1 circuits in any order you want. It also enables you to assign the order of T1/E1s within a DS3/E3. A DACS also enables you to perform T1-to-E1 format conversion. Most DACSs support console, Telnet, and Simple Network Management Protocol (SNMP) for configuration, maintenance, performance monitoring, and administration.

Channel Bank

Channel banks are devices implemented at a CO (public exchange) that convert analog signals from home and business users into digital signals to be carried over higher-speed lines between the CO and other exchanges. The analog signal is converted into a digital signal that transmits at a 64-kbps rate. The 64-kbps signal is multiplexed with other DS0 signals on the same line using TDM techniques to higher T1/E1 levels. Channel banks offer foreign exchange office (FXO), foreign exchange subscriber (FXS), special access office (SAO), dial pulse originating (DPO), dial pulse terminating (DPT), equalized transmission only (ETO), transmission only (TO), and pulse link repeater (PLR) facilities.

Summary

TDM is a common way to send multiple signals or streams of information over a medium at the same time in the form of a single, complex signal, and then recover the separate signals at the receiving end. There are various stages involved with analog signal generation and reception. These include analog-to-digital conversion, signal filtering, sampling, quantization, encoding, and μ-law and A-law coding for PCM transmission. TDM signaling methods, such as CAS and CCS, are used in the T-carrier as well as E-carrier systems. The North American T-carrier supports 24 * 64 kbps in DS single-framing format and also supports D4 and ESF multiframing formats. The EMEA, Asia Pacific, and Latin American E-carrier supports 32 * 64 kbps E1 FAS as well as E1 MFAS and E1 CRC error checking.

ISDN is a digital system that allows voice and data to be transmitted simultaneously using end-to-end digital connectivity. ISDN allows multiple digital channels to be transmitted simultaneously over the same wiring infrastructure used for analog lines. Two kinds of channels are defined in ISDN. The B channel or bearer channel carries user traffic, whereas the D channel or data channel carries CCS signaling data.

Various TDM network elements are used to build a TDM infrastructure. These include repeaters, CSUs/DSUs, digital access and cross-connect systems (DACS), and channel banks. The TDM network elements integrate to form the DLC, which supports various TDM architectures and topologies.

This chapter includes the following sections:

- **A Brief History of Fiber-Optic Communications**—This section discusses the history of fiber optics, from the optical semaphore telegraph to the invention of the first clad glass fiber invented by Abraham Van Heel. Today more than 80 percent of the world's long-distance voice and data traffic is carried over optical-fiber cables.

- **Fiber-Optic Applications**—Telecommunications applications of fiber-optic cable are widespread, ranging from global networks to desktop computers.

- **The Physics Behind Fiber Optics**—This section discusses the physics behind the operation of fiber-optic cables.

- **Optical-Cable Construction**—This section discusses fiber-optic cable construction. Fiber-optic cables are constructed of three types of materials: glass, plastic, and plastic-clad silica (PCS).

- **Propagation Modes**—There are two main modes of fiber-optic propagation: multimode and single mode. These two modes perform differently with respect to both attenuation and chromatic dispersion.

- **Fiber-Optic Characteristics**—Fiber-optic system characteristics include linear and nonlinear characteristics. Linear characteristics include attenuation and interference. Nonlinear characteristics include single-phase modulation (SPM), cross-phase modulation (XPM), four-wave mixing (FWM), stimulated Raman scattering (SRS), and stimulated Brillouin scattering (SBS).

- **Fiber Types**—This section discusses various multimode and single-mode fiber types currently used for premise, metro, aerial, submarine, and long-haul applications.

- **Fiber-Optic Cable Termination**—Removable and reusable optical termination in the form of metal and plastic connectors plays a vital role in an optical system.

- **Splicing**—Seamless permanent or semipermanent optical connections require fibers to be spliced. Fiber-optic cables might have to be spliced together for a number of reasons.

- **Physical-Design Considerations**—When designing a fiber-optic cable plant, you must consider many factors. First and foremost, the designer must determine whether the cable is to be installed for an inside-plant (ISP) or outside-plant (OSP) application.

- **Fiber-Optic Communications System**—This section discusses the end-to-end fiber-optic system.

- **Fiber Span Analysis**—Optical loss, or total attenuation, is the sum of the losses of each individual component between the transmitter and receiver. Loss-budget analysis is the calculation and verification of a fiber-optic system's operating characteristics.

CHAPTER 3

Fiber-Optic Technologies

A Brief History of Fiber-Optic Communications

Optical communication systems date back to the 1790s, to the optical semaphore telegraph invented by French inventor Claude Chappe. In 1880, Alexander Graham Bell patented an optical telephone system, which he called the Photophone. However, his earlier invention, the telephone, was more practical and took tangible shape. The Photophone remained an experimental invention and never materialized. During the 1920s, John Logie Baird in England and Clarence W. Hansell in the United States patented the idea of using arrays of hollow pipes or transparent rods to transmit images for television or facsimile systems.

In 1954, Dutch scientist Abraham Van Heel and British scientist Harold H. Hopkins separately wrote papers on imaging bundles. Hopkins reported on imaging bundles of unclad fibers, whereas Van Heel reported on simple bundles of clad fibers. Van Heel covered a bare fiber with a transparent cladding of a lower refractive index. This protected the fiber reflection surface from outside distortion and greatly reduced interference between fibers.

Abraham Van Heel is also notable for another contribution. Stimulated by a conversation with the American optical physicist Brian O'Brien, Van Heel made the crucial innovation of cladding fiber-optic cables. All earlier fibers developed were bare and lacked any form of cladding, with total internal reflection occurring at a glass-air interface. Abraham Van Heel covered a bare fiber or glass or plastic with a transparent cladding of lower refractive index. This protected the total reflection surface from contamination and greatly reduced cross talk between fibers. By 1960, glass-clad fibers had attenuation of about 1 decibel (dB) per meter, fine for medical imaging, but much too high for communications. In 1961, Elias Snitzer of American Optical published a theoretical description of a fiber with a core so small it could carry light with only one waveguide mode. Snitzer's proposal was acceptable for a medical instrument looking inside the human, but the fiber had a light loss of 1 dB per meter. Communication devices needed to operate over much longer distances and required a light loss of no more than 10 or 20 dB per kilometer.

By 1964, a critical and theoretical specification was identified by Dr. Charles K. Kao for long-range communication devices, the 10 or 20 dB of light loss per kilometer standard. Dr. Kao also illustrated the need for a purer form of glass to help reduce light loss.

In the summer of 1970, one team of researchers began experimenting with fused silica, a material capable of extreme purity with a high melting point and a low refractive index.

Corning Glass researchers Robert Maurer, Donald Keck, and Peter Schultz invented fiber-optic wire or "optical waveguide fibers" (patent no. 3,711,262), which was capable of carrying 65,000 times more information than copper wire, through which information carried by a pattern of light waves could be decoded at a destination even a thousand miles away. The team had solved the decibel-loss problem presented by Dr. Kao. The team had developed an SMF with loss of 17 dB/km at 633 nm by doping titanium into the fiber core. By June of 1972, Robert Maurer, Donald Keck, and Peter Schultz invented multimode germanium-doped fiber with a loss of 4 dB per kilometer and much greater strength than titanium-doped fiber. By 1973, John MacChesney developed a modified chemical vapor-deposition process for fiber manufacture at Bell Labs. This process spearheaded the commercial manufacture of fiber-optic cable.

In April 1977, General Telephone and Electronics tested and deployed the world's first live telephone traffic through a fiber-optic system running at 6 Mbps, in Long Beach, California. They were soon followed by Bell in May 1977, with an optical telephone communication system installed in the downtown Chicago area, covering a distance of 1.5 miles (2.4 kilometers). Each optical-fiber pair carried the equivalent of 672 voice channels and was equivalent to a DS3 circuit. Today more than 80 percent of the world's long-distance voice and data traffic is carried over optical-fiber cables.

Fiber-Optic Applications

The use and demand for optical fiber has grown tremendously and optical-fiber applications are numerous. Telecommunication applications are widespread, ranging from global networks to desktop computers. These involve the transmission of voice, data, or video over distances of less than a meter to hundreds of kilometers, using one of a few standard fiber designs in one of several cable designs.

Carriers use optical fiber to carry plain old telephone service (POTS) across their nationwide networks. Local exchange carriers (LECs) use fiber to carry this same service between central office switches at local levels, and sometimes as far as the neighborhood or individual home (fiber to the home [FTTH]).

Optical fiber is also used extensively for transmission of data. Multinational firms need secure, reliable systems to transfer data and financial information between buildings to the desktop terminals or computers and to transfer data around the world. Cable television companies also use fiber for delivery of digital video and data services. The high bandwidth provided by fiber makes it the perfect choice for transmitting broadband signals, such as high-definition television (HDTV) telecasts.

Intelligent transportation systems, such as smart highways with intelligent traffic lights, automated tollbooths, and changeable message signs, also use fiber-optic-based telemetry systems.

Another important application for optical fiber is the biomedical industry. Fiber-optic systems are used in most modern telemedicine devices for transmission of digital diagnostic images. Other applications for optical fiber include space, military, automotive, and the industrial sector.

The Physics Behind Fiber Optics

A fiber-optic cable is composed of two concentric layers, called the core and the cladding, as illustrated in Figure 3-1. The core and cladding have different refractive indices, with the core having a refractive index of n1, and the cladding having a refractive index of n2. The index of refraction is a way of measuring the speed of light in a material. Light travels fastest in a vacuum. The actual speed of light in a vacuum is 300,000 kilometers per second, or 186,000 miles per second.

Figure 3-1 *Cross Section of a Fiber-Optic Cable*

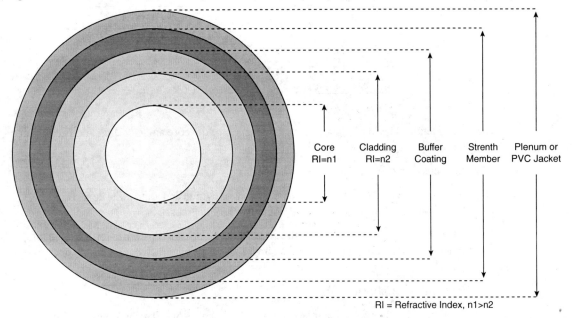

The index of refraction is calculated by dividing the speed of light in a vacuum by the speed of light in another medium, as shown in the following formula:

Refractive index of the medium = [Speed of light in a vacuum/Speed of light in the medium]

The refractive index of the core, n1, is always greater than the index of the cladding, n2. Light is guided through the core, and the fiber acts as an optical waveguide.

Figure 3-2 shows the propagation of light down the fiber-optic cable using the principle of total internal reflection. As illustrated, a light ray is injected into the fiber-optic cable on the left. If the light ray is injected and strikes the core-to-cladding interface at an angle greater than the critical angle with respect to the normal axis, it is reflected back into the core. Because the angle of incidence is always equal to the angle of reflection, the reflected light continues to be reflected. The light ray then continues bouncing down the length of the fiber-optic cable. If the angle of incidence at the core-to-cladding interface is less than the critical angle, both reflection and refraction take place. Because of refraction at each incidence on the interface, the light beam attenuates and dies off over a certain distance.

Figure 3-2 *Total Internal Reflection*

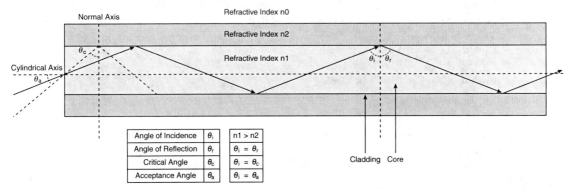

The critical angle is fixed by the indices of refraction of the core and cladding and is computed using the following formula:

$$\theta_c = \cos^{-1}(n2/n1)$$

The critical angle can be measured from the normal or cylindrical axis of the core. If $n1 = 1.557$ and $n2 = 1.343$, for example, the critical angle is 30.39 degrees.

Figure 3-2 shows a light ray entering the core from the outside air to the left of the cable. Light must enter the core from the air at an angle less than an entity known as the acceptance angle (θ_a):

$$\theta_a = \sin^{-1}[(n1/n0)\sin(\theta_c)]$$

In the formula, n0 is the refractive index of air and is equal to one. This angle is measured from the cylindrical axis of the core. In the preceding example, the acceptance angle is 51.96 degrees.

The optical fiber also has a numerical aperture (NA). The NA is given by the following formula:

$$NA = \sin\theta_a = \sqrt{(n1^2 - n2^2)}$$

From a three-dimensional perspective, to ensure that the signals reflect and travel correctly through the core, the light must enter the core through an acceptance cone derived by rotating the acceptance angle about the cylindrical fiber axis. As illustrated in Figure 3-3, the size of the acceptance cone is a function of the refractive index difference between the core and the cladding. There is a maximum angle from the fiber axis at which light can enter the fiber so that it will propagate, or travel, in the core of the fiber. The sine of this maximum angle is the NA of the fiber. The NA in the preceding example is 0.787. Fiber with a larger NA requires less precision to splice and work with than fiber with a smaller NA. Single-mode fiber has a smaller NA than MMF.

Figure 3-3 *Acceptance Cone*

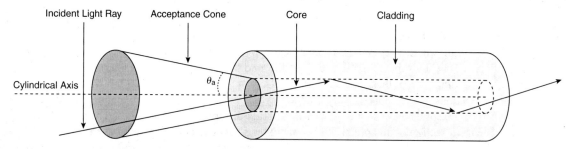

Performance Considerations

The amount of light that can be coupled into the core through the external acceptance angle is directly proportional to the efficiency of the fiber-optic cable. The greater the amount of light that can be coupled into the core, the lower the bit error rate (BER), because more light reaches the receiver. The attenuation a light ray experiences in propagating down the core is inversely proportional to the efficiency of the optical cable because the lower the attenuation in propagating down the core, the lower the BER. This is because more light reaches the receiver. Also, the less chromatic dispersion realized in propagating down the core, the faster the signaling rate and the higher the end-to-end data rate from source to destination. The major factors that affect performance considerations described in this paragraph are the size of the fiber, the composition of the fiber, and the mode of propagation.

Optical-Power Measurement

The power level in optical communications is of too wide a range to express on a linear scale. A logarithmic scale known as decibel (dB) is used to express power in optical communications.

The wide range of power values makes decibel a convenient unit to express the power levels that are associated with an optical system. The gain of an amplifier or attenuation in fiber is expressed in decibels. The decibel does not give a magnitude of power, but it is a ratio of the output power to the input power.

$$\text{Loss or gain} = 10\log 10(\text{POUTPUT}/\text{PINPUT})$$

The decibel milliwatt (dBm) is the power level related to 1 milliwatt (mW). Transmitter power and receiver dynamic ranges are measured in dBm. A 1-mW signal has a level of 0 dBm.

Signals weaker than 1 mW have negative dBm values, whereas signals stronger than 1 mW have positive dBm values.

$$\text{dBm} = 10\log 10(\text{Power(mW)}/1(\text{mW}))$$

Optical-Cable Construction

The core is the highly refractive central region of an optical fiber through which light is transmitted. The standard telecommunications core diameter in use with SMF is between 8 μm and 10 μm, whereas the standard core diameter in use with MMF is between 50 μm and 62.5 μm. Figure 3-4 shows the core diameter for SMF and MMF cable. The diameter of the cladding surrounding each of these cores is 125 μm. Core sizes of 85 μm and 100 μm were used in early applications, but are not typically used today. The core and cladding are manufactured together as a single solid component of glass with slightly different compositions and refractive indices. The third section of an optical fiber is the outer protective coating known as the *coating*. The coating is typically an ultraviolet (UV) light-cured acrylate applied during the manufacturing process to provide physical and environmental protection for the fiber. The buffer coating could also be constructed out of one or more layers of polymer, nonporous hard elastomers or high-performance PVC materials. The coating does not have any optical properties that might affect the propagation of light within the fiber-optic cable. During the installation process, this coating is stripped away from the cladding to allow proper termination to an optical transmission system. The coating size can vary, but the standard sizes are 250 μm and 900 μm. The 250-μm coating takes less space in larger outdoor cables. The 900-μm coating is larger and more suitable for smaller indoor cables.

Figure 3-4 *Optical-Cable Construction*

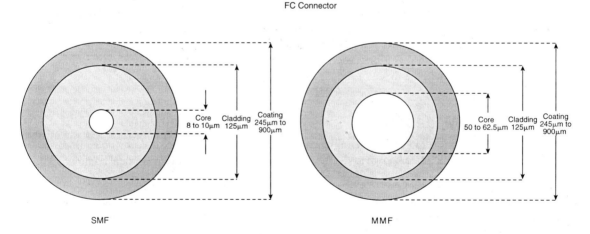

Fiber-optic cable sizes are usually expressed by first giving the core size followed by the cladding size. Consequently, 50/125 indicates a core diameter of 50 microns and a cladding diameter of 125 microns, and 8/125 indicates a core diameter of 8 microns and a cladding diameter of 125 microns. The larger the core, the more light can be coupled into it from the external acceptance angle cone. However, larger-diameter cores can actually allow in too much light, which can cause receiver saturation problems. The 8/125 cable is often used when a fiber-optic data link operates with single-mode propagation, whereas the 62.5/125 cable is often used in a fiber-optic data link that operates with multimode propagation.

Three types of material make up fiber-optic cables:

- Glass
- Plastic
- Plastic-clad silica (PCS)

These three cable types differ with respect to attenuation. Attenuation is principally caused by two physical effects: absorption and scattering. Absorption removes signal energy in the interaction between the propagating light (photons) and molecules in the core. Scattering redirects light out of the core to the cladding. When attenuation for a fiber-optic cable is dealt with quantitatively, it is referenced for operation at a particular optical wavelength, a window, where it is minimized. The most common peak wavelengths are 780 nm, 850 nm, 1310 nm, 1550 nm, and 1625 nm. The 850-nm region is referred to as the *first window* (as it was used initially because it supported the original LED and detector technology). The 1310-nm region is referred to as the *second window*, and the 1550-nm region is referred to as the *third window*.

Glass Fiber-Optic Cable

Glass fiber-optic cable has the lowest attenuation. A pure-glass, fiber-optic cable has a glass core and a glass cladding. This cable type has, by far, the most widespread use. It has been the most popular with link installers, and it is the type of cable with which installers have the most experience. The glass used in a fiber-optic cable is ultra-pure, ultra-transparent, silicon dioxide, or fused quartz. During the glass fiber-optic cable fabrication process, impurities are purposely added to the pure glass to obtain the desired indices of refraction needed to guide light. Germanium, titanium, or phosphorous is added to increase the index of refraction. Boron or fluorine is added to decrease the index of refraction. Other impurities might somehow remain in the glass cable after fabrication. These residual impurities can increase the attenuation by either scattering or absorbing light.

Plastic Fiber-Optic Cable

Plastic fiber-optic cable has the highest attenuation among the three types of cable. Plastic fiber-optic cable has a plastic core and cladding. This fiber-optic cable is quite thick. Typical dimensions are 480/500, 735/750, and 980/1000. The core generally consists of polymethylmethacrylate (PMMA) coated with a fluropolymer. Plastic fiber-optic cable was pioneered principally for use in the automotive industry. The higher attenuation relative to glass might not be a serious obstacle with the short cable runs often required in premise data networks. The cost advantage of plastic fiber-optic cable is of interest to network architects when they are faced with budget decisions. Plastic fiber-optic cable does have a problem with flammability. Because of this, it might not be appropriate for certain environments and care has to be taken when it is run through a plenum. Otherwise, plastic fiber is considered extremely rugged with a tight bend radius and the capability to withstand abuse.

Plastic-Clad Silica (PCS) Fiber-Optic Cable

The attenuation of PCS fiber-optic cable falls between that of glass and plastic. PCS fiber-optic cable has a glass core, which is often vitreous silica, and the cladding is plastic, usually a silicone elastomer with a lower refractive index. PCS fabricated with a silicone elastomer cladding suffers from three major defects. First, it has considerable plasticity, which makes connector application difficult. Second, adhesive bonding is not possible. And third, it is practically insoluble in organic solvents. These three factors keep this type of fiber-optic cable from being particularly popular with link installers. However, some improvements have been made in recent years.

NOTE For data center premise cables, the jacket color depends on the fiber type in the cable. For cables containing SMFs, the jacket color is typically yellow, whereas for cables containing MMFs, the jacket color is typically orange. For outside plant cables, the standard jacket color is typically black.

Multifiber Cable Systems

Multifiber systems are constructed with strength members that resist crushing during cable pulling and bends. The outer cable jackets are OFNR (riser rated), OFNP (plenum rated), or LSZH (low-smoke, zero-halogen rated). The OFNR outer jackets are composed of flame-retardant PVC or fluoropolymers. The OFNP jackets are composed of plenum PVC, whereas the LSZH jackets are halogen-free and constructed out of polyolefin compounds. Figure 3-5 shows a multiribbon, 24-fiber, ribbon-cable system. Ribbon cables are extensively used for inside plant and datacenter applications. Individual ribbon subunit cables use the MTP/MPO connector assemblies. Ribbon cables have a flat ribbon-like structure that enables installers to save conduit space as they install more cables in a particular conduit.

Figure 3-5 *Inside Plant Ribbon-Cable System*

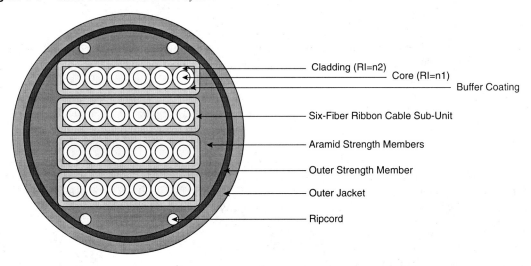

Figure 3-6 shows a typical six-fiber, inside-plant cable system. The central core is composed of a dielectric strength member with a dielectric jacket. The individual fibers are positioned around the dielectric strength member. The individual fibers have a strippable buffer coating. Typically, the strippable buffer is a 900-μm tight buffer. Each individual coated fiber is surrounded with a subunit jacket. Aramid yarn strength members surround the individual subunits. Some cable systems have an outer strength member that provides protection to the entire enclosed fiber system. Kevlar is a typical material used for constructing the outer strength member for premise cable systems. The outer jacket is OFNP, OFNR, or LSZH.

Figure 3-6 *Cross Section of Inside-Plant Cables*

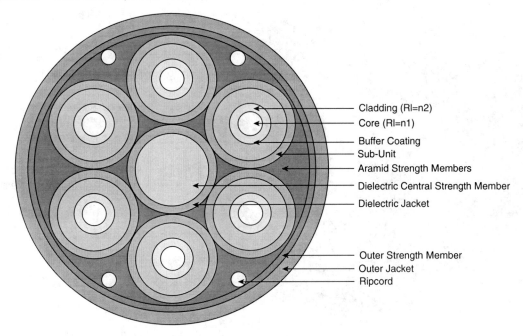

Figure 3-7 shows a typical armored outside-plant cable system. The central core is composed of a dielectric with a dielectric jacket or steel strength member. The individual gel-filled subunit buffer tubes are positioned around the central strength member. Within the subunit buffer tube, six fibers are positioned around an optional dielectric strength member. The individual fibers have a strippable buffer coating. All six subunit buffer tubes are enclosed within a binder that contains an interstitial filling or water-blocking compound. An outer strength member, typically constructed of aramid Kevlar strength members encloses the binder. The outer strength member is surrounded by an inner medium-density polyethylene (MDPE) jacket. The corrugated steel armor layer between the outer high-density polyethylene (HDPE) jacket, and the inner MDPE jacket acts as an external strength member and provides physical protection. Conventional deep-water submarine cables use dual armor and a special hermetically sealed copper tube to protect the fibers from the effects of deep-water environments. However, shallow-water applications use cables similar to those shown in Figure 3-7 with an asphalt compound interstitial filling.

Figure 3-7 *Cross Section of an Armored Outside-Plant Cable*

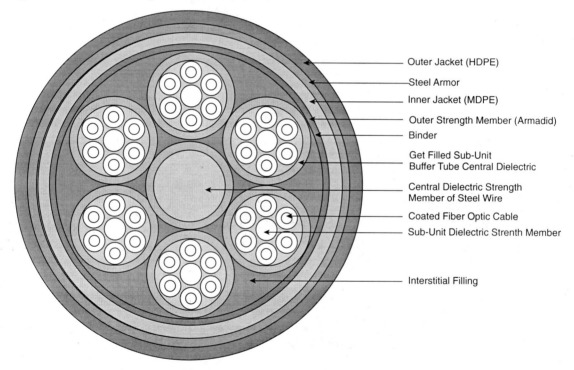

Propagation Modes

Fiber-optic cable has two propagation modes: multimode and single mode. They perform differently with respect to both attenuation and time dispersion. The single-mode fiber-optic cable provides much better performance with lower attenuation. To understand the difference between these types, you must understand what is meant by "mode of propagation."

Light has a dual nature and can be viewed as either a wave phenomenon or a particle phenomenon that includes photons and solitons. Solitons are special localized waves that exhibit particle-like behavior. For this discussion, let's consider the wave mechanics of light. When the light wave is guided down a fiber-optic cable, it exhibits certain modes. These are variations in the intensity of the light, both over the cable cross section and down the cable length. These modes are actually numbered from lowest to highest. In a very simple sense, each of these modes can be thought of as a ray of light. For a given fiber-optic cable, the number of modes that exist depends on the dimensions of the cable and the variation of the indices of refraction of both core and cladding across the cross section. The various modes include multimode step index, single-mode step index, single-mode dual-step index, and multimode graded index.

Multimode Step Index

Consider the illustration in Figure 3-8. This diagram corresponds to multimode propagation with a refractive index profile that is called *step index*. As you can see, the diameter of the core is fairly large relative to the cladding. There is also a sharp discontinuity in the index of refraction as you go from core to cladding. As a result, when light enters the fiber-optic cable on the left, it propagates down toward the right in multiple rays or multiple modes. This yields the designation multimode. As indicated, the lowest-order mode travels straight down the center. It travels along the cylindrical axis of the core. The higher modes, represented by rays, bounce back and forth, going down the cable to the left. The higher the mode, the more bounces per unit distance down to the right.

Figure 3-8 *Multimode Step Index*

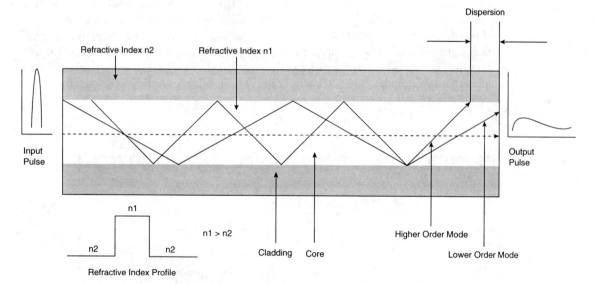

The illustration also shows the input pulse and the resulting output pulse. Note that the output pulse is significantly attenuated relative to the input pulse. It also suffers significant time dispersion. The reasons for this are as follows. The higher-order modes, the bouncing rays, tend to leak into the cladding as they propagate down the fiber-optic cable. They lose some of their energy into heat. This results in an attenuated output signal. The input pulse is split among the different rays that travel down the fiber-optic cable. The bouncing rays and the lowest-order mode, traveling down the center axis, are all traversing paths of different lengths from input to output. Consequently, they do not all reach the right end of the fiber-optic cable at the same time. When the output pulse is constructed from these separate ray components, the result is chromatic dispersion.

Fiber-optic cable that exhibits multimode propagation with a step index profile is thereby characterized as having higher attenuation and more time dispersion than the other propagation candidates. However, it is also the least costly and is widely used in the premises environment. It is especially attractive for link lengths up to 5 kilometers. It can be fabricated either from glass, plastic, or PCS. Usually, MMF core diameters are 50 or 62.5 μm. Typically, 50-μm MMF propagates only 300 modes as compared to 1100 modes for 62.5-μm fiber. The 50-μm MMF supports 1 Gbps at 850-nm wavelengths for distances up to 1 kilometer versus 275 meters for 62.5-μm MMF. Furthermore, 50-μm MMF supports 10 Gbps at 850-nm wavelengths for distances up to 300 meters versus 33 meters for 62.5-μm MMF. This makes 50-μm MMF the fiber of choice for low-cost, high-bandwidth campus and multitenant unit (MTU) applications.

Single-Mode Step Index

Single-mode propagation is illustrated in Figure 3-9. This diagram corresponds to single-mode propagation with a refractive index profile that is called *step index*. As the figure shows, the diameter of the core is fairly small relative to the cladding. Because of this, when light enters the fiber-optic cable on the left, it propagates down toward the right in just a single ray, a single mode, which is the lowest-order mode. In extremely simple terms, this lowest-order mode is confined to a thin cylinder around the axis of the core. The higher-order modes are absent.

Figure 3-9 *Single-Mode Step Index*

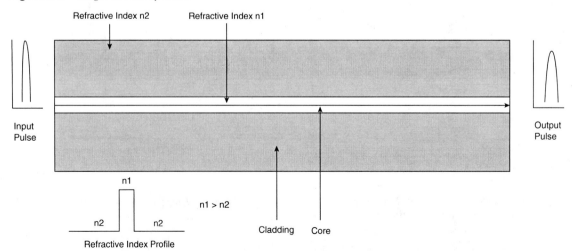

Consequently, extremely little or no energy is lost to heat through the leakage of the higher modes into the cladding, because they are not present. All energy is confined to this single, lowest-order mode. Because the higher-order mode energy is not lost, attenuation is not significant. Also, because the input signal is confined to a single ray path, that of the lowest-order mode, very little chromatic dispersion occurs. Single-mode propagation exists only above a certain specific wavelength called the *cutoff wavelength*.

The cutoff wavelength is the smallest operating wavelength when SMFs propagate only the fundamental mode. At this wavelength, the second-order mode becomes lossy and radiates out of the fiber core. As the operating wavelength becomes longer than the cutoff wavelength, the fundamental mode becomes increasingly lossy. The higher the operating wavelength is above the cutoff wavelength, the more power is transmitted through the fiber cladding. As the fundamental mode extends into the cladding material, it becomes increasingly sensitive to bending loss. Comparing the output pulse and the input pulse, note that there is little attenuation and time dispersion. Lower chromatic dispersion results in higher bandwidth. However, single-mode fiber-optic cable is also the most costly in the premises environment. For this reason, it has been used more with metropolitan- and wide-area networks than with premises data communications. Single-mode fiber-optic cable has also been getting increased attention as local-area networks have been extended to greater distances over corporate campuses. The core diameter for this type of fiber-optic cable is exceedingly small, ranging from 8 microns to 10 microns. The standard cladding diameter is 125 microns.

SMF step index fibers are manufactured using the outside vapor deposition (OVD) process. OVD fibers are made of a core and cladding, each with slightly different compositions and refractive indices. The OVD process produces consistent, controlled fiber profiles and geometry. Fiber consistency is important, to produce seamless spliced interconnections using fiber-optic cable from different manufacturers. Single-mode fiber-optic cable is fabricated from silica glass. Because of the thickness of the core, plastic cannot be used to fabricate single-mode fiber-optic cable. Note that not all SMFs use a step index profile. Some SMF variants use a graded index method of construction to optimize performance at a particular wavelength or transmission band.

Single-Mode Dual-Step Index

These fibers are single-mode and have a dual cladding. Depressed-clad fiber is also known as *doubly clad fiber*. Figure 3-10 corresponds to single-mode propagation with a refractive index profile that is called *dual-step index*. A depressed-clad fiber has the advantage of very low macrobending losses. It also has two zero-dispersion points and low dispersion over a much wider wavelength range than a singly clad fiber. SMF depressed-clad fibers are manufactured using the inside vapor deposition (IVD) process. The IVD or modified chemical vapor deposition (MCVD) process produces what is called *depressed-clad fiber* because of the shape of its refractive index profile, with the index of the glass adjacent to the core depressed. Each cladding has a refractive index that is lower than that of the core. The inner cladding a the lower refractive index than the outer cladding.

Figure 3-10 *Single-Mode Dual-Step Index*

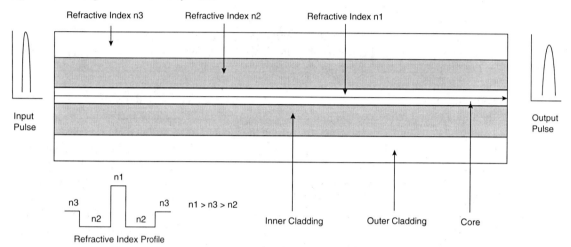

Multimode Graded Index

Multimode graded index fiber has a higher refractive index in the core that gradually reduces as it extends from the cylindrical axis outward. The core and cladding are essentially a single graded unit. Consider the illustration in Figure 3-11. This corresponds to multimode propagation with a refractive index profile that is called *graded index*. Here the variation of the index of refraction is gradual as it extends out from the axis of the core through the core to the cladding. There is no sharp discontinuity in the indices of refraction between core and cladding. The core here is much larger than in the single-mode step index case previously discussed. Multimode propagation exists with a graded index. As illustrated, however, the paths of the higher-order modes are somewhat confined. They appear to follow a series of ellipses. Because the higher-mode paths are confined, the attenuation through them due to leakage is more limited than with a step index. The time dispersion is more limited than with a step index; therefore, attenuation and time dispersion are present, but limited.

In Figure 3-11, the input pulse is shown on the left, and the resulting output pulse is shown on the right. When comparing the output pulse and the input pulse, note that there is some attenuation and time dispersion, but not nearly as much as with multimode step index fiber-optic cable.

Figure 3-11 *Multimode Graded Index*

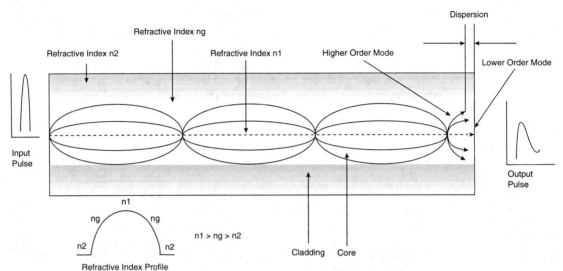

Fiber-optic cable that exhibits multimode propagation with a graded index profile is characterized as having levels of attenuation and time-dispersion properties that fall between the other two candidates. Likewise, its cost is somewhere between the other two candidates. Popular graded index fiber-optic cables have core diameters of 50, 62.5, and 85 microns. They have a cladding diameter of 125 microns—the same as single-mode fiber-optic cables. This type of fiber-optic cable is extremely popular in premise data communications applications. In particular, the 62.5/125 fiber-optic cable is the most popular and most widely used in these applications. Glass is generally used to fabricate multimode graded index fiber-optic cable.

Fiber-Optic Characteristics

Optical-fiber systems have many advantages over metallic-based communication systems. These advantages include interference, attenuation, and bandwidth characteristics. Furthermore, the relatively smaller cross section of fiber-optic cables allows room for substantial growth of the capacity in existing conduits. Fiber-optic characteristics can be classified as linear and nonlinear. Nonlinear characteristics are influenced by parameters, such as bit rates, channel spacing, and power levels.

Interference

Light signals traveling via a fiber-optic cable are immune from electromagnetic interference (EMI) and radio-frequency interference (RFI). Lightning and high-voltage interference is also eliminated. A fiber network is best for conditions in which EMI or RFI interference is heavy or safe operation free from sparks and static is a must. This desirable property of fiber-optic cable makes it the medium of choice in industrial and biomedical networks. It is also possible to place fiber cable into natural-gas pipelines and use the pipelines as the conduit.

Linear Characteristics

Linear characteristics include attenuation, chromatic dispersion (CD), polarization mode dispersion (PMD), and optical signal-to-noise ratio (OSNR).

Attenuation

Several factors can cause attenuation, but it is generally categorized as either intrinsic or extrinsic. Intrinsic attenuation is caused by substances inherently present in the fiber, whereas extrinsic attenuation is caused by external forces such as bending. The attenuation coefficient α is expressed in decibels per kilometer and represents the loss in decibels per kilometer of fiber.

Intrinsic Attenuation

Intrinsic attenuation results from materials inherent to the fiber. It is caused by impurities in the glass during the manufacturing process. As precise as manufacturing is, there is no way to eliminate all impurities. When a light signal hits an impurity in the fiber, one of two things occurs: It scatters or it is absorbed. Intrinsic loss can be further characterized by two components:

- Material absorption
- Rayleigh scattering

Material Absorption   Material absorption occurs as a result of the imperfection and impurities in the fiber. The most common impurity is the hydroxyl (OH-) molecule, which remains as a residue despite stringent manufacturing techniques. Figure 3-12 shows the variation of attenuation with wavelength measured over a group of fiber-optic cable material types. The three principal windows of operation include the 850-nm, 1310-nm, and 1550-nm wavelength bands. These correspond to wavelength regions in which attenuation is low and matched to the capability of a transmitter to generate light efficiently and a receiver to carry out detection.

Figure 3-12 *Attenuation Versus Wavelength*

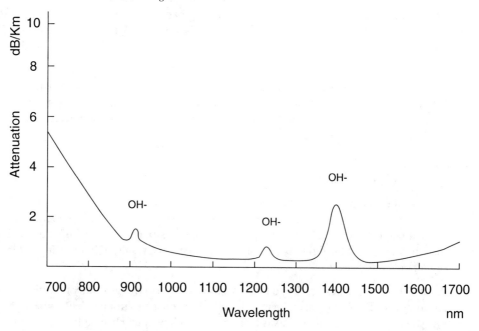

The OH- symbols indicate that at the 950-nm, 1380-nm, and 2730-nm wavelengths, the presence of hydroxyl radicals in the cable material causes an increase in attenuation. These radicals result from the presence of water remnants that enter the fiber-optic cable material through either a chemical reaction in the manufacturing process or as humidity in the environment. The variation of attenuation with wavelength due to the *water peak* for standard, single-mode fiber-optic cable occurs mainly around 1380 nm. Recent advances in manufacturing have overcome the 1380-nm water peak and have resulted in zero-water-peak fiber (ZWPF). Examples of these fibers include SMF-28e from Corning and the Furukawa-Lucent OFS AllWave. Absorption accounts for three percent to five percent of fiber attenuation. This phenomenon causes a light signal to be absorbed by natural impurities in the glass and converted to vibration energy or some other form of energy such as heat. Unlike scattering, absorption can be limited by controlling the amount of impurities during the manufacturing process. Because most fiber is extremely pure, the fiber does not heat up because of absorption.

Rayleigh Scattering As light travels in the core, it interacts with the silica molecules in the core. Rayleigh scattering is the result of these elastic collisions between the light wave and the silica molecules in the fiber. Rayleigh scattering accounts for about 96 percent of attenuation in optical fiber. If the scattered light maintains an angle that supports forward travel within the core, no attenuation occurs. If the light is scattered at an angle that does not support continued forward travel, however, the light is diverted out of the core and attenuation occurs. Depending on the incident angle, some portion of the light propagates forward and the other part deviates out of the propagation path and escapes from the fiber core. Some scattered light is reflected back toward the light source. This is a property that is used in an optical time domain reflectometer (OTDR) to test fibers. The same principle applies to analyzing loss associated with localized events in the fiber, such as splices.

Short wavelengths are scattered more than longer wavelengths. Any wavelength that is below 800 nm is unusable for optical communication because attenuation due to Rayleigh scattering is high. At the same time, propagation above 1700 nm is not possible due to high losses resulting from infrared absorption.

Extrinsic Attenuation

Extrinsic attenuation can be caused by two external mechanisms: macrobending or microbending. Both cause a reduction of optical power. If a bend is imposed on an optical fiber, strain is placed on the fiber along the region that is bent. The bending strain affects the refractive index and the critical angle of the light ray in that specific area. As a result, light traveling in the core can refract out, and loss occurs.

A macrobend is a large-scale bend that is visible, and the loss is generally reversible after bends are corrected. To prevent macrobends, all optical fiber has a minimum bend radius specification that should not be exceeded. This is a restriction on how much bend a fiber can withstand before experiencing problems in optical performance or mechanical reliability.

The second extrinsic cause of attenuation is a microbend. Microbending is caused by imperfections in the cylindrical geometry of fiber during the manufacturing process. Microbending might be related to temperature, tensile stress, or crushing force. Like macrobending, microbending causes a reduction of optical power in the glass. Microbending is very localized, and the bend might not be clearly visible on inspection. With bare fiber, microbending can be reversible.

Chromatic Dispersion

Chromatic dispersion is the spreading of a light pulse as it travels down a fiber. Light has a dual nature and can be considered from an electromagnetic wave as well as quantum perspective. This enables us to quantify it as waves as well as quantum particles. During the propagation of light, all of its spectral components propagate accordingly. These spectral components travel at different group velocities that lead to dispersion called *group velocity dispersion (GVD)*. Dispersion resulting from GVD is termed *chromatic dispersion* due to its wavelength dependence. The effect of chromatic dispersion is pulse spread.

As the pulses spread, or broaden, they tend to overlap and are no longer distinguishable by the receiver as 0s and 1s. Light pulses launched close together (high data rates) that spread too much (high dispersion) result in errors and loss of information. Chromatic dispersion occurs as a result of the range of wavelengths present in the light source. Light from lasers and LEDs consists of a range of wavelengths, each of which travels at a slightly different speed. Over distance, the varying wavelength speeds cause the light pulse to spread in time. This is of most importance in single-mode applications. Modal dispersion is significant in multimode applications, in which the various modes of light traveling down the fiber arrive at the receiver at different times, causing a spreading effect. Chromatic dispersion is common at all bit rates. Chromatic dispersion can be compensated for or mitigated through the use of dispersion-shifted fiber (DSF). DSF is fiber doped with impurities that have negative dispersion characteristics. Chromatic dispersion is measured in ps/nm-km. A 1-dB power margin is typically reserved to account for the effects of chromatic dispersion.

Polarization Mode Dispersion

Polarization mode dispersion (PMD) is caused by asymmetric distortions to the fiber from a perfect cylindrical geometry. The fiber is not truly a cylindrical waveguide, but it can be best described as an imperfect cylinder with physical dimensions that are not perfectly constant. The mechanical stress exerted upon the fiber due to extrinsically induced bends and stresses caused during cabling, deployment, and splicing as well as the imperfections resulting from the manufacturing process are the reasons for the variations in the cylindrical geometry.

Single-mode optical fiber and components support one fundamental mode, which consists of two orthogonal polarization modes. This asymmetry introduces small refractive index differences for the two polarization states. This characteristic is known as *birefringence*. Birefringence causes one polarization mode to travel faster than the other, resulting in a difference in the propagation time, which is called the *differential group delay (DGD)*. DGD is the unit that is used to describe PMD. DGD is typically measured in picoseconds. A fiber that acquires birefringence causes a propagating pulse to lose the balance between the polarization components. This leads to a stage in which different polarization components travel at different velocities, creating a pulse spread as shown in Figure 3-13. PMD can be classified as first-order PMD, also known as DGD, and second-order PMD (SOPMD). The SOPMD results from dispersion that occurs because of the signal's wavelength dependence and spectral width.

PMD is not an issue at low bit rates but becomes an issue at bit rates in excess of 5 Gbps. PMD is noticeable at high bit rates and is a significant source of impairment for ultra-long-haul systems. PMD compensation can be achieved by using PMD compensators that contain dispersion-maintaining fibers with degrees of birefringence in them. The introduced birefringence negates the effects of PMD over a length of transmission. For error-free transmission, PMD compensation is a useful technique for long-haul and metropolitan-area networks running at bit rates greater than 10 Gbps. Note in Figure 3-13 that the DGD is the difference between Z_1 and Z_2. The PMD value of the fiber is the mean value over time or frequency of the DGD and is represented as ps/$\sqrt{}$km. A 0.5-dB power margin is typically reserved to account for the effects of PMD at high bit rates.

Figure 3-13 *Polarization Mode Dispersion*

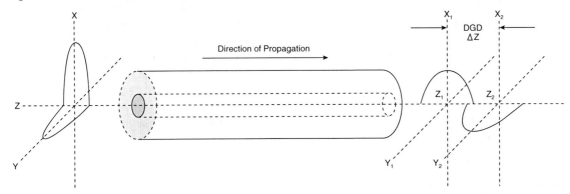

Polarization Dependent Loss

Polarization dependent loss (PDL) refers to the difference in the maximum and minimum variation in transmission or insertion loss of an optical device over all states of polarization (SOP) and is expressed in decibels. A typical PDL for a simple optical connector is less than .05 dB and varies from component to component. Typically, the PDL for an optical add/drop multiplexer (OADM) is around 0.3 dB. The complete polarization characterization of optical signals and components can be determined using an optical polarization analyzer.

Optical Signal-to-Noise Ratio

The optical signal-to-noise ratio (OSNR) specifies the ratio of the net signal power to the net noise power and thus identifies the quality of the signal. Attenuation can be compensated for by amplifying the optical signal. However, optical amplifiers amplify the signal as well as the noise. Over time and distance, the receivers cannot distinguish the signal from the noise, and the signal is completely lost. Regeneration helps mitigate these undesirable effects before they can render the system unusable and ensures that the signal can be detected at the receiver. Optical amplifiers add a certain amount of noise to the channel. Active devices, such as lasers, also add noise. Passive devices, such as taps and the fiber, can also add noise components. In the calculation of system design, however, optical amplifier noise is considered the predominant source for OSNR penalty and degradation.

OSNR is an important and fundamental system design consideration. Another parameter considered by designers is the Q-factor. The Q-factor, a function of the OSNR, provides a qualitative description of the receiver performance. The Q-factor suggests the minimum signal-to-noise ratio (SNR) required to obtain a specific BER for a given signal. OSNR is measured in decibels. The higher the bit rate, the higher the OSNR ratio required. For OC-192 transmissions, the OSNR should be at least 27 to 31 dB compared to 18 to 21 dB for OC-48.

Nonlinear Characteristics

Nonlinear characteristics include self-phase modulation (SPM), cross-phase modulation (XPM), four-wave mixing (FWM), stimulated Raman scattering (SRS), and stimulated Brillouin scattering (SBS).

Self-Phase Modulation

Phase modulation of an optical signal by itself is known as *self-phase modulation (SPM)*. SPM is primarily due to the self-modulation of the pulses. Generally, SPM occurs in single-wavelength systems. At high bit rates, however, SPM tends to cancel dispersion. SPM increases with high signal power levels. In fiber plant design, a strong input signal helps overcome linear attenuation and dispersion losses. However, consideration must be given to receiver saturation and to nonlinear effects such as SPM, which occurs with high signal levels. SPM results in phase shift and a nonlinear pulse spread. As the pulses spread, they tend to overlap and are no longer distinguishable by the receiver. The acceptable norm in system design to counter the SPM effect is to take into account a power penalty that can be assumed equal to the negative effect posed by XPM. A 0.5-dB power margin is typically reserved to account for the effects of SPM at high bit rates and power levels.

Cross-Phase Modulation

Cross-phase modulation (XPM) is a nonlinear effect that limits system performance in wavelength-division multiplexed (WDM) systems. XPM is the phase modulation of a signal caused by an adjacent signal within the same fiber. XPM is related to the combination (dispersion/effective area). CPM results from the different carrier frequencies of independent channels, including the associated phase shifts on one another. The induced phase shift is due to the *walkover* effect, whereby two pulses at different bit rates or with different group velocities walk across each other. As a result, the slower pulse sees the walkover and induces a phase shift. The total phase shift depends on the net power of all the channels and on the bit output of the channels. Maximum phase shift is produced when bits belonging to high-powered adjacent channels walk across each other.

XPM can be mitigated by carefully selecting unequal bit rates for adjacent WDM channels. XPM, in particular, is severe in long-haul WDM networks, and the acceptable norm in system design to counter this effect is to take into account a power penalty that can be assumed equal to the negative effect posed by XPM. A 0.5-dB power margin is typically reserved to account for the effects of XPM in WDM fiber systems.

Four-Wave Mixing

FWM can be compared to the intermodulation distortion in standard electrical systems. When three wavelengths ($\lambda1$, $\lambda2$, and $\lambda3$) interact in a nonlinear medium, they give rise to a fourth wavelength (λ4), which is formed by the scattering of the three incident photons, producing the fourth photon. This effect is known as *four-wave mixing (FWM)* and is a fiber-optic characteristic that affects WDM systems.

The effects of FWM are pronounced with decreased channel spacing of wavelengths and at high signal power levels. High chromatic dispersion also increases FWM effects. FWM also causes interchannel cross-talk effects for equally spaced WDM channels. FWM can be mitigated by using uneven channel spacing in WDM systems or nonzero dispersion-shifted fiber (NZDSF). A 0.5-dB power margin is typically reserved to account for the effects of FWM in WDM systems.

Stimulated Raman Scattering

When light propagates through a medium, the photons interact with silica molecules during propagation. The photons also interact with themselves and cause scattering effects, such as stimulated Raman scattering (SRS), in the forward and reverse directions of propagation along the fiber. This results in a sporadic distribution of energy in a random direction.

SRS refers to lower wavelengths pumping up the amplitude of higher wavelengths, which results in the higher wavelengths suppressing signals from the lower wavelengths. One way to mitigate the effects of SRS is to lower the input power. In SRS, a low-wavelength wave called *Stoke's wave* is generated due to the scattering of energy. This wave amplifies the higher wavelengths. The gain obtained by using such a wave forms the basis of Raman amplification. The Raman gain can extend most of the operating band (C- and L-band) for WDM networks. SRS is pronounced at high bit rates and high power levels. The margin design requirement to account for SRS/SBS is 0.5 dB.

Stimulated Brillouin Scattering

Stimulated Brillouin scattering (SBS) is due to the acoustic properties of photon interaction with the medium. When light propagates through a medium, the photons interact with silica molecules during propagation. The photons also interact with themselves and cause scattering effects such as SBS in the reverse direction of propagation along the fiber. In SBS, a low-wavelength wave called *Stoke's wave* is generated due to the scattering of energy. This wave amplifies the higher wavelengths. The gain obtained by using such a wave forms the basis of Brillouin amplification. The Brillouin gain peaks in a narrow peak near the C-band. SBS is pronounced at high bit rates and high power levels. The margin design requirement to account for SRS/SBS is 0.5 dB.

Fiber Types

This section discusses various MMF and SMF types currently used for premise, metro, aerial, submarine, and long-haul applications. The International Telecommunication Union (ITU-T), which is a global standardization body for telecommunication systems and vendors, has standardized various fiber types. These include the 50/125-μm graded index fiber (G.651), Nondispersion-shifted fiber (G.652), dispersion-shifted fiber (G.653), 1550-nm loss-minimized fiber (G.654), and NZDSF (G.655).

Multimode Fiber with a 50-Micron Core (ITU-T G.651)

The ITU-T G.651 is an MMF with a 50-μm nominal core diameter and a 125-μm nominal cladding diameter with a graded refractive index. The attenuation parameter for G.651 fiber is typically 0.8 dB/km at 1310 nm. The main application for ITU-T G.651 fiber is for short-reach optical transmission systems. This fiber is optimized for use in the 1300-nm band. It can also operate in the 850-nm band.

Nondispersion-Shifted Fiber (ITU-T G.652)

The ITU-T G.652 fiber is also known as standard SMF and is the most commonly deployed fiber. This fiber has a simple step-index structure and is optimized for operation in the 1310-nm band. It has a zero-dispersion wavelength at 1310 nm and can also operate in the 1550-nm band, but it is not optimized for this region. The typical chromatic dispersion at 1550 nm is high at 17 ps/nm-km. Dispersion compensation must be employed for high-bit-rate applications. The attenuation parameter for G.652 fiber is typically 0.2 dB/km at 1550 nm, and the PMD parameter is less than 0.1 ps/$\sqrt{}$km. An example of this type of fiber is Corning SMF-28.

Low Water Peak Nondispersion-Shifted Fiber (ITU-T G.652.C)

The legacy ITU-T G.652 standard SMFs are not optimized for WDM applications due to the high attenuation around the water peak region. ITU G.652.C-compliant fibers offer extremely low attenuation around the OH peaks. The G.652.C fiber is optimized for networks where transmission occurs across a broad range of wavelengths from 1285 nm to 1625 nm. Although G.652.C-compliant fibers offer excellent capabilities for shorter, unamplified metro and access networks, they do not fully address the needs for 1550-nm transmission. The attenuation parameter for G.652 fiber is typically 0.2 dB/km at 1550 nm, and the PMD parameter is less than 0.1 ps/$\sqrt{}$km. An example of this type of fiber is Corning SMF-28e.

Dispersion-Shifter Fiber (ITU-T G.653)

Conventional SMF has a zero-dispersion wavelength that falls near the 1310-nm window band. SMF shows high dispersion values over the range between 1500 nm and 1600 nm (third window band). The trend of shifting the operating transmission wavelength from 1310 nm to 1550 nm initiated the development of a fiber type called *dispersion-shifted fiber (DSF)*. DSF exhibits a zero-dispersion value around the 1550-nm wavelength where the attenuation is minimum. The DSFs are optimized for operating in the region between 1500 to 1600 nm. With the introduction of WDM systems, however, channels allocated near 1550 nm in DSF are seriously affected by noise induced as a result of nonlinear effects caused by FWM. This initiated the development of NZDSF. Figure 3-14 illustrates the dispersion slope of DSF with respect to SMF and NZDSF. G.53 fiber is rarely deployed any more and has been superseded by G.655.

Figure 3-14 *Fiber Dispersion Slopes*

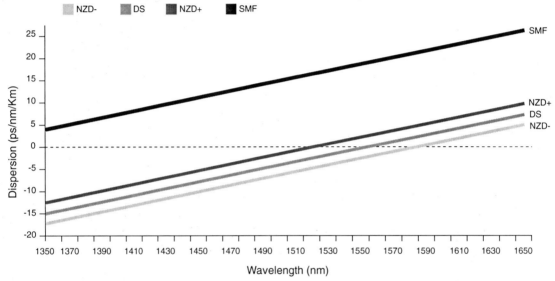

1550-nm Loss-Minimized Fiber (ITU-T G.654)

The ITU-T G.654 fiber is optimized for operation in the 1500-nm to 1600-nm region. This fiber has a low loss in the 1550-nm band. Low loss is achieved by using a pure silica core. ITU-T G.654 fibers can handle higher power levels and have a larger core area. These fibers have a high chromatic dispersion at 1550 nm. The ITU G.654 fiber has been designed for extended long-haul undersea applications.

Nonzero Dispersion Shifted Fiber (ITU-T G.655)

Using nonzero dispersion-shifted fiber (NZDSF) can mitigate nonlinear characteristics. NZDSF fiber overcomes these effects by moving the zero-dispersion wavelength outside the 1550-nm operating window. The practical effect of this is to have a small but finite amount of chromatic dispersion at 1550 nm, which minimizes nonlinear effects, such as FWM, SPM, and XPM, which are seen in the dense wavelength-division multiplexed (DWDM) systems without the need for costly dispersion compensation. There are two fiber families called nonzero dispersion (NZD+ and NZD–), in which the zero-dispersion value falls before and after the 1550-nm wavelength, respectively. The typical chromatic dispersion for G.655 fiber at 1550 nm is 4.5 ps/nm-km. The attenuation parameter for G.655 fiber is typically 0.2 dB/km at 1550 nm, and the PMD parameter is less than 0.1 ps/√km. The Corning LEAF fiber is an example of an enhanced G.655 fiber with a 32 percent larger effective area. Figure 3-14 illustrates the dispersion slope of NZDSF with respect to SMF and DSF.

Fiber-Optic Cable Termination

There are many types of optical connectors. The one you use depends on the equipment you are using it with and the application you are using it on. The connector is a mechanical device mounted on the end of a fiber-optic cable, light source, receiver, or housing. The connector allows the fiber-optic cable, light source, receiver, or housing to be mated to a similar device. The connector must direct light and collect light and must be easily attached and detached from equipment. A connector marks a place in the premises fiber-optic data link where signal power can be lost and the BER can be affected by a mechanical connection. Of the many different connector types, those for glass fiber-optic cable and plastic fiber-optic cable are discussed in this chapter. Other considerations for terminations are repeatability of connection and vibration resistance. Physical termination density is another consideration. Commonly used fiber-optic connectors are discussed in the following subsections and are shown in Figure 3-15.

Figure 3-15 *Fiber-Optic Connectors*

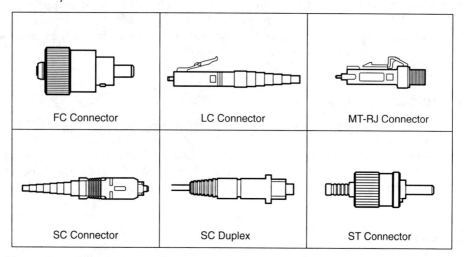

FC Connectors

These connectors are used for single-mode and multimode fiber-optic cables. FC connectors offer extremely precise positioning of the fiber-optic cable with respect to the transmitter's optical source emitter and the receiver's optical detector. FC connectors feature a position locatable notch and a threaded receptacle. FC connectors are constructed with a metal housing and are nickel-plated. They have ceramic ferrules and are rated for 500 mating cycles. The insertion loss for matched FC connectors is 0.25 dB. From a design perspective, it is recommended to use a loss margin of 0.5 dB or the vendor recommendation for FC connectors.

SC Connectors

SC connectors are used with single-mode and multimode fiber-optic cables. They offer low cost, simplicity, and durability. SC connectors provide for accurate alignment via their ceramic ferrules. An SC connector is a push-on, pull-off connector with a locking tab. Typical matched SC connectors are rated for 1000 mating cycles and have an insertion loss of 0.25 dB. From a design perspective, it is recommended to use a loss margin of 0.5 dB or the vendor recommendation for SC connectors.

ST Connectors

The ST connector is a keyed bayonet connector and is used for both multimode and single-mode fiber-optic cables. It can be inserted into and removed from a fiber-optic cable both quickly and easily. Method of location is also easy. ST connectors come in two versions: ST and ST-II. These are keyed and spring-loaded. They are push-in and twist types. ST connectors are constructed with a metal housing and are nickel-plated. They have ceramic ferrules and are rated for 500 mating cycles. The typical insertion loss for matched ST connectors is 0.25 dB. From a design perspective, it is recommended to use a loss margin of 0.5 dB or the vendor recommendation for ST connectors.

LC Connectors

LC connectors are used with single-mode and multimode fiber-optic cables. The LC connectors are constructed with a plastic housing and provide for accurate alignment via their ceramic ferrules. LC connectors have a locking tab. LC connectors are rated for 500 mating cycles. The typical insertion loss for matched LC connectors is 0.25 dB. From a design perspective, it is recommended to use a loss margin of 0.5 dB or the vendor recommendation for LC connectors.

MT-RJ Connectors

MT-RJ connectors are used with single-mode and multimode fiber-optic cables. The MT-RJ connectors are constructed with a plastic housing and provide for accurate alignment via their metal guide pins and plastic ferrules. MT-RJ connectors are rated for 1000 mating cycles. The typical insertion loss for matched MT-RJ connectors is 0.25 dB for SMF and 0.35 dB for MMF. From a design perspective, it is recommended to use a loss margin of 0.5 dB or the vendor recommendation for MT-RJ connectors.

MTP/MPO Connectors

MTP/MPO connectors are used with single-mode and multimode fiber-optic cables. The MTP/MPO is a connector manufactured specifically for a multifiber ribbon cable. The MTP/MPO single-mode connectors have an angled ferrule allowing for minimal back reflection, whereas the multimode connector ferrule is commonly flat. The ribbon cable is flat and appropriately named due to its flat ribbon-like structure, which houses fibers side by side in a jacket. The typical insertion loss for matched MTP/MPO connectors is 0.25 dB. From a design perspective, it is recommended to use a loss margin of 0.5 dB or the vendor recommendation for MTP/MPO connectors.

Splicing

Fiber-optic cables might have to be spliced together for a number of reasons—for example, to realize a link of a particular length. Another reason might involve *backhoe fade*, in which case a fiber-optic cable might have been ripped apart due to trenching work. The network installer might have in his inventory several fiber-optic cables, but none long enough to satisfy the required link length. Situations such as this often arise because cable manufacturers offer cables in limited lengths—usually 1 to 6 km. A link of 10 km can be installed by splicing several fiber-optic cables together. The installer can then satisfy the distance requirement and avoid buying a new fiber-optic cable. Splices might be required at building entrances, wiring closets, couplers, and literally any intermediate point between a transmitter and receiver.

Connecting two fiber-optic cables requires precise alignment of the mated fiber cores or spots in a single-mode fiber-optic cable. This is required so that nearly all the light is coupled from one fiber-optic cable across a junction to the other fiber-optic cable. Actual contact between the fiber-optic cables is not even mandatory.

There are two principal types of splices: fusion and mechanical. Fusion splices use an electric arc to weld two fiber-optic cables together. The process of fusion splicing involves using localized heat to melt or fuse the ends of two optical fibers together. The splicing process begins by preparing each fiber end for fusion. Fusion splicing requires that all protective coatings be removed from the ends of each fiber. The fiber is then cleaved using the score-and-break method. The quality of each fiber end is inspected using a microscope. In fusion splicing, splice loss is a direct function of the angles and quality of the two fiber-end faces.

The basic fusion-splicing apparatus consists of two fixtures on which the fibers are mounted with two electrodes. An inspection microscope assists in the placement of the prepared fiber ends into a fusion-splicing apparatus. The fibers are placed into the apparatus, aligned, and then fused together. Initially, fusion splicing used nichrome wire as the heating element to melt or fuse fibers together. New fusion-splicing techniques have replaced the nichrome wire with carbon dioxide (CO_2) lasers, electric arcs, or gas flames to heat the fiber ends, causing them to fuse together. Arc fusion splicers can splice single fibers or 12- and 24-fiber-count ribbon fibers at the same time. The small size of the fusion splice and the development of automated fusion-splicing machines have made electric arc fusion one of the most popular splicing techniques in commercial applications. The splices offer sophisticated, computer-controlled alignment of fiber-optic cables to achieve losses as low as 0.02 dB.

Splices can also be used as optical attenuators if there is a need to attenuate a high-powered signal. Splice losses of up to 10.0 dB can be programmed and inserted into the cable if desired. This way, the splice can act as an in-line attenuator with the characteristic nonreflectance of a fusion splice. Typical fusion-splice losses can be estimated at 0.02 dB for loss-budget calculation purposes. Mechanical splices are easily implemented in the field, require little or no tooling, and offer losses of about 0.5 to 0.75 dB.

Physical-Design Considerations

Many factors must be considered when designing a fiber-optic cable plant. First and foremost, the designer must determine whether the cable is to be installed for an inside-plant (ISP) or outside-plant (OSP) application. The answer to this question usually determines whether a loose buffer or a tight buffer cable will be used. An important factor in fiber-optic cable design and implementation is consideration of the cable's *minimum bend radius* and tensile loading. There are two kinds of submarine cable systems: shallow-water and deep-water systems.

Tight Buffer Versus Loose Buffer Cable Plants

Tight buffer or tight tube cable designs are typically used for ISP applications. Each fiber is coated with a buffer coating, usually with an outside diameter of 900 μm. Tight buffer cables have the following cable ratings:

- **OFNR**—Optical fiber, nonconductive riser rated
- **OFNP**—Optical fiber, nonconductive plenum rated
- **LSZH**—Low smoke, zero halogen rated

The type of ISP tight buffer cable selected usually depends on the application, environment, and building code. Loose-buffer or loose-tube cables mean that the fibers are placed loosely within a larger plastic tube. Usually 6 to 12 fibers are placed within a single tube. These tubes are filled with a gel compound that protects the fibers from moisture and physical stresses that may be experienced by the overall cable. Loose buffer designs are used for OSP applications such as underground installations, lashed or self-supporting aerial installations, and other OSP applications. These cables require additional cleaning, including the removal of the protective compounds when the fibers are to be terminated. Loose-tube cable designs include multifiber armored and non-armored cable systems.

Bend Radius and Tensile Loading

An important consideration in fiber-optic cable installation is the cable's minimum *bend radius*. Bending the cable farther than its minimum bend radius might result in increased attenuation or even broken fibers. Cable manufacturers specify the minimum bend radius for cables under tension and long-term installation. The ANSI TIA/EIA-568B.3 standard specifies a bend radius of 1.0 inch under no pull load and 2.0 inches when subject to tensile loading up to the rated limit.

For ISP cable other than two-fiber and four-fiber, the standard specifies 10× the cable's outside diameter under no pull load and 15× the cable's outside diameter when subject to tensile load. Cable tensile load ratings, also called cable *pulling tensions* or *pulling forces*, are specified under short-term and long-term conditions. The short-term condition represents a cable during installation and it is not recommended that this tension be exceeded. The long-term condition represents an installed cable subjected to a permanent load for the life of the cable. Typical loose-tube cable designs have a short-term (during installation) tensile rating of 600 pounds (2700 N) and a long-term (post installation) tensile rating of 200 pounds (890 N).

Submarine Cable Systems

Shallow-water systems are similar to their armored loose-buffered terrestrial counterparts, whereas deep-water submarine cables use a special hermetically sealed copper tube to protect the fiber from the effects of deep-water environments. Deep-water and submarine cables also have dual armor and an asphalt compound that is used to fill interstitial spaces and add negative buoyancy. In addition to the significant external physical forces that might be encountered in a submarine environment, the other major concern is the effect of hydrogen on the performance of the optical fiber in cables used in such applications.

The effect of hydrogen on fiber performance depends on specific system characteristics. System attributes include fiber type, system operating wavelength, and cable design and installation method. Hydrogen can chemically react with dopants, such as phosphorus, to produce irreversible absorption peaks, resulting in a significant increase in the attenuation coefficient across various wavelength ranges. This phenomenon, also known as the *Type 1 hydrogen effect*, occurred primarily in early optical-fiber designs that used a phosphorus dopant. Unlike early phosphorus fibers, current fibers using germania dopants are not susceptible to Type 1 hydrogen effects.

The second hydrogen effect arises from the propensity for molecular hydrogen to diffuse readily through most other materials. When diffused into glass optical fiber, hydrogen creates distinct absorption peaks at certain wavelengths. The most predominant of these occurs at 1240 nm and 1380 nm. The tails of these peaks can extend out, depending on the hydrogen concentration, affecting the optical performance at 1310 nm and 1550 nm. Unlike the Type 1 effect, the effect created by molecular hydrogen is reversible and is known as the *Type 2 hydrogen effect*. The major sources are typically understood to be the corrosion of the metal armoring and the presence of bacteria. Proper span design must take into consideration hydrogen safety margins for submarine applications. The attenuation coefficient is proportional to the water depth role because as depth increases, the partial pressure of hydrogen increases, resulting in an increase in the amount of interstitial hydrogen that can be present in the fiber.

Fiber-Optic Communications System

As depicted in Figure 3-16, information (voice, data, and video) from the source is encoded into electrical signals that can drive the transmitter. The fiber acts as an optical waveguide for the photons as they travel down the optical path toward the receiver. At the detector, the signals undergo an optical-to-electrical (OE) conversion, are decoded, and are sent to their destination.

Figure 3-16 *Fiber-Optic Communication System*

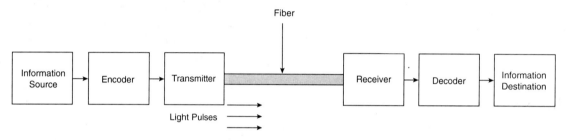

Transmitter

The transmitter component of Figure 3-16 serves two functions. First, it must be a source of the light launched into the fiber-optic cable. Second, it must modulate this light to represent the binary data that it receives from the source. A transmitter's physical dimensions must be compatible with the size of the fiber-optic cable being used. This means that the transmitter must emit light in a cone with a cross-sectional diameter of 8 to 100 microns; otherwise, it cannot be coupled into the fiber-optic cable. The optical source must be able to generate enough optical power so that the desired BER can be met over the optical path. There should be high efficiency in coupling the light generated by the optical source into the fiber-optic cable, and the optical source should have sufficient linearity to prevent the generation of harmonics and intermodulation distortion. If such interference is generated, it is extremely difficult to remove. This would cancel the interference resistance benefits of the fiber-optic cable. The optical source must be easily modulated with an electrical signal and must be capable of high-speed modulation; otherwise, the bandwidth benefits of the fiber-optic cable are lost. Finally, there are the usual requirements of small size, low weight, low cost, and high reliability. The transmitter is typically pulsed at the incoming frequency and performs a transducer electrical-to-optical (EO) conversion. Light-emitting diodes (LEDs) or vertical cavity surface emitting lasers (VCSELs) are used to drive MMF systems, whereas laser diodes are used to drive SMF systems. Two types of light-emitting junction diodes can be used as the optical source of the transmitter. These are the LED and the laser diode (LD). LEDs are simpler and generate incoherent, lower-power light. LEDs are used to drive MMF. LDs generate coherent, higher-power light and are used to drive SMF.

Figure 3-17 shows the optical power output, P, from each of these devices as a function of the electrical current input, I, from the modulation circuitry. As the figure indicates, the LED has a relatively linear P-I characteristic, whereas the LD has a strong nonlinearity or threshold effect. The LD can also be prone to kinks when the power actually decreases with increasing input current. LDs have advantages over LEDs in the sense that they can be modulated at very high speeds, produce greater optical power, and produce an output beam with much less spatial width than an LED. This gives LDs higher coupling efficiency to the fiber-optic cable. LED advantages include a higher reliability, better linearity, and lower cost.

Figure 3-17 *LED and LD P-I Characteristics*

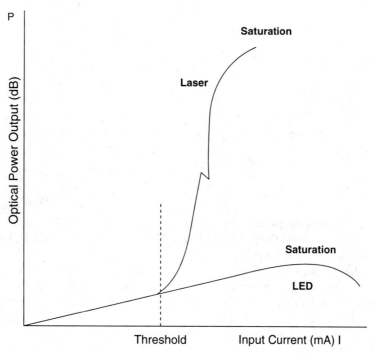

A key difference between the optical output of an LED and a LD is the wavelength spread over which the optical power is distributed. The spectral width, σ, is the 3-dB optical power width (measured in nanometers or microns). The spectral width impacts the effective transmitted signal bandwidth. A larger spectral width takes up a larger portion of the fiber-optic cable link bandwidth. Figure 3-18 shows the spectral width of the two devices. The optical power generated by each device is the area under the curve. The spectral width is the half-power spread. An LD always has a smaller spectral width than an LED. The specific value of the spectral width depends on the details of the diode structure and the semiconductor material. However, typical values for an LED are around 40 nm for operation at 850 nm and 80 nm at 1310 nm. Typical values for an LD are 1 nm for operation at 850 nm and 3 nm at 1310 nm.

Figure 3-18 *LED and LD Spectral Widths*

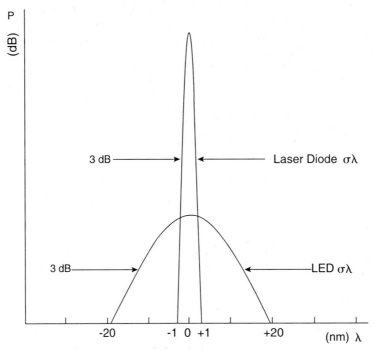

Other transmitter parameters include packaging, environmental sensitivity of device characteristics, heat sinking, and reliability. With either an LED or LD, the transmitter package must have a transparent window to transmit light into the fiber-optic cable. It can be packaged with either a fiber-optic cable pigtail or with a transparent plastic or glass window. Some vendors supply the transmitter with a package having a small hemispherical lens to help focus the light into the fiber-optic cable. Packaging must also address the thermal coupling for the LED or LD. A complete transmitter module can consume more than 1 watt, which could result in significant heat generation. Plastic packages can be used for lower-speed and lower-reliability applications. However, high-speed and high-reliability transmitters need metal packaging with built-in fins for heat sinking.

There are several different schemes for carrying out the modulation function. These include intensity modulation (IM), frequency shift keying (FSK), phase shift keying (PSK), and polarization modulation (PM). Within the context of a premise fiber-optic data link, the only one really used is IM. IM is used universally for premise fiber-optic data links because it is well matched to the operation of both LEDs and LDs. The carrier that each of these sources produces is easy to modulate with this technique. Passing current through them operates both of these devices. The amount of power that they radiate (sometimes referred to as the *radiance*) is proportional to this current. In this way, the optical power takes the shape of the input current. If the input current is the waveform m(t) representing the binary information stream, the resulting optical signal looks like bursts of optical signal when m(t) represents a 1 and the absence of optical signal when m(t) represents a 0. This is also known as *direct modulation* of the LED or LD.

Receiver

Figure 3-19 shows a schematic of an optical receiver. The receiver serves two functions: It must sense or detect the light coupled out of the fiber-optic cable and convert the light into an electrical signal, and it must demodulate this light to determine the identity of the binary data that it represents. The receiver performs the OE transducer function.

Figure 3-19 *Schematic of an Optical Receiver*

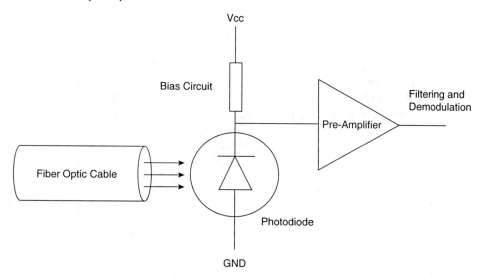

A receiver is generally designed with a transmitter. Both are modules within the same package. The light detection is carried out by a photodiode, which senses light and converts it into an electrical current. However, the optical signal from the fiber-optic cable and the resulting electrical current will have a small amplitude. Consequently, the photodiode circuitry must be followed by one or more amplification stages. There might even be filters and equalizers to shape and improve the information-bearing electrical signal.

The receiver schematic in Figure 3-19 shows a photodiode, bias resistor circuit, and a low-noise pre-amp. The output of the pre-amp is an electrical waveform version of the original information from the source. To the right of this pre-amp is an additional amplification, filters, and equalizers. All of these components can be on a single integrated circuit, a hybrid, or discretely mounted on a printed circuit board.

The receiver can incorporate a number of other functions, such as clock recovery for synchronous signaling, decoding circuitry, and error detection and recovery. The receiver must have high sensitivity so that it can detect low-level optical signals coming out of the fiber-optic cable. The higher the sensitivity, the more attenuated signals it can detect. It must have high bandwidth or a fast rise time so that it can respond fast enough and demodulate high-speed digital data. It must have low noise so that it does not significantly impact the BER of the link and counter the interference resistance of the fiber-optic cable transmission medium.

There are two types of photodiode structures: positive intrinsic negative (PIN) and the avalanche photodiode (APD). In most premise applications, the PIN is the preferred element in the receiver. This is mainly due to fact that it can be operated from a standard power supply, typically between 5 and 15V. APD devices have much better sensitivity. In fact, APD devices have 5 to 10 dB more sensitivity. They also have twice the bandwidth. However, they cannot be used on a 5V printed circuit board. They also require a stable power supply, which increases their cost. APD devices are usually found in long-haul communication links and can increasingly be found in metro-regional networks (because APDs have decreased in cost).

The demodulation performance of the receiver is characterized by the BER that it delivers to the user. The sensitivity curve indicates the minimum optical power that the receiver can detect compared to the data rate, to achieve a particular BER. The sensitivity curve varies from receiver to receiver. The sensitivity curve considers within it the SNR parameter that generally drives all communications-link performance. The sensitivity depends on the type of photodiode used and the wavelength of operation. Figure 3-20 shows sensitivity curve examples.

Figure 3-20 *Receiver Sensitivity Curves*

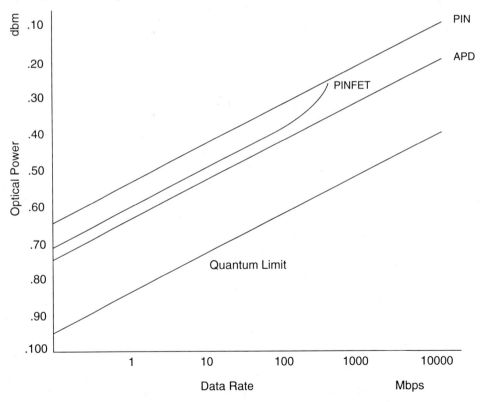

Receiver Sensitivities BER = 10-9, λ = 1.55 micrometer

The quantum limit curve serves as a baseline reference. In a sense it represents optimum performance on the part of the photodiode in the receiver—that is, performance in which there is 100 percent efficiency in converting light from the fiber-optic cable into an electric current for demodulation. All other sensitivity curves are compared to the quantum limit.

Fiber Span Analysis

Span analysis is the calculation and verification of a fiber-optic system's operating characteristics. This encompasses items such as fiber routing, electronics, wavelengths, fiber type, and circuit length. Attenuation and nonlinear considerations are the key parameters for loss-budget analysis. Before implementing or designing a fiber-optic circuit, a span analysis is recommended to make certain the system will work over the proposed link. Both the passive and active components of the circuit have to be included in the loss-budget calculation. Passive loss is made up of fiber loss, connector loss, splice loss, and losses involved with couplers or splitters in the link. Active components are system gain, wavelength, transmitter power, receiver sensitivity, and dynamic range.

Nonlinear effects occur at high bit rates and power levels. These effects must be mitigated using compensators, and a suitable budget allocation must be made during calculations.

The overall span loss, or *link budget* as it is sometimes called, can be determined by using an optical meter to measure true loss or by computing the loss of system components. The latter method considers the loss associated with span components, such as connectors, splices, patch panels, jumpers, and the optical safety margin. The safety margin sets aside 3 dB to compensate for component aging and repair work in event of fiber cut. Adding all of these factors to make sure their sum total is within the maximum attenuation figure ensures that the system will operate satisfactorily. Allowances must also be made for the type of splice, the age and condition of the fiber, equipment, and the environment (including temperature variations).

NOTE Considerations for temperature effects associated with most fibers usually yield ±1 dB that could be optionally included in optical loss-budget calculations.

Transmitter Launch Power

Power measured in dBm at a particular wavelength generated by the transmitter LED or LD used to launch the signal is known as the *transmitter launch power*. Generally speaking, the higher the transmitter launch power, the better. However, one must be wary of receiver saturation, which occurs when the received signal has a very high power content and is not within the receiver's dynamic range. If the signal strength is not within the receiver's dynamic range, the receiver cannot decipher the signal and perform an OE conversion. High launch powers can offset attenuation, but they can cause nonlinear effects in the fiber and degrade system performance, especially at high bit rates.

Receiver Sensitivity and Dynamic Range

Receiver sensitivity and dynamic range are the minimum acceptable value of received power needed to achieve an acceptable BER or performance. Receiver sensitivity takes into account power penalties caused by use of a transmitter with worst-case values of extinction ratio, jitter, pulse rise times and fall times, optical return loss, receiver connector degradations, and measurement tolerances. The receiver sensitivity does not include power penalties associated with dispersion or with back reflections from the optical path. These effects are specified separately in the allocation of maximum optical path penalty. Sensitivity usually takes into account worst-case operating and end-of-life (EOL) conditions. Receivers have to cope with optical inputs as high as –5 dBm and as low as –30 dBm. Or stated differently, the receiver needs an optical dynamic range of 25 dB.

Power Budget and Margin Calculations

To ensure that the fiber system has sufficient power for correct operation, you need to calculate the span's power budget, which is the maximum amount of power it can transmit. From a design perspective, worst-case analysis calls for assuming minimum transmitter power and minimum receiver sensitivity. This provides for a margin that compensates for variations of transmitter power and receiver sensitivity levels.

Power budget (P_B) = Minimum transmitter power (P_{TMIN}) – Minimum receiver sensitivity (P_{RMIN})

You can calculate the span losses by adding the various linear and nonlinear losses. Factors that can cause span or link loss include fiber attenuation, splice attenuation, connector attenuation, chromatic dispersion, and other linear and nonlinear losses. Table 3-1 provides typical attenuation characteristics of various kinds of fiber-optic cables. Table 3-2 provides typical insertion losses for various connectors and splices. Table 3-3 provides the margin requirement for nonlinear losses along with their usage criteria. For information about the actual amount of signal loss caused by equipment and other factors, refer to vendor documentation.

Span loss (P_S) = (Fiber attenuation * km) + (Splice attenuation * Number of splices) + (Connector attenuation * Number of connectors) + (In-line device losses) + (Nonlinear losses) + (Safety margin)

Table 3-1 *Typical Fiber-Attenuation Characteristics*

Mode	Material	Refractive Index Profile	λ (nm)	Diameter (μm)	Attenuation (dB/km)
Multimode	Glass	Step	800	62.5/125	5.0
Multimode	Glass	Step	850	62.5/125	4.0
Multimode	Glass	Graded	850	62.5/125	3.3
Multimode	Glass	Graded	850	50/125	2.7

Table 3-1 *Typical Fiber-Attenuation Characteristics (Continued)*

Mode	Material	Refractive Index Profile	λ (nm)	Diameter (μm)	Attenuation (dB/km)
Multimode	Glass	Graded	1310	62.5/125	0.9
Multimode	Glass	Graded	1310	50/125	0.7
Multimode	Glass	Graded	850	85/125	2.8
Multimode	Glass	Graded	1310	85/125	0.7
Multimode	Glass	Graded	1550	85/125	0.4
Multimode	Glass	Graded	850	100/140	3.5
Multimode	Glass	Graded	1310	100/140	1.5
Multimode	Glass	Graded	1550	100/140	0.9
Multimode	Plastic	Step	650	485/500	240
Multimode	Plastic	Step	650	735/750	230
Multimode	Plastic	Step	650	980/1000	220
Multimode	PCS	Step	790	200/350	10
Single-mode	Glass	Step	650	3.7/80 or 125	10
Single-mode	Glass	Step	850	5/80 or 125	2.3
Single-mode	Glass	Step	1310	9.3/125	0.5
Single-mode	Glass	Step	1550	8.1/125	0.2
Single-mode	Glass	Dual Step	1550	8.1/125	0.2

Table 3-2 *Component Loss Values*

Component	Insertion Loss
Connector Type	
SC	0.5 dB
ST	0.5 dB
FC	0.5 dB
LC	0.5 dB
MT-RJ	0.5 dB
MTP/MPO	0.5 dB
Splice	
Mechanical	0.5 dB
Fusion	0.02 dB
Fiber patch panel	2.0 dB

NOTE Typical multimode connectors have insertion losses between 0.25 dB and 0.5 dB, whereas single-mode connectors that are factory made and fusion spliced onto the fiber cable will have losses between 0.15 dB and 0.25 dB. Field-terminated single-mode connectors can have losses as high as 1.0 dB.

Table 3-3 *Reference Margin Values*

Characteristic	Loss Margin	Bit Rate	Signal Power
Dispersion margin	1 dB	Both	Both
SPM margin	0.5 dB	High	High
XPM margin (WDM)	0.5 dB	High	High
FWM margin (WDM)	0.5 dB	Both	High
SRS/SBS margin	0.5 dB	High	High
PMD margin	0.5 dB	High	Both

The next calculation involves the power margin (P_M), which represents the amount of power available after subtracting linear and nonlinear span losses (P_S) from the power budget (P_B). A P_M greater than zero indicates that the power budget is sufficient to operate the receiver. The formula for power margin (P_M) is as follows:

Power margin (P_M) = Power budget (P_B) − Span loss (P_S)

To prevent receiver saturation, the input power received by the receiver, after the signal has undergone span loss, must not exceed the maximum receiver sensitivity specification (P_{RMAX}). This signal level is denoted as (P_{IN}). The maximum transmitter power (P_{TMAX}) must be considered as the launch power for this calculation. The span loss (P_S) remains constant.

Input power (P_{IN}) = Maximum transmitter power (P_{TMAX}) − Span loss (P_S)

The design equation

Input power (P_{IN}) <= Maximum receiver sensitivity (P_{RMAX})

must be satisfied to prevent receiver saturation and ensure system viability. If the input power (P_{IN}) is greater than the maximum receiver sensitivity (P_{RMAX}), passive attenuation must be considered to reduce signal level and bring it within the dynamic range of the receiver.

Case 1: MMF Span Analysis

Consider the fiber-optic system shown in Figure 3-21 operating at OC-3 (155 Mbps). The minimum optical transmitter launch power is −12.5 dBm, and the maximum optical transmitter launch power is −2 dBm at 1310 nm. The minimum receiver sensitivity is −30 dBm, and the maximum receiver sensitivity is −3 dBm at 1310 nm. The example assumes inclusion of two patch panels in the path, two mechanical splices, with the system operating over 2 km of graded index 50/125-μm multimode fiber-optic cable. Refer to Tables 3-1, 3-2, and 3-3 for appropriate attenuation, component, and nonlinear loss values.

Figure 3-21 *MMF Span Analysis*

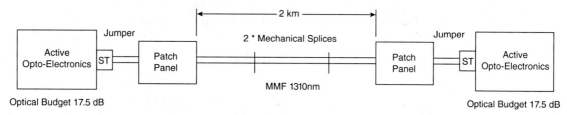

The system operates at 155 Mbps or approximately 155 MHz. At such bit rates, there is no need to consider SPM, PMD, or SRS/SBS margin requirements. Because the link is a single-wavelength system, there is no need to include XPM or FWM margins. However, it is safe to consider the potential for a degree of chromatic dispersion, because chromatic dispersion occurs at all bit rates. The span analysis and viability calculations over the link are computed as follows.

Component	dB Loss
Minimum transmitter launch power (P_{TMIN})	–12.5 dBm
Minimum receiver sensitivity (P_{RMIN})	–30 dBm
Power Budget (P_B) = ($P_{TMIN} - P_{RMIN}$)	**17.5 dB**

Component	dB Loss
MMF graded index 50/125-μm cable at 1310 nm (2 km * 0.7 dB/km)	1.4 dB
ST connectors (2 * 0.5 dB/connector)	1 dB
Mechanical splice (2 * 0.5 dB/splice)	1 dB
Patch panels (2 * 2 dB/panel)	4 dB
Dispersion margin	1 dB
Optical safety and repair margin	3 dB
Total Span Loss (P_S)	**11.4 dB**

Power margin (P_M) = Power budget (P_B) – Span loss (P_S)

P_M = 17.5 dB – 11.4 dB

P_M = 6.1 dB > 0 dB

88   Chapter 3: Fiber-Optic Technologies

In the preceding example, notice that the 11.4-dB total span loss is well within the 17.5-dB power budget or maximum allowable loss over the span.

To prevent receiver saturation, the input power received by the receiver, after the signal has undergone span loss, must not exceed the maximum receiver sensitivity specification (P_{RMAX}). This signal level is denoted as (P_{IN}). The maximum transmitter power (P_{TMAX}) must be considered as the launch power for this calculation. The span loss (P_S) remains constant.

Input power (P_{IN}) = Maximum transmitter power (P_{TMAX}) − Span loss (P_S)

$P_{IN} = -2 - 11.4$

$P_{IN} = -13.4$ dBm

−13.4 dBm (P_{IN}) <= −3 dBm

This satisfies the receiver sensitivity design equation and ensures viability of the optical system at an OC-3 rate over 2 km without the need for amplification or attenuation.

Case 2: SMF Span Analysis

Consider the fiber-optic system in Figure 3-22 operating at OC-192 (9.953 Gbps). The minimum optical transmitter launch power is −7.5 dBm, and the maximum optical transmitter launch power is 0 dBm at 1550 nm. The minimum receiver sensitivity is −30 dBm, and the maximum receiver sensitivity is −3 dBm at 1550 nm. The example assumes inclusion of two patch panels in the path, four fusion splices, with the system operating over 25 km of step index 8.1/125-μm SMF cable. Refer to Tables 3-1, 3-2, and 3-3 for appropriate attenuation, component, and nonlinear loss values.

Figure 3-22 *SMF Link-Budget Example*

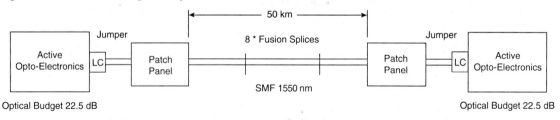

The system is operating at 9.953 Gbps or approximately 10 GHz. At such high bit rates, SPM, PMD, and SRS/SBS margin requirements must be taken into consideration. Also consider the potential for a degree of chromatic dispersion. Because the link is a single-wavelength system, there is no need to include XPM or FWM margins. The link loss and viability calculations over the link are computed as follows.

Component	dB Loss
Minimum transmitter launch power (P_{TMIN})	−7.5 dBm
Minimum receiver sensitivity (P_{RMIN})	−30 dBm
Power Budget (P_B) = (P_{TMIN} − P_{RMIN})	**22.5 dB**

Component	dB Loss
SMF step index 8.1/125-μm cable at 1550 nm (50 km * 0.2 dB/km)	10 dB
LC connectors (2 * 0.5 dB/connector)	1.0 dB
Fusion splices (8 * 0.02 dB/splice)	0.16 dB
Patch panels (2 * 2 dB/panel)	4 dB
Dispersion margin	1 dB
SPM margin	0.5 dB
PMD margin	0.5 dB
SRS/SBS margin	0.5 dB
Optical safety and repair margin	3 dB
Total Span Loss (P_S)	**20.66 dB**

Power margin (P_M) = Power budget (P_B) − Span loss (P_S)

P_M = 22.5 dB − 20.66 dB

P_M = 1.84 dB > 0 dB

In the example, notice that the 20.66-dB total span loss is well within the 22.5-dB power budget or maximum allowable loss over the span. To prevent receiver saturation, the input power received by the receiver, after the signal has undergone span loss, must not exceed the maximum receiver sensitivity specification (P_{RMAX}). This signal level is denoted as (P_{IN}). The maximum transmitter power (P_{TMAX}) must be considered as the launch power for this calculation. The span loss (P_S) remains constant.

Input power (P_{IN}) = Maximum transmitter power (P_{TMAX}) − Span loss (P_S)

P_{IN} = 0 − 20.66 dBm

P_{IN} = −20.66 dBm

−20.66 dBm (P_{IN}) <= −3 dBm (P_{RMAX})

This satisfies the receiver sensitivity design equation and ensures viability of the optical system at an OC-192 rate over 50 km without the need for amplification or attenuation. Note, however, that this example has not considered dispersion calculations or dispersion compensation. Dispersion compensation units insert their own loss component into the overall span.

NOTE In the preceding example, various margins for nonlinear effects were included in the span loss calculation. This is not necessary if the maximum power on the SMF is kept below +10 dBm to avoid nonlinear effects on the transmission signal. For dispersion-compensated spans, the maximum power on the dispersion compensation module (DCU) must be kept below +4 dBm to avoid nonlinear effects on DCU.

Summary

Fiber optics have become the industry standard for the terrestrial transmission of telecommunication information. Fiber optics will continue to be a major player in the delivery of broadband services. Carriers use optical fiber to carry POTS service across their nationwide networks. Today more than 80 percent of the world's long-distance traffic is carried over optical-fiber cables. Telecommunications applications of fiber-optic cable are widespread, ranging from global networks to desktop computers. These involve the transmission of voice, data, and video over distances of less than a meter to hundreds of kilometers, using one of a few standard fiber designs in one of several cable designs. Carriers use optical fiber to carry analog phone service. Cable television companies also use fiber for delivery of digital video services. Intelligent transportation systems and biomedical systems also use fiber-optic transmission systems. Optical cable is also the industry standard for subterranean and submarine transmission systems.

The principle of total internal reflection is used to propagate light signals. Light is guided through the core, and the fiber acts as an optical waveguide. SMF and MMF cables are constructed differently. MMF has a larger core diameter as compared to SMF. There are two types of propagation for fiber-optic cable: multimode or single mode. These modes perform differently with respect to both attenuation and time dispersion. SMF cable provides better performance than MMF cable. The three primary propagation modes include multimode step index, single-mode step index, and multimode graded index propagation.

In an optical communications system, information from the source is encoded into electrical signals that can drive the transmitter. The transmitter consists of an LED or laser and is pulsed at the incoming frequency. The transmitter performs an EO conversion. The fiber acts as an optical waveguide. At the detector, the signals undergo an OE conversion, are decoded, and are sent to their destination. Fiber-optic system characteristics include attenuation, interference, and bandwidth characteristics. Fiber-optic systems are also secure from data tapping, and tampering can be detected far more easily than with metallic-based transmission medium or free-space propagation. Furthermore, the relatively smaller cross section of fiber-optic cables allows room for substantial growth in the capacity of existing conduits. Attenuation characteristics can be classified as intrinsic and extrinsic. Intrinsic attenuation occurs because of substances inherently present in the fiber, whereas extrinsic attenuation occurs because of external influences such as bending.

Decibel loss at the connector interface is directly proportional to the alignment accuracy and rigidity of the connector. Many types of optical connectors are in use. The one you use depends on the equipment you use it with and the application you use it on. Seamless permanent or semipermanent optical connections require fibers to be spliced. Fiber-optic cables might have to be spliced together for any of a number of reasons. One reason is to realize a link of a particular length. Connecting two fiber-optic cables requires precise alignment of the mated fiber cores or spots in a single-mode fiber-optic cable. This is required so that nearly all the light is coupled from one fiber-optic cable across a junction to the other fiber-optic cable. The two main splicing techniques in use are mechanical and fusion splicing.

Optical loss, or total attenuation, is the sum of the losses of each individual component between a transmitter and receiver. Loss-budget analysis is the calculation and verification of a fiber-optic system's operating characteristics. This encompasses items such as fiber routing, electronics, wavelengths, fiber type, and circuit length. Attenuation and nonlinear fiber characteristics are the key parameters for fiber span analysis. Transmitter launch power, receiver sensitivity, and the dynamic range of the receiver are crucial numbers used in span analysis.

This chapter includes the following sections:

- **The Need for Wavelength-Division Multiplexing**—This section discusses the original drivers for the development of wavelength-division multiplexing (WDM) technology, including the need for sheer bandwidth.

- **Wavelength-Division Multiplexing**—WDM is the process of multiplexing wavelengths of different frequencies onto a single fiber. This operation creates many virtual fibers, each capable of carrying a different signal. WDM fundamentals are discussed in this section.

- **Coarse Wavelength-Division Multiplexing**—This section discusses coarse wavelength-division multiplexing (CWDM). CWDM systems are suited for the short-haul transport of data, voice, video, storage, and multimedia services.

- **Dense Wavelength-Division Multiplexing**—Dense wavelength-division multiplexing (DWDM) systems are suited for the short-haul as well as long-haul transport of data, voice, video, storage, and multimedia services. This section discusses DWDM systems.

- **The ITU Grid**—The ITU grid for DWDM and CWDM is discussed in this section. The ITU-T established a set of standards for telecommunications that drives most optical DWDM systems today.

- **Wavelength-Division Multiplexing Systems**—This section discusses various WDM network elements. WDM optical networks consist of network elements, such as transmitters, multiplexers and demultiplexers, amplifiers, optical-fiber media, and receivers.

- **WDM Characteristics and Impairments to Transmission**—This section discusses WDM characteristics, such as WDM bit error rate (BER), the Q-factor, and forward error correction (FEC).

- **Dispersion and Compensation in WDM**—This section discusses chromatic and polarization dispersion encountered at higher speeds. The DWDM system must compensate for chromatic and polarization dispersion to support higher speeds.

CHAPTER 4

Wavelength-Division Multiplexing

The Need for Wavelength-Division Multiplexing

The power of the Internet and the World Wide Web resides in its content. Retrieval of high-quality content from application servers, such as web servers, video servers, and e-commerce sites, in the shortest possible time has driven the *need for speed* for individual and corporate end users alike. Residential customers, small- and medium-sized businesses, and even large businesses are demanding affordable high-speed access services, such as xDSL and cable modem access, with or without service level agreement (SLA) guarantees. Larger enterprise customers continue to push for high-speed, managed multiservice IP virtual private networks (VPNs) with strict quality of service (QoS) as well as storage-area networks (SANs). Applications, such as high-definition television (HDTV) to the home, would also up the ante for cable companies to revisit the hybrid fiber-coax (HFC) model and consider a technology that would provide additional bandwidth, without having to lay new fiber. With increased aggregation at the access layer, the need arises for bandwidth at the distribution and the core of the network. This exponential growth has fueled the need for extremely scalable high-bandwidth core technologies. Technology has seen the limits of bandwidth and transmission speeds over traditional TDM media and head-end systems. Traditional networks have been built using a combination of circuit-switched TDM technology along with a TDM-capable Synchronous Optical Network/Synchronous Digital Hierarchy (SONET/SDH) infrastructure. TDM and SONET are essentially serial time-division multiplexed technologies that have finite limits in terms of bandwidth due to constraints, such as frame size, framing rate, clock speed, and device opto-electronics.

The original driver for the development of WDM technology was the need for sheer bandwidth. This requirement translated into a tangible need to pull additional terrestrial fiber-optic cable via congested conduits or, worse still, in the case of intercontinental links, laying additional submarine fiber-optic cable. The infrastructure and construction costs associated with the deployment of large-scale fiber plants were and continue to be prohibitively high. In some cases, city and local government ordinances prevent service providers from trenching streets and laying new fiber. In short, WDM can be applied wherever there is a need for fiber relief. WDM technology was initially expensive to engineer, deploy, and manage, which restricted the initial market deployment. Many WDM manufacturers have addressed these limitations by providing point-and-click network provisioning tools, network design modeling tools, and various operational enhancements.

Wavelength-Division Multiplexing

Wavelength-division multiplexing (WDM) is the process of multiplexing wavelengths of different frequencies onto a single fiber. This operation creates many virtual fibers, each capable of carrying a different signal. Figure 4-1 shows a schematic of a bidirectional WDM system. This system has *n* service interfaces and *n* wavelengths transmitted in either direction over a single fiber. Each wavelength operates at a different frequency.

Figure 4-1 *WDM Schematic*

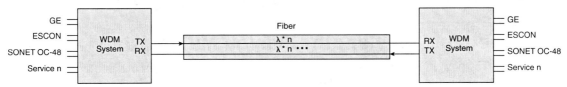

WDM uses wavelengths to transmit data parallel-by-bit or serial-by-character, which increases the capacity of the fiber by assigning incoming optical signals to specific frequencies (wavelengths) within a designated frequency band and then multiplexing the resulting signals out onto one fiber. Each signal can be carried at a different rate (OC-3/STM-1, OC-48/STM-16, and so on) and in a different format (SONET/SDH, ATM, data, and so on). This can increase the capacity of existing networks without the need for expensive recabling and tremendously reduce the cost of network infrastructure upgrades. WDM supports point-to-point, ring, and mesh topologies. Existing fiber in a SONET/SDH fiber plant can be easily migrated to WDM. Most WDM systems support standard SONET/SDH short-reach optical interfaces to which any SONET/SDH-compliant client device can attach. Long-haul WDM topologies are typically point to point. Perhaps the biggest reason for implementing WDM is the deployment speed of bandwidth service delivery. It is much easier to add a wavelength than to trench and add new fiber.

Four kinds of WDM systems are available:

- Metro WDM (<200 km)
- Long-haul or regional WDM (200 km to 800 km)
- Extended long-haul WDM (800 km to 2000 km)
- Ultra-long-haul WDM (>2000 km)

In today's regional and long-haul WDM systems, the user service interfaces are most often OC-48/STM-16 interfaces. In addition, other interfaces commonly supported include Ethernet, Fast Ethernet, Gigabit Ethernet, 10 Gigabit Ethernet, ESCON, Sysplex Timer and Sysplex Coupling Facility Links, and Fibre Channel. On the client side, there can be SONET/SDH terminals, add/drop multiplexers (ADMs), ATM switches, and routers. By converting incoming optical signals into the precise ITU-standard wavelengths to be multiplexed, transponders are currently a key determinant of the openness of WDM systems. WDM is an application option whenever there is a need for additional bandwidth, without the requirement to install additional fiber along the way. A cost model must be created to analyze and determine the financial feasibility of purchasing new end-equipment WDM opto-electronics versus the cost of leasing or laying additional fiber between the endpoints.

Wavelength-Division Multiplexing Fundamentals

WDM can be considered a form of frequency-division multiplexing (FDM) coupled with timed-division multiplexing (TDM), as depicted in Figure 4-2. The exact relationship between a WDM wavelength and frequency is determined from the equation $c = \lambda * f$; where c is the speed of light in a vacuum ($3 * 10^8$ m/s), λ is the wavelength measured in a vacuum; and f is the frequency. Frequency is standardized (rather than wavelength) because it is independent of the transmission medium, such as fiber or air. In WDM systems, the wavelength is measured in nanometers (nm) and the frequency is measured in gigahertz (GHz). Light travels considerably slower in a denser medium. The speed of light in glass is approximately $2 * 10^8$ m/s.

Figure 4-2 *WDM Depicted as a Combination of FDM and TDM*

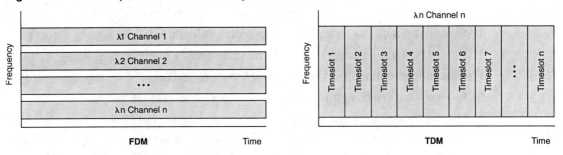

Various frequencies of light can travel down a single fiber, and each frequency can constitute a channel. However, transmitting light of various frequencies without some kind of clocking and pulsing isn't of much use in a digital communications system. Breaking down each lambda into time slots with framing and proper clocking mechanisms (similar to SONET/SDH) enables us to transmit information in the form of timed and framed pulses over a wavelength. Now, imagine a single wavelength capable of carrying an OC-192/STM-64 or roughly 10 Gbps worth of information. If we inject 80 lambdas over the same fiber, its bandwidth potential increases by a factor of 80, and the fiber will be able to carry up to

800 Gbps worth of information over a single fiber. In full-duplex mode, the resulting bandwidth would be 1.60 Tbps.

A few constraints are imposed by the laws of physics, in terms of injecting so many wavelengths into a fiber core. The diameter of the fiber core (8 μm for SMF), laser spectral width, channel spacing, the spectrum of light that can be used (C, L, or S bands), and linear and nonlinear impairments. The power of WDM resides in its potential to carry massive quantities of data, up to terabit levels. By varying the TDM signaling and framing, WDM can be customized to carry various kinds of application traffic from simple SONET/SDH to Gigabit Ethernet as well as SAN protocols, such as ESCON, FICON, and Fiber channel. The optical frequency bands used with various WDM systems are as follows:

- **O-band (original)**—A range from 1260 nm to 1360 nm
- **E-band (extended)**—A range from 1360 nm to 1460 nm
- **S-band (short wavelength)**—A range from 1460 nm to 1530 nm
- **C-band (conventional)**—A range from 1530 nm to 1565 nm
- **L-band (long wavelength)**—A range from 1565 nm to 1625 nm
- **U-band (ultra-long wavelength)**—A range from 1625 nm to 1675 nm

Standard SMF (ITU G.652) is recommended for use with O-band WDM systems. Low-water-peak fiber (ITU G.652.C) is recommended for use with E-band WDM systems, and nonzero dispersion-shifted fiber (ITU G.655) is recommended for use with S-, C-, and L-band WDM systems.

Unidirectional WDM

Unidirectional WDM systems multiplex a number of wavelengths for transmission in one direction on a single fiber. For example, signals at various wavelengths in the C-band are multiplexed together for transmission over a single fiber. The receiver receives multiplexed wavelengths on a separate fiber. The end-WDM device is responsible for demultiplexing the wavelengths and feeding them to the appropriate receiver. Figure 4-3 shows unidirectional WDM. Unidirectional WDM systems are very common with cable providers who transmit multicast traffic to downstream receiving stations.

Bidirectional WDM

A bidirectional WDM system transmits and receives multiple wavelengths over the same fiber. For example, signals at various wavelengths in the 1550-nm band are multiplexed together for transmission over a single fiber. At the same time, separate wavelengths in the 1550-nm band are also received over the same fiber. The end-WDM device is responsible for multiplexing and demultiplexing the wavelengths from and to their respective transmitters and receivers. Figure 4-4 shows a bidirectional WDM system.

Figure 4-3 *Unidirectional WDM*

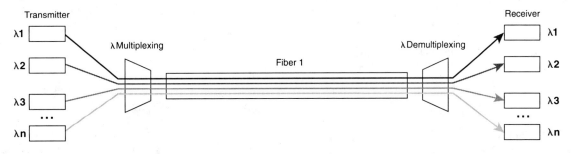
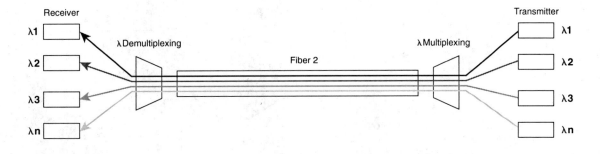

Figure 4-4 *Bidirectional WDM*

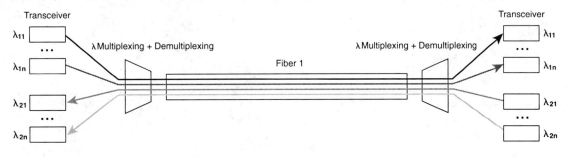
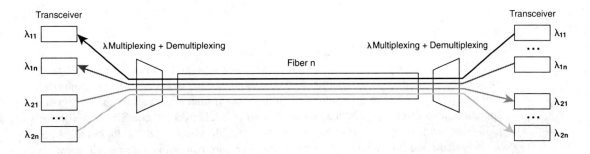

Various techniques are used to achieve full-duplex bidirectional transmission over a single fiber. Basically, the counter-propagating signals on the same fiber have to be separated by using suitable devices. Figure 4-5 shows three methods used to achieve bidirectional transmission.

Figure 4-5 *Bidirectional WDM Techniques*

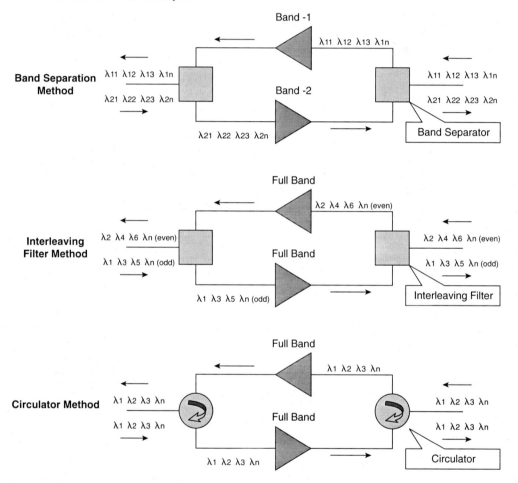

Band-Separation Method

In this method, the transmitted channels are divided in two or four groups known as *sub-bands*, traveling in opposite directions. Sub-bands are separated and combined by optical interleavers inserted in line along the transmission medium. To prevent the adjacent bands from interfering with each other along the transmission fiber and to allow for easier band separation, a spectral gap known as a *guard-band* is left between them. The need for the guard-band leads to inefficient utilization of the available spectral bandwidth in bidirectional WDM systems and fundamentally limits the number of transmissible channels. Typically, the number of wavelengths supported by the band-separation method is 32.

Interleaving-Filter Method

The interleaving technique uses wavelength-interleaving filters at each end of the span. As shown in Figure 4-5, interleaved channels are used in both directions of transmission. Even channels travel east to west, whereas odd channels travel west to east. As a consequence, channel spacing for wavelengths traveling in the same direction has to be doubled. However, the interleaving filters have a high insertion loss that contributes to higher system losses.

Circulator Method

In this technique, the same wavelengths are transmitted in both directions of propagation. To separate transmit and receive direction at any node, optical circulators are used. A *circulator* is a multiport device that allows signals to propagate in certain directions based on the port that the signal came from. The circulator essentially acts as an isolator that allows only unidirectional propagation.

Channel Spacing

The minimum wavelength separation between two different channels multiplexed on a fiber is known as *channel spacing*. Channel spacing ensures that neighboring channels do not overlap, causing power coupling between one channel and its neighbor. Channel spacing is a function of the precision of a laser. The more precise the tuning, the lower the channel spacing required. For example, a 100-GHz-spaced laser typically has one-half the preciseness of a 50-GHz-spaced laser. The precision of the laser has a linear relationship with the cost of the laser. The spacing that can be used is affected by the existing fiber characteristics. Fiber characteristics fall in the linear and nonlinear domains. Examples of linear fiber impairments would be that of attenuation and dispersion, whereas examples of nonlinear fiber impairments include four-wave mixing, cross-phase modulation, SRS, and SBS.

Another factor that affects channel spacing is the optical amplifier's capability to amplify the channel range. The closer the wavelengths are placed, the more important it is to ensure that the centers are identifiable from other signals on the same fiber. Channel-spacing values range from 200 GHz (1.6 nm), 100 GHz (0.8 nm), 50 GHz (0.4 nm), 25 GHz (0.2 nm), up to 12.5 GHz (0.1 nm). The ITU has published a wavelength grid as part of International Telecommunication Union Telecommunication Standardization Sector (ITU-T) G.694.1 and G.694.2 to provide interoperable standards for companies to work from.

To understand the effects of channel spacing on bandwidth capacity, consider a sample operating frequency band between 1505 nm and 1625 nm. This band provides a 120-nm spectrum. In theory, a fiber should be able to carry 150 wavelengths with 100-GHz spacing between wavelengths, 300 wavelengths with 50-GHz spacing, 600 wavelengths with 25-GHz spacing, or 1200 wavelengths with 12.5-GHz spacing. Assuming that each wavelength operates at 40 Gbps, it provides a theoretical maximum of 6 Tbps with 100-GHz spacing, 12 Tbps with 50-GHz spacing, 24 Tbps with 25-GHz spacing, or 48 Tbps with 12.5-GHz spacing.

Coarse Wavelength-Division Multiplexing

Coarse wavelength-division multiplexing (CWDM) systems are suited for the short-haul transport of data, voice, video, storage, and multimedia services. CWDM systems are ideally suited for fiber infrastructures with fiber spans that are 50 km or less and that don't need signal regeneration or the presence of optical amplifiers. The WDM laser bit rate directly determines the capacity of the wavelength and is responsible for converting the incoming electrical data signal into a wavelength.

CWDM systems use lasers that have a bit rate of up to 2.5 Gbps (OC-48/STM-16) and can multiplex up to 18 wavelengths. This provides a maximum of 45 Gbps over a single fiber. The transmitting laser and receiving detector are typically integrated into a single assembly called a *transceiver*.

Figure 4-6 shows a CWDM schematic. CWDM systems are characterized by a channel spacing of 20 nm or 2500 GHz as specified by the ITU standard G.694.2. The CWDM grid is defined in terms of wavelength separation. This grid is made up of 18 wavelengths defined within the range 1270 nm to 1610 nm. CWDM transceivers support the use of lower-cost distributed feedback (DFB) lasers that don't need any external cooling. These lasers are characterized with a drift of about 6 nm over a temperature range of 0 to 70 degrees Celsius. This, coupled with laser variations of up to +/−3 nm, yields a total wavelength variation of approximately 12 nm. The 20-nm channel spacing or guard-band between wavelengths provides adequate tolerance for the +/−12-nm wavelength drift. CWDM transceivers typically use highly sensitive avalanche photodiodes (APD) as receivers. CWDM gigabit interface converters (GBICs) support ITU standard G.694.2 wavelengths are increasingly being used with CWDM systems as a lower-cost alternative to conventional CWDM transceivers.

Figure 4-6 *Coarse WDM*

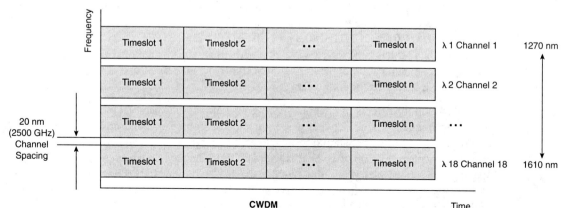

Dense Wavelength-Division Multiplexing

Dense wavelength-division multiplexing (DWDM) systems are suited for the short-haul and the long-haul transport of data, voice, video, storage, and multimedia services. DWDM systems are ideally suited in the metro or long-haul core where capacity demands are extremely high. These higher-capacity demands result from the aggregation of services received from multiple customers at the enterprise edge. In such a case, the service provider is faced with the option of obtaining permits, retrenching, and installing new fiber versus obtaining DWDM equipment and lighting up wavelengths. If more than 18 wavelengths are required during the planned life cycle of the equipment to meet the future capacity expectations, a DWDM system should be considered versus a CWDM system.

Typical DWDM systems use lasers that have a bit rate of up to 10 Gbps (OC-192/STM-64) and can multiplex up to 240 wavelengths. This provides a maximum of 2.4 Tbps over a single fiber. Newer DWDM systems will be able to support 40-Gbps wavelengths with up to 300 channels, resulting in 12 Tbps of bandwidth over a single fiber. DWDM transceivers consume more power and dissipate much more heat than CWDM transceivers. This creates a requirement for DWDM cooling subsystems.

Figure 4-7 shows a DWDM schematic. Metro DWDM systems deployed today typically use 100-GHz or 200-GHz frequency spacing. DWDM common spacing can be 200, 100, 50, 25, or 12.5 GHz with a channel count reaching up to 300 or more channels at distances of several thousand kilometers with amplification and regeneration along such a route. As specified by the ITU standard G.694.1, DWDM systems are characterized by channel spacing of 50 or 100 GHz. The ITU DWDM frequency grid is anchored to 193.1 THz. DWDM systems have a significantly finer granularity between wavelengths (100-GHz typical spacing) versus their CWDM counterparts. ITU grid DWDM products operate in the C-band between 1530 and 1565 nm or L-band between 1565 and 1625 nm. It must be noted that not all fiber plants deployed in the past can be used for DWDM transmission, because most DWDM equipment currently uses the C-band or L-band window. Legacy fiber was optimized for transmission in the O-band (1310-nm band) window. All fiber should be characterized and tested before deploying a DWDM infrastructure.

Figure 4-7 *Dense WDM*

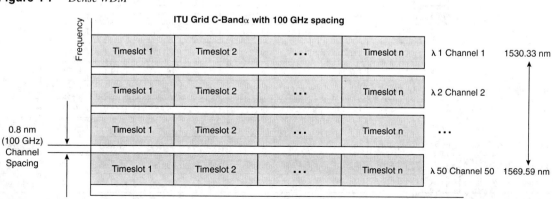

The ITU Grid

The ITU-T established a set of standards for telecommunications that drives all optical WDM systems today. All land-based DWDM systems follow this standard. The ITU-T has been engaged in standardization initiatives to enable international interoperability of various WDM systems. The introduction of the ITU-T G.694.1 frequency raster has made it easier to integrate WDM with older but more standard SONET systems. The ITU G.694.1 and G.694.2 standard replaces the earlier G.692 specification. The G.694.1 recommendation specifies a frequency grid for DWDM applications. The frequency grid, anchored to 193.1 THz or 1552.52 nm, supports a variety of channel spacing of 12.5 GHz (0.1 nm), 25 GHz (0.2 nm), 50 GHz (0.4 nm), and 100 GHz (0.8 nm). Most commercial DWDM systems deployed today typically utilize 50-GHz and 100-GHz frequency spacing. Table 4-1 illustrates the ITU-T G.694.1 grid for DWDM systems with 50-GHz (0.4-nm) and 100-GHz (0.8-nm) spacing between wavelengths.

Table 4-1   *ITU G.694.1 Grid*

L-Bandα		L-Bandβ		C-Bandα		C-Bandβ		S-Bandα		S-Bandβ	
100-GHz Grid		50-GHz Offset		100-GHz Grid		50-GHz Offset		100-GHz Grid		50-GHz Offset	
THz	nm	THz	nm	THz	nm	THz	nm	THz	nm	THz	nm
186.00	1611.79	186.05	1611.35	191.00	1569.59	191.05	1569.18	196.00	1529.55	196.05	1529.16
186.10	1610.92	186.15	1610.49	191.10	1568.77	191.15	1568.36	196.10	1528.77	196.15	1528.38
186.20	1610.06	186.25	1609.62	191.20	1567.95	191.25	1567.54	196.20	1527.99	196.25	1527.60
186.30	1609.19	186.35	1608.76	191.30	1567.13	191.35	1566.72	196.30	1527.22	196.35	1526.83
186.40	1608.33	186.45	1607.90	191.40	1566.31	191.45	1565.90	196.40	1526.44	196.45	1526.05
186.50	1607.47	186.55	1607.04	191.50	1565.50	191.55	1565.09	196.50	1525.66	196.55	1525.27
186.60	1606.60	186.65	1606.17	191.60	1564.68	191.65	1564.27	196.60	1524.89	196.65	1524.50
186.70	1605.74	186.75	1605.31	191.70	1563.86	191.75	1563.45	196.70	1524.11	196.75	1523.72
186.80	1604.88	186.85	1604.46	191.80	1563.05	191.85	1562.64	196.80	1523.34	196.85	1522.95
186.90	1604.03	186.95	1603.60	191.90	1562.23	191.95	1561.83	196.90	1522.56	196.95	1522.18
187.00	1603.17	187.05	1602.74	192.00	1561.42	192.05	1561.01	197.00	1521.79	197.05	1521.40
187.10	1602.31	187.15	1601.88	192.10	1560.61	192.15	1560.20	197.10	1521.02	197.15	1520.63
187.20	1601.46	187.25	1601.03	192.20	1559.79	192.25	1559.39	197.20	1520.25	197.25	1519.86
187.30	1600.60	187.35	1600.17	192.30	1558.98	192.35	1558.58	197.30	1519.48	197.35	1519.09
187.40	1599.75	187.45	1599.32	192.40	1558.17	192.45	1557.77	197.40	1518.71	197.45	1518.32
187.50	1598.89	187.55	1598.47	192.50	1557.36	192.55	1556.96	197.50	1517.94	197.55	1517.55
187.60	1598.04	187.65	1597.62	192.60	1556.55	192.65	1556.15	197.60	1517.17	197.65	1516.78
187.70	1597.19	187.75	1596.76	192.70	1555.75	192.75	1555.34	197.70	1516.40	197.75	1516.02
187.80	1596.34	187.85	1595.91	192.80	1554.94	192.85	1554.54	197.80	1515.63	197.85	1515.25
187.90	1595.49	187.95	1595.06	192.90	1554.13	192.95	1553.73	197.90	1514.87	197.95	1514.49
188.00	1594.64	188.05	1594.22	193.00	1553.33	193.05	1552.93	198.00	1514.10	198.05	1513.72

Table 4-1 ITU G.694.1 Grid (Continued)

L-Band α 100-GHz Grid		L-Band β 50-GHz Offset		C-Band α 100-GHz Grid		C-Band β 50-GHz Offset		S-Band α 100-GHz Grid		S-Band β 50-GHz Offset	
THz	nm	THz	nm	THz	nm	THz	nm	THz	nm	THz	nm
188.10	1593.79	188.15	1593.37	193.10	1552.52	193.15	1552.12	198.10	1513.34	198.15	1512.96
188.20	1592.95	188.25	1592.52	193.20	1551.72	193.25	1551.32	198.20	1512.58	198.25	1512.19
188.30	1592.10	188.35	1591.68	193.30	1550.92	193.35	1550.52	198.30	1511.81	198.35	1511.43
188.40	1591.26	188.45	1590.83	193.40	1550.12	193.45	1549.72	198.40	1511.05	198.45	1510.67
188.50	1590.41	188.55	1589.99	193.50	1549.32	193.55	1548.91	198.50	1510.29	198.55	1509.91
188.60	1589.57	188.65	1589.15	193.60	1548.51	193.65	1548.11	198.60	1509.53	198.65	1509.15
188.70	1588.73	188.75	1588.30	193.70	1547.72	193.75	1547.32	198.70	1508.77	198.75	1508.39
188.80	1587.88	188.85	1587.46	193.80	1546.92	193.85	1546.52	198.80	1508.01	198.85	1507.63
188.90	1587.04	188.95	1586.62	193.90	1546.12	193.95	1545.72	198.90	1507.25	198.95	1506.87
189.00	1586.20	189.05	1585.78	194.00	1545.32	194.05	1544.92	199.00	1506.49	199.05	1506.12
189.10	1585.36	189.15	1584.95	194.10	1544.53	194.15	1544.13	199.10	1505.74	199.15	1505.36
189.20	1584.53	189.25	1584.11	194.20	1543.73	194.25	1543.33	199.20	1504.98	199.25	1504.60
189.30	1583.69	189.35	1583.27	194.30	1542.94	194.35	1542.54	199.30	1504.23	199.35	1503.85
189.40	1582.85	189.45	1582.44	194.40	1542.14	194.45	1541.75	199.40	1503.47	199.45	1503.10
189.50	1582.02	189.55	1581.60	194.50	1541.35	194.55	1540.95	199.50	1502.72	199.55	1502.34
189.60	1581.18	189.65	1580.77	194.60	1540.56	194.65	1540.16	199.60	1501.97	199.65	1501.59
189.70	1580.35	189.75	1579.93	194.70	1539.77	194.75	1539.37	199.70	1501.21	199.75	1500.84
189.80	1579.52	189.85	1579.10	194.80	1538.98	194.85	1538.58	199.80	1500.46	199.85	1500.09
189.90	1578.69	189.95	1578.27	194.90	1538.19	194.95	1537.79	199.90	1499.71	199.95	1499.34
190.00	1577.86	190.05	1577.44	195.00	1537.40	195.05	1537.00	200.00	1498.96	200.05	1498.59
190.10	1577.03	190.15	1576.61	195.10	1536.61	195.15	1536.22	200.10	1498.21	200.15	1497.84
190.20	1576.20	190.25	1575.78	195.20	1535.82	195.25	1535.43	200.20	1497.46	200.25	1497.09
190.30	1575.37	190.35	1574.95	195.30	1535.04	195.35	1534.64	200.30	1496.72	200.35	1496.34
190.40	1574.54	190.45	1574.13	195.40	1534.25	195.45	1533.86	200.40	1495.97	200.45	1495.60
190.50	1573.71	190.55	1573.30	195.50	1533.47	195.55	1533.07	200.50	1495.22	200.55	1494.85
190.60	1572.89	190.65	1572.48	195.60	1532.68	195.65	1532.29	200.60	1494.48	200.65	1494.11
190.70	1572.06	190.75	1571.65	195.70	1531.90	195.75	1531.51	200.70	1493.73	200.75	1493.36
190.80	1571.24	190.85	1570.83	195.80	1531.12	195.85	1530.72	200.80	1492.99	200.85	1492.62
190.90	1570.42	190.95	1570.01	195.90	1530.33	195.95	1529.94	200.90	1492.25	200.95	1491.88

α = ITU 100-GHz grid
β = 50-GHz offset from the ITU 100 GHz grid

The G.694.2 recommendation specifies a wavelength grid for CWDM applications. The ITU has set a global standard for metro optical networks that will expand the use of CWDM in metropolitan networks. The wavelength plan contained in ITU-T Recommendation G.694.2 has a 20-nm channel spacing to accommodate lasers that have high spectral width and/or large thermal drift. Recommendation G.694.2 provides a grid for CWDM with 18 wavelengths defined within the range 1270 nm to 1610 nm spaced by 20 nm (2500 GHz). The CWDM ITU-T G.694.2 grid is illustrated in Table 4-2.

Table 4-2 *ITU G.694.2 Grid*

Center Wavelength (nm)
1270
1290
1310
1330
1350
1370
1390
1410
1430
1450
1470
1490
1510
1530
1550
1570
1590
1610

NOTE Currently, many existing Cisco DWDM devices operate in the C-band and L-band with spacing of 100 GHz using the older ITU grid standard G.692 and Telcordia GR-2918-CORE, issue 2. However, note that the G.692 wavelengths are a subset of the G.694 standard.

Wavelength-Division Multiplexing Systems

WDM optical networks consist of five main components:

- Transmitters
- Optical multiplexers and demultiplexers

- Amplifiers
- Optical-fiber media
- Receivers

The various WDM components are integrated to form WDM systems. This section introduces a typical WDM node and then proceeds to explain the various subsystems and their relevant technologies in greater detail. As illustrated in Figure 4-8, a WDM node consists of a multiplexer-demultiplexer section, a switching section, and a local interface section. The local interface section consists of transponders, which are further broken down into optical sources, optical detectors, and complex electronic circuitry. The multiplexer and demultiplexer sections consist of optical multiplexers and demultiplexers. The switching section typically has an array of O-E-O (opto-electro-opto) or O-O-O (optical-to-optical) switches in add/drop configuration or cross-connect configuration. A WDM node can typically add, drop, or pass through wavelengths. This is facilitated by O-E-O or O-O-O switch fabrics. At the ingress of each node, the composite WDM signal is pre-amplified by using optical amplifiers. The amplified signal is then fully demultiplexed and switched locally. The switching operation is followed by a multiplex section, which regroups the individual wavelengths into a composite signal.

Figure 4-8 *WDM System*

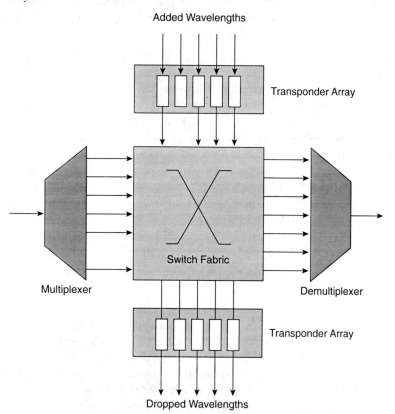

Transmitters

Transmitters format electrical bits to optical pulses and are frequency-specific. Lasers are used as optical sources for launching modulated data into an optical fiber. Lasers have a distinct property whereby they can emit a narrow beam of light with a narrow spectral width, while having a high-output optical power. Transmitters use narrow-band lasers with a narrow spectral width to generate the optical pulses. Transmission is in the infrared band and must be very tightly controlled to generate the correct wavelength. The laser transmitter expects certain environmental parameters for proper operation and regulated sources of electrical power.

Semiconductor materials, such as silicon, germanium, indium phosphide, and gallium arsenide, can be converted to P-type or N-type materials by doping the material (adding an impurity) with electrons to create an N-type material or by extracting free electrons from the material to create a P-type material. A P-N junction diode is formed by fusing the P-type material with N-type material at a boundary interface. Semiconductor lasers are based on the optical properties of a P-N junction. As the current crosses the P-N junction, free electrons absorb a quantum of energy and jump into an excited state. After a period of time, these excited electrons, which have absorbed the excess energy and have risen to a higher excited state, drop back to the original state by emitting the excess absorbed energy in the form of photons. This random oscillation of electrons from a lower energy level to a higher energy level and the subsequent emission of photonic radiation (light) is called *spontaneous emission*, in the sense that it has a random phase and frequency distribution.

Spontaneous emission cannot sustain optical communication for the simple reason of its low power and wide spectra of emission. Now, if an external photon is bombarded onto the electrons that are in the excited state, the electrons would fall from the excited state to the ground state, emitting photons that have the same frequency (as well as phase) as the incident-bombarded photon. This kind of emission is called *stimulated emission* because of the external stimulus involved in the emissive process. To sustain a stimulated emission of this kind, it is necessary to establish *population inversion*, in which the number of electrons in the higher state (excited) should be greater than the number of electrons in the lower (stable) state. One possible way to achieve population inversion is to have multiple energy levels. The cut-in point, at which stimulated emission is the dominant emission in the system, is called the *lasing threshold*. As soon as population inversion is established, the system exhibits an optical gain. By placing the P-N junction inside a cavity that consists of reflecting walls, optical feedback can be achieved that ensures the oscillatory function of the P-N junction and its continued operation as a laser.

Distributed Feedback Lasers

Distributed feedback lasers are constructed by placing the P-N junction in a cavity that has fully reflecting walls on all but one side and a partial reflector on the remaining side. Optical feedback is achieved by inserting a Bragg grating (corrugated surface) within the cavity as shown in Figure 4-9. A Bragg grating is made of a section of dry-etched indium gallium arsenide phosphide or a piece of silica that has holographic modifications to create periodic changes in its refractive index. Light traveling through the Bragg grating is refracted and then reflected back slightly, usually occurring at one particular wavelength. The reflected wavelength is known as the *Bragg resonance wavelength*. This optical feedback is called *distributed feedback* (DFB) due to its diverse occurrence in the cavity. The feedback is throughout the length of the cavity and is essential for maintaining the lasing threshold. The feedback wave adds in phase to the emitted radiation, which is due to the electrons dropping from the excited state to the ground state. The P-type material and N-type substrate are typically constructed out of indium phosphide. The waveguide could be constructed out of indium gallium arsenide (InGaAs) or a polymide. The waveguide layers help constrict the photon beam within a narrow path.

Figure 4-9 *Distributed Feedback Laser*

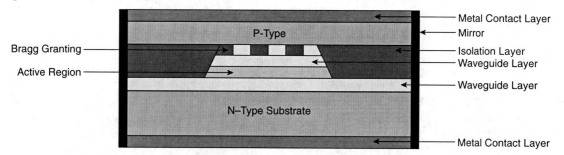

DFB lasers are edge emitting lasers currently used for both CWDM and DWDM transmitters. DBF lasers are also used with tunable optical sources. For metro applications, direct current modulation is preferred for CWDM and DWDM lasers to minimize the cost of the transmitter. This is in contrast to regional and long-haul applications, for which higher-cost externally modulated DWDM lasers are often used to avoid frequency chirping effects caused by direct modulation, associated dispersion, nonlinear impairments, and transmission-distance limitations.

CWDM-based DFB lasers are normally not cooled. Typical DWDM-based DFB lasers must be cooled to isolate the laser from the ambient temperature changes that would cause their wavelength to drift outside the DWDM filter pass-band. Each laser includes a thermister to monitor the laser temperature. In the case of fixed-wavelength lasers, it is often the thermister that is calibrated during manufacture to set the precise ITU wavelength of a laser. The calibrated thermister increases the manufacturing yield by allowing a wider variation in the nominal wavelength of the laser than the DWDM filter pass-band would suggest. In the case of 50-GHz, 25-GHz, and 12.5-GHz spaced lasers, a thermister enables coarse temperature control, but an external wavelength locker is required for finer wavelength control.

Distributed Bragg Reflector Lasers

Distributed Bragg reflector (DBR) lasers use a similar principle of operation to DFB lasers. DBR lasers extend the feedback that is associated through the grating through the entire region of the cavity. The Bragg grating now extends to the mirrored walls, thus enhancing tunability. Wavelength tunability can be achieved by varying the grating periods outside the active region of the P-N junction. The change of current over the grating also changes the Bragg's wavelength and the associated feedback. In this way, a DBR laser can be tuned across several nanometers relatively quickly. DBR lasers are used with tunable optical sources. Figure 4-10 illustrates the schematic of an edge emitting DBR laser.

Figure 4-10 *Distributed Bragg Reflector Laser*

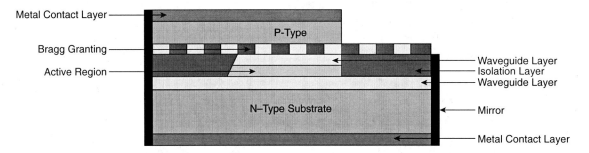

DBR and DFB lasers are temperature dependent and need temperature-controlling elements for stable uniform operations. The temperature-controlling elements add a significant cost to the laser transmitters used in C-band applications.

Tunable Lasers

DWDM systems need separate lasers mounted on cards or modules that are configured for a specific ITU-T wavelength. This means that carriers have to keep spares readily on hand that are wavelength-specific. This results in huge inventories of spare lasers and their associated modules that are vendor-specific. DFB and DBR lasers can be considered as first-generation tunable lasers, because they can be configured to two or possibly four different wavelengths. However, tunable laser modules can transmit many wavelengths across a wide spectrum as compared to the DFB and DBR lasers widely used today. DFB arrayed lasers have their tuning range limited by the number of elements in the array and require 1 to 10 milliseconds (ms) of tuning time.

Mechanical Tunable Lasers

Mechanical tunable lasers use micro-electro-mechanical systems (MEMS) as their wavelength actuators. The Fabry Perot (FP) cavity laser emits wavelengths that are a function of the cavity length (FP cavity) between the highly reflective walls of the cavity. Varying the length between the walls of the cavity can change the resonant frequency. MEMS actuators can physically vary the length between the cavity walls resulting in a variation of the output wavelength. The physical distance between cavity walls is carefully calibrated to emitted ITU-T wavelengths. In this way, variation in voltage across the MEMS actuators results in modification of the emitted wavelengths. Figure 4-11 illustrates the schematic of a mechanical tunable laser using an FP cavity with the air gap between the polished surfaces of two fibers. The MEMS actuator mechanism is physically attached to the fiber on one side. The MEMS actuator provides the linear motion required to vary the output wavelength. Mechanical lasers have a tuning range of 10 to 20 nm and require 100 to 500 ms of tuning time.

Figure 4-11 *Mechanical Tunable Laser*

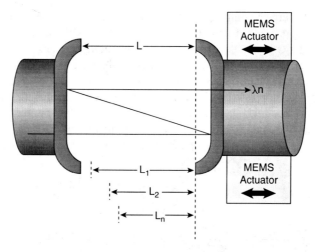

External Cavity Tunable Lasers

External cavity tunable lasers include collimating optics, mirrors, and an external grating. Commercial external cavity tunable laser modules also include an integrated thermal tuner, a gain chip, an isolator, and a number of flexures that precisely align the laser to the fiber before it is mounted in the package. External cavity tunable lasers have a typical output power in excess of 20 milliwatts (mW). Generally, the more widely tunable the laser, the less its transmit power. Newer DWDM systems are being designed with external cavity tunable lasers. External cavity tunable lasers can switch from one wavelength to another in the order of nanoseconds, and this makes them potential candidates for building optical switches that could route traffic on a wavelength basis. Figure 4-12 shows a schematic of an external cavity tunable laser using an external grating and collimating optics. The specific ITU wavelength can be tuned by means of the integrated thermal tuner.

Figure 4-12 *External Cavity Tunable Laser*

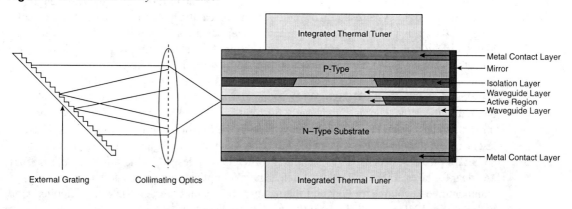

Vertical Cavity Surface Emitting Lasers

Vertical cavity surface emitting lasers (VCSELs) are a type of semiconductor diode laser whose cavity is perpendicular to the wafer plane. A VCSEL emits light perpendicular to the plane of P-N junction, unlike conventional semiconductor lasers. As formed on the wafer, the cavity of a VCSEL is complete, requiring no cleaving of facets or any other mechanical intrusion into the device. A VCSEL cavity contains distributed Bragg gratings that stabilize the wavelength and suppress the side modes. This design provides the advantages of low-cost manufacturing and high spectral performance. The VCSEL contains an active region and mirrors. A sandwich of active regions between the mirrors is created by stacking the subcomponents vertically on top of each other. Figure 4-13 shows a schematic of a VCSEL.

Figure 4-13 *Vertical Cavity Surface Emitting Laser*

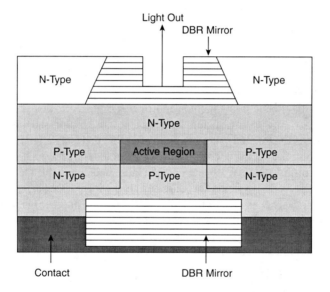

The cavity length of VCSELs is very short and typically one to three times the wavelengths of the emitted light. As a result, in a single pass of the cavity, a photon has a small chance of triggering a stimulated emission event at low carrier densities. Therefore, VCSELs require highly reflective mirrors to be efficient, necessitating the requirement of distributed Bragg reflectors. Most VCSEL devices use quantum wells within the cavity. For fiber-optic communication, VCSEL lasers use indium gallium arsenide phosphide (InGaAsP) for wavelengths of 850 and 1310 nm. When a small current is applied across the device, light is emitted in the active region of the laser. This light is reflected back and forth between the mirrors, while a fraction of the light leaks through the mirror to form the laser beam. These lasers are efficient and have voltage requirements due to the high gain and small volume of the VCSEL structures. Because light is emitted vertically with respect to the wafer plane, VCSELs are ideal for laser array applications, such as parallel optical interconnects. They are now manufactured in volume for Gigabit Ethernet and 10 Gigabit Ethernet WDM LAN applications with 850-nm/1310-nm and single-mode or multimode options.

> **NOTE** VCSELs are capable of fast direct modulation speeds up to 2.5 Gbps. This means that no additional external components, such as modulators, are required in the transmitter construction.

Chirp

Chirp is an abrupt change of the center wavelength of a laser, caused by laser instability. When the modulator pulses the laser, a difference in the refractive index of the laser output can cause chirp in a WDM system. Chirp is the phenomenon of the rising edge of a pulse having a slightly different frequency than the falling edge. Chirp can be defined as the change in frequency of the transmitted optical signal with respect to time. In semiconductor lasers, the frequency of a pulse shifts to a shorter frequency from the original frequency due to chirp. Nonlinear effects can also introduce chirp in optical communication systems. You can reduce the chirping effect that results from lasing by using external modulators. Chirp usually occurs with a value of +1 GHz to –1 GHz. Each laser transmits coherent light at a different center frequency for each λ. Chirp can be provisioned to match the system input requirements on many WDM systems. On systems that allow changes, the technician can adjust the chirp value to support the network requirement; generally, the technician can report only the presence and degree of chirp.

Modulators

Modulation can be achieved by superimposing a serial data stream onto a carrier signal by altering one of the parameters of the carrier signal with respect to a change in the data stream. In an optical network, data is modulated onto the light that a laser emits. Modulators modulate the laser signal by either pulsing it off and on or by changing the phase of the signal so that it carries information. There are two main techniques for modulation: direct and external modulation. In semiconductor lasers, the frequency of a pulse shifts to a shorter frequency from the original frequency due to an effect called chirp. Chirp results from the sudden change in density of electron-hole pairs in the active region of a laser diode, on application of a voltage. Chirp influences the refractive index of the material and also increases the temperature of the active region. Chirping effects are more pronounced in direct modulators. However, chirping that results from the lasing effect can be reduced by using external modulators.

Direct Modulation

The direct modulator performs an electrical-to-optical (EO) conversion, and the parallel-to-serial converter converts parallel data to a serial bit stream. The electrical bit stream is used to pulse the laser resulting in a series of light pulses launched into the fiber. In the direct modulation technique, serialized data is directly coupled with the laser drive current. Direct modulation uses the return-to-zero (RZ) modulation format, and the laser diode switches between on and off for a logical 1 and logical 0, respectively. Direct modulation has severe drawbacks at high data rates. It typically cannot be used at bit rates that are greater than 2.5 Gbps. Direct modulation creates nonlinearity, such as self-phase modulation (SPM), and also increases the laser chirp. Directly modulated lasers are limited by distance and are suited for metro applications. Figure 4-14 illustrates the schematic of a directly modulated laser with a feedback circuit. The feedback circuit controls the voltage and biasing current across the laser to maintain constant pulse amplitude.

Figure 4-14 *Direct Modulation*

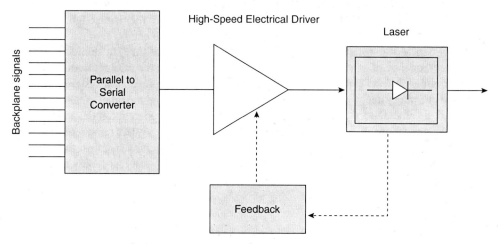

External Modulation

In an externally modulated system, a DC-biased laser producing a continuous wave (CW) feeds an external modulator, which modulates the CW signal into desired optical bit streams. Externally modulated lasers are more stable and are used frequently in WDM systems. Most commercial systems contain the laser diode and the modulator as a single unit. In external modulation, the laser output power is generally modulated in an external cavity. External modulation avoids nonlinearity and excessive chirp. Typically, two kinds of modulators are available: electro-absorption modulators (EAMs) and Mach-Zehnder interferometer (MZI) modulators. Lithium niobate-based MZIs are more common.

EAM lasers have the advantage of size over their MZI counterparts because they are much smaller than MZIs. The single biggest advantage of having external modulators is the reduced chirp, which means that the signal occupies less bandwidth. Typically, MZI modulators are known to have twice the bit rate as their bandwidth. This also means that the spacing between adjacent channels in a WDM system can be greatly reduced. Most external modulators use the nonreturn-to-zero (NRZ) modulation format. Return-to-zero (RZ) and carrier-suppressed return-to-zero (CS-RZ) modulation formats are also used in WDM systems. The RZ format is preferred in long-haul and ultra-long-haul systems. Figure 4-15 illustrates a schematic of an externally modulated system.

Figure 4-15 *External Modulation*

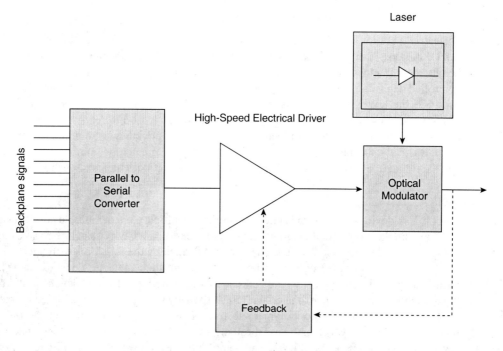

NOTE It is a common practice for enterprise engineers and technicians to peek into the end of a fiber connector to detect a *red* light that would indicate an active *link*. However, note that DWDM and other long-haul optical systems use the invisible infrared band. There is the risk of self-inflicted damage to the technician's eyes by looking directly at these lasers. DWDM lasers are usually Class I lasers, and that means that enough light power is present to cause eye damage or blindness if the person exposed looks directly into a fiber end. Most modern lasers have automatic laser safety (ALS) systems that prevent such an occurrence.

Optical Multiplexers and Multiplexers

Optical multiplexers combine discrete wavelengths for launch into the fiber, whereas demultiplexers separate the discrete wavelengths. Many technologies are used in optical multiplexing, including the following:

- Thin film filters
- Fiber Bragg gratings (FBGs)
- Arrayed waveguide gratings (AWGs)
- FP cavity filter
- Acousto optical tunable filters

- Mach-Zehnder interferometers
- Interleavers, periodic filters, and frequency slicers

Dielectric thin film filter (TFF) technology is used to manufacture low-cost CWDM and DWDM filters. The desired filter characteristics (center wavelength, channel bandwidth, ripple height, insertion loss, skirt width, and adjacent channel isolation) are all achieved through controlled deposition of optical layers of different refractive index dielectric material on a glass substrate. For DWDM applications, the same TFF technology has been refined and has become an attractive, low-cost alternative to FBGs. As a result, the cost differences between CWDM and DWDM filters is now determined by the number of optical layers required to implement them. Basically, the narrower the wavelength spacing, the greater the number of optical layers and the tighter the accuracy needed to meet the thickness requirements of each layer.

Thin Film Filter

The TFF is a device used in some optical networks to multiplex and demultiplex optical signals. TFFs are similar to cavity filters in the sense that the resonant cavity selects the wavelengths that are allowed to traverse through. The cavity is formed by the thin films with interfaces that act as reflectors. The wavelength or group of wavelengths that is selected depends on the length of this cavity. TFFs are typically made of quarter wavelength ($\lambda/4n$)-thick layers of alternating high- and low-refractive index materials. TFFs use many ultra-thin layers of dielectric material coating deposited on a glass or polymer substrate. This substrate can be made to let only photons of a specific wavelength pass through, while all others are reflected. By integrating several of these components, you can then demultiplex several wavelengths. Figure 4-16 shows a TFF schematic with n wavelengths.

Figure 4-16 *Thin Film Filter*

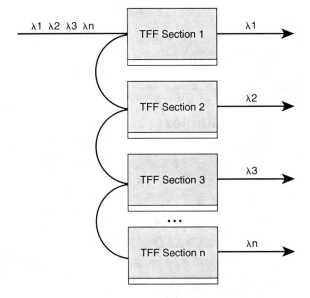

The first TFF section passes wavelength 1 and reflects 2, 3, and 4 to the second, which then passes 2 and reflects 3 and 4. This allows for demultiplexing or multiplexing of optical signals. Typically there are more than 200 layers required for a 100-GHz filter design, as used in metro DWDM products, whereas there are only 50 layers in a 20-nm filter (as used in metro CWDM products).

Fiber Bragg Grating

A Bragg grating is made of a small section of fiber that has been modified by exposure to ultraviolet radiation to create periodic changes in the refractive index of the fiber. The result, shown in Figure 4-17, is that light traveling through the Bragg grating is refracted and then reflected back slightly, usually occurring at one particular wavelength.

Figure 4-17 *Fiber Bragg Grating*

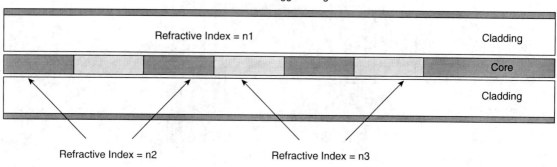

The reflected wavelength, known as the *Bragg resonance wavelength*, depends on the amount of refractive index change that has been applied to the Bragg grating fiber and this also depends on how distantly spaced these changes to refraction are. Gratings can be etched onto a fiber in numerous ways. One popular method is to use photosensitivity of doped germanium in fiber and etch a grating pattern by exposing the photosensitive fiber to alternating intensities of ultraviolet light. The FBG can be characterized by low-loss (0.1-dB) and low-channel cross talk. FBGs find applications in most WDM systems, such as channel-drop elements, dispersion-compensation devices, and filters.

FBGs have a range of 10 nm and require a tuning time of 1 to 10 seconds with mechanical tuning mechanisms.

Arrayed Waveguide

An AWG device consists of multiple waveguides of different lengths converging at the same point(s). Signals passing through each of these waveguides interfere with signals passing through their neighboring waveguides at the converging point(s). The interference could be either constructive or destructive depending on the net phase difference between the signal and its interfering counterpart(s). In the transmit direction, the AWG mixes individual wavelengths from different lines etched into the AWG substrate (the base material that supports the waveguides) into one etched line called the *output waveguide*, thereby acting as a multiplexer. In the opposite direction, the AWG can demultiplex the composite λs onto individual etched lines. Usually one AWG is for transmit and a second one is for receive. Figure 4-18 illustrates the AWG demultiplexer in the receiver.

Figure 4-18 *Arrayed Waveguide Demultiplexer*

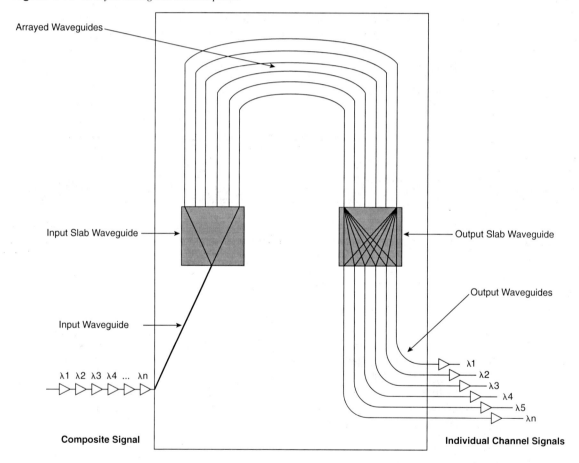

The AWG can replace multiple Bragg gratings; each Bragg grating supports only one wavelength and occupies the same physical space as an 8-λ AWG. Multiple Bragg gratings also cost more than a single AWG. For some applications, AWG offers a higher channel capacity at a lower cost per channel with a smaller footprint. This results in fewer components and provides for component integration in multiplexers and switching systems. AWG tunable filters have a tuning range of 40 nm. The tuning time of AWG tunable filters is approximately 10 ms. AWG tunable filters use thermo-optic tuning mechanisms.

Fabry Perot Cavity Filter

An FP cavity is constructed out of two reflective surfaces separated by a refractive medium. The two reflective surfaces have a reflectivity that is a function of the operating wavelength, and the cavity acts as a resonant cavity for particular wavelengths. As illustrated in Figure 4-19, the reflectivity can be changed for different resonant wavelengths. The reflective surfaces are connected to an electromechanical transducer, and the distance between the reflective surfaces can be changed by varying the current to the transducer. Applying a voltage to the transducer mechanically shifts the mirrors closer or apart. The change in length is a function of the wavelength. FP cavity mirrors are typically about 150 to 200 μm apart, and common cavities are constructed by using the air gap between the polished surfaces of two fibers. In such construction, the air gap acts as the refractive medium and the two polished surfaces act as the reflective surfaces.

Figure 4-19 *Fabry Perot Cavity Filter*

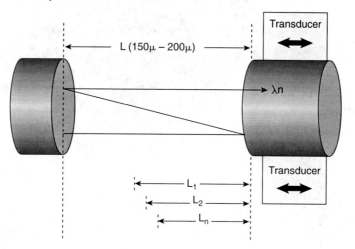

Acousto Optical Tunable Filter

Acousto optical tunable filters (AOTFs) create a series of ultrasound waves inside a tellurium dioxide (TeO2) crystal waveguide forming an acoustic grating. As illustrated in Figure 4-20, the ultrasound waves are generated by applying an RF signal to the crystal. Ultrasound waves are essentially longitudinal waves with propagation based on formation of compression and rarefaction zones, unlike the crest and troughs in a transverse wave (light wave). These compressions and rarefactions are equivalent to regions of high- and low-refractive index. Light passing through such a disturbance has the same effect as passing through a grating. The interaction of light with the acoustic waves is called the *photon-phonon interaction* and results because of an effect known as the *photo-elastic effect*. A photon-phonon interaction can easily be understood as collision under energy conservation. AOTF can be fabricated best by using

TeO$_2$ or lithium niobate (LiNbO$_3$) waveguides, producing small polarization-independent filters. AOTFs are characterized by a tuning range of 250 nm covering both C- and L-bands. Tuning times of AOTF are very low, in the range of several microseconds. AOTFs can also be used in optical router implementations.

Figure 4-20 *Acoustic Optical Tunable Filter*

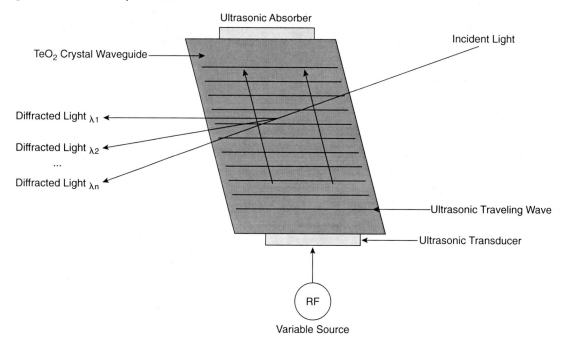

Mach-Zehnder Interferometers

A Mach-Zehnder interferometer (MZI) consists of two passive couplers connected in tandem. The couplers are connected to the two arms of the MZI. The voltage can be varied across the MZI arm to change the coupling ratio between the two arms. As illustrated in Figure 4-21, the couplers are equally balanced and the input power is equally split between the two arms. The signals in the two arms interact with each other twice. The first coupler (A) divides the signal into two. The two propagating signals can be made to obtain different phase shifts by varying the lengths of the two arms. The signals, upon interfering with each other at the second coupler (B), might have constructive or destructive interference. The constructive or destructive interference blocks one wavelength. MZIs can be built on planar silica, indium phosphide, or lithium niobate substrates. MZIs have a tuning range of 4 nm and have tuning times in the order of tens of nanoseconds.

Figure 4-21 *Mach-Zehnder Interferometer*

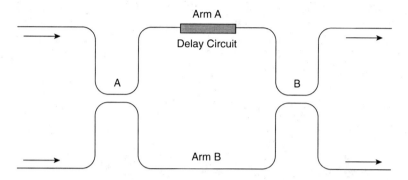

Couplers, Circulators, and Isolators

Couplers are passive bidirectional devices that could have multiple inputs and output ports. The purpose of a coupler is to fuse the cores of the n input fibers to the cores of m output fibers, so as to create an optical power transfer device. In practice, 2×2 couplers are most common and are known as 3-dB couplers because of their 3-dB insertion loss. Couplers are used for monitoring WDM ports and for passively adding channels into a fiber. They are also used as optical hubs in passive optical networks (PONs). Figure 4-22 (a) illustrates the schematic of a 2×2 coupler.

A *circulator* is a multiport device that allows signals to propagate in certain directions based on the port that the signal came from. The circulator essentially acts as an isolator that allows only unidirectional propagation. As illustrated in Figure 4-22 (b), a circulator typically supports three ports. A signal entering port 1 is directed to port 2, whereas the signal entering port 2 is directed to port 3. The connection between port 3 and port 1 is generally not available. The circulator uses polarization beam splitters to separate wavelengths. If a WDM system multiplexer is connected to port 1, all channels will exit from port 2 and travel down the single transmission fiber toward the next node. In contrast, the WDM signal coming from the other node will enter the circulator from port 2 and will exit from node 3, in turn connected to the demultiplexing stage. The circulator has a port-to-port loss of approximately 0.7 dB, leading to a total span loss of 1.4 dB when also considering the circulator at the other end. In addition, circulator port isolation is not infinite. As a result, a small fraction of the signal entering from port 1 and directed to port 2 will leak into port 3, leading to some cross talk with the signal entering from port 2 and directed to port 3. The circulator and isolator both operate on the principles of light polarization.

An *isolator* is constructed out of polarizers and Faraday rotators, as illustrated in Figure 4-22 (c). A *polarizer* is a device that allows light to pass through only if it is polarized in a certain phase.

Only the incident light that matches the phase of the polarizer passes through. This light is now subjected to a Faraday rotator, which rotates the state of polarization (SOP) by 45 degrees. A further rotation of 45 degrees by the second rotator makes the output state of polarization at the end of the second rotator 90 degrees as compared to the original input state (SOP). If this light is reflected back, it is blocked by the polarizer, because its SOP is 90 degrees out of phase with that of the polarizer.

Figure 4-22 *Couplers, Circulators, and Isolators*

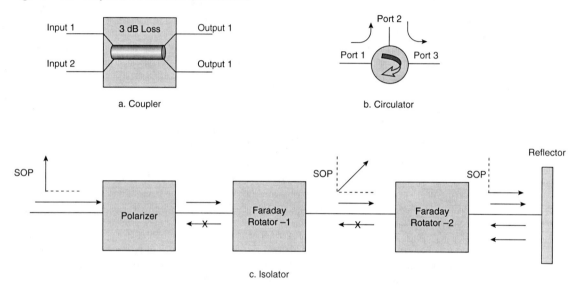

Periodic Filters, Frequency Slicers, and Interleavers

Periodic filters, frequency slicers, and interleavers are devices that can share the same functions and are usually used together. Stage 1 is a kind of periodic filter, such as an AWG. Stage 2 is representative of a frequency slicer on its input—in this instance, another AWG—and an interleaver function on the output, provided by six Bragg gratings. Six λs are received at the input to the AWG, which then breaks the signal down into odd λ and even λ.

The odd λs and even λs go to their respective Stage 2 frequency slicers and then are delivered by the interleaver in the form of six discrete interference-free optical channels for end-customer use. By splitting a DWDM spectrum into multiple complementary sets of periodic spectra, the combined devices can create hierarchical suites of wavelengths for more complex wavelength routing and switching. Figure 4-23 illustrates the schematic of the combined devices.

Figure 4-23 *Combined Devices*

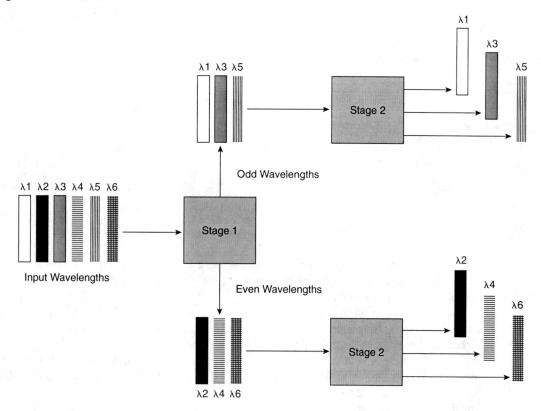

Amplifiers

Pre-amplifiers boost signal pulses at the transmit side, whereas postamplifiers boost signal pulses at the receive side. In-line amplifiers (ILA) are placed at different distances from the source to provide recovery of the signal before it is degraded by loss. Amplifiers are defined as type 1R, 2R, or 3R:

- **1R**—Re-amplify
- **2R**—Re-amplify and reshape
- **3R**—Re-amplify, reshape, and retime

Expansion of WDM networks to greater distances and/or more nodes requires the insertion of either a repeater or an amplifier. Repeaters can provide 3R regeneration to overcome loss and dispersion limitations, whereas amplifiers provide 1R regeneration to overcome optical power-loss limitations only. The 1R device amplifies only the signal received. A 2R device provides amplification and reshaping of the waveform to provide some data recovery. The 3R device provides amplification and reshaping and requires a time source so that it can provide retiming for the transponder. 3R devices include 1R and 2R as well as 3R. Figure 4-24 illustrates network generation using 1R, 2R, and 3R devices.

Figure 4-24 *Network Regeneration*

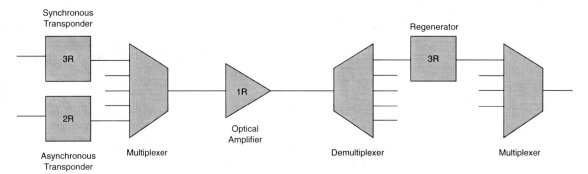

For WDM systems, it is not sufficient to regenerate a single wavelength. It might be necessary to regenerate all wavelengths at appropriate points in the optical network before the optical signal degrades too much. Such multiwavelength regenerators have 3R regeneration bit rates programmed for each channel at the time of connection establishment. Note that a multiwavelength 3R repeater is not practical in a high-capacity DWDM network due to the higher number of wavelengths to regenerate and/or the size, power, and cost associated with each wavelength regenerated. This is why C- and L-band optical amplifiers are preferred for regional and long-haul DWDM networks. For loss-limited, WDM metro networks with a high number of nodes, an alternative lower-cost solution is a simple silicon optical amplifier (SOA). Small, low-cost SOAs cover the O-, E-, S-, C-, and L-bands and can be used for CWDM and for DWDM applications.

SOAs include rare-earth elements, such as the following, to make rare-earth-doped fibers into optical amplifiers:

- Tellurium, which is a compound of tellurite and oxygen (TeO_2)
- Thulium, which is a compound of thulium and fluoride (TmF_3)

Other optical amplifiers include the following:

- **PDFA**—Pracydymium-doped fiber amplifier (1310–1380 nm)
- **EDFA**—Erbium-doped fiber amplifier (1530–1565 nm)
- **GS-EDFA**—Gain-shifted EDFA (1565–1625 nm)
- **EDTFA**—Tellurium-based gain-shifted TDFA (1530–1610 nm)
- **GS-TDFA**—Gain-shifted thulium-doped fiber amplifier (1490–1530 nm)
- **TDFA**—Thulium-doped fluoride-based fiber amplifier (1450–1490 nm)
- **RFA**—Raman fiber amplifier (1420–1650 nm or more)

Erbium-Doped Fiber Amplifiers

EDFAs provide the gain mechanism for DWDM amplification. DWDM systems use erbium amplifiers because they work well and are very efficient as amplifiers in the 1530- to 1565-nm range. Only a few parts per billion of erbium are needed for doping. Light is pumped in at around 980 nm and/or 1480 nm (pump laser diode) to excite the erbium ions, which then amplify the incoming C-band wavelengths from the source. Figure 4-25 illustrates a schematic of an EDFA.

Figure 4-25 *Single-Stage EDFA Schematic*

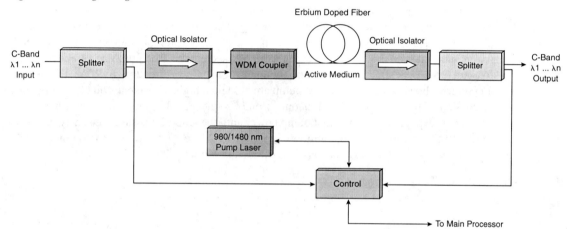

EDFAs use an active medium, which is essentially a few meters of rare-earth-doped special fiber in which the power exchange between a pump signal and the transmission signal takes place. The optical amplification process is intrinsically unidirectional. The unidirectional behavior is further enhanced by the presence of optical isolators that are analogous to an electronic diode and only allow the signals to flow in one direction. Optical isolators are required to prevent optical back reflections from the transmission fiber to enter the amplifying medium and disturb the amplification process. The EDFA is the last active component in the DWDM system on the transmit side (post amplifier). On the receive side, the pre-amplifier (a receive EDFA) is the first active component. Different C- and L-band amplifiers are required because EDFA must be optimized for either C-band or L-band amplification. L-band EDFAs are doped with erbium and with ytterbium ions:

- High pump power with short EDFA fiber is used for C-band amplifiers.
- Medium pump power with long EDFA fiber is used for L-band amplifiers.

Increased pumping power can enhance the gain of a signal. A pumping efficiency of 1 dB/mW can be achieved with a 980-nm pump providing a WDM gain of 30 dB over all channels. EDFAs can be made to operate with pumping in the same and in opposite directions of the signal. When a pump signal is in the same direction of the WDM signal, it is known as *forward pumping*. When a pump signal is in the opposite direction of the WDM signal, it is known as *reverse pumping*. *Bidirectional pumping* is achieved when two pumps are applied in the forward and reverse direction at the same time.

Raman Fiber Amplifiers

Raman fiber amplifiers (RFAs) use the effects of stimulated Raman scattering (SRS). SRS is a type of nonlinear scattering that results in broadband amplification of multiple optical channels. Raman amplification occurs when a lower wavelength pump signal is made to propagate through a fiber. The pump signal creates a wide-band Stoke's wave that transfers energy and amplifies the multiple channels in a WDM system. The gain spectra for Raman amplification is quite broad (150 to 200 nm) and covers the entire operating S-, C-, L-, and U-bands.

In Raman amplification, the pump photon loses its energy to create another photon at a lower frequency (higher wavelength) with lower energy. The difference in energy creates optical phonons, which are absorbed by the medium. The phono-to-phonon interaction is responsible for creating the gain or amplification of the input wavelengths. The pumping can be done in forward (co-propagating) and in reverse (counter-propagating) direction. In RFAs, the signal is amplified only when both signal and pump are at logical high. This means that the pump must be continuously run to amplify all incoming pulse sequences. The amount of amplification achieved is exponential to the input pump power until the amplifier is saturated and gain begins to stabilize. Typically, Raman amplifiers can produce 20- to 35-dB gain with 800 mW to 1W pump power. A schematic of an RFA is illustrated in Figure 4-26.

Figure 4-26 *Raman Fiber Amplifier Schematic*

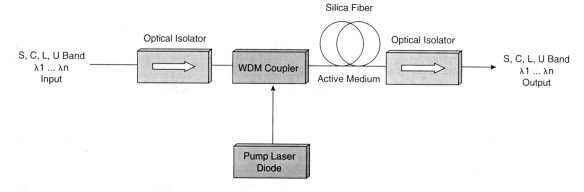

Hybrid and Distributed Amplifiers

RFAs have a relatively flat gain profile over an operating band and a lower noise figure (NF) as compared to EDFAs. This results in a better optical signal-to-noise ratio (OSNR) for an optical signal if we use Raman amplifiers rather than EDFAs. However, Raman amplifiers have certain disadvantages compared to EDFAs. They need high-input pump power, which can cause severe nonlinear impairments, such as self-phase modulation (SPM) and cross-phase modulation (XPM). In addition, they are generally more expensive than EDFAs. The low-noise and wide-operating bandwidth of RFAs and low pump power requirements of C-band and L-band EDFAs can be combined into hybrid amplifiers that can be applied in long-haul and ultra-long-haul networks. Optical signals can be amplified at various points throughout the network. This kind of amplification is called *distributed amplification* and can be achieved by injecting pumps at various sections of the fiber-optic cable or by ensuring a large effective length for Raman amplification. Distributed amplification can be achieved for both Raman amplifiers and EDFAs. In addition, bidirectional pumping helps flatten the gain profile. Table 4-3 compares the various amplifier types.

Table 4-3   *Optical Amplifier Comparison*

	Amplifier Type		
Parameter	**EDFA**	**RAMAN**	**SOA**
Gain	~30 dB	~20–25 dB	~10–20 dB
Output Power	High	High	Low
Input Power	Moderate	High	High
Cross Talk	Low	Low	Very high
Gain Tilt	High	Low	High
Application	Metro, long haul	Long haul, ultra long haul	Short haul, single channel

Optical-Fiber Media

Optical-fiber transmission media is used to carry optical pulses. Many different kinds of fiber are currently in use. Optical-fiber systems and their characteristics are examined in great detail in an earlier chapter. See Chapter 3, "Fiber-Optic Technologies," for details on fiber-optic cable types and their characteristics.

Receivers

Receivers detect optical pulses and convert them back into electrical bits. Receivers typically use photodiodes to convert energy from photons back into electrons. Figure 4-27 illustrates a WDM receiver schematic. In contrast to standard single-protocol receivers, the receivers used in multichannel CWDM and DWDM systems require larger bandwidths that can capture all the specified bit rates and protocols.

Figure 4-27   *WDM Receiver Schematic*

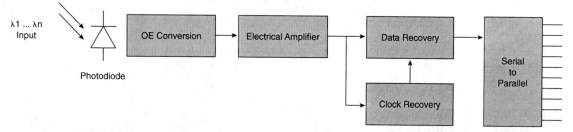

Photodetectors need to meet stringent requirements to achieve desirable performance. Requirements include high sensitivity to a wide range of wavelengths used for transmission in the various O-, E-, S-, C-, L-, and U-operating bands. In addition, other requirements include low noise characteristics, extremely low sensitivity to temperature variations, low cost, and an extended operating life. Even though several types of photodetectors are available, semiconductor-based photodetectors or photodiodes are used exclusively for optical communications. The two widely used receiver photodiodes in use are as follows:

- **PIN photodiodes**—Simplest and fastest
- **Avalanche photodiodes (APDs)**—Slower, but are much more sensitive to light as compared to PIN photodiodes

Wide-band trans-impedance amplifiers (TIAs) are integrated with the detectors for maximum receiver sensitivity. Regeneration is then provided using either 2R or 3R techniques. The latter is nearly always used for the higher data rates to maximize receiver sensitivity and transmission distances.

PIN Photodiode

The PIN photodiode or P-type intrinsic N-type diode is a modified P-N junction diode that has a slightly doped intrinsic material embedded in between the P-N junction. This increases the depletion width of the P-N junction. The depletion region is the region in the P-N junction that is formed by some of the electrons from the N-type material moving over and depleting the holes in the P-type material, thereby creating a region of neutral charge, upon condition of reverse bias. A high reverse-biased voltage is applied across the PIN diode so that the intrinsic region is completely depleted. When light (photons) is incident on a semiconductor material, electrons in the valence band absorb the light. As a result of this absorption, the photons transfer their energy and excite electrons from the valance band to the conduction band, leaving holes in the valance band. After the application of voltage across the depletion region, the formed electron-hole pairs induce an electric current flow known as *photocurrent* in an external circuit. Each electron-hole pair generates one electron flow. Figure 4-28 illustrates the schematic of a PIN photodiode.

Figure 4-28 *PIN Photodiode*

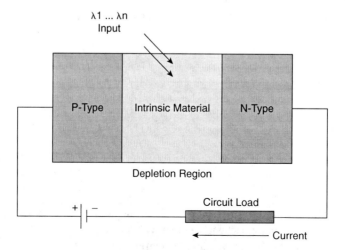

Avalanche Photodiode

APDs have an increased sensitivity as compared to PIN diodes. When light is absorbed by a PIN photodiode, only a single electron-hole pair is generated per photon. The sensitivity of the detectors can be increased if more electrons are generated. This means that the optical signal can travel longer and lower optical power signals can be detected. APDs use a high electric field that is applied across the photodiode to excite more electrons from the valence band to the conduction band. This, in turn, results in more electron-hole pairs being generated. These secondary electron-hole pairs that are generated by the preceding process, in turn, produce more electron-hole pairs. This process of multiplication of electron-hole pairs is known as *avalanche multiplication*. The schematic of an APD is illustrated in Figure 4-29. The benefit of the APD detectors is a 9- to 10-dB improvement in receiver sensitivity over PIN photodiodes.

Figure 4-29 *Avalanche Photodiode*

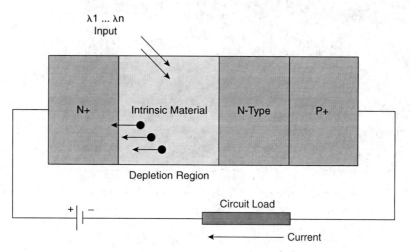

WDM Characteristics and Impairments to Transmission

The bit error rate (BER) is a ratio of error bits to total transmitted bits. Typical values are 10^{-12} BER for SONET and 10^{-15} for next-generation long-haul transport equipment. The value 10^{-15} is 1 error bit in 10^{15} bits, which equates to one error in 11.6 days for a 10-Gbps signal. The eye pattern in Figure 4-30 is a visual depiction an OC-192/STM-64 waveform. It consists of the waveform for each wavelength overlaid on one screen. The eye pattern display allows for quick verification of signals that meet performance specifications. In the display, the 1 signals are above the center point, and the 0 signals are below the center point. An eye pattern is a *sampling oscilloscope* display in which a pseudorandom optical data signal from a optical receiver is repetitively sampled and applied to the vertical input, while the optical signaling rate is used to trigger the horizontal sweep. System performance information can be derived by analyzing the display. An open eye pattern corresponds to minimal signal distortion. Distortion of the signal waveform due to intersymbol interference and noise appears as closure of the eye pattern.

Figure 4-30 *Eye Pattern*

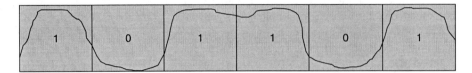

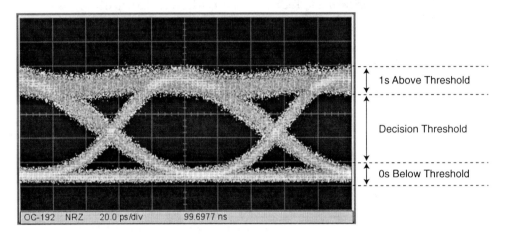

The Q-factor is a measure of how noisy a pulse is for diagnostic purposes. The sampling oscilloscope will typically generate a report that shows what the Q-factor number is as opposed to the ideal Q-factor. The Q-factor can also be determined from system OSNR values. A larger number in the result means that the pulse is relatively free from noise. The arrows show the desired points that are the farthest apart. They show that the eye is open as much as possible and indicate that the data can be recovered easily with low effects from noise.

Forward Error Correction

Forward error correction (FEC) is used to support higher capacity by increasing the channel count and longer transmission distances by improving the BER. FEC makes the system more robust in respect to errors. The FEC code bytes are used at the end of a transmitted frame by the receiving system to find and correct errors. The two main kinds of FEC used in optical transmission are in-band and out-of-band. In-band is sometimes called *simple* FEC.

In-Band FEC

FEC is a form of coding the source data to improve the RX data BER by correcting recovered bit errors in the demodulator output. SONET in-band FEC (IB-FEC) inserts 39 bits of FEC overhead into unused section and line overhead locations. The IB-FEC code specified in the

T1X1.5 is the ANSI T1.105.08 standard. IB-FEC uses the Bose-Chaudhuri-Hocquenghem-3 (BCH-3) (4359, 4320) codes. The BCH-3 codes are described as (N,K), where N is the total number of symbols per codeword, K is the number of information symbols, and R $(N - R)$ is the number of check symbols. This means that the ANSI T1.105.08 (4359,4320) specification provides for 4359 symbols per codeword along with 4320 information symbols and 39 check symbols. IB-FEC provides a correction service to the SONET line layer. The ANSI T1.105.08 standard provides for 24 bits of burst error correction with an interleaving depth of 8. The coding efficiency is 99 percent.

Out-of-Band FEC

Out-of-band FEC (OOB-FEC) improves transmission quality and provides network monitoring and supervision capabilities.

OOB-FEC is typically used for DWDM systems. FEC bytes are added on top of the signal to be carried. Adding OOB-FEC to an OC-192/STM-64 signal changes the signal from 9.953 Gbps to 10.7 Gbps for 10-Gbps SONET transport, resulting in 6 percent overhead added outside the normal signal envelope. OOB-FEC yields concurrent $5 * 10^{-15}$ BER to 10^{-15} BER performance. The effect of approximately 7-dB optical system gain, depending on OSNR and other impairments on the DWDM route, can be achieved. The 7-dB gain is not an actual power gain, rather it is an improvement in the OSNR. This can permit greater distance between sites on the optical span or increased channel count without impacting the optical span budget. FEC provides a method to overcome some of the limitations inherent in DWDM over long-haul networks. Constraints founded in physics principles and the fiber itself can cause suboptimal long-haul DWDM performance and limit capacity. Dispersion, attenuation, and modulation are just some of the occurrences that can impact both the integrity and span of a light signal. OOB-FEC counters these occurrences by compensating for signal noise caused by dispersion, attenuation, and modulation and by improving the overall performance of the signal. OOB-FEC yields a significant improvement in OSNR margins over a non-FEC solution.

OOB-FEC uses a coding scheme that adds redundancy to transmitted data streams, enabling the network to identify and correct corrupted bits and reduce the overall BER. OOB-FEC increases the WDM channel count by supporting 25-GHz channel spacing and compensating for nonlinear effects. It also increases span length and the number of spans. The Cisco OOB-FEC solution is based on Reed-Solomon error-correction algorithms, conforming to the ITU-T G.975 industry standard, and translates to an average gain of 7 dB compared with non-FEC solutions. Two ITU specifications recommend FEC in transmission systems: ITU-T G.709 and ITU-T G.975. ITU-T G.709 defines the network node interface for the optical transport network operating at 2.5, 10, and 40 Gbps, and describes a "wrapper" approach that incorporates a simple framing structure and an FEC section. Specifically intended for use in optical-fiber submarine cable systems, ITU-T G.975 recommends FEC in systems operating at 2.5 Gbps and higher.

OOB-FEC also provides monitoring and supervisory capabilities. With OOB-FEC, it is possible to detect link-section degradation well in advance of service degradation, thereby enabling service providers to guarantee the agreed levels of QoS. Figure 4-31 shows an OOB-FEC schematic. Reed-Solomon codes are described as (N,K), where N is the total number of symbols per codeword, K is the number of information symbols, and R $(N - R)$ is the number of check symbols. For example, the Reed-Solomon codes used in ITU-T G.709 and G.975 are both (255,239)—that is, each consists of 255 total symbols, 239 information symbols, and 16 check symbols. Reed-Solomon codes treat errors on a symbol basis; therefore, a symbol that has all bits in error is as easy to detect and correct as a symbol that has a single bit in error. Reed-Solomon codes let the system detect and correct one error symbol for every two check symbols. Interleaving data from different codewords improves the efficiency of Reed-Solomon codes because the effect of burst errors is shared across many codewords. Both ITU-T G.709 and G.975 specify interleaving as part of the transport frame, which is useful for increasing the error correction and has the additional benefit of making the hardware implementation easier. ITU-T G.709 and G.795 standards provide for a burst error correction of 1024 bits with an interleaving depth of 16 accomplished with 16 parallel encoders/decoders. The coding efficiency is 93.72 percent.

Figure 4-31 *OOB-FEC Schematic*

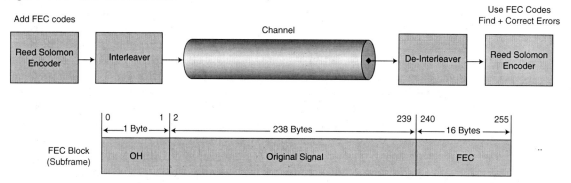

Optical Signal-to-Noise Ratio

Noise is observed in systems that include optical amplification. The noise results from the presence of amplified spontaneous emission (ASE) that is generated within the amplification process and is relatively broadband. The OSNR specifies the ratio of the net signal power to the net noise power. If a signal and noise are both amplified, the system OSNR indicates the quality of the signal by calculating this ratio. System design based on OSNR is an important fundamental design tool. Better OSNR is indicated by higher numbers. One of the consequences of having a low OSNR is that no matter how strong the signal presented to a good receiver, errors will be associated with the presence of the ASE. The exact level of BER for a given OSNR depends upon the receiver. The OSNR is measured in decibels. The OSNR can be determined using the following equation:

$$OSNR = 10 \log_{10}(P_s/P_n)$$

P_s = Power level of the signal (dB)

P_n = Power level of the noise (dB)

OSNR is not just limited to optical amplifier-based networks. Other active and passive devices can also add noise and create an OSNR-limited system design problem. Active devices, such as lasers and amplifiers, also add noise. Passive devices, such as taps and the fiber, can add components of noise. In the calculation of system design, however, optical amplifier noise is considered the predominant source for OSNR penalty and degradation. In WDM links with cascaded amplifier stages, the amplifier noise (ASE) becomes a serious issue. The end-to-end system OSNR degrades because of the cumulative effect of noise figures of each amplifier stage. A figure of merit for optical amplifiers is the noise figure (NF). The NF is the ratio of input OSNR to the output OSNR in an optical amplifier. The NF calculation is used to assist in the design of the routes needed in a DWDM network and OSNR calculations. The NF is measured in decibels. The NF can be determined using the following equation:

$$NF = SNR_{INPUT}/SNR_{OUTPUT}$$

NOTE Raman amplifiers are effective for WDM signals because of their broad flattened gain profiles. They also have lower NFs compared to EDFAs and can be used at higher bit rates. This results in better OSNR for an optical signal if Raman amplifiers are used rather than EDFAs. However, Raman amplifiers have higher pump power requirements that could create nonlinear impairments. In addition, stimulated Raman scattering can cause four-wave mixing, creating harmonic peaks and severe cross talk. Typically, Raman amplifiers cost more than EDFAs. The designer must evaluate the benefits of both amplifier types and take into consideration a cost-benefit analysis before selecting an amplifier type.

The OSNR of a single amplifier stage is determined by the following equation:

$$OSNR = (P_{IN})/(NF_{STAGE} h \nu \nabla f)$$

P_{IN} = Amplifier input power (dB)

NF_{STAGE} = NF for the amplifier stage (dB)

h = Planck's constant (6.6260×10^{-34})

ν = Optical frequency constant (193 THz)

∇f = Bandwidth constant that measures the NF (0.1 nm)

For N stages of amplification, the OSNR is determined by the following equation:

$$1/OSNR = (1/OSNR_1) + (1/OSNR_2) + \cdots + (1/OSNR_N)$$

For an N-stage amplifier system, each amplifier compensates for the span loss of the previous span. We can approximate the final OSNR for a N-stage system with the following assumptions:

- The NF of every amplifier is the same, assuming uniformity of products.
- The span loss is the same for all spans.
- The noise is nonpolarized.

In a multispan system, it is good design practice to consider the highest span loss (worst case) and use its value in OSNR calculations. The final OSNR for an N-stage amplification system is determined by the following equation:

$$OSNR_{FINAL} = P_{IN} - P_S - NF - 10Log_{10}(N) - 10Log_{10}(h\nu\nabla f)$$

P_{IN} = Amplifier input power (dB)

P_S = Total span loss (dB)

NF = NF of the amplifier (dB)

N = Number of spans

h = Planck's constant ($6.6260 * 10^{-34}$)

v = Optical frequency constant (193 THz)

∇f = Bandwidth constant that measures the NF (0.1 nm or 12.5 GHz)

Substituting the values for Planck's constant (h), v and ∇f in the previous equation, we get the following equation:

$OSNR_{FINAL} = P_{IN} + 58 - P_S - NF - 10Log_{10}N$

P_{IN} = Amplifier input power (dB)

P_S = Total span loss (dB)

NF = NF of the amplifier (dB)

N = Number of spans

OSNR Calculation

Consider the WDM system schematic shown in Figure 4-32. Determine the final OSNR for the system and the signal power at the receiver. The amplifier gain at each stage is 21 dB with an NF of 6 dB. The transmitter and receiver specification is shown on the drawing. The attenuation coefficient of the fiber in use is 0.2. The splice and connector loss for each span is represented as L.

Figure 4-32 *OSNR Calculation*

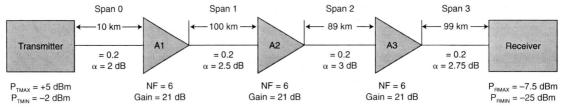

Solution:

From the drawing, you can deduct that there are four spans ($N = 4$). Notice in Figure 4-32 that the spans have varying splice and connector losses. From a design perspective, we consider the highest splice and connector loss value ($L = 3$ dB). We also consider the longest span length (100 km) to determine the highest attenuation per span.

Attenuation per span = $\alpha * 100 = 0.2 * 100 = 20$ dB

Total loss per span $P_S = 20$ dB + 3 dB = 23 dB

To determine the value of the amplifier input power (P_{IN}), we need to consider the value of P_{TMIN} (–2 dBm).

We can use the equation to determine the value of the final system OSNR as follows:

$OSNR_{FINAL} = P_{IN} + 58 - P_S - NF - 10Log_{10}N$

P_{IN} = Amplifier input power = -2 dB

P_S = Total span loss (dB) = 23 dB

NF = NF of the amplifier = 6 dB

N = Number of spans = 4

$OSNR_{FINAL} = -2 + 58 - 23 - 6 - 10Log_{10}(4)$

$OSNR_{FINAL} = 21$ dB

The signal power at the receiver is determined as follows:

$P_{RX} = P_{IN}$ − Loss (Span 0) + Gain (A1) − Loss (Span 1) + Gain (A2) − Loss (Span 2) + Gain (A3) − Loss (Span 3)

Loss (Span 0) = $(\alpha * 10) + 2 = (0.2 * 10) + 2 = 4$ dB

Loss (Span 1) = $(\alpha * 100) + 2.5 = (0.2 * 100) + 2.5 = 22.5$ dB

Loss (Span 2) = $(\alpha * 89) + 3 = (0.2 * 89) + 3 = 20.8$ dB

Loss (Span 3) = $(\alpha * 99) + 2.75 = (0.2 * 99) + 2.75 = 22.55$ dB

$P_{RX} = -2 - 4 + 21 - 22.5 + 21 - 20.8 + 21 - 22.55$

$P_{RX} = -8.85$ dB

P_{RX} is within the dynamic range of the receiver ($P_{RMIN} < P_{RX} < P_{RMAX}$) and the system is viable.

Dispersion and Compensation in WDM

Linear characteristics include attenuation, chromatic dispersion (CD), and polarization mode dispersion (PMD), whereas nonlinear characteristics include cross-phase modulation (XPM), four-wave mixing (FWM), stimulated Raman scattering (SRS), and stimulated Brillouin scattering (SBS). Chapter 3 covers these characteristics. Note that nonlinear characteristics can also be mitigated by using nonzero dispersion shifted fiber (NZDSF). NZDSF fiber overcomes these nonlinear effects by moving the zero-dispersion wavelength outside the 1550-nm operating window. The practical effect of this is to have a small but finite amount of CD at 1550 nm, which minimizes nonlinear effects, such as FWM, SPM, and XPM, which are seen in the DWDM systems without the need for dispersion compensation.

This section discusses chromatic and polarization mode dispersion and covers compensation techniques. There are two main kinds of dispersion; the most common is called *chromatic dispersion* and is routinely compensated for by DWDM systems for proper operation. The effects of PMD are much more subtle and are difficult to compensate for in real networks. Some dispersion is required in WDM networks, because it keeps down cross talk by minimizing stimulated Brillouin scattering. Newer fibers have just enough dispersion to eliminate cross talk. However, they still need a degree of dispersion compensation. The velocity of propagation of light depends on wavelength. The degradation of light waves is caused by the various spectral components present within the wave, each traveling at its own velocity. This phenomenon is called *dispersion*. Several types of dispersion exist, two of which include CD and PMD. CD is common at all bit rates. PMD is comparatively effective only at high bit rates. Waveguide and material dispersion are forms of CD, whereas PMD is a measure of differential group delay of the different polarization profiles of the optical signal. However, note that properly engineered spans safely work within the allowed tolerances for chromatic and polarization impairments.

Chromatic Dispersion

During the propagation of light, all of its spectral components propagate along the fiber at different group velocities. This phenomenon leads to dispersion called *group velocity dispersion (GVD)*. Due to the difference in velocities experienced by various spectral components, the output pulse is time scattered and dispersed in the time domain. GVD is also referred to as *chromatic dispersion*. CD results in a broadening of the input signal as it travels down the length of the fiber. CD is a primary cause of concern in high-bit-rate (> 2.5 Gbps) single-mode WDM systems. Dispersion in an optical pulse creates pulse broadening such that the pulse spreads into the time domain slots of the other pulses. This is known as *intersymbol interference (ISI)*. This leads to severe distortion of the pulses—that is, they become indistinguishable at the receiver. Figure 4-33 illustrates the effects of CD. Each signal has a spectrum that is affected by dispersion. The faster the rate of transmission, the greater the effects of dispersion. Different fibers have different CD characteristics and have been engineered specifically for this reason. The primary capability of CD is directly related to the concentricity of the glass and the size of the effective area. Standard SMF fiber has an average of 18 ps/nm-km of dispersion in the C-band.

Figure 4-33 *Effects of Chromatic Dispersion*

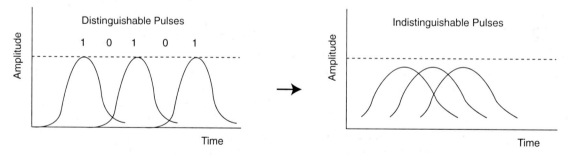

For example, the CD of a fiber at 1550 nm is 18 ps/nm-km. This means that a 500-km system generates 18 ps/nm-km * 500 km = 9000 ps/nm of dispersion. Typically, a 10-Gbps transmitter and fiber system can tolerate about 1200 ps/nm of dispersion. This means that the system is not feasible at the 500-km length.

Chromatic Dispersion Compensation

The DWDM system must compensate for CD to support higher speeds. Dispersion limits the transmission length of an optical link, especially at high data rates. Methods are available to compensate for dispersion. This section discusses dispersion-compensation techniques.
A dispersion-compensation unit (DCU) is a device that has the opposite CD effect as the transmission fiber and provides negative dispersion in ps/nm. CD can be compensated for using

either postcompensation or precompensation techniques. Postcompensation techniques are implemented in receivers with adjoining circuitry that can reduce the amount of dispersion that a signal undergoes. Precompensation is the reverse of postcompensation. In precompensation, the pulses are compensated for dispersion even before they are transmitted through the fiber. The pulses are chirped in such a way that the effects of the fiber channel are not sufficient to disperse the pulse out of its intended time slot. GVD in optical fibers can be compensated for easily by using high-dispersion fibers. The second fiber that is used is the high-dispersion fiber, which usually has a strong dispersion profile that is opposite to the dispersion profile of the first fiber, thereby yielding almost zero dispersion over the entire channel. The problem of having dispersion-compensating fibers is the high-attenuation coefficients of these dispersion fibers. Various technologies are available that can compensate for all wavelengths in a band or for each wavelength. Compensating for all wavelengths greatly reduces the cost of compensation. Per-band compensation is used in some DWDM products. Typically, the higher the dispersion compensated for by providing a negative ps/nm value, the higher the insertion loss of the DCU. For example, a DCU compensating at 350 ps/nm might have a 3-dB insertion loss, whereas a DCU compensating at 1150 ps/nm might have a 6-dB insertion loss.

The various technologies used in DCUs include the following:

- Dispersion-compensating fiber.
- Fiber Bragg gratings.
- High-order mode devices.
- Virtual image phase array (VIPA) is a free-space dispersion device.

The dispersion limit of the system depends on many factors:

- Receiver tolerance.
  - Depends on pulse type; NRZ is better than RZ pulses.
  - Depends on data rate.
  - Depends on transmitter chirp.
- Variability of fiber dispersion.
  - Dispersion of fiber has a range of values.
  - Some temperature dependence occurs.
- Variability of DCU values.
  - Includes slope of dispersion and manufacturing tolerances.
- Variability of components in network elements.
- Receiver dispersion tolerance must be less than the sum of dispersion variances.

Dispersion-compensating fiber (DCF) has been successfully deployed to address dispersion issues in the past. However, DCF is not suitable for next-generation systems operating over a larger number of channels at 10 Gbps or higher. VIPAs use a combination of three-dimensional mirrors and lenses to compensate for dispersion. The main drawback to this approach is the high insertion loss of up to 10 dB.

For long-haul and ultra-long-haul WDM systems operating at 10 Gbps and 40 Gbps, dispersion management across the whole transmission band through accurate dispersion slope matching of each fiber span or through residual slope compensation at the receiver is necessary. In addition, slope matching and dynamically tunable dispersion compensation is needed. Fiber Bragg gratings (FBGs) are an excellent candidate for DCUs. Figure 4-34 (a) shows a schematic of an FBG DCU, and Figure 4-34 (b) shows the typical placement of DCUs on a span.

Figure 4-34 *Chromatic Dispersion Compensation*

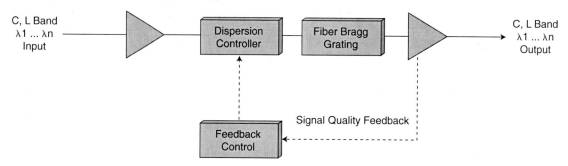

a. Dispersion Compensation Unit

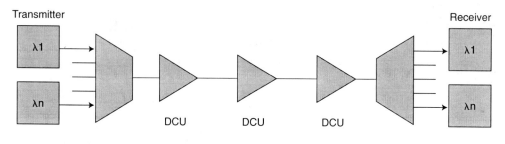

b. DCU Placement

Dispersion compensators are needed at different points along the optical network depending on system requirements. Dispersion compensators can be used periodically across an optical link to compensate for dispersion accumulation. Dispersion compensators are typically inserted at the midstage access of a two-stage amplifier. Dispersion slope mismatch compensators can be employed to compensate for residual dispersion accumulation along multispans of fiber. Dispersion compensators can also take the form of a dynamically tunable device used in the fine-tuning of dispersion on a per-channel basis or for dynamically compensating sub-bands.

Polarization Mode Dispersion

Electromagnetic light exhibits linear polarization, whereby some light rays propagate vertically, whereas others propagate horizontally. As light propagates through a fiber, the wave constantly interacts with the medium. This interaction leads to a condition in which the individual components are no longer equal in magnitude and direction, which in turn leads to PMD. Light, which takes different paths within the fiber, will have polarization differences resulting in dispersion. The fiber core is not a perfect cylinder. This gives rise to birefringence within the fiber that causes the pulse to spread. Ultimately, the amplitude of the light pulses is reduced and a pulse spread occurs as a result of PMD at high bit rates. Figure 4-35 shows the effects of PMD.

Figure 4-35 *Polarization Mode Dispersion*

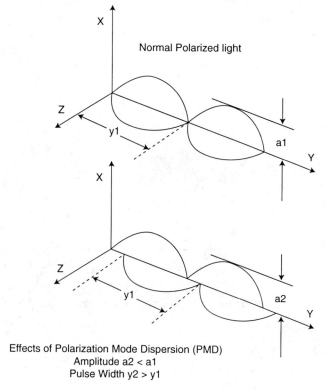

Effects of Polarization Mode Dispersion (PMD)
Amplitude a2 < a1
Pulse Width y2 > y1

PMD was not considered in early fiber manufacturing because of the limited impairments that PMD represented at the lower data rates prevalent at that time. Later, as faster data transmission rates became practical, various manufacturers began to provide solutions that helped manage the PMD effects. (For instance, the outside vapor deposition method produced low-PMD fibers [Corning].) Manufacturing methods have improved, and now fibers can be manufactured that have low PMD. The PMD standard, Standard Reference Materials (SRM) 2518, published by the National Institute of Standards and Technology (NIST), states 0.5 ps of PMD per the square root of the fiber length in kilometers as the proven PMD management interface:

$$0.5 \text{ ps}/\sqrt{km}$$

The new fiber types have less than 0.5 PMD. For example, 10-Gbps signals with 10 ps of PMD tolerance, derived from the preceding formula, would exhibit a range of about 400 km; 40 Gbps with 2.5 ps tolerance has an effective range of 25 km. (PMD compensation is required.)

Research is underway to make even lower PMD fibers. LEAF fiber, with 0.1 ps/√km, allows theoretical distances of up to 10,000 km at 10 Gbps or 625 km at 40 Gbps.

Polarization Mode Dispersion Compensation

PMD compensation techniques include dispersion-maintaining fibers that are constructed by intentionally introducing degrees of birefringence in them that negate the effects of PMD over a length of transmission. A PMD compensator (PMDC) also compensates for PMD. The PMDC device illustrated in Figure 4-36 has a feedback control mechanism. The compensator applies the opposite amount of PMD as that produced by the physical attributes of the fiber network itself. PMD compensation is performed at the receiver.

Figure 4-36 *PMD Compensation*

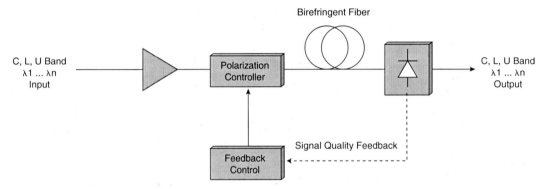

PMDCs must consider the following conditions in their dynamic PMDC operation:

- Signal rate
- Noise accumulation limit of amplifiers
- CD compensation limit
- PMD compensation limit

Summary

This chapter discussed WDM, CWDM, DWDM, the ITU grid, WDM systems, WDM characteristics, and impairments to transmission and to dispersion and compensation in WDM systems. One of the original drivers for the development of WDM technology was the need for sheer bandwidth. This requirement translated into a tangible need to pull additional terrestrial fiber-optic cable. WDM provides additional bandwidth, without having to lay new fiber. WDM is the process of multiplexing signals of different wavelengths onto a single fiber. This operation creates many virtual fibers, each capable of carrying a different signal. WDM uses wavelengths to transmit data parallel-by-bit or serial-by-character, which increases the capacity of the fiber by assigning incoming optical signals to specific frequencies within a designated frequency band and then multiplexing the resulting signals out onto one fiber. WDM can be considered a form of FDM coupled with TDM.

CWDM systems are suited for the short-haul transport of data, voice, video, storage, and multimedia services. CWDM systems are ideally suited for fiber infrastructures with fiber spans that need fewer wavelengths and don't need signal regeneration or the presence of optical amplifiers. CWDM systems use lasers that have a bit rate of up to 2.5 Gbps (OC-48/STM-16) and can multiplex up to 16 wavelengths. This provides a maximum of 40 Gbps over a single fiber. DWDM systems are suited for the short-haul and for the long-haul transport of data, voice, video, storage, and multimedia services. DWDM systems are ideally suited for fiber infrastructures with fiber spans that might need signal regeneration or the presence of optical amplifiers. If more than 16 wavelengths are required during the planned life cycle of the equipment to meet the future capacity expectations, a DWDM system should be considered versus a CWDM system. DWDM systems use lasers that have a bit rate of up to 40 Gbps (OC-768/STM-256) and can multiplex up to 32 wavelengths. This provides a maximum of 320 Gbps over a single fiber.

The ITU-T established a set of standards for telecommunications that drives all optical DWDM systems today. Systems are based on an absolute reference to 193.10 THz, which corresponds to a wavelength of 1552.52 nm with individual wavelengths spaced in steps of 50 GHz, or a wavelength step of 0.41 nm from the reference. All land-based DWDM systems follow this standard. The ITU-T has been engaged in standardization initiatives to enable international interoperability of various WDM systems. The introduction of the ITU-T G.694.1 frequency raster has made it easier to integrate WDM with older but more standard SONET systems. The ITU G.694.1 and G.694.2 standards replace the earlier G.692 specification.

WDM optical networks consist of network elements, such as transmitters, multiplexers and demultiplexers, amplifiers, optical-fiber media, and receivers. Transmitters change electrical bits to optical pulses and are frequency-specific. Tunable lasers are widely used as DWDM transmitter sources. Tunable lasers combine multiple individual lasers onto one piece of silicon. Optical multiplexers combine discrete wavelengths for launch into the fiber, whereas demultiplexers separate the discrete wavelengths. Thin film filters, fiber Bragg gratings, arrayed waveguides, periodic filters, frequency slicers, and interleavers are widely used WDM multiplexers and demultiplexers. Erbium-doped fiber amplifiers and Raman amplifiers provide the gain mechanism for WDM amplification. Receivers detect optical pulses and convert them back into electrical bits. WDM BER is a ratio of error bits to total signals that are below the center point. The Q-factor is a measure of how noisy a pulse is for diagnostic purposes. FEC is used to support higher capacity and longer transmission distances by improving the BER. Impairments on the DWDM route cause changes in OSNR. Such impairments can be compensated for by proper use of pre-emphasis. To optimize optical signals for transmission, equalization is used to adjust the signal across the complete network. Some dispersion is required in WDM networks, because it keeps down cross talk by minimizing stimulated Brillouin scattering. Newer fibers have just enough dispersion to eliminate cross talk. However, they still need a degree of dispersion compensation. There are two kinds of dispersion; the most common is called chromatic dispersion and is routinely compensated for by DWDM systems for proper operation. The effects of PMD are much more subtle and are difficult to compensate for in real networks. The DWDM system must compensate for chromatic and polarization dispersion to support higher speeds.

This chapter includes the following sections:

- **SONET Integration of TDM signals**—This section describes Synchronous Optical Network (SONET) integration of TDM signals.
- **SONET Electrical and Optical Signals**—The differences between the terms OC-N and STS-N are discussed in this section.
- **SONET Layers**—The four SONET optical interface layers—path, line, section, and photonic—are described in this section.
- **SONET Framing**—SONET framing of the Synchronous Transport Signal level 1 (STS-1) is described in this section with a description of the SONET envelope, payload, and path overhead.
- **SONET Transport Overhead**—SONET provides substantial overhead information, allowing simpler multiplexing and greatly expanded operations, administration, maintenance, and provisioning (OAM&P) capabilities. The various SONET section and line overhead bytes are explained in this section in great detail.
- **SONET Alarms**—SONET alarms are defined as anomalies, defects, and failures. This section describes various SONET alarms.
- **Virtual Tributaries**—The STS SPE can be subdivided into smaller structures, known as virtual tributaries (VTs). This section describes the various VT types, such as VT1.5, VT2, VT3, and VT6. This section also describes VT groups, the VT superframe, and VT path overhead.
- **SONET Multiplexing**—Time-division multiplexed (TDM) plesiochronous digital hierarchy (PDH) services can be natively accepted by SONET systems. This section covers various SONET multiplexing schemes.
- **SONET Network Elements**—Various SONET network elements (NEs) are covered in this section including section-, line-, and path-terminating equipment.
- **SONET Topologies**—This section covers SONET topologies and protection architectures that center around network survivability and 50-millisecond or lower service restoration.
- **SONET Protection Architectures**—SONET protection mechanisms are discussed in this section. It covers automatic protection switching (APS) as well as 1+1, 1:1, and 1:N protection architectures.
- **SONET Ring Architectures**—This section discusses SONET ring architectures and describes their inner workings.
- **SONET Network Management**—SONET NEs need OAM&P support to be managed by carriers and service providers. This section describes SONET NE management.

CHAPTER 5

SONET Architectures

SONET Integration of TDM Signals

The SONET format allows different types of signal formats to be transmitted over the fiber-optic cable. It allows adding or dropping signals within a single multiplexer. Communication between various localized networks is complex due to differences in digital signal hierarchies, encoding techniques, and multiplexing strategies. For example, the DS1 signals with alternate mark inversion (AMI) encoding consist of 24 voice signals and 1 framing bit per frame, with a rate of 1.544 Mbps. It robs a bit from an 8-bit byte for signaling and therefore has a rate of 56 kbps per channel. However, with B8ZS encoding, every bit is used for transmission providing a data rate of 64 kbps per channel. The CEPT-1 (E1) signal consist of 30 voice signals and two channels for framing and signaling with a rate of 2.048 Mbps. Communications between TDM networks previously described require complicated multiplexing, demultiplexing, coding, and decoding to convert a signal from one format to another.

To solve this signal-conversion problem, SONET standardizes the rates, framing format, signaling, and termination between SONET equipment. As illustrated in Figure 5-1, SONET multiplexer equipment accepts various native TDM signal formats and multiplexes (adds) these signals without conversion. These signals can be demultiplexed (dropped) at any node or intermediate node.

Figure 5-1 *SONET Integration of TDM*

The SONET standards define the bit rate, format, physical layer, NE architectural features, and network operational criteria of SONET networks. The base standard for SONET is the T1.105 ANSI Telecommunications standard. The various ANSI SONET standards are as follows:

- **ANSI T1.105: SONET**—Basic description including multiplex structure, rates, and formats

- **ANSI T1.105.01: SONET**—APS
- **ANSI T1.105.02: SONET**—Payload mappings
- **ANSI T1.105.03: SONET**—Jitter at network interfaces
- **ANSI T1.105.03a: SONET**—Jitter at network interfaces (DS1 supplement)
- **ANSI T1.105.03b: SONET**—Jitter at network interfaces (DS3 wander supplement)
- **ANSI T1.105.04: SONET**—Data communication channel (DCC) protocol and architectures
- **ANSI T1.105.05: SONET**—Tandem connection maintenance
- **ANSI T1.105.06: SONET**—Physical layer specifications
- **ANSI T1.105.07: SONET**—Sub-STS-1 interface rates and formats specification
- **ANSI T1.105.09: SONET**—NE timing and synchronization
- **ANSI T1.119: SONET**—OAM&P (communications)
- **ANSI T1.119.01: SONET**—OAM&P communications protection switching fragment

The Telcordia FR-SONET-17 family of requirements (FR) contains the generic requirements (GR) criteria that provide the Telcordia framework of broadband transport network requirements relative to SONET technology. The SONET generic requirements are as follows:

- GR-1042 Generic Requirements for Operations Interfaces Using OSI Tools
- GR-1230 SONET Bidirectional Line-Switched Ring Equipment Generic Criteria
- GR-1250 Generic Requirements for Synchronous Optical Network (SONET) File Transfer
- GR-1365 SONET Private Line Service Interface Generic Criteria for End Users
- GR-1374 SONET Inter-Carrier Interface Physical Layer Generic Criteria for Carriers
- GR-1400 SONET Dual-Fed Unidirectional Path Switch Ring (UPSR) Equipment Generic Criteria
- GR-253 Synchronous Optical Network (SONET): Common Generic Criteria
- GR-2891 SONET Digital Cross-Connect Systems with ATM Functionality—Generic Criteria
- GR-2918 Dense Wavelength-Division Multiplexing with Digital Tributaries for Use in Metropolitan Applications—Common Physical Layer Generic Criteria
- GR-2950 Information Model for Wideband and Broadband Digital Cross-Connect Systems (DCS)
- GR-2955 Generic Requirements for Hybrid SONET/ATM Element Management
- GR-2979 Common Generic Requirements for Optical Add/Drop Multiplexers Systems
- GR-2996 Generic Criteria for SONET Digital Cross-Connect Systems
- GR-496 SONET Add-Drop Multiplexer (SONET ADM) Generic Criteria

Plesiochronous Digital Hierarchy

The plesiochronous digital hierarchy (PDH) includes all ITU TDM hierarchies, including the T-carrier and E-carrier systems. Signals are considered to be plesiochronous if their *significant instants occur at nominally the same rate*, with any variations constrained to specified limits. Plesiochronous network timing signals can suffer phase shift, jitter, and wander. For the most part, however, they stay within the predefined limits.

Asynchronous TDM transmission systems are free running with each terminal in the network running on its own clock. In digital transmission systems, clocking maintains network synchronization. Asynchronous systems use terminal internal oscillators or trunk lines to provide clocking. Because these clocks are not synchronized, large variations occur in the clock rate of free-running systems that affect the signal bit rate. For example, a DS3 signal specified at 44.736 Mbps + 20 parts per million (ppm) can produce a variation of up to 1789 bps between one incoming DS3 and another. Asynchronous multiplexing uses multiple stages. Signals, such as asynchronous DS1s, are multiplexed, and extra bits are added using bit-stuffing mechanisms to account for the variations of each individual stream. The extra bits are then combined with framing bits to form a DS2 stream. Asynchronous DS2s are multiplexed, and extra bits are added using bit-stuffing mechanisms to account for the variations of each individual stream, and the extra bits are then combined with framing bits to form a DS3 stream. DS3s are multiplexed up to higher rates in the same manner. Demultiplexing the asynchronous streams provides access to the individual DS-N signals.

In a synchronous system such as SONET, the average frequency of all clocks in the system is the same. Every clock can be traced back to a highly stable reference supply. In synchronous systems, there are several types of stratum clock sources. Typically, a Stratum 3 or Stratum 3E is used as the internal reference source. An internal clock is inadequate for long-term clocking due to wander effects of the free-running clock. This is precisely why Stratum 2 or higher sources are used externally for clocking the SONET network. Stratum 2 clock sources are not part of the NE due to costs associated with production. Stratum 3E is the preferred clocking source for NEs. The STS-1 rate remains at a nominal 51.84 Mbps, allowing many synchronous STS-1 signals to be stacked together when multiplexed without any bit stuffing. Thus, the STS-1s are easily accessed at a higher STS-N rate. Low-speed synchronous virtual tributary (VT) signals are also simple to interleave and transport at higher rates. At low speeds, DS1s are transported by synchronous VT1.5 signals at a constant rate of 1.728 Mbps. Single-step multiplexing up to STS-1 requires no bit stuffing, and VTs are easily accessed. SONET systems are never truly synchronous, especially when the SONET payload must be filled and emptied by plesiochronous T-carrier equipment. Payload pointers accommodate differences in the reference source frequencies and phase wander and prevent frequency differences during synchronization failures. When SONET links are deployed with T-carrier or E-carrier sources and destinations, or when two carriers operating mid-span meet with SONET, payload pointers become critical in keeping the receivers informed about the SPE's location in the SONET frame.

SONET Electrical and Optical Signals

The term OC-N refers to the optical carrier SONET transmission characteristics of an Nth level transmission link, whereas STS-N refers to the electrical SONET transmission characteristics of an Nth level transmission link where $1 <= N <= 768$. SONET frames are created, multiplexed, demultiplexed, and managed electrically. As shown in Figure 5-2, user data is sent to the STS-1 multiplexing device that adds the path overhead (POH) to create the SPE. The transport overhead (TOH) is then added to create the complete SONET STS-1 frame. All of this happens in the electrical domain. Multiple electrical STS-1 frames could be multiplexed at this stage. The electrical STS-N is then fed to an optical transmitter over an STSX-N interface. The STSX-N interface is an interdevice interface for electrical-to-optical (EO) systems.

Figure 5-2 *SONET Electrical and Optical Signals*

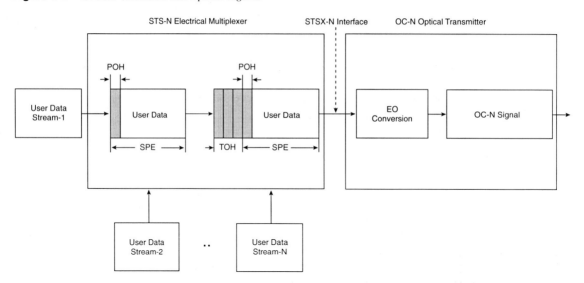

The STSX-1 interface rating permits a distance of 450 feet or 137 meters, whereas the STSX-3 interface rating permits a distance of 225 feet or 69 meters between the electrical and optical devices. The transmitter performs the EO conversion of the STS-N signal and converts the STS-N to an OC-N. The OC-N signal is scrambled before being launched over the fiber-optic media. The Synchronous Digital Hierarchy (SDH) does not make the distinction between electrical and optical nomenclature. The term STM-N is used for both electrical and optical signals.

SONET Synchronization

Network timing between SONET devices is an integral part of maintaining accurate information transmitted over a SONET network. In the earlier days of networking, the method used in timing was asynchronous. In asynchronous timing, each switch runs its own clock. In synchronous timing, switches can use a single common clock to maintain timing. This single common clock is referred to as a *primary reference source (PRS)* or *master clock*. Synchronous timing maintains better accuracy than asynchronous timing because it uses a single common clocking scheme to maintain timing. Five methods are typically used in obtaining synchronous timing in a SONET network:

- Internal timing from a free-running onboard internal oscillator (Stratum 3E or Stratum 3 clock).
- Line timing derived from a scrambled incoming SONET signal on a high-speed interface.
- Through timing derived from signal input, but clocks the signal in a different outbound direction.
- Loop timing is similar to line timing, but is used with customer premises equipment (CPE) or terminal multiplexer equipment rather than ADMs.
- External timing from an external source such as a Stratum 1 global positioning system (GPS) or atomic clock source.

The SONET network is organized with a master-slave relationship with clocks of the higher-level nodes feeding timing signals to clocks of the lower-level nodes. All nodes can be traced up to a primary reference source, a Stratum 1 GPS or atomic clock with extremely high stability and accuracy. Lower-stratum or less-stable clocks are adequate to support the lower nodes. The internal clock of a SONET terminal can derive its timing signal from a building-integrated timing supply (BITS) used by switching systems and other equipment. Thus, this terminal serves as a master for other SONET nodes, providing timing on its outgoing OC–N signal. Other SONET nodes operate in a slave mode called *loop timing*, with their internal clocks timed by the incoming OC–N signal. Current standards specify that a SONET network must be able to derive its timing from a Stratum 3 or higher clock.

Table 5-1 shows the long-term accuracy requirements within the stratum clock hierarchy, which ranges from Stratum 1 through 4. Although synchronous networks display accurate timing, some variations can occur between different network devices or between networks. This difference is known as *phase variation*. Phase variations are defined as jitter or wander. *Jitter* is defined as short-term phase variations above 10 Hz. *Wander* is defined as long-term phase variations below 10 Hz. In digital networks, jitter and wander are handled by buffers found in the interfaces within different network devices. One example is a slip buffer. The slip buffer is used to handle frequency differences between read and write operations. To prevent against write operations happening faster than read operations, read operations are handled at a slightly higher rate. On a periodic basis, read operations are paused, while a bit is stuffed into a stream to account for any timing differences between the read and write operations. This bit-stuffing scheme is referred to as a *controlled slip* (frame). The stratum clocking accuracy timing requirements are also defined in Table 5-1 for wander, bit slips, and controlled slips. The timing accuracy requirements increase as the stratum hierarchy increases.

Table 5-1 *Stratum Clock Hierarchy*

Stratum	Minimum Free-Run Accuracy	Wander (0.12 μs increment)	Bit Slips	Controlled Frame Slips
1	$+/- 1 * 10^{-11}$	3.3 hrs	18 hrs	20.6 wks
2	$+/- 1.6 * 10^{-8}$	7.5 sec	41 sec	2.17 hrs
3	$+/- 4.6 * 10^{-6}$	26 ms	140 ms	27 sec
4	$+/- 32 * 10^{-6}$	4 ms	20 ms	3.9 sec

SONET Layers

SONET has four optical interface layers, which are depicted in Figure 5-3 and are as follows:

- Path
- Line
- Section
- Photonic

Figure 5-3 *SONET Layers*

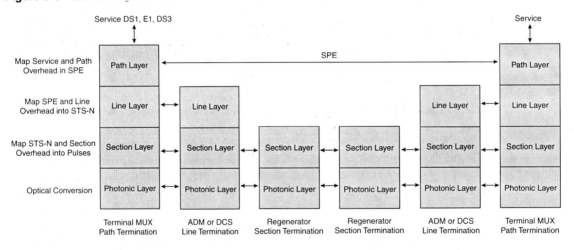

Path Layer

The path layer deals with the transport of services between the path-terminating equipment (PET) NEs. The main function of the path layer is to map the signals into a format required by the line layer. Its functions include reading, interpreting, and modifying the POH for performance and APS.

Path-Terminating Equipment (PTE)

The STS PTE is an NE that multiplexes or demultiplexes the STS payload. As an example, STS-1 PTE assembles 28 1.544-Mbps DS1 signals and inserts POH to form a 51.84-Mbps STS-1 signal. Figure 5-4 shows the PTE.

Figure 5-4 *SONET Termination Equipment*

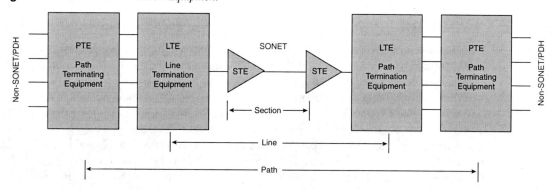

Line Layer

The line layer deals with the transport of the path layer payload and its overhead across the physical medium. The main function of the line layer is to provide synchronization and to perform multiplexing for the path layer. Its functions include protection switching, synchronization, multiplexing, line maintenance, and error monitoring.

Line-Terminating Equipment (LTE)

The line-terminating equipment (LTE) is the NE that originates or terminates the line signal. Figure 5-4 shows the LTE.

Section Layer

The section layer deals with the transport of an STS-N frame across the physical medium. Its main functions are framing, scrambling, error monitoring, and section maintenance.

Section-Terminating Equipment (STE)

The section-terminating equipment (STE) is the NE that can acts as a termination device or a regenerator. The element is able to access, modify, originate, or terminate its overhead. Figure 5-4 shows the STE.

Photonic Layer

The photonic layer mainly deals with the transport of bits across the physical fiber medium. Its main function is the conversion between STS-N and OC-N signals. Its functions include wavelength launching, pulse shaping, and modulation of power levels.

SONET Framing

The STS level 1 (STS-1) is the basic building block of SONET optical interfaces with a rate of 51.84 Mbps. SDH does not have an STS-1 equivalent, because the STM-1 frame is functionally equivalent to the STS-3c. The STS consists of two parts: the STS payload and the STS overhead used for signaling and protocol information. The term *payload* refers to the user data within the STS-N SONET frame. Payloads could be sub-STS-1 level, such as T1 or E1 payloads. As shown in Figure 5-5, the SONET envelope is the portion of the frame that carries the payload and path overhead. The POH and payload constitute the SPE. The SPE floats within the SONET envelope and could begin anywhere within the envelope from columns 4 to 90 of the STS-1. This is due to the fact that the SPE must be frame synchronous with the customer's payload, and that payload could be derived from a TDM system. Multiplexing of various customer payloads onto the synchronous SONET system could lead to phase differences.

Figure 5-5 *SONET Framing*

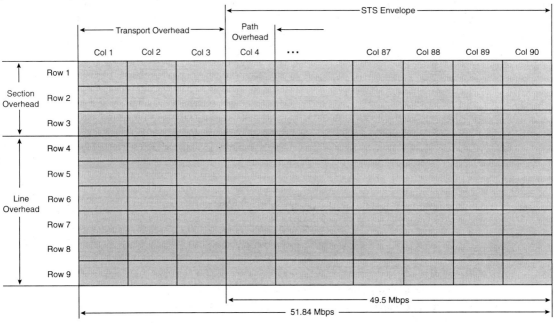

SONET adopts a frame rate of 8000 frames per second. As shown in Figure 5-5, each two-dimensional frame has 9 rows and 90 columns of bytes with a total of 810 bytes (9 rows * 90 columns = 810 bytes). The SONET line rate is synchronous and is flexible enough to support many different signals. The STS-1/OC-1 line rate is 51.84 Mbps that accommodates 28 DS1 signals or 1 DS3 signal. The first 3 columns constitute the transport overhead of 27 bytes. (9 rows * 3 columns = 27 bytes.)

The SONET STS-1 frame has an actual line rate of 51.84 Mbps that can be computed as follows:

Line rate = 90 columns * 9 rows * 8 bits/byte * 8000 Frames/sec = 51.84 Mbps

Columns 4 to 90 constitute the SPE that provides an actual data rate of 50.112 Mbps. The actual data rate of the SPE can be computed as follows:

Data Rate = 87 columns * 9 rows * 8 bits/byte * 8000 frames/sec = 50.112 Mbps

The STS-1 frame is transmitted starting from the first byte in row 1, column 1, to the last byte in row 9, column 90. The most significant bit (MSB) of a byte is transmitted first. After the 90th byte of row 1 is transmitted, the byte in the first column of row 2 is transmitted and so on. Higher line rates are obtained by synchronous multiplexing the lower line rates. Standard OC-N rates have values of N as 1, 3, 12, 48, 192, and 768. Table 5-2 shows line rates for various values of N.

Table 5-2 *OC-N Line Rates*

Optical Level	Electrical Level	Line Rate (Mbps)	Payload Rate (Mbps)	Overhead Rate (Mbps)	SDH Equivalent
OC-1	STS-1	51.840	50.112	1.728	None
OC-3	STS-3	155.520	150.336	5.184	STM-1
OC-9	STS-9	466.560	451.008	15.552	STM-3
OC-12	STS-12	622.080	601.344	20.736	STM-4
OC-18	STS-18	933.120	902.016	31.104	STM-6
OC-24	STS-24	1244.160	1202.688	41.472	STM-8
OC-36	STS-36	1866.240	1804.032	62.208	STM-13
OC-48	STS-48	2488.320	2405.376	82.944	STM-16
OC-96	STS-96	4976.640	4810.752	165.888	STM-32
OC-192	STS-192	9953.280	9621.504	331.776	STM-64
OC-768	STS-768	39813.120	38486.016	1327.104	STM-256

As the line rates increase, the percentage of overhead increases and in turn the useful payload capacity decreases. The additional overheads are used for control, parity, stuffing, alarm, and signaling. For SONET STS-N or OC-N, the ratio of overhead to payload (1.728 Mbps : 50.112 Mbps) remains constant at 3.455 regardless of the value of N.

SONET SPE

Figure 5-6 depicts the STS-1 SPE that occupies the STS-1 envelope. The STS-1 SPE consists of 783 bytes and can be depicted as an 87-column by 9-row structure. Column 1 contains 9 bytes, designated as the STS POH. Columns 30 and 59 are not used for payload, but they are designated as the fixed-stuff columns. The 756 bytes in the remaining 84 columns are designated as the STS-1 payload capacity.

Figure 5-6 *SONET SPE*

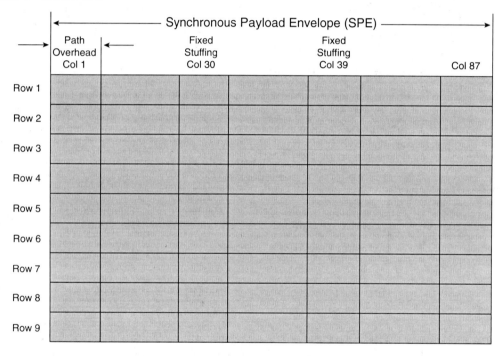

The STS-1 SPE can begin anywhere in the STS-1 envelope capacity, as shown in Figure 5-7. Typically, it begins in one STS-1 frame and ends in the next. The STS payload pointer contained in the transport overhead designates the location of the byte in which the STS-1 SPE begins. The STS POH is associated with each payload and is used to communicate various kinds of information from the point at which a payload is mapped into the STS-1 SPE to the point at which it is delivered.

The STS-1 transport overhead houses a pointer (H1 and H2 bytes) that allows the SPE to be separated from the transport overhead. The pointer is just an offset value that points to the byte where the SPE begins. Figure 5-7 shows the typical case of the SPE overlapping two STS-1 frames. To maintain synchronization, if there are any frequency or phase variations between the STS-1 frame and its SPE, the pointer values are increased or decreased accordingly. This allows the receiving SONET equipment to align with the incoming transmission stream and to identify the start of the SPE. Otherwise, the buffers would be needed to delay the sending of the SPE until the start of the next frame. These buffers would be quite large to support STS-N frame sizes. Buffer memory is relatively inexpensive these days. During the early days of SONET development, however, the cost of memory was a consideration in the design of SONET NEs and their implementation. H1 and H2 bytes are also used to indicate concatenation and STS path alarm indication signals (AIS-Ps). The H3 pointer action byte is used for frequency justification in each and every STS-1 signal of an STS-N frame. It is used to compensate for the timing variations. Depending on the pointer value, the byte is used to adjust the input buffers.

Figure 5-7 *SPE Positioning Across STS-1 Frames*

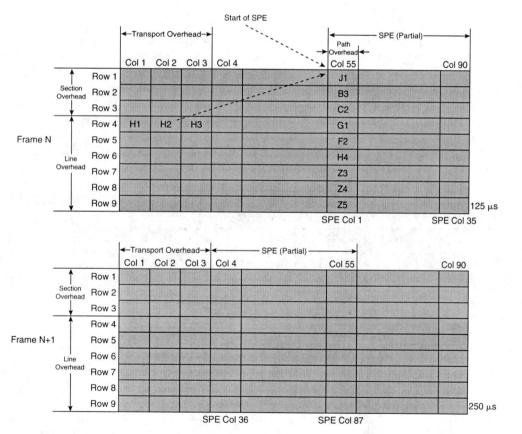

STS-N Framing

A SONET STS-N frame contains N STS-1s. An STS-N frame is created by byte interleaving columns of STS-1 frames.

An STS-N frame is a specific sequence of N * 810 bytes. The transport overhead of the individual STS-1 frames is aligned before interleaving. However, the associated STS SPEs are not required to be aligned because each STS-1 has a payload pointer to indicate the location of the SPE or to indicate concatenation. The new larger STS-N frames are transmitted at a rate of 8000 frames per second. The STS-Ns are still considered N-independent frame streams, and each frame has its own set of payload pointers. All user information must fit within an STS-1 frame SPE, and the frames are multiplexed before transmission.

STS-1 frames can also be concatenated to form an STS-Nc structure. Because individual frames are concatenated, one SPE is present and the STS-Nc frame has only one set of payload pointers. STS-1s can be concatenated up to STS-3c. Beyond STS-3, concatenation is done in multiples of STS-3c. Figure 5-8 graphically illustrates the difference between an STS-3 and an STS-3c frame structure. There are three sets of overhead bytes and three column interleaved payloads in each STS-3 frame, whereas there is only one set of overhead bytes in an STS-3c concatenated frame. The concatenation is indicated by putting special values in the H1/H2 pointers known as *concatenation indicators*. The receivers are able to detect the concatenation value. Because there is only one SPE, only one set of POH bytes is needed that indicates the first column of the SPE. Typically, STS-Nc concatenated frame structures are used to transport ATM cells.

Figure 5-8 *STS-3 and STS-3c Frame Structures*

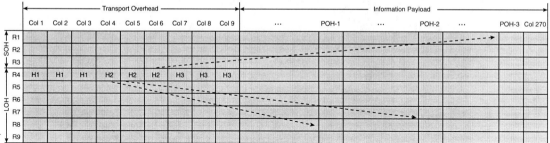

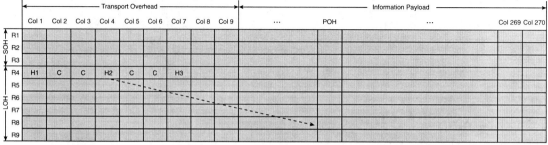

Consider an STS-N frame with N=3. The STS-3 frame consists of 9 rows and is sent 8000 times per second. The STS-3 overhead and SPE columns are multiplied by a factor of 3. The resultant STS-3 frame is 270 columns wide (3 * 90), of which the first 9 columns are the TOH (3 * 3)

and the remaining 261 (3 * 87) are the payload capacity. The entire STS-3 frame is 2430 bytes (9 rows * 270 columns). The line rate for the OC-3 can thus be computed as follows:

OC-3 line rate = 2430 bytes/frame * 8000 frames/sec * 8 bits/byte = 155.52 Mbps

Multiplying the OC-1 line rate of 51.84 Mbps by 3 can also derive the line rate. Now, do the math with STS-192. Consider an STS-N frame with N=192. The STS-192 overhead and SPE columns are multiplied by a factor 192. The resultant STS-192 frame is 17,280 columns wide (192 * 90), of which the first 576 columns are the TOH (192 * 3) and the remaining 16,704 (192 * 87) are the payload capacity. The entire STS-192 frame is 155,520 bytes (9 rows * 17,280 columns). The line rate for the OC-192 can thus be computed as follows:

OC-192 line rate = 155,520 bytes/frame * 8000 frames/sec * 8 bits/byte = 9.953 Gbps

SONET Transport Overhead

SONET provides substantial overhead information, allowing simpler multiplexing and greatly expanded OAM&P capabilities. The SONET TOH, as shown in Figure 5-9, has two layers that include the section overhead and line overhead. Section overhead is used for communications between adjacent NEs, such as regenerators, whereas line overhead is used for the STS-N signal between STS-N multiplexers. You should make a genuine attempt to comprehend this section because a large part of understanding SONET includes understanding the purpose and use of the overhead bytes.

Figure 5-9 *SONET Transport Overhead*

		Transport Overhead				Path Overhead
		Col 1	Col 2	Col 3		Col N
Section Overhead	Row 1	Framing A1	Framing A2	Trace/Growth J0/Z0		Trace J1
	Row 2	BIP-8 B1	Orderwire E1	User F1		BIP-8 B3
	Row 3	Data Com D1	Data Com D2	Data Com D3		Signal Label C2
Line Overhead	Row 4	Pointer H1	Pointer H2	Pointer Action H3		Path Status G1
	Row 5	BIP-8 B2	APS K1	APS K2		User Channel F2
	Row 6	Data Com D4	Data Com D5	Data Com D6		Indicator H4
	Row 7	Data Com D7	Data Com D8	Data Com D9		Growth/DQDB Z3
	Row 8	Data Com D10	Data Com D11	Data Com D12		Growth Z4
	Row 9	Sync/Growth S1/Z1	REI-L/Growth M0 or M1/Z2	Orderwire E2		Tandem Z5

SONET Section Overhead

The section overhead (SOH) of the SONET STS-1 frame consists of 9 bytes found in the first 3 rows of columns 1, 2, and 3 of the SONET frame. The SOH is responsible for performance monitoring of the STS-N signal, carrying local orderwire communications, carrying OAM&P information over the DCCs, and for framing. The various bytes of the SOH are shown in Figure 5-10 and described in the next section.

Figure 5-10 *SONET Section Overhead*

	Col 1	Col 2	Col 3
Row 1	Framing A1	Framing A2	Trace Growth J0/Z0
Row 2	BIP-8 B1	Orderwire E1	User F1
Row 3	Data Com D1	Data Com D2	Data Com D3

(Transport Overhead; Section Overhead spans Rows 1–3)

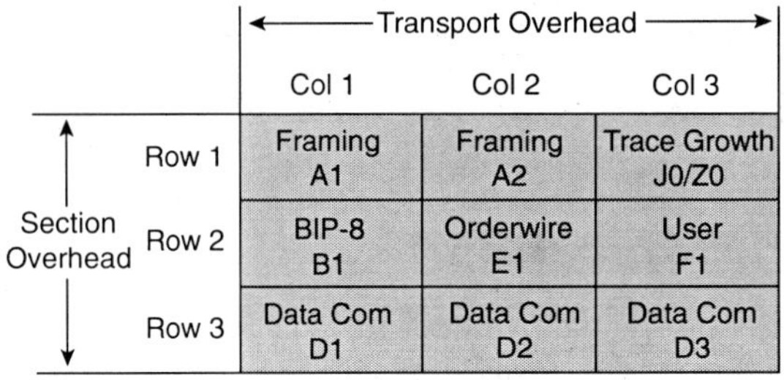

Framing Bytes (A1, A2)

In Figure 5-10, A1 and A2 are the two framing bytes dedicated to each STS-1 that indicate the beginning of an STS-1 frame. The A1, A2 byte pattern is F628 hex. The A1, A2 byte pattern of F628 hex is never scrambled. When four consecutive errored framing patterns have been received, an out-of-frame (OOF) condition is declared. Alarms are discussed later in this chapter. When two consecutive error-free framing patterns have been received, an in-frame condition is declared.

Section Trace or Growth Byte (J0/Z0)

The J0 section trace byte is used to trace the origin of an STS-1 frame over the SONET network. In the case of multiple STS-1s in an STS-N, the J0 byte is defined only for the first STS-1, because all remaining STS-1s would have originated from the same device. The J0 byte is used to indicate growth as the Z0 byte for the remaining STS-1s.

The section trace byte used to be known as the STS-1 ID C1 byte and was assigned a number in an STS-N frame according to the order of its appearance. The C1 byte of the first STS-1 signal in an STS-N frame is set to 1; the second STS-1 signal is 2, and so on. The C1 byte is assigned before byte interleaving and remains within the STS-1 until de-interleaving. Currently, many SONET systems use the J0 byte as a C1 byte.

BIP-8 Byte (B1)

The bit-interleaved parity (BIP-8) B1 byte is allocated from the first STS-1 of an STS-N for section-error monitoring. Using a BIP-8 code with even parity, the value of the B1 byte is computed over all bits of the previous STS-N frame after scrambling. The parity is then inserted into the B1 field of the current frame before scrambling. Each piece of section equipment calculates the B1 byte of the current STS-N frame and compares it with the B1 byte received from the first STS-1 of the next STS-N frame. If the B1 bytes match, there is no error. If the B1 bytes do not match and the threshold is reached, the alarm indicator is set.

Orderwire Byte (E1)

The E1 byte is allocated from the first STS-1 of an STS-N frame as local orderwire channel for voice communications. This is the SOH orderwire used as the local orderwire. The orderwire byte of a SONET frame at 8000 frames per seconds provides a 64-kbps voice communications channel that could be used by engineers or technicians to plug in a butt set (no pun intended) to communicate with each other during equipment installation or troubleshooting.

User Byte (F1)

The F1 byte is set for user purposes. It is passed from one section level to another and is terminated at all section equipment. The F1 byte is used by some service providers for their own customized management applications. Many vendor proprietary EM applications use the F1 field for configuration and management communications. This provides individual vendors with flexibility at the cost of interoperability.

DCC Bytes (D1, D2, and D3)

D1, D2, and D3 are the SONET DCC bytes allocated from the first STS-1 of an STS-N frame. This 192-kbps message channel can be used for alarms, maintenance, control, monitoring, administration, and communication needs between two section-terminating devices. The use of the DCC bytes must be in compliance with the TMN standards. The DCC is sometimes referred to as the *embedded operations channel (EOC)* in SONET and the *embedded communications channel (ECC)* in SDH.

NOTE Scrambling and descrambling techniques are necessary to make the data appear random and guarantee a degree of user privacy and security. The scrambling is done after the multiplexing step, but before the SOH byte insertion and the EO conversion. Therefore the SOH bytes A1, A2, and the C1 (J0/Z0) bytes are *not* scrambled. The scrambler-exclusive ORs the STS-N frame starting from the byte just after the Nth C1 byte with a 127-bit sequence.

Line Overhead

The line overhead (LOH) of the SONET STS-1 frame consists of 18 bytes found in rows 4 to 9 of columns 1, 2, and 3 of the SONET frame. The LOH is processed by all SONET NEs except for regenerators or repeaters. The LOH is responsible for performance monitoring of the individual STS-1s, carrying express orderwire communications information, data channels for OAM&P, pointers to indicate the start of the SPE, protection switching information, a line alarm indication signal (AIS), and line far-end receive failure (FERF) indication. The various LOH bytes are shown in Figure 5-11 and described in the next section.

Figure 5-11 *SONET Line Overhead*

	Col 1	Col 2	Col 3
Row 4	Pointer H1	Pointer H2	Trace Growth J0/Z0
Row 5	BIP-8 B2	APS K1	APS K2
Row 6	Data Com D4	Data Com D5	Data Com D6
Row 7	Data Com D7	Data Com D8	Data Com D9
Row 8	Data Com D10	Data Com D11	Data Com D12
Row 9	Sync/Growth S1/Z1	REI-L/Growth M0 or M1/Z2	Orderwire E2

Pointer Bytes (H1, H2)

H1 and H2 are known as *pointer bytes*. H1 and H2 are used to indicate the offset in the bytes between the pointer bytes themselves and the first byte of the STS SPE, in each STS-1 signal of an STS-N frame. Jitter and timing differences can cause the start of the SPE to move by 1 byte forward or backward within the information payload. The pointer bytes inform the receiver of the correct position of the SPE at all times. The pointers are used to align the STS-1 SPE in an STS-N signal as well as to perform frequency justification. The first pointer byte contains the actual pointer to the SPE, and the following pointer bytes contain the linking indicators.

Pointer Action Byte (H3)

The pointer action byte is used for frequency justification in each and every STS-1 signal of an STS-N frame. It is used to compensate for the timing variations. Depending on the pointer value, the byte is used to adjust the input buffers. The SPE matrix contains 9 rows with 87 columns, providing a full payload of 783 bytes. As a result of timing jitter, more than 783 bytes might be ready to be transmitted in an SPE within a 125 μs time period. If the excess bits were less than 8 bits, they would be buffered and transmitted as the first bits of the next frame. When a full byte (8 bits) has been accumulated in the buffer, however, the pointer action byte is used to carry the byte. This mechanism is called *negative timing justification*. Positive timing justification is used when 783 bytes are not available in the device buffer to fill the SPE. Many SONET implementations lock the value of the H1 and H2 bytes to 20A (hex), which points to row 1, column 4, or the beginning of the SPE in the next SONET frame.

BIP-8 Byte (B2)

The BIP-8 B2 byte in each of the STS-1 signals of an STS-N frame is used for line-error monitoring. It is similar to the B1 byte in the SOH. The B2 byte uses BIP-8 code with even parity. It contains the result from the computation of all the bits of LOH and STS-1 envelope capacity of the previous STS-1 frame before scrambling. The SOH is not included in the computation, reflecting the hierarchical nature of the SONET overhead. The LOH places the BIP-8 into the B2 field of the current frame before it is scrambled. These byte values are defined for all STS-1s of an STS-N signal.

APS Bytes (K1, K2)

The K1 and K2 bytes are used for APS signaling between line-level entities for line-level bidirectional APS. These byte values are defined only for the first STS-1 of an STS-N signal. The K1 and K2 bytes communicate APS and switching commands between LTE NEs. K1 and K2 bytes are also used for line-switched SONET protection mechanisms. The K2 byte also carries line-level alarm messages. Table 5-3 describes the use of the K1 and K2 bytes.

Table 5-3 *K1 and K2 Bytes*

Byte	Bits	Value	Function
K1	1 to 4	0000	No request
K1	1 to 4	0001	Do not revert
K1	1 to 4	0010	Reverse request
K1	1 to 4	0011	Not used
K1	1 to 4	0100	Exercise
K1	1 to 4	0101	Not used
K1	1 to 4	0110	Wait to restore
K1	1 to 4	0111	Not used
K1	1 to 4	1000	Manual switch
K1	1 to 4	1001	Not used
K1	1 to 4	1010	SD low priority
K1	1 to 4	1011	SD high priority
K1	1 to 4	1100	SF low priority

continues

Table 5-3 *K1 and K2 Bytes (Continued)*

Byte	Bits	Value	Function
K1	1 to 4	1101	SF high priority
K1	1 to 4	1110	Forced switch
K1	1 to 4	1111	Lock out of protection
K1	5 to 8	nnnn	Indicates the number of the channel requested
K2	1 to 4	nnnn	Selects channel number
K2	5	0	1+1 protection architecture
K2	5	1	1:N protection architecture
K2	6 to 8	000	Future use
K2	6 to 8	001	Future use
K2	6 to 8	010	Future use
K2	6 to 8	011	Future use
K2	6 to 8	100	Unidirectional provisioning
K2	6 to 8	101	Bidirectional provisioning
K2	6 to 8	110	RDI-L
K2	6 to 8	111	AIS-L

DCC Bytes (D4 to D12)

The LOH DCC bytes, from D4 to D12 (9 bytes), are allocated for LOH data communications and should be considered as one 576-kbps message-based channel. These bytes can be used for alarms, maintenance, control, monitoring, administration, and communication between two section-terminating devices. The LOH DCC uses the same protocol used in the SOH DCC. OAM&P messages can be internally generated or acquired from an external source. The LOH DCC bytes are defined only for the first STS-1 of an STS-N signal. The D4 to D12 bytes of the rest of the STS-N frame are not defined. The D4 to D12 bytes are being proposed in the OIF and G.ASON forums for use in signaling and communications between NEs.

Synchronization Status Byte (S1)

The S1 byte is used to transport the synchronization status of the SONET NE. The S1 byte is used by NEs to select the best clocking source among many timing sources to avoid timing loops within the network. Bits 1 to 4 of the S1 byte are not currently defined. Most vendors use only bits 5 to 8 of the S1 byte. The S1 byte is defined for the first STS-1 of an STS-N signal. In the remaining STS-1s of the STS-N, this byte is defined as the growth (Z1) byte reserved for future functions not yet defined by SONET standards.

REI-L Byte (M0 or M1)

The M0 byte is used for single STS-1 frames, whereas the M1 byte is used with a channelized STS-N structure containing N STS-1 frames. The M0 byte is used for a line-level remote error (REI-L) function formerly known as the line *far-end block error (FEBE)*. The M0 byte conveys the LOH BIP-8 error count back to the source. Bits 1 to 4 of the M0 byte are not currently defined. Most vendors use only bits 5 to 8 of the M0 byte. The M1 byte performs a function similar to the M0 byte for STS-N frames. The M1 byte is used with STS-N frame structures equal to or higher than STS-3 with minor variations above the STS-48 level. Many SONET implementations do not use the REI-L byte and define this byte as the growth (Z2) byte reserved for future functions not yet defined by SONET standards.

Growth Bytes (Z1, Z2)

The Z1 and Z2 growth bytes are set aside for functions not yet defined by SONET standards.

Orderwire Byte (E2)

The E2 byte is a 64-kbps voice channel similar to the SOH E1. The E2 byte is allocated for orderwire between line entities. This byte is defined only for STS-1 number 1 of an STS-N signal. The E2 byte is the LOH orderwire and is used as the express orderwire.

SONET Path Overhead

The POH is assigned to and transported with the SONET payload. It is created by the PTE as part of the SPE until the payload is demultiplexed at the terminating path equipment. The SPE contains nine bytes of POH that form a column in the SPE. Because the SPE can float within the STS-1 frame payload area, the position of the POH column floats as well. The POH supports the following four classes of operations:

- **Class A**—Payload-independent functions required by all payload types
- **Class B**—Mapping-dependent functions not required by all payload types
- **Class C**—Application-specific functions
- **Class D**—Undefined functions reserved for future use

The STS POH is responsible for performance monitoring of the STS SPE, signal labeling (equipped or unequipped), carrying the path status, and path trace. Path-level overhead is carried end to end and is added to DS1 signals when they are mapped into VTs and for STS-1 payloads that travel end to end. The various POH bytes are shown in Figure 5-12 and described in the next section.

Figure 5-12 *SONET Path Overhead*

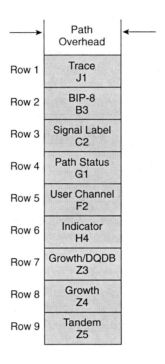

STS Path Trace Byte (J1)

The J1 byte (Class A) is used by the receiving terminal to verify the path connection. The J1 byte transmits a repeated 64-byte string to the receiving PTE that enables it to verify its connection to the device originating the SPE. Null characters are transmitted in the absence of a source message. The J1 field could be used to transmit IP addresses or E.164 addresses of CPE equipment. Carriers use the J1 byte to carry common language location identifier (CLLI) code to identify carrier central offices (COs).

BIP-8 Byte (B3)

The B3 byte (Class A) is assigned for path-error monitoring. The B3 byte's function is analogous to that of the SOH and LOH BIP-8 fields. The B3 byte is calculated over all bits of the previous STS SPE before scrambling using BIP-8 code with even parity. The SOH and LOH are excluded from the parity check, reflecting the hierarchical nature of SONET overhead.

STS Path Signal Label Byte (C2)

The signal label C2 byte (Class A) is assigned to indicate the construction of the STS SPE. It informs network equipment about the contents of the SPE. SONET permits the interleaving of STS frames containing various data types, such as TDM voice, ATM, and IP. The C1 byte informs the receiver about the contents of the SPE. Table 5-4 shows predefined values of the C2 byte.

Table 5-4 *SONET C2 Byte Values*

C2 Byte	Hex	SPE Content
0000 0000	00	Unequipped signal
0000 0001	01	Equipped signal—nonspecific payload
0000 0010	02	Floating VT mode
0000 0011	03	Locked VT mode
0000 0100	04	Asynchronous mapping for DS3
0001 0010	12	Asynchronous mapping for 139.264-Mbps DS4NA
0001 0011	13	Mapping for ATM
0001 0100	14	Mapping for MAN DQDB IEEE 802.6
0001 0101	15	Asynchronous mapping for FDDI
0001 0110	16	Mapping for IP inside PPP with scrambling
1100 1111	CF	Mapping for IP inside PPP without scrambling/HDLC
1111 1110	FE	0.181 Test signal mapping ITU-T G.707
1111 1111	FF	AIS-P

Path Status Byte (G1)

The G1 byte (Class A) notifies the originating end of the path performance and status of the entire duplex path. It is assigned to carry back path-terminating status and performance information to the originating PTE. The G1 byte carries two maintenance signals: the REI-P and the RDI-P. The B3 error count, known as the *path remote error indicator (REI-P)*, is carried in bits 1 to 4. The *path remote defect indicator (RDI-P)* is carried in bits 5 to 7. Bit 8 is not defined. This allows the complete duplex path to be monitored from either end. Table 5-5 shows predefined values of the C2 byte.

Table 5-5 *G1 Byte Values*

Byte	Bits	Value	Function
G1	1 to 4	NNNN	STS path REI function
G1	5 to 7	STS path RDI function. Trigger and interpretation	
G1	5 to 7	000	No remote defect
G1	5 to 7	001	No remote defect

continues

Table 5-5 *G1 Byte Values (Continued)*

Byte	Bits	Value	Function
G1	5 to 7	010	PLM-P, LCD-P remote payload defect
G1	5 to 7	011	No remote defect
G1	5 to 7	100	AIS-P, LOP-P remote defect
G1	5 to 7	101	AIS-P, LOP-P remote server defect
G1	5 to 7	110	UNEQ-P, TIM-P remote connectivity defect
G1	5 to 7	111	AIS-P, LOP-P remote defect
G1	8	N	Undefined

Path User Channel Byte (F2)

The F2 byte (Class C) is allocated for user communications between path elements. Some SONET implementations use the F2 byte to carry layer management information generated by distributed queue, dual-bus (DQDB) networks.

Indicator Byte (H4)

The H4 byte (Class C) provides a generalized multiframe indicator for payload. The H4 byte is used when a frame is organized into various mappings that determine the structure of a portion of the SPE carrying user data. VTs are an example of a very common mapping. With VTs, up to 28 DS1 circuits can be carried over a single STS-1 frame, with each DS1 mapping to a VT. In DQDB networks, the H4 byte is used to carry link status information.

Z3 to Z5 Bytes

Z3 to Z5 are growth bytes (Class D) reserved for future use. The Z3 byte is used to carry additional layer management information in DQDB networks. The Z4 POH byte is used in proprietary Cisco optical transport mechanisms to increase the BLSR node count on a ring from 16 to 32.

Tandem Connection Byte (Z5)

The Z5 byte has been defined for use as a tandem connection maintenance channel and a path-level DCC by ANSI. The path-level DCC performs a similar OAM&P function as the line- and section-level DCCs.

SONET Alarms

The SONET overhead described in the previous section provides a variety of management information and other functions including error performance monitoring; pointer adjustment information; path status and trace; section trace; remote defect, error, and failure indications; signal labels; data flag indications; DCC information; APS control; orderwire information; and synchronization status messages. A majority of the overhead information is involved with alarm and in-service monitoring of the particular SONET sections.

SONET alarms are defined as anomalies, defects, and failures. An *anomaly* is the smallest discrepancy that can be observed between the actual and desired characteristics of an item. The occurrence of a single anomaly does not constitute an interruption in the ability to perform a required function. If the density of anomalies reaches a level at which the ability to perform a required function has been interrupted, it is termed a *defect*. Defects are used as an input for performance monitoring, control of consequent actions, and determination of fault cause. The inability of a function to perform a required action persisted beyond the maximum time allocated is termed a *failure*. Various SONET alarm anomalies, defects, and failures are described in the next section.

Loss-of-Signal (LOS) Alarm

The loss-of-signal (LOS) alarm is raised when the synchronous signal (STS-STS-N) level drops below the threshold at which a bit error rate (BER) of 1 in 10^3 is predicted. This can be the result of excessive attenuation of the signal, receiver saturation, cable cut, or equipment fault. The LOS state clears when two consecutive framing patterns are received and no new LOS condition is detected. The LOS alarm condition is configurable in certain SONET NEs.

Out-of-Frame (OOF) Alignment

The OOF state occurs when four or five consecutive SONET frames are received with invalid (errored) framing patterns (A1 and A2 bytes). The maximum time to detect OOF is 625 microseconds. OOF state clears when two consecutive SONET frames are received with valid framing patterns.

Loss-of-Frame (LOF) Alignment

The LOF state occurs when the OOF state exists for a specified time in milliseconds. LOF state clears when an in-frame condition exists continuously for a specified time in milliseconds.

Loss-of-Pointer (LOP)

The LOP state occurs when N consecutive invalid pointers are received or N consecutive new data flags (NDFs) are received (other than in a concatenation indicator), where N = 8, 9, or 10.

LOP state clears when three equal valid pointers or three consecutive AIS indications are received. The LOP can also be identified as the STS path loss of pointer (SP-LOP) or VT path loss of pointer (VP-LOP).

Alarm Indication Signal (AIS)

The AIS is an all-1s characteristic or adapted information signal. It is generated to replace the normal traffic signal when that signal contains a defect condition to prevent consequential downstream failures being declared or alarms being raised. The AIS can be identified as the line alarm indication signal (AIS-L), STS path alarm indication signal (SP-AIS), or the VT path alarm indication signal (VP-AIS).

Remote Error Indication (REI)

The REI is an indication returned to a transmitting node (source) that an errored block has been detected at the receiving node (sink). This indication was formerly known as FEBE. The REI can also be identified as the line remote error indication (REI-L), STS path remote error indication (REI-P), or the VT path remote error indication (REI-V).

Remote Defect Indication (RDI)

The RDI is a signal returned to the transmitting terminating equipment on detection of a LOS, LOF, or AIS defect. RDI was previously known as FERF. The RDI can also be identified as the line remote defect indication (RDI-L), STS path remote defect indication (RDI-P), or VT path remote defect indication (RDI-V).

Remote Failure Indication (RFI)

A failure is a defect that persists beyond the maximum time allocated to the transmission system protection mechanisms. When this situation occurs, an RFI is sent to the far end and initiates a protection switch if this function has been enabled. The RFI can be identified as the line remote failure indication (RFI-L), STS path remote failure indication (RFI-P), or the VT path remote failure indication (RFI-V).

B1 Error

Parity errors evaluated by byte B1 (BIP-8) of an STS-N are monitored. If any of the eight parity checks fail, the corresponding block is assumed to be in error.

B2 Error

Parity errors evaluated by byte B2 (BIP-24 * N) of an STS-N are monitored. If any of the N * 24 parity checks fail, the corresponding block is assumed to be in error.

B3 Error

Parity errors evaluated by byte B3 (BIP-8) of a VT-N (N = 3, 4) are monitored. If any of the eight parity checks fail, the corresponding block is assumed to be in error.

BIP-2 Error

Parity errors contained in bits 1 and 2 (BIP-2: bit-interleaved parity 2) of byte V5 of a VT-N (N = 11, 12, 2) are monitored. If any of the two parity checks fail, the corresponding block is assumed to be in error.

Loss of Sequence Synchronization (LSS)

Sequence synchronization is considered to be lost and resynchronization is started if the BER is greater than or equal to 0.20 during an integration interval of 1 second.

Virtual Tributaries

The STS SPE can be subdivided into smaller structures, known as *virtual tributaries (VTs)*. VTs are used for transporting and switching payloads smaller than the STS-1 rate. All services below the DS3, such as DS1 signals, are transported in the VT structure. The various sub-DS3 TDM payloads that can be carried over VTs are shown in Table 5-6. An individual VT containing a DS1 can be added or dropped without multiplexing or demultiplexing the entire STS-1. This improved accessibility improves switching and grooming at VT or STS levels. The STS-1 payload can be subdivided into VTs that are synchronous signals used to transport lower-speed transmissions. There are four sizes of VTs: VT1.5 (1.728 Mbps), VT2 (2.304 Mbps), VT3 (3.456 Mbps), and VT6 (6.912 Mbps). The various VT types are shown in Figure 5-13 and described in Table 5-6.

Table 5-6   *Virtual Tributary Types*

VT Type	VT Bit Rate	Size	TDM Payload
1.5	1.728 Mbps	9 rows, 3 columns	DS1 (1.544 Mbps)
2	2.304 Mbps	9 rows, 4 columns	E1 (2.048 Mbps)
3	3.456 Mbps	9 rows, 6 columns	DS1C (3.152 Mbps)
6	6.912 Mbps	9 rows, 12 columns	DS2 (6.312 Mbps)

Figure 5-13 *VT Types and Structures*

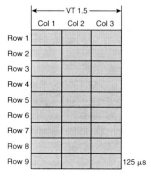
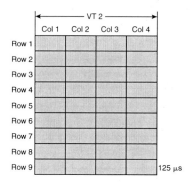
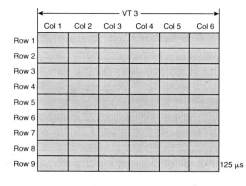
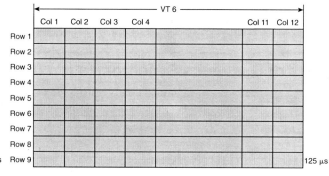

VT Groups

To accommodate a mix of different VT types within an STS-1 SPE, VTs are grouped together into VT groups. An STS-1 SPE that carries VTs is divided into 7 VT groups, with each virtual group (VTG) consuming 12 columns of the STS-1 SPE. Each VTG is 108 bytes long (9 rows * 12 columns). Simple math reveals that only 84 columns (7 groups * 12 columns per group) are used for carrying VT information within the 87-column SPE. The three missing columns include the POH column and two other columns (SPE columns 30 and 59) that are used for fixed elements. As shown in Table 5-6, the numbers of columns in each of the different VT types are factors of 12. Each VTG can contain only one type of VT and would contain one of the following combinations: 4 VT1.5s with 3 columns per VT1.5, 3 VT2s with 4 columns per VT2, 2 VT3s with 6 columns per VT3, or 1 VT6 with 12 columns per VT6.

Within an STS-1 SPE, there can be a mix of the different VTGs. For example, an STS-1 SPE may contain five VT1.5 groups and two VT6 groups, for a total of seven VTGs. An SPE can carry a mix of any of the seven groups. VTGs have no overhead or pointers and are used only to organize the different VTs within an STS-1 SPE.

Virtual Tributaries

The VT columns within a VTG are not placed in consecutive columns within that group. The columns of the individual VTs within the VTG are interleaved by column. The intent of interleaving VTs within a VTG is to reduce the delay on isochronous voice traffic. This is illustrated in Figure 5-14 for the VT1.5 structure. As shown in the figure, four VT1.5s are interleaved within a VTG. The four VT1.5s in the figure have been labeled VT1.5 (A), VT1.5 (B), VT1.5 (C), and VT1.5 (D). VT1.5 (A) has three columns (A1, A2, and A3) that map into VTG columns 1, 5, and 9. VT1.5 (B) has three columns (B1, B2, and B3) that map into VTG columns 2, 6, and 10. VT1.5 (C) has three columns (C1, C2 and C3) that map into VTG columns 3, 7, and 11, and VT1.5 (D) has three columns (D1, D2 and D3) that map into VTG columns 4, 8, and 12.

Figure 5-14 *VT1.5 Interleaving Within a VTG*

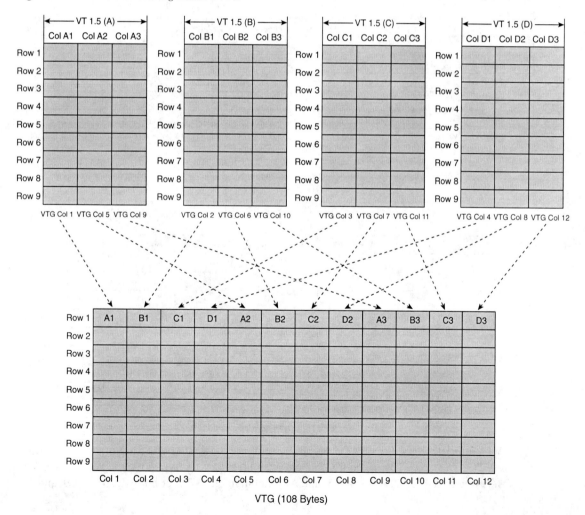

168 Chapter 5: SONET Architectures

As shown in Figure 5-15, three VT2s are interleaved within a VTG. The three VT2s in the figure have been labeled VT2 (A), VT2 (B), and VT2 (C). VT2 (A) has four columns (A1, A2, A3, and A4) that map into VTG columns 1, 4, 7 and 10. VT2 (B) has four columns (B1, B2, B3, and B4) that map into VTG columns 2, 5, 8, and 11; and VT2 (C) has four columns (C1, C2, C3, and C4) that map into VTG columns 3, 6, 9, and 12.

Figure 5-15 *VT2 Interleaving Within a VTG*

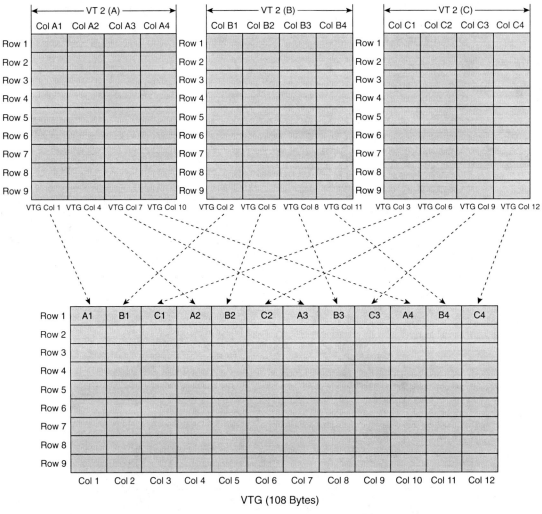

VT3 interleaving within a VTG is shown in Figure 5-16. Two VT3s are interleaved within a VTG. The two VT3s in the figure have been labeled VT3 (A) and VT3 (B). VT3 (A) has six columns (A1, A2, A3, A4, A5, and A6) that map into VTG columns 1, 3, 5, 7, 9, and 11.

VT3 (B) has six columns (B1, B2, B3, B4, B5, and B6) that map into VTG columns 2, 4, 6, 8, 10, and 12.

Figure 5-16 *VT3 Interleaving Within a VTG*

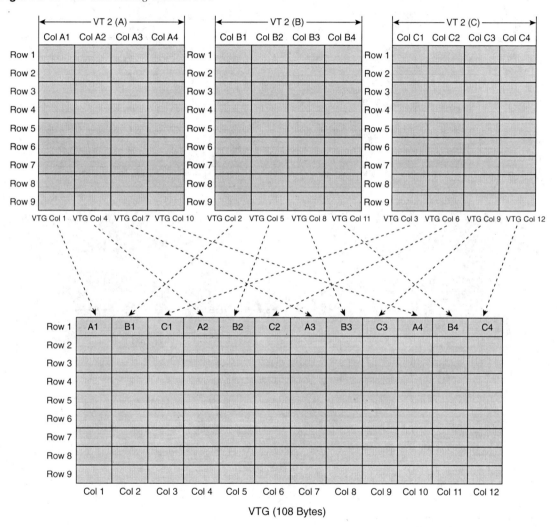

VT6 interleaving within a VTG is shown in Figure 5-17. There is a one-to-one mapping between a VT6 and the VTG. VT6 (A) has 12 columns (A1, A2, A3, A4, A5, A6, A7, A8, A9, A10, A11, and A12) that map into VTG columns 1, 2, 3, 4, 5, 6, 7, 8, 9, 10, 11, and 12. Table 5-7 indicates the VT capacity for each VT type within an STS-1 SPE.

Figure 5-17 *VT6 Interleaving Within a VTG*

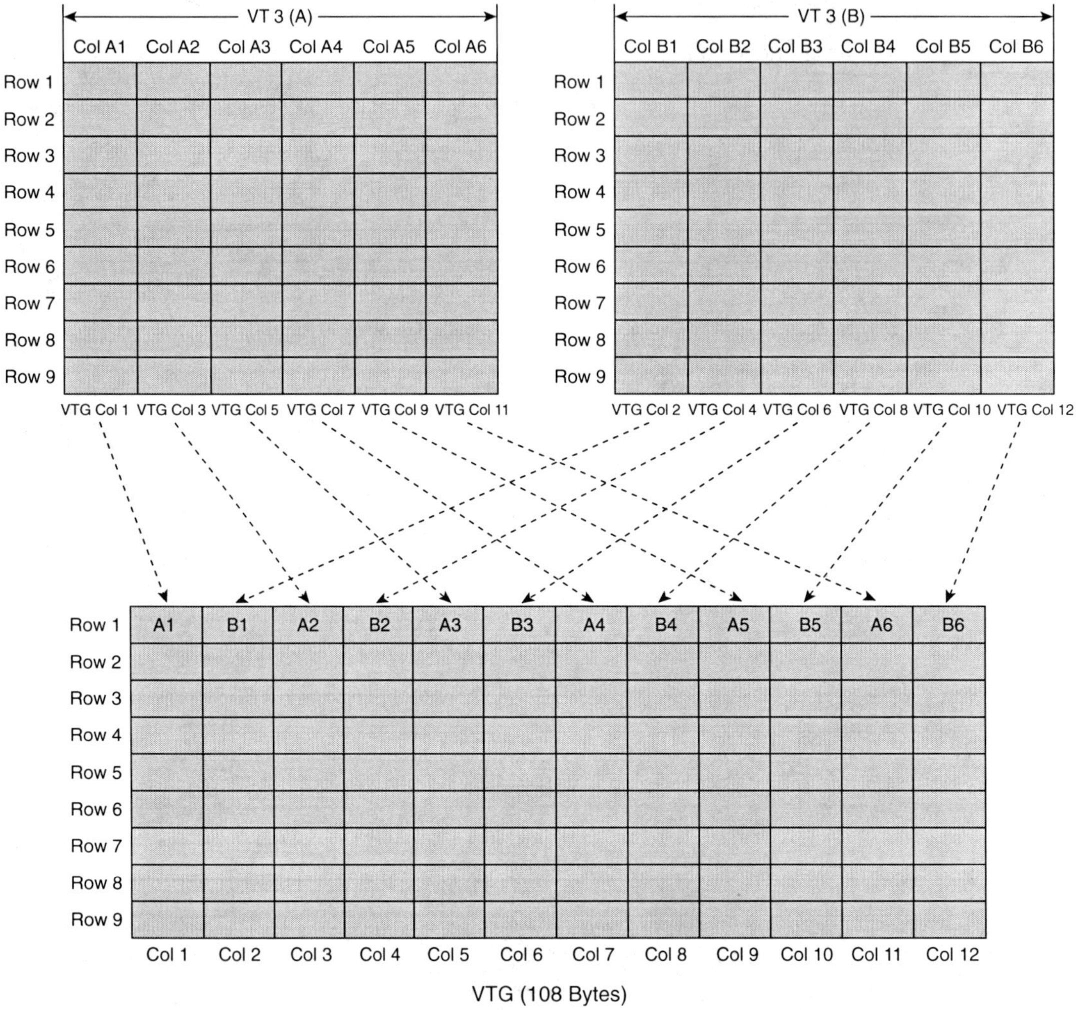

Table 5-7 *Virtual Tributary Capacity*

VT Type	Payload Signal	Signals/VTG	Signals/STS-1
VT1.5	DS1	4	28
VT2	E1	3	21
VT3	DS1C	2	14
VT6	DS2	1	7

VTG columns are not placed in consecutive columns within an STS-1 SPE. The VTG columns are interleaved by column within the STS-1 SPE. The interleaving of VTGs within an SPE is shown in Figure 5-18. Seven VTGs numbered VTG1 through VTG7 are interleaved by column within the SPE. The first column of the SPE is reserved for the POH and is not used. SPE columns 30 and 59 are reserved for fixed stuff bytes that are ignored by the receiver. The first column of VTG1 occupies column 2 of the SPE, and the first column of VTG2 occupies column 3 of the SPE and so on. The VTG column assignments skip over columns 30 and 59. Note that an SPE assigned for VTG transmission by its signal label cannot carry any other traffic type. However, the value of the signal label can change by SPE.

Figure 5-18 *VTG Interleaving Within an SPE*

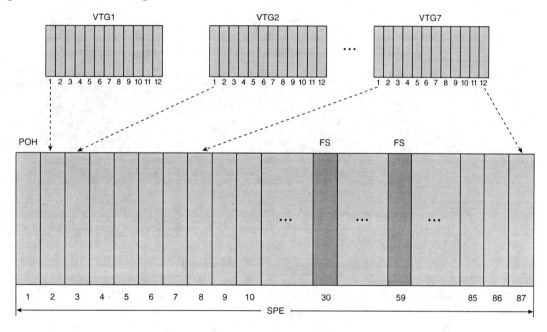

VT Superframe

DS1 frames are organized as a series of 12 frames known as the *D4 superframe* or as a series of 24 frames known as the *D5 extended superframe (ESF)*. It is important to preserve the D4 or ESF structure during transmission over SONET. In addition to the division of VTs into VTGs, a 500-microsecond structure called a *VT superframe* is defined for VTs. As shown in Figure 5-19, the VT superframe, also known as the *VT multiframe*, is defined as four consecutive VT SPEs. Four consecutive 125-microsecond frames of the VT-structured STS-1 SPE are organized into a 500-microsecond superframe. The SPEs are numbered 00, 01, 10, and 11 in binary.

Figure 5-19 *VT Superframe and SPE*

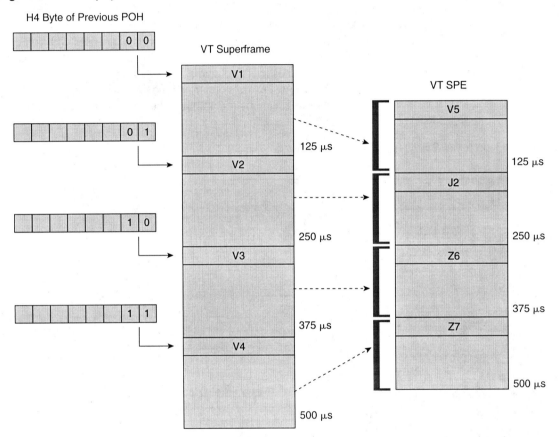

The first VT overhead (VTOH) byte in each of the four SPEs is V1, V2, V3, and V4. The phase of the VT superframe is indicated by the H4 VT multiframe indicator byte in the previous STS POH. A value of 00 in bits 7 and 8 of the H4 byte in the previous STS POH indicates the presence of the first VT multiframe SPE and V1 byte in the current SPE. Similarly, a value of 01, 10, and 11 in bits 7 and 8 of the H4 byte in the previous STS POH indicates the presence of the second, third, or fourth VT multiframe and V2, V3, or V4 byte respectively in the current SPE. The V1, V2, V3, and V4 act as VT multiframe pointer bytes. The VT payload pointers provide flexible and dynamic alignment of the VT SPE within the VT envelope capacity, independent of other VT SPEs. The VT envelope capacity, and therefore the size of the VT SPE, is different for each VT size. Each VT SPE contains 4 bytes of VT POH (V5, J2, Z6, and Z7), and the remaining bytes constitute the VT payload capacity, which is different for each VT.

For the VT1.5 superframe, each VT1.5 SPE has 3 columns and 9 rows that constitute 27 bytes. Subtracting the V1, V2, V3, or V4 VTOH byte from the 27-byte VT1.5 leaves 26 bytes per VT1.5 SPE. Now, 24 of these bytes are for the DS1 inside the VT1.5, so this leaves 2 other bytes per SPE for other functions. Because there are 4 SPEs in a VT multiframe, a total of 8 bytes in the VT1.5 are available for additional overhead. The 4 VT overhead (VTOH) bytes are V5, J2, Z6, and Z7. The position of the VTOH bytes can be anywhere in the VT1.5 superframe, except for the initial byte position that is reserved for the V1, V2, V3, or V4 byte in each SPE. Table 5-8 indicates the VT superframe payload capacity for each of the VT types.

Table 5-8 *VT Superframe Payload Capacities*

	VT1.5	VT2	VT3	VT6
VT SPE 00	25 bytes	34 bytes	52 bytes	106 bytes
VT SPE 01	25 bytes	34 bytes	52 bytes	106 bytes
VT SPE 10	25 bytes	34 bytes	52 bytes	106 bytes
VT SPE 11	25 bytes	34 bytes	52 bytes	106 bytes
Total Payload Capacity	100 bytes per VT superframe	136 bytes per VT superframe	208 bytes per VT superframe	424 bytes per VT superframe

VT Path Overhead

VT POH contains four evenly distributed POH bytes per VT SPE, starting at the first byte of the VT SPE. VT POH provides communication between the point of creation of a VT SPE and its point of disassembly.

The VT POH and bit assignments are described in Table 5-9. Four bytes (V5, J2, Z6, and Z7) are allocated for VT POH. The first byte of a VT SPE (the byte in the location pointed to by the VT payload pointer) is the V5 byte, whereas the J2, Z6, and Z7 bytes occupy the corresponding locations in the subsequent 125-microsecond frames of the VT superframe. The V5 byte provides the same functions for VT paths that the B3, C2, and G1 bytes provide for STS paths (namely, error checking, signal label, and path status).

Table 5-9 *VT Path Overhead*

Byte	Bits	Value	Function
V5	1 to 2	NN	Allocated for error performance monitoring.
V5	3	N	
V5	4	N	
V5	5 to 7		Allocated for a VT path signal label to indicate the content of the VT SPE.
V5	5 to 7	000	Unequipped.
V5	5 to 7	001	Equipped—Non-specific payload.
V5	5 to 7	010	Asynchronous mapping.
V5	5 to 7	011	Bit synchronous mapping (no longer valid for DS1).

continues

Table 5-9 *VT Path Overhead (Continued)*

Byte	Bits	Value	Function
V5	5 to 7	100	Byte synchronous mapping.
V5	5 to 7	101	Unassigned.
V5	5 to 7	110	Unassigned.
V5	5 to 7	111	Unassigned.
V5	8	N	Allocated for a VT RDI-V signal.
J2	1 to 8		VT path trace identifier—Used to support the end-to-end monitoring of a path.
Z6	1 to 8		The Z6 byte is known as N2 in the SDH standard and is allocated to provide a lower-order tandem connection monitoring (LO-TCM) function.
Z7			The Z7 byte is known as K4 in the SDH standard.
Z7	1 to 4		Allocated for APS signaling and protection at the lower-order path level.
Z7	5 to 7		Bits 5–7 are used in combination with V5 bit 8 for ERDI-V.
Z7	8		Reserved.

SONET Multiplexing

Various TDM PDH services can be natively accepted by SONET systems. These signals are mapped into the SONET payload envelope through service adapters. All inputs are eventually converted to a base format of a synchronous STS-1 signal, with the exception of concatenated signals. As shown in Figure 5-20, DS1 inputs are mapped into VT1.5 SPEs and consequently into VT1.5s. Other low-speed signals, such as E1, DS1C, and DS2, are also mapped into their respective VTs. The VTs combine to form a virtual group (VG). Seven VGs are then mapped into an STS-1 SPE. The STS-SPE is then provided with its appropriate overhead and converted into an electrical STS-1. Several synchronous STS-1s are then multiplexed together in either a single- or two-stage process to form an electrical STS-N signal. STS-N multiplexing is performed along with byte interleaving in such a way that the low-speed signals are visible. The signal is then scrambled. Scrambling is the process of encoding digital 1s and 0s onto a line in such a way that provides an adequate number for a 1s density requirement. The ANSI standard for T1 transmission requires an average density of 1s of 12.5 percent. The primary reason for enforcing a 1s density requirement is for timing recovery or network synchronization. However, other factors, such as automatic-line-build-out (ALBO), equalization, and power usage are also affected by 1s density.

NOTE The A1 and A2 framing alignment bytes are used to ensure that the beginning of a SONET frame can be detected by the equipment. A1 is always set to 11110110, and A2 is always 00101000. These values are never scrambled during the frame-scrambling process.

No additional signal processing occurs, except a direct conversion from an electrical STS-N to an optical OC-N signal.

Figure 5-20 *SONET Multiplexed Hierarchy*

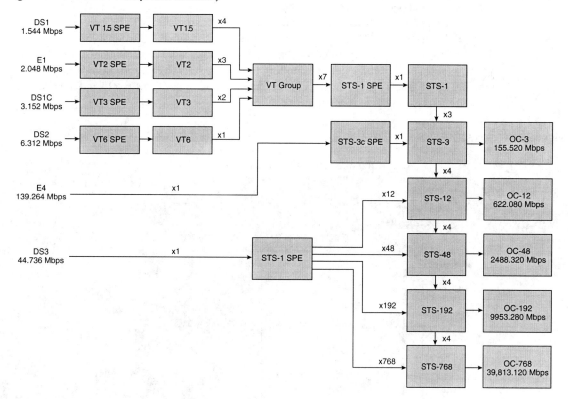

SONET Network Elements

Devices that implement SONET transmission are defined as SONET NEs. These NEs include section-, line-, and path-terminating equipment. The SONET NEs integrate to form the digital loop carrier (DLC) supporting various SONET architectures and topologies.

Regenerator

The regenerator is an STE that regenerates attenuated signals. A regenerator or amplifier is needed when, due to the long distance between multiplexers, the signal level in the fiber gets attenuated and becomes too low to drive a receiver. The regenerator clocks itself off of the received signal and replaces the SOH bytes before retransmitting the signal. The LOH, payload, and POH are not altered. The regenerator is sometimes called a *repeater*. Regenerators can be cascaded to extend the reach of the optical-fiber system. However, nonlinear impairments must be taken into consideration while cascading regenerators.

Terminal Multiplexer

The terminal multiplexer (TM) is a PTE that can concentrate or aggregate DS1s, DS3s, E1s, E3s, STS-Ns, and STM-Ns. An implementation with two TMs represents a SONET link with a section, line, and path all in one link. The schematic of a TM is shown in Figure 5-21. Various PDH signals, such as DS1, E1, DS3, are mapped to their associated SONET electrical payloads in the TM. For example, DS1 signals are mapped to VT1.5s, and DS3 signals are mapped to STS-1 SPEs. An EO conversion takes place, and the OC-N signals are launched into the fiber. The reverse happens during signal reception. In practice, the TM Is an ADM operating in terminal mode. The TM is analogous to the channel bank in the TDM world and allows lower-speed user access to the SONET network.

Figure 5-21 *SONET Terminal Multiplexer*

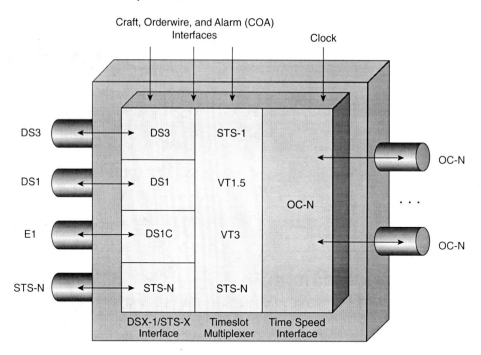

Add/Drop Multiplexer

An ADM is a PTE that can multiplex or demultiplex various signals to or from an OC-N signal. At an add/drop site, only those signals that need to be accessed are dropped or inserted. The remaining traffic continues through the NE without requiring special pass-through units or other signal processing. Signals can also be looped or terminated at the ADM. Figure 5-22 shows a schematic of a TM.

Figure 5-22 *SONET Add/Drop Multiplexer*

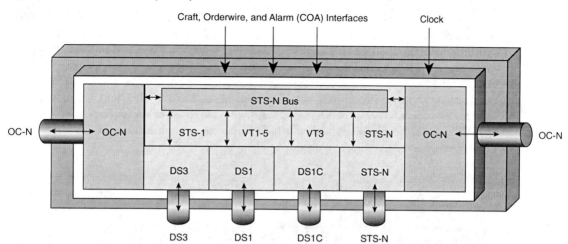

Various PDH signals and STS-N signals are mapped to their associated SONET electrical payloads in the TM. The electrical payloads are then mapped into STS-N signals based on the line rate of the OC-N transmission. The STS-N signals then undergo an EO conversion and are launched directionally onto the fiber based on the ring type and protection mechanism. An ADM can be deployed at a terminal site or any intermediate location for consolidating traffic from widely separated locations. Several ADMs can also be configured as a survivable ring. SONET enables add, drop, and pass-through capability: A signal that terminates at one node, is duplicated (repeated), and is then sent to the next and subsequent nodes. ADMs can be used to interconnect SONET rings. In such a case, the ADMs are referred to as *matching nodes*. In ring-survivability applications, drop and pass-through capability provides alternate routing for traffic passing through interconnecting rings in a matched node configuration. If the connection cannot be made through one of the nodes, the signal is repeated and passed along an alternate route to the destination node.

Single-stage ADMs can multiplex or demultiplex one or more tributary signals into or from an STS-N signal. It can be used in terminal sites, intermediate sites, or hub configurations. At an intermediate add/drop site, it can drop lower-rate signals to be transported on different facilities, or it can add lower-rate signals into the higher-rate STS-N signal. The rest of the traffic just continues straight through. In multinode distribution applications, one transport channel can carry traffic between multiple distribution nodes. Although NEs are compatible at the OC-N level, they can differ in features from vendor to vendor. SONET does not restrict manufacturers to providing a single type of product, nor require them to provide all types. For example, one vendor might offer an ADM with access at DS1 only, whereas another might offer simultaneous access at DS1 and DS3 rates. The next-generation multiservice platforms have their roots in the ADM. Multiservice platforms carry data services over SONET and also integrate dense wavelength-division multiplexing (DWDM) capabilities within the box.

Broadband Digital Cross-Connect

The broadband digital cross-connect (BDCS) can make two-way cross-connections at the DS3, STS-1, and STS-Nc levels. It can interface various SONET signals and DS3s. The BDCS is the synchronous equivalent of a DS3 digital access and cross-connect system (DACS) and supports hub-network architectures. The BDCS accesses the STS-1 signals and switches at the STS-1 level. It accepts optical signals and allows overhead to be maintained for integrated OAM&P purposes. As shown in Figure 5-23, the BDCS can be used as a SONET hub, where it can be used for grooming STS-1s, for broadband restoration purposes, or for routing traffic. The BDCS can also bridge and roll payload and facilities, make unidirectional or one-way connections, and support interconnection to older SONET NEs.

Figure 5-23 *Broadband Digital Cross-Connect*

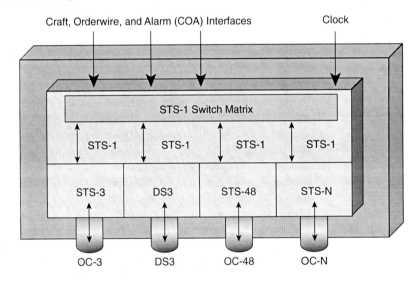

One major difference between a cross-connect and an ADM is that a cross-connect can be used to interconnect a much larger number of STS-1s. The BDCS can be used for grooming, consolidation, or segregation of STS-1s or for broadband traffic management. For example, it can be used to segregate high-bandwidth from low-bandwidth traffic and send it separately to a high-bandwidth switch and a low-bandwidth switch.

Wideband Digital Cross-Connect

The wideband digital cross-connect (WDCS) is a digital cross-connect that terminates SONET and DS3 signals and performs VT and DS1-level cross-connections. The WDCS accepts optical OC-N signals as well as STS-1s, DS1s, and DS3s. In a WDCS, the switching is done at the VT level as it cross-connects the constituent VTs between STS-N terminations.

The WDCS accepts various optical carrier rates, accesses the VT-level signals, and switches at this level. Figure 5-24 shows a schematic of a WDCS.

Figure 5-24 *Wideband Digital Cross-Connect*

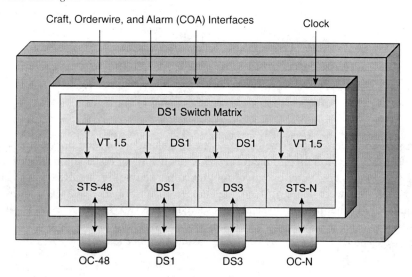

The WDCS is suitable for DS1-level grooming applications at hub locations. One major advantage of the WDCS is that less demultiplexing and multiplexing is required because only the required tributaries are accessed and switched. Because SONET is synchronous, the low-speed tributaries are visible and accessible within the STS-1 signal. Therefore, the required tributaries can be accessed and switched without demultiplexing, which isn't possible with existing TDM digital cross-connects. The WDCS also cross-connects the constituent DS1s between DS3 terminations, as well as between DS3 and DS1 terminations. The features of the WDCS make it useful in several applications. Because the WDCS can automatically cross-connect VTs and DS1s, it can be used as a network management system and for grooming at hub locations.

Digital Loop Carrier

The integrated digital loop carrier (IDLC), which consists of intelligent remote digital terminals (RDTs), and digital switch elements called *integrated digital terminals (IDTs)*, which are connected by a digital line. The IDLCs are designed to more efficiently integrate DLC systems with existing digital switches. A DLC is intended for service in the CO or a controlled environment vault (CEV) that belongs to the carrier. A schematic of a DLC is shown in Figure 5-25.

Figure 5-25 *Digital Loop Carrier*

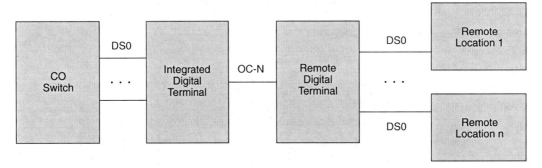

From a voice perspective, the DLC may be considered a concentrator of low-speed services before they are brought into the local CO for switching or distribution. The concentration can be achieved using a combination of multiplexing and intelligent GR.303 concentration. If concentration were not performed, the number of subscribers that a CO could serve would be limited by the number of physical lines served by the CO. The DLC itself is actually a system of multiplexers and switches designed to perform concentration from the remote terminals to the CO. The DLC devices are referred to as *remote fiber terminals (RFTs)* by several optical vendors.

Multiservice Provisioning Platforms (MSPPs)

Multiservice provisioning platforms (MSPPs) combine the functionality of the SONET ADM, BDCS, and WDCS NEs in a single platform. The platform also integrates Ethernet services and DWDM functionality in the box. The MSPP provides TDM solutions with interfaces, such as DS1, DS3, EC1, and data solutions with 10/100/1000 Ethernet solutions along with OC-3 to OC-192 optical transport bit rates including integrated DWDM wavelengths. The MSPP supports various SONET topologies including ring, linear point-to point, linear add/drop, star, and hybrid topologies. The protection and restoration choices include unidirectional path-switched ring (UPSR), two-fiber and four-fiber bidirectional line-switched ring (BLSR), 1+1 APS, unprotected span, and the Cisco Path Protected Mesh Networking (PPMN). The MSPP devices are easily configurable using integrated graphical user interface (GUI)-based tools and are fully manageable using element management systems. The MSPP devices are New Equipment Building Standards (NEBS) level 3 as well as OSMINE compliant and provide carrier-class reliability, with high MTBF (mean time between failure) and low MTTR (mean time to repair) figures. The Cisco ONS MSPP is discussed in great length and detail in later chapters.

SONET Topologies

SONET topologies center around survivability of the network with 50 ms or lower rapid recovery. Various topologies can be configured using either ADMs or DCSs. It is interesting to note that SONET ring topologies are often used in North America, whereas Europe, Asia, and

Latin America mainly rely on SDH-based ring as well as meshed network topologies. Protection switching is mainly performed by the intelligent ADMs or DCSs.

Point-to-Point Topology

Point-to-point topologies are created by connecting two SONET PTEs back to back over dark fiber. As depicted in Figure 5-26, a point-to-point deployment involves two PTE ADMs or TMs linked by fiber with or without an STE regenerator in the link. The TM could act as a DS1 concentrator and transport the DS1s over an OC-N link. The synchronous island can coexist within an asynchronous domain. Point-to-point topologies are extremely popular to provide storage-area network (SAN) interconnectivity between datacenters. Albeit, SAN protocols, such as ESCON, FICON, or Fibre Channel, use 1.25- or 2.125-Gbps channels over DWDM systems. The system always terminates the link using a next-generation TM, such as an MSPP.

Figure 5-26 *Point-to-Point Topology*

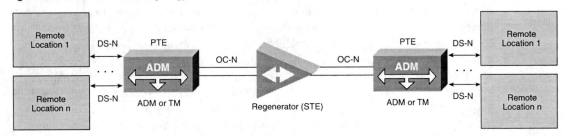

Point-to-point topologies use 1:N protection mechanisms in which one standby (protect) link is used to protect N active (working) links. Maximum protection is obtained by using a 1:1 ratio or 1+1 topology. Under normal working conditions, the working path is used. In the event of failure, however, the protect path is used with a switchover time less than 50 ms. Ideally, the protect path or fiber must use diverse physical routing to achieve maximum redundancy.

Point-to-Multipoint Topology

A point-to-multipoint architecture implements the adding and dropping of circuits along the path. The SONET ADM is a unique NE specifically designed for this task. As shown in Figure 5-27, the ADM is typically placed along a SONET link to facilitate adding and dropping of tributary or STS channels at intermediate points in the network. The point-to-multipoint topology is also referred to as a *linear add/drop architecture*. This topology is implemented for medium- and long-haul linear SONET architectures, where the service provider might have to add or drop circuits on the way to the destination. Conventional TDM systems would need complex demultiplexing, multiplexing, and cross-connections to achieve similar functionality at much lower bandwidths. As with point-to-point topologies, point-to-multipoint topologies also use 1:N protection mechanisms in which one standby (protect) link is used to protect N active (working) links. As before, maximum protection is obtained by using a 1:1 ratio or 1+1 topology.

Figure 5-27 *Point-to-Multipoint Topology*

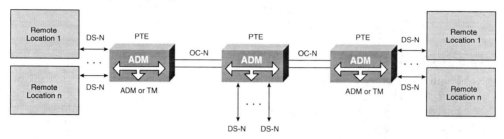

Hub Topology

The hub network topology is a scalable architecture that uses PTE devices in a hub-and-spoke configuration. It accommodates unexpected growth and changes as compared to point-to-point network topologies. As shown in Figure 5-28, the hub is implemented as a WDCS or BDCS that concentrates traffic at a central site and allows easy reprovisioning of the circuits. In a WDCS implementation, two or more ADMs and a wideband cross-connect switch allows cross-connecting services at the tributary level. In a BDCS implementation, two or more ADMs and a BDCS switch allows cross-connecting at both the SONET STS and the tributary level. An MSSP could replace the BDCS and subtending ADMs at the hub. The hub architecture is implemented in situations where the fiber is laid out in a hub-and-spoke topology. As with the linear topologies, hub topologies also use 1:N protection mechanisms in which one standby (protect) link is used to protect N active (working) links, with maximum protection obtained by using a 1:1 ratio or 1+1 topology.

Figure 5-28 *Hub Topology*

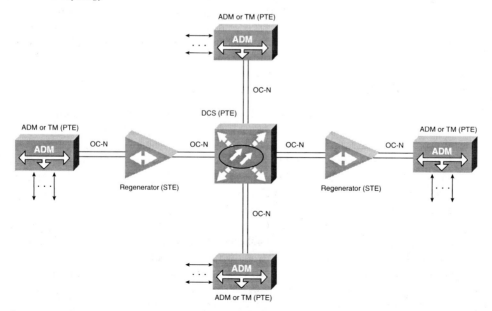

Ring Topology

By far the most popular metro topology for SONET is the ring architecture. SONET rings provide 50-ms and lower restoration times as well as robust protection mechanisms. The SONET building block for a ring architecture is the ADM or MSPP. Multiple ADMs can be daisy chained in a ring configuration for either bidirectional or unidirectional traffic, as shown in Figure 5-29. Access and core rings can be interconnected using ADMs as matching nodes. The main advantage of the ring topology is its survivability and fast restoration or healing time. If a fiber cable is cut, the ADMs have the intelligence to reroute the affected services via an alternate path through the ring without interruption. The demand for survivable services, diverse routing of fiber facilities, and flexibility to rearrange services to alternate serving nodes with automatic restoration have made rings the most popular metro access and core SONET architecture. Rings use advanced protection mechanisms and protocols, such as APS, UPSR (two-fiber), BLSR (two-fiber), and BLSR (four-fiber). UPSR protection topologies normally use a 1:N protection mechanism in which one standby (protect) link is used to protect the active (working) links. Ring protection mechanisms is explained in greater detail in a later section.

Figure 5-29 *Ring Topology*

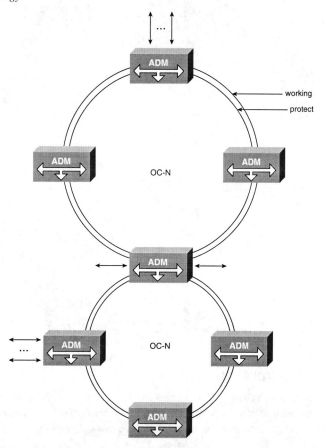

Mesh Topology

Meshed networks refer to any number of sites arbitrarily connected together with at least one loop. Mesh topologies are created using a hybrid combination of multiple point-to-point links and/or two-fiber UPSR links. As shown in Figure 5-30, meshed networks refer to any number of sites arbitrarily connected together with at least one loop. Mesh topologies provide maximum redundancy with multiple rerouting options for the SONET ADMs or DCS devices. Next-generation MSPPs leverage protection protocols, such as PPMN, by using advanced Dijkstra's algorithmic procedures for network discovery and route recalculation on fiber or node failure. Such information is carried in the SONET DCC channels and is used to build Dijkstra's tables and routing tables on startup or link state change. As soon as all the nodes on a network are turned up, they begin the process of autodiscovery. Within minutes, each node has a full description and status of the other nodes and connections throughout the network. Creating an overlaying TDM or Ethernet circuit is then easily accomplished by just specifying the source and destination node.

Figure 5-30 *Mesh Topology*

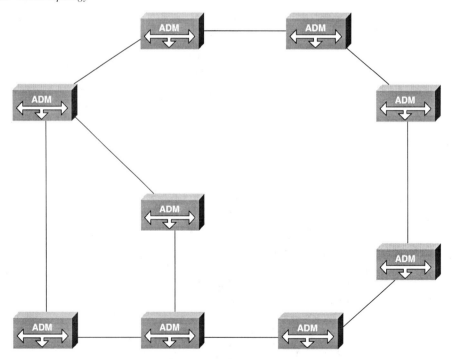

Fiber Routing and Diversity

Various fiber routing schemes help achieve redundancy in SONET networks. Trunks are the main fiber cables that can carry hundreds of fiber strands owned by carriers. Laterals are the fiber cables from the customer premises to the nearest splice point on the cable trunk. Generally laterals are used exclusively by the customer. Within cities, laterals can be as short as a few meters or could extend several kilometers in suburban and rural areas. The minimum size of a lateral is usually 12 strands. But even though a lateral can have 12 strands, only 2 or 4 of those strands can be spliced to dedicated fibers on the trunk. Most fiber-provisioning companies provide additional spare strands on the trunk to which the customer can connect to at a later time, depending on requirements. Building entrances and termination panels are the facilities within the customer's premise for the termination of the fiber.

Aerial installation on existing poles is the most cost-effective installation method and offers moderate reliability. Most regulatory bodies have well-established rules and procedures for licensed carriers and fiber installers to access existing utility and telephone poles. Installation in the existing conduit is the next-best option. Many regulatory bodies require carriers to install extra conduits accessible by any other licensed carrier or fiber installer. As with poles, regulators have set prices for the cost of access to this conduit. Jet fiber is another approach to fiber deployment. In this case, the fiber provider installs *microconduit* rather than fiber. When a customer requires a fiber pair, it is blown into the microconduit using high-pressure air. The advantage of this approach is that far fewer splices are required and the fiber can be blown all the way into the customer premises.

If there are no existing conduits or poles, commonly referred to as *support structures*, traditional trenching deployment can be used including the following:

- Direct-bury, fiber-plough techniques
- In-the-groove technologies, where a very narrow grove is cut into the existing road bed
- Sewer systems, where robotic systems install fiber in storm or sanitary sewers using either specialized cable or stainless steel tubing
- Gas pipeline systems, where the fiber is installed in active gas pipelines
- Directional boring methods

Note that individual fiber cables are sheathed, shielded, and armored. Various conduits can share a feeder into a building. As shown in Figure 5-31, a single feeder into the building with a single conduit from a CO is the most basic fiber routing topology. A degree of redundancy can be achieved by using diverse conduit routing from the CO point of presence with a single feeder into the building. The single feeder is the weakest link in such a physical fiber routing topology because it could be susceptible to backhoe digging or trenching.

True diverse routing can be obtained by having multiple feeders into the building with diverse conduit routing to the CO. Such a form of routing is shown in Figure 5-32. Dual COs with diverse conduit routing and diverse feeders provide the ultimate diversity for 1:1 protection schemes.

NOTE Twenty-five feet of facilities separation can qualify a fiber plant to possess route diversity from a construction perspective.

186 Chapter 5: SONET Architectures

Figure 5-31 *Fiber Routing with Conduit Diversity*

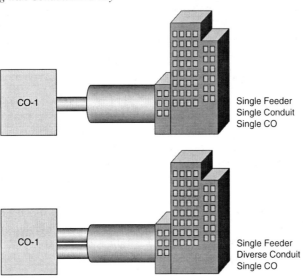

Figure 5-32 *Fiber Diversity*

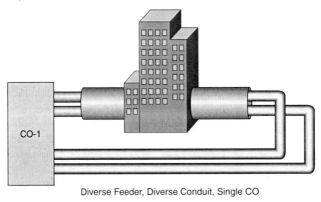

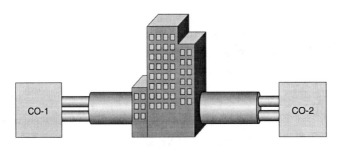

SONET Protection Architectures

This section discusses various SONET protection architectures and mechanisms. This includes APS, linear, and ring protection architectures. SONET rings use path-switching or line-switching, self-healing techniques. The Telcordia GR-253-CORE specification requires SONET protection architectures to restore within 50 ms. For example, SONET defines a maximum switch time of 50 ms for a BLSR ring with no extra traffic and less than 1200 km of fiber. The specification actually states 60 ms, with 10 ms for discovery of the problem and 50 ms to perform the switch, but this is commonly referred to as just 50 ms. However, most SONET networks are much smaller and simpler than this example and even faster. Y-cable switchover times are much higher and are typically in the order of 250 ms. SONET protection switching is invoked if there is an LOS, LOF, or even signal degradation, such as BER exceeding a pre-configured limit. Protection switching can also be initialized manually to allow for routine maintenance or testing without having to take the circuits out of service. Protection implies that a backup resource has been established for recovery if the primary resource fails, and restoration implies re-establishing the end-to-end path based on resources available after the failure. SONET protection architectures support nonrevertive and revertive protection mechanisms. With nonrevertive protection, after the original working line has been restored to its proper operating condition, the system will not revert to use the original line as the working path. Instead, it continues to use what was originally the protection line as the working line. With revertive protection, the system reverts to the original line after restoration. Revertive protection is most common with 1:N protection schemes. 1:1 protection can use revertive or nonrevertive protection mechanisms.

Automatic Protection Switching

APS is a standard devised to provide for link recovery in the case of failure. Link recovery is made possible by having SONET devices with two sets of fiber. One set (transmit and receive) is used for working traffic, and the other set (transmit and receive pair) is used for protection. The working and protect fibers are routed over diverse physical paths to be effective. The fibers used for protection may or may not carry a copy of the working traffic depending on how protection has been configured (1:N or 1+1). APS protection can be configured for linear or ring topologies. Each type of topology has specific choices of 1:1, 1:N, or 1+1 protection. These can further be configured with unidirectional or bidirectional switching mechanisms.

1+1 Protection

In a 1+1 protection architecture, the head-end signal is continuously bridged at the STS-N level to working and protection equipment so that an identical payload is transmitted over a separate fiber pair to the tail-end working and protection equipment. At the tail end, the working and protection signals are continuously monitored independently for failures. The tail-end equipment selects between either the working or the protection signals. Because of the continuous head-end bridge, the 1+1 architecture does not permit unprotected extra traffic over the span. Figure 5-33 shows a schematic of 1+1 protection. The signal is bridged on both paths and the tail-end receiver can select between the two signals based on local criteria. Switchover to the alternative signal is a local decision that does not require coordination with any other nodes. Because the signal is always available on both paths, the time to switch to the protected path is very short. The receiver could detect failure or signal degradation and perform an immediate switch. Recovery times of much less than 50 ms are possible. For 1+1 protection to be truly effective, the fiber plant must provide diverse routing.

Figure 5-33 *1+1 Protection*

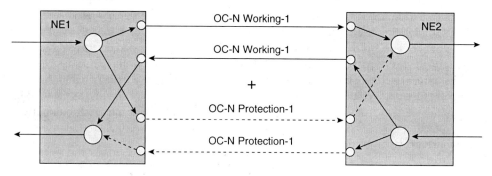

1:1 and 1:N Protection

In a 1:1 or 1:N protection architecture, traffic is carried on the working line until a failure occurs. The protection line is invoked if the working line fails. A schematic of 1:1 and 1:N protection is shown in Figure 5-34. In 1:N protection, one protection path is established as a backup for N working paths. If a failure occurs, an APS protocol is used to switchover the traffic to the protection facility. Figure 5-34 shows that the Nth multiple working path has failed and its signal has been bridged to the reserved protection path. Any one of the multiple working paths could be bridged to the protection path, but only one path can be protected at a time. If a failure that affects multiple working lines occurs, a priority mechanism must be configured to identify which line gets protection. 1:N protection supports a revertive option that allows the traffic to automatically switch back to the original working fibers after restoration with preconfigured wait time called *Wait To Restore (WTR)*. In 1:1 protection, there must be one protect line for every working line. Because the protect line does not actually carry traffic when not in use, it is possible, for the protect line to carry other nonpriority traffic called extra traffic. The extra traffic is dropped if protection is invoked. In 1:N protection schemes, a single line can protect one or more working lines.

Figure 5-34 *1:1 and 1:N Protection*

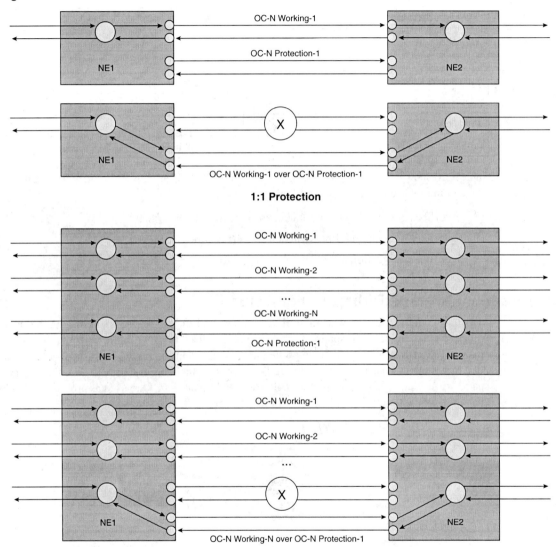

SONET Ring Architectures

This section examines SONET unidirectional and bidirectional ring architectures and examines the differences between two-fiber and four-fiber SONET rings. A comparison is also made between line (ring) switching versus path (span) switching. SONET provides for three attributes with two choices each, as shown in the Table 5-10.

Table 5-10 *SONET Ring Types*

SONET Attribute	Value
Fibers per link	Two-fiber
	Four-fiber
Signal direction	Unidirectional
	Bidirectional
Protection switching	Line switching
	Path switching

Table 5-10 shows various SONET ring configurations that differ in at least one major attribute. The commonly used ring types and topologies are as follows:

- Two-fiber unidirectional path-switched rings (two-fiber UPSRs)
- Two-fiber bidirectional line-switched rings (two-fiber BLSRs)
- Four-fiber bidirectional line-switched rings (four-fiber BLSRs)

Unidirectional Versus Bidirectional Rings

In a unidirectional ring, the working traffic is routed over the clockwise spans around the ring, while the counterclockwise spans are protection spans used to carry traffic when the working spans fail. Consider the two-fiber ring schematic presented in Figure 5-35. Traffic from NE1 to NE2 would traverse span 1 in a clockwise flow, while traffic from NE2 to NE1 would traverse span 2, span 3, and span 4 in a clockwise flow as well. Spans 5, 6, 7 are used as protection spans and carry production traffic, when one of the working clockwise spans fail.

Figure 5-35 *Unidirectional Versus Bidirectional Rings*

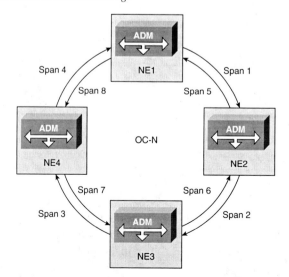

Bidirectional traffic flows can be illustrated using the schematic of Figure 5-35. In a bidirectional ring, traffic from NE1 to NE2 would traverse span 1 in a clockwise flow. However, traffic from NE2 to NE1 would traverse span 5 in a counterclockwise fashion. If the links between NE1 and NE2 were to fail, traffic between NE1 and NE2 would use the spans between NE2-NE3, NE3-NE4, and NE4-NE1.

Two-Fiber Versus Four-Fiber Rings

Unidirectional and bidirectional systems both implement two-fiber and four-fiber systems. Most commercial unidirectional systems are two-fiber, whereas bidirectional systems implement both two-fiber and four-fiber infrastructures. A two-fiber OC-N unidirectional system with two nodes is illustrated in Figure 5-36. Fiber span 1 carries N working channels eastbound, and fiber span 5 carries N protection channels westbound. For example, an OC-48 system would carry 48 working STS-1s eastbound from NE1 to NE2, while carrying 48 separate protection STS-1s westbound from NE2 to NE1. The SONET transport and POH bytes are carried on both the working and protection fiber spans.

Figure 5-36 *Two-Fiber Unidirectional Ring*

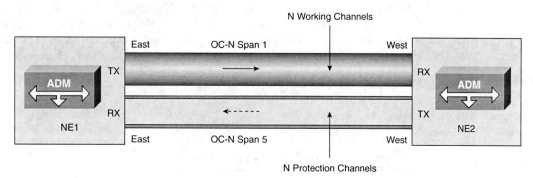

A two-fiber OC-N bidirectional system with two nodes is illustrated in Figure 5-37. On each fiber, a maximum of half the bandwidth or number of channels are defined as working channels, and the other half are defined as protection channels. Fiber span 1 carries (N/2) working channels and (N/2) protection channels eastbound, and fiber span 5 carries (N/2) working channels and (N/2) protection channels westbound. For example, an OC-48 system would carry 24 working STS-1s and 24 protection STS-1s eastbound from NE1 to NE2, while carrying 24 working STS-1s and 24 protection STS-1s westbound from NE2 to NE1. Each fiber has a set of SONET transport and POH bytes for the working and protection channels.

Figure 5-37 *Two-Fiber Bidirectional Ring*

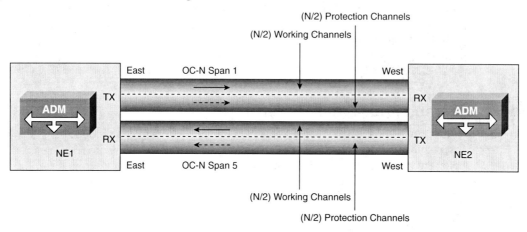

A four-fiber OC-N bidirectional system with two nodes is shown in Figure 5-38. Fiber pair span 1A and 5A carries N working channels full duplex east and westbound, while fiber pair span 1B and 5B carries N protection channels full duplex east and westbound. For example, an OC-48 system would carry 48 working STS-1s eastbound from NE1 to NE2 as well as 48 working STS-1s westbound from NE2 to NE1. The same system would also carry 48 protection STS-1s eastbound from NE1 to NE2 as well as 48 protection STS-1s westbound from NE2 to NE1. A set of SONET transport and POH bytes is dedicated either to working or protection channels for the four-fiber ring.

Figure 5-38 *Four-Fiber Bidirectional Ring*

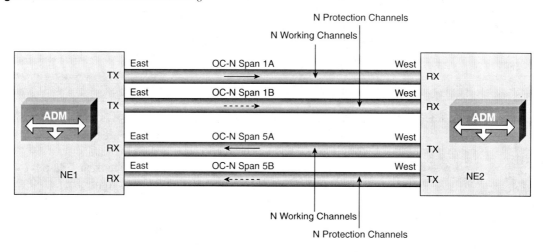

As can be seen for an OC-N fiber system, two-fiber UPSR provides N * STS-1s, and two-fiber BLSR provides (N/2) * STS-1s in either direction. Four-fiber BLSR, on the other hand, provides N * STS-1s in either direction. Usually two segment failures will cause a network failure or outage on a two-fiber ring of either type. However, four-fiber systems with diverse routing can suffer multiple failures and still function. Four-fiber systems are widely used for rings spanning large geographical areas or when the traffic being carried on the network is very critical, such as voice traffic or government operations (FAA or DoD, for instance).

SONET rings are limited to 16 nodes per ring according to GR.253-CORE specifications, because the K1/K2 bytes that define the source and destination node were defined with only 4 bits each. However, vendors have implemented proprietary mechanisms that use unused bytes from other fields in the SONET header to extend the limit on the number of nodes. For example, the Cisco ONS 15454 uses 4 bits from the K1/K2 fields and 2 additional bits of the K3 byte in BLSR configurations. The K3 byte also carries information on the K1/K2 bytes. Out of the 2 K3 bits, 1 bit is used to define the source and the other bit is used to define the destination node. The K3 byte is actually the Z4 growth byte found in the POH. Use of the 2 additional K3 bits increases the node count to 32 NEs per BLSR ring. In such a case, however, if the span had to pass through third-party equipment, the K3 byte would need to be remapped to an unused SONET overhead byte, such as the Z2, E2, or F1 byte.

Path and Line Switching

Path switching works by restoring working channels at a level below the entire OC-N capacity in a single protection operation. This means that levels lower than an OC-N, such as STS-1s or VT1.5s, can be restored the event of failure. Path switching is shown in Figure 5-39. Live protected user traffic is always sent on the working fiber. However, a copy of the protected traffic is also transmitted over the protection fiber. The receiver constantly senses the signal level of both the working and protection fibers. In the event of a fiber cut or signal degradation on the working fiber, the receiver switches to the incoming signal available on the protection fiber. All unprotected traffic is dropped for the duration of the outage. Path switching is mostly implemented on two-fiber unidirectional rings.

Figure 5-39 *Path Switching*

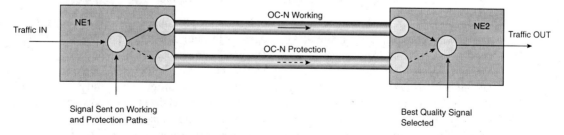

Line switching works by restoring all working channels of the entire OC-N capacity in a single protection operation. The protection channels or fiber are idle while the ring operates normally. Line switching is shown in Figure 5-40. Live protected user traffic is always sent on the working channels or fiber. In the event of fiber or node failure, the protected traffic is switched to the protection channels or fiber at both ends of the span. Channels within the *line*

are switched this way, which is why it is called *line switching*. In the event of failure, all unprotected traffic being transmitted on the protection link or protected channels is dropped. This is called *protection channel access (PCA)*, and the traffic carried this way is called *extra traffic*. Carriers typically discount unprotected PCA bandwidth, thereby allowing customers to maintain a more cost-effective network without having to pay for a five-nines service level agreement (SLA). Line-switching systems are able to restore service within 50 ms. Line switching is mostly implemented on two-fiber and four-fiber bidirectional rings.

Figure 5-40 *Line Switching*

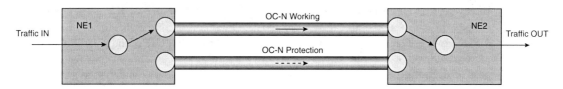

Dual-Ring Interconnect

The dual-ring interconnect (DRI) architecture allows subtending rings sharing traffic to be resilient from a *matching node* failure perspective. As shown in Figure 5-41, a DRI topology uses two interconnecting matching nodes (DRI node 3 and DRI node 4) to connect the two OC-N rings. If one of the interconnected nodes were to fail, traffic will be routed through the surviving DRI node. The benefit to the service provider is that continuous network operation is maintained even though a node failure has occurred. The DRI topology provides an extra level of path protection between rings. In a DRI configuration, traffic is dropped and continued at the interconnecting nodes to eliminate single points of failure. Each ring protects against failures within itself using path-switched and/or line-switched protection mechanisms, while DRI provides protection against failures at the interconnections. DRI cannot provide protection if both DR1 node 3 and DRI node 4 experience simultaneous failure.

As shown in Figure 5-41, a signal input at node 1 destined for node 7 is bridged east and west. The downstream primary eastbound signal passes through node 2 and arrives at the DRI node 3. At DRI node 3, a duplicate copy of the signal is dropped and transmitted to DRI node 4. Similarly, the downstream secondary westbound signal passes through node 5 and arrives at the DRI node 4. At DRI node 4, a duplicate copy of the signal is dropped and transmitted to DRI node 3. The downstream path selector at node 3 always selects the primary downstream signal during steady-state normal operation. However, the downstream path selector at node 4 always selects the secondary downstream signal during steady-state normal operation. The primary downstream signal at node 3 is then continued and transmitted to node 6 on ring 2 that acts as a pass-through node and transmits the signal to node 7. Similarly, the secondary downstream signal at node 4 is then continued and transmitted to node 8 on ring 2 that acts as a pass-through node and transmits the signal to node 7.

Figure 5-41 *Dual-Ring Interconnect*

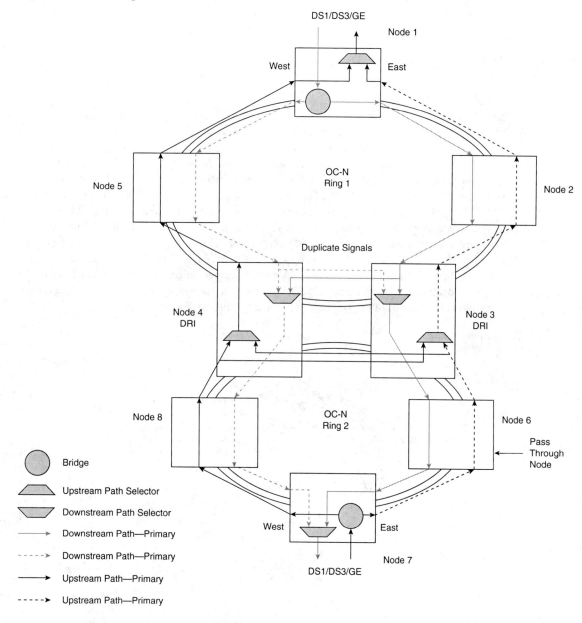

Node 7 receives two copies of the downstream signal (primary and secondary). However, the path selector in node 7 always selects the primary downstream signal during steady-state normal operation. A similar process takes place with the primary and secondary upstream signal. Let us assume that DRI node 3 fails. In such an event, the path selector at node 7 will switch to the secondary downstream signal. The upstream traffic is not affected because the primary upstream path is through the surviving node 4.

Unidirectional Path Switched Rings

A UPSR is a survivable, closed-loop, transport architecture that protects against fiber cuts and node failures by providing duplicate, geographically diverse paths for each circuit. UPSR provides dual fiber paths around the ring. Working traffic flows clockwise in one direction, and protection traffic flows counterclockwise in the opposite direction. If fiber or node failure occurs in the working traffic path, the receiving node switches to the path coming from the opposite direction. Because each traffic path is transported around the entire ring, UPSRs are best suited for networks where traffic concentrates at one or two locations and is not widely distributed. UPSR capacity is equal to its bit rate. This means that an OC-N UPSR ring will always provide N * STS-1s of capacity. Services can originate and terminate on the same core UPSR ring or can be passed across a matching node to an access ring for transport to the service-terminating location. Figure 5-42 shows a basic UPSR configuration. This drawing can also be used to explain basic two-fiber UPSR operation with its various subtleties. The schematic illustrates the operation of a two-fiber OC-12 ring using UPSR as its protection mechanism. The outer Fiber 1 is the working fiber that carries traffic in a clockwise direction. The inner Fiber 2 is the protection fiber that would carry a *copy* of the working traffic in a counterclockwise direction.

Figure 5-42 *Two-Fiber UPSR*

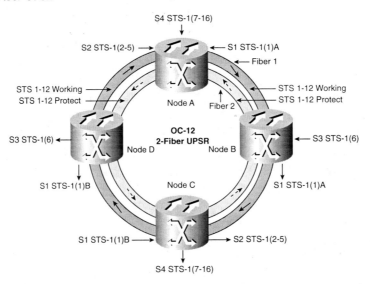

If node A sends a signal S1 to node B, the working signal travels on the working traffic path to node B. The same signal is also sent on the protect traffic path from node A to node B, via nodes D and C. For node B to reply to node A, the signal uses the working STS-1 path around the ring via nodes C and D. Note that signal S1 or STS-1(1)A is the first STS-1 of the OC-12. Signal S1 consumes the entire STS-1 around the ring. Therefore, it is *not* possible for another signal, such as S1 STS-1(1)B, to be transmitted between node C and node D using UPSR protection. However, STS-1(1)B could be transmitted *unprotected* between nodes C and D. This circuit would be dropped in event of UPSR protection being invoked.

Signal S2 is added at node A and dropped at node C. Signal S2 contains STS-1(2–5) or [STS-1, number 2 to number 5 of the OC-12]. This also means that STS-1(2–5) cannot be used for adds or drops at other nodes on the ring. For node C to reply to node A, the signal uses the working STS-1 path around the ring via node D. Signal S3 STS-1(6) is added at node B and dropped at node D, effectively blocking any other adds or drops for STS-1(6) at any of the other nodes. For node D to reply to node B, the signal uses the working STS-1 path around the ring via node A. Finally, signal S4 is added at node A and dropped at node C. Signal S4 contains STS-1(7–12), which means that STS-1(7–12) cannot be used for adds or drops at other nodes on the ring. For node C to reply to node A, the signal uses the working STS-1 path around the ring via node D.

As shown in Figure 5-43, if a fiber break occurs on the working Fiber 1 between node A and node B, node B switches its active receiver to the protect signal coming through node C. Signals S1 and S3 would be received on the protection fiber for the duration of the outage. The switchover would happen within the 50-ms SONET restoration time. Signals S2 and S4 would be received on node C via node D on the protection fiber. If there were a fiber cut on the protection Fiber 2, however, the system would continue operating without any disruption. The element management system would detect the fiber cut and report the LOS on the protection fiber. Repairs could be performed on the Fiber 2 without service interruption.

Figure 5-43 *Two-Fiber UPSR Protection*

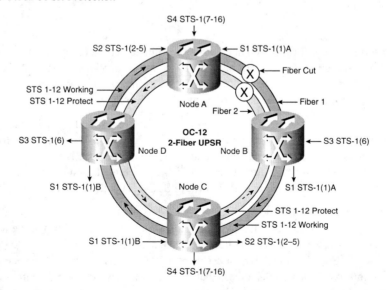

Asymmetrical Delay

As shown in Figure 5-42, any signal from node A to node B traverses a single span. When node B has to reply to node A, however, the signal has to traverse multiple spans via node C and node D. In the case of small metropolitan rings, this does not create any issues. For large transcontinental rings, however, a finite delay occurs that could affect voice or data applications. In the case of voice applications, the cumulative delay should not exceed 100 ms. So long as the asymmetric delay does not exceed 100 ms, the human user would not perceive any delay. In the case of data, transport layer windowing comes into play. With asymmetric delay, two end hosts might experience a 40-ms round-trip delay. One host might perceive a 5-ms inbound delay with a 35-ms outbound delay. It would be the exact opposite for the other host. Issues occur when one data application tries to adjust its window size for a 20/20-ms split in delay, while data arrives early (5 ms). The host on the other end experiences exactly the opposite effect when adjusting its window size for a 20/20-ms split, with data arriving late (25 ms).

Bidirectional Line-Switched Rings

BLSR uses bidirectional line-switched protection mechanisms. BLSR is commonly implemented on two-fiber as well as four-fiber systems. BLSR nodes can terminate traffic that is fed from either side of the ring and are suited for distributed node-to-node traffic applications such as interoffice networks and access networks. BLSRs allow bandwidth to be reused around the ring and can carry more traffic than a network with traffic flowing through one central hub. BLSR supports nonrevertive and revertive protection mechanisms.

Two-Fiber BLSR

In a two-fiber BLSR ring, each fiber carries working and protection STS-1s. In an OC-12 BLSR, as shown in Figure 5-44, however, STSs 1 through 6 carry the working traffic, and STSs 7 through 12 are reserved for protection. Working traffic travels clockwise in one direction on one fiber and counterclockwise in the opposite direction on the second fiber.

In Figure 5-44, signal S1 STS-1(1)A added at node A, destined for a drop at node B, typically will travel on Fiber 1, unless that fiber is full, in which case circuits will be routed on Fiber 2 through node C and D. Traffic from node A to node C (or node B to node D), can be routed on either fiber, depending on circuit-provisioning requirements and traffic loads. For node B to reply to node A, the signal uses the working STS-1(1) path on Fiber 2.

Signal S2 STS-1(2–5) added at node A, destined for a drop at node C, typically will travel on Fiber 1 via node B, unless that fiber is full, in which case the circuit will be routed on Fiber 2 via node D. For node C to reply to node A, the signal uses the working STS-1(2–5) path on Fiber 2 via node B. Signal S3 STS-1(6) added at node B, destined for a drop at node D, typically will travel on Fiber 1 via node C, unless that fiber is full, in which case the circuit will be routed on Fiber 2 via node A. For node D to reply to node B, the signal uses the working STS-1(6) path on Fiber 2 via node C. It is quite apparent that only STS-6 worth of bandwidth can be configured on a two-fiber OC-12 BLSR. This is not entirely true. Unlike UPSR, the provisioning of STS-1(1) does not consume the entire first STS-1 of the OC-12 around the ring. Bandwidth is reusable. This can be exemplified by looking at S1 STS-1(1)B in Figure 5-44, which can be provisioned between node C and D. With careful bandwidth-capacity planning, BLSR can be quite efficient.

Figure 5-44 *Two-Fiber BLSR*

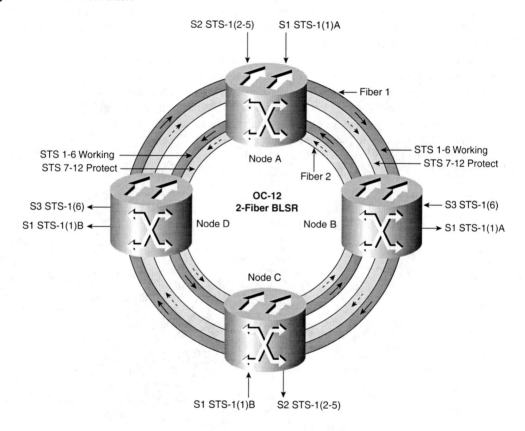

NOTE The bidirectional bandwidth capacity of two-fiber BLSR is the OC-N rate divided by two, multiplied by the number of nodes in the ring, minus the number of pass-through STS-1 circuits.

The SONET K1 and K2 bytes carry the information that governs BLSR protection switching. Each BLSR node monitors the K bytes to determine when to switch the SONET signal to an alternate physical path. The K bytes communicate failure conditions and actions taken between nodes in the ring. If a break occurs on one fiber, working traffic targeted for a node beyond the break switches to the protect bandwidth on the second fiber. The traffic travels in reverse direction on the protect bandwidth until it reaches its destination node. At that point, traffic is switched back to the working bandwidth.

As shown in Figure 5-45, if a break occurs in Fiber 1 between node A and node B, signal S1 STS-1(1)A that would normally travel between nodes A and B using STS-1(1) of Fiber 1 would line switch to STS-1(7) of Fiber 2 and reach node B via nodes D and C for the duration of the outage. The switchover would happen within the 50-ms SONET restoration time. Signal S2 STS-1(2–5) added at node A and destined for node C would also be affected. S2 would be line switched to STS-1(8–11) of Fiber 2 and would reach node C via node D. Signal S3 STS-1(6) would not be affected. Now consider the case where Fiber 1 is intact and there is a break in Fiber 2 between nodes A and B. In such a case, the return path for signal S1 STS-1(1)A between node B and node A is lost. A line switch would occur and signal STS-1(1)A would switch to STS-1(7) of Fiber 1 and reach node A via nodes C and D. The return path for signal S2 STS-1(2–5) between node C, destined for node A, would also be affected. Node C would transmit signal S2 STS-1(2–5) back to node A over Fiber 2. However, the fiber cut on Fiber 2 (between nodes A and B), detected by node B, would cause all return traffic to node A to be line switched to STS-1(8–11) of Fiber 1 and retransmitted to node A via nodes C and D.

Figure 5-45 *Two-Fiber BLSR Protection*

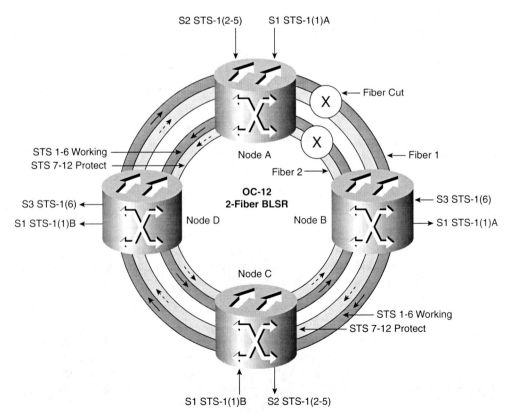

Finally, consider a case of a dual fiber cut of both Fiber 1 and Fiber 2 between nodes A and B. In such a case, signal S1 STS-1(1)A added at node A and destined for node B would be line switched to STS-1(7) of Fiber 2 and would reach node B via nodes D and C. The return path for signal S1 STS-1(1)A between node B and node A would line switch to STS-1(7) of Fiber 1 and reach node A via nodes C and D. Signal S2 STS-1(2–5) added at node A and destined for node C would be line switched to STS-1(8–11) of Fiber 2 and would reach node C via node D. Node C would transmit the return signal S2 STS-1(2–5) back to node A over Fiber 2. However, the fiber cut on Fiber 2 (between nodes A and B), detected by node B, would cause all return traffic to node A to be line switched to STS-1(8–11) of Fiber 1 and retransmitted to node A via nodes C and D. All unprotected traffic carried over the protection STS-1s is dropped in event of a line switch.

BLSR Node Failure

BLSR restoration gets complex in the event of a node failure. BLSR uses a protection scheme called *shared protection*. Shared protection is required because of the construction of the BLSR ring and the reuse of STSs around the ring. This creates a situation in which the STSs on a protection fiber cannot be guaranteed to protect traffic from a specific working STS. Shared protection, which provides BLSR its capability to reuse bandwidth, brings with it additional problems when a node failure occurs in a BLSR ring. Let us examine the BLSR schematic in Figure 5-46. This schematic illustrates a complete failure of node D. Let us trace the path of signal S3 as it gets added on to node B with a destination node D. Signal S3 STS-1(6) gets sent out to node D on Fiber 1 and proceeds to node C. Node C has sensed an LOS from failed node D and reroutes S3 on to Fiber 2 as signal S3 STS-1(12). Signal S3 passes via node B and arrives at node A. Because node A cannot deliver this traffic to node D, however, it places S3 on Fiber 1 as S3 STS-1(6). This signal gets dropped off at node B, because STS-1(6) already has a connection from node A to node B (signal S4). This event results in the traffic being delivered to the wrong node and is called a *misconnection*. In some situations, it is possible that bridging traffic after a node failure could also lead to a misconnection.

BLSR misconnections can be avoided by using the squelching mechanism. The squelching feature uses automatically generated squelch maps that require no manual record-keeping to maintain. Each node maintains squelch tables to know which connections need to be squelched in the event of a node failure. The squelch table contains a list of inaccessible nodes. Any traffic received by a node for the inaccessible node is never placed on the fiber and is removed if discovered.

Squelching involves sending the STS path alarm indication signal (SP-AIS) in all channels that normally terminated in the failed node rather than real traffic. The misconnection is avoided by the insertion of an AIS path by nodes A and C into channel STS-1(6). In an AIS path, all the bits belonging to that path are set to 1 so that the information carried in that channel is invalidated. This way, node B is informed about the error condition of the ring, and a misconnection is prevented. Misconnection can only occur in BLSR when a node is cut off and traffic happens to be terminate on that node from both directions on the same channel (STS). In some implementations, the path trace might also be used to avoid this problem. If node B monitors the path trace byte, it will recognize that it has changed after the misconnection. This change should be sufficient indication that a fault has occurred and that traffic should not be terminated.

Figure 5-46 *BLSR Node Failure*

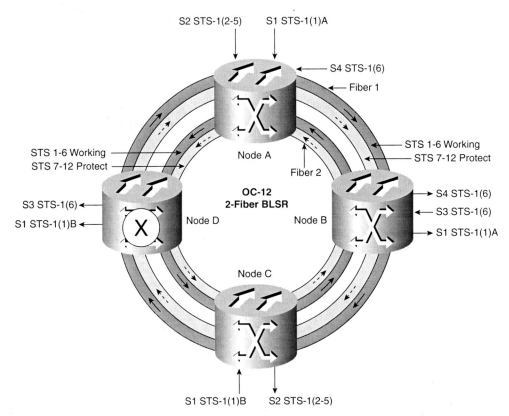

Four-Fiber BLSR

Four-fiber BLSRs double the bandwidth of Two-fiber BLSRs. As shown in Figure 5-47, two fibers are allocated for working traffic and two fibers are allocated for protection signal. S1 from node A to node B would use STS-1(1) of the working Fiber 1, and the return path from node B to node A would use STS-1(1) of the working Fiber-3. Signal S2 added at node A, destined for node C, would use STS-1(2–5) of the working Fiber 1 and would use STS-1(2–5) of the working Fiber-3 for its return path from node C to node A.

Signal S3 added at node B, destined for node D, would use STS-1(6) of the working Fiber 1 via node C and would use STS-1(6) of the working Fiber-3 for its return path from node D to node B, via node C. Signal S4 added at node A, destined for node C, would use STS-1(7–12) of the working Fiber 1 and would use STS-1(7–12) of the working Fiber-3 for its return path from node C to node A.

Four-fiber BLSR allows path (span) switching as well as line (ring) switching, thereby increasing the reliability and flexibility of traffic protection. Path (span) switching occurs when a working span fails. Traffic switches to the protect fibers between the nodes and then returns to the working fibers. Multiple span switches can occur at the same time. Line (ring) switching occurs when a span switch cannot recover traffic, such as when both the working and protect fibers fail on the same span. In a line (ring) switch, traffic is routed to the protect fibers throughout the full ring.

Figure 5-47 *Four-Fiber BLSR*

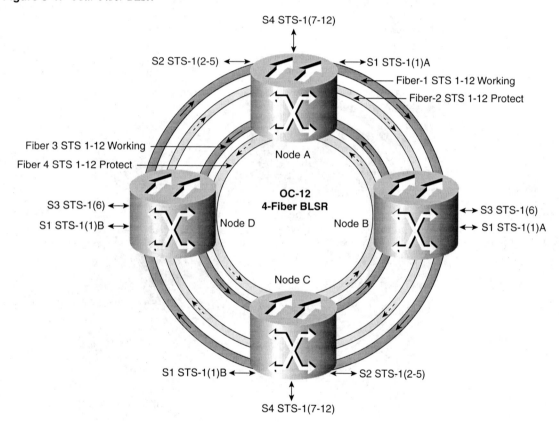

As shown in Figure 5-48, if the working fiber pair between node A and B fails, all working traffic between these nodes is shunted onto the protection fiber pair. Any unprotected traffic mapped between other nodes on the ring is unaffected by this outage.

Signal S1 from node A to node B would use STS-1(1) of protection Fiber 2, and the return path from node B to node A would use STS-1(1) of protection Fiber-4. Signal S2 added at node A, destined for node C, would use STS-1(2–5) of protection Fiber 2 between node A and B, after which it would revert to STS-1(2–5) of the working Fiber 1 between node B and C. Signal S2 would use STS-1(2–5) of the working Fiber-3 for its return path from node C to node B, after which it would use STS-1(2–5) of protection Fiber-4 between nodes B and A. Signal S3 would be unaffected. However, Signal S4 added at node A, destined for node C, would use STS-1(7–12) of protection Fiber 2 between node A and node B, after which it would revert to STS-1(7–12) of the working Fiber 1 between nodes B and C. Signal S4 would use STS-1(7–12) of the working Fiber-3 for its return path from node C to node B, after which it would use STS-1(7–12) of protection Fiber-4 between nodes B and A. Four-fiber BLSR ring switching is shown in Figure 5-49. If both fiber pairs between node A and B fail, all working traffic between these nodes is wrapped onto the protection fiber pairs. Any unprotected traffic mapped between other nodes on the ring is preempted and dropped because all the protection pairs are used during the outage.

Figure 5-48 *Four-Fiber BLSR Span Switch*

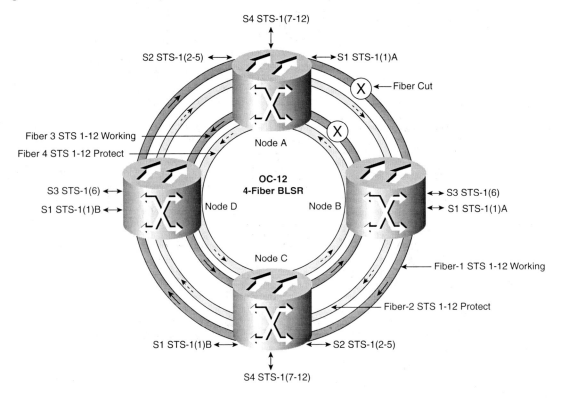

Signal S1 from node A to node B would use STS-1(1) of protection Fiber-4 via nodes D and C, and the return path from node B to node A would use STS-1(1) of protection Fiber 2 via node C and D. Signal S2 added at node A, destined for node C, would use STS-1(2–5) of protection Fiber-4 via node D. On its return path from node C to node A, signal S2 would use STS-1(2–5) of the working Fiber-3 between node C to node B. Node B would cause a wrap and switch the traffic to STS-1(2–5) of protection Fiber 2 for a drop at node A via nodes C and D. Signal S3 would be unaffected. Signal S4 added at node A, destined for node C, would use STS-1(7–12) of protection Fiber-4 via node D. On its return path from node C to node A, signal S4 would use STS-1(7–12) of the working Fiber-3 between node C to node B. Node B would cause a wrap and switch the traffic to STS-1(7–12) of protection Fiber 2 for a drop at node A via nodes C and D.

Figure 5-49 *Four-Fiber BLSR Ring Switch*

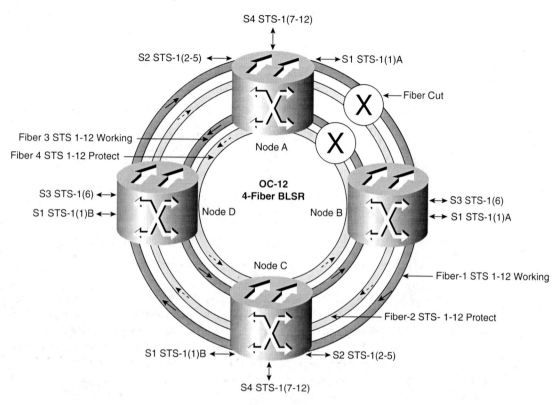

SONET Network Management

SONET NEs need OAM&P support to be managed by carriers and service providers. The OAM&P of an NE is the task of its EM. EMs are device-specific and vary by vendor. In a typical service provider environment, there could be multiple EMs. The integration of the various EMs (along with fault management (FM), performance management (PM), accounting management (AM), security management (SM), configuration management (CM), trouble ticketing, and billing applications) is the function of the operations support system (OSS). Multiple OSS systems that manage the data communications network (DCN) constitute the Telecommunication Management Network (TMN). The TMN has been standardized by the ITU-T under Recommendation M.3010. Telcordia has defined the network management architecture for SONET operational communications in the GR-253 standard. According to GR-253, SONET devices can be remotely managed through the use of in-band management channels in the SOH and LOH, known as the *section data communications channel (SDCC) and line data communications channel (LDCC)*. Figure 5-50 illustrates an intercarrier TMN model. OSS-1 is operated by Carrier-1 and OSS-N is operated by Carrier-N. The OSS accesses the DCN via a gateway network element (GNE).

Figure 5-50 *OSS and TMN Schematic*

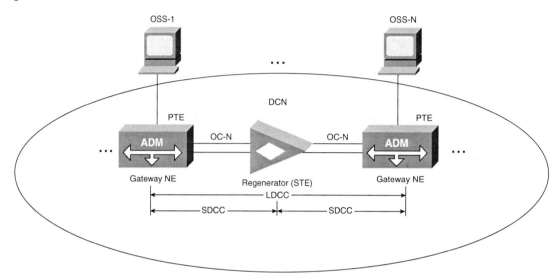

The SDCC and LDCC channels can transport operations and management messages that let OSS systems comply with the TMN specification. However, many SONET equipment vendors have established proprietary EM schemes and there is little interoperability between vendors in the use of these bytes. The SDCC bytes are D1 through D3 in the SOH. These 3 bytes provide a 192-kbps communications channel. The LDCC bytes are D4 through D12 in the LOH. The 9 LDCC bytes provide a 576-kbps communications channel. Most SONET systems use SDCC bytes for management purposes and don't use the LDCC bytes alone. The Cisco Transport Manager (CTM) enables service providers to manage their Cisco SONET and optical transport devices collectively under one management system. Cisco also uses a craft tool and EMS for comprehensive SONET and optical transport management called Cisco Transport Controller (CTC). Cisco has developed its management application based on an IP stack coupled with an Open Shortest Path First (OSPF)-based topology discovery mechanism. Furthermore, Cisco ONS devices can tunnel their SDCC bytes through an ONS network. The bytes are tunneled by copying the SDCC bytes into three of the LDCC bytes. Because there are three SDCC bytes and nine LDCC bytes, the ONS devices have the capability to transport traffic from three different SONET networks simultaneously across any given span. The SDCC bytes are restored as they leave the ONS network, thus permitting interoperability with non-ONS networks. The Cisco EMS supports Transaction Language 1 (TL-1), Common Object Request Broker Architecture (CORBA), and SNMP for OAM&P purposes.

Summary

Communication between various localized networks is complex due to differences in digital signal hierarchies, encoding techniques, and multiplexing strategies. The SONET format allows different types of signal formats to be transmitted over the fiber-optic cable. The PDH includes all ITU TDM hierarchies including the T-carrier and E-carrier systems. OC-N refers to the optical SONET transmission characteristics of an Nth level transmission link, and STS-N refers to the electrical SONET transmission characteristics of an Nth level transmission link. The four SONET optical interface layers are the path, line, section, and the photonic layers. Path termination, line termination, and section termination occur at these various layers.

SONET framing of the STS level 1 (STS-1) and the STS-N signal can occur as discrete or concatenated modes. The STS-1 signal consists of the transport overhead, POH, and the payload envelope. The SPE floats within the STS-1 envelope. SONET provides substantial overhead information, allowing simpler multiplexing and greatly expanded OAM&P capabilities. The SONET TOH consists of the line and section overhead, and the SONET POH is assigned to and transported with the SONET payload. SONET alarms are defined as anomalies, defects, and failures. An anomaly is the smallest discrepancy that can be observed between the actual and desired characteristics of an item. The occurrence of a single anomaly does not constitute an interruption in the ability to perform a required function. If the density of anomalies reaches a level at which the ability to perform a required function has been interrupted, it is termed a defect. The STS SPE can be subdivided into smaller structures, known as VTs. VTs are used for transporting and switching payloads smaller than the STS-1 rate. The various VT types include VT1.5, VT2, VT3, and Vt6. VT structures can be grouped into VT groups. The VT superframe consists of four VT SPEs.

SONET NEs include section-, line-, and path-terminating equipment. The SONET NEs integrate to form the DLC supporting various SONET architectures and topologies. Some of these NEs include regenerators, TMs, ADMs, BDCS, WDCS, and the IDLC, which consists of intelligent remote digital terminals and digital switch elements called integrated digital terminals. SONET topologies and protection architectures center around network survivability and sub-50-ms service restoration. SONET topologies include the linear ADM point-to-point, point-to-multipoint, hub, ring, and mesh topology. The SONET ADM and DCS are NEs used to build such topologies. The ring and mesh topologies offer various survivability and protection mechanisms. Fiber routing and diversity schemes also provide redundancy. Several ring configurations and protection mechanisms provide various levels of redundancy. Path switching works by restoring working channels at a level below the entire OC-N capacity in a single protection operation, whereas line switching works by restoring all working channels of the entire OC-N capacity in a single protection operation. Two-fiber UPSR, two-fiber BLSR, and four-fiber BLSR rings are some of the commonly used ring protection mechanisms. SONET NEs need OAM&P support to be managed by carriers and service providers. The OAM&P of an NE is the task of its EM. EMs are device specific and vary by vendor. In a typical service provider environment, there could be multiple EMs. The integration of the various EMs along with the integrated NMS constitutes the OSS system.

This chapter includes the following sections:

- **SDH Integration of TDM Signals**—This section describes Synchronous Digital Hierarchy (SDH) integration of time-division multiplexed (TDM) signals. Communication between various localized networks is complex due to differences in digital signal hierarchies, encoding techniques, and multiplexing strategies.
- **SDH Layers**—The four SDH layers (path, multiplex section, regenerator section, and photonic layer) are explained in this section. Path termination, multiplex section termination, and regenerator section termination equipment are also described.
- **SDH Multiplexing**—TDM plesiochronous digital hierarchy (PDH) services can be natively accepted by SDH systems.
- **SDH Framing**—SDH framing of the STM-1 is described in this section. The derivation of various STM-N rates and STM-1 frame creation is also described in this section.
- **SDH Transport Overhead**—The various SDH regenerator section, multiplex section, and path overhead bytes are explained in this section in great detail.
- **SDH Alarms**—SDH alarms are defined as anomalies, defects, and failures. This section describes various SDH alarms.
- **SDH Higher-Level Framing**—This section describes higher-level STM-N frames, such as the STM-4, STM-16, STM-64, and STM-256.
- **SDH Network Elements**—Various SDH network elements are covered in this section including the regenerator section, multiplex section, and path terminating equipment.
- **SDH Topologies**—This section covers SDH topologies and protection architectures that center around network survivability and 50-millisecond or lower service restoration.
- **SDH Protection Architectures**—SDH protection mechanisms are discussed in this section. It covers automatic protection switching (APS) as well as 1+1, 1:1, and 1:N protection architectures.
- **SDH Ring Architectures**—This section discusses SDH ring architectures and describes their inner workings.
- **SDH Network Management**—SDH network elements (NEs) need OAM&P support to be managed by carriers and service providers. This section describes SDH NE management.

CHAPTER 6

SDH Architectures

SDH Integration of TDM Signals

The Synchronous Digital Hierarchy (SDH) has evolved as a result of standardization proposals in the International Telecommunications Union—Telecommunications Services Sector (ITU—TS) for an international optical interface standard for carriers and features for improved network management. The SDH format allows different types of signal formats to be transmitted over the fiber-optic cable. It allows adding or dropping signals within a single multiplexer. Communication between various localized networks is complex due to differences in digital signal hierarchies, encoding techniques, and multiplexing strategies. Communications between time-division multiplexed (TDM) networks previously described require complicated multiplexing, demultiplexing, coding, and decoding to convert a signal from one format to another. To solve this signal-conversion problem, SDH standardizes the rates, framing format, signaling, and termination between SDH equipment. As illustrated in Figure 6-1, SDH multiplexer equipment accepts various native TDM signal formats and multiplexes (adds) these signals without conversion. These signals can be demultiplexed (dropped) at any node or intermediate node.

Figure 6-1 *SDH Integration of TDM*

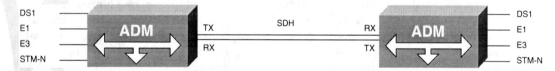

The ITU-T SDH standards define the bit rate, format, physical layer, network element (NE) architectural features, and network operational criteria of SDH networks. The various ITU-T SDH standards are as follows:

- **ITU-T Recommendation G.691**—Optical Interfaces for Single-Channel SDH Systems with Optical Amplifiers, and STM-64 Systems
- **ITU-T Recommendation G.707**—Network Node Interface for the Synchronous Digital Hierarchy (SDH)
- **ITU-T Recommendation G.781**—Structure of Recommendations on Equipment for the Synchronous Digital Hierarchy (SDH)

- **ITU-T Recommendation G.782**—Types and Characteristics of Synchronous Digital Hierarchy (SDH) Equipment
- **ITU-T Recommendation G.783**—Characteristics of Synchronous Digital Hierarchy (SDH) Equipment Functional Blocks
- **ITU-T Recommendation G.803**—Architecture of Transport Networks Based on the Synchronous Digital Hierarchy (SDH)
- **ITU-T Recommendation G.813**—Timing Characteristics of SDH Equipment Slave Clocks (SEC)
- **ITU-T Recommendation G.825**—The Control of Jitter and Wander Within Digital Networks Which Are Based on the Synchronous Digital Hierarchy (SDH)
- **ITU-T Recommendation G.826**—Error Performance Parameters and Objectives for International, Constant Bit Rate Digital Paths at or Above the Primary Rate
- **ITU-T Recommendation G.831**—Management Capabilities of Transport Networks Based on Synchronous Digital Hierarchy (SDH)
- **ITU-T Recommendation G.957**—Optical Interfaces for Equipment and Systems Relating to the Synchronous Digital Hierarchy (SDH)
- **ITU-T Recommendation G.958**—Digital Line Systems Based on the Synchronous Digital Hierarchy (SDH) for Use on Optical-Fiber Cables
- **ITU-T Recommendation I.432**—B-ISDN User-Network Interface Physical Layer Specification Criteria

Plesiochronous Digital Hierarchy (PDH)

The plesiochronous digital hierarchy (PDH) includes all ITU TDM hierarchies, including the T-carrier and E-carrier systems. Signals are considered to be plesiochronous if their *significant instants occur at nominally the same rate*, with any variations constrained to specified limits. Plesiochronous network timing signals can suffer phase shift, jitter, and wander. For the most part, however, they stay within the predefined limits. PDH is the conventional multiplexing technology for network transmission systems. PDH is an older transmission standard defined by ITU-T G.702, where transmission rates are independent but kept within a closely defined range. PDH refers to the three digital communication standards that are based on the fundamental concepts of TDM. As illustrated in Table 6-1, these include the T-N carrier system in the North American digital hierarchy (NADH), the E-N carrier system in the European digital hierarchy (EDH), and the J-N carrier system in the Japanese digital hierarchy (JDH). All PDH systems share the base rate of 64 kbps that corresponds to a pulse code modulation (PCM) voice channel; however, the rate hierarchies and overhead and bit-stuffing schemes vary. Line rates range from 1.544 Mbps (NADH T1), 2.048 Mbps (EDH E1), and all the way up to 274.176 Mbps (NADH T4). PDH systems, such as

T1 and E1, are widely used for Public Switched Telephone Networks (PSTNs) and TDM access into frame- or cell-based data networks.

Table 6-1 *Plesiochronous Digital Hierarchy*

Level	Voice Channels	NADH Mbps	EDH Mbps	JDH Mbps
DS0/E0/J0	1	0.064	0.064	0.064
DS1/J1	24	1.544	—	1.544
E1	30	—	2.048	—
DS1C/J1C	48	3.152	—	3.152
DS2/J2	96	6.312	—	6.312
E2	120	—	8.448	—
E3/J3	480	—	34.368	32.064
DS3	672	44.736	—	—
DS3C	1344	89.472	—	—
J3C	1440	—	—	95.728
E4	1920	—	139.264	—
DS4	4032	274.176	—	—
J4	5760	—	—	397.2
E5	7680	—	565.148	—

PDH networks are unable to identify individual channels in a higher-order bit stream, such as E3 or a DS3. For example, to recover a 64-kbps channel from a 139.264 Mbps E4, it would be necessary to demultiplex the signal all the way down to the 2.048-Mbps level before the location of the 64-kbps channel can be identified. PDH networks also lack the in-band capacity for network management. This led to the development of proprietary PDH network management. There is no standardized definition of PDH bit rates greater than 274.176 Mbps and the various PDH hierarchies used worldwide led to interoperability issues. Fortunately, the SDH standard has succeeded in its goal of international interoperability and interworking of various regional PDH signals by providing a common base transport system. SDH is an international standard and supports the European Telecommunications Standards Institute (ETSI)–defined PDH bit rates of 2/34/140 Mbps.

Asynchronous TDM transmission systems are free running with each terminal in the network running on its own clock. In digital transmission systems, clocking maintains network synchronization. Asynchronous systems use terminal internal oscillators or trunk lines to provide clocking. Because these clocks are not synchronized, large variations occur in the clock rate of free-running systems that affect the signal bit rate.

SDH Synchronization

Network timing between SDH devices is an integral part of maintaining accurate information transmitted over an SDH network. In the earlier days of networking, the method used in timing was asynchronous. In asynchronous timing, each switch runs its own clock. In synchronous timing, switches can use a single common clock to maintain timing. This single common clock is referred to as a primary reference source (PRS) or master clock. Synchronous timing maintains better accuracy than asynchronous timing because it uses a single common clocking scheme to maintain timing. Five methods are typically used in obtaining synchronous timing in an SDH network:

- Internal timing from a free-running onboard internal oscillator (Stratum 3E or Stratum 3 clock).
- Line timing derived from a scrambled incoming SDH signal on a high-speed interface.
- Through timing derived from a signal input, but clocks the signal in a different outbound direction.
- Loop timing is similar to line timing, but is used with customer premises equipment (CPE) or terminal multiplexer equipment rather than ADMs.
- External timing from an external source such as a Stratum 1 global positioning system (GPS) or an atomic clock.

The SDH network is organized with a master-slave relationship with clocks of the higher-level nodes feeding timing signals to clocks of the lower-level nodes. All nodes can be traced up to a primary reference source, a Stratum 1 GPS or atomic clock with extremely high stability and accuracy. Lower-stratum or less-stable clocks are adequate to support the lower nodes. The internal clock of an SDH terminal can derive its timing signal from a building-integrated timing supply (BITS) used by switching systems and other equipment. Thus, this terminal serves as a master for other SDH nodes, providing timing on its outgoing OC–N signal. Other SDH nodes operate in a slave mode called *loop timing*, with their internal clocks timed by the incoming OC–N signal. Current standards specify that an SDH network must be able to derive its timing from a Stratum 3 or higher clock.

Table 6-2 shows the long-term accuracy requirements within the stratum clock hierarchy that ranges from Stratum 1 through 4. Although synchronous networks display accurate timing, some variations can occur between different network devices or between networks. This difference is known as *phase variation*. Phase variations are defined as jitter or wander. *Jitter* is defined as short-term phase variations above 10 Hz, and *wander* is defined as long-term phase variations below 10 Hz. In digital networks, jitter and wander are handled by buffers found in the interfaces within different network devices. One example is a slip buffer. The slip buffer is used to handle frequency differences between read and write operations. To prevent write operations from happening faster than read operations, read operations are handled at a slightly higher rate. On a periodic basis, read operations are paused, while a bit is stuffed into a stream to account for any timing differences between the read and write operations. This bit-stuffing scheme is referred to as a *controlled slip (frame)*. The stratum clocking accuracy timing requirements are also defined in Table 6-2 for wander, bit slips, and controlled slips. The timing accuracy requirements increase as the stratum hierarchy increases.

Table 6-2 *Stratum Clock Hierarchy*

Stratum	Minimum Free-Run Accuracy	Wander (0.12 μs increment)	Bit Slips	Controlled Frame Slips
1	+/– 1 x 10^{-11}	3.3 hrs	18 hrs	20.6 wks
2	+/– 1.6 x 10^{-8}	7.5 sec	41 sec	2.17 hrs
3	+/– 4.6 x 10^{-6}	26 ms	140 ms	27 sec
4	+/– 32 x 10^{-6}	4 ms	20 ms	3.9 sec

SDH Electrical and Optical Signals

The term Synchronous Transport Module-N (STM-N) refers to the SDH transmission characteristics of an Nth-level transmission link. Unlike SONET, which uses the term OC-N to refer to the optical carrier and the term STS-N to refer to the electrical signal level, SDH uses the term STM-N to indicate the optical as well as electrical signal level. The basic unit of transmission in SDH is 155.52 Mbps (STM-1). This offers partial compatibility and interworking with SONET at the sub-STM-1 levels. However, interworking for alarms and performance management is generally not possible between SDH and SONET systems, unless proprietary byte-mapping methods are used. Higher SDH rates are defined as integer multiples of STM-1 in an N * 4 sequence. This offers rates from STM-1 up to STM-256, as indicated in Table 6-3. The upper rate limits are set by technology rather than by lack of standards as was the case with PDH.

Table 6-3 *STM-N Line Rates*

SDH Level	Line Rate (Mbps)	Payload Rate (Mbps)	Overhead Rate (Mbps)	SONET Equivalent
STM-0	51.840	50.112	1.728	STS-1
STM-1	155.520	150.336	5.184	STS-3
STM-3	466.560	451.008	15.552	STS-9
STM-4	622.080	601.344	20.736	STS-12
STM-6	933.120	902.016	31.104	STS-18
STM-8	1244.160	1202.688	41.472	STS-24
STM-13	1866.240	1804.032	62.208	STS-36

continues

Table 6-3 *STM-N Line Rates (Continued)*

SDH Level	Line Rate (Mbps)	Payload Rate (Mbps)	Overhead Rate (Mbps)	SONET Equivalent
STM-16	2488.320	2405.376	82.944	STS-48
STM-32	4976.640	4810.752	165.888	STS-96
STM-64	9953.280	9621.504	331.776	STS-192
STM-256	39813.120	38486.016	1327.104	STS-768

Each STM-N interface rate contains SDH overhead information to support a range of facilities and a payload capacity for traffic. The overhead and payload areas can be fully or partially filled. In the SDH multiplexing process, payloads are layered into lower-order (LO) and higher-order (HO) virtual containers, each including a range of overhead functions for management and error monitoring. Transmission is then supported by the attachment of further layers of overheads. SDH management layer communications are transported within dedicated data communications channel (DCC) time slots inside the interface rate.

SDH Layers

SDH has four optical interface layers, which are depicted in Figure 6-2 and are as follows:

- Path layer
- Multiplex section layer
- Regenerator section layer
- Photonic layer

Figure 6-2 *SDH Layers*

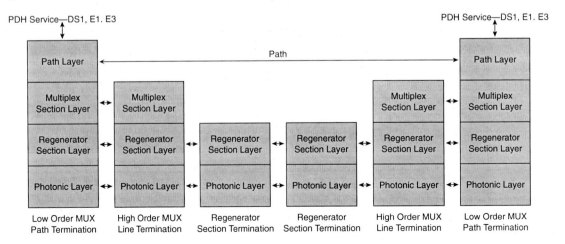

Path Layer

The path is a logical connection between the point at which the standard SDH frame format for the signal at a given rate is assembled and the point at which the standard SDH frame format for the signal is disassembled. The path overhead (POH) is first on, last off, so the alarm and error information contained within this layer represents end-to-end status. The POH is added or stripped at this layer.

Path-Terminating Equipment (PTE)

The SDH path-terminating equipment (PTE) is an NE that multiplexes or demultiplexes the VC-N payload. Virtual containers (VC-Ns) are discussed in more detail later. The POH is generated where the lower-rate, non-SDH signal enters the SDH network by a PTE, such as an ADM. The POH is removed when the payload exits the SDH network by a PTE, such as an ADM. Low-order multiplexers, wideband cross-connect systems, and subscriber loop access systems are examples of SDH PTE devices. Figure 6-3 shows the PTE.

Figure 6-3 *PTE, MSTE, and RSTE Devices*

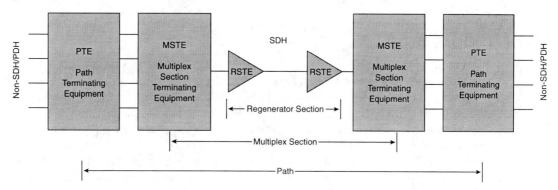

Multiplex Section

The multiplex section is defined as the transmission medium, together with the associated equipment, required to provide the means of transporting information between two consecutive NEs. One of the NEs originates the line signal, and the other NE terminates the line signal in either direction. The multiplex section overhead (MSOH) bytes are added or stripped at this layer.

Multiplex Section-Terminating Equipment (MSTE)

An SDH MSTE is an NE that originates and/or terminates STM-N signals. The MSOH is generated by the MSTE. The MSOH is used for communication and synchronization between major nodes and for the monitoring of errors. The high-order multiplexer, optical-line terminal, and broadband cross-connect systems are examples of SDH MSTE devices. Figure 6-3 shows the MSTE.

Regenerator Section Layer

The regenerator section in SDH is the portion of a transmission facility, including terminating points, between a terminating NE and a regenerator, or two regenerators. The regenerator section overhead (RSOH) bytes are added or stripped at this layer.

Regenerator Section-Terminating Equipment (RSTE)

Regenerator section-terminating equipment (RSTE) is the NE that regenerates an STM-N signal for long-haul transport. The RSTE can originate, access, modify, or terminate the RSOH, or it can perform a combination of these actions. Figure 6-3 shows the RSTE.

Photonic Layer

The photonic layer mainly deals with the transport of bits across the physical fiber medium. Its main function is the conversion between STM-N signals and light pulses on the fiber media. Its functions include wavelength launching, pulse shaping, and modulation of power levels.

SDH Multiplexing

It is very important to understand the SDH multiplexing hierarchy to truly understand and appreciate SDH. Multiplexing follows a rigid hierarchy in SDH. From an extremely high-level perspective, it is safe to say that low-level PDH signals are mapped to an SDH entity known as a container (C). The C is mapped along with POH bytes to form another entity known as a lower-order virtual container (VC). The lower-order VCs are aligned with tributary unit (TU) pointers to form entities known as *tributary units (TUs)*. The TUs are multiplexed to form tributary unit groups (TUGs). The TUGs are further multiplexed to form higher-order VCs. These higher-order VCs are aligned with fixed byte-stuffing and administration units (AU) to form administration units (AUs). The AUs are further multiplexed to form administrative unit groups (AUGs). The AUGs are finally multiplexed along with RSOH and MSOH bytes to form the STM-N signal.

There are variations to the flow just described as you will notice in the subsequent discussion. Before you read on, you need to understand a few simple terms. The multiplexing principles of SDH use the following terms:

- **Mapping**—A process used when tributaries are adapted into VCs by adding justification bits and POH information.
- **Aligning**—This process takes place when a pointer is included in a TU or an AU, to allow the first byte of the VC to be located.
- **Multiplexing**—This process is used when multiple lower-order path layer signals are adapted into a higher-order path signal, or when the higher-order path signals are adapted into a multiplex section.

- **Stuffing**—As the tributary signals are multiplexed and aligned, some spare capacity has been designed into the SDH frame to provide enough space for all the various tributary rates. Therefore, at certain points in the multiplexing hierarchy, this space capacity is filled with fixed stuffing bits that carry no information, but are required to fill up the particular frame.

PDH traffic signals to be mapped into SDH are by definition continuous. Each PDH signal is mapped to an SDH container (C). The Cs are mapped to VCs. The purpose of this function is to create a uniform VC payload by using bit stuffing to bring all inputs to a common bit rate ready for synchronous multiplexing.

There are two kinds of Cs: lower-order tributaries (Cs) and higher-order tributaries (Cs). Lower-order Cs (C-Nx, where N = 1, 2, and x = 1, 2) typically accommodate PDH signals, such as DS1, E1, and other PDH signals up to the DS2/J2 level. Higher-order Cs (C-N, where N = 3, 4) typically accommodate PDH signals, such as E3, DS3, and other PDH signals up to the E4 level. Similarly, there are two kinds of VCs: lower-order VCs and higher-order VCs. Lower-order VCs (VC-Nx, where N = 1, 2 and x = 1, 2) typically accommodate PDH signals, such as DS1, E1, and other PDH signals up to the DS2/J2 level. Higher-order VCs (VC-N, where N = 3, 4) typically accommodate PDH signals, such as E3, DS3, and other PDH signals up to the E4 level. Various VCs ranging from VC-11 to VC-4 are covered by the SDH hierarchy. Currently, the ITU-T has defined six SDH Cs/VCs. The six VCs defined are VC-11, VC-12, VC-2, VC-3 (E3/J3/ISDN H31), VC-3 (DS3/ISDN H32), and VC-4. Table 6-4 lists the various SDH VC levels.

Table 6-4 *VC Levels*

Low-Order VC Level	VC Bit Rate (Mbps)	PDH Level	NADH (Mbps)	EDH (Mbps)	JDH (Mbps)
VC-11	1.728	DS1/J1/ISDN H11	1.544	—	1.544
VC-12	2.304	E1/ISDN H12	—	2.048	—
VC-2	6.912	DS2/J2	6.312	—	6.312
High-Order VC Level		**PD Level**	**NADH (Mbps)**	**EDH (Mbps)**	**JDH (Mbps)**
VC-3	48.960	E3/J3/ISDN H31	—	34.368	32.064
VC-3	48.960	DS3/ISDN H32	44.736	—	—
VC-4	150.336	E4/ISDN H4	—	139.264	—

Through the use of pointers and offset values, VCs can be carried in the SDH payload as independent data packages. VCs are used to transport lower-speed tributary signals. VCs can start at any point within the STM-1 frame. The start location of the J1 POH byte is indicated by the AU pointer byte values. VCs can also be concatenated to provide more capacity.

Table 6-5 illustrates the various SDH paths and associated overhead. Higher levels of the synchronous hierarchy are formed by byte interleaving the payloads from a number N of STM–1 signals and then adding a transport overhead of size N times that of an STM–1 and filling it with new management data and pointer values as appropriate. Before transmission, the STM–N signal has scrambling applied overall to randomize the bit sequence for improved transmission performance. A few bytes of overhead are left unscrambled to simplify subsequent demultiplexing.

Table 6-5 *SDH Multiplexing Overhead*

Path	SDH Overhead
C-N to VC-N	C-N + VC-N POH = VC-N (N = 11, 12, 2, 3, or 4)
VC-N to TU-N	VC-N + TU-N pointers = TU-N (N = 11, 12, 2, or 3)
VC-N to AU-N	VC-N + AU-N pointers = AU-N (N = 3 or 4)
AUG-N to STM-N	AUG-N + RSOH + MSOH = STM-N

SDH Multiplexing of E1 Signals

Low-level E1 (2.048-Mbps) signals are mapped to the C-12 container. As illustrated in Figure 6-4, the C-12 container gets mapped with VC-12 POH bytes into a lower-order VC-12 virtual container. The VC-12 along with TU-12 pointers gets aligned into a TU-12 tributary unit. The TU-12 gets multiplexed (x3) to a TUG-2, which means that three TU-12s are multiplexed into a TUG-2. The TUG-2 gets multiplexed (x7) into a higher-order VC-3. This VC-3 gets aligned with AU-3 pointers to form an AU-3. The AU-3 can get directly multiplexed (x1) to form an STM-0 signal or get multiplexed (x3) to form an AUG-1. The AUG-1 gets multiplexed (x1) along with MSOH and RSOH bytes to form an STM-1.

SDH Multiplexing of DS1 Signals

Low-level DS1 (1.544-Mbps) signals are mapped to the C-11 container. As illustrated in Figure 6-4, the C-11 container gets mapped with VC-11 POH bytes into a lower-order VC-11 virtual container. The VC-11 along with TU-11 pointers gets aligned with a TU-11 tributary unit. The TU-11 gets multiplexed (x4) to a TUG-2, which means that four TU-11s are multiplexed into a TUG-2. In the case of the VC-11, however, there is an alternate path. The VC-11 could also get aligned along with TU-12 pointers to form a TU-12 tributary unit, in which case the TU-12 would get multiplexed (x3) to a TUG-2. The TUG-2 gets multiplexed (x7) into a higher-order VC-3. This VC-3 gets aligned with AU-3 pointers to form an AU-3. The AU-3 can get directly multiplexed (x1) to form an STM-0 signal or get multiplexed (x3) to form an AUG-1. The AUG-1 gets multiplexed (x1) along with MSOH and RSOH bytes to form an STM-1.

Figure 6-4 *SDH Multiplexing*

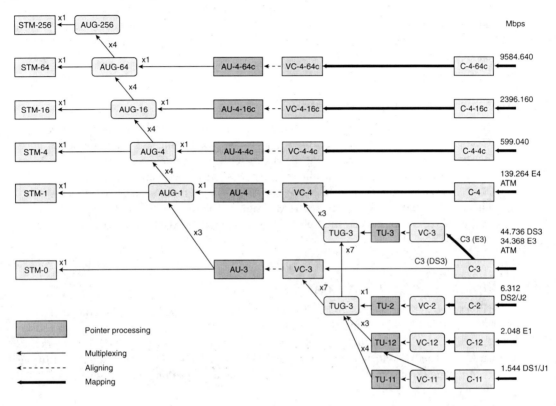

SDH Multiplexing of DS2 Signals

Low-level DS2 (6.312-Mbps) signals are mapped to the C-2 container. As illustrated in Figure 6-4, the C-2 container gets mapped with VC-2 POH bytes into a lower-order VC-2 virtual container. The VC-2 along with TU-2 pointers gets aligned with a TU-2 tributary unit. The TU-2 gets multiplexed (x1) to a TUG-2. The TUG-2 gets multiplexed (x7) into a higher-order VC-3. This VC-3 gets aligned with AU-3 pointers to form an AU-3. The AU-3 could get directly multiplexed (x1) to form an STM-0 signal or get multiplexed (x3) to form an AUG-1. The AUG-1 gets multiplexed (x1) along with MSOH and RSOH bytes to form an STM-1.

SDH Multiplexing of E3 Signals

High-level E3 (34.368-Mbps) signals are mapped to the C-3 container. However, the C-3 mapping for E3 takes place differently as compared to the C-3 mapping for DS3. This is illustrated in Figure 6-4. The E3 gets mapped to a C-3 that in turn gets mapped along with VC-3 POH bytes into a VC-3. This VC-3 is aligned along with TU-3 pointers to form a TU-3. The TU-3 gets multiplexed into a TUG-3. The TUG-3 gets multiplexed (x3) into a higher-order VC-4, which in turn gets aligned with AU-4 pointers to form an AU-4. This AU-4 gets multiplexed (x1) into an AUG-1 that can get multiplexed (x4) into a higher-order AUG-4 or multiplexed (x1) along with MSOH and RSOH bytes to form an STM-1.

SDH Multiplexing of DS3 Signals

High-level DS3 (44.736-Mbps) signals are also mapped to the C-3 container. However, the C-3 mapping for DS3 takes place differently as compared to the C-3 mapping for E3. This is illustrated in Figure 6-4. The DS3 gets mapped to the C-3 container that gets directly mapped with the VC-3 POH bytes to the higher-order VC-3 virtual container. The VC-3 is aligned along with AU-3 pointers into an AU-3. The AU-3 can get multiplexed (x3) into an AUG-1 or directly multiplexed (x1) to form an STM-0 signal. The AUG-1 can get further multiplexed (x4) to an AUG-4 or directly multiplexed (x1) along with MSOH and RSOH bytes to form an STM-1.

SDH Multiplexing of E4 Signals

High-level signals, such as an E4 (139.264 Mbps), are mapped to the C-4 container. As illustrated in Figure 6-4, the C-4 container along with the VC-4 POH bytes is mapped to a higher-order VC-4 virtual container. The VC-4 is aligned along with AU-4 pointers into an AU-4. The AU-4 then gets multiplexed (x1) to an AUG-1. The AUG-1 could get further multiplexed (x4) to an AUG-4 or directly multiplexed (x1) along with MSOH and RSOH bytes to form an STM-1.

SDH Framing

SDH adopts a frame rate of 8000 frames per second. The STM-1 frame is the basic transmission format for SDH. As illustrated in Figure 6-5, each two-dimensional frame has 9 rows and 270 columns of bytes with a total of 2430 bytes (9 rows * 270 columns = 2430 bytes). The STM-1 frame consists of an transport overhead plus a virtual container (VC-4) capacity. The first 9 columns of each frame make up the transport overhead, and the remaining 261 columns make up the virtual container (VC-4). The VC-4 plus the nine pointer bytes (H1 * 3, H2 * 3, H3 * 3) in row 4 is known as the administrative unit (AU-4). The first column of the VC-4 is the POH, as indicated in Figure 6-5. Actually, the POH floats within the STM-1 frame and can occupy any column from column 10 to column 270 of the STM-1 frame. The STM-1 can accommodate a single VC-4 or three VC-3s. The AU-N pointer bytes indicate the position of the VC-4 POH or VC-3 POHs within the STM-1 frame. VCs can have any phase alignment within the AU, and this alignment is indicated by the AU pointer in row 4. Within the transport overhead, the first three rows are used for the RSOH, and the last five rows are used for the MSOH.

Figure 6-5 *SDH Framing*

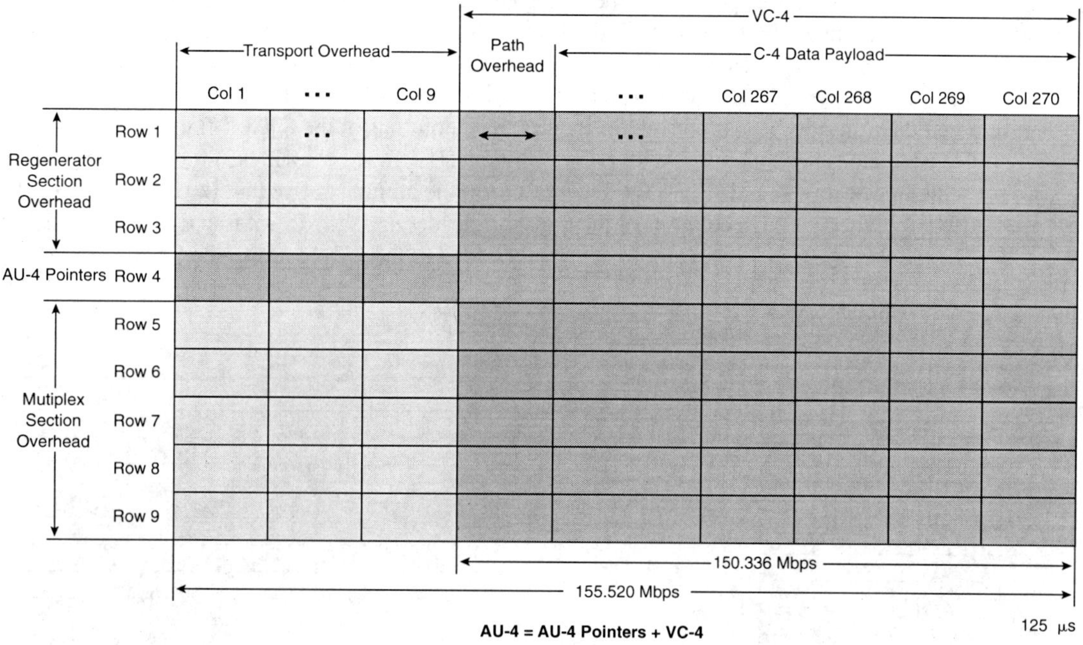

The SDH line rate is synchronous and is flexible enough to support many different signals. The SDH STM-1 line rate of 155.520 Mbps can be computed as follows:

Line rate = 270 columns * 9 rows * 8 bits/byte * 8000 frames/sec = 155.520 Mbps

Columns 10 to 270 constitute the VC with the POH that provides a payload data rate of 150.336 Mbps. The actual data rate of the payload can be computed as follows:

Data rate = 261 columns * 9 rows * 8 bits/byte * 8000 frames/sec = 150.336 Mbps.

The STM-1 frame is transmitted starting from the first byte in row 1, column 1, to the last byte in row 9, column 270. The most significant bit (MSB) of a byte is transmitted first. After the 270th byte of row 1 is transmitted, the byte in the first column of row 2 is transmitted and so on. Higher line rates are obtained by synchronous multiplexing the lower line rates. Standard STM-N rates have values of N as 1, 4, 16, 64, and 256. Table 6-3 indicates line rates for various values of N. As the line rates increase, the percentage of overhead increases, and in turn the useful payload capacity decreases. The additional overheads are used for control, parity, stuffing, alarm, and signaling. For SDH STM-N, the ratio of overhead to payload remains constant regardless of the value of N.

STM-1 Frame Creation

The STM-1 frame is the SDH unit that gets translated into optical pulses and launched onto the fiber. This section considers an example that examines the construction of an STM-1 frame, beginning with a C-4 container. The formation of an STM-1, beginning with a C-4 is illustrated in Figure 6-6. As seen previously, the E4 signal (139.264 Mbps) maps to a C-4 container. The C-4 container along with the E4 payload adds the VC-7 POH bytes to form a higher-order VC-4 virtual container. As shown in Figure 6-6, the VC-4 consists of the C-4 payload along with the VC-4 POH bytes. The POH is typically the first column of a VC-4. It is important to remember that the transport header constitutes the first nine columns of the STM-1 frame. The VC-4 is aligned along with the nine AU-4 pointers bytes (H1 * 3, H2 * 3, H3 * 3) found in the first nine columns of row 4 in the transport header to form the administrative unit (AU-4). The RSOH found in the first three columns of the transport header and the MSOH found in columns 5, 6, 7, 8, and 9 of the transport overhead are finally added to the AU-4 to form the STM-1 frame.

Figure 6-6 *STM-1 Frame Creation*

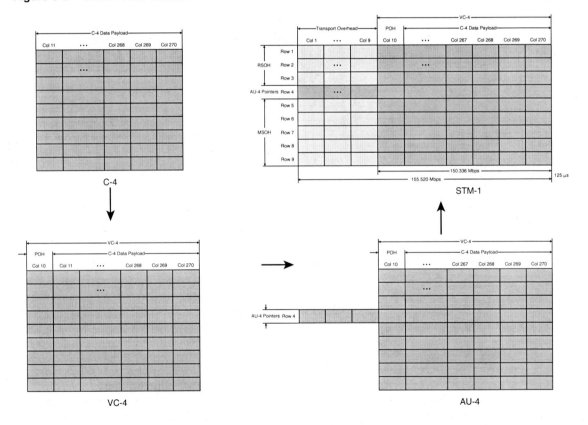

SDH Tributaries

SDH TUs are analogous to SONET VTs. A TU-11 has 3 columns multiplied by 9 rows that can be used to transport a DS1. A TU-12 has 4 columns multiplied by 9 rows that can be used to transport an E1, and the TU-2 has 12 columns multiplied by 9 rows that can be used to transport a DS2. To accommodate a mix of different TU types within a VC-4, the TUs are grouped together. A VC-4 that carries TUs is divided into three TUG-3s, each of which can contain seven TUG-2s or a single TU-3. There can be a mix of the different TUGs. For example, the first TUG-3 could contain 12 TU-12s and 3 TU-2s, making a total of 7 TUG-2 groups. The TU groups have no overhead or pointers because they are just a way of multiplexing and organizing the different TUs within the VC-4 of an STM-1. The columns in a TUG are not consecutive within the VC, but are byte interleaved column by column with respect to the other TUGs. The first TUG-2 group within a TUG-3, called group 1, is found in every seventh column, skipping columns 1 and 2 of the TUG-3, and starting with column 3. The TU columns within a group are not placed in consecutive columns within that group. The columns of the individual TUs within the TUG are byte interleaved as well. Each TUG occupies 12 columns of the VC-4. As a result, a TUG could contain one of the following combinations:

- 4 TU-11s (with 4 columns per TU-11)
- 3 TU-12s (with 4 columns per TU-12)
- 1 TU-2 (with 12 columns per TU-2)

TU Multiframe

In the floating TU mode, four consecutive 125-microsecond frames of the VC-4 are combined into one 500-microsecond structure, called a *TU multiframe*. The 500-microsecond multiframe is aligned to the 125-microsecond VC-4. The occurrence of the TU multiframe and its phase is indicated in the VC-N POH by the multiframe indicator byte (H4). A value of XXXXXX00 in the multiframe indicator byte indicates that the next STM frame contains the first frame in the TU multiframe. A value of XXXXXX01 in the multiframe indicator byte indicates that the next VC-4 contains the second frame in the TU multiframe and so on. Only the last 2 bits of the H4 byte have a value of 0 or 1 assigned. The first 6 bits are unassigned and are denoted by the X. The TUs also contain payload pointers to allow for flexible and dynamic alignment of the VC. In this case, the TU pointer value indicates the offset from the TU to the first byte of the VC. TU pointers allow AU and TU payloads to differ in phase with respect to each other and the network while still allowing AUs and TUs to be synchronously multiplexed. The TU multiframe overhead consists of 4 bytes: V1, V2, V3, and V4. Each of these 4 bytes is located in the first byte of the respective TU frame in the TU multiframe. The payload pointers V1 and V2 indicate the start of the payload within the multiframe. The V3 byte provides a 64-kbps channel for a payload pointer movement opportunity, and the V4 byte is reserved. The remaining bytes in the TU multiframe define the TU container capacity, which carries the VC, and the POH. The container capacity differs for the different TU types because their size varies according to the number of columns in each type.

SDH Pointers

SDH provides payload pointers to permit differences in the phase and frequency of the VCs (VC-N) with respect to the STM-N frame. Lower-order pointers are also provided to permit phase differences between VC-11/VC-12/VC-2 and the higher-order VC-3/VC-4. On a frame-by-frame basis, the payload pointer indicates the offset between the VC payload and the STM-N frame by identifying the location of the first byte of the VC in the payload. In other words, the VC is allowed to float within the STM-1 frame capacity. To make this possible, within each STM-N frame, there is a pointer, known as the *VC payload pointer*, that indicates where the actual payload container starts. For a VC-4 payload, this pointer (AU-4 pointer) is located in columns 1 and 4 of the fourth row of the transport overhead. The bytes H1 and H2 can be viewed as one value. The pointer value indicates the offset in bytes from the pointer to the first byte of the VC, which is the J1 byte. Because the transport overhead bytes are not counted, and starting points are at 3-byte increments for a VC-4 payload, the possible range is as follows:

Total STM-1 bytes − Transport overhead bytes = Pointer value range

For example:

(2430 − 81)/3 = 783 valid pointer positions

This means that the value of the pointer has a range of 0 to 782. If the VC-4 payload pointer has a value of 0, for example, the VC-4 begins in the byte adjacent to the H3 byte of the overhead. If the payload pointer has a value of 87, the VC-4 begins in the byte adjacent to the K2 byte of the overhead in the next row. The pointer value, which is a binary number, is carried in bits 7 through 16 of the H1-H2 pointer word. The first 4 bits of the VC-4 payload pointer make provision for indicating a change in the VC, and thus an arbitrary change in the value of the pointer. These 4 bits, the N bits, are known as the *new data flag*. The VC pointer value that accompanies the new data flag will indicate the new offset. If there is a difference in phase or frequency, the pointer value is adjusted. To accomplish this, a process known as *byte stuffing* is used. The VC payload pointer indicates where in the container capacity a VC starts, and the byte-stuffing process allows dynamic alignment of the VC in case it slips in time. When the data rate of the VC is slow in relation to the rate of the STM-1 frame, a noninformational byte is stuffed in right behind the H3 byte, allowing the alignment of the container to slip back in time. This is known as *positive stuffing* or *positive pointer justification*. When the data rate of the VC is fast in relation to the rate of the STM-1 frame, however, actual data is written in the H3 byte, allowing the alignment of the container to move forward in time. This is known as *negative stuffing* or *negative pointer justification*.

SDH Transport Overhead

SDH provides substantial overhead information, allowing simpler multiplexing and greatly expanded OAM&P capabilities. The SDH transport overhead, as illustrated in Figure 6-7, has two layers that include the RSOH and MSOH. The RSOH is used for communications between adjacent NEs, such as regenerators, whereas the multiplex section overhead is used for the STM-N signaling between STM-N multiplexers. The reader must make a genuine attempt to comprehend this section because a large part of understanding SDH includes understanding the purpose and use of the overhead bytes.

Figure 6-7 *SDH Transport Overhead*

		Col 1	Col 2	Col 3	Col 4	Col 5	Col 6	Col 7	Col 8	Col 9
RSOH	Row 1	Framing A1	Framing A1	Framing A1	Framing A2	Framing A2	Framing A2	Trace J0	Reserved for National Use	Reserved for National Use
	Row 2	BIP-8 B1	Media Dependent	Media Dependent	Orderwire E1	Media Dependent	Reserved	User F1	Reserved for National Use	Reserved for National Use
	Row 3	Data Com D1	Media Dependent	Media Dependent	Data Com D2	Media Dependent	Reserved	Data Com D3	Reserved	Reserved
AU-4 Pointers		Pointer H1	Pointer H1	Pointer H1	Pointer H2	Pointer H2	Pointer H2	Pointer Action H3	Pointer Action H3	Pointer Action H3
MSOH	Row 5	BIP-8 B2	BIP-8 B2	BIP-8 B2	APS K1	Reserved	Reserved	APS K2	Reserved	Reserved
	Row 6	Data Com D4	Reserved	Reserved	Data Com D5	Reserved	Reserved	Data Com D6	Reserved	Reserved
	Row 7	Data Com D7	Reserved	Reserved	Data Com D8	Reserved	Reserved	Data Com D9	Reserved	Reserved
	Row 8	Data Com D10	Reserved	Reserved	Data Com D11	Reserved	Reserved	Data Com D12	Reserved	Reserved
	Row 9	Sync S1	Reserved	Reserved	Reserved	Reserved	REI-L M1	Orderwire E2	Reserved for National Use	Reserved for National Use

SDH Regenerator Section Overhead

The RSOH contains only the information required by the NEs located at both ends of a section. This might be two regenerators, a piece of line-terminating equipment and a regenerator, or two pieces of line-terminating equipment. The RSOH is found in the first three rows of Columns 1 through 9 of the STM-1 frame, as shown in Figure 6-7.

Framing Bytes (A1, A2)

These 2 bytes indicate the beginning of the STM-N frame. The A1, A2 bytes are unscrambled. A1 has the binary value 11110110, and A2 has the binary value 00101000. The frame-alignment word of an STM-N frame is composed of (3 * N) A1 bytes followed by (3 * N) A2 bytes.

Regenerator Section Trace Byte (J0)

The J0 byte is used to transmit a section access point identifier so that a section receiver can verify its continued connection to the intended transmitter. The coding of the J0 byte is the same as for J1 and J2 bytes. This byte is defined only for STM-1 number 1 of an STM-N signal.

RS Bit-Interleaved Parity Code (BIP-8) Byte (B1)

The B1 byte is an even parity code byte used to check for transmission errors over a regenerator section. Its value is calculated over all bits of the previous STM-N frame after scrambling, and then placed in the B1 byte of STM-1 before scrambling. Therefore, this byte is defined only for STM-1 number 1 of an STM-N signal.

RS Orderwire Byte (E1)

The E1 byte is allocated to be used as a local orderwire channel for voice communication between regenerators.

RS User Channel Byte (F1)

The F1 byte is set aside for the user's purposes; it can be read and/or written to at each section-terminating equipment in that line.

RS Data Communications Channel (DCC) Bytes (D1, D2, D3)

The 3 DCC bytes form a 192-kbps message channel providing a message-based channel for operations, administration, and maintenance (OAM) between pieces of section-terminating equipment. The channel can be used from a central location for control, monitoring, administration, and other communication needs.

Reserved Bytes

These bytes, which are located at various positions in the RSOH of an STM-N signal (N > 1), are reserved for future international standardization.

AU Pointers

The value carried in the pointer bytes (H1, H2) designates the location of the VC-4 frame within the STM-N frame. The last 10 bit (7 through 16) carries the pointer value, which is a binary number ranging from 0 to 782. The H3 pointer action byte is used for frequency justification. Depending on the pointer value, this byte is used to adjust the fill-input buffers. It carries valid information only in the event of negative justification; otherwise, it is not defined.

SDH Multiplex Section Overhead

The MSOH contains the information required between the multiplex section termination equipment at each end of the multiplex section. The MSOH is found in rows 5 to 9 of columns 1 through 9 of the STM-1 frame, as shown in Figure 6-7.

Bit-Interleaved Parity Code (BIP-24) Byte (B2)

This bit-interleaved parity (BIP) N * 24 code is used to determine whether a transmission error has occurred over a multiplex section. The B2 byte uses even parity, and is calculated on the overall bits of the MSOH and the STM-N frame of the previous STM-N frame before scrambling. The value is placed in the three B2 bytes of the MSOH before scrambling. These bytes are provided for all STM-1 signals in an STM-N signal.

Automatic Protection Switching Bytes (K1, K2)

The K1 and K2 bytes are used for multiplex section protection (MSP) signaling between multiplex level entities for bidirectional APS and for communicating alarm indication signal (AIS) and remote defect indication (RDI) conditions. The multiplex section remote defect indication (MS-RDI) is used to return an indication to the transmit end that the received end has detected an incoming section defect or is receiving MS-AIS. MS-RDI is generated by inserting a "110" code in positions 6, 7, and 8 of the K2 byte before scrambling. The various values of the K1 and K2 bytes are illustrated in Table 6-6.

Table 6-6 *K1 and K2 Bytes*

Byte	Bits	Value	Function
K1	1 to 4	0000	No request
K1	1 to 4	0001	Do not revert
K1	1 to 4	0010	Reverse request
K1	1 to 4	0011	Not used
K1	1 to 4	0100	Exercise
K1	1 to 4	0101	Not used
K1	1 to 4	0110	Wait to restore
K1	1 to 4	0111	Not used
K1	1 to 4	1000	Manual switch
K1	1 to 4	1001	Not used
K1	1 to 4	1010	Signal Degrade - Low priority
K1	1 to 4	1011	Signal Degrade - High priority
K1	1 to 4	1100	Signal Fail - Low priority
K1	1 to 4	1101	Signal Fail - High priority
K1	1 to 4	1110	Forced switch
K1	1 to 4	1111	Lock out of protection
K1	5 to 8	nnnn	Indicates the number of the channel requested
K2	1 to 4	nnnn	Selects channel number
K2	5	0	1+1 protection architecture
K2	5	1	1:N protection architecture
K2	6 to 8	000	Future use

continues

Table 6-6 *K1 and K2 Bytes (Continued)*

Byte	Bits	Value	Function
K2	6 to 8	001	Future use
K2	6 to 8	010	Future use
K2	6 to 8	011	Future use
K2	6 to 8	100	Unidirectional provisioning
K2	6 to 8	101	Bidirectional provisioning
K2	6 to 8	110	MS-RDI
K2	6 to 8	111	MS-AIS

Data Communications Channel (DCC) Bytes (D4 to D12)

The D4 to D12 bytes form a 576-kbps message channel from a central location for OAM information (control, maintenance, remote provisioning, monitoring, administration, and other communication needs).

Synchronization Status Message Byte (S1)

Bits 5 to 8 of the S1 byte are used to carry the synchronization messages. Table 6-7 illustrates the assignment of bit patterns to the four synchronization levels agreed to within ITU-T. The other values are reserved.

Table 6-7 *S1 Byte Values*

S1 Bits (5–8)	Synchronization Message
0000	Quality Unknown
0010	G.811 PRC
0100	SSU-A (G.812 transit)
1000	SSU-B (G.812 local)
1011	G.813 Option 1 Synchronous Equipment Timing Clock (SEC)
1111	Do not use for synchronization

MS Remote Error Indication Byte (M1)

The M1 byte of an STM-1 or the first STM-1 of an STM-N is used for an MS layer remote error indication (MS-REI). Bits 2 to 8 of the M1 byte are used to carry the error count of the interleaved bit blocks that the MS BIP-24xN has detected to be in error at the far end of the section. This value is truncated at 255 for STM-N > 4.

MS Orderwire Byte (E2)

The E2 orderwire byte provides a 64-kbps channel between multiplex entities for an express orderwire. It is a voice channel for use by technicians and can be accessed at multiplex section terminations.

Reserved Bytes

These bytes, which are located at various positions in the MSOH of an STM-N signal (N > 1), are reserved for future international standardization.

SDH High-Order Path Overhead

The SDH POH is assigned to and transported with the VC from the time it's created by path-terminating equipment until the payload is demultiplexed at the termination point in a piece of path-terminating equipment. The POH is found in rows 1 to 9 of the first column of the VC-4 or VC-3. The higher-order POH for an STM-1 frame is shown in Figure 6-8.

Figure 6-8 *SDH Path Overhead*

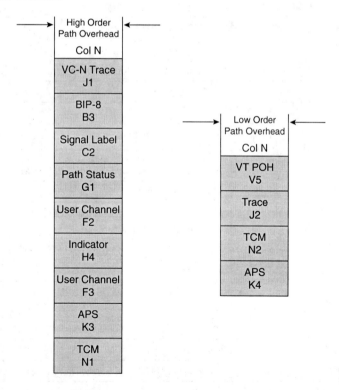

Higher-Order VC-N Path Trace Byte (J1)

The J1 user-programmable byte repetitively transmits a 15-byte, E.64 format string plus 1-byte CRC-7. A 64-byte free-format string is also permitted for this access point identifier. This allows the receiving terminal in a path to verify its continued connection to the intended transmitting terminal.

Path Bit-Interleaved Parity Code (Path BIP-8) Byte (B3)

The B3 is an even parity code byte used to determine whether a transmission error has occurred over a path. Its value is calculated over all the bits of the previous VC before scrambling and placed in the B3 byte of the current frame.

Path Signal Label Byte (C2)

The C2 byte specifies the mapping type in the VC-N. The standard binary values for C2 are shown in Table 6-8.

Table 6-8 *SDH C2 Byte Values*

C2 Byte	Hex	VC-N Content
0000 0000	00	Unequipped signal
0000 0001	01	Equipped signal – Nonspecific payload
0000 0010	02	TUG structure
0000 0011	03	Locked TU-N
0000 0100	04	Asynchronous mapping for E3 (34.368 Mbps) or DS3 (44.736 Mbps) into C-3
0001 0010	12	Asynchronous mapping for E4 (139.264 Mbps) into C-4
0001 0011	13	Mapping for ATM
0001 0100	14	Mapping for metropolitan-area network (MAN) distributed queue dual-bus (DQDB) IEEE 802.6
0001 0101	15	Asynchronous mapping for FDDI (ISO 9314)
0001 0110	16	Mapping for IP inside PPP with scrambling
0001 0111	17	Mapping of simple data link (SDL) with SDH self-synchronizing scrambler
0001 1000	18	Mapping of HDLC/LAP-S framed signals
0001 1001	19	Mapping of SDL with set-reset scrambler
0001 1010	1A	Mapping of 10-Gbps Ethernet frames (IEEE 802.3)

Table 6-8 *SDH C2 Byte Values (Continued)*

C2 Byte	Hex	VC-N Content
1100 1111	CF	Mapping for IP inside PPP without scrambling/HDLC
1110 0001	E1	Reserved for national use
Up to	—	—
1111 1100	FC	Reserved for national use
1111 1110	FE	Test signal, O.181 specific mapping
1111 1111	FF	VC-AIS

Path Status Byte (G1)

The G1 byte is used to convey the path-terminating status and performance back to the originating path-terminating equipment. The bidirectional path in its entirety can be monitored, from either end of the path. Byte G1 is allocated to convey back to a VC-4-Xc/VC-4/VC-3 trail-termination source the status and performance of the complete trail. Bits 5 to 7 can be used to provide an enhanced RDI with additional differentiation between the payload defect (PLM), server defects (AIS, LOP), and connectivity defects (TIM, UNEQ). Table 6-9 lists the G1 codes.

Table 6-9 *G1 Byte Values*

G1 Bits 5–7	Interpretation
000	No remote defect
001	No remote defect
010	E-RDI payload defect
011	No remote defect
100	E-RDI server defect
101	101 Remote E-RDI server defect
110	110 Remote E-RDI connectivity defect
111	111 Remote E-RDI server defect (Note 1)

Path User Channel Byte (F2)

The F2 byte is used for user communication between path elements.

Position and Sequence Indicator Byte (H4)

The H4 byte provides a multiframe and sequence indicator for virtual VC-3/4 concatenation and a generalized position indicator for payloads. In the latter case, the content is payload specific. For example, H4 can be used as a multiframe indicator for a VC-2/1 payload. For mapping of DQDB in VC-4, the H4 byte carries the slot boundary information and the link status signal (LSS). Bits 1 through 2 are used for the LSS code as described in the IEEE 802.6 standard. Bits 3 through 8 form the slot offset indicator. The slot offset indicator contains a binary number indicating the offset in octets between the H4 octet and the first slot boundary following the H4 octet. The valid range of the slot offset indicator value is 0 to 52. A received value of 53 to 63 corresponds to an error condition.

Path User Channel Byte (F3)

The F3 byte is allocated for communication purposes between path elements and is payload-dependent.

APS Signaling Byte (K3)

The K3 bits 1 through 4 are allocated for protection at the VC-4/3 path levels. K3 bits 5 through 8 are allocated for future use. These bits have no defined value and can be used in vendor proprietary implementations. The receiver is required to ignore their content unless programmed to do so.

Network Operator Byte (N1)

The N1 byte is allocated to provide a higher-order tandem connection monitoring (HO-TCM) function. N1 is allocated for TCM for a contiguous concatenated VC-4, VC-4, and VC-3 levels.

SDH Low-Order Path Overhead

The SDH low-order POH consists of 4 bytes. The 4 bytes (V5, J2, N2, and K4) are allocated to the VC-2/VC-1 POH. The 4 bytes of the lower-order POH are described in the next section.

VT Path Overhead Byte (V5)

The V5 byte is the first byte of the multiframe, and its position is indicated by the TU-2/TU-1 pointer. The V5 byte provides the functions of error checking, signal label, and path status of the VC-2/VC-1 paths. The bit assignments for the V5 byte and the byte-by-byte lower-order POH is shown in Table 6-10.

Table 6-10 *V5 Byte Values*

V5 Bit	Description
Bits 1–2	Allocated for error performance monitoring. A BIP-2 scheme is specified. It includes POH bytes, but excludes V1, V2, V3, and V4.
Bit 3	A VC-2/VC-1 path LP-REI that is set to 1 and sent back toward a VC-2/VC-1 path originator if one or more errors are detected by the BIP-2; otherwise, it is set to 0.

Table 6-10 *V5 Byte Values (Continued)*

V5 Bit	Description	
Bit 4	A VC-2/VC-1 path LP-RFI. This bit is set to 1 if a failure is declared; otherwise, it is set to 0. A failure is a defect that persists beyond the maximum time allocated to the transmission system protection mechanisms.	
Bits 5–7	Provides a VC-2/VC-1 signal label.	
	000	Unequipped or supervisory unequipped.
	001	Equipped – nonspecific.
	010	Asynchronous.
	011	Bit synchronous.
	100	Byte synchronous.
	101	Reserved for future use.
	110	Test signal, O.181 specific mapping.
	111	VC-AIS.
Bit 8	Set to 1 to indicate a VC-2/VC-1 path LP-RDI; otherwise, it is set to 0.	

Path Trace Byte (J2)

The J2 byte is used to repetitively transmit a lower-order access path identifier so that a path-receiving terminal can verify its continued connection to the intended transmitter. A 16-byte frame is defined for the transmission of path access point identifiers. This 16-byte frame is identical to the 16-byte frame of the J1 and J0 bytes.

Network Operator Byte (N2)

The N2 byte is allocated to provide a lower-order tandem connection monitoring (LO-TCM) function. N2 is allocated for TCM for the VC-2, VC-12, and VC-11 level.

APS Signaling Byte (K4)

The K4 bits 1 through 4 are allocated for protection at the VC-2/1 path levels. K4 bits 5 through 7 are allocated for future use. These bits have no defined value and can be used in vendor proprietary implementations. The receiver is required to ignore their content unless programmed to do so. Bit 8 is reserved for future use and has no defined value.

SDH and SONET Interworking

It is of significant importance that Europe, Middle East, and Africa (EMEA), Asia-Pacific, and Latin American SDH networks be interoperable with their North American Synchronous

Optical Network (SONET) counterparts. Especially in the view that the continental United States terminates a large percentage of the global voice and Internet data traffic. The SDH payload types are compared with the SONET payload types in Table 6-11. As can be seen from the table, SDH and SONET are compatible in terms of their payload types. Table 6-12 illustrates compatible SDH and SONET payload mappings. If you have noticed, E3 is not compatible with the SONET hierarchy and is missing from the table. However, many vendors map the individual E3 DS0s to a DS3 signal using TDM digital access and cross-connect system (DACS) equipment, before transmitting the DS3 over a SONET STS-1 signal.

NOTE There are 2 bits known as the "ss" bits (size bits) defined in the SDH AU-3 or AU-4 pointer byte H1 (bits 5 and 6). SDH uses these bits to define the structure of some payloads. Typically in SDH, these bits are set to 10 for AU-3/AU-4. When SDH equipment is interfaced with SONET equipment, receiving SONET equipment ignores these bits, because they are used to indicate payload mapping information, which is communicated via other fields, such as the C2 byte. SONET equipment, typically sets the ss bits to 00 before transmission. The ss bit value 11 is reserved and not normally used. Some older SDH ADMs set the value of the ss-bit to 01.

Table 6-11 *SDH and SONET Payload Type Comparison*

C2 Byte	Hex	SDH VC-N Content	SONET SPE Content
0000 0000	00	Unequipped signal	Unequipped signal
0000 0001	01	Equipped signal – Nonspecific payload	Equipped signal – Nonspecific payload
0000 0010	02	TUG structure	Floating VT mode
0000 0011	03	Locked TU-N	Locked VT mode
0000 0100	04	Asynchronous mapping for E3 (34.368 Mbps) or DS3 (44.736 Mbps) into C-3	Asynchronous mapping for DS3
0001 0010	12	Asynchronous mapping for E4 (139.264 Mbps) into C-4	Asynchronous mapping for 139.264-Mbps DS4NA
0001 0011	13	Mapping for ATM	Mapping for ATM
0001 0100	14	Mapping for MAN DQDB IEEE 802.6	Mapping for MAN DQDB IEEE 802.6
0001 0101	15	Asynchronous mapping for FDDI (ISO 9314)	Asynchronous mapping for FDDI

Table 6-11 *SDH and SONET Payload Type Comparison (Continued)*

C2 Byte	Hex	SDH VC-N Content	SONET SPE Content
0001 0110	16	Mapping for IP inside PPP with scrambling	Mapping for IP inside PPP with scrambling
0001 0111	17	Mapping of SDL with SDH self-synchronizing scrambler	—
0001 1000	18	Mapping of HDLC/LAP-S framed signals	—
0001 1001	19	Mapping of SDL with set-reset scrambler	—
0001 1010	1A	Mapping of 10-Gbps Ethernet frames (IEEE 802.3)	—
1100 1111	CF	Mapping for IP inside PPP without scrambling/HDLC	Mapping for IP inside PPP without scrambling/HDLC
1110 0001	E1	Reserved for national use	—
Up to	—	—	—
1111 1100	FC	Reserved for national use	—
1111 1110	FE	Test signal, O.181 specific mapping	Test signal, O.181 mapping ITU-T G.707
1111 1111	FF	VC-AIS	AIS-P

Table 6-12 *Compatible SDH and SONET Payload Mappings*

Payload	SDH	SONET
DS1	VC-11/VC-12	VT1.5
E1	VC-12	VT2
DS2	VC-2	VT6
DS3	VC-3	STS-1
ATM	VC-3/VC-4	STS-1/STS-3c

SDH Alarms

The SDH overhead described in the previous section provides a variety of management information and other functions including error performance monitoring, pointer information, path status and trace, path trace, section trace, remote defect, error, failure indications, signal labels, data flag indications, DCC information, APS control, orderwire information, and synchronization status messages. A majority of the overhead information is involved with alarm and in-service monitoring of the particular SDH sections.

SDH alarms are defined as anomalies, defects, and failures. An *anomaly* is the smallest discrepancy that can be observed between the actual and desired characteristics of an item. The occurrence of a single anomaly does not constitute an interruption in the ability to perform a required function. If the density of anomalies reaches a level at which the ability to perform a required function has been interrupted, it is termed a *defect*. Defects are used as an input for performance monitoring, control of consequent actions, and determination of fault cause. The inability of a function to perform a required action beyond the maximum time allocated is termed a *failure*.

Error performance monitoring in the SDH is based on bit-interleaved-parity (BIP) checks calculated on a frame-by-frame basis. These BIP checks are inserted in the RSOH, MSOH, and POH. In addition, higher-order path-terminating equipment (HO PTE) and lower-order path-terminating equipment (LO PTE) produce remote error indications (REIs) based on errors detected in the HO path and LO path BIP respectively. The REI signals are sent back to the equipment at the originating end of a path. Various SDH alarm anomalies, defects, and failures are described in the next section. Figure 6-9 illustrates the various SDH maintenance interactions, alarms and the overhead bytes used to carry the alarms and maintenance information.

Figure 6-9 *SDH Maintenance Interactions*

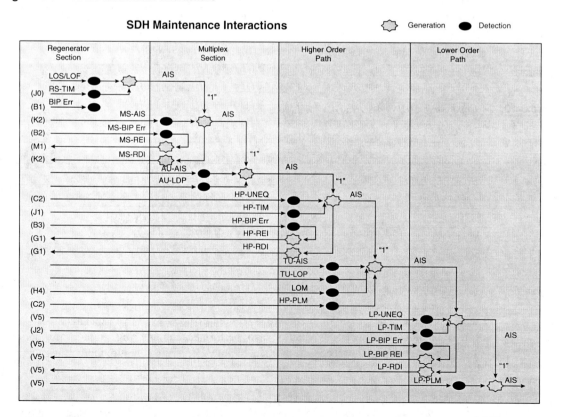

Loss of Signal (LOS)

The loss-of-signal (LOS) alarm is raised when the synchronous signal (STS–N) level drops below the threshold at which a bit error rate (BER) of 1 in 10^3 is predicted. This can be the result of excessive attenuation of the signal, receiver saturation, cable cut, or equipment fault. The LOS state clears when two consecutive framing patterns are received and no new LOS condition is detected. The LOS alarm condition is configurable in certain SDH NEs.

Out-Of-Frame (OOF) Alignment

The out-of-frame (OOF) state occurs when four or five consecutive SDH frames are received with invalid (errored) framing patterns (A1 and A2 bytes). The maximum time to detect OOF is 625 microseconds. OOF state clears when two consecutive SDH frames are received with valid framing patterns.

Loss-Of-Frame (LOF) Alignment

The LOF state occurs when the OOF state exists for a specified time in milliseconds. The LOF state clears when an in-frame condition exists continuously for a specified time in milliseconds. The time for detection and clearance is normally 3 milliseconds.

Loss-Of-Pointer (LOP)

The loss-of-pointer (LOP) state occurs when N consecutive invalid pointers are received or N consecutive new data flags (NDFs) are received (other than in a concatenation indicator), where N = 8, 9, or 10. LOP state clears when three equal valid pointers or three consecutive AIS indications are received. The LOP can be identified as an administrative unit loss of pointer (AU-LOP) or tributary unit loss of pointer (TU-LOP).

Alarm Indication Signal (AIS)

The AIS is an all-1s characteristic or adapted information signal. It is generated to replace the normal traffic signal when that signal contains a defect condition to prevent consequential downstream failures being declared or alarms being raised. The AIS can be identified as a multiplex section alarm indication signal (MS-AIS), administrative unit alarm indication signal (AU-AIS), or tributary unit alarm indication signal (TU-AIS).

Remote Error Indication (REI)

The REI is an indication returned to a transmitting node (source) that an errored block has been detected at the receiving node (sink). This indication was formerly known as *far-end block error (FEBE)*. The REI can be identified as a multiplex section remote error indication (MS-REI), higher-order path remote error indication (HP-REI), or lower-order path remote error indication (LP-REI).

Remote Defect Indication (RDI)

The RDI is a signal returned to the transmitting terminating equipment on detection of a LOS, LOF, or AIS defect. RDI was previously known as FERF. The RDI can be identified as a multiplex section remote defect indication (MS-RDI), higher-order path remote defect indication (HP-RDI), or lower-order path remote defect indication (LP-RDI).

Remote Failure Indication (RFI)

A failure is a defect that persists beyond the maximum time allocated to the transmission system protection mechanisms. When this situation occurs, an RFI is sent to the far end and initiates a protection switch if this function has been enabled. The RFI can be identified as a lower-order path remote failure indication (LP-RFI).

B1 Error

Parity errors evaluated by byte B1 (BIP–8) of an STM–N are monitored. If any of the eight parity checks fail, the corresponding block is assumed to be in error.

B2 Error

Parity errors evaluated by byte B2 (BIP–24 * N) of an STM–N are monitored. If any of the N * 24 parity checks fail, the corresponding block is assumed to be in error.

B3 Error

Parity errors evaluated by byte B3 (BIP–8) of a VC–N (N = 3, 4) are monitored. If any of the eight parity checks fail, the corresponding block is assumed to be in error.

BIP–2 Error

Parity errors contained in bits 1 and 2 (BIP-2) of byte V5 of a VC-M (M = 11, 12, 2) are monitored. If any of the two parity checks fail, the corresponding block is assumed to be in error.

Loss of Sequence Synchronization (LSS)

Sequence synchronization is considered to be lost, and resynchronization is started if the bit error ratio is greater than or equal to 0.20 during an integration interval of 1 second.

SDH Higher-Level Framing

Higher-level STM-N frames can be simplistically perceived as *4 multiples of a basic STM-1. As illustrated in Figure 6-10, an STM-4 is constructed by byte-interleaved multiplexing of 4 STM-1s into a frame that is 9 rows by 1080 columns wide. The STM-4 signal has a line rate of 622.080 Mbps (4 * 155.520 Mbps). The four STM-1s (STM-1(1), STM-1(2), STM-1(3), and STM-1(4)) are frame aligned before multiplexing. Frame alignment is achieved by ensuring that the first 12 bytes of the STM-4 signal are A1 framing bytes drawn from STM-1(1), the next 3 from STM-1(2), then 3 from STM-1(3), and finally 3 from STM-1(4). The 12 A1 framing bytes are followed by 12 A2 framing bytes that are obtained from the 4 STM-1s in a process, similar to the way that the A1 bytes were obtained. As shown in Figure 6-10, there are 12.

Figure 6-10 *STM-4 and STM-16 Frames*

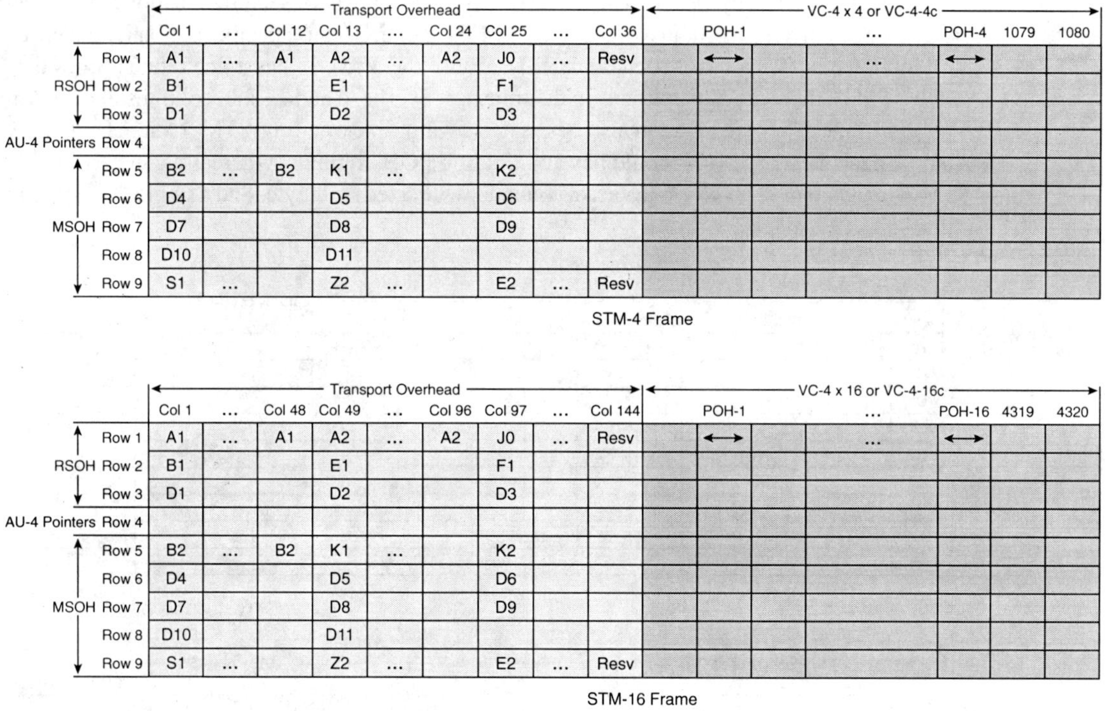

A1 bytes are followed by 12 A2 bytes, 4 J0 bytes, and 8 bytes reserved for national use. There are also 12 B2 bytes present in row 5 of the STM-4 frame. Because the illustration is condensed, all J0 national bytes have not been depicted. Because the STM-4 has a line rate 4 times that of the STM-1, there are 36 columns in the transport overhead and 1080 columns in the entire STM-4 frame. A nonconcatenated STM-4 would have four floating POH columns, whereas a concatenated STM-4c would have a VC-4-4c virtual container with fixed stuff bytes and a single POH column.

The STM-16 is constructed by byte-interleaved multiplexing 16 STM-1s into a frame that is 9 rows by 4320 columns wide. The STM-16 signal has a line rate of 2488.320 Mbps (16 * 155.520 Mbps). The 16 STM-1s are frame aligned before multiplexing. As shown in Figure 6-10, there are 48 A1 bytes followed by 48 A2 bytes, 16 J0 bytes, and 32 bytes reserved for national use. There are also 48 B2 bytes present in row 5 of the STM-16 frame. Because the illustration is condensed, all J0 national bytes have not been depicted. Because the STM-16 has a line rate 16 times that of the STM-1, there are 144 columns in the transport overhead and 4320 columns in the entire STM-4 frame. The STM-16 can also be created by multiplexing four STM-4 signals directly. A nonconcatenated STM-16 would have 16 floating POH columns, whereas a concatenated STM-16c would have a VC-4-16c virtual container with fixed stuff bytes and a single POH column.

The STM-64 is constructed by byte-interleaved multiplexing 64 STM-1s into a frame that is 9 rows by 17,280 columns wide. The STM-64 signal has a line rate of 9953.280 Mbps (64 * 155.520 Mbps). The 64 STM-1s are frame aligned before multiplexing. As shown in Figure 6-11, there are 192 A1 bytes followed by 192 A2 bytes, 64 J0 bytes, and 128 bytes reserved for national use. There are also 192 B2 bytes present in row 5 of the STM-64 frame. Because the illustration is condensed, all J0 national bytes have not been depicted. Because the STM-64 has a line rate 64 times that of the STM-1, there are 576 columns in the transport overhead and 17,280 columns in the entire STM-64 frame. The STM-64 can also be created by multiplexing four STM-16 signals directly. A nonconcatenated STM-64 would have 64 floating POH columns, whereas a concatenated STM-64c would have a VC-4-64c virtual container with fixed stuff bytes and a single POH column.

Figure 6-11 *STM-64 and STM-256 Frames*

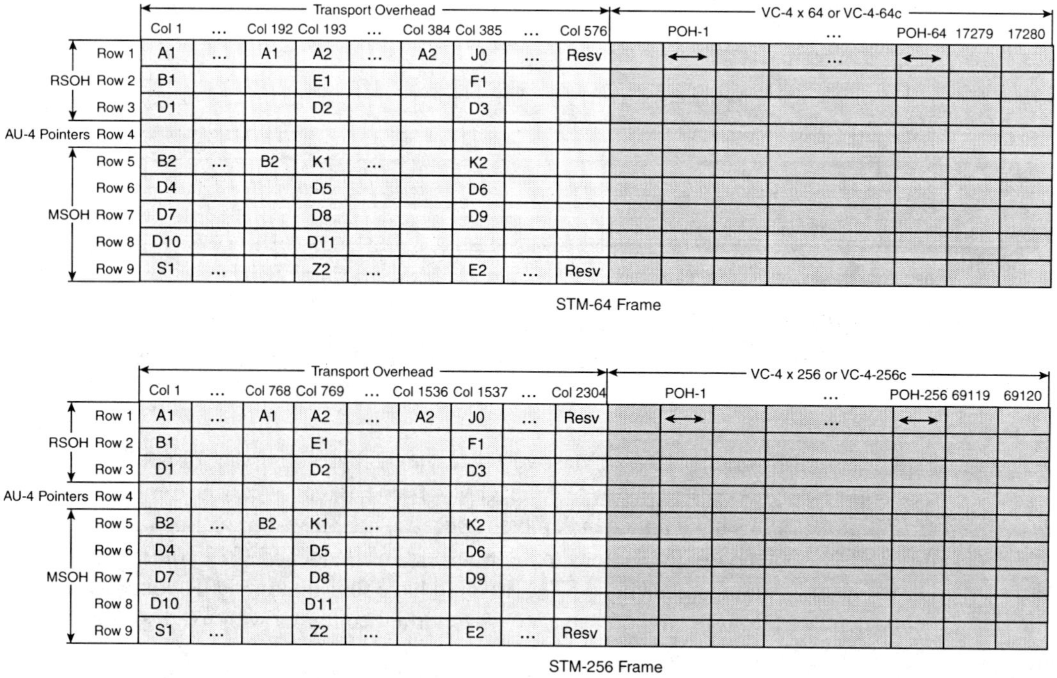

The STM-256 is constructed by byte-interleaved multiplexing 256 STM-1s into a frame that is 9 rows by 69,120 columns wide. The STM-256 signal has a line rate of 39813.120 Mbps (256 * 155.520 Mbps). The 256 STM-1s are frame aligned before multiplexing. As shown in Figure 6-11, there are 768 A1 bytes followed by 768 A2 bytes, 256 J0 bytes, and 512 bytes reserved for national use. There are also 768 B2 bytes present in row 5 of the STM-256 frame. Because the illustration is condensed, all J0 national bytes have not been depicted. Because the STM-256 has a line rate 256 times that of the STM-1, there are 2304 columns in the transport overhead and 69,120 columns in the entire STM-256 frame. The STM-256 can also be created by multiplexing four STM-64 signals directly. A nonconcatenated STM-256 would have 256 floating POH columns, whereas a concatenated STM-256c would have a VC-4-256c virtual container with fixed stuff bytes and a single POH column.

SDH Network Elements

Devices that implement SDH transmission are defined as SDH NEs. These NEs include regenerator section, multiplex section, and path-terminating equipment. The SDH NEs integrate to form the digital loop carrier (DLC) supporting various SDH architectures and topologies.

Regenerator

The regenerator is an RSTE that regenerates attenuated signals. A regenerator or amplifier is needed when, due to the long distance between multiplexers, the signal level in the fiber gets attenuated and becomes too low to drive a receiver. The regenerator is sometimes called a *repeater*. Regenerators can be cascaded to extend the reach of the optical-fiber system. However, nonlinear impairments must be taken into consideration while cascading regenerators. The regenerator clocks itself off of the received signal and replaces the RSOH bytes before retransmitting the signal. The MSOH, POH, and payload are not altered.

Terminal Multiplexer

The terminal multiplexer (TM) is a path-terminating element (PTE) that can concentrate or aggregate DS1s, DS3s, E1s, E3s, and STM-Ns. An implementation with two TMs represents an SDH link with a regenerator section, multiplex section, and path all in one link. The schematic of a TM is represented in Figure 6-12. Various PDH signals, such as DS1, E1, and E3 levels, are mapped to their associated SDH electrical payloads in the TM. For example, DS1 signals are mapped to VC-11s, E1 signals are mapped to VC-12s, and E3 signals are mapped to VC-3 virtual containers. An electrical-to-optical (EO) conversion takes place after the multiplexing, and the STM-N signals are launched into the fiber. The reverse happens during signal reception. In practice, the TM is an ADM operating in terminal mode. The TM is analogous to the channel bank in the TDM world and allows lower-speed user access to the SDH network.

Figure 6-12 *SDH Terminal Multiplexer*

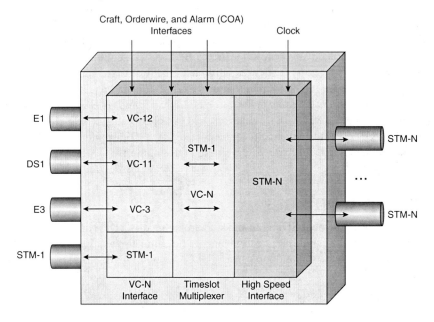

Add/Drop Multiplexer

An ADM is a PTE that can multiplex or demultiplex various signals to or from an STM-N signal. At an add/drop site, only those signals that need to be accessed are dropped or inserted. The remaining traffic continues through the NE without requiring special pass-through units or other signal processing. Signals can also be looped or terminated at the ADM. Figure 6-13 shows a schematic of a TM.

Figure 6-13 *SDH ADM*

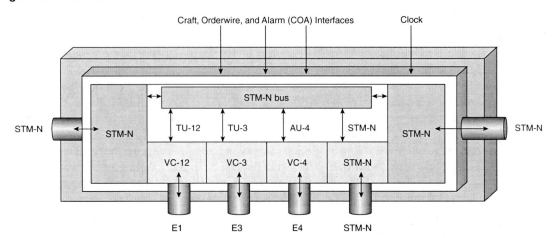

Various PDH signals are mapped to their associated SDH VC-Ns in the TM. The multiplexed payloads are then mapped into STM-N signals based on the line rate of the STM-N transmission. The STM-N signals then undergo an EO conversion and are launched directionally onto the fiber based on the ring type and protection mechanism. An ADM can be deployed at a terminal site or any intermediate location for consolidating traffic from widely separated locations. Several ADMs can also be configured as a survivable ring. SDH enables add, drop, and pass-through capability where a signal that terminates at one node is duplicated (repeated) and is then sent to the next and subsequent nodes. ADMs can be used to interconnect SDH rings. In such a case, the ADMs are referred to as *matching nodes*. In ring survivability applications, drop and pass-through capability provides alternate routing for traffic passing through interconnecting rings in a matched node configuration. If the connection cannot be made through one of the nodes, the signal is repeated and passed along an alternate route to the destination node. Signals can also be hair-pinned at intermediate ADMs by connecting ports externally.

Single-stage ADMs can multiplex or demultiplex one or more tributary signals into or from an STM-N signal. It can be used in terminal sites, intermediate sites, or hub configurations. At an intermediate add/drop site, it can drop lower-rate signals to be transported on different facilities, or it can add lower-rate signals into the higher-rate STS-N signal. The rest of the traffic just continues straight through. In multinode distribution applications, one transport channel can carry traffic between multiple distribution nodes. Although NEs are compatible at the STM-N level, they can differ in features from vendor to vendor. SDH does not restrict manufacturers to providing a single type of product, nor require them to provide all types. For example, one vendor might offer an ADM with access at E1 only, whereas another might offer simultaneous access at E1 and E3 rates. The next-generation multiservice platforms have their roots in the ADM. Multiservice platforms carry data services over SDH and also integrate DWDM capabilities within the box.

Broadband Digital Cross-Connect

The broadband digital cross-connect (BDCS) can make two-way cross-connections at the E3, E4, and STM-N levels. It can interface various high-level PDH and SDH signals, such as E3, DS3, E4, and others. The BDCS is the synchronous equivalent of a DACS and supports hub-network SDH architectures. The BDCS accesses the high-level signals and switches at the AU-4 level. It accepts optical signals and allows overhead to be maintained for integrated OAM&P purposes. As illustrated in Figure 6-14, the BDCS can be used as an SDH hub, where it can be used for grooming VC-3s, for broadband restoration purposes, or for routing traffic. The BDCS can also bridge and roll payload and facilities, make unidirectional or one-way connections, and support interconnection to older SDH NEs.

One major difference between a cross-connect and an ADM is that a cross-connect can be used to interconnect a much larger number of STM-Ns. The BDCS can be used for grooming, consolidation, segregation of STM-1s, or for broadband traffic management. For example, it can be used to segregate high-bandwidth from low-bandwidth traffic and send them separately to a high-bandwidth switch and a low-bandwidth switch.

Figure 6-14 *Broadband Digital Cross-Connect*

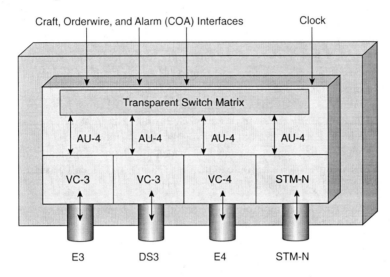

Wideband Digital Cross-Connect

The wideband digital cross-connect (WDCS) is a digital cross-connect that terminates high-level SDH and E3 signals; it also performs low-level E1 and DS1-level cross-connections. In a WDCS, the switching is done at the TU-12 level as it cross-connects the constituent VCs between STM-N terminations. The WDCS accepts various optical carrier rates, accesses the TU-12 level signals, and switches at this level. Figure 6-15 shows a schematic of a WDCS.

Figure 6-15 *Wideband Digital Cross-Connect*

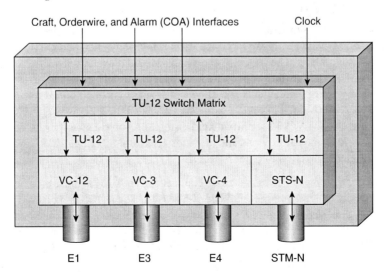

The WDCS is suitable for E1-level grooming applications at hub locations. One major advantage of the WDCS is that less demultiplexing and multiplexing is required because only the required tributaries are accessed and switched. Because SDH is synchronous, the low-speed tributaries are visible and accessible within the STM-1 signal. Therefore, the required tributaries can be accessed and switched without demultiplexing, which isn't possible with existing TDM digital cross-connects. The WDCS also cross-connects the constituent E1s between E3 terminations, as well as between E3 and E1 terminations. The features of the WDCS make it useful in several applications. Because the WDCS can automatically cross-connect VCs and E1s, it can be used as a network management system and for grooming at hub locations.

Digital Loop Carrier

The integrated digital loop carrier (IDLC) consists of intelligent remote digital terminals (RDTs) and digital switch elements called *integrated digital terminals (IDTs)*, which are connected by a digital line. The IDLCs are designed to more efficiently integrate DLC systems with existing digital switches. A DLC is intended for service in the central office (CO) or a controlled environment vault (CEV) that belongs to the carrier. Figure 6-16 shows a schematic of a DLC.

Figure 6-16 *Digital Loop Carrier*

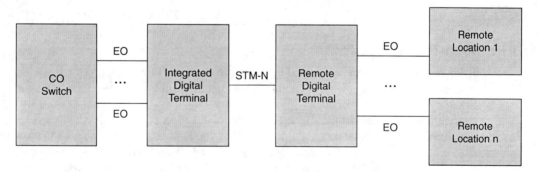

From a voice perspective, the DLC might be considered a concentrator of low-speed services before they are brought into the local CO for switching or distribution. The concentration can be achieved using a combination of multiplexing and intelligent GR.303 concentration. If concentration were not performed, the number of subscribers that a CO could serve would be limited by the number of physical lines served by the CO. The DLC itself is actually a system of multiplexers and switches designed to perform concentration from the remote terminals to the CO. The DLC devices are referred to as *remote fiber terminals (RFTs)* by several optical vendors.

Multiservice Provisioning Platforms (MSPPs)

Multiservice provisioning platforms (MSPPs) combine the functionality of the SDH ADM, BDCS, and WDCS NEs in a single platform. The platform also integrates Ethernet services and DWDM functionality in the box. The MSPP provides TDM solutions with interfaces such as E1,

E3, EC1, and data solutions with 10/100/1000 Ethernet solutions along with STM-1 to STM-64 optical transport bit rates including integrated DWDM wavelengths. The MSPP supports various SDH topologies, including ring, linear point-to point, linear add/drop, star, and hybrid topologies. The protection and restoration choices include two-fiber subnetwork connection protection (SNCP), two-fiber and four-fiber multiplex section-shared protection rings (MS-SPRings), 1+1 APS, unprotected span, and the Cisco Path Protected Mesh Networking (PPMN). The MSPP devices are easily configurable using integrated graphical user interface (GUI)-based tools and are fully manageable using element management systems. The MSPP devices are New Equipment Building Standards (NEBS) level 3 as well as OSMINE compliant and provide carrier-class reliability, with high mean time between failure (MTBF) and low mean time to repair (MTTR) figures. The Cisco ONS MSPP is discussed in great length and detail in later chapters.

SDH Topologies

SDH topologies center around survivability of the network with 50 ms or lower rapid recovery. Various topologies can be configured using either SDH ADMs or DCSs. It is interesting to note that SONET ring topologies are often used in North America, whereas Europe, Asia, and Latin America mainly rely on SDH-based ring as well as meshed network topologies. Protection switching is mainly performed by the intelligent ADMs or DCSs.

SDH Point-to-Point Topology

Point-to-point topologies are created by connecting two SDH PTEs back to back over dark fiber. As depicted in Figure 6-17, a point-to-point deployment involves two PTE ADMs or TMs linked by fiber with or without an STE regenerator in the link. The TM can act as an E1 concentrator and transport the E1s over an STM-N link. The synchronous island can coexist within an asynchronous domain. Point-to-point topologies are extremely popular to provide storage-area network (SAN) interconnectivity between datacenters. Albeit, SAN protocols, such as ESCON, FICON, and Fibre Channel, use 1.25- or 2.125-Gbps channels over WDM systems. The system always terminates the link using a next-generation device, such as an MSPP with WDM capability.

Figure 6-17 *SDH Point-to-Point Topology*

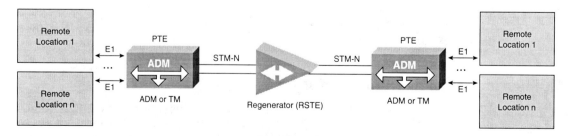

Point-to-point topologies use 1:N protection mechanisms in which one standby (protect) link is used to protect N active (working) links. Maximum protection is obtained by using a 1:1 ratio or 1+1 topology. Under normal working conditions, the working path is used. In the event of a failure, however, the protect path is used with a switchover time less than 50 ms. Ideally, the protect path or fiber must use diverse physical routing to achieve maximum redundancy.

SDH Point-to-Multipoint Topology

A point-to-multipoint architecture implements the adding and dropping of circuits along the path. The SDH ADM is a unique NE specifically designed for this task. As illustrated in Figure 6-18, the ADM is typically placed along an SDH link to facilitate adding and dropping of tributary or STM-N channels at intermediate points in the network. The point-to-multipoint topology is also referred to as *linear add/drop architecture*. This topology is implemented for medium- and long-haul linear SDH architectures, where the service provider might have to add or drop circuits on the way to the destination. Conventional TDM systems would need complex demultiplexing, multiplexing, and cross-connections to achieve similar functionality at much lower bandwidths. As with point-to-point topologies, point-to-multipoint topologies also use 1:N protection mechanisms in which one standby (protect) link is used to protect N active (working) links. As before, maximum protection is obtained by using a 1:1 ratio or 1+1 topology.

Figure 6-18 *SDH Point-to-Multipoint Topology*

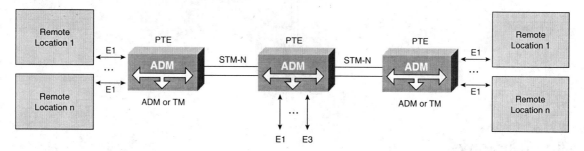

SDH Hub Topology

The hub network topology is a scalable architecture that uses PTE devices in a hub-and-spoke configuration. It accommodates unexpected growth and changes as compared to point-to-point network topologies. As illustrated in Figure 6-19, the hub is implemented as a WDCS or BDCS that concentrates traffic at a central site and allows easy reprovisioning of the circuits. In a WDCS implementation, two or more ADMs and a wideband cross-connect switch allows cross-connecting services at the tributary level. In a BDCS implementation, two or more ADMs and a BDCS switch allows cross-connecting at both the SDH STS and the tributary level. An MSSP could replace the BDCS and subtending ADMs at the hub. The hub architecture is implemented in situations where the fiber is laid out in a hub-and-spoke topology. As with the linear topologies, hub topologies also use 1:N protection mechanisms in which one standby (protect) link is used to protect N active (working) links, with maximum protection obtained by using 1:1 or 1+1 protection.

Figure 6-19 *SDH Hub Topology*

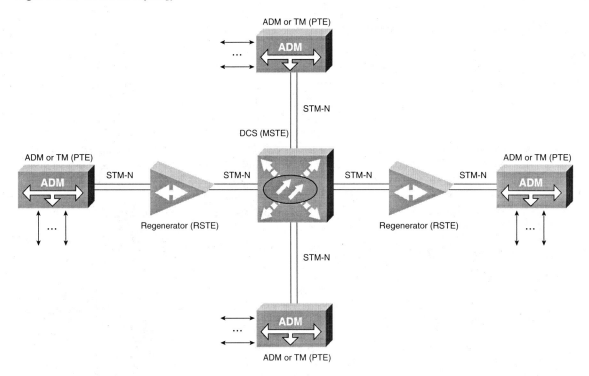

Ring Topology

SDH rings provide 50 ms and lower restoration times as well as robust protection mechanisms. The SDH building block for ring architecture is the ADM or MSPP. Multiple ADMs can be daisy-chained in a ring configuration for either bidirectional or unidirectional traffic flow as illustrated in Figure 6-20. Access and core rings can be interconnected using ADMs as matching nodes. The main advantage of the ring topology is its survivability and fast restoration or healing time. If a fiber cable is cut, the ADMs have the intelligence to reroute the affected services via an alternate path through the ring without interruption. The demand for survivable services, diverse routing of fiber facilities, and flexibility to rearrange services to alternate serving nodes with automatic restoration have made rings the most popular metro access and core SDH architecture. Rings use advanced protection mechanisms and protocols, such as APS, Subnetwork Dependent Conversion Protocol (SNCP) two-fiber, MS-SPRing two-fiber, and MS-SPRing four-fiber. SNCP protection topologies normally use a 1:N protection mechanism in which one standby (protect) link is used to protect the active (working) links. Ring protection mechanisms are explained in greater detail in a later section.

Figure 6-20 *SDH Ring Topology*

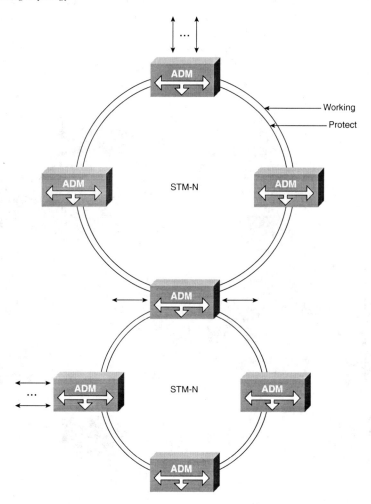

Mesh Topology

Meshed networks refer to any number of sites arbitrarily connected together with at least one loop. Mesh topologies are created using a hybrid combination of multiple point-to-point links and/or two-fiber SNCP links. As illustrated in Figure 6-21, meshed networks refer to any number of sites arbitrarily connected together with at least one loop. Mesh topologies provide maximum redundancy with multiple rerouting options for the SDH ADMs or DCS devices. Next-generation MSPPs leverage protection protocols, such as Path Protected Mesh Networking (PPMN), by using advanced Dijkstra's algorithmic procedures for network discovery and route recalculation on fiber or node failure. Such information is carried in the SDH DCC channels and is used to build Dijkstra's tables and routing tables on startup or link-state change. As soon as all the nodes on a network are turned up, they begin the process of autodiscovery. Within minutes, each node has a full description and status of the other nodes and connections throughout the network. Creating an overlaying TDM or Ethernet circuit is then easily accomplished by specifying the source and destination node.

Figure 6-21 *SDH Mesh Topology*

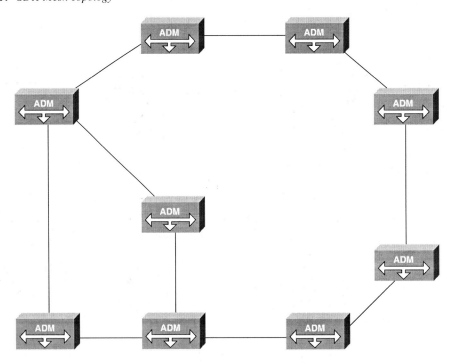

Fiber Routing and Diversity

Various fiber routing schemes help achieve redundancy in SDH networks. Trunks are the main fiber cables that can carry hundreds of fiber strands owned by carriers. Laterals are the fiber cables from the customer premises to the nearest splice point on the cable trunk. Generally, laterals are used exclusively by the customer. Within cities, laterals can be as short as a few meters or could extend several kilometers in suburban and rural areas. The minimum size of a lateral is usually 12 strands. However, even though a lateral might have 12 strands, only 2 or 4 of those strands can be spliced to dedicated fibers on the trunk. Most fiber provisioning companies provide additional spare strands on the trunk to which the customer can connect to at a later time, depending on requirements. Building entrances and termination panels are the facilities within the customer's premise for the termination of the fiber.

Aerial installation on existing poles is the most cost-effective installation method and offers moderate reliability. Most regulatory bodies have well-established rules and procedures for licensed carriers and fiber installers to access existing utility and telephone poles. Installation in existing conduits is the next-best option. Many regulatory bodies require carriers to install extra conduits accessible by any other licensed carrier or fiber installer. As with poles, regulators have set prices for the cost of access to this conduit. Jet fiber is another approach to fiber deployment. In this case, the fiber provider installs *microconduit* rather than fiber. When a customer requires a fiber pair, it is blown into the microconduit using high-pressure air. The advantage of this approach is that far fewer splices are required and the fiber can be blown all the way into the customer premises.

If there are no existing conduits or poles, commonly referred to as *support structures*, traditional trenching deployment can be used, including the following:

- Direct-bury fiber-plough techniques
- In-the-groove technologies, where a very narrow grove is cut into the existing road bed
- Sewer systems, where robotic systems install fiber in storm or sanitary sewers using either specialized cable or stainless steel tubing
- Gas pipeline systems, where the fiber is installed in active gas pipelines

Note that individual fiber cables are sheathed, shielded, and armored. Various conduits can share a feeder into a building. As shown in Figure 6-22, a single feeder into the building with a single conduit from a CO is the most basic fiber routing topology. A degree of redundancy can be achieved by using diverse conduit routing from the CO point of presence with a single feeder into the building. The single feeder is the weakest link in such a physical fiber routing topology, because it could be susceptible to backhoe digging or trenching.

True diverse routing can be obtained by having multiple feeders into the building with diverse conduit routing to the CO. Such a form of routing is presented in Figure 6-23. Dual COs with diverse conduit routing and diverse feeders provide the ultimate diversity for 1:1 protection schemes.

NOTE Ten meters of facilities separation can qualify a fiber plant to possess route diversity from a construction perspective.

Figure 6-22 *Fiber Routing with Conduit Diversity*

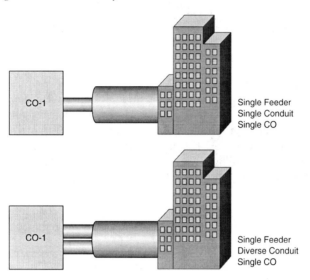

Figure 6-23 *Fiber Diversity*

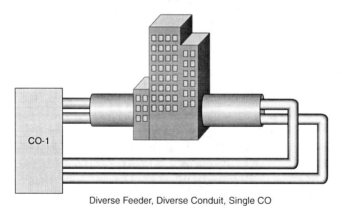

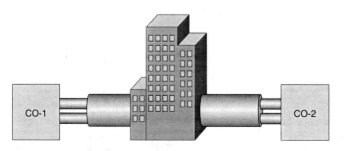

SDH Protection Architectures

This section discusses various SDH protection architectures and mechanisms. This includes APS, linear, and ring protection architectures. SDH rings use path-switching or line-switching, self-healing techniques. SDH defines a maximum switch time of 50 ms for an MS-SPRing ring with no extra traffic and less than 1200 km of fiber. The specification actually states 60 ms, with 10 ms for discovery of the problem and 50 ms to perform the switch, but this is commonly referred to as just 50 ms. However, most SDH networks are much smaller and simpler than this example and switch even faster. Y-cable switchover times are much higher and are typically in the order of 250 ms. SDH protection switching is invoked if there is an LOS, LOF, or even signal degradation, such as the BER exceeding a preconfigured limit. Protection switching can also be initialized manually to allow for routine maintenance or testing without having to take the circuits out of service. Protection implies that a backup resource has been established for recovery if the primary resource fails, and restoration implies re-establishing the end-to-end path based on resources available after the failure. SDH protection architectures support nonrevertive and revertive protection mechanisms. With nonrevertive protection, after the original working line has been restored to its proper operating condition, the system will not revert to use the original line as the working path. Instead, it continues to use what was originally the protection line as the working line. With revertive protection, the system reverts to the original line after restoration. Revertive protection is most common with 1:N protection schemes. 1:1 protection can use revertive or nonrevertive protection mechanisms.

Automatic Protection Switching

APS is a standard devised to provide for link recovery in the case of failure. Link recovery is made possible by having SDH devices with two sets of fiber. One set (transmit and receive) is used for working traffic, and the other set (transmit and receive pair) is used for protection. The working and protect fibers are routed over diverse physical paths to be effective. The fibers used for protection may or may not carry a copy of the working traffic depending on how protection has been configured (1:N or 1+1). APS protection can be configured for linear or ring topologies. Each type of topology has specific choices of 1:1, 1:N, or 1+1 protection. These can further be configured with unidirectional or bidirectional switching mechanisms. Only the multiplex section in SDH is protected by the APS mechanism. The multiplex section protection mechanisms are coordinated by the K1 and K2 bytes in the MSOH. Path protection is managed at a higher level by network management functions. In SDH, the transmission is protected on optical sections from the point at which the MSOH is inserted to the point where the MSOH is terminated. The K1 and K2 bytes in the MSOH of the STM-1 signal carry a multiplex section protection (MSP) protocol used to coordinate protection switching between the near end and the far end. Protection switching is initiated as a result of signal failure, signal degradation, or in response to commands from a local craft terminal or a remote network manager.

1+1 Protection

In a 1+1 protection architecture, the head-end signal is continuously bridged at the VC-4/STM-1 level to working and protection equipment so that an identical payload is transmitted over a separate fiber pair to the tail-end working and protection equipment. At the tail end, the working and protection signals are continuously monitored independently for failures. The tail-end equipment selects between either the working or the protection signals. Because of the continuous head-end bridge, the 1+1 architecture does not permit unprotected extra traffic over the span. Figure 6-24 shows a schematic of 1+1 protection. The signal is bridged on both paths, and the tail-end receiver can select between the two signals based on local criteria. Switchover to the alternative signal is a local decision that does not require coordination with any other nodes. Because the signal is always available on both paths, the time to switch to the protected path is very short. The receiver could detect failure or signal degradation and perform an immediate switch. Recovery times of much less than 50 ms are possible. For 1+1 protection to be truly effective, the fiber plant must provide diverse routing.

Figure 6-24 *1+1 Protection*

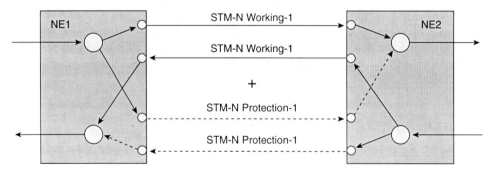

1:1 and 1:N Protection

In a 1:1 or 1:N protection architecture, traffic is carried on the working line until a failure occurs. The protection line is invoked if the working line fails. A schematic of 1:1 and 1:N protection is illustrated in Figure 6-25. In 1:N protection, one protection path is established as a backup for N working paths. If a failure occurs, an APS protocol is used to switch over the traffic to the protection facility. Figure 6-25 shows that the Nth multiple working path has failed and its signal has been bridged to the reserved protection path. Any one of the multiple working paths could be bridged to the protection path, but only one path can be protected at a time. If a failure that affects multiple working lines occurs, a priority mechanism must be configured to identify which line gets protection. 1:N protection supports a revertive option that allows the traffic to automatically switch back to the original working fibers after restoration with preconfigured wait time called *Wait To Restore (WTR)*. In 1:1 protection, there must be one protect line for every working line. Because the protect line does not actually carry traffic when not in use, it is possible for the protect line to carry other nonpriority traffic called *extra traffic*. The extra traffic is dropped if protection is invoked. In 1:N protection schemes, a single line can protect one or more working lines.

Figure 6-25 *1:1 and 1:N Protection*

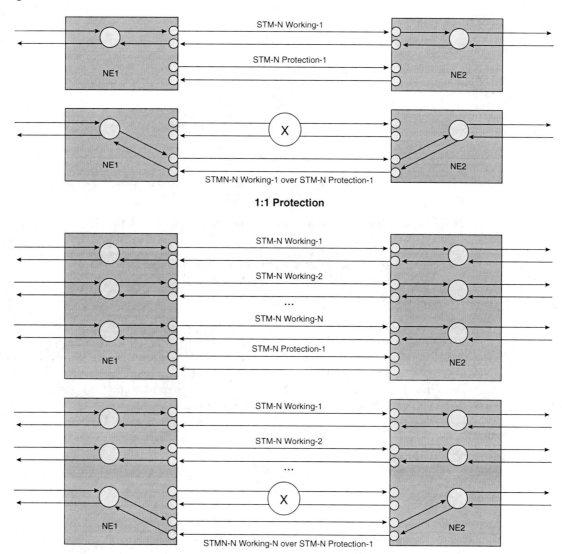

SDH Ring Architectures

This section examines SDH unidirectional and bidirectional ring architectures and examines the differences between two-fiber and four-fiber SDH rings. A comparison is also made between multiplex section (ring) switching versus path (span) switching. SDH provides for three attributes with two choices each, as illustrated in Table 6-13.

256 Chapter 6: SDH Architectures

Table 6-13 *SDH Ring Types*

SDH Attribute	Value
Fibers per link	2-fiber
	4-fiber
Signal direction	Unidirectional
	Bidirectional
Protection switching	Multiplex section switching
	Path switching

Table 6-13 shows various SDH ring configurations that differ in at least one major attribute. The commonly used ring types and topologies are as follows:

- Two-fiber subnetwork connection protection ring (two-fiber SNCP)
- Two-fiber multiplex section-shared protection ring (two-fiber MS-SPRing)
- Four-fiber multiplex section-shared protection ring (four-fiber MS-SPRing)

Unidirectional Versus Bidirectional Rings

In a unidirectional ring, the working traffic is routed over the clockwise spans around the ring, and the counterclockwise spans are protection spans used to carry traffic when the working spans fail. Consider the two-fiber ring schematic presented in Figure 6-26. Traffic from NE1 to NE2 traverses span 1 in a clockwise flow, and traffic from NE2 to NE1 traverses span 2, span 3, and span 4 in a clockwise flow as well. Spans 5, 6, 7 are used as protection spans and carry production traffic when one of the working clockwise spans fail.

Figure 6-26 *Unidirectional Versus Bidirectional Rings*

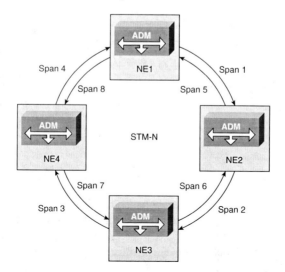

Bidirectional traffic flows can also be illustrated using the schematic of Figure 6-26. In a bidirectional ring, traffic from NE1 to NE2 would traverse span 1 in a clockwise flow. However, traffic from NE2 to NE1 would traverse span 5 in a counterclockwise fashion. If the links between NE1 and NE2 were to fail, traffic between NE1 and NE2 would use the spans between NE2-NE3, NE3-NE4, and NE4-NE1.

Two-Fiber Versus Four-Fiber Rings

Unidirectional and bidirectional systems both implement two-fiber and four-fiber systems. Most commercial unidirectional systems, such as SNCP, are two-fiber systems, whereas bidirectional systems, such as MS-SPRing, implement both two-fiber and four-fiber infrastructures. A two-fiber STM-N unidirectional system with two nodes is illustrated in Figure 6-27. Fiber span 1 carries N working channels eastbound, and fiber span 5 carries N protection channels westbound. For example, an STM-16 system would carry 16 working VC-4s eastbound from NE1 to NE2, while carrying 16 separate protection VC-4s westbound from NE2 to NE1. The SDH transport and POH bytes are carried on both working and protection fiber spans.

Figure 6-27 *Two-Fiber Unidirectional Ring*

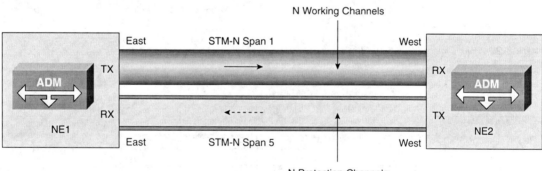

A two-fiber STM-N bidirectional system with two nodes is illustrated in Figure 6-28. On each fiber, a maximum of half the bandwidth or number of channels are defined as working channels, and the other half are defined as protection channels. Fiber span 1 carries (N/2) working channels and (N/2) protection channels eastbound, and fiber span 5 carries (N/2) working channels and (N/2) protection channels westbound. For example, an STM-16 system would carry eight working VC-4s and eight protection VC-4s eastbound from NE1 to NE2, while carrying eight working VC-4s and eight protection VC-4s westbound from NE2 to NE1. Each fiber has a set of SDH transport and POH bytes for the working and protection channels.

A four-fiber STM-N bidirectional system with two nodes is shown in Figure 6-29. Fiber pair span 1A and 5A carry N working channels full duplex east and westbound, while fiber pair span 1B and 5B carry N protection channels full duplex east and westbound. For example, an STM-16 system would carry 16 working VC-4s eastbound from NE1 to NE2 as well as 16 working VC-4s westbound from NE2 to NE1. The same system would also carry 16 protection VC-4s eastbound from NE1 to NE2 as well as 16 protection VC-4s westbound from NE2 to NE1. A set of SDH transport and POH bytes is dedicated either to working or protection channels for the four-fiber ring.

Figure 6-28 *Two-Fiber Bidirectional Ring*

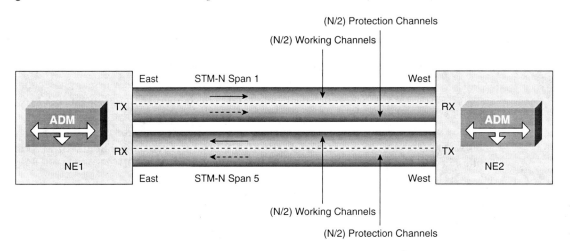

Figure 6-29 *Four-Fiber Bidirectional Ring*

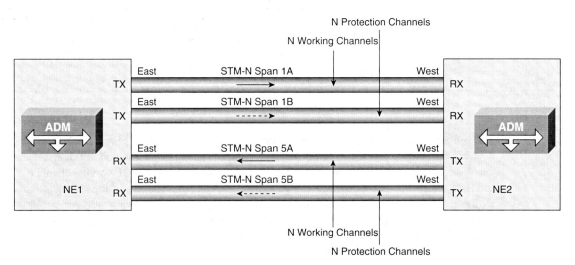

As can be seen for an STM-N fiber system, two-fiber SNCP provides N * VC-4s, whereas two-fiber MS-SPRing provides (N/2) * VC-4s in either direction. Four-fiber MS-SPRing, on the other hand, provides N * VC-4s in either direction. Usually two segment failures will cause a network failure or outage on a two-fiber ring of either type. However, four-fiber systems with diverse routing can suffer multiple failures and still function. Four-fiber systems are widely used for rings spanning large geographical areas or when the traffic being carried on the network is mission critical.

SDH rings are limited to 16 nodes per ring because the K1/K2 bytes that define the source and destination node were defined with only 4 bits each. However, vendors have implemented proprietary mechanisms that use unused bytes from other fields in the SDH header to extend the limit on the number of nodes. For example, the Cisco ONS 15454 SDH uses 4 bits from the K1/K2 fields and 2 additional bits of the K3 byte in MS-SPRing configurations. The K3 byte also carries information on the K1/K2 bytes. Out of the 2 K3 bits, 1 bit is used to define the source and the other bit is used to define the destination node. Use of the 2 additional K3 bits increases the node count to 32 NEs per MS-SPRing ring. In such a case, however, if the span has to pass through third-party equipment, the K3 byte needs to be remapped to an unused SDH overhead byte, such as the E2 or F1 byte.

Path and Multiplex Section Switching

Path switching works by restoring working channels at a level below the entire STM-N capacity in a single protection operation. This means that levels lower than an STM-N, such as VC-3s, VC-12, or VC-11s, can be restored in the event of a failure. Path switching is shown in Figure 6-30. Live protected user traffic is always sent on the working fiber. However, a copy of the protected traffic is also transmitted over the protection fiber. The receiver constantly senses the signal level of both the working and protection fibers. In the event of a fiber cut or signal degradation on the working fiber, the receiver switches to the incoming signal available on the protection fiber. All unprotected traffic is dropped for the duration of the outage. Path switching is mostly implemented on two-fiber SNCP rings.

Figure 6-30 *Path Switching*

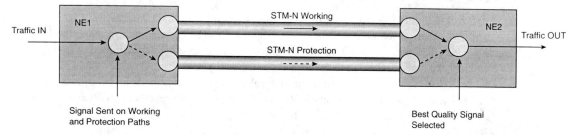

MS switching works by restoring all working channels of the entire STM-N capacity in a single protection operation. The protection channels or fiber are idle while the ring operates normally. MS switching is shown in Figure 6-31. Live protected user traffic is always sent on the working channels or fiber. In the event of a fiber or node failure, the protected traffic is switched to the protection channels or fiber at both ends of the span. Channels within the MS are switched this way, which is why it is called line switching. In the event of a failure, all unprotected traffic being transmitted on the protection link or protected channels is dropped.

This is called *protection channel access (PCA)*, and the traffic carried this way is called *extra traffic*. Carriers typically discount unprotected PCA bandwidth, thereby enabling customers to maintain a more cost-effective network without having to pay for a five-nines service level agreement (SLA). Line-switching systems are able to restore service within 50 ms. Line switching is mostly implemented on two-fiber and four-fiber bidirectional rings.

Figure 6-31 *Multiplex Section Switching*

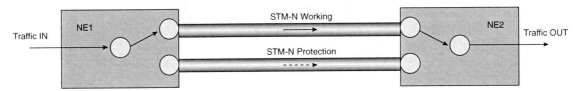

Dual-Ring Interconnect

The dual-ring interconnect (DRI) architecture allows subtending rings sharing traffic to be resilient from a matching node failure perspective. As shown in Figure 6-32, a DRI topology uses two interconnecting matching nodes, DRI node 3 and DRI node 4, to connect the two STM-N rings. If one of the interconnected nodes fails, traffic is routed through the surviving DRI node. The benefit to the service provider is that continuous network operation is maintained even though a node failure has occurred. The DRI topology provides an extra level of path protection between rings. In a DRI configuration, traffic is dropped and continued at the interconnecting nodes to eliminate single points of failure. Each ring protects against failures within itself using path-switched and/or MS-switched protection mechanisms, whereas DRI provides protection against failures at the interconnections. DRI cannot provide protection if both DRI node 3 and DRI node 4 experience simultaneous failure.

As shown in Figure 6-32, a signal input at node 1 destined for node 7 is bridged east and west. The downstream primary eastbound signal passes through node 2 and arrives at the DRI node 3. At DRI node 3, a duplicate copy of the signal is dropped and transmitted to DRI node 4. Similarly, the downstream secondary westbound signal passes through node 5 and arrives at the DRI node 4. At DRI node 4, a duplicate copy of the signal is dropped and transmitted to DRI node 3. The downstream path selector at node 3 always selects the primary downstream signal during steady-state normal operation. However, the downstream path selector at node 4 always selects the secondary downstream signal during steady-state normal operation. The primary downstream signal at node 3 is then continued and transmitted to node 6 on ring 2 that acts as a pass-through node and transmits the signal to node 7. Similarly, the secondary downstream signal at node 4 is then continued and transmitted to node 8 on ring 2 that acts as a pass-through node and transmits the signal to node 7.

Node 7 receives two copies of the downstream signal (primary and secondary). However, the path selector in node 7 always selects the primary downstream signal during steady-state normal operation. A similar process takes place with the primary and secondary upstream signal. Suppose that DRI node 3 fails. In such an event, the path selector at node 7 will switch to the secondary downstream signal. The upstream traffic is not affected, because the primary upstream path is through the surviving node 4.

Figure 6-32 *Dual-Ring Interconnect*

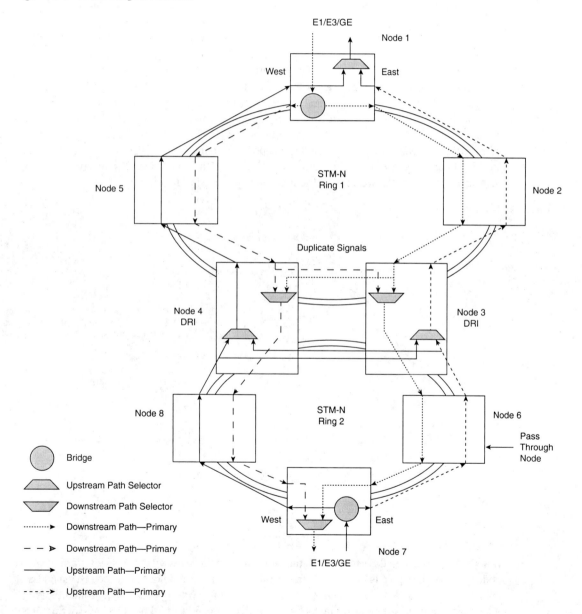

Subnetwork Connection Protection Rings

An SNCP ring is a survivable, closed-loop, transport architecture that protects against fiber cuts and node failures by providing duplicate, geographically diverse paths for each circuit. SNCP provides dual fiber paths around the ring. Working traffic flows clockwise in one direction, and protection traffic flows counterclockwise in the opposite direction. If fiber or node failure occurs in the working traffic path, the receiving node switches to the path coming from the opposite direction. Because each traffic path is transported around the entire ring, SNCPs are best suited for networks where traffic concentrates at one or two locations and is not widely distributed. SNCP capacity is equal to its bit rate. This means that an STM-N SNCP ring will always provide N * VC-4s of capacity. Services can originate and terminate on the same core SNCP ring, or can be passed across a matching node to an access ring for transport to the service-terminating location. Figure 6-33 shows a basic SNCP configuration. This drawing can also be used to explain basic two-fiber SNCP operation with its various subtleties. The schematic illustrates the operation of a two-fiber STM-16 ring using SNCP as its protection mechanism. The outer Fiber 1 is the working fiber that carries traffic in a clockwise direction. The inner Fiber 2 is the protection fiber that carries a *copy* of the working traffic in a counterclockwise direction.

Figure 6-33 *Two-Fiber SNCP*

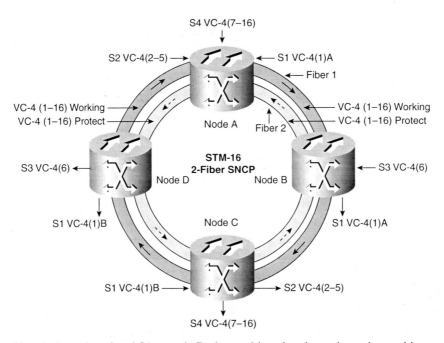

If node A sends a signal S1 to node B, the working signal travels on the working traffic path to node B. The same signal is also sent on the protect traffic path from node A to node B, via nodes D and C. For node B to reply to node A, the signal uses the working VC-4 path around the ring via nodes C and D. Note that signal S1 or VC-4(1)A is the first VC-4 of the STM-16. Signal S1 consumes the entire VC-4 around the ring. Therefore, it is *not* possible for another signal, such as S1 VC-4(1)B, to be transmitted between node C and node D using SNCP protection. However, VC-4(1)B could be transmitted *unprotected* between nodes C and D. This circuit would be dropped in the event SNCP protection is invoked.

Signal S2 is added at node A and dropped at node C. Signal S2 contains VC-4(2–5) or [VC-4, number 2 to number 5 of the STM-16]. This also means that VC-4(2–5) cannot be used for adds or drops at other nodes on the ring. For node C to reply to node A, the signal uses the working VC-4 path around the ring via node D. Signal S3 VC-4(6) is added at node B and dropped at node D, effectively blocking any other adds or drops for VC-4(6) at any of the other nodes. For node D to reply to node B, the signal uses the working VC-4 path around the ring via node A. Finally, signal S4 is added at node A and dropped at node C. Signal S4 contains VC-4(7–16); that means VC-4(7–16) cannot be used for adds or drops at other nodes on the ring. For node C to reply to node A, the signal uses the working VC-4 path around the ring via node D.

As shown in Figure 6-34, if a fiber break occurs on the working Fiber 1 between node A and node B, node B switches its active receiver to the protect signal coming through node C. Signals S1 and S3 would be received on the protection fiber for the duration of the outage. The switchover would happen within the 50-ms SDH restoration time. Signals S2 and S4 would be received on node C via node D on the protection fiber. If there were a fiber cut on the protection Fiber 2, however, the system would continue operating without any disruption. The element management system would detect the fiber cut and report the LOS on the protection fiber. Repairs could be performed on the Fiber 2 without service interruption.

Figure 6-34 *Two-Fiber SNCP Protection*

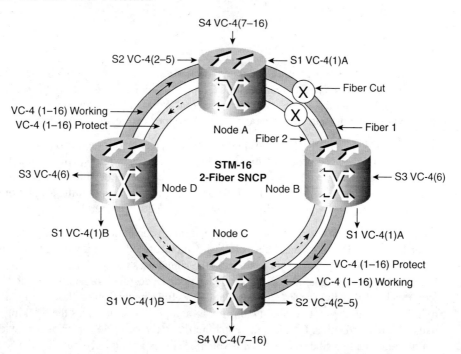

Asymmetrical Delay

As shown in Figure 6-33, any signal from node A to node B traverses a single span. When node B has to reply to node A, however, the signal has to traverse multiple spans via node C and node D. In the case of small metropolitan rings, this does not create any issues. However, for large transcontinental rings, a finite delay could affect voice or data applications. In the case of voice applications, the cumulative delay should not exceed 100 ms. So long as the asymmetric delay does not exceed 100 ms, the human user would not perceive any delay. In the case of data, transport layer windowing comes into play. With asymmetric delay, two end hosts might experience a 40-ms round-trip delay. One host might perceive a 5-ms inbound delay with a 35-ms outbound delay. It would be the exact opposite for the other host. Issues occur when one data application tries to adjust its window size for a 20/20-ms split in delay, while data keeps arriving early (5 ms). The host on the other end experiences exactly the opposite effect, when adjusting its window size for a 20/20-ms split, with data arriving late (25 ms).

Multiplex Section–Shared Protection Rings

MS-SPRing uses bidirectional multiplex section–switched protection mechanisms. MS-SPRing is commonly implemented on two-fiber as well as four-fiber systems. MS-SPRing nodes can terminate traffic that is fed from either side of the ring and are suited for distributed node-to-node traffic applications, such as interoffice networks and access networks. MS-SPRings allow bandwidth to be reused around the ring and can carry more traffic than a network with traffic flowing through one central hub. MS-SPRing supports nonrevertive and revertive protection mechanisms.

Two-Fiber MS-SPRing

In a two-fiber MS-SPRing ring, each fiber carries working and protection VC-3s. In an STM-16 MS-SPRing, as shown in Figure 6-35, for example, VC-4s 1 through 8 carry the working traffic, and VC-4s 9 through 16 are reserved for protection. Working traffic travels clockwise in one direction on one fiber and counterclockwise in the opposite direction on the second fiber.

In Figure 6-35, signal S1 VC-4(1)A added at node A, destined for a drop at node B, typically will travel on Fiber 1, unless that fiber is full (in which case, circuits will be routed on Fiber 2 through nodes C and D). Traffic from node A to node C (or node B to node D) can be routed on either fiber, depending on circuit-provisioning requirements and traffic loads. For node B to reply to node A, the signal uses the working VC-4(1) path on Fiber 2.

Signal S2 VC-4(2–5) added at node A, destined for a drop at node C, typically will travel on Fiber 1 via node B, unless that fiber is full (in which case, the circuit will be routed on Fiber 2 via node D). For node C to reply to node A, the signal uses the working VC-4(2–5) path on Fiber 2 via node B. Signal S3 VC-4(6) added at node B, destined for a drop at node D, typically will travel on Fiber 1 via node C, unless that fiber is full (in which case, the circuit will be routed on Fiber 2 via node A). For node D to reply to node B, the signal uses the working VC-4(6) path on Fiber 2 via node C. It is quite apparent that only VC-4 * 8 worth of bandwidth can be configured on a two-fiber STM-16 MS-SPRing. This is not entirely true. Unlike SNCP, the provisioning of VC-4(1) does not consume the entire first VC-4 of the STM-16 around the ring. Bandwidth is reusable, as shown by S1 VC-4(1)B in Figure 6-35, and can be provisioned between nodes C and D. With careful bandwidth-capacity planning, MS-SPRing could be quite efficient.

Figure 6-35 *Two-Fiber MS-SPRing*

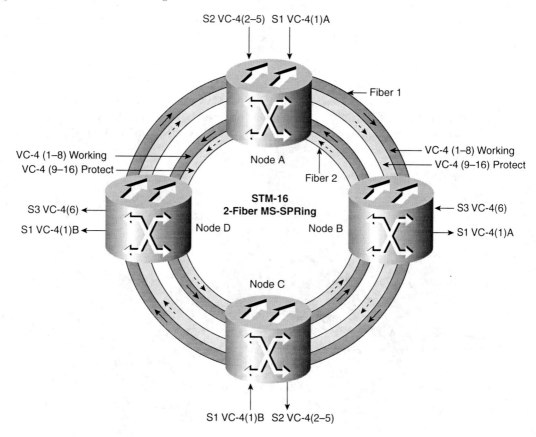

| NOTE | The bidirectional bandwidth capacities of two-fiber MS-SPRings is the STM-N rate divided by two, multiplied by the number of nodes in the ring, minus the number of pass-through VC-4 circuits. |

The SDH K1 and K2 bytes carry the information that governs MS-SPRing protection switching. Each MS-SPRing node monitors the K bytes to determine when to switch the SDH signal to an alternate physical path. The K bytes communicate failure conditions and actions taken between nodes in the ring. If a break occurs on one fiber, working traffic targeted for a node beyond the break switches to the protect bandwidth on the second fiber. The traffic travels in reverse direction on the protect bandwidth until it reaches its destination node. At that point, traffic is switched back to the working bandwidth.

As shown in Figure 6-36, if a break occurs in Fiber 1 between node A and node B, signal S1 VC-4(1)A that would normally travel between node A and B using VC-4(1) of Fiber 1 would MS switch to VC-4(9) of Fiber 2 and reach node B via nodes D and C for the duration of the outage. The switchover would happen within the 50-ms SDH restoration time. Signal S2 VC-4(2–5) added at node A and destined for node C would also be affected. S2 would be MS switched to VC-4(10–13) of Fiber 2 and would reach node C via node D. Signal S3 VC-4(6) would not be affected. Now consider the case where Fiber 1 is intact and there is a break in Fiber 2 between nodes A and B. In such a case, the return path for signal S1 VC-4(1)A between node B and node A is lost. An MS switch would occur and signal VC-4(1)A would switch to VC-4(9) of Fiber 1 and reach node A via nodes C and D. The return path for signal S2 VC-4(2–5) between node C, destined for node A, would also be affected. Node C would transmit signal S2 VC-4(2–5) back to node A over Fiber 2. However, the fiber cut on Fiber 2 (between nodes A and B), detected by node B, would cause all return traffic to node A to be MS switched to VC-4(10–13) of Fiber 1 and retransmitted to node A via nodes C and D.

Figure 6-36 *Two-Fiber MS-SPRing Protection*

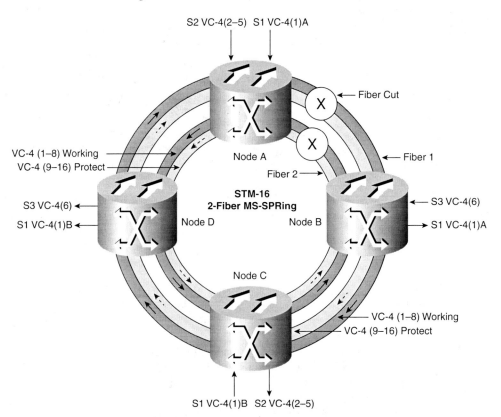

Finally, consider a case of a dual fiber cut of both Fiber 1 and Fiber 2 between nodes A and B. In such a case, signal S1 VC-4(1)A added at node A and destined for node B would be MS switched to VC-4(9) of Fiber 2 and would reach node B via nodes D and C. The return path for signal S1 VC-4(1)A between node B and node A would MS switch to VC-4(9) of Fiber 1 and reach node A via nodes C and D. Signal S2 VC-4(2-5) added at node A and destined for node C would be MS switched to VC-4(10–13) of Fiber 2 and would reach node C via node D. Node C would transmit the return signal S2 VC-4(2–5) back to node A over Fiber 2. However, the fiber cut on Fiber 2 (between node A and B), detected by node B, would cause all return traffic to node A to be MS switched to VC-4(10–13) of Fiber 1 and retransmitted to node A via nodes C and D. All unprotected traffic carried over the protection VC-4s is dropped in the event of an MS switch.

MS-SPRing Node Failure

MS-SPRing restoration gets complex in the event of a node failure. MS-SPRing uses a protection scheme called shared protection. Shared protection is required because of the construction of the MS-SPRing ring and the reuse of VC-4s around the ring. This creates a situation in which the VC-4s on a protection fiber cannot be guaranteed to protect traffic from a specific working VC-4. Shared protection, which provides MS-SPRing its capability to reuse bandwidth, brings with it additional problems when a node failure occurs in a MS-SPRing ring. Consider the MS-SPRing schematic in Figure 6-37. This schematic shows a complete failure of node D. Trace the path of signal S3 as it gets added on to node B with a destination node D. Signal S3 VC-4(6) gets sent out to node D on Fiber 1 and proceeds to node C. Node C has sensed an LOS from failed node D and reroutes S3 on to Fiber 2 as signal S3 VC-4(12). Signal S3 passes via node B and arrives at node A. However, because node A cannot deliver this traffic to node D, it places S3 on Fiber 1 as S3 VC-4(6). This signal gets dropped off at node B, because VC-4(6) already has a connection from node A to node B (signal S4). This event results in the traffic being delivered to the wrong node and is called a *misconnection*. In some situations, it is possible that bridging traffic after a node failure could also lead to a misconnection.

MS-SPRing misconnections can be avoided by using the squelching mechanism. The squelching feature uses automatically generated squelch maps that require no manual record-keeping to maintain. Each node maintains squelch tables to know which connections need to be squelched in the event of a node failure. The squelch table contains a list of inaccessible nodes. Any traffic received by a node for the inaccessible node is never placed on the fiber and is removed if discovered.

Squelching involves sending the AIS in all channels that normally terminated in the failed node rather than real traffic. The misconnection is avoided by the insertion of an AIS path by nodes A and C into channel VC-4(6). In an AIS path, all the bits belonging to that path are set to 1 so that the information carried in that channel is invalidated. This way, node B is informed about the error condition of the ring, and a misconnection is prevented. Misconnection can occur only in MS-SPRing when a node is cut off and traffic happens to be terminated on that node from both directions on the same channel (VC-4). In some implementations, the path trace might also be used to avoid this problem. If node B monitors the path trace byte, it will recognize that it has changed after the misconnection. This change should be a sufficient indication that a fault has occurred, and that traffic should not be terminated.

Figure 6-37 *MS-SPRing Node Failure*

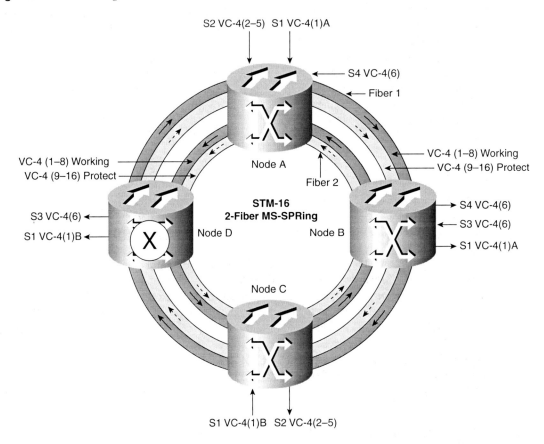

Four-Fiber MS-SPRing

Four-fiber MS-SPRings double the bandwidth of two-fiber MS-SPRings. As shown in Figure 6-38, two fibers are allocated for working traffic and two fibers are allocated for protection. Signal S1 from node A to node B would use VC-4(1) of the working Fiber 1, and the return path from node B to node A would use VC-4(1) of the working Fiber-3. Signal S2, added at node A and destined for node C, would use VC-4(2–5) of the working Fiber 1, and would use VC-4(2–5) of the working Fiber-3 for its return path from node C to node A.

Signal S3, added at node B and destined for node D, would use VC-4(6) of the working Fiber 1 via node C, and would use VC-4(6) of the working Fiber-3 for its return path from node D to node B, via node C. Signal S4, added at node A and destined for node C, would use VC-4(7–12) of the working Fiber 1, and would use VC-4(7–12) of the working Fiber-3 for its return path from node C to node A.

Figure 6-38 *Four-Fiber MS-SPRing*

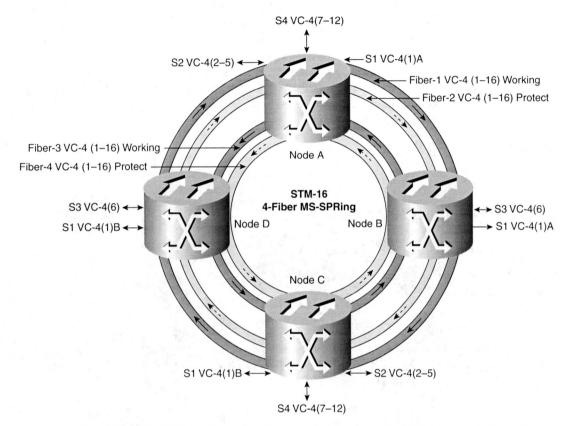

Four-fiber MS-SPRing allows path (span) switching as well as MS (ring) switching, thereby increasing the reliability and flexibility of traffic protection. Path (span) switching occurs when a working span fails. Traffic switches to the protect fibers between the nodes and then returns to the working fibers. Multiple span switches can occur at the same time. MS (ring) switching occurs when a span switch cannot recover traffic, such as when both the working and protect fibers fail on the same span. In an MS (ring) switch, traffic is routed to the protect fibers throughout the full ring.

As shown in Figure 6-39, if the working fiber pair between node A and B fails, all working traffic between these nodes is shunted onto the protection fiber pair. Any unprotected traffic mapped between other nodes on the ring is unaffected by this outage.

Signal S1 from node A to node B would use VC-4(1) of protection Fiber 2, and the return path from node B to node A would use VC-4(1) of protection Fiber-4. Signal S2, added at node A and destined for node C would use VC-4(2–5) of protection Fiber 2 between node A and B, after which it would revert to VC-4(2–5) of the working Fiber 1 between nodes B and C. Signal S2 would use VC-4(2–5) of the working Fiber-3 for its return path from node C to node B, after which it would use VC-4(2–5) of protection Fiber-4 between nodes B and A. Signal S3 would be unaffected. However, signal S4, added at node A and destined for node C, would use VC-4(7–12) of protection Fiber 2 between node A and node B, after which it would revert to VC-4(7–12) of the working Fiber 1 between nodes B and C. Signal S4 would use VC-4(7–12) of the working Fiber-3 for its return path from node C to node B, after which it would use VC-4(7–12) of protection Fiber-4 between nodes B and A. Four-fiber MS-SPRing ring switching is shown in Figure 6-40. If both fiber pairs between node A and B fail, all working traffic between these nodes is wrapped onto the protection fiber pairs. Any unprotected traffic mapped between other nodes on the ring is preempted and dropped because all the protection pairs are used during the outage.

Figure 6-39 *Four-Fiber MS-SPRing Span Switch*

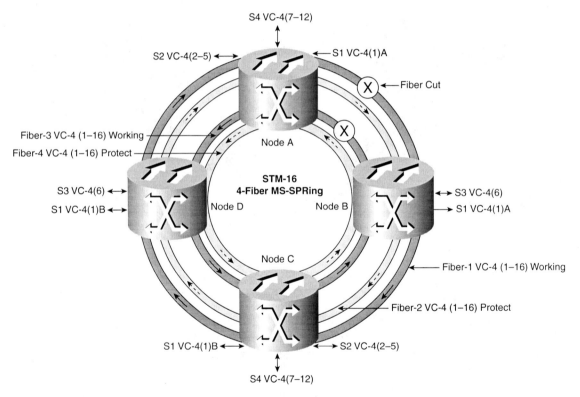

Signal S1 from node A to node B would use VC-4(1) of protection Fiber-4 via nodes D and C, and the return path from node B to node A would use VC-4(1) of protection Fiber 2 via nodes C and D. Signal S2, added at node A and destined for node C, would use VC-4(2–5) of protection Fiber-4 via node D. On its return path from node C to node A, signal S2 would use VC-4(2–5) of the working Fiber-3 between node C to node B. Node B would cause a wrap and switch the traffic to VC-4(2–5) of protection Fiber 2 for a drop at node A via nodes C and D. Signal S3 would be unaffected. Signal S4, added at node A and destined for node C, would use VC-4(7–12) of protection Fiber-4 via node D. On its return path from node C to node A, signal

S4 would use VC-4(7–12) of the working Fiber-3 between node C to node B. Node B would cause a wrap and switch the traffic to VC-4(7–12) of protection Fiber 2 for a drop at node A via nodes C and D.

Figure 6-40 *Four-Fiber MS-SPRing Ring Switch*

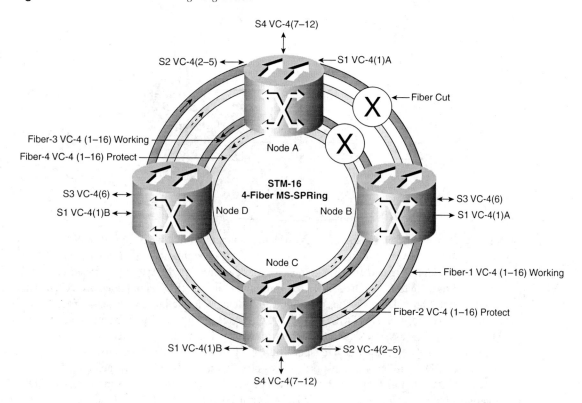

SDH Network Management

SDH NEs need OAM&P support to be managed by carriers and service providers. The OAM&P of an NE is the task of its EM. EMs are device-specific and vary by vendor. In a typical service provider environment, there could be multiple EMs. The integration of the various EMs along with fault management (FM), performance management (PM), accounting management (AM), security management (SM), configuration management (CM), and trouble ticketing and billing applications is the function of the OSS. Multiple OSS systems that manage the data communications network (DCN) constitute the TMN. The TMN has been standardized by the ITU-T under Recommendation M.3010. SDH devices can be remotely managed through the use of in-band management channels in the RSOH and MSOH, known as DCCs. Figure 6-41 shows an intercarrier TMN model. OSS-1 is operated by Carrier 1 and OSS-N is operated by Carrier N. The OSS accesses the DCN via a gateway network element (GNE).

Figure 6-41 *OSS and TMN Schematic*

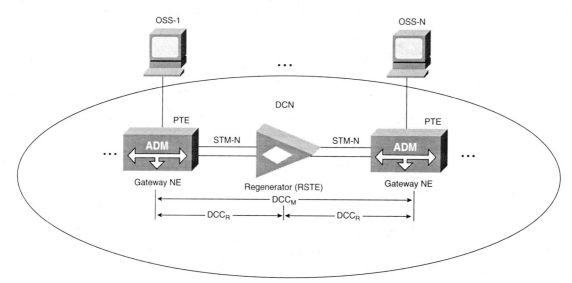

The DCC channels can transport operations and management messages that let OSS systems comply with the TMN specification. However, many SDH equipment vendors have established proprietary element management schemes, and there is little interoperability between vendors in the use of these bytes. The DCC bytes, D1 through D3 in the RSOH, are known as DCC_R. The 3 DCC_R bytes provide a 192-kbps communications channel. The DCC bytes, D4 through D12 in the MSOH, are known as DCC_M. The 9 DCC_M bytes provide a 576-kbps communications channel. Most SDH systems use DCC_R bytes for management purposes and don't use the DCC_M bytes by themselves. The Cisco Transport Manager (CTM) enables service providers to manage their Cisco SDH and optical transport devices collectively under one management system. Cisco also uses a craft tool and element management system (EMS) for comprehensive SDH and optical transport management called Cisco Transport Controller (CTC). Cisco has developed its management application based on an IP stack coupled with an Open Shortest Path First (OSPF)-based topology discovery mechanism. Furthermore, Cisco ONS devices can tunnel their DCC_R bytes through an ONS network. The bytes are tunneled by copying the DCC_R bytes into 3 of the DCC_M bytes. Because there are 3 DCC_R bytes and 9 DCC_M bytes, the ONS devices have the capability to transport traffic from 3 different SDH networks simultaneously across any given span. The DCC_R bytes are restored as they leave the ONS network, thus permitting interoperability with non-ONS networks. The Cisco EMS supports Transaction Language 1 (TL-1), Common Object Request Broker Architecture (CORBA), and SNMP for OAM&P purposes.

Summary

Communication between various localized networks is complex due to differences in digital signal hierarchies, encoding techniques, and multiplexing strategies. The SDH format allows different types of signal formats to be transmitted over the fiber-optic cable. The PDH includes all ITU TDM hierarchies including the E-carrier and T-carrier systems. STM-N refers to the electrical and optical SDH transmission characteristics of an Nth-level transmission link. The four SDH optical interface layers are the path, multiplex section (MS), regenerator section (RS), and the photonic layers. Path termination, multiplex section termination, and regenerator section termination occur at these various layers.

The STM-1 signal consists of the transport overhead, POH, and the payload container. SDH provides substantial overhead information, allowing simpler multiplexing and greatly expanded OAM&P capabilities. The SDH transport overhead consists of the MSOH, RSOH, and AU-N pointer bytes. SDH alarms are defined as anomalies, defects, and failures. An anomaly is the smallest discrepancy that can be observed between the actual and desired characteristics of an item. The occurrence of a single anomaly does not constitute an interruption in the ability to perform a required function. If the density of anomalies reaches a level at which the ability to perform a required function has been interrupted, it is termed a defect.

SDH NEs include RS-, MS-, and path-terminating equipment. The SDH NEs integrate to form the DLCs supporting various SDH architectures and topologies. Some of these NEs include regenerators, TMs, ADMs, BDCS, WDCS, and the IDLC, which consists of intelligent remote digital terminals and digital switch elements called integrated digital terminals. SDH topologies and protection architectures center around network survivability and sub-50-ms service restoration. SDH topologies include the linear ADM point-to-point, point-to-multipoint, hub, ring, and mesh topology. The SDH ADM and DCS are NEs used to build such topologies. The ring and mesh topologies offer various survivability and protection mechanisms. Fiber routing and diversity schemes also provide redundancy. Several ring configurations and protection mechanisms provide various levels of redundancy.

Path switching works by restoring working channels at a level below the entire STM-N capacity in a single protection operation. MS switching works by restoring all working channels of the entire STM-N capacity in a single protection operation. Two-fiber SNCP, two-fiber MS-SPRing, and four-fiber MS-SPRings are some of the commonly used ring protection mechanisms. SDH NEs need OAM&P support to be managed by carriers and service providers. The OAM&P of an NE is the task of its EM. EMs are device-specific and vary by vendor. In a typical service provider environment, there could be multiple EMs. The integration of the various EMs along with the integrated NMS constitutes the OSS.

This chapter includes the following sections:

- **Ethernet Services**—This section discusses the ubiquitous nature of Ethernet and examines the need to carry Ethernet end to end in a consistent packet format from start to finish throughout the entire transport path. This would eliminate the need for additional layers of protocol and synchronization that result in extra costs and complexities. This section also compares various metropolitan-area network (MAN) technologies, such as Gigabit Ethernet (GE) and multiservice provisioning platforms (MSPPs). It provides a fair comparison of both technologies along with an introduction to the shared packet ring (SPR) and resilient packet ring (RPR) features that can be implemented on Synchronous Optical Network/Synchronous Digital Hierarchy (SONET/SDH) MSPP devices.

- **Ethernet over SONET/SDH**—SPR or RPR technologies can be leveraged to carry Ethernet services over SONET/SDH. SPRs are built by concatenating SONET STS-1s or SDH VC-3s. This section also introduces various Ethernet over SONET/SDH encapsulation schemes, such as Ethernet over SONET/SDH using ANSI T1X1.5 147R1 Generic Framing Procedure (GFP) headers, Ethernet over Packet over SONET/SDH using ITU-T x.86 Link Access Procedure (LAPS) headers, and Ethernet over Multiprotocol Label Switching (MPLS).

- **Shared Packet Ring**—SPR calls for the provisioning of optical transport pipes at Layer 1.5 between Ethernet switch cards present in the ONS nodes. Ethernet switches in the various nodes can be daisy chained to form an SPR. This section discusses SPR architecture with respect to the Cisco implementation of SPR. The various design specific features of the E-Series and ML-Series cards are explained in great detail that will provide the reader with an idea as to design constraints involved while designing SPRs with Cisco equipment.

- **Resilient Packet Ring**—This section discusses the IEEE 802.17 standard for RPR implementation in great detail. Data is carried in packets rather than over time-division multiplexed (TDM) circuits. RPR networks retain many of the performance characteristics, such as protection, low latency, and jitter of SONET/SDH. RPR architectures are highly scalable, very reliable, and easy to manage in comparison to legacy point-to-point topologies. RPR combines the advantages of both SONET/SDH and Ethernet and allows the support of the newer services while simultaneously supporting traditional carrier-class features, such as resiliency, restoration, and quality of service (QoS).

CHAPTER 7

Packet Ring Technologies

Ethernet Services

Ethernet services are ubiquitous in nature and could someday replace ATM as the technology of choice, thereby changing the utopian adage of "ATM everywhere" to "Ethernet everywhere." IPv4 and/or IPv6 would obviously ride on Ethernet fulfilling the requirement as a routable protocol that could provide corporate intranet connectivity as well as full compatibility with the public Internet. Ethernet has become the leading contender for packet transport in the metro. As an alternative to circuit-oriented technology, such as SONET/SDH and ATM, optical Gigabit Ethernet (GE) technology, which is capable of supporting fiber spans of more than 50 miles, is a good match for carrying IP data traffic. Because nearly all IP packets begin and end their trip across the Internet as Ethernet frames, carrying data in a consistent packet format from start to finish throughout the entire transport path eliminates the need for additional layers of protocol and synchronization that result in extra costs and complexities. In addition to efficient handling of IP packets, Ethernet has the advantages of familiarity, simplicity, and low cost.

Currently, many technologies are available to build broadband MANs and WANs, such as ATM, GE, Dynamic Packet Transport (DPT), Packet over SONET (PoS), multiservice provisioning platform (MSPP), and so forth. The two leading technology contenders are GE and MSPP. The issue becomes a choice of technology, wherein each technology has its own merits and demerits. Most incumbent local exchange carriers (ILECs) and inter-exchange carriers (IXCs) have a huge SONET/SDH installed base with dark fiber laid out in rings, which does not lend itself very well to the implementation of MANs using GE, for example. Furthermore, the choice of GE as a MAN technology forces the use of Voice over IP (VoIP) for voice transport or, worse still, forcing TDM over IP. GE also has inherent deficiencies, including Ethernet unfairness of bandwidth allocation for downstream switches in a ring topology and a dependency on the Spanning Tree Protocol (STP) algorithm or its variations for loop prevention and protection. Ethernet also suffers from poor jitter and latency control.

The move toward Layer 3 virtual private networks (VPNs) will lead to continued use of technologies such as MPLS and Internet Protocol Security (IPSec). (Refer to the Cisco Press title *Advanced MPLS Design and Implementation* for a detailed discussion and comparison of various Layer 2 and Layer 3 VPN technologies.) Within the metro space, however, the paradigm has changed a bit. Most service providers offer Ethernet services in increments of 10, 100, and 1000 bits per second (bps). The core technologies that can meet such bandwidth requirements include Gigabit switches and optical MSPPs.

The optical MSPP devices, such as the Cisco ONS 15000 family of products, offer service providers the solid reliability of SDH/SONET along with transport of literally any kind of service including 100-Mbps Ethernet, GE, TDM, and so on. They can also carry VoIP over Ethernet for future voice-packet applications. In terms of scalability, the MSPP devices offer dense wavelength-division multiplexing (DWDM), whereby bandwidth can be scaled to multiples of OC-192 (currently 32 * OC-192 or 320 Gbps). Data can be carried using conventional SONET/SDH-TDM or over SPRs or RPRs. Voice can be carried over SONET/SDH-TDM or over IP-Ethernet using differentiated service codepoint (DSCP) for class of service (CoS). Customer VPNs can be created using double VLAN tagging, which scales up to 16 million customer VPNs. The MSPP has emerged as the technology of choice for service provider and large enterprise network applications in which there is a requirement to carry GE services, 10/100 Ethernet services, and TDM with full SONET/SDH redundancy and oversubscription for voice and for data. The MSPP technology is tried, proven, and it works. MSPP systems scale from 1.5-Mbps to 10-Gbps transport in a single shelf. The Cisco ONS 15000 MSPP systems can also be DWDM enabled for an even more efficient use of the fiber plant, with bandwidth scalability up to 640 Gbps and beyond. Pure GE switches cannot provide the massive bandwidth, SONET/SDH redundancy, voice, and TDM services with hard QoS guarantees that carriers depend upon to honor their service level agreement (SLA) commitments.

MSPP technology uses SONET/SDH as a Layer 1.5 transport mechanism to carry TDM traffic and Ethernet traffic with specific switching functions and VLAN capability. The technology supports traditional SONET/SDH circuits and RPR or RPR-like functionality (SPRs) for transporting Ethernet over SONET/SDH. VPNs can be created using 802.1Q tagging or TDM pipes. The MSPP network can also offer double VLAN tagging (tag stacking), which increases the maximum number of configurable VLANs or VPNs to 16,777,216 versus only 4096 for conventional GE solutions. SPR uses a combination of SONET/SDH bidirectional line-switched ring (BLSR)/multiplex section-shared protection ring (MS-SPRing) or unidirectional path-switched ring (UPSR)/subnetwork connection protection (SNCP) technology along with conventional 802.1D bridging or per-VLAN Fast Spanning Tree Protocol (IEEE 802.1w) to provide a stable Layer 1.5 and Layer 2 infrastructure. VLANs provide the VPN partitioning mechanism for customer-closed user groups.

Ethernet over SONET/SDH

MSPPs use SPRs or RPRs to carry Ethernet data services over SONET/SDH. SPRs are built by concatenating SONET STS-1s or SDH VC-3s. For example, an OC-48 UPSR SONET ring has 48 protected STS-1s. Out of these 48 STS-1s, 12 of them could be concatenated to form an SPR-12 that would interconnect the 802.1D bridges on the MSPP nodes to each other. The remaining 36 STS-1s are free to be used for TDM traffic or for creating other SPRs. RPR technology uses a new Layer 2 protocol for the MAN and WAN and combines the advantages of both SONET/SDH and Ethernet.

Ethernet over SONET/SDH Encapsulation

There are three types of Ethernet over SONET/SDH encapsulations:

- Ethernet over SONET/SDH using ANSI T1X1.5 147R1 Generic Framing Procedure (GFP) headers
- Ethernet over Packet-over-SONET/SDH using ITU-T x.86 LAPS headers
- Ethernet-over-MPLS. Transport is achieved using STS mapping or SPR technology

Cisco supports Ethernet over Packet over SONET/SDH using ITU-T x.86 LAPS for the ONS 15000 series MSPP. Transport is achieved using STS mapping or SPR technology. Ethernet can be transported over SONET/SDH using one of the two possible mechanisms or a combination of both:

- SPRs
- RPRs

Shared Packet Ring

MSPPs support SPRs to provide Ethernet and packet transport over a SONET/SDH infrastructure. The implementation of this technology varies from vendor to vendor. The Cisco implementation of SPR calls for the provisioning of optical transport pipes at Layer 1.5 between Ethernet switch cards present in the various ONS nodes. At present Cisco supports 10-Mbps, 100-Mbps, and Gigabit Ethernet on the Ethernet modules. This technology allows the provisioning of bandwidth on the SONET/SDH ring for packet transport by statistically multiplexing Ethernet traffic onto a shared packet ring (circuit) that each MSPP node can access. This section discusses SPR implementation for the Cisco ONS nodes.

To understand SPR, the reader must understand the concept of logical SONET/SDH STS-1/VC-3 ports. Each card has logical or "virtual" interfaces that are mapped to SONET/SDH optical interfaces for transport with other services between network elements. Each Ethernet card used in the ONS MSPP has physical Ethernet interfaces and logical SONET STS-1 or SDH VC-3 ports. The STS-1/VC-3 ports are logical and are implemented via means of backplane cross-connections to the STS/VC-3 matrix. The ONS ML-Series Ethernet cards have up to 48 logical SONET/SDH STS-1/VC-3 ports on its back end. This means that the card can support up to STS-24c/VC-4-8c eastward and up to STS-24c/VC-4-8c westward. As illustrated in Figure 7-1, the Ethernet card in each ONS node can be perceived as a switch that has local Ethernet electrical interfaces that would connect to customer premises equipment (CPE), such as routers, and logical STS-N/VC-N ports that connect the switches to other Ethernet switches present in other ONS nodes.

In this way, it is possible to daisy chain the Ethernet cards in the various ONS nodes to form an SPR. The SPR shown in Figure 7-1 actually consists of four separate bidirectional Ethernet circuits.

The first circuit is built between nodes 1 and 2, and the second circuit is built between nodes 2 and 3. The third circuit is built between nodes 3 and 4, and the last circuit is provisioned between nodes 4 and 1. The concatenation granularity for an SPR is STS-1/VC-3. The circuits in an SPR span are $N * $ STS-1/VC-3. Traffic can be differentiated within the SPR using VLANs. SPRs are built by daisy chaining ONS nodes with Ethernet cards around a ring. The SONET/SDH links serve only to act as an optical transport mechanism. The Ethernet cards essentially act as Layer 2 Ethernet bridges. The CPE side ports are assigned to VLANs that are typically assigned to customers.

Figure 7-1 *SPR Schematic*

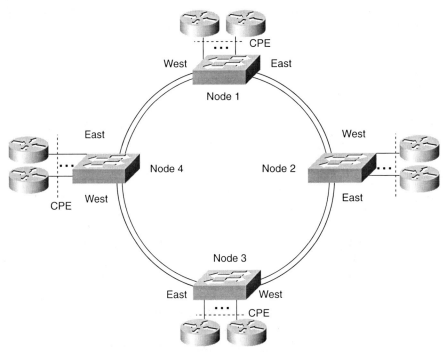

Ethernet uses VLAN identifiers to create closed user groups. Customers will be assigned VLAN IDs as a VPN mechanism for keeping their data secure over the carrier MSPP infrastructure. As shown in Figure 7-2, the 12 bits of VLAN identifier (802.1Q) or tag permits a theoretical maximum of 4096 VLANs. However, using the mechanism of 24-bit double tagging, a theoretical maximum of 16,777,216 VLANs is possible. The ONS 15000 series MSPPs work with Ethernet devices that do and do not support IEEE 802.1Q tagging. The ONS 15000 series MSPPs support VLANs that provide private network service across a SONET/SDH backbone. You can define specific Ethernet ports and SONET/SDH STS channels as a VLAN group. VLAN groups isolate subscriber traffic from users outside the VLAN group and keep "outside" traffic from "leaking" into the VPN. Each IEEE 802.1Q VLAN represents a different logical network, and multiple VLANs can be trunked over the SPR.

Figure 7-2 *VLAN Identifiers and Double Tagging*

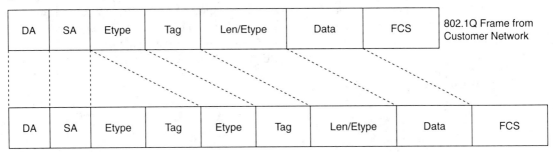

DA = Destination Address
SA = Source Address
Len/Etype = Length/EtherType
FCS = Frame Check Sequence

The SPR Ethernet cards can be modeled as Ethernet bridges, as shown in Figure 7-3. The bridges clearly indicate the possibility of bridging loops. Loops are prevented by running an instance of the STP protocol. The ONS 15000 series MSPP nodes use the IEEE 802.1D standard to provide STP. STP detects and eliminates network loops; the ONS 15000 series MSPP uses STP internally and externally. Internally, it detects multiple circuit paths between any two network ports and blocks ports until only one path exists. The single path eliminates possible bridge loops.

Figure 7-3 *SPR Bridge and STP Model*

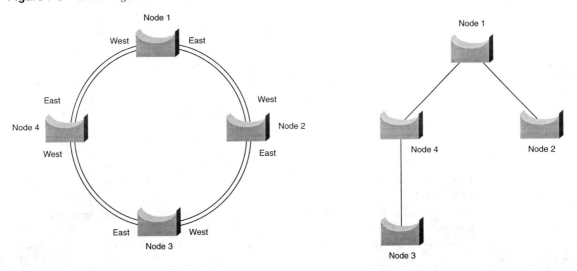

As shown in Figure 7-4, you can enable STP externally at the Ethernet port level to allow parallel connections to external CPE networking equipment. STP will allow only one loop-free CPE-to-CPE link to be established at any given time. You can disable STP protection on a circuit-by-circuit basis on the ONS Ethernet. However, note that the ONS runs STP on the optical ports by default. The electrical CPE-side STP can be configured, but only if needed.

Figure 7-4 *External SPR Bridge and STP Model*

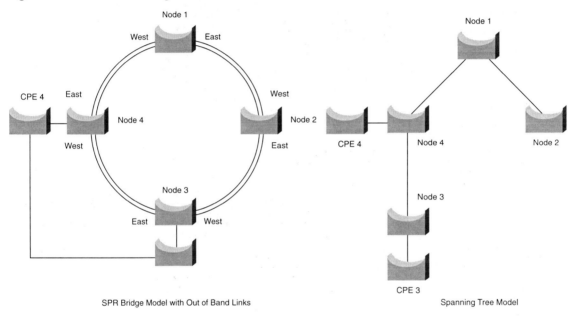

The ONS 15454 MSPPs have various kinds of Ethernet cards. The main genres include the E-Series, ML-Series, and G-Series cards. Chapter 8, "Multiservice SONET and SDH Platforms," discusses the various ONS platform architectures and card types in detail. From an STP perspective, however, the entire ONS node is perceived as a single bridge, even though it might contain multiple E-Series cards. For ML-Series cards, however, each card is accounted for as a discrete bridge, and STP uses each and every ML-Series card for its STP computation.

NOTE It is important to remember that in the case of the E-Series cards, the Timing and Control 2 (TCC2) common card actually runs the spanning-tree engine with a single instance of STP per node. Each E-Series card plugged into the chassis maintains only a copy of its Layer 2 forwarding table. In the case of the ML-Series cards, however, the individual cards maintain copies of the Layer 2 forwarding table and they run an instance of STP.

The ONS ML-Series Ethernet cards implement transparent bridging along with the enhanced functionality of Ethernet switching. Transparent bridges are so named because their presence and operation are transparent to network hosts. When transparent bridges are powered on, they learn the network's topology by analyzing the source address of incoming frames from all attached networks. If a bridge sees a frame arrive on port 1 from host A, for example, the bridge concludes that host A can be reached through the network connected to port 1. Through this process, transparent bridges build a table. The bridge uses its table as the basis for traffic forwarding. When a frame is received on one of the bridge's interfaces, the bridge looks up the frame's destination address in its internal table. If the table contains an association between the destination address and any of the bridge's ports other than the port on which the frame was received, the frame is forwarded out the indicated port. If no association is found, the frame is flooded to all ports except the inbound port. The transparent-bridge algorithm fails when multiple paths of bridges exist between any two LAN elements in the network. STP designates a loop-free subset of the network's topology by placing those bridge ports (that if active would create loops) into a standby (blocking) condition. Blocked bridge ports are activated in the event of primary link failure, providing a new path through the network. In this regard, STP does double duty as a Layer 2 protection mechanism.

On the ONS 15454s, per-VLAN STP is supported. STP is always active on the optical line interfaces by default. This can be turned *off* if needed. On the user-interface side, STP can be provisioned on a per-port basis to allow for redundant connections if required. One of the issues associated with STP is the length of time associated with network reconvergence in the event of a connection failure. An STP convergence could take 30 to 50 seconds when circuits are first provisioned, and 30 to 50 seconds in the event of a span failure (according to the 802.1D STP specification). By using SONET/SDH layer protection along with STP (such as UPSR/SNCP or BLSR/MS-SPRing), however, you can eliminate data loss during a failover. The 50ms switch time is sufficient to allow traffic to be redirected and communication to be maintained between nodes, such that the STP process is not aware that a failure has occurred.

The ML-Series uses 802.1w to greatly increase the convergence of the STP protocol. For example, BPDU frames are sent immediately in the event of link-failure detection to initiate STP on adjacent and downstream-upstream nodes. In addition, forwarding ports continue to forward, in broadcast mode when necessary, to reduce the impact of STP convergence. Finally, STP processes in one ring do not cause STP processes in other rings, even if the VLANs traverse rings.

SPR Design Constraints

Apart from understanding SPR as described earlier, it is also important to understand the ONS equipment available to build an SPR. This section examines the various subtleties and caveats that one would encounter while designing an SPR using Cisco ONS 15454 equipment. I have considered the ONS 15454 as the design platform, because the 15454 is the most widely deployed MSPP with maximum versatility at the metro edge and core. The 15454 operates with the following common cards: the XC, XCVT, and XC10G cross-connect fabric cards. It also uses the TCC2 common card. The ONS 15454 chassis and card architecture are discussed in greater detail in Chapter 8. The ONS 15454 uses E100T-12 and E100T-G cards for Ethernet (10 Mbps) and Fast Ethernet (100 Mbps). Each card provides twelve 10/100BASE-T autosensing Ethernet interfaces. The ports autoconfigure to operate at either half or full duplex and can determine whether to enable or disable flow control. These ports can also be configured manually.

The E100T-G card operates with the XC, XCVT, and XC10G cross-connect cards. The E100T-12 operates with the XC or XCVT and is incompatible with the XC10G. The ONS 15454 uses the E1000-2 and E1000-2-G cards for GE. These cards provide two GE ports with full-duplex operation. The E1000-2-G card operates with the XC, XCVT, and XC10G cross-connect cards. The E1000-2 operates with the XC and XCVT and is incompatible with the XC10G.

As shown in Figure 7-5, the E-Series card has 12 logical SONET/SDH STS-1/VC-3 ports on its back end. This means that the E-Series card can support up to STS-6c/VC-4-2c eastward and up to STS-6c/VC-4-2c westward. The E-Series cards have 12 PoS ports that have the flexibility to be grouped as the following SPRs:

- 6 groups * STS-1/VC-3 (east) + 6 groups * STS-1/VC-3 (west)
- 2 groups * STS-3c/VC-4 (east) + 2 groups * STS-3c/VC-4 (west)
- 1 group * STS-6c/VC-4-2c (east) + 1 Group * STS-6c/VC-4-2c (west)

Figure 7-5 *E-Series Card Ports*

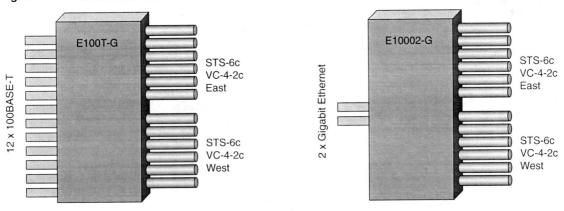

The ML-Series cards, on the other hand, do not have such flexibility and can build only two PoS groups per ML-Series card. As illustrated in Figure 7-6, the ML-Series card has 48 logical SONET/SDH STS-1/VC-3 ports on its back end. The ML-Series card can support up to STS-24c/VC-4-8c eastward and up to STS-24c/VC-4-8c westward.

Figure 7-6 *ML-Series Card Ports*

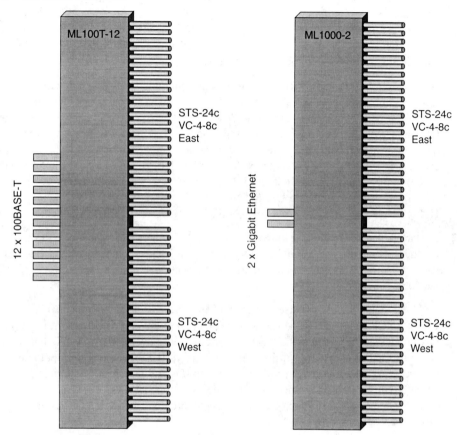

The ML-Series card can build the following SPR combinations of circuits:

- 1 group * STS-1/VC-3 (east) + 1 group * STS-1/VC-3 (west)
- 1 group * STS-3c/VC-4 (east) + 1 group * STS-3c/VC-4 (west)
- 1 group * STS-6c/VC-4-2c (east) + 1 group * STS-6c/VC-4-2c (west)
- 1 group * STS-9c/VC-4-3c (east) + 1 group * STS-9c/VC-4-3c (west)
- 1 group * STS-12c/VC-4-4c (east) + 1 group * STS-12c/VC-4-4c (west)
- 1 group * STS-24c (east) + 1 Group * STS-24c (west)

The ONS 15454 uses the ML100T-12 for 10/100-Mbps Ethernet services. Each card provides twelve 10/100BASE-T autosensing Ethernet interfaces. The ports autoconfigure to operate at either half or full duplex and can determine whether to enable or disable flow control. These ports can also be configured manually. The ML100T-12 card operates with the XC, XCVT, and XC10G cross-connect cards. The ONS 15454 uses the ML1000-2 card for GE service. These cards provide two GE ports with full-duplex operation. The ML1000-2 card operates with the XC, XCVT, and XC10G cross-connect cards.

As shown in Figure 7-7, SPRs are built by daisy chaining ONS Ethernet cards around a ring. In the example, a circuit size of STS-6c/VC-4-2c east-west is used. One could also build an Ethernet circuit of size STS-24c/VC-4-8c east-west across the ring daisy chaining between ONS node/ML-Series cards around a ring. The SONET/SDH links serve only to act as an optical transport mechanism. The E- or ML-Series cards essentially act as Layer 2 Ethernet bridges. ONS SPRs are built on SONET/SDH UPSR/SNCP (protected), SONET/SDH UPSR/SNCP (unprotected), or BLSR/MS-SPRing circuits. ONS nodes have up to 96 provisionable STS-1/VC-3s on unprotected UPSR/SNCP, 48 provisionable STS-1/VC-3s on protected UPSR/SNCP, and up to 24 provisionable STS-1/VC-3s on BLSR/MS-SPRing.

At startup, the ONS nodes use the SONET/SDH DCC channel to transmit OSPF-like information to other ONS nodes present in the network and form a Dijkstra's tree with the node as the root node of the tree. This way, each node knows the location of other nodes within the SONET/SDH network.

The ONS data network can use 802.1D STP as a Layer 2 protection mechanism while using UPSR/SNCP, or use BLSR/MS-SPRing protection at Layer 1.5 with or without STP. The ONS nodes can build a UPSR/SNCP network using a conventional SONET/SDH ring topology with core and tributary rings or in a mesh topology. BLSR/MS-SPRing topologies use core and tributary ring topologies with unique ring IDs and nodal identification. Ethernet or TDM circuits are built over the SONET/SDH UPSR or BLSR Layer 1.5 network. Ethernet data circuits can be built as protected or unprotected circuits. However, TDM circuits must be protected.

The ONS 15000 series MSPP supports single-card and multicard EtherSwitches. When you provision single-card EtherSwitch, each Ethernet card is a single switching entity within the MSPP. This option allows STS-12c/VC-4-4c of bandwidth between two Ethernet circuit points. Single-card EtherSwitch on the E-Series supports one STS-12c/VC-4-4c, two STS-6c/VC-4-2c, four STS-3c/VC-4, or twelve STS-1/VC-3 circuits. In the case of ML-Series cards, the single-card EtherSwitch supports two groups, from STS-1/VC-3s all the way up to two STS-24c/VC-4-8cs.

When you provision multicard EtherSwitch, two or more Ethernet cards within the same chassis can act as a single Layer 2 switch. In the case of E-Series Ethernet cards, the multicard EtherSwitch supports one STS-6c/VC-4-2c SPR, two STS-3c/VC-4 SPRs, or six STS-1/VC-3 SPRs. The bandwidth of the single switch formed by the Ethernet cards matches the bandwidth of the provisioned Ethernet circuit up to STS-6c/VC-5-2c worth of bandwidth. Single-card mode provides greater provisioning flexibility and control options.

Figure 7-7 *STS-6c/VC-4-2c SPR*

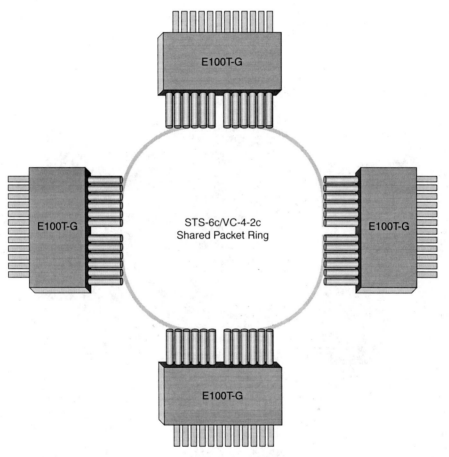

In the case of a matching node (a node that connects the collector ring to the core), as shown in Figure 7-8, there would be two E-Series or ML-Series cards present. This example shows two E1000-2-G cards present. Each E-Series or ML-Series card presents itself as an isolated discrete Ethernet bridge. This means that external connectivity, via means of a jumper or an external Layer 2 bridge, is needed to provide a contiguous data path from the collector ring to the core ring.

Figure 7-8 *SPRs with a Matching Node*

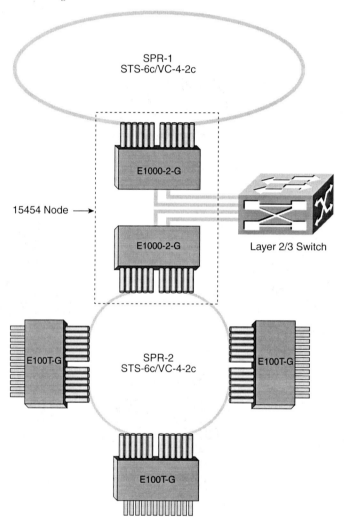

NOTE ONS 15454 E-Series SPR configuration and an example is presented in Chapter 11, "Ethernet, IP, and RPR over SONET and SDH." The E-Series interface integrates Ethernet switching in a SONET/SDH network along with double-tagged VLANs that can be leveraged as a highly scalable VPN mechanism.

Resilient Packet Ring

RPR is a MAC layer, ring-based protocol that combines the intelligence of IP routing and statistical multiplexing with the bandwidth efficiencies and resiliency of optical rings. RPR networks consist of two counter-rotating fiber rings that are fully utilized for transport at all times for superior fiber utilization, unlike SONET/SDH-based networks. RPR permits more efficient use of bandwidth using statistical multiplexing. RPR eliminates the need for manual provisioning, because the architecture lends itself to the implementation of automated provisioning. Moreover, there is no need for channel provisioning as each ring member can communicate with every other member based on the MAC address. RPR also protects existing investment in fiber and other transmission infrastructures. Because most current metro-area fiber is ring-based, RPR will best utilize existing fiber facilities. Moreover, apart from dark fiber, RPR can also operate over SONET/SDH ADMs or WDM equipment, allowing smooth and efficient migration. RPR also provides two priority queues at the transmission level, which allow the delivery of delay- and jitter-sensitive applications, such as voice and video.

SPR technology is a precursor to true RPR. SPR possesses certain inherent deficiencies that limit the scalability of the SPR solution. At every node on the SPR ring, a router or switch will process each packet, which can be time-consuming for large rings. As a result, Ethernet will have trouble meeting the jitter and latency requirements for voice and video. Conventional SONET/SDH has implemented improvements, such as virtual concatenation and LCAS (link capacity adjustment scheme), to suit data applications. However, SONET/SDH transport creates point-to-point circuits that are not particularly suited for data applications. SONET/SDH also reserves bandwidth for every source on the ring and prevents nodes from claiming unused bandwidth.

Standards-based RPR (IEEE 802.17) can be achieved using a software-firmware upgrade to the ML-Series Ethernet modules in the ONS 15000 series MSPPs. RPR is a fiber-based ring network architecture. Data is carried in packets rather than over TDM circuits. RPR networks retain many of the performance characteristics, such as protection, low latency, and low jitter of SONET/SDH. RPR architectures are highly scalable, very reliable, and easy to manage in comparison to legacy point-to-point topologies. RPR achieves a loop-free topology across the SONET/SDH rings with rapid reconvergence on ring break. RPR supports autodiscovery of other RPR network elements on the ring. New RPR nodes announce themselves to their direct neighbors with control messages and distribute changes in their settings or topologies.

Because RPR has been implemented as a Layer 2 technology that runs on top of existing Layer 1 and 1.5 technologies, service providers and carriers can add RPR equipment to their existing infrastructure without disrupting existing services. RPR traffic can travel over SONET/SDH networks and can be added to networks without affecting current SONET/SDH TDM or voice traffic. RPR rides on Ethernet's Layer 1 and can encapsulate Ethernet frames within an RPR frame. However, most Ethernet devices expect Ethernet to be the Layer 1 and the Layer 2 protocol of the frame, and won't carry RPR. Instead, the RPR equipment must be the core of the network, with Ethernet devices feeding services in to or out of the RPR network.

The following are some of the features of IEEE 802.17 RPR:

- Support for dual counter-rotating ring topology
- Full compatibility with IEEE's 802 architecture as well as 802.1D, 802.1Q, and 802.1F
- Protection mechanism with sub-50-ms failover
- Destination stripping of packets
- Adoption of existing physical layer media

RPR combines the advantages of both SONET/SDH and Ethernet and allows the support of the newer services while simultaneously supporting traditional carrier-class features, such as resiliency, restoration, and QoS. RPR resilience is on par with SONET/SDH achieved with sub-50ms protection switching, whereas bandwidth efficiency is achieved by delivering packet services rather than circuits. Bandwidth management gives the flexibility to oversubscribe the total ring bandwidth with a greater number of users for certain (nonguaranteed) services and temporarily reclaim reserved bandwidth from idle nodes. RPR lets traffic be provisioned bidirectionally and utilizes bandwidth traditionally set aside for SONET/SDH protection. RPR strips data from the ring when it reaches its destination, leaving spare bandwidth to be reused (spatial reuse). Bandwidth is consumed only on fiber spans between the transmitting and receiving nodes. This stripping yields a fourfold increase of the effective bandwidth over the equivalent SONET/SDH network. For multicast packets, one packet is circulated to multiple nodes. This is more efficient than the flooding with multiple packets by Ethernet. Each RPR ring supports up to 255 nodes. Matching nodes perform the RPR routing between multiple rings.

Dynamic Packet Transport

Cisco also has a pre-RPR variant known as *Dynamic Packet Transport (DPT)* that is supported on most 12000 GSR products. DPT was built around the Spatial Reuse Protocol (SRP) that allows effective use of idle bandwidth. SRP is also a part of the RPR specification. DPT is a resilient, ring-based technology that is optimized for packet-based traffic. DPT is a proprietary protocol developed by Cisco, but DPT is also published as an informational RFC that vendors can use to create their own DPT. For example, Riverstone has a line of DPT cards that work with the Cisco proprietary DPT. Because DPT is such a successful technology, the IEEE decided to ratify the RPR 802.17 standard based on it. The DPT frame format is closely aligned to that of Ethernet. DPT uses SONET/SDH framing for backward compatibility with legacy transport infrastructures, but is physical (PHY) layer agnostic and can run directly over fiber without the need for SONET/SDH framing. SONET/SDH framing allows DPT to be transported over legacy SONET/SDH infrastructures. DPT uses the SRP protocol to reuse unused ring bandwidth. It also implements a fairness algorithm so that bandwidth is shared equally between DPT nodes. DPT also has QoS features.

DPT combines the intelligence of IP with the bandwidth efficiencies and protection of optical rings. DPT is a Layer 2 protocol with optical protection schemes that increase efficiency over

SONET networks. DPT is based on SONET BLSR technology without the need to reserve half of the ring bandwidth. As illustrated in Figure 7-9, a DPT ring consists of dual counter-rotating fiber rings referred to as the *inner* and *outer rings*. There is no working or protect facility as there is in SONET rings. Both rings support data simultaneously. Whenever a data packet is transmitted on one of the rings, control packets are sent in the other direction to share the bandwidth-usage information around the ring. Both rings support data and control packets. The control and data packets from the same source always flow in opposite directions, but each ring can transport both data and control traffic. DPT is optimized for transporting TCP/IP traffic. Other Layer 3 technologies are not supported, but they might be tunneled through the DPT ring. DPT supports sub-50-ms SONET restoration times. DPT borrows its frame format from the IEEE 802.3, while using many of the ring-based concepts of SONET. The bandwidth allocation is not fixed in DPT. The system tries to use all available bandwidth as much as possible using statistical TDM. DPT is available on the following Cisco platforms:

- Cisco 12000 series routers
- Cisco 10720 router
- Cisco 7200 VXR and 7500 series routers
- ONS 15194 IP Transport Concentrator

Figure 7-9 *RPR/DPT Ring*

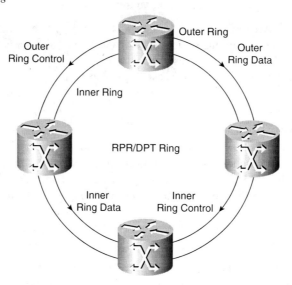

DPT by design is a physical layer agnostic protocol. Although it can run over any technology, the current framer uses SONET framing so that DPT interfaces can be internetworked with the large number of deployed SONET networks. DPT does not require a SONET infrastructure. DPT devices might be connected directly to each other over dark fiber leased from a service provider. DWDM can be leveraged with DPT, as long as the transponders support the same SONET line rate as the DPT interfaces. Similar to PoS, the SONET framing allows DPT to be used anywhere SONET framing is supported. DPT implementations support the following physical media:

- Dark fiber
- DWDM transponder-based networks
- SONET networks

RPR Operation

RPR consists of at least two counter-rotating fiber rings in which multiple nodes share the bandwidth without the requirement of provisioning circuits. Figure 7-9 shows an RPR. The nodes on the ring can automatically negotiate for bandwidth among themselves via a fairness control algorithm. Each station has a topology map of the ring and can send data on the optimal ring toward its destination. To transmit, each station looks up its topology map to help it select a ring based on the least-hop count and distance metric to a destination node. Each RPR node will process only frames addressed to it. Other frames are just transmitted through the RPR MAC. Both rings can be used to carry working traffic. The protection algorithm avoids failed spans to protect against fiber or station failure.

The RPR Media Access Control (MAC) protocol defines the manner in which available bandwidth can be utilized by transmitting stations. The MAC protocol also defines how a station would react to congestion or outright collisions on the media. Finally, the MAC regulates access to the media by buffering and prioritizing packets onto the media.

RPR Topology Discovery

RPR supports a topology-discovery mechanism. Topology messages are broadcast from each station to the other stations on the ring. Each station constructs a topology map, containing information about the location, capabilities, and health of other nodes on the ring. Messages are generated periodically and upon the detection of change in the local status. When an RPR node starts up, its topology map contains information about the local node only. The node listens for broadcast messages from other RPR nodes. The node also broadcasts its latest topology information periodically or when a change has been detected.

A new node initializes itself after joining a ring and broadcasting its topology message. Upon detecting a change, the other nodes on the ring send their topology messages. A node discovers its immediate neighbors by receiving messages that have traveled only one hop. If a node is removed or a fiber span fails, the nodes adjacent to the failure record the status in their topology maps. Protection messages are sent. All RPR nodes then update their topology map to reflect the change in connectivity.

The topology-discovery message contains information about the node sending it. The node's capabilities, such as its capability to receive jumbo frames (MTU > 1500, up to 9218 bytes) and its capability to perform wrap protection, are transmitted in the topology-discovery messages. In addition, extended capabilities are carried in the Type-Length-Value (TLV) fields of the extended topology message. The TLV frame offers a flexible format in which to deliver detailed information about the capabilities of a station.

The TLV values transmitted in the extended topology fields include the following information:

- Station fairness weight per ring
- Total Class A reserved bandwidth per ring
- Multicast groups assigned to a station
- Address of adjacent neighbors per ring
- Individual bandwidth reserved at each hop per ring
- Vendor ID and specific information

RPR CoS

RPR offers multiple classes of service for packets on or entering the ring. New services are easily accommodated on a single network by having separate classes for latency- and jitter-sensitive traffic, committed information rate (CIR), and best-effort traffic. There are three classes of service defined for RPR traffic:

- **High priority (Class A)**—CIR services. This service supports guaranteed bandwidth. Class A service reserves exclusive network bandwidth. It is meant for low-jitter and latency-sensitive traffic. It is not subject to the bandwidth-sharing algorithm, and the bandwidth cannot be used by any other traffic, even if the bandwidth is idle. High priority can be provisioned up to 50-percent link capacity. High-priority traffic is guaranteed to be delivered even during a fiber break, because RPR can use both rings concurrently. If a fiber outage or break occurs, the designer must ensure that enough bandwidth is available on the other ring, because priority traffic configured on the second ring takes precedence.

- **Medium priority (Class B)**—Reserves bandwidth; however, unused bandwidth can be used by other medium- or low-priority traffic. Medium-priority assignments are suitable for time-sensitive and bursty traffic. Bandwidth would be allocated during periods of burst, while being available during nonbursty periods. Ethernet LAN traffic can be assigned as medium-priority traffic.

- **Low priority (Class C)**—Utilizes unused bandwidth and cannot reserve bandwidth. Low-priority traffic can be provisioned up to 100-percent link capacity. The bandwidth-sharing algorithm manages this CoS, and bandwidth is dynamically negotiated between the multiple RPR nodes on the ring. Best-effort residential Internet services are a typical application for low-priority service.

RPR Fairness Algorithm

RPR implements a weighted fairness routine that shares unused bandwidth among nodes. Because low-priority traffic can theoretically occupy any unused portion of the fiber bandwidth, a node on the network could consume all unused bandwidth and starve the other RPR nodes. The RPR fairness algorithm prevents such occurrences. The fairness algorithm is implemented during fiber outages and during normal operation. If medium- or low-priority traffic exceeds what the node is capable of transmitting, the node invokes the weighted fairness algorithm. Each RPR node is assigned with a weight that determines its transmission priority. Any RPR node can send fairness messages, but when congestion occurs at a higher priority node, it sends a message to the other RPR nodes on the ring to throttle back their traffic so that its traffic gets priority over the others. When a fairness request is received, nodes sending data through the requesting node will queue or drop packets marked fairness eligible (FE) in their header. The FE value of high-priority traffic is ignored, because high-priority traffic is never dropped. When the congestion at a RPR node eases, other RPR nodes are allowed to access available bandwidth.

RPR Bandwidth Management

The 802.17 RPR standard supports the SRP protocol and statistical multiplexing of bursty traffic through the use of a fairness algorithm. By having both rings carry working traffic, no fiber wasted is wasted. Instead, it is reserved only to serve as backup when a ring failure occurs. By stripping unicast packets at their destination, RPR networks can achieve "spatial reuse" or bandwidth multiplication. Spatial reuse increases the capacity of simultaneous conversations on each ringlet. RPR also has an efficient mechanism for sending multicast packets. Unlike meshed topologies, where many packet copies need to be sent to cover multiple paths, multiple RPR stations can share a single packet.

Spatial Reuse

The destination node removes packets from the ring. This enables the reuse of bandwidth by other nodes on the RPR, thereby increasing bandwidth utilization. The effective capacity of the RPR can be upwardly scaled, depending on the specific topology and traffic pattern. For example, adjacent nodes could fully utilize the ring span between them, while other nodes could use other parts of the ring without interference.

Through the statistical multiplexing approach, RPR systems can dynamically allocate bandwidth per service in real time, which is in turn used for efficient statistical multiplexing of bursty services over the shared media ring. To do this, stations advertise information in real time about current loads so that other stations can determine how much traffic they can send through that station.

The statistical multiplexing scheme, which uses a fairness algorithm, enforces fair behavior of the stations sharing the ring. When a congestion threshold has been crossed, the adjacent station advertises its fair usage to the other stations. Stations that get this message adjust their usage according to their respective weighted fair usage. Multiple congestion points can be handled over the ring, allowing for stations that do not use the congested span(s) to maximize the available resources.

The ring topology is used to maximize utilization of network resources. As opposed to an alternative hub-and-spoke-based technology, RPR ring traffic doesn't have to be backhauled to a hub for switching and aggregation. Instead, RPR's bandwidth-management scheme enables an operator to effectively aggregate the whole ring traffic "locally," and thus saves the cost of backhauling unaggregated traffic to a distant hub. In a comparable hub-and-spoke architecture only the distant hub has "awareness" to the entire ring traffic to enable effective aggregation.

RPR Traffic Protection and Rerouting

In the event of a fiber break, RPR offers two methods of redundancy: the wrap mechanism and the steer mechanism. Each connection created during ring provisioning is specified as wrap or steer:

- **Wrap method**—This mechanism does not require a lot of CPU processing at the node. Traffic is wrapped and transmitted to RPR nodes on either side of the break and to the opposite ring, on to its destination. Traffic is rerouted around a break without considering its destination. Essentially, traffic proceeds as normal until a fault is detected. When the fault is detected, the traffic is looped back around a ring to the nearest RPR node, from where it is rerouted around the ring. Wrap places lower requirements on the RPR nodes but causes more traffic delays. Such delays could affect time-sensitive traffic, such as voice and video. However, the wrap mechanism suffers from lower packet loss during protection switching.

- **Steer method**—This method is the default mechanism for many vendors. The steer method requires higher RPR node intelligence and CPU processing at the source node. In the event of a fiber break, the source RPR node determines the best path to the destination and places the traffic on the proper ring, ignoring the ring the traffic was originally provisioned for. The source RPR node must make a determination as to the location of the destination with respect to the fiber break. The traffic must then be transmitted over that fiber link considering the priority of each packet and the available bandwidth. The steer method minimizes jitter and the number of hits on traffic during protection switching.

The RPR 802.17 standard supports a unified protection mechanism known as the *selective wrap independent steer (SWIS)* mechanism. The SWIS mechanism enables coexistence of the wrap and steer mechanisms, while enabling the selection of either protection mechanism during the provisioning phase. The RPR specification calls for the steer mechanism to be the default operating condition. Typically, if the network operates the wrap mechanism, it is recommended that future provisioning consider the wrap mechanism.

To enable coexistence of the steer and wrap mechanisms, a Type-Value field is defined in the RPR header. RPR nodes mark their sourced frames with either "normal data" or Steer Only Data (SOD), which indicate the desired behavior to be applied to the frame by the node detecting the defect. The RPR node that detects the defect can wrap the frame if it is marked normal data or discard it, if it is marked SOD. The node also transmits alarm indications to the other nodes in the ring. Other nodes in the RPR can use the alarm indication to perform a steer operation. When the ring failure is cleared, the node will get out of wrap mode and continue normal operation.

RPR Protection Hierarchy

The protection mechanism can handle simultaneous failures. The more severe event takes precedence over less severe events. For example, protection may be removed from a span with a signal degrade event after a signal fail occurs elsewhere on the ring.

RPR failure states, in order of decreasing severity, are as follows:

- **Forced Switch**—An operator initiated command to force a protection event on an interface
- **Signal fail**—A protection event caused by signal failure from the physical media or RPR keepalive message failure
- **Signal Degrade**—A protection event caused by excessive bit errors
- **Manual Switch**—Similar to forced switch but of lower priority
- **Wait To Restore**—A configurable delay timer to restore a link after a fault has been cleared

RPR Media Access Control

RPR is essentially a Layer 2 protocol and has its own MAC header that encapsulates higher-level data. The 802.17 RPR MAC uses three frame formats:

- **Data**—Data frames carry user data in its payload field. RPR supports jumbo frames containing a maximum of 9218 bytes.
- **Control**—Control frames carry control messages other than fairness information, such as topology and protection messages.
- **Fairness**—Fairness control frames carry fairness control messages to communicate bandwidth requirements between RPR nodes.

RPR Data Frame

The RPR Data MAC header consists of 24 bytes. Figure 7-10 shows the RPR Data header.

Figure 7-10 *RPR Data Frame*

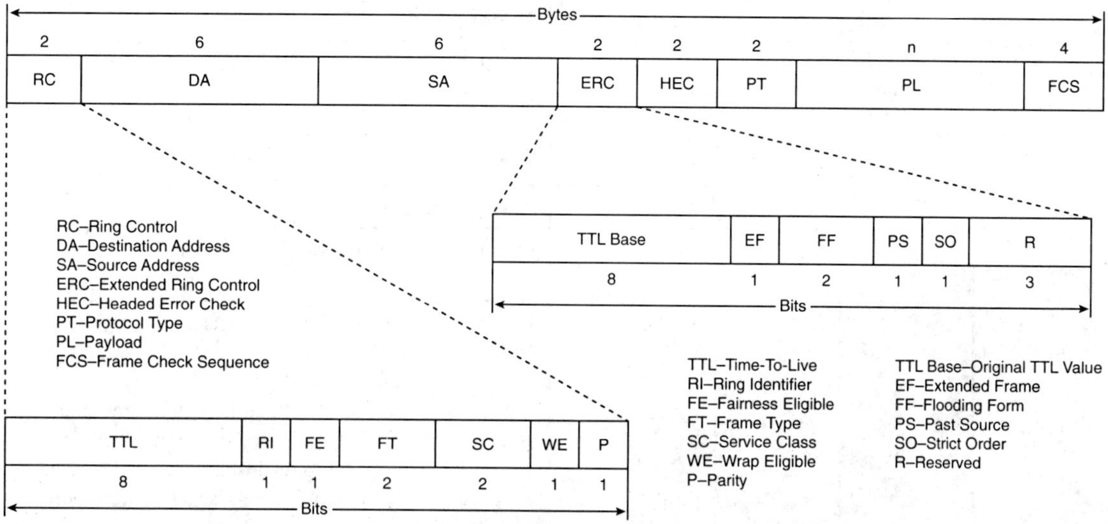

```
|<-------------------------------- Bytes --------------------------------->|
|  2  |    6    |    6    |  2  |  2  |  2  |    n    |  4  |
| RC  |   DA    |   SA    | ERC | HEC | PT  |   PL    | FCS |
```

RC–Ring Control
DA–Destination Address
SA–Source Address
ERC–Extended Ring Control
HEC–Headed Error Check
PT–Protocol Type
PL–Payload
FCS–Frame Check Sequence

```
| TTL Base | EF | FF | PS | SO | R |
|    8     | 1  | 2  | 1  | 1  | 3 |
|<----------------- Bits ----------------->|
```

```
| TTL | RI | FE | FT | SC | WE | P |
|  8  | 1  | 1  | 2  | 2  | 1  | 1 |
|<----------------- Bits ----------------->|
```

TTL–Time-To-Live
RI–Ring Identifier
FE–Fairness Eligible
FT–Frame Type
SC–Service Class
WE–Wrap Eligible
P–Parity

TTL Base–Original TTL Value
EF–Extended Frame
FF–Flooding Form
PS–Past Source
SO–Strict Order
R–Reserved

The Ring Control (RC) and Extended Ring Control (ERC) fields contain subfields that perform various control tasks for the RPR nodes and network. Table 7-1 describes the major fields and subfields.

Table 7-1 *RPR Data Frame Fields*

Field		Length	Description
RC		2 bytes	Ring Control field.
	TTL	8 bits	Time-To-Live.
	RI	1 bit	Ring ID—0 for inner ring, 1 for outer ring.
	FE	1 bit	Fairness Eligible—The FE flag indicates whether the fairness algorithm needs to be invoked.
	FT	2 bits	Frame Type—The FT field indicates whether the frame contains user data, fairness requests, or control data for other RPR nodes.
	SC	2 bits	Service Class—The value of the SC field determines the packet's priority on the network. High, medium, or low.
	WE	1 bit	Wrap Eligible—The WE flag indicates whether the frame is eligible for rerouting in the event of a fiber break.
	P	1 bit	Parity.

continues

Table 7-1 *RPR Data Frame Fields (Continued)*

Field		Length	Description
DA		6 bytes	48-bit destination MAC address that identifies the destination RPR node.
SA		6 bytes	48-bit source MAC address that identifies the source RPR node.
ERC		2 bytes	Extended Ring Control field.
	TTL Base	8 bits	Original TTL value (not decremented).
	EF	1 bit	Extended Frame—This flag indicates whether the frame is a Base Data frame or an Extended Data frame. The Base Data frame is used by all data traffic that travels from source to destination on the same ring. If data needs to travel from one ring to another subtending ring to get to its destination, an Extended Data frame, which includes the original source address and final destination address after the HEC, is used.
	FF	2 bits	Flooding Form—The two FF bits indicate whether data is flooded in a unidirectional manner, bidirectional manner, or not at all.
	PS	1 bit	Past Source—The PS bit is used during a wrap function to indicate the frame has traveled past the source address on its way back to the destination.
	SO	1 bit	Strict Order—The SO field is used when frames need to be kept in order.
	R	3 bits	Reserved (all 0s).
HEC		2 bytes	Header Error Check—16-bit CRC to protect the RPR MAC header fields, SA, and DA.
PT		2 bytes	Protocol Type—When the value of the PT field is less than 1535, it indicates the length of the frame. If the value is equal to or greater than 1536, it indicates the MAC client protocol. The value of the PT field is designated by the IEEE Type Field Register. The protocol bytes determine type or length, but never both.
PL		Variable	Payload—This field contains higher-level user data.
FCS		4 bytes	32-bit Frame Check Sequence.

RPR Control Frame

The RPR Control MAC header consists of 22 bytes. Figure 7-11 shows the RPR Control header.

Figure 7-11 *RPR Control Frame*

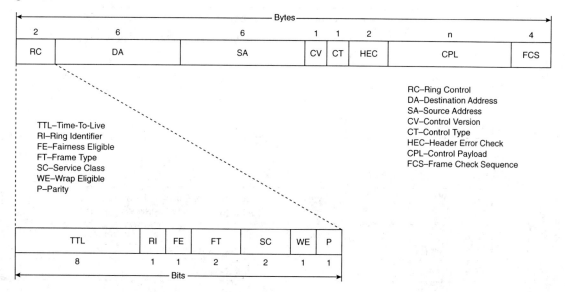

The Ring Control (RC) field contains subfields that perform various control tasks for the RPR nodes and network. Table 7-2 describes the major fields and subfields.

Table 7-2 *RPR Control Frame Fields*

Field		Length	Description
RC		2 bytes	Ring Control field.
	TTL	8 bits	Time-To-Live.
	RI	1 bit	Ring ID—0 for inner ring, 1 for the outer ring.
	FE	1 bit	Fairness Eligible—The FE flag indicates whether the fairness algorithm needs to be invoked.
	FT	2 bits	Frame Type—The FT field indicates whether the frame contains user data, fairness requests, or control data for other RPR nodes.
	SC	2 bits	Service Class—The value of the SC field determines the packet's priority on the network. High, medium, or low.
	WE	1 bit	Wrap Eligible—The WE flag indicates whether the frame is eligible for rerouting in the event of a fiber break.
	P	1 bit	Parity.

continues

Table 7-2 *RPR Control Frame Fields (Continued)*

Field	Length	Description
DA	6 bytes	48-bit destination MAC address that identifies the destination RPR node.
SA	6 bytes	48-bit source MAC address that identifies the source RPR node.
CV	1 byte	Control Version—The CV byte indicates the version of the control algorithm.
CT	1 byte	Control Type—The CT byte indicates the type of control frame. 1=Topology, 2=Protection, and 3=OAM are currently defined.
HEC	2 bytes	Header Error Check—16-bit CRC to protect the RPR MAC header fields, SA, and DA.
CPL	Variable	Control Payload—This field contains control payload information.
FCS	4 bytes	32-bit checksum.

RPR Fairness Control Frame

The RPR Fairness Control MAC header consists of 12 bytes. Figure 7-12 shows the RPR FC header.

Figure 7-12 *RPR FC Frame*

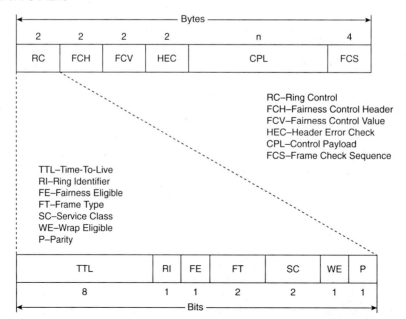

The Ring Control (RC) field contains subfields that perform various control tasks for the RPR nodes and network. Table 7-3 describes the major fields and subfields.

Table 7-3 *RPR Fairness Frame Control Fields*

Field		Length	Description
RC		2 bytes	Ring Control field.
	TTL	8 bits	Time-To-Live.
	RI	1 bit	Ring ID—0 for the inner ring, 1 for the outer ring.
	FE	1 bit	Fairness Eligible—The FE flag indicates whether the fairness algorithm needs to be invoked.
	FT	2 bits	Frame Type—The FT field indicates whether the frame contains user data, fairness requests, or control data for other RPR nodes.
	SC	2 bits	Service Class—The value of the SC field determines the packet's priority on the network. High, medium, or low.
	WE	1 bit	Wrap Eligible—The WE flag indicates whether the frame is eligible for rerouting in the event of a fiber break.
	P	1 bit	Parity.
FCH		2 bytes	Fairness Control Header—The FCH field contains information about the message, such as single choke versus multichoke messages.
FCV		2 bytes	Fairness Control Value—The FCV field carries the advertised fair rate.
HEC		2 bytes	Header Error Check—16-bit CRC to protect the RPR MAC header fields, SA, and DA.
CPL		Variable	Control Payload—This field contains fairness control payload information.
FCS		4 bytes	32-bit checksum.

RPR MAC Operation

The primary purpose of the RPR MAC is to provide access to the SONET/SDH ring media. Each station on the ring contains two MACs to communicate over the two fiber rings. As illustrated in Figure 7-13, CPE devices will communicate via Ethernet interfaces to the routing and switching engine of the RPR node. The RSE will then communicate data and control information, such as topology and protection messages, to the MAC control sublayer.

Figure 7-13 *RPR MAC Schematic*

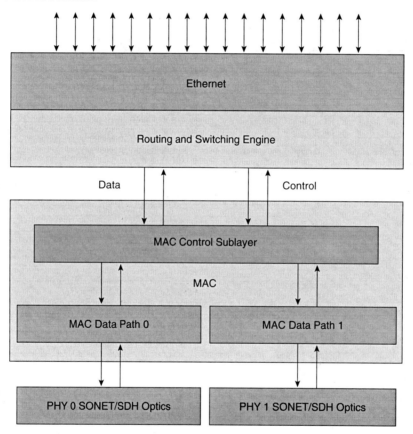

The MAC control sublayer will then frame the data as RPR frames and select the appropriate MAC for transmission over the physical layer. RPR frames received by either MAC will be routed to the RSE in the case of a destination address match. Other packets will be placed back on the ring to transit toward other RPR nodes on the ring.

As illustrated in Figure 7-14, upon frame reception, the MAC determines which frames to receive and where to deliver them. Frames can be delivered to the client interface, MAC control sublayer, or the transit buffer. Multicast and matching unicast packets destined for the station client or its control sublayer are received and processed by the local RPR node, whereas multicast and nonmatching unicast are placed in the transit buffer to continue around the ring.

Dual-transit-queue implementations place the packet in the appropriate queue. The RPR MAC transmits high-priority and reserved traffic from the client before transmitting unreserved traffic. Dual-queue implementations have both primary (PTQ) and secondary (STQ) transit queues. The PTQ is used only for high-priority traffic. The size of the PTQ is on the same order as in the single-queue implementation. The STQ is a larger queue capable of holding many frames. Class B and Class C frames can be temporarily delayed in the STQ, while other packets are transmitted from the client. The STQ is not allowed to overflow. When the STQ is almost full, frames in the STQ are transmitted before client-sourced frames. This condition also triggers the fairness algorithm to request more bandwidth. The transmit order for the dual-queue implementation is as follows:

1. Frames in the primary transit queue (PTQ)
2. Frames in the STQ (only when the STQ is near full)
3. Control frames
4. Frames from the client
5. Frames in the STQ

Figure 7-14 *RPR MAC Detailed Block Schematic*

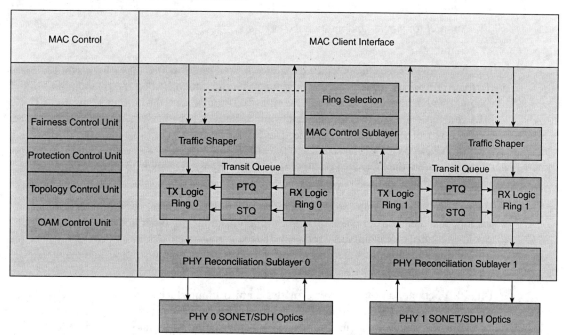

During frame transmission, the MAC or the client can select the ring on which to transmit. The topology information can be used to select the ring and set the TTL. During wrap or steering protection events, packets can be redirected to the other ring to avoid failed spans. Transmit rate controls are placed on each of the classes of traffic. Class A and Class B shapers ensure bandwidth up to their provisioned values. Class C traffic is shaped according to the fair rate determined by the fairness control unit (FCU).

A token-bucket algorithm is used to shape the Class A, Class B, and Class C traffic. The purpose of the traffic shapers is to smooth the delivery of each class of traffic and prevent large bursts of any particular class. The client receives permission for each class of traffic and ring separately.

The FCU ensures that the fairness eligible bandwidth is shared (Classes B1 and C) among the stations on the ring. Each 802.17 MAC has an FCU that is responsible for the following:

- Fast response
- High-bandwidth utilization
- Scalability
- Bandwidth reclamation
- Source-based weighted fairness
- Support for single- and multichoke clients
- Stability

The FCU provides the clients with a dynamic allowed rate at which they can transmit fairness-eligible traffic. The stations on the ring negotiate this rate based on the total amount of uncommitted bandwidth available on the ring. To communicate the current status, the 802.17 FCU generates and receives fairness control messages. It generates messages to advertise the current usage of station and its fair rate. The FCU also receives fairness messages from which it determines ring congestion status and the allowed rate for the RPR node. With the received rate information, the FCU also configures the MAC transmit shapers to comply with the bandwidth limitations.

The protection control unit provides protection for both station and span failures. A state machine and database are maintained via communication with other nodes on the ring. The topology control unit maintains a topology database and state machine and communicates this information with other stations on the ring. The operations, administration, and maintenance (OAM) control unit provides configuration and fault status functions.

This PHY reconciliation sublayer provides a uniform interface between the MAC and the physical layer. Because multiple physical interfaces can be defined for RPR, a separate reconciliation sublayer is defined for each PHY, translating its IO into a consistent format for the MAC. The standard currently defines such interfaces for both SONET/SDH and Ethernet physical layers. Two reconciliation sublayers are defined for SONET/SDH physical interfaces.

The SONET/SDH reconciliation sublayer (SRS) defines an HDLC-like encapsulation for RPR frames with a SONET payload and is similar to Packet over SONET/SDH encapsulation. The GFP reconciliation sublayer (GRS) defines an encapsulation mechanism using Generic Framing Protocol (GFP). Specifications for SPI Level 3 (8- and 32-bit) and SPI Level 4 Phase 1 and 2 interfaces have been defined. For Ethernet, the GE reconciliation sublayer (GERS) defines an interface between the RPR MAC and the GMII interface. The 10GE reconciliation sublayer (XGERS) defines interfaces between the RPR MAC and the XGMII and XAUI interfaces.

RPR OAM and Layer Management

The OAM functions of the RPR MAC offer three management services to the client:

- Configuration management
- Fault management
- Performance management

Messages are exchanged between stations. Echo/response messages are used for connectivity monitoring and path localization between stations. The echo frame can request the ringlet, CoS, and protection mode of the response. A station sending an echo request receives a response from the desired node. Flush messages are used to help prevent the misordering of packets on a ring during protection switches. A station can send a flush frame around the ring for any service class. The frame circulates around the ring and arrives back at the source. Afterward, this station is sure that all of its other packets on the ring (for that service class and higher) have also circulated the ring and the station is free to switch ringlets. For management, the RPR MAC defines a consistent interface to a Management Information Base (MIB). This interface allows the management entity to "set" and "get" parameters for configuration, topology, and protection.

NOTE The ML-Series interface integrates multilayer switching in a SONET/SDH network along with line-rate RPR. Chapter 11 covers the ONS 15454 ML-Series RPR configuration and provides an example.

Summary

There is a need to carry Ethernet end to end in a consistent packet format from start to finish throughout the entire transport path. This would eliminate the need for additional layers of protocol and synchronization that result in extra costs and complexities. Currently, many technologies are available to build broadband MANs and WANs, such as ATM, GE, DPT, PoS, MSPP, and so on. The two leading technology contenders are GE and MSPP. The issue becomes a choice of technology, wherein each technology has its own merits and demerits. Most ILECs and IXCs have a huge SONET/SDH installed base with dark fiber laid out in rings, which does not lend itself very well to the implementation of MANs using GE, for example. Furthermore, the choice of GE as a MAN technology forces the use of VoIP for voice transport or, worse still, forcing TDM over IP. GE also has inherent deficiencies that include Ethernet unfairness of bandwidth allocation for downstream switches in a ring topology as well as dependency on the STP algorithm or its variations for loop prevention and protection. Ethernet also suffers from poor jitter and latency control.

MSPP technology uses SONET/SDH as a Layer 1.5 transport mechanism to carry TDM traffic and Ethernet traffic with specific switching functions and VLAN capability. The technology supports traditional SONET/SDH STS-N circuits as well as RPR or RPR-like functionality (SPRs) for transporting Ethernet over SONET/SDH. VPNs can be created using 802.1Q tagging or TDM pipes. The MSPP network can also offer double VLAN tagging, which increases the maximum number of configurable VLANs or VPNs to 16,777,216 versus only 4096 for conventional GE solutions. SPR uses a combination of SONET/SDH BLSR or UPSR technology along with conventional 802.1D bridging or per-VLAN FSTP (IEEE 802.1w) to provide a stable Layer 1.5 and Layer 2 infrastructure. VLANs provide the VPN-partitioning mechanism for customer-closed user groups.

MSPPs support SPRs to provide Ethernet and packet transport over a SONET/SDH infrastructure. The implementation of this technology varies from vendor to vendor. The Cisco implementation of SPR calls for the provisioning of optical transport pipes at Layer 1.5 between Ethernet switch cards present in the various ONS nodes. At present, Cisco supports 10-Mbps, 100-Mbps, and Gigabit Ethernet on the Ethernet modules. This technology allows the provisioning of bandwidth on the SONET/SDH ring for packet transport by statistically multiplexing Ethernet traffic onto an SPR (circuit) that each MSPP node can access. Each Ethernet card used in the ONS MSPP has physical Ethernet interfaces as well as logical SONET/SDH STS-1/VC-3 ports. The STS-1/VC-3 ports are logical and are implemented via means of backplane cross-connections to the STS matrix. The ONS Ethernet cards have up to 48 logical SONET/SDH STS-1/VC-3 ports on its back end. The logical PoS ports can be daisy chained to PoS ports on other MSPPs to form an SPR.

RPR networks retain many of the performance characteristics, such as protection, low latency, and jitter of SONET/SDH. RPR architectures are highly scalable, very reliable, and easy to manage in comparison to legacy point-to-point topologies. RPR combines the advantages of both SONET/SDH and Ethernet and allows the support of the newer services while simultaneously supporting traditional carrier-class features such as resiliency, restoration, and QoS. RPR resilience is on par with SONET/SDH achieved with sub-50ms protection switching, whereas bandwidth efficiency is achieved by delivering packet services rather than circuits. Bandwidth management gives the flexibility to oversubscribe the total ring bandwidth with a greater number of users for certain (nonguaranteed) services and temporarily reclaim reserved bandwidth from idle nodes. RPR lets traffic be provisioned bidirectionally and uses bandwidth traditionally set aside for SONET/SDH protection and strips data from the ring when it reaches its destination, leaving spare bandwidth to be reused (spatial reuse). Bandwidth is consumed only on fiber spans between the transmitting and receiving nodes. This stripping yields a fourfold increase of the effective bandwidth over the equivalent SONET/SDH network.

This chapter includes the following sections:

- **Next-Generation SONET and SDH Platforms**—The Cisco Complete Optical Multiservice Edge and Transport solution, known as COMET, is an architecture that offers an end-to-end broadband solution. The COMET solution addresses photonics, protection, protocols, provisioning, and service enablement for service provider and enterprise applications in the edge, metro, core, and long haul. This section describes the various ONS series of flexible and modular platforms that are designed to scale metro, central office, and long-haul networks while extending the geographical reach of dense wavelength-division multiplexing (DWDM) infrastructures. The entire ONS range of products can be managed via the ONS element management system. Cisco Transport Controller (CTC) is a GUI-based craft interface and console for the ONS series of MSxPs. Cisco Transport Manager (CTM) is an integrated element management system designed to provision, manage, and monitor all edge, metro, and core MSxP platforms.

- **ONS 15400 Series of Optical Platforms**—This section describes the 15400 series of optical platforms including the Cisco ONS 15454 SONET multiservice provisioning platform (MSPP).

- **Cisco Transport Controller (CTC)**—Cisco Transport Controller (CTC) is a graphical-user interface (GUI)-based element management tool that can be used for OAM&P of various ONS 15000 MSxPs. This section describes the CTC craft tool. CTC is typically used as a GUI-based craft tool during the deployment and implementation of an ONS-based network.

- **Cisco Transport Manager (CTM)**—Cisco Transport Manager (CTM) is a carrier-class element management system (EMS) that enables you to operate, administer, manage, and provision the entire range of ONS 15000 network elements. This section describes CTM and its applications. CTM is a superset of CTC and has a similar look and feel of the CTC GUI tool.

CHAPTER 8

Multiservice SONET and SDH Platforms

Next-Generation SONET and SDH Platforms

With the massive increase in inter-exchange carrier (IXC) core bandwidth, metropolitan-area transport has become the stress point of the entire public telecommunications infrastructure. The Internet and other forms of data communications continue to drive residential and business demand for higher bandwidth access. The metropolitan-area network (MAN) providers, such as incumbent local-exchange carriers (ILECs), competitive local-exchange carriers (CLECs), and cablecos, are caught in between and are responsible for metro transport and last mile access. However, ILECs and CLECs that offer voice services depend on their SONET/SDH networks, and would prefer a unified multiservice voice and data infrastructure that could leverage their existing facilities and fiber plant.

The transition of SONET and SDH from a TDM infrastructure to a truly multiservice technology platform cannot be achieved without proper vendor support and implementation of proposed technology drafts, standards, and customer demands. Many incumbent vendors have stuck to their tunnel vision of SONET/SDH as a TDM transport base, and have added Ethernet over SONET/SDH functionality in a bid to tout their products as multiservice. True visions of multiservice broadband cannot be achieved without considering *last mile* access. Some vendors have focused on core technologies and have not paid much attention to the functional interoperability and bandwidth bottleneck of the last mile interface with network backbones. What is needed is a seamless, end-to-end interoperable broadband solution from the network edge to the core with long-haul capability. The Cisco Complete Optical Multiservice Edge and Transport solution (COMET) is an architecture that offers an end-to-end broadband solution. The COMET solution addresses photonics, protection, protocols, provisioning, and service enablement for service provider and enterprise applications in the edge, metro, core, and long haul. The ONS series of flexible and modular platforms are designed to scale metro, central office (CO), and long-haul networks while extending the geographical reach of DWDM infrastructures. The entire ONS range of products can be managed via the ONS element management system. Cisco Transport Controller (CTC) is a GUI-based craft interface and console for the ONS series of MSPPs. CTC is used for initial provisioning and for OAM&P of smaller networks, whereas the Cisco Transport Manager (CTM) is an integrated element management system designed to provision, manage, and monitor all edge, metro, and core MSxP platforms. In the optical world, the terms *east* and *west* typically signify an eastbound or westbound direction of the fiber with respect to the NE. The terms *north* and *south* are usually associated with the carrier's operations support system (OSS). In this regard, the term *northbound* would indicate an element management link or circuit to an element manager, such as CTC or CTM, with respect to the NE. Northbound traffic would indicate OAM&P traffic from the NE to the element manager. The term *southbound* would indicate a link or circuit from an element manager to an NE with respect to the element manager. Southbound traffic would indicate OAM&P traffic from the element manager to the NE.

ONS 15100 Series

The ONS 15100 series products are IP-aware optical systems ideally suited for scaling, managing, and extending Packet over SONET (POS) and resilient packet ring (RPR) networks. The Cisco ONS 15104 Optical Regenerator regenerates and retransmits OC-48 or STM-16 optical signals over distances up to 80 Km, whereas the ONS 15194 IP Transport Concentrator enables the implementation of logical rings while using a physical fiber hub-and-spoke infrastructure. These devices also support bidirectional wavelength division multiplexing (BWDM).

ONS 15200 Series

The Cisco ONS 15200 series of products are metro DWDM platforms that interoperate with the ONS 15454 and ONS 15327 MSPPs. The Cisco ONS 15200 series includes the family of ONS 15216 platforms. The ONS 15216 platform provides DWDM and optical-filtering capability to multiplex and demultiplex wavelengths launched by MSPPs, such as the ONS 15454 and ONS 15327.

The ONS 15216-OADM is an optical add/drop multiplexer that provides ITU wavelength filtering for sites along an optical DWDM ring. The ONS 15216-EDFA is a C-band erbium-doped fiber amplifier that provides optical amplification along with constant gain control, gain flatness, transient suppression, and a high signal-to-noise ratio. The ONS 15216 Optical Performance Manager (OPM) provides optical monitoring of individual ITU wavelengths at any location within a DWDM network. The ONS 15216-OPM provides spectral measurement of DWDM, which includes wavelength, power, and optical signal-to-noise ratio (OSNR) on a per-channel basis. The ONS 15216-DCU is a dispersion compensation unit assembly that provides compensation for the accumulated chromatic dispersion effect in line and terminal sites of optical telecommunications systems. DWDM channels use the optical supervisory channel. The ONS 15216 OSC-1510 provides an out-of-band management solution for all DWDM platforms that have Ethernet communication ports to enable an exchange of management information. The ONS 15216 family of products also includes the FlexLayer 15216. The ONS 15216 FlexLayer is an asymmetric DWDM system focused on unidirectional video-on-demand (VoD) applications. These applications are unique in that they require only one channel to be bidirectional. The other channels are unidirectional. The current ONS 15216 R2.1 channel plan supports 32 channels with 100-GHz spacing.

ONS 15300 Series

The ONS 15300 series is comprised of access and aggregation optical platforms. These MSPPs are typically used on access or collector rings to aggregate and groom customer traffic for transport to a core or backbone optical network. The 15300 series includes the ONS 15327, 15302, and the ONS 15305. The ONS 15302 is an access device for use in SDH networks. The ONS 15302 combines Ethernet and TDM traffic inside an SDH STM-1 frame structure for transport over an SDH network. The ONS 15305 integrates the functions of an SDH add/drop multiplexer (ADM) and Layer 2 switch fabric to combine TDM (E1, E3/T3, STM-1, STM-4, STM-16) and Ethernet (10/100BASE-T, Gigabit Ethernet) services in an SDH link up to STM-16 bandwidth. It implements a full nonblocking 64 × 64 cross-connect matrix with VC-12/VC-3 and VC-4 granularity. The ONS 15327 is a multiservice ADM that combines SONET/SDH transport along with IP, Ethernet, and TDM services. The ONS 15327 aggregates data, voice, and video services for transport over a SONET/SDH infrastructure. The platform's card slots support SONET/SDH, TDM, IP, and Ethernet.

NOTE The following subsection introduces various ONS 15000 platforms. However, a greater emphasis has been placed on the architecture and design specifications of the ONS 15454 SONET/SDH MSPP. The ONS 15454 SONET/SDH is also equipped with DWDM capabilities and is the most versatile and widely used MSPP in the metro access and core space. The next section addresses the SONET and SDH versions of the ONS 15454 in greater detail.

ONS 15500 Series

The ONS 15500 series is designed for carrying storage, data, and legacy applications over a metro optical DWDM infrastructure. The ONS 15500 series includes the ONS 15530 DWDM multiservice aggregation platform, ONS 15540 ESPx DWDM transport platform, and ONS 15501 EDFA optical amplifier. The ONS 15530 offers service-aggregation capabilities for services, such as enterprise system connection (ESCON), Fibre Channel, fiber connection (FICON), and Gigabit Ethernet. The ONS 15540 ESPx system with its external cross-connect functionality and cable management system supports high-density transparent services, such as 1-Gbps and 2-Gbps Fibre Channel, Gigabit Ethernet, and SONET/SDH. It also supports configuration flexibility, both in terms of client transponder placement in the chassis as well as direct ITU wavelength insertion from the ONS 15454 or ONS 15530. It also permits cross-connection of individual wavelengths within a node.

The ONS 15530 and ONS 15540 ESPx are designed to work together as a combination. They can, however, operate independently, if desired. Both platforms are designed for storage-area networking (SAN), server and storage clustering, and mirroring applications. Optical network designers working with applications, such as IBM Geographically Dispersed Parallel Sysplex (GDPS) and EMC Symmetrix Remote Data Facility (SRDF), should consider using the ONS 15530 and ONS 15540 ESPx.

ONS 15600 Series

The ONS 15600 series is highly scalable multiservice ADM that complements the ONS 15454 and ONS 15327 MSPP. The Cisco ONS 15600 MSSP combines the functionality of multiple metro systems including SONET/SDH multiplexers and digital cross-connect network elements (NEs) in one platform. The ONS 15600 has been designed to scale metro architectures that require a large number of rings with various protection schemes, such as a mix of unidirectional path-switched ring (UPSR) and bidirectional line-switched ring (BLSR), or a mix of subnetwork connection protection (SNCP) and multiplexed section protection ring (MS-SPRing). The 15600 scales from a single shelf (320 Gbps) to a multishelf (5.12 Tbps) in-service and provides full, nonblocking STS-1/VC-N cross-connections. The 15600s are not responsible for service termination and are positioned to act as matching nodes between metro collector and core rings. The Cisco ONS 15600 MSSP provides complete integration of metro core and edge networks for service provisioning and network management. It supports all metro topologies, such as point to point, linear add/drop, UPSR rings, BLSR rings, and mesh. The Cisco ONS 15600 MSSP provides greater than five-nines reliability, or 99.9995-percent uptime and delivers scalability of up to 320 Gbps of traffic in a single shelf, with three shelves per rack.

ONS 15800 Series

The Cisco ONS 15800 DWDM systems are American National Standards Institute (ANSI) and European Telecommunications Standards Institute (ETSI) DWDM platforms that are designed for long-haul and extended-long-haul applications. The primary application for the 15800 platform is for point-to-point fiber relief. The Cisco ONS 15808 is the mainstay of the Cisco DWDM 15800 series DWDM platforms. Long-haul transmission is generally considered to range from spans of 0 to 600 km, although use and applications vary and this range can be extended beyond 600 km. Extended long-haul transmission is generally considered to range from spans of 600 to 2000 km, although use and applications vary and this range can be extended beyond 2000 km.

For long-haul applications up to 600 km, the ONS 15808 supports up to 80 channels of 10 Gbps per channel with 50-GHz channel spacing in the C-band. It also supports an in-service upgrade to L-band for 80 more channels of 10 Gbps each. This provides 160 channels of 10 Gbps each with a resultant bandwidth of 1.6 terabits per second. For extended-long-haul applications up to 2000 to 2250 km, the ONS 15808 supports 40 channels of 10 Gbps per channel or 400 Gbps in the L-band using 50-GHz channel spacing, Raman amplifiers, and forward error correction (FEC). The ONS 15808 rack density is currently 36 channels per rack. The ONS 15808 also supports 10 Gigabit Ethernet LAN PHY. In terms of scalability, the ONS 15808 architecture is designed to scale up to 300 channels with a 40-Gbps rate per channel, essentially providing up to 12 Tbps of bandwidth per ONS 15080 system, albeit with shorter distances.

The 15808 platform can be deployed in a variety of topologies including point to point, ring, and mesh. The ONS 15808's open architecture allows the platform to be a building block of both legacy TDM voice networks and Greenfield pure-IP networks.

ONS 15400 Series of Optical Platforms

The 15400 series of optical platforms includes the Cisco ONS 15454 SONET multiservice provisioning platform. The ONS 15454 is the most versatile and widely deployed MSPP that supports transport of TDM and Ethernet services over SONET and SDH. The 15454 MSPP provides carrier-class service provisioning at the network edge. The Cisco ONS 15454 multiservice transport platform (MSTP) provides carrier-class multiservice delivery via metro DWDM. In addition to integrating DWDM and 32-channel OC-192/STM-64 capability onto the service platform, this product also scales optical transport from 10s of kilometers to 100s of kilometers. There are three ONS 15400 series platforms:

- ONS 15454 SONET MSPP
- ONS 15454 SDH MSPP
- ONS 15454 MSTP

ONS 15454 SONET MSPP

The ONS 15454 SONET MSPP is a multiservice ADM that combines SONET transport, DWDM, along with multiservice IP, Ethernet, and TDM services. The platform supports bit rates up to $32 \times$ OC-192/STM-64 and offers integrated bandwidth management. The ONS 15454 supports all metro topologies, such as point to point, linear add/drop, rings, and mesh. The platform supports multiple service interfaces including DS1, DS3, DS3 transmux, EC1/STS-1, 10/100/1000-Mbps Ethernet, and optical OC-3/OC-3c, OC-12/OC-12c, OC-48/OC-48c, and OC-192 interfaces. The ONS 15454 also serves as a distributed bandwidth manager with support for STS-1 and VT1.5 bandwidth management, packet switching, cell transport, and 3/1 and 3/3 transmux functionality.

The ONS 15454 enables optical network designers to terminate multiple rings or linear systems on a single chassis as well as mix and match the service interfaces that enable a unified data, voice, and video multiservice network. The ONS 15454 supports UPSR rings, two- and four-fiber BLSR rings, unidirectional and bidirectional linear ADM, automatic protection switching (APS), path-protected mesh networking (PPMN), and spanning-tree protection mechanisms.

Layer 3 IP functionality is achieved by adding the ML-Series Ethernet cards. The ML-Series cards provide IP switching along with support for static routing, RIPv2, Enhanced Interior Gateway Routing Protocol (EIGRP), Open Shortest Path First (OSPF), Border Gateway Protocol (BGP), Intermediate System-to-Intermediate System (IS-IS), Hot Standby Routing Protocol (HSRP), Virtual Routing and Forwarding (VRF-Lite), as well as quality of service (QoS) and multicast support.

The ONS 15454 supports the Cisco Transport Controller (CTC) tool, which enables initial provisioning and OAM&P support. CTC permits A-to-Z circuit provisioning, auto NE discovery with network topology, and custom bandwidth management, enabling service providers to design transport networks around subscriber needs rather than around equipment limitations. The ONS 15454 also supports Cisco Transport Manager (CTM). CTM is the carrier-class element management system that provides advanced capabilities in the functional management areas of configuration, faults, performance, and security for Cisco optical NEs.

The ONS 15454 SONET Platform

The ONS 15454 SONET MSPP is a NEBS-compliant shelf assembly that contains 17 card (module) slots, a backplane interface, a fan-tray assembly, a front panel with an LCD, and alarm indicators. ONS 15454 has 17 card slots numbered 1 to 17. All slots are card ready, meaning that when you plug in a card it automatically boots up and becomes ready for service. The ONS 15454 houses five types of cards: common control, alarm interface, electrical, optical, and Ethernet. The common control cards include the timing and control cards (TCC2), cross-connect cards (XCVT or XC10G), and the alarm interface controller card (AIC). Figure 8-1 shows the ONS 15454 chassis with its air filter, fan tray, fiber routing panels, and card slots.

Figure 8-1 *ONS 15454 Chassis*

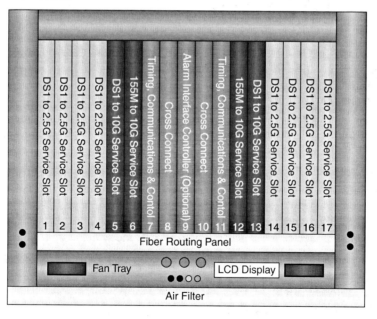

The ONS 15454 electrical cards require electrical interface assemblies (EIAs) installed in the rear of the chassis to provide the cable connection points for the shelf assembly, whereas optical and Ethernet cards have frontal faceplate connections rather than backplane connections. The Building Integrated Timing Supply (BITS) timing connector, LAN, alarms, modem, and craft port use backplane wire-wrap pins.

ONS 15454 Timing and Control (TCC2) Card

The TCC2s control the main processing functions of the ONS 15454. TCC2s also have an RS-232 craft port for CTC. The TCC2s combine timing, control, and switching functions, including the following:

- System initialization
- Provisioning
- Alarm reporting
- Maintenance
- Diagnostics
- IP address detection and resolution
- Timing
- SONET data communications channel (DCC) termination
- System fault detection

The CRIT, MAJ, MIN, and REM alarm LEDs on the TCC2 faceplate indicate whether a critical, major, minor, or remote alarm is present anywhere on the ONS 15454 or on a remote node in the network.

The node name, configuration database, IP address, and system software are stored in the TCC card's NVRAM, which allows quick recovery if power or card failures occur. System software can be upgraded without affecting traffic on the ONS 15454 if dual TCC cards are used. The upgrade takes place first on the standby TCC. The system verifies that the upgrade is successful and switches from the active TCC card running the older release to the upgraded standby TCC card running the newer release. After the switch, the second TCC card undergoes the upgrade. The TCC then loads new software to each of the installed line (traffic) cards.

ONS 15454 Cross-Connect Cards

The cross-connect card is the central switching element in the ONS 15454. The ONS 15454 offers two cross-connect cards: the XCVT and XC10G. Circuit cross-connect information is provisioned using CTC or TL1. The TCC2 then establishes the proper internal cross-connect information and relays the setup information to the cross-connect card. The two cross-connect cards used by the ONS 15454 are as follows:

- **XCVT card**—The XCVT card establishes connections and performs time-division switching (TDS) at the STS-1 and VT1.5 level between ONS 15454 traffic cards. The switch matrix on the XCVT card consists of 144 STS-1 bidirectional ports and adds a VT matrix that can manage up to 336 bidirectional VT1.5s. The VT1.5-level signals can be cross-connected, dropped, or rearranged.

- **XC10G card**—The XC10G card supports STS-192 signal rates. The switch matrix on the XC10G consists of 576 STS-1 bidirectional ports, and its VT matrix can manage up to 336 bidirectional VT1.5s. The XC10G is required to operate the OC-192 card, OC-48 AS card, G1K-4 Gigabit Ethernet card, ML-Series Ethernet cards, and the four-port OC-12 card. The XC10G card is also required for operation of the SDH version of the ONS 15454.

ONS 15454 Alarm Interface Controller

The optional alarm interface controller (AIC-I) card provides user-provisionable alarm capability and supports local and express orderwire. Orderwire is a 64-kbps channel in the SONET header that installation technicians can use to communicate with other technicians in-band. The orderwire ports are standard RJ-11 receptacles. The AIC-I card provides input/output alarm contacts for user-defined alarms and controls, also known as environmental alarms. The AIC-I provides up to 16 input contacts, or 12 input contacts and 4 input/output contacts. The AIC-I cards use the backplane wire-wrap field to make the physical connections. An optional alarm expansion panel (AEP) can be installed on the backplane of the AIC-I. The AEP provides 32 input contacts and 16 output contacts.

ONS 15454 Electrical Cards

The electrical cards provide the ports and physical interfaces for TDM signal ingress and egress. Slots 1 to 6 and 12 to 17 host electrical cards. Each card has faceplate LEDs showing active, standby, or alarm status, and you can also obtain the status of all electrical card ports using the LCD screen on the ONS 15454 an-tray assembly. The EC1, DS1, DS3, DS3E, and DS3XM electrical cards require EIAs to provide the cable connection points for the shelf assembly. In most cases, EIAs are ordered with the ONS 15454 and come pre-installed on the backplane. The ONS 15454 EIAs include Bayonet Neill Concelman (BNC), high-density BNC, AMP Champ, and server management bus (SMB) cable connectors. The various electrical cards used by the ONS 15454 are as follows:

- **DS1-14 and DS1N-14 cards**—The DS1-14 card provides 14 Telcordia-compliant, GR-499 DS1 ports. Each port operates at 1.544 Mbps over a 100-ohm twisted-pair copper cable. The DS1-14 card can function as a working or protect card in 1:1 protection schemes and as a working card in 1:N protection schemes. The DS1N-14 card is identical to the DS1-14 and operates as a protect card in a 1:N protection group. The traffic from an entire DS1-14 card can be grouped and mapped to a single STS-1. Individual DS1 ports can be mapped to a VT1.5.

- **DS3-12 and DS3N-12 cards**—The DS3-12 card provides 12 Telcordia-compliant, GR-499 DS3 ports per card. Each port operates at 44.736 Mbps over a single 75-ohm 728A or equivalent coaxial span. The DS3-12 card operates as a working or protect card in 1:1 protection schemes and as a working card in a 1:N protection scheme. The DS3N-12 card is identical to the DS3-12 and operates as a protect card in a 1:N protection group.

- **DS3-12E and DS3N-12E cards**—The DS3-12E and DS3N-12E are enhanced DS3 cards. The 12-port DS3-12E and DS3N-12E cards provide enhanced performance-monitoring functions. By monitoring additional overhead in the DS3 frame, subtle network degradations are detected. The DS3N-12E can operate as the protect card in a 1:N protection group. The DS3N-12E card can function only as the protect card for one other DS3-12E card.
- **DS3XM-6 card**—The DS3XM-6 card, commonly referred to as a *transmux card*, provides six Telcordia-compliant, GR-499-CORE M13 multiplexing functions. The DS3XM-6 converts six framed DS3 network connections to 28 × 6 (or 168) VT1.5s.
- **EC1-12 card**—The EC1-12 card provides 12 Telcordia-compliant, GR-253 STS-1 ports per card. Each port operates at 51.840 Mbps over a single 75-ohm 728A or equivalent coaxial span.

  The EC1-12 terminates the twelve selected working STS-1 signals from the backplane. The EC1-12 maps each of the 12 received EC1 signals into 12 STS-1s with visibility into the SONET path overhead.

ONS 15454 Optical Cards

ONS 15454 optical cards have SC fiber connectors on the card faceplate. The optical cards reside in slots 1 to 6 and 12 to 17. However, the OC-12 cards can only reside in slots 1 to 4 and 14 to 17, whereas the OC-48 and OC-192 cards can reside in slots 5, 6, 12, and 13. The OC-48 any-slot (AS) cards can reside in the same slots as all other optical cards when used in conjunction with the XC10G. Without the XC10G, the AS cards work only in slots 5, 6, 12, and 13. You can provision an optical card as a drop card or span card in a linear ADM (1+1), UPSR or BLSR protection scheme. However, OC-3 cards cannot be used in a BLSR.

Each card faceplate has three card-level LED indicators. When illuminated, the red FAIL LED represents a hardware problem; the amber SF LED represents a signal failure or condition such as a loss of frame or a high bit error rate, and the green ACT LED indicates that the card is operational. The various optical cards used by the ONS 15454 are as follows:

- **OC3 IR/STM1 SH 1310-4 card**—The OC3 IR 4/STM1 SH 1310 card provides four intermediate or short-range 1310-nm OC-3/STM-1 ports compliant with ITU-T G.707, ITU-T G.957, and Telcordia GR-253-CORE. The port operates at 155.52 Mbps over a single-mode fiber (SMF) span. Each card supports VT and nonconcatenated or concatenated payloads at the STS-1 or STS-3c signal levels. This card provides port-to-port protection.
- **OC3 IR/STM1 SH 1310-8 card**—The OC3 IR/STM1 SH 1310-8 card provides eight intermediate or short-range OC-3/STM-1 ports compliant with ITU-T G.707, ITU-T G.957, and Telcordia GR-253-CORE. Each port operates at 155.52 Mbps over an SMF span. The card supports VT and nonconcatenated or concatenated payloads at the STS-1 or STS-3c signal levels.

- **OC12 IR/STM4 SH 1310 card**—The OC12 IR/STM4 SH 1310 provides one intermediate or short-range 1310-nm OC-3/STM-1 port compliant with ITU-T G.707, ITU-T G.957, and Telcordia GR-253-CORE. The port operates at 622.08 Mbps over an SMF span and supports VT and nonconcatenated or concatenated payloads at STS-1, STS-3c, STS-6c, or STS-12c signal levels.

- **OC12 IR/STM4-4 1310 card**—The OC12 IR/STM4-4 1310 card provides four intermediate or short-range 1310-nm wavelength OC-12/STM-4 ports compliant with ITU-T G.707, ITU-T G.957, and Telcordia GR-253-CORE. Each port operates at 622.08 Mbps over an SMF span. The card supports VT and nonconcatenated or concatenated payloads at the STS-1, STS-3c, STS-6c, or STS-12c signal levels.

- **OC12 LR/STM4 LH 1310 card**—The OC12 LR/STM4 LH 1310 card provides one long-range 1310-nm wavelength OC-12/STM-4 port per card compliant with ITU-T G.707, ITU-T G.957, and Telcordia GR-253-CORE. The port operates at 622.08 Mbps over an SMF span. The card supports VT and nonconcatenated or concatenated payloads at STS-1, STS-3c, STS-6c, or STS-12c signal levels.

- **OC12 LR/STM4 LH 1550 card**—The OC12 LR/STM4 LH 1550 card provides one long-range 1550-nm wavelength OC-12/STM-4 port compliant with the ITU-T G.707, ITU-T G.957, and Telcordia GR-253-CORE. The port operates at 622.08 Mbps over an SMF span. The card supports VT and nonconcatenated or concatenated payloads at STS-1, STS-3c, STS-6c, or STS-12c signal levels.

- **OC12 IR/STM4 SH 1310-4 card**—The OC12 IR/STM4 SH 1310-4 card provides four intermediate or short-range 1310-nm wavelength OC-12/STM-4 ports compliant with the ITU-T G.707, ITU-T G.957, and Telcordia GR-253-CORE. Each port operates at 622.08 Mbps over an SMF span. The card supports VT and nonconcatenated or concatenated payloads at the STS-1, STS-3c, STS-6c, or STS-12c signal levels.

    The OC12 IR/STM4 SH 1310-4 card supports 1+1 unidirectional or bidirectional protection switching. You can provision protection on a per-port basis. The OC12 IR/STM4 SH 1310-4 detects loss of signal (LOS), loss of frame (LOF), loss of pointer (LOP), alarm indication signal-line (AIS-L), and remote defect indication-line (RDI-L) conditions. The card also counts section and line bit-interleaved parity (BIP) errors.

- **OC48 IR 1310 card**—The OC48 IR 1310 card provides one intermediate-range, Telcordia-compliant, GR-253 1310-nm wavelength OC-48 port per card. The port operates at 2.49 Gbps over an SMF span. The card supports VT and nonconcatenated or concatenated payloads at STS-1, STS-3c, STS-6c, STS-12c, or STS-48c signal levels. The OC-48 IR 1310-nm card is restricted to slots 5, 6, 12, and 13 of the chassis, regardless of the cross-connect card installed.

- **OC48 IR/STM16 SH AS 1310 card**—The OC48 IR/SMT16 SH AS 1310 card provides the same capability as the OC48 IR 1310 card, but it can be installed in any traffic slot if used with an XC10G card. However, this card is restricted to slots 5, 6, 12, and 13 when used with an XC or XCVT cross-connect card.

- **OC48 LR 1550 card**—The OC48 LR 1550 card provides one long-range, Telcordia-compliant, GR-253 1550-nm wavelength OC-48 port per card. The port operates at 2.49 Gbps over an SMF span. The card supports VT and nonconcatenated or concatenated payloads at STS-1, STS-3c, STS-6c, STS-12c, or STS-48c signal levels. The OC-48 LR 1550-nm card is restricted to slots 5, 6, 12, and 13 of the chassis, regardless of the cross-connect card installed.

- **OC48 LR/STM16 LH AS 1550 card**—The OC48 IR/SMT16 SH AS 1550 card provides the same capability as the OC48 LR 1550 card, but it can be installed in any traffic slot if used with an XC10G card. However, this card is restricted to slots 5, 6, 12, and 13 when used with an XC or XCVT cross-connect card.

- **OC48 ELR/STM16 EH 100 GHz DWDM cards**—Thirty-seven distinct OC48 ELR/STM16 EH 100-GHz cards provide the ONS 15454 DWDM channel plan. Each card has one OC-48/STM-16 port that complies with Telcordia GR-253-CORE, ITU-T G.692, and ITU-T G.958. The port operates at 2.49 Gbps over an SMF span. The card carries VT, concatenated, and nonconcatenated payloads at STS-1, STS-3c, STS-6c, STS-12c, or STS-48c signal levels. The OC-48 ELR/STM16 EH 100-GHz DWDM card is restricted to slots 5, 6, 12, and 13 of the chassis, regardless of the cross-connect card installed.

  Nineteen of the cards operate in the blue band with spacing of 100 GHz on the ITU grid. The other 18 cards operate in the red band with spacing of 100 GHz on the ITU grid. These cards are also designed to interoperate with the Cisco ONS 15216 DWDM and ONS 15454 MSTP solutions.

- **OC192 SR/STM64 IO 1310 card**—The OC192 SR/STM64 IO 1310 card provides one short-reach OC-192/STM-64 port in the 1310-nm wavelength range, compliant with ITU-T G.707, ITU-T G.957, and Telcordia GR-253-CORE. The port operates at 9.95328 Gbps over unamplified distances up to 2 km. The card supports VT and nonconcatenated or concatenated payloads. The OC192 SR/STM64 IO 1310 card is restricted to slots 5, 6, 12, and 13 of the chassis and can be used only with the XC10G cross-connect card.

- **OC192 IR/STM64 SH 1550 card**—The OC192 IR/STM64 SH 1550 card provides one intermediate-reach OC-192/STM-64 port in the 1550-nm wavelength range, compliant with ITU-T G.707, ITU-T G.957, and Telcordia GR-253-CORE. The port operates at 9.95328 Gbps over unamplified distances up to 40 km with SMF-28 fiber limited by loss and/or dispersion. The card supports VT and nonconcatenated or concatenated payloads. The OC192 IR/STM64 SH 1550 card is restricted to slots 5, 6, 12, and 13 of the chassis and can be used only with the XC10G cross-connect card.

- **OC192 LR/STM64 LH 1550 card**—The OC192 LR/STM64 LH 1550 card provides one long-range, Telcordia-compliant, GR-253 OC-192/STM-64 port per card, compliant with ITU-T G.707, ITU-T G.957, and Telcordia GR-253-CORE. The port operates at 9.95 Gbps over an SMF span up to 60 km. The card supports VT and nonconcatenated or concatenated payloads at STS-1, STS-3c, STS-6c, STS-12c, STS-48c, or STS-192c signal levels. The OC192 LR/STM64 LH 1550 card is restricted to slots 5, 6, 12, and 13 of the chassis and can be used only with the XC10G cross-connect card.

- **OC192 LR/STM64 LH ITU DWDM card**—Sixteen distinct OC192/STM64 ITU 100-GHz DWDM cards comprise the ONS 15454 DWDM channel plan. The OC192 LR/STM64 LH ITU 15xx.xx card provides one long-reach STM-64/OC-192 port per card, compliant with ITU-T G.707, ITU-T G.957, and Telcordia GR-253-CORE. The port operates at 9.95328 Gbps over unamplified distances up to 60 km with standard SMF-28 non-dispersion-shifted fiber (NDSF) fiber. With other fiber types, the transmission distance might be limited due to optical loss or dispersion penalties. This card is restricted to slots 5, 6, 12, and 13 of the chassis and must be used only with the XC-10G cross-connect card.

ONS 15454 Ethernet Cards

The ONS 15454 Ethernet cards eliminate the need for external Ethernet aggregation equipment and provide efficient transport and coexistence of traditional TDM traffic with packet-switched data traffic. Multiple Ethernet cards installed in an ONS 15454 can operate independently or act as a single switch, known as EtherSwitch, supporting a variety of SONET port configurations. The ONS 15454 supports the E-Series, G-Series, and ML-Series cards. The E-Series cards support full Layer 2 switching, whereas the G-Series cards do not provide a switching function on the card. The G-Series cards perform only the Ethernet-to-SONET mapping function. The ML-Series cards perform full multilayer Layer 2/3 switching and routing.

Three gigabit interface converter (GBIC) or SFP modules can be used with the ONS 15454 Ethernet cards, as follows:

- The IEEE 1000BASE-SX-compliant, 850-nm, optical module (SX) is designed for multimode fiber and distances of up to 722 feet (220 meters) on 62.5-micron fiber and up to 1804 feet (550 meters) on 50-micron fiber.
- The IEEE 1000BASE-LX-compliant, 1300-nm, optical module (LX) is designed for single-mode fiber and distances of up to 6.2 miles (10 kilometers).
- The IEEE 1000BASE-ZX-compliant, 1550-nm, optical module (ZX) is designed for single-mode fiber and distances of up to 50 miles (80 kilometers).

The ONS 15454 Ethernet cards are as follows:

- **E100T-G card**—The ONS 15454 uses E100T-G cards for 10/100-Mbps Ethernet. Each card provides 12 switched, IEEE 802.3-compliant, 10/100BASE-T Ethernet interfaces. Each port autoconfigures itself for speed, duplex, and flow control. You can also manually configure Ethernet ports. Each E100T-G card supports wire-speed, Layer 2 Ethernet switching between its Ethernet ports. The IEEE 802.1Q tag and port-based VLANs logically isolate customer traffic. Priority queuing is also supported to provide multiple classes of service. The E100T-G card operates with the XC, XCVT, and XC10G cross-connect cards.
- **E1000-2-G card**—The ONS 15454 uses the E1000-2-G cards for 1000-Mbps Gigabit Ethernet. Each card provides two ports of IEEE-compliant, 1000-Mbps interfaces for high-capacity customer LAN interconnections. Each port autoconfigures itself for half or full duplex and flow control. The E1000-2-G card supports SX and LX GBICs. The E100T-G card operates with the XC, XCVT, and XC10G cross-connect cards.

- **G1K-4 card**—The G1K-4 card provides four ports of IEEE-compliant, 1000-Mbps interfaces. Each interface supports full-duplex operation for a maximum bandwidth of 1 Gbps or 2 Gbps bidirectional per port, and 2.5 Gbps or 5 Gbps bidirectional per card. Each port autoconfigures itself for half or full duplex and flow control. The card operates on Layer 1 and can be used only for point-to-point circuits. The G1K-4 card performs an Ethernet-to-SONET mapping function with an OC-48c of maximum SONET bandwidth per G1K-4 card. The G1K-4 card supports three types of standard Cisco GBIC modules: SX, LX, and ZX.
- **ML100T-12 card**—The ML100T-12 card provides 12 ports of IEEE 802.3-compliant 10/100 interfaces. Each interface supports full-duplex operation for a maximum bandwidth of 200 Mbps per port and 2.488 Gbps per card. Each port autoconfigures itself for speed, duplex, and flow control. The card provides high-throughput, low-latency packet switching of Ethernet-encapsulated traffic (IP and other Layer 3 protocols) across a SONET network while using the inherent self-healing capabilities of SONET protection. Each ML100T-12 card supports wire-speed, Layer 2 Ethernet switching between its Ethernet ports. The IEEE 802.1Q tag and port-based VLANs logically isolate customer traffic. Priority queuing is also supported to provide multiple classes of service. An ONS 15454 with an XC10G card can host the card in any traffic card slot, but an ONS 15454 with an XC or XCVT can host the card only in slots 5, 6, 12, and 13.
- **ML1000-2 card**—The ML1000-2 Gigabit Ethernet card provides two ports of IEEE-compliant, 1000-Mbps interfaces. Each interface supports full-duplex operation for a maximum bandwidth of 2 Gbps per port and 4 Gbps per card. Each port autoconfigures itself for half or full duplex and flow control. The card provides high-throughput, low-latency packet switching of Ethernet-encapsulated traffic (IP and other Layer 3 protocols) across a SONET network while using the inherent self-healing capabilities of SONET protection. The ML1000-2 card uses SFP modules for the optical interfaces. The card supports two types of standard SFP modules: SX and LX. An ONS 15454 with an XC10G card can host the card in any traffic card slot, but an ONS 15454 with an XC or XCVT can host the card only in slots 5, 6, 12, and 13.

ONS 15454 Storage Area Network (SAN) Cards

The ONS 15454 SL-Series card provides Fibre Channel/Fiber Connection (FICON) transport services over SONET or the ONS 15454 optical transport platform. The SL-Series card is a single-slot card that offers four client ports, each supporting 1.0625 or 2.125 Gbps Fibre Channel or FICON. This provides 16 protected, full line-rate 1-Gbps Fibre Channels (each requiring two OC-192 rings) or 40 subrate Fibre Channels per network element. The SL-Series card can be deployed in any of the four high-speed (STS-48) slots in the 2.5-Gbps ONS 15454 shelves. The SL-Series uses GBIC optical modules for the client-side interfaces.

The SL-Series card Uses Generic Framing Procedure-Transparent (GFP-T) to deliver low-latency SAN transport in compliance with the ITU-T G.7041 GFP-T for SONET encapsulation specification. The SL-Series card takes advantage of SONET resiliency schemes offered by the ONS 15454. The SL-Series offers subrate capability and VCAT. These features not only optimize the use of the SONET bandwidth but also allow the service provider to offer Fibre Channel services at increments of 51.84 Mbps.

The SL-Series uses the buffer-to-buffer credits supported by the connected Fibre Channel switching devices to overcome distance limitations in 1-2 Gbps line-rate SAN extension applications. Furthermore, distance extension functions via R_RDY spoofing enables the SL-Series to serve as an integrated Fibre Channel extension device, obviating the need for external SAN extension devices. The SL-Series solution supports distances up to 2800 km.

ONS 15454 Card Protection

The ONS 15454 provides 1:1 and 1:N electrical protection and 1+1 optical protection methods. The ONS 15454 also supports protection switching and protection channel access (PCA).

Electrical Protection

The ONS 15454 supports 1:N protection, which allows a single card to protect several working cards. A DS1N-14 card provides protection for up to five DS1-14 cards, and a DS3N-12/DS3N-12E card protects up to five DS3-12/DS3-12E cards. However, the standard DS1-14 card and DS3-12 card can provide 1:1 protection only. The 1:N protection mechanism operates only at the DS1 and DS3 levels. The 1:N protect cards must match the levels of their working cards. For example, a DS1N-14 protects only DS1-14 or other DS1N-14 cards, and a DS3N-12 protects only DS3-12 or other DS3N-12 cards. 1:N cards have added circuitry to act as the protection card in a 1:N protection group. Otherwise, the card is identical to the standard card and can serve as a normal working card. 1:1 protection in the ONS 15454 supports revertive and nonrevertive switching.

Optical Protection

The ONS 15454 supports 1+1 protection to create redundancy for optical cards and spans. With 1+1 protection, one optical port can protect another optical port. In any two optical slots, a single working card and a single dedicated protect card of the same type can be paired for protection. If the working port fails, the protect port takes over. 1+1 span protection can be either revertive or nonrevertive.

Multiport cards support port-to-port protection. However, the ports on the protect card must support the corresponding ports on the working card. (That is, port 1 on the protect card supports port 1 on the working card.)

Protection Switching

The ONS 15454 SDH supports revertive and nonrevertive, unidirectional or bidirectional switching for optical signals. 1:N electrical protection is always revertive and bidirectional.

1:1 electrical protection is also bidirectional but provides a revertive or nonrevertive option. The ONS 15454 supports APS. APS can be configured as revertive or nonrevertive. When a failure occurs, APS switches the signal from the working card to the protect card. However, nonrevertive APS switching does not revert the traffic to the working card automatically when the working card reverts to active status. However, when a failure is cleared, revertive APS switching automatically switches the signal back to the working card after the provisionable revertive time period has elapsed.

When a failure occurs to a signal that is provisioned as bidirectional, both the transmit and receive signals are switched away from the point of failure (the port or card). A unidirectional signal switches only the failure direction, either transmit or receive.

ONS 15454 SDH MSPP

The ONS 15454 SDH is a multiservice ADM that combines SDH transport, DWDM, along with multiservice IP, Ethernet, and TDM services. The platform supports bit rates up to $32 \times$ STM-64 and offers integrated bandwidth management. The ONS 15454 SDH supports all metro topologies, such as point to point, linear add/drop, rings, and mesh. The platform supports multiple service interfaces including E1, E3, DS3i-N-12, 10/100/1000-Mbps Ethernet and optical STM-1, STM-4, STM-16, and STM-64 interfaces. The ONS 15454 SDH also serves as a distributed bandwidth manager with nonblocking cross-connect capacity and packet-/frame-switching functionality for VC-4-Nc, VC-3-Nc, and VC-12-Nc bandwidth levels.

The ONS 15454 SDH enables optical network designers to terminate multiple rings or linear systems on a single chassis and to mix and match the service interfaces that enable a unified data, voice, and video multiservice network. The ONS 15454 SDH supports SNCP rings, two- and four-fiber MS-SPRing rings, unidirectional and bidirectional linear circuits, enhanced SNCP mesh, and spanning-tree protection mechanisms.

Layer 3 IP functionality is achieved by adding the ML-Series Ethernet cards. The ML-Series cards provide IP switching along with support for static routing, RIPv2, EIGRP, OSPF, BGP, IS-IS, HSRP, VPN routing and forwarding (VRF-lite), and QoS and multicast support. The ONS 15454 SDH supports the Cisco Transport Controller (CTC) tool, which enables initial provisioning and OAM&P support. CTC permits A-to-Z circuit provisioning, auto NE discovery with network topology, and custom bandwidth management, enabling service providers to design transport networks around subscriber needs rather than around equipment limitations.

The ONS 15454 SDH Platform

The ONS 15454 is a CE Mark-compliant device. It has a lower- and upper-shelf assembly. The lower-shelf assembly contains 17 card (module) slots, a backplane interface, a fan-tray assembly, a front panel with an LCD, and alarm indicators. ONS 15454 has 17 card slots numbered 1 to 17. All slots are card ready, meaning that when you plug in a card it automatically boots up and becomes ready for service. The ONS 15454 SDH houses five types of cards: common control, alarm interface, electrical, optical, and Ethernet.

The common control cards include the advanced timing communications and control card (TCC2), the 10 Gigabit cross-connect card (XC10G), the international cross-connect, 10 Gigabit AU3/AU4 high-capacity tributary card (XC-VXL-10G), and the international cross-connect, 2.5 Gigabit AU3/AU4 high-capacity tributary card (XC-VXL-2.5G). Slot 9 is reserved for the alarm interface controller-international card (AIC-I). The BITS timing connector, LAN, alarms, modem, and craft port use backplane wire-wrap pins.

The upper-shelf assembly has 12 front-mounted electrical connector (FMEC) slots numbered 18 to 29. FMECs provide serial ports, LAN ports, a modem connection for future use, electrical connections, redundant power supplies, timing connections, and alarm connections for the AIC-I card. Figure 8-2 shows the ONS 15454 SDH chassis with its air filter, fan tray, fiber routing panels, FMEC slots, and card slots.

Figure 8-2 *ONS 15454 SDH Chassis*

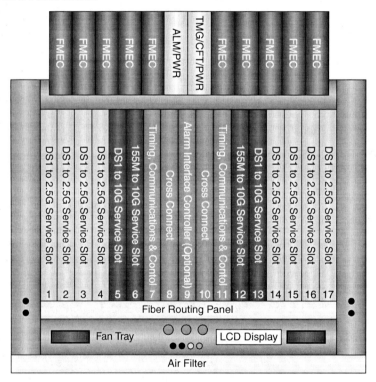

ONS 15454 SDH Timing and Control (TCC2) Card

The TCC2s control the main processing functions of the ONS 15454 SDH. The TCC2s also have an RS-232 craft port for CTC. The TCC2 combines timing, control, and switching functions including the following:

- System initialization
- Provisioning

- Alarm reporting
- Maintenance
- Diagnostics
- IP address detection and resolution
- Timing (Stratum 3 ITU-T G.813)
- SDH data communications channel (DCC) termination
- System fault detection

The CRIT, MAJ, MIN, and REM alarm LEDs on the TCC2 faceplate indicate whether a critical, major, minor, or remote alarm is present anywhere on the ONS 15454 or on a remote node in the network.

The node name, configuration database, IP address, and system software are stored in the TCC2 card's NVRAM, which allows quick recovery if power or card failures occur. System software can be upgraded without affecting traffic on the ONS 15454 SDH if dual TCC2 cards are used. The upgrade takes place first on the standby TCC2. The system verifies that the upgrade is successful and switches from the active TCC2 card running the older release to the upgraded standby TCC2 card running the newer release. After the switch, the second TCC2 card undergoes the upgrade. The TCC2 then loads new software to each of the installed line (traffic) cards.

ONS 15454 SDH Cross-Connect (XC) Cards

The cross-connect card is the central switching element in the ONS 15454 SDH. The ONS 15454 SDH offers three cross-connect cards: the XC10G, XC-VXL-10G, and XC-VXL-2.5G. Circuit cross-connect information is provisioned using CTC. The TCC2 then establishes the proper internal cross-connect information and relays the setup information to the cross-connect card. The various cross-connect cards used by the ONS 15454 SDH are as follows:

- **XC10G card**—The XC10G card cross-connects standard VC-4, VC-4-4c, VC-4-16c, and VC-4-64c signal rates and the nonstandard VC-4-2c, VC-4-3c, and VC-4-8c signal rates providing a maximum of 384 × 384 VC-4 cross-connections. Any VC-4 on any port can be connected to any other port, meaning that the virtual circuit (VC) cross-connection capacity is nonblocking. The XC10G card manages up to 192 bidirectional VC-4 cross-connects.

- **XC-VXL-10G card**—The international cross-connect 10 gigabit AU3/AU4 high-capacity tributary card (XC-VXL-10G) cross-connects E1, E3, DS3, STM-1, STM-4, and STM-16 signal rates. The XC-VXL-10G provides a maximum of 384 × 384 VC-4 nonblocking cross-connections, 384 × 384 VC-3 nonblocking cross-connections, or 2016 × 2016 VC-12 nonblocking cross-connections. It is designed for 10-Gbps solutions. The XC-VXL-10G card manages up to 192 bidirectional STM-1 cross-connects, 192 bidirectional E3 or DS3 cross-connects, or 1008 bidirectional E1 cross-connects. The TCC2 assigns bandwidth to each slot on a per STM-1 basis. The XC-VXL-10G card works with the TCC2 card to maintain connections and set up cross-connects within the system.

- **XC-VXL-2.5G card**—The international cross connect 2.5-gigabit AU3/AU4 high-capacity tributary card (XC-VXL-2.5G) cross-connects E1, E3, DS3, STM-1, STM-4, STM-16, and STM-64 signal rates. The XC-VXL-2.5G provides a maximum of 192 × 192 VC-4 nonblocking cross-connections, 384 × 384 VC-3 nonblocking cross-connections, or 2016 × 2016 VC-12 nonblocking cross-connections. It is designed for 2.5-Gbps solutions.

   The XC-VXL-2.5G card manages up to 192 bidirectional STM-1 cross-connects, 192 bidirectional E3 or DS3 cross-connects, or 1008 bidirectional E1 cross-connects. The TCC2 assigns bandwidth to each slot on a per STM-1 basis. The XC-VXL-2.5G card works with the TCC2 card to maintain connections and set up cross-connects within the system.

AIC-I Card

The optional AIC-I card provides customer-defined alarm I/O and supports user data, data communications, and local and express orderwire channels. The AIC-I card provides I/O alarm contact closures. You can define up to 16 external alarm inputs and 4 external, user-configurable, alarm inputs/outputs. The physical connections are made using the MIC-A/P card. The alarms are defined using CTC and TL1. LEDs on the front panel of the AIC-I indicate the status of the alarm lines. Orderwire is a 64-kbps channel in the SDH header that installation technicians can use to communicate with other technicians in-band. The ONS 15454 SDH supports up to four orderwire channel terminations per shelf. The orderwire ports are standard RJ-11 receptacles.

The AIC-I allows simultaneous use of both local (section overhead signal) and express (line overhead channel) orderwire channels on a SONET/SDH ring or particular optics facility. Express orderwire also allows communication via regeneration sites when the regenerator is not a Cisco device. Orderwire is provisioned using CTC. The various channels used by the AIC-I card are as follows:

- **User data channel**—The user data channel (UDC) features a dedicated 64-kbps data channel (F1 byte) between two nodes in an ONS 15454 SDH network. Each AIC-I card provides two UDCs, UDC-A and UDC-B, through separate RJ-11 connectors on the front of the AIC-I. Each UDC can be routed to an individual optical interface in the ONS 15454 SDH system. All of the UDCs are provisioned using CTC.

- **Data communication channel/generic communication channel**—The data communications channel/generic communication channel (DCC/GCC) features a dedicated data channel of 5712 kbps (D4 to D12 bytes) between two nodes in an ONS 15454 SDH network. Each AIC-I card provides two DCC/GCCs, DCC-A and DCC-B, through separate RJ-45 connectors on the front of the AIC-I. Each DCC/GCC can be routed to an individual optical interface in the ONS 15454 SDH system. Each node that contains TCC2 cards can terminate up to 32 section DCCs. DCC connections cannot be provisioned if DCC tunneling is configured on the span.

ONS 15454 SDH FMECs

Front-mounted electrical connections (FMECs) provide serial ports, LAN ports, a modem connection for future use, electrical connections, redundant power supplies, timing connections, and alarm connections for the AIC-I card. The ONS 15454 SDH upper-shelf assembly has 12 FMEC slots numbered 18 to 29. FMEC slots 18 to 22 support the electrical cards in slots 1 to 5 of the lower shelf. FMEC slots 25 to 29 support the electrical cards in slots 13 to 17 of the lower shelf. FMEC slot 23 is used for the MIC-A/P alarm and power card. FMEC slot 24 supports the MIC-C/T/P timing, craft, and power card. The E1-75/120, which also provides FMEC, is not installed in an FMEC slot. Instead, it is mounted directly on the rack. The various FMECs cards used by the ONS 15454 SDH are as follows:

- **E1 FMEC**—The E1 FMEC card enables you to terminate unbalanced E1 interfaces. The card provides FMEC for 14 ITU-G.703-compliant E1 ports.
- **DS1/E1 FMEC**—The DS1/E1 FMEC card enables you to terminate balanced E1 interfaces. The card provides FMEC for 14 ITU-compliant, G.703 E1 ports.
- **E3/DS3 FMEC**—The E3/DS3 FMEC card enables you to terminate E3 interfaces. The card provides FMEC for 12 ITU-G.703-compliant E3 or DS3 ports.
- **BLANK FMEC**—The BLANK card covers empty FMEC slots and fulfills EMC requirements.
- **MIC-A/P FMEC**—The MIC-A/P FMEC card provides system power and alarm connections for the TCC2 and AIC-I cards. The MIC-A/P card also stores manufacturing and inventory data. For proper system operation, the MIC-A/P FMEC card must be installed in the shelf.
- **MIC-C/T/P FMEC**—The MIC-C/T/P FMEC card provides front-panel access for the Timing A and Timing B connectors, two system power connectors at −48V, and two standard eight-pin modular LAN connectors for each TCC2 card. For proper system operation, the MIC-C/T/P FMEC card must be installed in the shelf.
- **FMEC E1-120NP**—The ONS 15454 SDH FMEC E1-120NP card provides FMEC for 42 ITU-compliant, G.703 E1 ports. Each FMEC E1-120NP card port features E1-level inputs supporting cable losses of up to 6 dB at 1024 kHz. With the FMEC E1-120NP card, each E1-42 port operates at 2.048 Mbps over a 120-ohm balanced interface. Twenty-one interfaces are led through one common Molex 96-pin LFH connector.
- **FMEC E1-120PROA**—The ONS 15454 SDH FMEC E1-120PROA card provides FMEC for 42 ITU-compliant, G.703 E1 ports. Each FMEC E1-120PROA card port features E1-level inputs supporting cable losses of up to 6 dB at 1024 kHz. With the FMEC E1-120PROA card, each E1-42 port operates at 2.048 Mbps over a 120-ohm balanced interface. Twenty-one interfaces are led through one common Molex 96-pin LFH connector.
- **FMEC E1-120PROB**—The ONS 15454 SDH FMEC E1-120PROB card provides front mount electrical connection for 42 ITU-compliant, G.703 E1 ports. Each FMEC E1-120PROB card port features E1-level inputs supporting cable losses of up to 6 dB at

1024 kHz. With the FMEC E1-120PROB card, each E1-42 port operates at 2.048 Mbps over a 120-ohm balanced interface. Twenty-one interfaces are led through one common Molex 96-pin LFH connector.

- **FMEC STM1E NP**—The ONS 15454 SDH FMEC STM1E NP card provides FMEC for 12 ITU-compliant, G.703 STM1E ports. Ports 9 to 12 can be switched to E4 rather than STM-1 (via CTC on the STM1E-12 card). With the FMEC STM1E NP card, each interface of an STM1E-12 card operates at 155.52 Mbps for STM-1 or 139.264 Mbps for E4 over a 75-ohm unbalanced coaxial 1.0/2.3 miniature coax connector. Each FMEC STM1E NP card interface features STM1-level inputs supporting cable losses of up to 12.7 dB at 78 MHz.

- **FMEC STM1E 1:1**—The ONS 15454 SDH FMEC STM1E 1:1 card provides FMEC for 12 ITU-compliant, G.703 STM1E ports. Ports 9 to 12 can be switched to E4 rather than STM-1 (via CTC on the STM1E-12 card). With the FMEC STM1E 1:1 card, each interface of an STM1E-12 card operates at 155.52 Mbps for STM-1 or 139.264 Mbps for E4 over a 75-ohm unbalanced coaxial 1.0/2.3 miniature coax connector. Each FMEC STM1E 1:1 card interface features STM1-level inputs supporting cable losses of up to 12.7 dB at 78 MHz.

  The FMEC STM1E 1:1 card is required if you want to use the 1:1 protection feature of the STM1E-12 card. You can also use the FMEC STM1E 1:1 for connection to two unprotected STM1E-12 cards.

- **FMEC STM1E 1:3**—The ONS 15454 SDH FMEC STM1E 1:3 card provides FMEC for 12 ITU-compliant, G.703 STM1E ports. Ports 9 to 12 can be switched to E4 rather than STM-1 (via CTC on the STM1E-12 card). With the FMEC STM1E 1:3 card, each interface of an STM1E-12 card operates at 155.52 Mbps for STM-1 or 139.264 Mbps for E4 over a 75-ohm unbalanced coaxial 1.0/2.3 miniature coax connector. Each FMEC STM1E 1:3 card interface features STM1-level inputs supporting cable losses of up to 12.7 dB at 78 MHz.

  The FMEC STM1E 1:3 card is required if you want to use the 1:3 protection feature of the STM1E-12 card. You can also use the FMEC STM1E 1:3 for connection to four unprotected STM1E-12 cards.

ONS 15454 SDH Electrical Cards

The electrical cards provide the ports and physical interfaces for TDM signal ingress and egress. Lower-shelf slots 1 to 5 and 13 to 17 host all electrical cards (E1-N-14, E3-12, DS3i-N-12, and E1-42). Each card has faceplate LEDs showing active, standby, or alarm status. You can also obtain the status of all electrical card ports using the LCD screen on the fan-tray assembly. The upper-shelf assembly has 12 FMEC slots numbered 18 to 22 and 25 to 29. These slots support the electrical cards in slots 1 to 5 and 13 to 17 of the lower shelf. Each FMEC slot supports the

NOTE Optical cards and Ethernet cards have connectors on the faceplate rather than FMEC connections. Some Ethernet cards require GBICs that plug into the card faceplate.

electrical card in a specific slot of the lower shelf. The various electrical cards used by the ONS 15454 SDH are as follows:

- **E1-N-14 card**—The E1-N-14 card provides 14 ITU-compliant, G.703 E1 ports. Each port operates at 2.048 Mbps over a 120-ohm twisted-pair copper cable with the DS1/E1 FMEC, or over a 75-ohm unbalanced coaxial cable with the E1 FMEC. The E1-N-14 card can be used as a working or protect card in 1:1 or 1:N protection schemes. If you use the E1-N-14 as a standard E1 card in a 1:1 protection group, you can install the E1-N-14 card in slots 1 to 6, 12 to 14, or 17. To use the 1:N functionality of the card, you must install an E1-N-14 card in slot 3 or slot 15.

- **E3-12 card**—The E3-12 card provides 12 ITU-compliant, G.703 E3 ports per card. Each port operates at 34.368 Mbps over a 75-ohm coaxial cable with the E3/DS3 FMEC.

  The E3-12 card can be used as a working or protect card in 1:1 protection schemes. When creating circuits, the E3-12 card must use port grouping with VC low-order path tunnels. Three ports form one port group. In one E3-12 card, for example, there are four port groups: Ports 1 to 3 comprise PG1, ports 4 to 6 comprise PG2, ports 7 to 9 comprise PG3, and ports 10 to 12 comprise PG4.

- **DS3i-N-12 card**—The DS3i-N-12 card provides 12 DS3 ports per card, compliant with ITU-T G.703, Telcordia GR-499, and ITU-T G.704. Each port operates at 44.736 Mbps over a 75-ohm coaxial cable with the E3/DS3 FMEC. The 12-port DS3i-N-12 card provides enhanced performance-monitoring functions. By monitoring additional overhead in the DS3 frame, subtle network degradations are detected.

  The DS3i-N-12 can operate as the protect card in a 1:N (N < 4) DS3 protection group. It can protect up to four working DS3i-N-12 cards. When creating circuits, the DS3i-N-12 card must use port grouping with VC low-order path tunnels.

- **E1-42 card**—The 42-port ONS 15454 SDH E1-42 card provides 42 ITU-compliant, G.703 E1 ports. Each port of the E1-42 card operates at 2.048 Mbps over a 120-ohm, twisted-pair copper cable. FMEC is done using the FMEC E1-120 NP card for unprotected operation, the FMEC E1-120PROA for 1:3 protection in the left side of the shelf, or the FMEC E1-120PROB for 1:3 protection in the right side of the shelf. If you need 75-ohm unbalanced interfaces, you must additionally use the E1-75/120 conversion panel.

- **STM1E-12 card**—The 12-port STM1E-12 card provides 12 ITU-compliant, G.703 STM-1 ports per card. Each port features ITU-T G.703-compliant inputs supporting cable losses of up to 12.7 dB at 78 MHz. The STM1E-12 card supports 1:1 protection and 1:3 protection. Ports 9 to 12 can be remotely switched to E4 rather than STM-1 using the CTC software. Each interface operates at 155.52 Mbps for STM-1 or 139.264 Mbps for E4 over a 75-ohm coaxial cable. In E4 mode, framed or unframed signal operation is possible.

ONS 15454 SDH Optical Cards

The optical cards, with the exception of the STM-64 cards, reside in slots 1 to 6 or 12 to 17. The STM-64 cards reside in slot 5, 6, 12, or 13. You can provision an optical card as a drop card or span card in a linear ADM (1+1), SNCP, or MS-SPRing protection scheme.

Each card faceplate has three card-level LED indicators. When illuminated, the red FAIL LED represents a hardware problem, the amber signal failure (SF) LED represents a signal failure or condition (for example, a loss of frame or a high bit error rate), and the green ACT LED indicates that the card is operational. ONS 15454 SDH optical cards have SC or LC fiber connectors on the card faceplate. The various optical cards used by the ONS 15454 SDH are as follows:

- **STM1 SH 1310 card**—The OC3 IR 4/STM1 SH 1310 card provides four intermediate- or short-range ports compliant with ITU-T G.707 and G.957. Each port operates at 155.52 Mbps over an SMF span. The card supports payloads at the STM-1 signal level on a per VC-4 basis.

- **STM1 SH 1310-8 card**—The OC3IR/STM1SH 1310-8 card provides eight intermediate- or short-range, ITU-T G.707- and G.957- compliant, SDH, STM-1 ports per card. Each port operates at 155.52 Mbps over an SMF span. The card supports payloads at the STM-1 signal level on a per VC-4 basis.

- **STM4 SH 1310 card**—The OC12 IR/STM4 SH 1310 card provides one intermediate- or short-range port per card, compliant with ITU-T G.707 an SMF and G.957. The port operates at 622.08 Mbps over an SMF span. The card supports concatenated or nonconcatenated VC-4 payloads.

- **STM4 SH 1310-4 card**—The OC12 IR/STM4 SH 1310-4 card provides four intermediate- or short-range ports per card, compliant with Telcordia GR-253 IR-1, Telcordia GR-2918-CORE, ITU-T G.707, and ITU-T G.957. Each port operates at 622.08 Mbps over an SMF span. The card supports concatenated or nonconcatenated VC-4 payloads.

- **STM4 LH 1310 card**—The OC12 LR/STM4 LH 1310 card provides one long-range port per card, compliant with ITU-T G.707 and G.957. The port operates at 622.08 Mbps over an SMF span. The card supports concatenated or nonconcatenated VC-4 payloads.

- **STM4 LH 1550 card**—The OC12 LR/STM4 LH 1550 card provides one long-range port per card, compliant with ITU-T G.707 and G.957. The port operates at 622.08 Mbps over an SMF span. The card supports concatenated or nonconcatenated VC-4 payloads.

- **STM16 SH AS 1310 card**—The OC48 IR/STM16 SH AS 1310 card provides one intermediate-range port per card, compliant with ITU-T G.707 and G.957. The port operates at 2.488 Gbps over an SMF span. The card supports concatenated or nonconcatenated VC-4 payloads.

- **STM16 LH AS 1550 card**—The OC48 IR/STM16 LH AS 1550 card provides one long-range port per card, compliant with ITU-T G.707 and G.957. The port operates at 2.488 Gbps over an SMF span. The card supports concatenated or nonconcatenated VC-4 payloads.

- **STM16 EH 100-GHz DWDM card**—Eighteen distinct OC48 ELR/STM16 EH 100-GHz DWDM cards provide the ONS 15454 SDH STM-16 DWDM channel plan. Although the ONS 15454 SDH uses 200-GHz spacing, these cards also work in 100-GHz-capable systems. Each STM-16 DWDM card provides one port per card, compliant with ITU-T G.692, G.707, G.957, and G.958. The port operates at 2.488 Gbps over an SMF span. The card supports concatenated or nonconcatenated VC-4 payloads.

Nine of the cards operate in the blue band with spacing of 100 GHz on the ITU grid. The other nine cards operate in the red band with spacing of 100 GHz on the ITU grid.

- **STM64 LH 1550 card**—The OC192 LR/STM64 LH 1550 card provides one long-range port per card, compliant with ITU-T G.707 and G.957. The port operates at 9.95 Gbps over unamplified distances of up to 80 km with different types of fiber, such as C-SMF or dispersion-compensated fiber limited by loss and/or dispersion. The card supports concatenated or nonconcatenated VC-4 payloads.

- **STM64 LH ITU DWDM card**—The OC192 LR/STM64 LH ITU 15xx.xx card provides one long-range, ITU-T G.707- and G.957-compliant, SDH, STM-64 port per card. The port operates at 9.95328 Gbps over unamplified distances up to 60 km with different types of fiber, such as C-SMF or dispersion-compensated fiber limited by loss and/or dispersion. The card supports concatenated or nonconcatenated payloads on a VC-4 basis, as well as VC-4, VC-3, and VC-12 payloads.

  The ONS 15454 SDH STM-64 DWDM channel plan is composed of eight distinct STM-64, ITU, 100-GHz DWDM cards. Four of the cards operate in the blue band with spacing of 100 GHz in the ITU grid (1534.25 nm, 1535.04 nm, 1535.82 nm, and 1536.61 nm). The other four cards operate in the red band with spacing of 100 GHz in the ITU grid (1550.12 nm, 1550.92 nm, 1551.72 nm, and 1552.52 nm).

ONS 15454 SDH Ethernet Cards

The ONS 15454 SDH Ethernet cards eliminate the need for external Ethernet aggregation equipment and provide efficient transport and coexistence of traditional TDM traffic with packet-switched data traffic. Multiple Ethernet cards installed in an ONS 15454 SDH can operate independently or act as a single switch known as EtherSwitch, supporting a variety of SONET port configurations. The ONS 15454 supports the E-Series, G-Series, and ML-Series cards. The E-Series cards support full Layer 2 switching, whereas the G-Series cards do not provide a switching function on the card. The G-Series cards perform only the Ethernet-to-SDH mapping function. The ML-Series cards perform full multilayer Layer 2/3 switching and routing.

Three GBIC and SFP modules can be used with the ONS 15454 Ethernet cards:

- The IEEE 1000BASE-SX compliant, 850-nm, optical module (SX) is designed for multimode fiber and distances of up to 722 feet (220 meters) on 62.5-micron fiber and up to 1804 feet (550 meters) on 50-micron fiber.

- The IEEE 1000BASE-LX compliant, 1300-nm, optical module (LX) is designed for single-mode fiber and distances of up to 6.2 miles (10 kilometers).

- The IEEE 1000BASE-ZX compliant, 1550-nm, optical module (ZX) is designed for single-mode fiber and distances of up to 50 miles (80 kilometers).

The ONS 15454 SDH Ethernet cards are as follows:

- **E100T-G card**—The ONS 15454 SDH uses E100T-G cards for 10/100-Mbps Ethernet. Each card provides 12 switched, IEEE 802.3-compliant, 10/100BASE-T Ethernet interfaces. Each port autoconfigures itself for speed, duplex, and flow control. You can also manually configure Ethernet ports. Each E100T-12 card supports wire-speed, Layer 2 Ethernet switching between its Ethernet ports. The IEEE 802.1Q tag and port-based VLANs logically isolate customer traffic. Priority queuing is also supported to provide multiple classes of service.

- **E1000-2-G card**—The ONS 15454 SDH uses the E1000-2-G cards for 1000-Mbps Gigabit Ethernet. Each card provides two ports of IEEE-compliant, 1000-Mbps interfaces for high-capacity customer LAN interconnections. Each port autoconfigures itself for half or full duplex and flow control. The E1000-2-G card supports SX and LX GBICs.

- **G1K-4 card**—The G1K-4 card provides four ports of IEEE-compliant 1000-Mbps interfaces. Each interface supports full-duplex operation for a maximum bandwidth of 1 Gbps or 2 Gbps bidirectional per port, and 2.5 Gbps or 5 Gbps bidirectional per card. Each port autoconfigures itself for half or full duplex and flow control. The G1K-4 card supports three types of standard Cisco GBIC modules: SX, LX, and ZX.

- **ML100T-12 card**—The ML100T-12 card provides 12 ports of IEEE 802.3-compliant 10/100 interfaces. Each interface supports full-duplex operation for a maximum bandwidth of 200 Mbps per port and 2.488 Gbps per card. Each port autoconfigures itself for speed, duplex, and flow control. The card provides high-throughput, low-latency packet switching of Ethernet-encapsulated traffic (IP and other Layer 3 protocols) across an SDH network while using the inherent self-healing capabilities of SDH protection. Each ML100T-12 card supports wire-speed, Layer 2 Ethernet switching between its Ethernet ports. The IEEE 802.1Q tag and port-based VLANs logically isolate customer traffic. Priority queuing is also supported to provide multiple classes of service.

- **ML1000-2 card**—The ML1000-2 Gigabit Ethernet card provides two ports of IEEE-compliant, 1000-Mbps interfaces. Each interface supports full-duplex operation for a maximum bandwidth of 2 Gbps per port and 4 Gbps per card. Each port autoconfigures itself for half or full duplex and flow control. The card provides high-throughput, low-latency packet switching of Ethernet-encapsulated traffic (IP and other Layer 3 protocols) across an SDH network while using the inherent self-healing capabilities of SDH protection. The ML1000-2 card uses SFP modules for the optical interfaces. The card supports two types of standard SFP modules: SX and LX.

ONS 15454 SDH Storage Area Network (SAN) Cards

The ONS 15454 SDH SL-Series card provides Fibre Channel/Fiber Connection (FICON) transport services over SDH for the ONS 15454 SDH optical transport platform. The SL-Series card is a single-slot card that offers four client ports, each supporting 1.0625 or 2.125 Gbps Fibre Channel or FICON. This provides 16 protected, full line-rate, 1-Gbps Fibre Channels (each requiring two STM-64 rings) or 40 subrate Fibre Channels per network element. The

SL-Series card can be deployed in any of the 12 interface slots within the ONS 15454 SONET SDH shelves. The SL-Series uses GBIC optical modules for the client-side interfaces.

The SL-Series card GFP-T is used to deliver low-latency SAN transport in compliance with the ITU-T G.7041 GFP-T for SONET and SDH encapsulation specification. The SL-Series card takes advantage of SDH resiliency schemes offered by the ONS 15454 SDH. The SL-Series offers subrate capability and VCAT. These features optimize the use of the SDH bandwidth.

The SL-Series uses the buffer-to-buffer credits supported by the connected Fibre Channel switching devices to overcome distance limitations in 1-2 Gbps line-rate SAN extension applications. Furthermore, distance extension functions via R_RDY spoofing enables the SL-Series to serve as an integrated Fibre Channel extension device, obviating the need for external SAN extension devices. The SL-Series solution supports distances up to 2800 km.

ONS 15454 SDH Card Protection

The ONS 15454 SDH provides 1:1 and 1:N electrical protection and 1+1 optical protection methods. The ONS 15454 SDH also supports protection switching and protection channel access (PCA).

Electrical Protection

In 1:1 protection, a working card is paired with a protect card of the same type. If the working card fails, the traffic from the working card switches to the protect card. When the failure on the working card is resolved, by default traffic automatically reverts to the working card. 1:N protection operates only at the E1 and DS3 levels. The 1:N protect cards must be the same speed as their working cards. For example, an E1-N-14 card protects E1-N-14 cards, and a DS3i-N-12 card protects DS3i-N-12 cards. Each side of the shelf assembly has only one card that protects all the cards on that side.

Optical Protection

The ONS 15454 SDH supports 1+1 protection to create redundancy for optical cards and spans. With 1+1 protection, one optical port can protect another optical port; therefore, in slots 5 to 6 or 12 to 13, there is a single working card and a single dedicated protect card of the same type. For example, two STM-16 cards can be paired for protection. If the working port fails, the protect port takes over. Several of the ONS 15454 SDH optical cards have multiple ports. Multiport cards support port-to-port protection. The ports on the protect card support the corresponding ports on the working card.

Protection Switching

The ONS 15454 SDH supports revertive and nonrevertive, unidirectional or bidirectional switching for optical signals. 1:N electrical protection is always revertive and bidirectional. 1:1

electrical protection is also bidirectional, but provides a revertive or nonrevertive option. The ONS 15454 SDH supports APS. APS can be configured as revertive or nonrevertive. When a failure occurs, APS switches the signal from the working card to the protect card. However, nonrevertive APS switching does not revert the traffic to the working card automatically when the working card reverts to active status. When a failure is cleared, however, revertive APS switching automatically switches the signal back to the working card after the provisionable revertive time period has elapsed. When a failure occurs to a signal that is provisioned as bidirectional, both the transmit and receive signals are switched away from the point of failure (the port or card). A unidirectional signal switches only the failure direction, either transmit or receive.

Protection Channel Access (PCA)

1:1 and 1+1 protection configurations use only a portion of ring bandwidth. The remaining bandwidth is idle until a switch occurs. PCA circuits run through the idle bandwidth on two-fiber and four-fiber MS-SPRings. Because they are considered to be lower priority, PCA circuits are preempted when a switch occurs to make room for the protected circuits. They are restored when the protected circuit is no longer needed. PCA circuits help you to use networks more efficiently by lowering overall network cost per bit, leveraging untapped network bandwidth, and sharing network costs across a larger service base. ONS 15454 SDH PCA circuits are compliant with Telcordia GR-1230-CORE Section 3.4.

ONS 15454 MSTP

The ONS 15454 multiservice transport platform (MSTP) is a multiservice ADM that combines SONET/SDH transport and DWDM with various broadband multiservice interfaces such as the following:

- Fibre Channel
- FICON
- ESCON
- Gigabit Ethernet
- 10 Gigabit Ethernet
- OC-192/STM-64
- OC-3/STM-1
- OC-12/STM-4
- OC-48/STM-16
- D1 video
- HDTV

Applications for the ONS 15454 MSTP include metro fiber relief, transport of SAN protocol and bearer traffic, server and storage clustering, as well as mirroring applications, traditional voice and data services over SONET/SDH, transport of broadband video and HDTV, and all possible broadband IP applications.

The ONS 15454 MSTP integrates various DWDM transmission elements and also includes the following capabilities:

- DWDM of up to 32 ITU-T wavelengths with 100-GHz spacing
- DWDM of up to 64 ITU-T wavelengths with 50-GHz spacing
- One- and four-band flexible OADM functionality
- One-, two-, and four-channel OAMD
- Optical EDFA pre-amplifier
- Optical EDFA booster amplifier
- Amplified OADM functionality
- Optical service channel modules (OSCMs)
- Optical OSCMs with combiner-separator
- Variable optical attenuation
- Optical power monitoring and equalization
- 10 Gigabit multirate transponder with wavelength tunability
- 4 × OC-48, 10 Gigabit transponder with wavelength tunability
- 100-Mbps to 2.5-Gbps multirate transponder with variable client SFP and wavelength tunability
- FE UDC

The Cisco ONS 15454 MSTP integrates full DWDM functionality within the box and reduces dependency on external devices, such as DWDM multiplexers, demultiplexers, OADMS, and SCMs. Table 8-1 shows the wavelengths launched by the ONS 15454 MSTP DWDM modules. These wavelengths fall in the 1530-nm to 1561-nm range.

Table 8-1 *ONS 15454 MSTP 32-Channel Wavelength Plan (100-GHz Spacing)*

λ 1 to 8 (nm)	λ 9 to 16 (nm)	λ 17 to 24 (nm)	λ 25 to 32 (nm)
1530.33	1538.19	1546.12	1554.13
1531.12	1538.98	1546.92	1554.94
1531.90	1539.77	1547.72	1555.75
1532.68	1540.56	1548.51	1556.55
1534.25	1542.14	1550.12	1558.17
1535.04	1542.94	1550.92	1558.98
1535.82	1543.73	1551.72	1559.79
1536.61	1544.53	1552.52	1560.61

The Cisco ONS 15454 MSTP supports the CTC graphical user interface (GUI) craft interface, allowing point-and-click, A-to-Z wavelength provisioning with nodal control. Integration with EMSs and OSSs is accomplished through the CTM element management system, which supports the entire COMET product line. The *MetroPlanner* DWDM optical design software simplifies network engineering and deployment by verifying optical network designs and creating automatic bill-of-material (BOM) documents. It also provides node configurations and fiber connections along with exportable optical setup files that speed up node turn-ups. The ONS 15454 MSTP supports the following node configurations:

- Terminal
- Hub
- Line amplifier
- Optical ADM amplified or unamplified

The ONS 15454 MSTP supports the following network topologies:

- Linear
- Open-ring single hub
- Open-ring multihub
- Closed-ring no hub

An important capability of the Cisco ONS 15454 MSTP is automatic optical power management. Traditional DWDM solutions require considerable manual interaction to turn up, manage, and upgrade DWDM networks. Through strategic optical power monitoring and variable optical attenuation (VOA), Cisco ONS 15454 MSTP software is able to dynamically monitor and control optical power, a critical operation in amplified DWDM networks. The Cisco ONS 15454 MSTP software reconstructs a model of the provisioned DWDM network. Software algorithms automatically provide network-wide optical power management by equalizing channels that are intrinsically unequal, adjusting for optical paths with different insertion losses (add, drop, express, and hitless paths) and maintaining constant power when wavelengths are added or dropped.

ONS 15454 MSTP Platform

The ONS 15454 MSTP is a NEBS-compliant shelf assembly that contains 17 card (module) slots, a backplane interface, a fan-tray assembly, a front panel with an LCD, and alarm indicators. The MSTP shelf assembly supports the SA-ANSI or SA-ETSI version of the 15454 chassis. ONS 15454 has 17 card slots numbered 1 to 17. All slots are card ready, meaning that when you plug in a card it automatically boots up and becomes ready for service. The ONS 15454 MSTP houses five types of cards that include common control, alarm interface, electrical, optical, and Ethernet. The common control cards include the TCC2, XCVT or XC10G, and the AIC. The ONS 15454 MSTP also houses all electrical and optical cards supported by the 15454 MSPP. In addition, the MSTP supports the following cards.

ONS 15454 MSTP Service Interfaces

The ONS 15454 service interface cards include client-side transponder and muxponder interfaces operating at various speeds over protected or unprotected DWDM lines. The various service interfaces used by the ONS 15454 MSTP are as follows:

- **MR-L1-xx.x/MR-1-xx.x card**—100-Mbps to 2.5-Gbps multirate transponder card, SFP client module slot, 100-GHz ITU wavelength, unprotected DWDM line with LC connector. There are eight modules in the MRP-L1 series with four-channel tunable lasers for a total of 32 channels.

- **MRP-L1-xx.x/MRP-1-xx.x card**—100-Mbps to 2.5-Gbps multirate transponder card, SFP client module slot, 100-GHz ITU wavelength, protected DWDM line with LC connector. There are eight modules in the MRP-L1 series with four-channel tunable lasers for a total of 32 channels.

- **10T-L1-xx.x/10T-xx.x card**—10-Gbps multirate transponder card, 10 Gigabit Ethernet LAN physical layer, 10 Gigabit Ethernet WAN physical layer, OC-192, STM-64, 100-GHz ITU wavelength (50-GHz laser stability) DWDM line with LC connector. There are 16 modules in the 10T-L1 series with two-channel tunable lasers for a total of 32 channels.

- **10M-L1-xx.x/10M-xx.x card**—4 × OC-48/STM-16 10-Gbps muxponder card, intermediate-reach, 1310-nm SFP client interface with LC connectors, OC-192/STM-64 100-GHz ITU wavelength (50-GHz laser stability) DWDM line with LC connector. There are 16 modules in the 10M-L1 series with two-channel tunable lasers for a total of 32 channels. This card can accommodate a maximum of four client SFP optics, and can scale from a minimum of one SFP to four SFPs in service.

ONS 15454 MSTP Optical Transmission Elements

The ONS 15454 optical transmission elements include optical service channel cards with integrated combiner and separators. These elements also include C-band pre-amplifiers, C-band amplifiers, multiplexers, demultiplexers, OADMs, and external 15216 dispersion-compensation units. The various optical transmission elements used by the ONS 15454 MSTP are as follows:

- **OSCM card**—Optical service channel card, 1510-nm, LC connector, and includes two 2-meter LC/LC fiber-optic cables

- **OSC-CSM card**—Optical service channel card with integrated combiner and separator, 1510-nm, LC connector, and includes two 2-meter LC/LC fiber-optic cables

- **OPT-PRE card**—Optical pre-amplifier, C-band, 64 channel ready, 50-GHz compatible, LC connector, midstage access for dispersion compensation, and includes two 2-meter 5-dB attenuated LC/LC fiber-optic cables

- **OPT-BST card**—Optical booster amplifier, C-band, 64 channel ready, 50-GHz compatible, LC connector, and includes two 2-meter LC/LC fiber-optic cables

- **32MUX-O card**—32-channel multiplexer card, 100-GHz, monitor port, 8-fiber multipath push-on (MPO) connector
- **32DMX-O card**—32-channel demultiplexer card, 100-GHz, 8-fiber MPO connector, and includes one 2-meter LC/LC fiber-optic cable
- **4MD-xx.x card**—Four-channel multiplexer and demultiplexer card that is used with band OADM cards, 100-GHz, LC connector, and includes two 2-meter LC/LC fiber-optic cables
- **AD-1C-xx.x card**—One-channel OADM, 100-GHz, LC connector, and includes two 2-meter LC/LC fiber-optic cables
- **AD-2C-xx.x card**—Two-channel OADM, 100-GHz, LC connector, and includes two 2-meter LC/LC fiber-optic cables
- **AD-4C-xx.x card**—Four-channel OADM, 100-GHz, LC connector, and includes two 2-meter LC/LC fiber-optic cables
- **AD-1B-xx.x card**—One-band OADM, 100-GHz, LC connector, and includes two 2-meter LC/LC fiber-optic cables
- **AD-4B-xx.x card**—Four-band OADM, 100-GHz, LC connector, and includes two 2-meter LC/LC fiber-optic cables

Cisco Transport Controller (CTC)

Cisco Transport Controller (CTC) is a GUI-based element management tool that can be used for operations, administration, management, and provisioning (OAM&P) of various ONS 15000 MSPPs, such as the ONS 15327, ONS 15454, and ONS 15600. CTC is typically used as a GUI-based craft tool during the deployment and implementation of an ONS-based network. CTC accepts graphical user input and converts the commands to Transaction Language 1 (TL1) commands accepted by the ONS NEs.

CTC software is pre-installed on the TCC2 common control card of the ONS 15454. If the software release of the MSPP needs to be upgraded, a new version of CTC must be installed on the TCC2. The craft tool PC or workstation needs a standard Java-enabled browser with Java Runtime Environment (JRE) installed on the machine. CTC is automatically downloaded as a Java applet from the TCC2 and installed on the PC or workstation after the PC or workstation logs in to the ONS 15454. You can automatically download the CTC software files to ensure your computer is running the same CTC software version as the TCC+ that you are accessing. The computer CTC software files are stored in the temporary directory designated by your computer's operating system. If the files are deleted, they are downloaded the next time you connect to an ONS 15454. Downloading the files takes 1 to 2 minutes. To connect to an ONS 15454, you enter the ONS 15454 IP address in the URL field of the web browser. You can also use TL1 commands to communicate with the Cisco ONS 15454 through VT100 terminals and VT100 emulation software, or you can Telnet to an ONS 15454 using TL1 port 3083. Figure 8-3 shows the CTC console.

Figure 8-3 *ONS 15454 CTC Console*

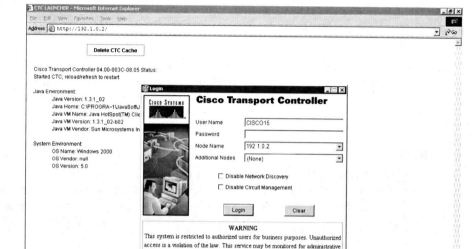

You can connect a PC to the ONS 15454 using the RJ-45 LAN port on the TCC+ or the wire-wrap LAN 1 pins on the ONS 15454 backplane. Each ONS 15454 must have a unique IP address that you use to access the ONS 15454. The address is displayed on the front-panel LCD. The initial IP address, 192.1.0.2, is the default address for ONS 15454 access and configuration. Each computer used to communicate with the ONS 15454 should have only one IP address. Each ONS 15454 comes preconfigured with a user account that has superuser or administrator privileges. The user ID for this account is CISCO15. The CISCO15 account is not assigned a password and cannot be deleted. This account can be used to create other users. It is recommended that a password be assigned to the CISCO15 account. To create one, click the **Provisioning > Security** tab after you log in and change the default password.

When you upgrade CTC software, the TCC stores the older CTC version as the protect CTC version, and the newer CTC release becomes the working version. You can view the software versions that are installed on an ONS 15454 by choosing the **Maintenance** tab followed by the **Software** subtab. Select these tabs in node view to display the software installed on one node. Select the tabs in network view to display the software versions installed on all network nodes.

The CTC provides three views of the ONS 15454/15600/15327 and ONS network:

- Node view displays all cards in a shelf and their administrative, manageable, and configurable options.

- Network view displays all the nodes in a SONET/SDH network as well as its administrative, manageable, and configurable options for the node you are logged in to.
- Card view shows all ports in a card and its administrative, manageable, and configurable options.

Various navigational methods are available within the CTC window to access views and perform management actions. You can double-click and right-click objects in the graphic area and move the mouse over nodes, cards, and ports to view pop-up status information.

CTC Node View

This is the very first *view* that appears when you log in to an ONS 15454. The node view shows a graphic of the ONS 15454 shelf and provides access to tabs and subtabs that you use to administer and provision the node. Figure 8-4 shows the node view and the various areas of the screen.

Figure 8-4 *CTC Node View*

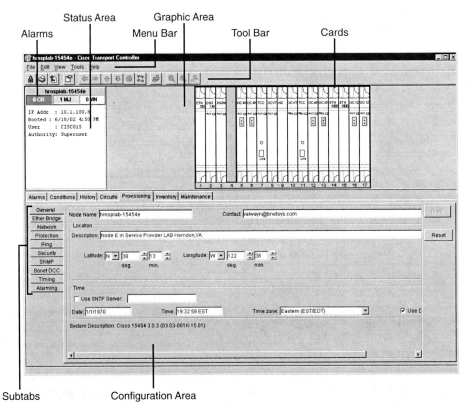

The login node is the first node displayed, and it is the *base view* for the session. Node view enables you to view and manage the ONS 15454 node you are logged in to. The status area shows the node name; IP address; session boot date and time; number of critical (CR), major (MJ), and minor (MN) alarms; the name of the current logged-in user, and the security level of the user. The graphic area of the CTC window depicts the ONS 15454 shelf assembly. The colors of the cards in the graphic reflect the real-time status of the physical card and slot, as shown in Table 8-2. The card graphics also indicate whether they are *active* or *standby*.

Table 8-2 *Node View Card Color and Status*

Card Color	Status
Gray	Slot is not provisioned and card not installed.
Violet	Slot is provisioned and card not installed.
White	Slot is provisioned and a functioning card is installed.
Yellow	Slot is provisioned and a minor alarm condition exists.
Orange	Slot is provisioned and a major alarm condition exists.
Red	Slot is provisioned and a critical alarm exists.

Moving the mouse pointer over cards in the graphic area causes pop-ups to display additional information about the card, including the card type; card status (active or standby); the number of critical, major, and minor alarms (if any); and the alarm profile used by the card. Right-clicking a card reveals a shortcut menu that you can use to open, reset, or delete a card. Right-click a *gray* slot to preprovision a card or to provision a slot before installing the card.

Ports can be assigned one of four states: OOS, IS, OOS_AINS, or OOS_MT. The color of the port in both card and node view indicates the port state. Table 8-3 shows the port colors and their states.

Table 8-3 *Node View Card Port Color and State*

Port Color	State	Description
Gray	OOS	Port is out of service; no signal will be transmitted.
Violet	OOS_AINS	Port is out of service, auto in service. The port will transmit a signal but will suppress alarms and allow loopbacks. The port will transition to IS when a signal is received for the amount of time specified in the AINS_SOAK field so long as no hardware-related alarms, such as EQPT or IMPROPRMVL, are present on the node.
Cyan	OOS_MT	Port is out of service, maintenance. The port will transmit a signal but alarms are suppressed and loopbacks are allowed. The port will not transition to IS until manually assigned by the user.
Green	IS	Port is in service. The port will transmit signal and display alarms; loopbacks are not allowed.

Node view presents various tabs. Each of these tabs has associated subtabs that provide information or configuration screens for various parameters. Table 8-4 lists the SONET and SDH tabs and subtabs available in node view.

Table 8-4 *Node View Tabs and Subtabs*

Tab	Description	Subtabs
Alarms	Lists current alarms (CR, MJ, and MN) for the node and updates them in real time	None
Conditions	Displays a list of standing conditions on the node.	None
History	Provides a history of node alarms including date, type, and severity of each alarm. The Session subtab displays alarms and events for the current session. The Node subtab displays alarms and events retrieved from a fixed-size log on the node.	Session, Node
Circuits	Create, delete, edit, and map circuits.	None
Provisioning	Provision the ONS 15454 node.	General, Ether Bridge, Network, Protection, BLSR or MS-SPRing, Security, SNMP, SONET DCC or SDH DCC, Timing, Alarm Behavior, Defaults Editor, and UCP
Inventory	Provides inventory information (part number, serial number, CLEI codes) for cards installed in the node. Enables you to delete and reset cards.	None
Maintenance	Performs maintenance tasks for the node.	Database, Ether Bridge, Protection, BLSR or MS-SPRing, Software, XC Cards, Overhead XConnect, Diagnostic, Timing, Audit, Routing Table, and RIP Routing Table Test Access

CTC Network View

The CTC network view displays all ONS nodes in the network with configurable options for the node you are logged in to. Network view enables you to view and manage ONS 15454s that have

DCC connections to the node that you logged in to and any login node groups you may have selected. Nodes with DCC connections to the login node will not display if you choose Exclude Dynamically Discovered Nodes in the Login dialog box. The CTC network view graphic area shown in Figure 8-5 displays a background image with colored ONS 15454 icons. The icon colors indicate the node status. Green lines show DCC connections between the nodes. Selecting a node or span in the graphic area displays information about the node and span in the status area.

Figure 8-5 *CTC Network View*

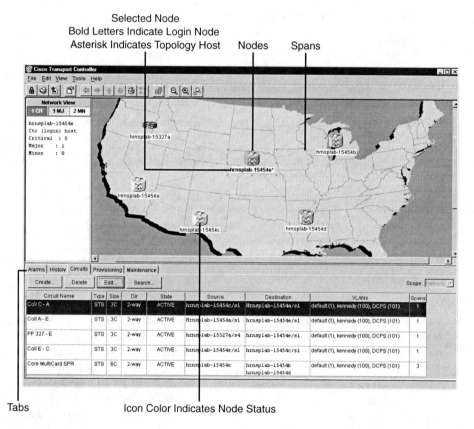

Depending on the state of current alarms, the nodes display various colors. Red indicates that a node has a critical alarm. Orange indicates that the node has a major alarm. Yellow indicates that the node has a minor alarm, and green indicates that the node is operationally functional without any alarms. If the node is gray, it is initializing. If the node is gray with an IP address, however, it means that the node is initializing or a problem exists with IP routing from node to CTC. Network view also indicates the number of various alarms present in the network. A superuser can configure CTC so that each user will see the same network view, or the user can create a custom view with maps. Table 8-5 lists the tabs and subtabs available in CTC network view.

Table 8-5 *Node View Card Port Color and State*

Tab	Description	Subtabs
Alarms	Lists current alarms (CR, MJ, and MN) for the network and updates them in real time.	None
Conditions	Displays a list of standing conditions on the network.	None
History	Provides a history of network alarms including date, type, and severity of each alarm.	None
Circuits	Create, delete, edit, filter and search for network circuits.	none
Provisioning	Provision security, alarm profiles, BLSR, and overhead circuits.	Security, Alarm Profiles, BLSR, Overhead Circuits
Maintenance	Displays network nodes status; displays working and protect software versions, and allows software to be downloaded.	Software

Card View

Card view provides access to individual ONS 15454 cards. This view provides a graphic of the card and provides access to tabs and subtabs that you use to manage card-specific maintenance and provisioning. A graphic of the selected card is shown in the graphic area. The status area displays the node name, slot, number of alarms, card type, equipment type, and either the card status (active or standby), card state (IS, OOS, OOS_AINS, or OOS_MT), or port state (IS, OOS). The information that is displayed and the actions you can perform depend on the card installed in the ONS chassis. CTC displays a card view for all ONS 15454 cards except the TCC2 and XC, XCVTs, XC10G, and XC-VXL cards. Figure 8-6 shows the CTC card view.

Card view provides access to the following tabs: Alarms, History, Circuits, Provisioning, Maintenance, Performance, and Conditions. However, the Performance tab is not displayed for the AIC card. The subtabs, fields, and information displayed under each tab depend on the card type selected.

Figure 8-6 *CTC Card View*

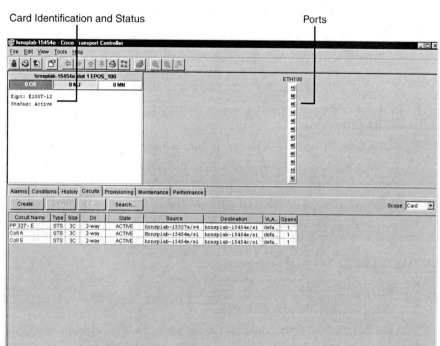

Cisco Transport Manager (CTM)

Cisco Transport Manager is an optical transport domain manager that delivers element and network management layer functionality for the Cisco ONS series of optical networking elements. CTM supports configuration, fault, performance, and security management functional areas. CTM also serves as a foundation for integration into a larger overall OSS environment.

CTM is an EMS that enables you to operate, administer, manage, and provision the entire range of ONS 15000 NEs. CTM provides the following features:

- Integrated IP, SONET, SDH, and DWDM OAM&P in a single scalable platform
- A Java-based GUI that provides similar screens on both Microsoft Windows and Sun Solaris client platforms
- User-defined Domain Explorer network views with *bubble-up*, alarm-severity propagation and drill-down capabilities to isolate fault conditions and service-delivery impact

- Geographic network maps and Explorer views that reflect the physical layout and configuration of the network
- Alarm Browser and alarm log views that provide a robust listing of all current and historical alarms and events
- A desktop-resident dashboard that provides alarm status for the CTM user's entire span of control with quick access to the Domain Explorer and the Alarm Browser
- Real-time network surveillance with configurable pop-up alarm and event notifications
- Real-time shelf views with full alarm and operational status indicators
- Automated configuration backup with manual restore capabilities, plus remote software download capability across the entire network domain
- Integrated A-to-Z automated circuit provisioning
- Performance-monitoring (PM) statistics collected across the SONET/SDH, TDM, DWDM, and Ethernet interfaces available for display or export

CTM includes the following components:

- Cisco Transport Manager
- Cisco Transport Manager GateWay

CTM provides standard fault, configuration, performance, and security management capabilities across the element and network management layers of the Telecommunications Management Network (TMN) reference architecture. The CTM client/server-based platform scales to manage up to 100 simultaneous client (user) sessions and up to 1000 NEs.

CTM GateWay is an architectural component that provides northbound EMS-to-network management system (NMS) interface mediation. CTM GateWay enables service providers to integrate CTM with their existing OSSs by using open, standard interfaces. CTM offers TL1, Simple Network Management Protocol (SNMP) trap forwarding, and Common Object Request Broker Architecture (CORBA) interface options for the CTM GateWay.

CTM is client/server-based, and the CTM server must be installed on a Sun hardware platform along with the Solaris operating system. The CTM server also needs Sun Java Runtime Environment (JRE) and an Oracle database engine. CTM uses CiscoView to configure and monitor certain ONS 15000 NEs. CiscoView is a server-based configuration tool that supports a device package for each supported ONS 15000 NE. There is also a version of CiscoView that is embedded in the hardware, which provides the same functionality as server-based CiscoView and has the same client requirements. CiscoView is automatically launched from CTM when you use certain CTM options. CTM first tries to launch server-based CiscoView. If that fails, CTM launches embedded CiscoView. If embedded CiscoView is not available, an error message displays. However, only server-based CiscoView is available for the ONS 15501. Figure 8-7 shows a schematic of a CTM system.

Figure 8-7 *CTM Schematic*

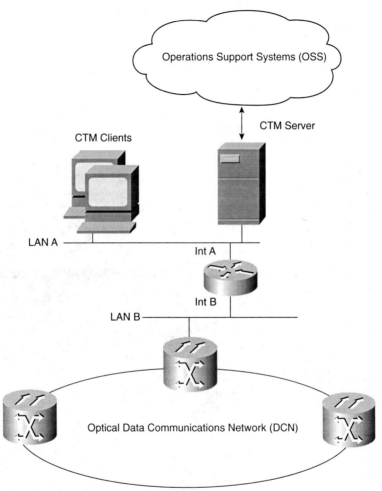

CTM Client

The CTM client software can run on the same machine as the CTM server or can be installed on a network-attached Sun workstation or a PC-based client. Check the documentation that comes along with your licensed copy of CTM for details on installation and platform-specific requirements. This section briefly describes how to use the CTM client, including how to start the client and descriptions of the major screens and functions.

You can run the CTM client on a Windows 2000/NT or Sun Solaris workstation. To start the CTM client in Windows 2000/NT, double-click the **CTM** desktop icon. To start the CTM client in Sun Solaris 2.8, enter the following command at the console: **ctmc-start**. Use the default username and password for the administrator. The username for this account in Release 3.2 or higher of CTM software is SysAdmin. Enter the CTM server host name or IP address and click **OK**. You need DNS enabled on the network if you enter a host name.

CTM Domain Explorer

The CTM Domain Explorer is the CTM *base* window and provides a logical view of the network plus alarm, connectivity, and operational status. Administrators use the Domain Explorer to create groups of NEs and organize the domain in a hierarchy. The Domain Explorer window is divided into two sections: the tree and the property sheets. The tree consists of a management domain, groups, and NEs, which appear in a hierarchical format. The top level of the hierarchy is the management domain, followed by groups and then NEs. You can drag and drop NEs to reposition them in the tree. Groups and NEs can exist in multiple locations in the tree. Figure 8-8 shows an example of the CTM Domain Explorer.

Figure 8-8 *CTM Domain Explorer*

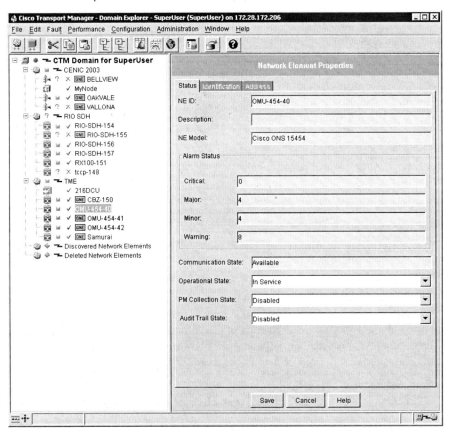

By default, the Domain Explorer contains the following groups, which are visible to administrators only:

- **Discovered NEs**—Contains NEs that have been automatically discovered by the CTM server. When an ONS 15454 NE is added as a gateway network element (GNE), for example, all ONS 15327 or ONS 15454 NEs that are DCC-connected to the GNE are automatically discovered. When a properly configured ONS 15540 NE is added, other interconnected ONS 15540 NEs are automatically discovered. Autodiscovered NEs are added to the Discovered NEs group, and CTM starts managing them automatically.

- **Deleted NEs**—Contains NEs that have been deleted. An NE appears in this group only when the last instance of that NE has been deleted.

- **Lost and Found**—The CTM client performs a minimal check at initialization to verify that the tree representation for the administrator's domain is valid. If there are any mismatches between the groups and NEs in the CTM domain and those in the administrator's domain, the mismatched NEs or groups are shown in the Lost and Found group in the Domain Explorer tree. If the administrator moves the NEs or groups in the Lost and Found group to another group and then clicks Refresh Data or restarts the CTM client, the Lost and Found group disappears.

CTM Subnetwork Explorer

The CTM Subnetwork Explorer window is similar in appearance and function to the Domain Explorer. A key difference is that the Subnetwork Explorer provides a single-level grouping of NEs based on subnetworks. A *subnetwork* is a set of NEs interconnected at a specific network layer (such as physical, section, line, and so on). For the ONS 15327 and ONS 15454, the subnetwork groups define the set of NEs for which you can create A-to-Z circuits.

The Subnetwork Explorer also shows the alarm, connectivity, and operational status of subnetworks and NEs. To open the Subnetwork Explorer, choose **File > Subnetwork Explorer** from the Domain Explorer window. Figure 8-9 shows an example of the CTM Subnetwork Explorer.

Figure 8-9 *CTM Subnetwork Explorer*

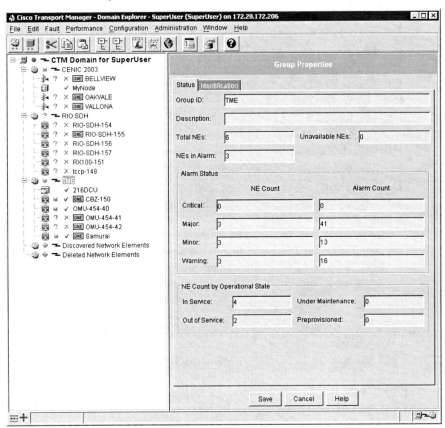

CTM Alarm Browser

The CTM Alarm Browser displays standing alarms and conditions in the managed domain. To display the Alarm Browser, choose an NE, group, subnetwork, or domain node. Then choose **Fault > Alarm Browser** from the Domain Explorer or Subnetwork Explorer. The Alarm Browser window lists conditions that are assigned a severity level of critical, major, minor, or warning. The Alarm Browser window also shows cleared alarms that are not acknowledged. Figure 8-10 shows an example of the CTM Alarm Browser.

CTM Node View

The CTM node view displays a graphical representation of the NE and its basic system information. To display the node view, choose an NE, and then choose **File > Node View** in the Domain Explorer or Subnetwork Explorer. Figure 8-11 shows an example of the CTM node view for an ONS 15252.

Cisco Transport Manager (CTM) 349

Figure 8-10 *CTM Alarm Browser*

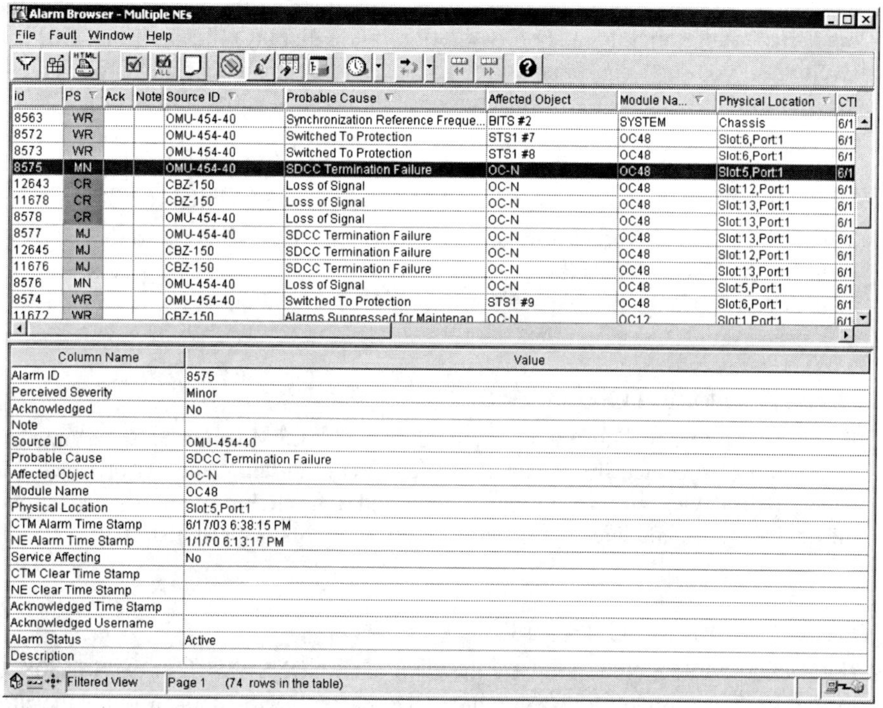

Figure 8-11 *CTM Node View for an ONS 15252*

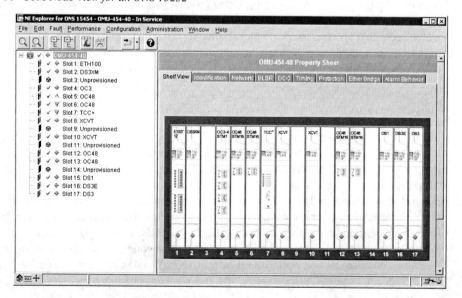

The node view shows the physical configuration of the system. Moving the cursor over the NE, its shelves, or Client Layer Interface Port (CLIPs) displays the current alarms for the NE, shelf, or CLIP. Double-clicking a shelf or CLIP module displays the shelf or CLIP in the NE Explorer. If you right-click a shelf or CLIP, a shortcut menu opens that you can use to display the shelf or CLIP in the Alarm Browser or in the NE inventory table.

Node View for the ONS 15327, ONS 15454, and ONS 15600

If you choose **File > Node View** for the ONS 15327, ONS 15454, and ONS 15600, CTM launches the CTC, which shows a shelf view of all line cards, alarm status, and navigation to all provisioning functions.

Node View for the ONS 15540

If you choose **File > Node View** for the ONS 15540, CTM launches CiscoView, which shows a continuously updated shelf view of the device configuration and performance. Moving the cursor to the device or one of its components and right-clicking displays menus for configuring the device, cards, and interfaces, and for monitoring real-time statistics.

Node View for the ONS 1580x

If you choose **File > Node View** for the ONS 15800, ONS 15801, or ONS 15808, CTM opens the NE Explorer window with a graphic of the selected ONS 1580x. Moving the cursor over the NE's racks, subracks, or line card modules displays the current alarms for the rack, subrack, or module. Double-clicking a rack, subrack, or module in the graphic displays the rack, subrack, or module in the NE Explorer. If you right-click the rack, a subrack, a module, or a shortcut menu opens that you can use to display the rack, subrack, or module in the Alarm Browser or in the NE inventory table. If the selected module supports performance-monitoring data collection and you right-click that module, it opens the 15-minute or 1-day PM table.

CTM Network Map

The CTM Network Map window enables you to see a geographical layout of the network. You can open the network map for an NE, group, subnetwork, or for the CTM domain. Choose a node in the Domain Explorer or Subnetwork Explorer tree, and then choose **File > Network Map** (or click the Open Network Map tool). The network map is organized into a multilevel hierarchy that corresponds to the structure of your Domain Explorer and Subnetwork Explorer trees when launched from the respective Explorer window. The network map hierarchy consists of management domains, subnetworks, groups, and NEs, which appear graphically. The network map enables you to visualize the structure of your network and customize node positions, node icons, and background map images. You can also create and modify links between distinct nodes. CTM automatically discovers links between various ONS NEs. You can create manual links in the network map; however, you cannot delete autodiscovered links. Figure 8-12 shows an example of a CTM network map.

Figure 8-12 *CTM Network Map*

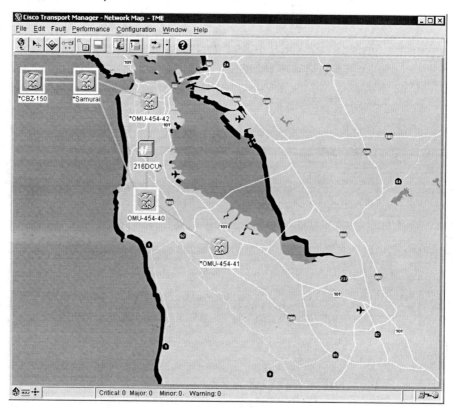

CTM NE Explorer

The CTM NE Explorer window displays service-provisioning information for the selected NE. The configuration information is retrieved through CORBA for the ONS 15327, ONS 15454, and 15600, through SNMP for the ONS 15200 and ONS 15540, and through TL1 for the ONS 1580x. To open the NE Explorer, choose an NE. Then choose **Configuration > NE Explorer** in the Domain Explorer or Subnetwork Explorer.

NE Explorer for the ONS 15200

If you choose **Configuration > NE Explorer** for the ONS 15200, CTM launches a web-based interface that runs in a Java-enabled browser. The screen is divided into two sections. The navigation area on the left side of the screen is used to view the ONS 15252 multichannel unit (MCU), ONS 15201 single-channel unit (SCU), modules installed in the network, events, and alarm logs. The display area on the right side of the screen shows information specific to each MCU, SCU, module, or log.

NE Explorer for the ONS 15327, ONS 15454, ONS and 15600

If you choose **Configuration > NE Explorer** for the ONS 15327, ONS 15600, or ONS 15454, CTM launches CTC. Service-provisioning information for the ONS 15327 and ONS 15454 is available through CTC. Figure 8-13 shows an example of a CTM NE Explorer for an ONS 15454.

Figure 8-13 *CTM NE Explorer for the ONS 15454*

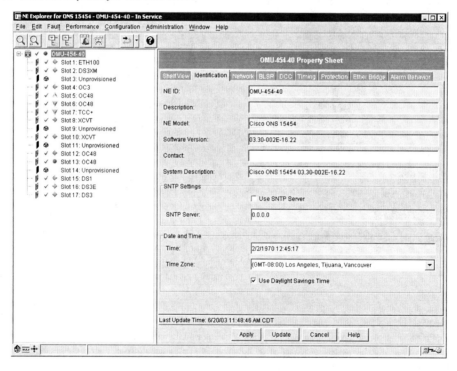

NE Explorer for the ONS 15540

If you choose **Configuration > NE Explorer** for the ONS 15540, CTM launches CiscoView, which shows a continuously updated shelf view of device configuration and performance. Moving the cursor to the device or one of its components and right-clicking displays menus for configuring the device, cards, and interfaces, and for monitoring real-time statistics.

NE Explorer for the ONS 1580x

If you choose **Configuration > NE Explorer** for the ONS 1580x, the window displayed by CTM consists of a tree on the left side and a property sheet on the right. The tree provides a hierarchical view and alarm status of the NE's physical racks, subracks, slots, and line cards. The property sheet shows information about the selected entity, such as the NE, rack, subrack, and slot.

CTM Control Panel

The CTM Control Panel window enables you to view and modify certain client and server configuration parameters. To view the Control Panel, choose **Administration > Control Panel** in the CTM Domain Explorer. The left side of the window displays the tree, which contains the different CTM functions and services. The right side of the window displays the property sheet that corresponds to the selected client or server component. Figure 8-14 shows an example of a CTM control panel.

Figure 8-14 *CTM Control Panel*

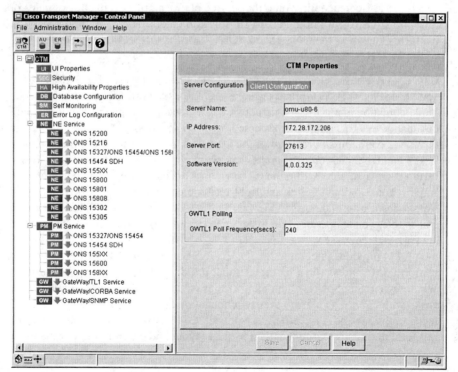

Summary

The Cisco Complete Optical Multiservice Edge and Transport solution, known as COMET, is an architecture that offers an end-to-end broadband solution. The COMET solution addresses photonics, protection, protocols, provisioning, and service–enablement for service provider and enterprise applications in the edge, metro, core, and long haul. Various ONS platforms are designed to scale metro, central office, and long-haul networks while extending the geographical reach of DWDM infrastructures. The entire ONS range of products can be managed via the ONS element management system. CTC is a GUI-based craft interface and console for the ONS series of MSPPs. CTC is used for initial provisioning and for OAM&P of smaller networks, whereas the CTM is an integrated element management system designed to provision, manage, and monitor all edge, metro and core MSxP platforms.

The ONS 15454 is the most versatile and widely deployed MSPP that supports transport of TDM and Ethernet services over SONET and SDH. The 15454 MSPP provides carrier-class service provisioning at the network edge. The Cisco ONS 15454 MSTP provides carrier-class multiservice delivery via metro DWDM. In addition to integrating DWDM and 32-channel OC-192/STM-64 capability onto the service platform, this product also scales optical transport from 10s of kilometers to 100s of kilometers. There are three ONS 15400 series platforms known as the ONS 15454 SONET MSPP, ONS 15454 SDH MSPP, and the ONS 15454 MSTP.

CTC is a GUI-based element management tool that can be used for OAM&P of various ONS 15000 MSPPs. CTC is typically used as a GUI-based craft tool during the deployment and implementation of an ONS-based network. CTC accepts graphical user input and converts the commands to TL1 commands accepted by the ONS NEs. CTC software is pre-installed on the TCC2 common control card of the ONS 15454. CTC provides three views of the ONS 15454 and ONS network: the node view, network view, and card view. Various navigational methods are available within the CTC window to access views and perform management actions. You can double-click and right-click objects in the graphic area and move the cursor over nodes, cards, and ports to view pop-up status information. CTM is a carrier-class EMS that enables you to operate, administer, manage, and provision the entire range of ONS 15000 network elements. CTM is a superset of CTC and has a similar look and feel of the CTC GUI tool. CTM provides integrated IP, SONET, SDH, and DWDM OAM&P in a single scalable platform. It features a Java-based GUI that provides similar screens on both Microsoft Windows and Sun Solaris client platforms as well as user-defined Domain Explorer network views with *bubble-up* alarm-severity propagation and drill-down capabilities to isolate fault conditions and service-delivery impact. CTM provides geographic network maps and Explorer views that reflect the physical layout and configuration of the network.

It also provides real-time shelf views with full alarm and operational status indicators. CTM includes automated configuration backup with manual restore capabilities, plus remote software download capability across the entire network domain. It performs integrated A-to-Z automated circuit provisioning and provides performance-monitoring statistics collected across the SONET/SDH, TDM, DWDM, and Ethernet interfaces available for display or export.

This chapter includes the following sections:

- **Provisioning the SONET MSPP**—This section introduces the provisioning aspect of the Synchronous Optical Network (SONET) multiservice provisioning platform (MSPP), with a reminder that provisioning needs a solid understanding of SONET theory.
- **Initial Provisioning Tasks**—Initial provisioning tasks must be completed, prior to the configuration of SONET and circuits. This section discusses the CTC craft interface and the setup of basic node information.
- **Provisioning of Protection Groups**—This section discusses the protection types that can be set up for ONS 15454 cards and ways to provision protection groups.
- **ONS 15454 Timing**—SONET timing parameters must be set for each ONS 15454. This section discusses SONET timing and the configuration of ONS timing. Each ONS 15454 independently accepts its timing reference from an external, line, or internal timing source.
- **Node Inventory**—This section discusses inventory information about cards installed in the ONS 15454 node, including part numbers, serial numbers, hardware revisions, and equipment types. The inventory feature provides a central repository that can be used to obtain information about installed cards, software, and firmware versions.
- **IP Networking of ONS nodes for OAM&P**—This section explains how to set up Cisco ONS 15454s in an IP operation, administration, maintenance, and provisioning (OAM&P) network.
- **UPSR Configuration**—Unidirectional path-switched ring (UPSR) provides duplicate fiber paths around the ring. This section discusses the installation of UPSR trunk cards, data communications channel (DCC) terminations, timing, and enabling of ports with respect to UPSR configuration.
- **BLSR Configuration**—This section discusses ONS bidirectional line-switched ring (BLSR) configuration.
- **Subtending Ring Configuration**—This section discusses subtending rings and their benefits in terms of optical network design. The ONS 15454 supports up to 32 SONET DCCs with the Timing and Control Card (TCC2).
- **Linear ADM Configurations**—This section discusses the configuration of ONS 15454 MSPPs as linear add/drop multiplexers (ADMs).
- **Path Protected Mesh Networking (PPMN)**—This section discusses path-protected mesh networks (PPMNs), which can include multiple ONS 15454 SONET topologies, such as a combination of linear ADM and UPSR, that extend the protection provided by a single UPSR to the meshed architecture of several interconnecting rings.
- **Circuit Provisioning**—Provisioning a UPSR, BLSR, linear ADM, or PPMN is not complete without assigning circuits for transport of user traffic. This section explains how to create and administer various types of Cisco ONS 15454 circuits and tunnels.

CHAPTER 9

Provisioning the Multiservice SONET MSPP

Provisioning the SONET MSPP

This book has so far covered substantial theory that serves as a buildup to the actual provisioning and implementation of the MSPP SONET network. It is important that you understand the theory behind SONET and TDM before jumping into provisioning mode. This chapter discusses the provisioning of the MSPP, SONET UPSR, BLSR, and TDM. Prior to the provisioning of TDM circuits, one must perform the basic node provisioning for the SONET infrastructure. The ONS 15454 is used as an example because it exemplifies most of the SONET provisioning aspects. The ONS 15454 is also the most deployed MSPP device in the metro edge and core. See Chapter 8, "Multiservice SONET and SDH Platforms," for a detailed description of the ONS 15454 architecture and its various optical, electrical, and common cards. This chapter covers initial node provisioning tasks and the provisioning of SONET circuits. Most of the ONS 15454 features are software dependent. This chapter covers Release 4.x of the ONS 15454 system software. The configuration commands, functions, and parameters are also supported across other ONS families, such as the ONS 15600 and ONS 15327.

Initial Provisioning Tasks

Before you begin your ONS node setup, ensure that you have the prerequisite information, including the node name, contact, location, current date, and time. If the ONS 15454 will be connected to an OAM&P network, you also need the IP address and subnet mask assigned to the node and the IP address of the default router. If Dynamic Host Configuration Protocol (DHCP) is used, you need the IP address of the DHCP server. If you intend to create card protection groups, you need the card protection scheme that will be used and what cards will be included in it. Proper network design requires an analysis of the requirements and a determination of the SONET protection topology that will be used for the network.

The ONS Craft Interface

Cisco Transport Controller (CTC) is a graphical user interface (GUI)-based craft tool that is used to provision the ONS 15454. Alternatively, the Transaction Language 1 (TL1) interface or Cisco Transport Manager (CTM) could be used. CTM is typically used in a network operations center (NOC) environment to configure and provision customer circuits. CTC is more suited for smaller networks and is used by engineers and technicians during deployment. CTC software is pre-installed on the TCC. If the CTC release is upgraded, new software needs to be installed on the TCC. CTC is downloaded from the TCC and installed on the PC or workstation automatically after the PC or workstation logs in to the ONS 15454. To connect to an ONS 15454, you enter the ONS 15454 IP address in the URL field of a web browser. Figure 9-1 shows the CTC console.

Figure 9-1   *ONS 15454 CTC Console*

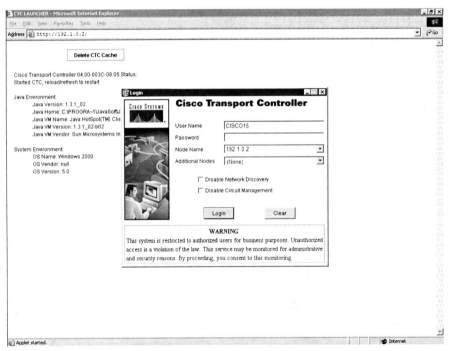

You can connect a PC to the ONS 15454 using the RJ-45 LAN port on the TCC+ or TCC2 or the LAN 1 pins on the ONS 15454 backplane. Each ONS 15454 must have a unique IP address that you use to access the ONS 15454. The address is displayed on the front-panel LCD. The initial IP address, 192.1.0.2, is the default address for ONS 15454 access and configuration. Each computer used to communicate with the ONS 15454 should have only one IP address. Each ONS 15454 comes preconfigured with a user account that has Superuser privileges. The user ID for this account is CISCO15. The CISCO15 account is not assigned a password. This account can be used to create other users. It is recommended that a password be assigned to the CISCO15 account. To create one, click the **Provisioning > Security** tab after you log in and change the CISCO15 password.

NOTE To use CTC with the ONS 15454 running Release 4.x system software, the PC or workstation must have a web browser with Java Runtime Environment (JRE) Release 1.3 or higher installed. If you want, you can also use TL1 commands to communicate with the Cisco ONS 15327 through VT100 terminals and VT100 emulation software, or you can Telnet to an ONS 15327 using TL1 port 3083.

NOTE When you upgrade CTC software, the TCC stores the older CTC version as the protect CTC version, and the newer CTC release becomes the working version. You can view the software versions that are installed on an ONS 15454 by choosing the Maintenance tab followed by the Software subtab. Choose these tabs in node view to display the software installed on one node. Choose the tabs in network view to display the software versions installed on all network nodes.

Set Up Basic Node Information

Setting basic information for each Cisco ONS 15454 node is one of the first provisioning tasks you perform. This information includes node name, location, contact, and timing. Completing the information for each node facilitates ONS 15454 management, particularly when the node is connected to a large ONS 15454 network. Complete the following steps for basic node setup. Figure 9-2 provides an example of basic node configuration.

Figure 9-2 *Basic Node Configuration*

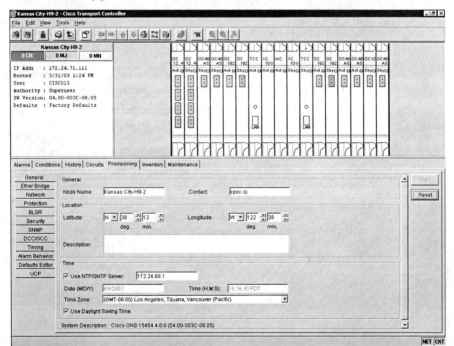

Step 1 Log in to the ONS 15454 node. The CTC node view is displayed.

Step 2 Click the **Provisioning > General** tab.

Step 3 Enter the following information:

- **Node Name**—Type a name for the node. For TL1 compliance, names must begin with an alphabetic character and have no more than 20 alphanumeric characters.

- **Contact**—Type the name of the node contact person and the phone number (optional).

- **Location**—Type the node location (for example, a city name or specific office location [optional]).

- **Latitude**—Enter the node latitude: N (north) or S (south), degrees, and minutes (optional).

- **Longitude**—Enter the node longitude: E (east) or W (west), degrees, and minutes (optional).

  CTC uses the latitude and longitude to position node icons on the network view map. (You can also position nodes manually by pressing Ctrl and dragging the node icon to a new location.) To convert a coordinate in degrees to degrees and minutes, multiply the number after the decimal by 60. For example, the latitude 29.36958 converts to 29 degrees, 22 minutes (.36958 * 60 = 22.1748, rounded to the nearest whole number).

- **Use SNTP Server**—When checked, CTC uses a Simple Network Time Protocol (SNTP) server to set the date and time of the node. Using an SNTP server ensures that all ONS 15454 network nodes use the same date and time reference. The server synchronizes the node's time after power outages or software upgrades. If you check Use SNTP Server, type the server's IP address in the next field. If you do not use an SNTP server, complete the Date, Time, and Time Zone fields. The ONS 15454 uses these fields for alarm dates and times. (CTC displays all alarms in the login node's time zone for cross-network consistency).

- **Date**—Type the current date.

- **Time**—Type the current time.

- **Time Zone**—Choose the time zone.

Step 4 Click **Apply**.

Initial Provisioning Tasks   361

Set Up Network Information

ONS 15454s mostly operate in network environments and need to be managed by an element manager, such as CTM, within an OSS system. The ONS 15454 ships with a default IP address of 192.1.0.2. Before you connect an ONS 15454 to other ONS 15454s or to a LAN, the default IP address must be changed to reflect a valid address of the IP subnet that the node would reside on. Any change of the ONS 15454 IP address causes the TCC cards to restart. Within IP networks, ONS 15454s often exist as hosts on management IP subnets. Complete the following steps for basic network setup. Figure 9-3 provides an example of IP network setup.

Figure 9-3 *IP Network Setup*

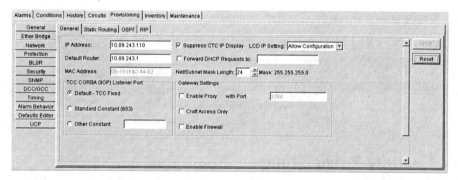

Step 1 From the CTC node view, click the **Provisioning > Network** tab.

Step 2 Enter the following information:

- **IP Address**—Type the IP address assigned to the ONS 15454 node.

- **Prevent LCD IP Configuration**—If checked, prevents the ONS 15454 IP address from being changed using the LCD front panel.

- **Default Router**—If the ONS 15454 must communicate with a device on a network to which the ONS 15454 is not connected, the ONS 15454 forwards the packets to the default router. Type the IP address of the default router in this field. If the ONS 15454 is not connected to a LAN, leave the field blank.

- **Subnet Mask Length**—If the ONS 15454 is part of a subnet, type the subnet mask length (decimal number representing the subnet mask length in bits) or click the arrows to adjust the subnet mask length. The subnet mask length is the same for all ONS 15454s in the same subnet.

NOTE   The MAC address is read only. It displays the ONS 15454 address as it is identified on the IEEE 802 Media Access Control (MAC) layer.

- **Forward DHCP Request To**—When checked, forwards Dynamic Host Configuration Protocol requests to the IP address entered in the Request To field. DHCP is a TCP/IP protocol that enables CTC computers to get temporary IP addresses from a server. If you enable DHCP, CTC computers that are directly connected to an ONS 15454 node can obtain temporary IP addresses from the DHCP server.

- **TCC CORBA (IIOP) Listener Port**—Sets a listener port to allow communication with the ONS 15454 through firewalls if needed.

Step 3 Click **Apply**.

Step 4 Click **Yes** in the Confirmation dialog box.

Both ONS 15454 TCC cards will reboot, one at a time.

NOTE You can change the ONS 15454 IP address, subnet mask, and default router address using the Slot, Status, and Port buttons on the front-panel LCD.

User and Security Provisioning

Up to 500 users can be added to one ONS 15454. Each ONS 15454 user can be assigned one of the following security levels:

- **Retrieve users**—Can retrieve and view CTC information, but cannot set or modify parameters
- **Maintenance users**—Can access only the ONS 15454 maintenance options
- **Provisioning users**—Can access provisioning and maintenance options
- **Superusers**—Can perform all the functions of the other security levels and can set names, passwords, and security levels for other users

Table 9-1 shows the actions that each user can perform in node view.

Table 9-1 *ONS 15454 Security Levels for Node View*

CTC Tab	Subtab	Actions	Retrieve	Maintenance	Provisioning	Superuser
Alarms	None	Synchronize alarms	Yes	Yes	Yes	Yes
Conditions	None	Retrieve	Yes	Yes	Yes	Yes
History	Session	Read only	Yes	Yes	Yes	Yes
	Node	Retrieve alarms/events	Yes	Yes	Yes	Yes

Table 9-1 *ONS 15454 Security Levels for Node View (Continued)*

CTC Tab	Subtab	Actions	Retrieve	Maintenance	Provisioning	Superuser
Circuits	None	Create/delete/edit/upgrade	No	No	Yes	Yes
		Path selector switching	No	Yes	Yes	Yes
		Search	Yes	Yes	Yes	Yes
		Switch retrieval	Yes	Yes	Yes	Yes
Provisioning	General	Edit	No	No	Yes	Yes
	Ether-Bridge	Spanning trees: Edit	No	No	Yes	Yes
		Thresholds: Create/delete	No	No	Yes	Yes
	Network	All	No	No	No	Yes
	Protection	Create/delete/edit	No	No	Yes	Yes
		Browse groups	Yes	Yes	Yes	Yes
	Ring	All (BLSR)	No	No	Yes	Yes
	Security	Create/delete	No	No	No	Yes
		Change password	Same user	Same user	Same user	All users
	SNMP	Create/delete/edit	No	No	No	Yes
		Browse trap destinations	Yes	Yes	Yes	Yes
	SONET DCC	Create/delete	No	No	No	Yes
	Timing	Edit	No	No	Yes	Yes
	Alarming	Edit	No	No	Yes	Yes
Inventory	None	Delete	No	No	Yes	Yes
		Reset	No	Yes	Yes	Yes

continues

Table 9-1 *ONS 15454 Security Levels for Node View (Continued)*

CTC Tab	Subtab	Actions	Retrieve	Maintenance	Provisioning	Superuser
Maintenance	Database	Backup/restore	No	No	No	Yes
	Ether-Bridge	Spanning-tree retrieve	Yes	Yes	Yes	Yes
		Spanning-tree clear/clear all	No	Yes	Yes	Yes
		MAC table retrieve	Yes	Yes	Yes	Yes
		MAC table clear/clear all		Yes	Yes	Yes
		Trunk utilization refresh	Yes	Yes	Yes	Yes
	Protection	Switch/lockout operations	No	Yes	Yes	Yes
	Ring	BLSR maintenance	No	Yes	Yes	Yes
	Software	Download/upgrade/activate/revert	No	No	No	Yes
	XC Cards	Protection switches	No	Yes	Yes	Yes
	Diagnostic	Retrieve/lamp test	No	Yes	Yes	Yes
	Timing	Edit	No	Yes	Yes	Yes
	Audit	Retrieve	Yes	Yes	Yes	Yes
	Routing Table	Read only	Yes	Yes	Yes	Yes

Each ONS 15454 user has a specified amount of time that he or she can leave the system idle before the CTC window is locked. The lockouts prevent unauthorized users from making changes. Higher-level users have shorter idle times, as shown in Table 9-2.

Table 9-2 *ONS 15454 User Idle Times*

Security Level	Idle Time
Superuser	15 minutes
Provisioning	30 minutes
Maintenance	60 minutes
Retrieve	Unlimited

You can perform ONS 15454 user management tasks from network or node view. In network view, you can add, edit, or delete users from multiple nodes at one time. If you perform user management tasks in node view, you can only add, edit, or delete users from that node. Figure 9-4 provides an example of user creation and setup.

Figure 9-4 *User Creation and Setup*

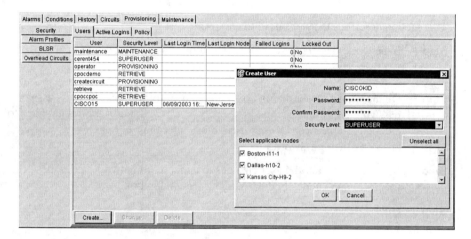

NOTE You must add the same username and password to each node the user will access.

Create New Users

This section describes the steps used to create new users:

Step 1 In network view, choose the **Provisioning > Security** tab.

Step 2 On the Security pane, click **Create**.

Step 3 In the Create User dialog box, enter the following:

- **Name**—Type the username.
- **Password**—Type the user password. The password must be a minimum of 6 and a maximum of 10 alphanumeric (a–z, A–Z, 0–9) and special characters (+, #, %), where at least 2 characters are nonalphabetic and at least 1 character is a special character
- **Confirm Password**—Type the password again to confirm.
- **Security Level**—Choose the user's security level.

Edit a User

This section describes the steps used to edit users:

Step 1 In network view, choose the **Provisioning > Security** tab.

Step 2 Click **Change**.

Step 3 On the Change User dialog box, edit the user information: name, password, password confirmation, and/or security level. (A Superuser does not need to enter an old password. Other users must enter their old passwords when changing their password.)

Step 4 If you do not want the user changes to apply to all network nodes, deselect the unchanged nodes in the Change Users dialog box.

Step 5 Click **OK**.

NOTE You cannot change the CISCO15 default Superuser name. Also, changed user permissions and access levels do not take effect until the user logs out of CTC and logs back in.

Delete a User

This section describes the steps used to delete a user:

Step 1 In network view, choose the **Provisioning > Security** tab.

Step 2 Click **Delete**.

Step 3 On the Delete User dialog box, enter the name of the user you want to delete.

Step 4 If you do not want to delete the user from all network nodes, deselect the nodes.

Step 5 Click **OK** and click **Apply**.

Provisioning of Protection Groups

The ONS 15454 provides several card protection methods. When you set up protection for ONS 15454 cards, you must choose between maximum protection and maximum slot availability. The highest protection reduces the number of available card slots, whereas the highest slot availability reduces the protection. Table 9-3 shows the protection types that you can set up for ONS 15454 cards.

Table 9-3 *Protection Types*

Type	Cards	Description
1:1	DS1 DS3 EC1-12 DS3XM-6	Pairs one working card with one protect card. Install the protect card in an odd-numbered slot and the working card in an even-numbered slot next to the protect slot toward the center—for example: protect in slot 1, working in slot 2; protect in slot 3, working in slot 4; protect in slot 15, working in slot 14.
1:N	DS1 DS3	Assigns one protect card for several working cards. The maximum is 1:5. Protect cards (DS1N-14, DS3N-12) must be installed in slots 3 or 15, and the cards they protect must be on the same side of the shelf. Protect cards must match the cards they protect. For example, a DS1N-14 can only protect DS1-14 or DS1N-14 cards. If a failure clears, traffic reverts to the working card after the reversion time has elapsed.
1+1	Any optical	Pairs a working optical port with a protect optical port. Protect ports must match the working ports. For example, port 1 of an OC-3 card can be protected only by port 1 of another OC-3 card. Cards do not need to be in adjoining slots.
Unprotected	Any	Unprotected cards can cause signal loss if a card fails or incurs a signal error. Because no card slots are reserved for protection, however, unprotected schemes maximize the service available for use on the ONS 15454. Unprotected is the default protection type.

Figure 9-5 provides an example of protection group creation.

Figure 9-5 *Protection Group Creation*

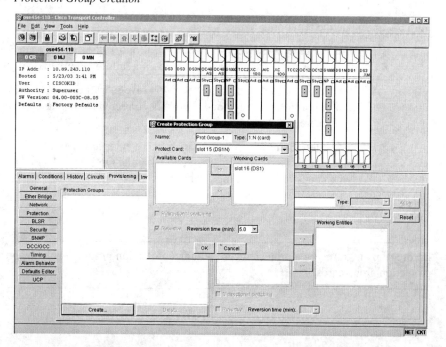

Create a Protection Group

This section describes the steps used to create a protection group:

Step 1 From the CTC node view, click the **Provisioning > Protection** tab.

Step 2 Under Protection Groups, click **Create**.

Step 3 In the Create Protection Group dialog box, enter the following:

(a) **Name**—Type a name for the protection group. The name can have up to 32 alphanumeric characters.

(b) **Type**—Choose the protection type (1:1, 1:N, or 1+1) from the drop-down menu. The protection selected determines the cards that are available to serve as protect and working cards. If you choose 1:N protection, for example, only DS-1N and DS-3N cards display.

(c) **Protect Card or Port**—Choose the protect card (if using 1:1 or 1:N) or protect port (if using 1+1) from the drop-down menu.

Based on these selections, a list of available working cards or ports displays under Available Cards or Available Ports.

Step 4 From the Available Cards or Available Ports list, choose the card or port that you want to be the working card or port (the card(s) or port(s) that will be protected by the card or port selected in Protect Cards or Protect Ports). Click the top arrow button to move each card/port to the Working Cards or Working Ports list.

Step 5 Complete the remaining fields:

- **Bidirectional Switching**—(Optical cards only) Click if you want both transmit and receive channels to switch if a failure occurs to one.

- **Revertive**—If checked, the ONS 15454 reverts traffic to the working card or port after failure conditions stay corrected for the amount of time entered in Reversion Time.

- **Reversion Time**—If Revertive is checked, enter the amount of time following failure-condition correction that the ONS 15454 should switch back to the working card or port.

Step 6 Click **OK**.

NOTE Before running traffic on a protected card within a protection group, enable the ports of all protection group cards.

Enable Ports

This section describes the steps used to enable a port:

Step 1 Log in to the node in CTC and display the card you want to enable in card view.

Step 2 Click the **Provisioning > Line** tab.

Step 3 Under the Status column, choose **In Service**.

Step 4 Click **Apply**.

Edit Protection Groups

This section describes the steps used to edit protection groups:

Step 1 From the CTC node view, click the **Provisioning > Protection** tab.

Step 2 In the Protection Groups section, choose a protection group.

Step 3 In the Selected Group section, edit the fields as appropriate.

Step 4 Click **Apply**.

Delete Protection Groups

This section describes the steps used to delete protection groups:

Step 1 From the CTC node view, click the **Maintenance > Protection** tab.

Step 2 Verify that working traffic is not running on the protect card:

(a) In the Protection Groups section, choose the group you want to delete.

(b) In the Selected Group section, verify that the protect card is in standby mode. If it is in standby mode, continue with Step 3. If it is active, complete Substep C.

(c) If the working card is in standby mode, manually switch traffic back to the working card. In the Selected Group pane, click the working card and then click Manual. Verify that the protect card switches to standby mode and the working card is active. If it does, continue with Step 3. If the protect card is still active, do not continue. Begin troubleshooting procedures or call technical support.

Step 3 From the node view, click the **Provisioning > Protection** tab.

Step 4 In the Protection Groups section, click a protection group.

Step 5 Click **Delete**.

ONS 15454 Timing

SONET timing parameters must be set for each ONS 15454. Each ONS 15454 independently accepts its timing reference from one of three sources:

- **External**—Building integrated timing supply (BITS) pins on the ONS 15454 backplane
- **Line**—An OC-N card installed in the ONS 15454 that is connected to a node receiving timing from a BITS source
- **Internal**—The internal ST3 clock on the TCC card

In typical ONS 15454 networks, one node is always set to external. The external node derives its timing from a BITS source wired to the BITS backplane pins. The BITS source, in turn, derives its timing from a primary reference source (PRS) such as a Stratum 1 clock or global-positioning system (GPS) signal. The other nodes are set to line. The line nodes derive timing from the externally timed node through the OC-N trunk cards. You can set three timing references for each ONS 15454. The first two references are typically two BITS-level sources, or two line-level sources optically connected to a node with a BITS source. The third reference is the internal clock provided on every ONS 15454 TCC card. This clock is a Stratum 3 (ST3). If an ONS 15454 becomes isolated, there is a slow and gradual return to ST3. After isolation, the ST1 equivalent is held for almost 24 hours according to SONET specification.

Synchronization Status Messaging (SSM) is a SONET protocol that communicates information about the quality of the timing source. SSM messages are carried in the S1 byte of the SONET line layer. They enable SONET devices to automatically choose the highest-quality timing reference and to avoid timing loops. SSM messages are either generation 1 or generation 2. Generation 1 is the first and most widely deployed SSM message set. Generation 2 is a newer version. If you enable SSM for the ONS 15454, consult your timing reference documentation to determine which message set to use. An example of provisioning of ONS 15454 timing is shown in Figure 9-6.

Figure 9-6 *Provisioning ONS Timing*

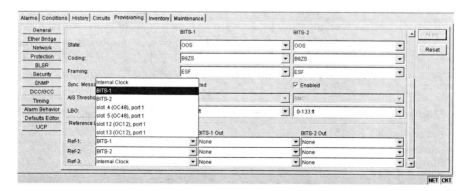

Provisioning ONS 15454 Timing

This section describes the steps used to provision timing on the ONS 15454:

Step 1 From the CTC node view, click the **Provisioning > Timing** tab.

Step 2 In the General Timing section, complete the following information:

- **Timing Mode**—Set to External if the ONS 15454 derives its timing from a BITS source wired to the backplane pins; set to Line if timing is derived from an OC-N card that is optically connected to the timing node. A third option, Mixed, enables you to set external and line timing references. (Because mixed timing may cause timing loops, Cisco does not recommend its use. Use this mode with care.)

- **SSM Message Set**—Choose the message set level supported by your network. If a generation 1 node receives a generation 2 message, the message will be mapped down to the next available Generation 1. For example, an ST3E message becomes an ST3.

- **Quality of RES**—If your timing source supports the reserved S1 byte, you set the timing quality here. (Most timing sources do not use RES.) Qualities display in descending quality order as ranges. For example, ST3<RES<ST2 means the timing reference is higher than a Stratum 3 and lower than a Stratum 2.

- **Revertive**—If checked, the ONS 15454 reverts to a primary reference source after the conditions that caused it to switch to a secondary timing reference are corrected.

- **Revertive Time**—If Revertive is checked, indicate the amount of time the ONS 15454 will wait before reverting back to its primary timing source.

Step 3 In the BITS Facilities section, complete the following information:

> **NOTE** The BITS Facilities section sets the parameters for your BITS1 and BITS2 timing references. Many of these settings are determined by the manufacturer of the timing source. If equipment is timed through BITS Out, you can set timing parameters to meet the requirements of the equipment.

- **State**—Set the BITS reference to IS (in service) or OOS (out of service). For nodes set to line timing with no equipment timed through BITS Out, set State to OOS. For nodes using external timing or line timing with equipment timed through BITS Out, set State to IS.

- **Coding**—Set to the coding used by your BITS reference to either B8ZS or AMI.

- **Framing**—Set to the framing used by your BITS reference to either ESF (Extended Super Frame) or SF (D4) (Super Frame). SSM is not available with Super Frame.
- **Sync Messaging**—Check to enable SSM.
- **AIS Threshold**—Sets the quality level where a node sends an alarm indication signal (AIS) from the BITS1 Out and BITS2 Out backplane pins. When a node times at or below the AIS threshold quality, AIS is sent (used when SSM is disabled or frame is SF).

Step 4 Under Reference Lists, complete the following information:

> **NOTE** Reference lists define up to three timing references for the node and up to six BITS Out references. BITS Out references define the timing references used by equipment that can be attached to the node's BITS Out pins on the backplane. If you attach equipment to BITS Out pins, you normally attach it to a node with line mode because equipment near the external timing reference can be directly wired to the reference.

- **NE Reference**—Enables you define three timing references (Ref 1, Ref 2, and Ref3). The node uses Reference 1 unless a failure occurs to that reference, (in which case, the node uses Reference 2). If that fails, the node uses Reference 3, which is typically set to Internal Clock. This is the Stratum 3 clock provided on the TCC card. The options displayed depend on the Timing Mode setting.
- **Timing Mode Set to External**—Options are BITS1, BITS2, and Internal Clock.
- **Timing Mode Set to Line**—Options are the node's working optical cards and Internal Clock. Choose the cards/ports that are directly or indirectly connected to the node wired to the BITS source (that is, the node's trunk cards). Set Reference 1 to the trunk card that is closest to the BITS source. If slot 5 is connected to the node wired to the BITS source, for example, choose slot 5 as Reference 1.
- **Timing Mode Set to Mixed**—Both BITS and optical cards are available, enabling you to set a mixture of external BITS and optical trunk cards as timing references.
- **BITS1 Out/BITS2 Out**—Define the timing references for equipment wired to the BITS Out backplane pins. Normally, BITS Out is used with line nodes, so the options displayed are the working optical cards. BITS1 and BITS2 Out are enabled as soon as BITS1 and BITS2 facilities are placed in service.

Setup of Internal Timing

If no BITS source is available, you can set up internal timing by timing all nodes in the ring from the internal clock of one node:

NOTE Internal timing is Stratum 3 and not intended for permanent use. All ONS 15454s should be timed to a Stratum 2 or better primary reference source.

Step 1 Log in to the node that will serve as the timing source.

Step 2 In CTC node view, click the **Provisioning > Timing** tab.

Step 3 In the General Timing section, enter the following:

- **Timing Mode**—Set to External.
- **SSM Message Set**—Set to Generation 1.
- **Quality of RES**—Set to DUS.
- **Revertive**—Is not relevant for internal timing. Leave at default value.
- **Revertive Time**—Default setting of 5 minutes is sufficient.

Step 4 In the BITS Facilities section, enter the following information:

- **State**—Set BITS1 and BITS2 to OOS.
- **Coding**—Not relevant for internal timing. Leave the field at its default value of B8ZS.
- **Framing**—Not relevant for internal timing. Leave the field at its default value of ESF.
- **Sync Messaging**—Checked.
- **AIS Threshold**—Not available.

Step 5 In the Reference Lists section, enter the following information:

- **NE Reference**:

  Ref1—Set to Internal Clock.

  Ref2—Set to Internal Clock.

  Ref3—Set to Internal Clock.

  BITS1 Out/BITS2 Out—Set to None.

Step 6 Click **Apply**.

Step 7 Log in to a node that will be timed from the node set up in Steps 1 through 4.

Step 8 In CTC node view, click the **Provisioning > Timing** tab.

Step 9 In the General Timing section, enter the same information as entered in Step 3, except for the following:

- **Timing Mode**—Set to Line.
- **NE Reference**:

    Ref1—Set to the OC-N trunk card with the closest connection to the node in Step 3.

    Ref2—Set to the OC-N trunk card with the next closest connection to the node in Step 3.

    Ref3—Set to Internal Clock.

Step 10 Click **Apply**.

Step 11 Repeat Steps 7 through 10 at each node that will be timed by the node in Step 3.

Node Inventory

The Inventory tab displays information about cards installed in the ONS 15454 node including part numbers, serial numbers, hardware revisions, and equipment types. The tab provides a central location to obtain information about installed cards, software, and firmware versions. Using the ONS 15454 export feature, you can export inventory data from ONS 15454 nodes into spreadsheet and database programs to consolidate ONS 15454 information for network inventory management and reporting. You can preprovision a slot before the card is installed by right-clicking the slot in node view and selecting a card type. The Inventory tab displays the following information about the cards installed in the ONS 15454:

- **Location**—The slot where the card is installed
- **Eqpt Type**—Equipment type the slot is provisioned for (for example, OC-12 or DS1)
- **Actual Eqpt Type**—The actual card that is installed in the slot (for example, OC-12 IR 4 1310, or DS1N-14)
- **HW Part #**—Card part number
- **HW Rev**—Card revision number
- **Serial #**—Card serial number unique to each card
- **CLEI Code**—Common Language Equipment Identifier code
- **Firmware Rev**—Revision number of the software used by the ASIC chip installed on the card

An example of ONS 15454 inventory is shown in Figure 9-7.

Figure 9-7 *ONS 15454 Inventory*

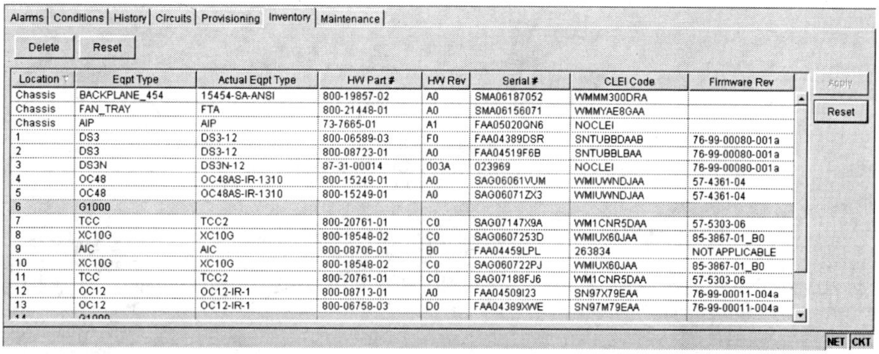

Location	Eqpt Type	Actual Eqpt Type	HW Part #	HW Rev	Serial #	CLEI Code	Firmware Rev
Chassis	BACKPLANE_454	15454-SA-ANSI	800-19857-02	A0	SMA06187052	WMMM300DRA	
Chassis	FAN_TRAY	FTA	800-21448-01	A0	SMA06156071	WMMYAE8GAA	
Chassis	AIP	AIP	73-7665-01	A1	FAA05020QN6	NOCLEI	
1	DS3	DS3-12	800-06589-03	F0	FAA04389DSR	SNTUBBDAAB	76-99-00080-001a
2	DS3	DS3-12	800-08723-01	A0	FAA04519F6B	SNTUBBLBAA	76-99-00080-001a
3	DS3N	DS3N-12	87-31-00014	003A	023969	NOCLEI	76-99-00080-001a
4	OC48	OC48AS-IR-1310	800-15249-01	A0	SAG06061VUM	WMIUVWNDJAA	57-4361-04
5	OC48	OC48AS-IR-1310	800-15249-01	A0	SAG06071ZX3	WMIUVWNDJAA	57-4361-04
6	G1000						
7	TCC	TCC2	800-20761-01	C0	SAG07147X9A	WM1CNR5DAA	57-5303-06
8	XC10G	XC10G	800-18548-02	C0	SAG06072S3D	WMIUX6QJAA	85-3867-01_B0
9	AIC	AIC	800-08706-01	B0	FAA04459LPL	263834	NOT APPLICABLE
10	XC10G	XC10G	800-18548-02	C0	SAG060722PJ	WMIUX6QJAA	85-3867-01_B0
11	TCC	TCC2	800-20761-01	C0	SAG07188FJ6	WM1CNR5DAA	57-5303-06
12	OC12	OC12-IR-1	800-08713-01	A0	FAA04509123	SN97X79EAA	76-99-00011-004a
13	OC12	OC12-IR-1	800-06758-03	D0	FAA04389XWE	SN97M79EAA	76-99-00011-004a

IP Networking of ONS Nodes for OAM&P

This section explains how to set up Cisco ONS 15454s in an IP operation, administration, maintenance, and provisioning (OAM&P) network. ONS 15454s can be connected in many different ways within an IP environment:

- They can be connected to LANs through direct connections or a router.

- IP subnetting can create ONS 15454 node groups, which enable you to provision non-DCC connected nodes in a network.

- Different IP functions and protocols can be used to achieve specific network goals. For example, Proxy Address Resolution Protocol (ARP) enables one LAN-connected ONS 15454 to serve as a gateway for ONS 15454s that are not connected to the LAN.

- You can create static routes to enable connections among multiple CTC sessions with ONS 15454s that reside on the same subnet but have different destination IP addresses.

- If ONS 15454s are connected to Open Shortest Path First (OSPF) networks, ONS 15454 network information is automatically communicated across multiple LANs and WANs.

CTC and ONS Nodes on the Same IP Subnet

A basic ONS 15454 LAN configuration with the ONS 15454s and CTC computer residing on the same 192.168.1.0 subnet is shown in Figure 9-8. All ONS 15454s connect to LAN A, and all ONS 15454s have DCC connections configured.

The SONET DCC connections need to be provisioned between the ONS nodes for UPSR, BLSR, or linear ADM topologies.

Figure 9-8 *CTC and ONS Nodes on the Same IP Subnet*

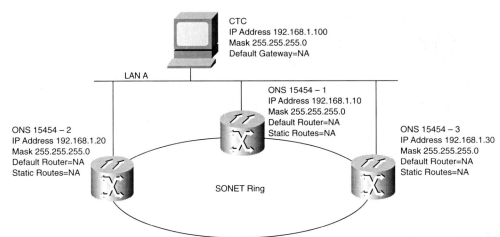

CTC and ONS Nodes on Separate IP Subnets

In this scenario, the CTC computer resides on subnet 192.168.1.0 and attaches to LAN A, as shown in Figure 9-9. The ONS 15454s reside on a different subnet (192.168.2.0) and attach to LAN B. A router connects LAN A to LAN B. The IP address of router interface A is set to 192.168.1.1, and the IP address of router interface B is set to 192.168.2.1.

Figure 9-9 *CTC and ONS Nodes on the Separate IP Subnets*

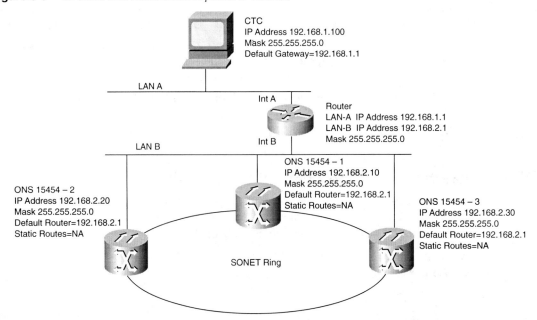

On the CTC computer, the default gateway is set to router interface A. If the LAN uses DHCP, the default gateway and IP address are assigned automatically. In this example, a DHCP server is not available.

Using Proxy ARP to Enable an ONS 15454 Gateway

In this scenario, ONS nodes 2 and 3 connect to ONS 15454 1 through the SONET DCC. Because all three ONS 15454s are on the same IP subnet, Proxy ARP enables ONS 15454 1 to serve as a gateway for ONS 15454s 2 and 3. This is shown in Figure 9-10.

Figure 9-10 *Using Proxy ARP to Enable an ONS Gateway*

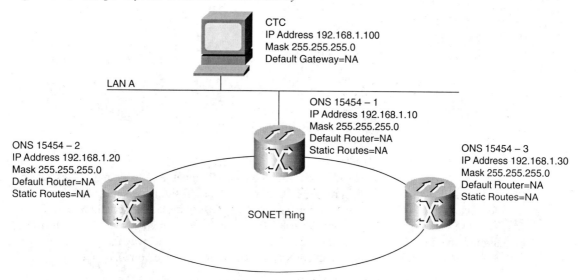

Proxy ARP enables one LAN-connected ONS 15454 to respond to the ARP request for ONS 15454s not connected to the LAN. ONS 15454 Proxy ARP requires no user configuration. For this to occur, the DCC-connected ONS 15454s must reside on the same IP subnet. When a LAN device, such as CTC, sends an ARP request to an ONS 15454 that is not connected to the LAN, the gateway ONS 15454 returns its MAC address to the LAN device. The LAN device then sends the datagram for the remote ONS 15454 to the MAC address of the proxy ONS 15454. The proxy ONS uses the DCC to communicate with the remote ONS devices.

Default Gateway on the CTC

In this scenario, ONS nodes 2 and 3 reside on separate subnets, 192.168.2.0 and 192.168.3.0, respectively. ONS node 1 and the CTC computer are on subnet 192.168.1.0. The network includes different subnets because Proxy ARP is not used. For the CTC computer to communicate with ONS nodes 2 and 3, ONS node 1 is entered as the default gateway on the CTC computer. This is shown in Figure 9-11.

Figure 9-11 *Default Gateway on the CTC*

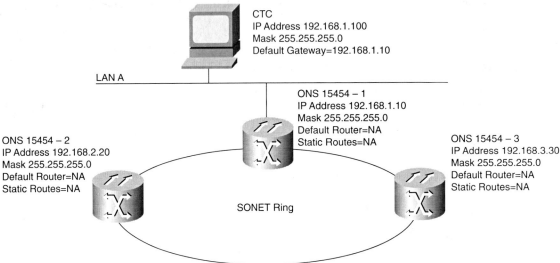

Static Route on the ONS Node

Static routes are used by ONS 15454s residing on one IP subnet to connect to CTC sessions on another subnet connected via a router. In Figure 9-12, CTC residing on subnet 192.168.1.0 connects to a router through interface A. The router in this example is not configured for OSPF. The ONS nodes residing on subnet 192.168.2.0 are connected through ONS node 1 to the router through interface B. Proxy ARP enables ONS 15454 node 1 as a gateway for ONS nodes 2 and 3. To connect to CTC computers on LAN A, a static route is created on ONS node 1.

The destination and subnet mask entries control access to the ONS 15454s:

- If a single CTC computer is connected to router, enter the complete CTC "host route" IP address as the destination with a subnet mask of **255.255.255.255**.
- If CTC computers on a subnet are connected to router, enter the destination subnet (in this example, **192.168.1.0**) and a subnet mask of **255.255.255.0**.
- If all CTC computers are connected to router, enter a destination of **0.0.0.0** and a subnet mask of **0.0.0.0**. Figure 9-13 shows an example.
- The IP address of router interface B is entered as the next hop, and the cost in number of hops from source to destination is 2.

Use the following steps to create a static route on the ONS node:

Step 1 Log in to the ONS 15454 and choose the **Provisioning > Network** tab.

Step 2 Click the **Static Routing** tab, and then click **Create**.

Figure 9-12 *Static Route with a Single CTC Destination*

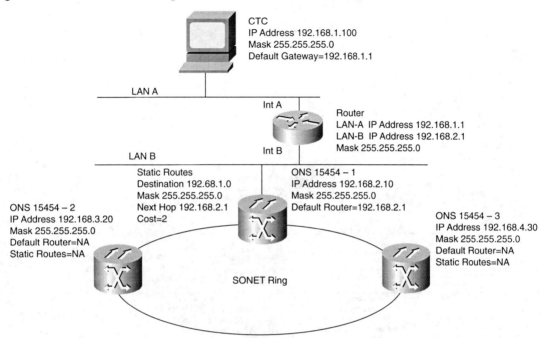

Step 3 In the Create Static Route dialog box. Enter the following:

- **Destination**—Enter the IP address of the computer running CTC. To limit access to one computer, enter the full IP address (in the example, **192.168.1.100**). To allow access to all computers on the 192.168.1.0 subnet, enter **192.168.1.0** and a subnet mask of **255.255.255.0**. You can enter a destination of 0.0.0.0 to allow access to all CTC computers that connect to the router.

- **Mask**—Enter a subnet mask. If the destination is a host route (single CTC computer), enter a 32-bit subnet mask (**255.255.255.255**). If the destination is a subnet, adjust the subnet mask accordingly (for example, **255.255.255.0**). If the destination is 0.0.0.0, enter a subnet mask of **0.0.0.0** to provide access to all CTC computers.

- **Next Hop**—Enter the IP address of the router port (in this example, **192.168.90.**1) or the node IP address if the CTC computer is connected to the node directly.

- **Cost**—Enter the number of hops between the ONS 15454 and the computer. In this example, the cost is two, one hop from the ONS 15454 to the router and a second hop from the router to the CTC workstation.

Step 4 Click **OK**. Verify that the static route displays in the Static Route window, or ping the node.

Figure 9-13 *Static Route with Multiple LAN Destinations*

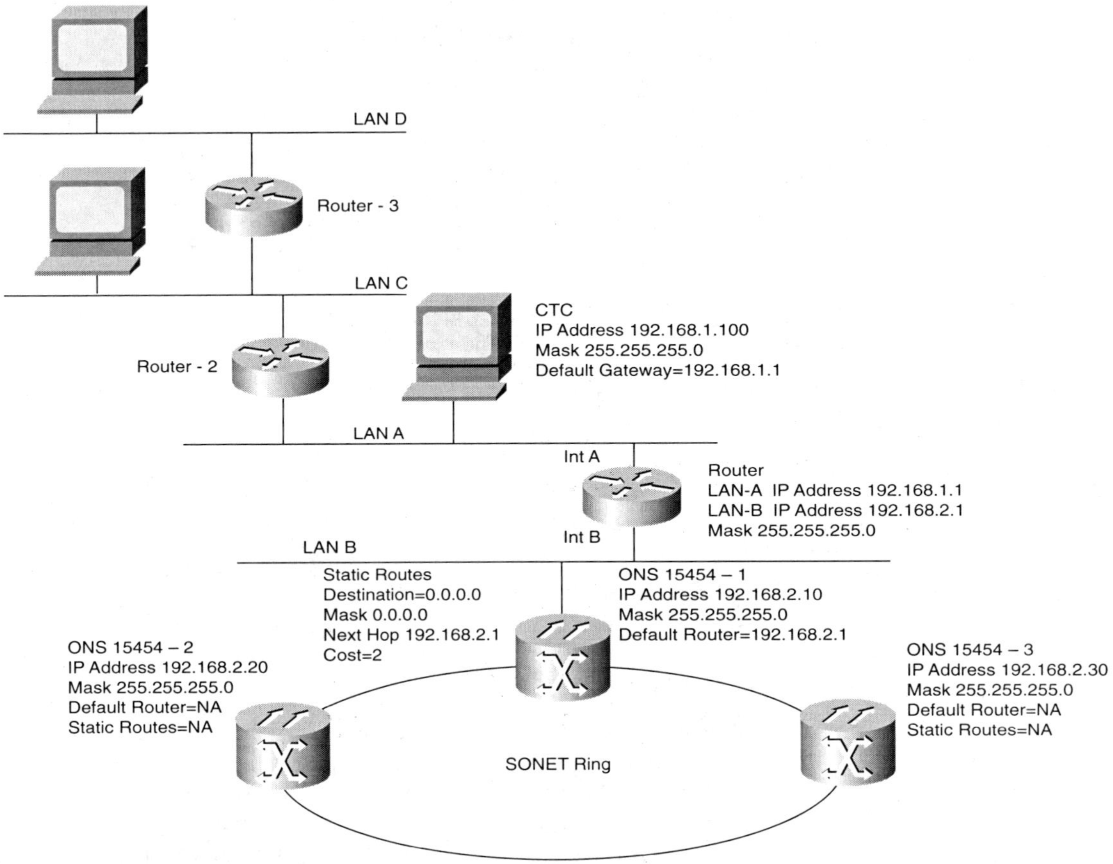

Static Routes for Multiple CTCs

This scenario illustrates the static routes used when multiple CTC computers need to access ONS nodes residing on the same subnet, as shown in Figure 9-14. In this scenario, CTC nodes 1 and 2 as well as all ONS nodes are on the same IP subnet 192.168.1.0. ONS node 1 and CTC 1 are attached to LAN A. ONS node 2 and CTC 2 are attached to LAN B. Static routes are added to ONS node 1 pointing to CTC 1, and to ONS node 2 pointing to CTC 2. The static routes are entered from the node's perspective.

ONS OSPF Configuration

ONS 15454s use the OSPF protocol in internal ONS networks for node discovery, circuit routing, and node management. You can enable OSPF on the ONS 15454s so that the ONS node topology is sent to OSPF routers on a LAN. Advertising the ONS 15454 network topology to LAN routers eliminates the need to manually enter static routes for ONS 15454 subnets.

Figure 9-15 shows the same network enabled for OSPF, and Figure 9-16 shows the same network without OSPF. Static routes must be manually added to the router for CTC computers on LAN A to communicate with ONS nodes 2 and 3 because these nodes reside on different subnets.

Figure 9-14 *Static Routes on the ONS Nodes*

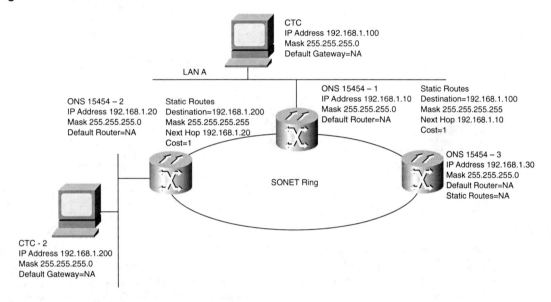

Figure 9-15 *ONS Configuration with OSPF Enabled*

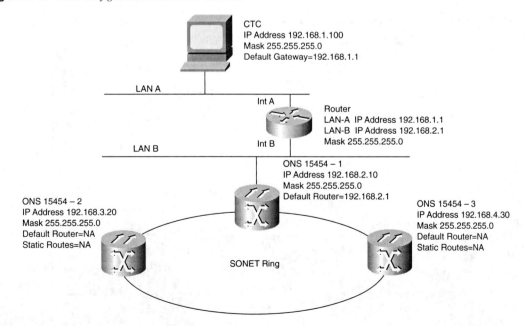

Figure 9-16 *ONS Configuration with OSPF Disabled*

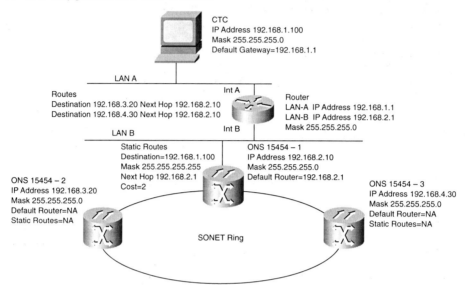

Use the following procedure to enable OSPF on each ONS 15454 node that you want included in the OSPF network topology. ONS 15454 OSPF settings must match the router OSPF settings, so you need to get the OSPF area ID, hello and dead intervals, and authentication key (if OSPF authentication is enabled) from the router to which the ONS 15454 network is connected before enabling OSPF. An example of OSPF provisioning is shown in Figure 9-17.

Figure 9-17 *OSPF Configuration for the ONS 15454*

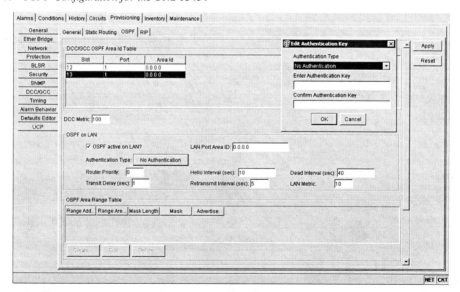

Step 1 Log in to the ONS 15454 node.

Step 2 In node view, choose the **Provisioning > Network > OSPF** tab.

Step 3 On the upper-left side, complete the following:

- **DCC OSPF Area ID**—Enter the number that identifies the ONS 15454s as a unique OSPF area. The OSPF area number can be an integer between 0 and 4,294,967,295, and it can take a form similar to an IP address. The number must be unique to the LAN OSPF area.

- **DCC Metric**—This value is normally unchanged. It sets a "cost" for sending packets across the DCC, which is used by OSPF routers to calculate the shortest path. This value should always be higher than the LAN metric. The default DCC metric is 100.

Step 4 In the OSPF on LAN area, complete the following:

- **OSPF Active on LAN**—When checked, enables ONS 15454 OSPF topology to be advertised to OSPF routers on the LAN. Enable this field on ONS 15454s that directly connect to OSPF routers.

- **Area ID for LAN Port**—Enter the OSPF area ID for the router port where the ONS 15454 is connected. (This number is different from the DCC area ID.)

Step 5 In the Authentication area, complete the following:

- **Type**—If the router where the ONS 15454 is connected uses authentication, choose **Simple Password**. Otherwise, choose **No Authentication**.

- **Key**—If authentication is enabled, enter the OSPF key (password).

Step 6 In the Priority and Intervals area, complete the following:

- The OSPF priority and intervals default to values most commonly used by OSPF routers. In the Priority and Interval area, verify that these values match those used by the OSPF router where the ONS 15454 is connected.

- **Router Priority**—Used to choose the designated router for a subnet.

- **Hello Interval (sec)**—Sets the number of seconds between OSPF hello packet advertisements sent by OSPF routers. The default is 10 seconds.

- **Dead Interval**—Sets the number of seconds that will pass while an OSPF router's packets are not visible before its neighbors declare the router down. The default is 40 seconds.

- **Transit Delay (sec)**—Indicates the service speed. The default is 1 second.

- **Retransmit Interval (sec)**—Sets the time that will elapse before a packet is resent. The default is 5 seconds.
- **LAN Metric**—Sets a cost for sending packets across the LAN. This value should always be lower than the DCC metric. The default is 10.

Step 7 In the OSPF Area Range Table area, complete the following:

Area range tables consolidate the information that is propagated outside an OSPF area border. One ONS 15454 in the ONS 15454 OSPF area is connected to the OSPF router. An area range table on this node points the router to the other nodes that reside within the ONS 15454 OSPF area.

To create an area range table, follow these substeps:

(a) Under OSPF Area Range Table, click **Create**.

(b) In the Create Area Range dialog box, enter the following:

- **Range Address**—Enter the area IP address for the ONS 15454s that reside within the OSPF area. If the ONS 15454 OSPF area includes nodes with IP addresses 10.10.20.100, 10.10.30.150, 10.10.40.200, and 10.10.50.250, the range address would be 10.10.0.0.
- **Range Area ID**—Enter the OSPF area ID for the ONS 15454s. This is either the ID in the DCC OSPF Area ID field or the ID in the Area ID for LAN Port field.
- **Mask Length**—Enter the subnet mask length. In the range address example, this is 16.
- **Advertise**—Check if you want to advertise the OSPF range table.

(c) Click **OK**.

Step 8 All OSPF areas must be connected to area 0. If the ONS 15454 OSPF area is not physically connected to area 0, use the following substeps to create a virtual link table that will provide the disconnected area with a logical path to Area 0:

(a) Under OSPF Virtual Link Table, click **Create**.

(b) In the Create Virtual Link dialog box, complete the following fields (OSPF settings must match OSPF settings for the ONS 15454 OSPF area):

- **Neighbor**—Enter the router ID of the area 0 router.
- **Transit Delay (sec)**—The service speed. The default is 1 second.
- **Hello Interval (sec)**—The number of seconds between OSPF hello packet advertisements sent by OSPF routers. The default is 10 seconds.

- **Auth Type**—If the router where the ONS 15454 is connected uses authentication, choose **Simple Password**. Otherwise, set it to **No Authentication**.

- **Retransmit Interval (sec)**—Sets the time that will elapse before a packet is resent. The default is 5 seconds.

- **Dead Interval (sec)**—Sets the number of seconds that will pass while an OSPF router's packets are not visible before its neighbors declare the router down. The default is 40 seconds.

(c) Click **OK**.

Step 9 After entering ONS 15454 OSPF area data, click **Apply**.

NOTE If you changed the area ID, the TCC cards will reset, one at a time.

UPSR Configuration

UPSRs provide duplicate fiber paths around the ring. Working traffic flows in one direction, and protection traffic flows in the opposite direction. If a problem occurs in the working traffic path, the receiving node switches to the path coming from the opposite direction. Please refer to Chapter 5, "SONET Architectures," for details on UPSR protection switching. CTC automates UPSR ring configuration. UPSR traffic is defined within the ONS 15454 on a circuit-by-circuit basis. If a path-protected circuit is not defined within a 1+1 or BLSR line protection scheme and path protection is available and specified, CTC uses UPSR as the default.

Because each traffic path is transported around the entire ring, UPSRs are best suited for networks where traffic concentrates at one or two locations and is not widely distributed. UPSR capacity is equal to its bit rate. Services can originate and terminate on the same UPSR, or they can be passed to an adjacent access or interoffice ring for transport to the service-terminating location.

To set up UPSR, you perform four basic procedures:

- Install the UPSR trunk cards.
- Create the DCC terminations.
- Configure the timing.
- Enable the ports.

After you enable the ports, you set up the UPSR circuits. UPSR signal thresholds—the levels that determine when the UPSR path is switched—are set at the circuit level. To create UPSR circuits, see the "Create UPSR Circuits" section.

Install the UPSR Trunk Cards

This section describes the steps used to install UPSR trunk cards:

Step 1 Install the OC-N cards that will serve as the UPSR trunk cards. You can install the OC-3, OC-12, and OC-48AS cards in any slot, but the OC-48 and OC-192 cards can be installed only in slots 5, 6, 12, or 13.

Step 2 Allow the cards to boot.

Step 3 Attach the fiber to the east and west UPSR ports at each node. To avoid errors, make the east port the farthest slot to the right and the west port the farthest left. Fiber connected to an east port at one node must plug into the west port on an adjacent node. Figure 9-18 shows fiber connections for a four-node UPSR with trunk cards in slot 5 (west) and slot 12 (east). Always plug the fiber plugged into the transmit (TX) connector of an OC-N card at one node into the receive (RX) connector of an OC-N card at the adjacent node. The card will display an SF LED if TX and RX fibers are mismatched.

Figure 9-18 *Four-Node UPSR Configuration*

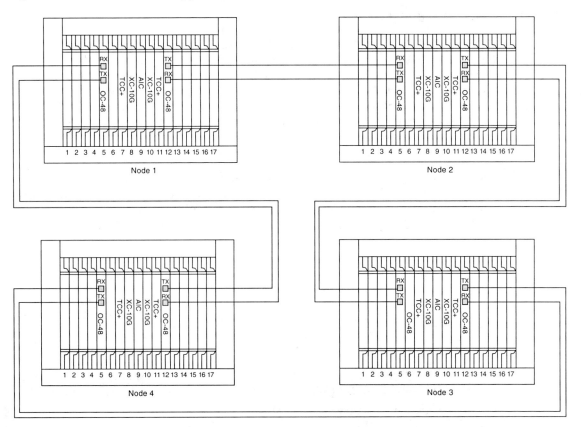

Configure the UPSR DCC Terminations

This section describes the steps used to configure UPSR DCC terminations:

Step 1 Log in to the first node that will be in the UPSR.

Step 2 Click the **Provisioning > SONET DCC** tab.

Step 3 In the SDCC Terminations section, click **Create**.

Step 4 In the Create SDCC Terminations dialog box, press Control and click the two slots/ports that will serve as the UPSR ports at the node. For example, slot 5 (OC-48)/port 1 and slot 12 (OC-48)/port 1.

> **NOTE** The ONS 15454 uses the SONET section layer DCC (SDCC) for data communications. It does not use the line DCCs. Line DCCs can be used to tunnel DCCs from third-party equipment across ONS 15454 networks. For procedures, see the "Creating DCC Tunnels" section.

Step 5 Click **OK**. The slots/ports display in the SDCC Terminations section.

Step 6 Complete Steps 2 through 5 at each node that will be in the UPSR.

An example of UPSR SONET DCC configuration is shown in Figure 9-19. After configuring the SONET DCC, set the timing for the node. For procedures, see the "Provisioning ONS 15454 Timing" section in this chapter. After configuring the timing, enable the UPSR ports, as described in the following procedure.

Figure 9-19 *SONET DCC Provisioning for UPSR*

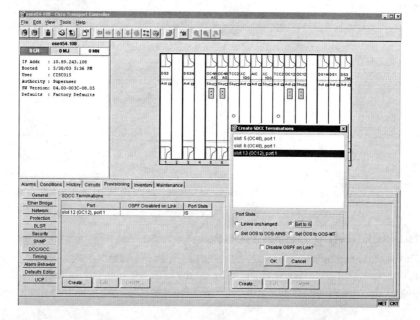

Enable the UPSR Ports

This section describes the steps used to enable UPSR ports:

Step 1 Log in to the first UPSR node.

Step 2 Double-click one of the cards that you configured as an SDCC termination.

Step 3 Click the **Provisioning > Line** tab.

Step 4 Under State, choose **In Service** for each port that you want enabled.

Step 5 Repeat Steps 2 through 4 for the second card.

Step 6 Click **Apply**.

The steps described previously describe UPSR configuration for a single node. An example is shown in Figure 9-20. Apply the same procedures to configure the additional nodes. To create circuits, see the "Creating Circuits and VT Tunnels" section.

Figure 9-20 *Enabling UPSR Ports for the ONS 15454*

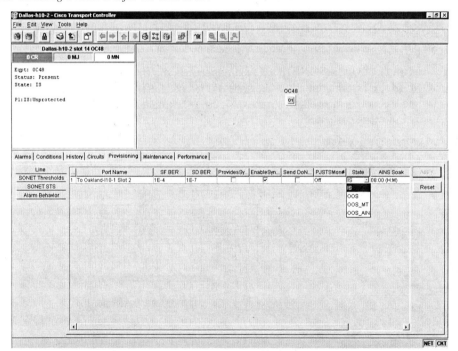

Adding and Removing UPSR Nodes

This section explains how to add and remove nodes in an ONS 15454 UPSR configuration. To add or remove a node, you switch traffic on the affected spans to route traffic away from the area of the ring where service will be performed. Use the span selector switch option to switch traffic from a UPSR span at different protection levels. The span selector switch option is useful when you need to reroute traffic from a UPSR span temporarily to add or drop nodes, perform maintenance, or perform other operations.

Switch UPSR Traffic

This section describes the steps used to switch UPSR traffic:

Step 1 Display the network view.

Step 2 Right-click the span that will be cut to add or delete a node and choose **Circuits** from the shortcut menu.

Step 3 On the Circuits on Span dialog box, choose the protection from the Switch all UPSR Circuits menu:

- **Clear** removes a previously set **switch** command.
- **Manual** switches the span if the new span is error free.
- **Force** forces the span to switch, regardless of whether the new span is error free.
- **Lockout** locks out or prevents switching to a highlighted span. (**Lockout** is available only when revertive traffic is enabled.)

> **NOTE** Force and Lockout commands override normal protective switching mechanisms. Applying these commands incorrectly can cause traffic outages.

Step 4 Click **Apply**.

Step 5 When the confirmation dialog box appears, click **OK** to confirm the protection switching. The column under Switch State changes to your chosen level of protection.

Step 6 Click **Close** after Switch State changes.

Add a UPSR Node

You can add only one node at a time. Perform these steps onsite and not from a remote location. This section describes the steps used to add a UPSR node:

Step 1 Log in to CTC and display the UPSR nodes in network view. Verify the following:

- All UPSR spans on the network map are green.
- No critical or major alarms (LOF, LOS, ASP, or ASL) are displayed on the Alarms tab.

- On the Conditions tab, no UPSR switches are active.
- At each physical UPSR node, all fibers are securely connected to the appropriate ports.
- If trouble is indicated (for example, a critical or major alarm exists), resolve the problem before proceeding.

Step 2 At the node that will be added to the UPSR, do the following:

(a) Verify that the OC-N cards are installed and fiber is available to connect to the other nodes.

(b) Run test traffic through the cards that will connect to the UPSR.

(c) Use the "UPSR Configuration" section to provision the new node.

Step 3 Log in to a node that will directly connect to the new node.

Step 4 Use the "Switch UPSR Traffic" procedure to initiate a force switch to switch traffic away from the span that will connect to the new node.

NOTE Traffic is not protected during a protection switch.

Step 5 Two nodes will connect directly to the new node; remove their fiber connections:

(a) Remove the east fiber connection from the node that will connect to the west port of the new node.

(b) Remove the west fiber connection from the node that will connect to the east port of the new node.

Step 6 Replace the removed fiber connections with connections from the new node.

Step 7 Log out of CTC, and then log back in.

Step 8 Display the network view. The new node should appear in the network map. Wait for a few minutes to allow all the nodes to appear.

Step 9 Click the **Circuits** tab and wait for all the circuits to appear, including spans. The affected circuit will display as incomplete.

Step 10 In the network view, right-click the new node and choose **Update Circuits with New Node** from the list of options. Wait for the confirmation dialog box to appear. Verify that the number of updated circuits displayed in the dialog box is correct.

Step 11 Click the **Circuits** tab and verify that no incomplete circuits are displayed. If incomplete circuits display, repeat Step 9.

Step 12 Use the "Switch UPSR Traffic" procedure to clear the protection switch.

Remove a UPSR Node

The following procedure is designed to minimize traffic outages while nodes are removed, but traffic will be lost when you delete and re-create circuits that passed through the removed node:

Step 1 Log in to CTC and display the UPSR nodes in network view. Verify the following:

- All UPSR spans on the network map are green.
- No critical or major alarms (LOF, LOS, ASP, or ASL) are displayed on the Alarms tab.
- On the Conditions tab, no UPSR switches are active.
- At each physical UPSR node, all fibers are securely connected to the appropriate ports.
- If trouble is indicated (for example, a critical or major alarm exists), resolve the problem before proceeding.

Step 2 Use the "Switch UPSR Traffic" procedure to initiate a force switch to switch traffic away from the node you remove. Initiate a force switch on all spans connected to the node you remove.

> **NOTE** Traffic is not protected during a forced protection switch.

Step 3 In the node that will be removed, delete circuits that originate or terminate in that node. (If a circuit has multiple drops, delete only the drops that terminate on the node you delete.)

(a) Click the **Circuits** tab.

(b) Choose the circuit(s) to delete. To choose multiple circuits, press the **Shift** or **Ctrl** key.

(c) Click **Delete**.

(d) Click **Yes** when prompted.

Step 4 From the node that will be deleted, remove the east and west span fibers. At this point, the node should no longer be a part of the ring.

Step 5 Reconnect the span fibers of the nodes remaining in the ring.

Step 6 Open the **Alarms** tab of each newly connected node and verify that the span cards are free of alarms. Resolve any alarms before proceeding.

Step 7 One circuit at a time, delete and re-create each circuit that passed through the deleted node on different STSs.

NOTE If the removed node was the BITS timing source, choose a new node as the BITS source or choose another node as the master timing node.

Step 8 Use the "Switch UPSR Traffic" procedure to clear the protection switch.

An example of UPSR circuit deletion is shown in Figure 9-21.

Figure 9-21 *UPSR Circuit Deletion*

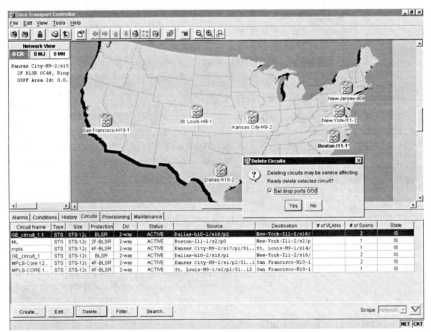

BLSR Configuration

The ONS 15454 can support two concurrent BLSRs in one of the following configurations:

- Two two-fiber BLSRs
- One two-fiber and one four-fiber BLSR

Each BLSR can have up to 32 ONS 15454 nodes. Because the working and protect bandwidths must be equal, you can create only OC-12 (two-fiber only), OC-48, or OC-192 BLSRs. Two-fiber BLSRs can support up to 24 ONS 15454s, but switch times are slightly longer for rings containing more than 16 nodes. BLSRs with 16 or fewer nodes will meet the GR-1230 sub-50-ms switch time requirement. However, four-fiber BLSRs can support only 16 nodes. In two-fiber BLSRs, each fiber is divided into *working* and *protect* bandwidths. In an OC-48 BLSR, for example, STSs 1 through 24 carry the working traffic, and STSs 25 through 48 are reserved for protection. Working traffic (STSs 1 through 24) travels in one direction on one fiber and in the opposite direction on the second fiber. The Cisco Transport Controller (CTC) circuit routing routines calculate the "shortest path" for circuits based on many factors including requirements set by the circuit provisioner, traffic patterns, and distance.

The SONET K1 and K2 bytes carry the information that governs BLSR protection switches. Each BLSR node monitors the K bytes to determine when to switch the SONET signal to an alternate physical path. The K bytes communicate failure conditions and actions taken between nodes in the ring. If a break occurs on one fiber, working traffic targeted for a node beyond the break switches to the protect bandwidth on the second fiber. The traffic travels in reverse direction on the protect bandwidth until it reaches its destination node. At that point, traffic is switched back to the working bandwidth.

Four-fiber BLSRs double the bandwidth of two-fiber BLSRs. Because they allow span switching as well as ring switching, four-fiber BLSRs increase the reliability and flexibility of traffic protection. Two fibers are allocated for working traffic and two fibers for protection. To implement a four-fiber BLSR, you must install four OC-48 or OC-48AS cards, or four OC-192 cards at each BLSR node. Four-fiber BLSRs provide span and ring switching. Span switching occurs when a working span fails and traffic switches to the protect fibers between the failed span and then returns to the working fibers. Multiple span switches can occur at the same time. Ring switching occurs when a span switch cannot recover traffic, such as when both the working and protect fibers fail on the same span. In a ring switch, traffic is routed to the protect fibers throughout the full ring.

The ONS 15454 uses the K3 overhead byte for BLSR automatic protection switching (APS) to allow an ONS 15454 BLSR to have more than 16 nodes. If a BLSR is routed through third-party equipment that cannot transparently transport the K3 byte, remap the BLSR extension byte on the trunk cards on each end of the span. It can be remapped to the Z2, E2, or F1 bytes on OC-48AS cards. If you remap the K3 byte, you must remap it to the same byte on each BLSR trunk card that connects to the third-party equipment. All other BLSR trunk cards should remain mapped to the K3.

NOTE Do not perform K3 byte remapping unless it is required to complete a BLSR that uses third-party equipment.

BLSR nodes can terminate traffic that is fed from either side of the ring and are suited for distributed node-to-node traffic applications, such as interoffice networks and access networks. BLSRs allow bandwidth to be reused around the ring and can carry more traffic than a network with traffic flowing through one central hub. A properly designed BLSR system can carry more traffic than a UPSR operating at the same OC-N rate. To set up a BLSR on the ONS 15454, you perform five basic procedures:

- Install the BLSR trunk cards.
- Create the BLSR DCC terminations.
- Enable the BLSR ports.
- Set up BLSR timing.
- Provision the BLSR.

394 Chapter 9: Provisioning the Multiservice SONET MSPP

Install the BLSR Trunk Cards

This section describes the steps used to install BLSR trunk cards:

Step 1 Install the OC-12, OC-48, OC-48AS, or OC-192 cards that will serve as the BLSR trunk cards. You can install the OC-12 and OC-48AS cards in any slot, but you can install the OC-48 and OC-192 cards only in high-speed slots 5, 6, 12, or 13.

Step 2 Allow the cards to boot.

Step 3 Attach the fiber to the east and west BLSR ports at each node. Plan your fiber connections and use the same card plan for all BLSR nodes. For example, make the east port the farthest slot to the right and the west port the farthest left. Plug fiber connected to an east port at one node into the west port on an adjacent node.

Figure 9-22 shows fiber connections for a two-fiber BLSR with trunk cards in slot 5 (west) and slot 12 (east). For four-fiber BLSRs, use the same east-west connection pattern for the working and protect fibers. Do not mix working and protect card connections. The BLSR will not function if working and protect cards are interconnected.

Figure 9-22 *Two-Fiber BLSR*

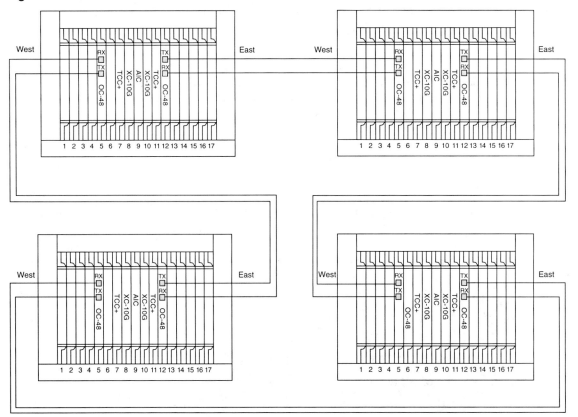

Figure 9-23 shows fiber connections for a four-fiber BLSR. Slot 5 (west) and slot 12 (east) carry the working traffic. Slot 6 (west) and slot 13 (east) carry the protect traffic.

Figure 9-23 *Four-Fiber BLSR*

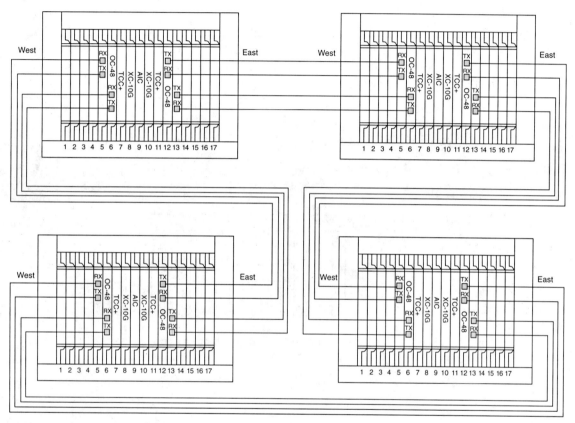

Create the BLSR DCC Terminations

This section describes the steps used to create BLSR DCC terminations:

Step 1   Log in to the first node that will be in the BLSR.

Step 2   Click the **Provisioning > SONET DCC** tab.

Step 3   In the SDCC Terminations section, click **Create**.

Step 4   In the Create SDCC Terminations dialog box, press **Ctrl** and click the two slots/ports that will serve as the BLSR ports at the node. For example, slot 5 (OC-48)/port 1 and slot 12 (OC-48)/port 1. For four-fiber BLSRs, provision the working cards, but not the protect cards, as DCC terminations.

Step 5 Click **OK**.

The slots/ports appear in the SDCC Terminations list.

Step 6 Complete Steps 2 through 5 at each node that will be in the BLSR.

NOTE The ONS 15454 uses the SONET section layer DCC (SDCC) for data communications. Because it does not use the line DCCs, the line DCCs are available for tunneling of DCCs from third-party equipment across ONS 15454 networks.

An example of SONET DCC provisioning for BLSR is shown in Figure 9-24.

Figure 9-24 *SONET DCC Provisioning for BLSR*

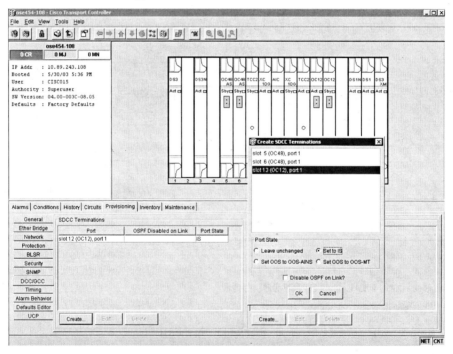

Enable the BLSR Ports

This section describes the steps used to enable BLSR ports:

Step 1 Log in to one of the nodes that will be in the BLSR.

Step 2 Double-click one of the OC-N cards that you configured as a DCC termination.

Step 3 Click the **Provisioning > Line** tab.

Step 4 Click **Status** and choose **In Service**.

Step 5 Click **Apply**.

Step 6 Repeat Steps 2 through 4 for the other optical card configured as a DCC termination.

Step 7 If you are configuring four-fiber BLSR, repeat Steps 2 through 4 for each protect card.

Step 8 Repeat Steps 2 through 5 at each node that will be in the BLSR.

NOTE After configuring the SONET DCC, set the timing for the node. For procedures, see the "Provisioning ONS 15454 Timing" section. After you configure the timing, you can provision the BLSR.

Remapping the K3 Byte

K3 byte remapping should be performed only when specifically required to run BLSRs through third-party equipment that cannot transparently transport the K3 byte. K3 bytes can be remapped only on OC-48AS cards. This section describes the steps used to remap the K3 byte:

Step 1 Log in to one of the nodes that connects to the third-party equipment.

Step 2 Double-click the OC-48AS card that connects to the third-party equipment.

Step 3 Click the **Provisioning > Line** tab.

Step 4 Click BLSR Ext byte and choose the alternate byte: **Z2**, **E2**, or **F1**.

Step 5 Click **Apply**.

Step 6 For four-fiber BLSRs, repeat Steps 2 through 5 for each protect card.

Step 7 Repeat Steps 2 through 5 at the node and card on the other end of the BLSR span.

Provision the BLSR

This section describes the steps used to provision the BLSR:

Step 1 Log in to a BLSR node.

Step 2 Choose the **Provisioning > Ring** tab.

Step 3 Click **Create**.

Step 4 In the Create BLSR dialog box, set the BLSR properties:

- **Ring Type**—Choose the BLSR ring type, either two-fiber or four-fiber.

- **Ring ID**—Assign a ring ID (a number between 0 and 9999). Nodes in the same BLSR must have the same ring ID.

- **Node ID**—Assign a node ID to the ONS node. The node ID identifies the node to the BLSR. Nodes in the same BLSR must have unique node IDs.

- **Ring Reversion**—Set the amount of time that will pass before the traffic reverts to the original working path. The default is 5 minutes. All nodes in a BLSR ring should have the same ring reversion setting, particularly if Never (nonrevertive) is selected.

- **West Port**—Assign the west BLSR port for the node from the pull-down menu. This is slot 5 in Figure 9-22.

- **East Port**—Assign the east BLSR port for the node from the pull-down menu. This is slot 12 in Figure 9-22.

The east and west ports must match the fiber connections and DCC terminations set up in the "Install the BLSR Trunk Cards" procedure and the "Create the BLSR DCC Terminations" procedure.

For four-fiber BLSRs, complete the following:

- **Span Reversion**—Set the amount of time that will pass before the traffic reverts to the original working path following a span reversion. The default is 5 minutes. Span reversions can be set to Never. If you set a ring reversion time, the times must be the same for both ends of the span. That is, if node A's west fiber is connected to node B's east port, the node A west span reversion time must be the same as the node B east span reversion time. To avoid reversion time mismatches, it is recommended that you use the same span reversion time throughout the ring.

- **West Protect**—Assign the west BLSR port that will connect to the west protect fiber from the pull-down menu. This is slot 6 in Figure 9-23.

- **East Protect**—Assign the east BLSR port that will connect to the east protect fiber from the pull-down menu. This is slot 13 in Figure 9-23.

Step 5 Click **OK**. An example for two-fiber BLSR creation is shown in Figure 9-25.

Figure 9-25 *BLSR Provisioning for the ONS 15454*

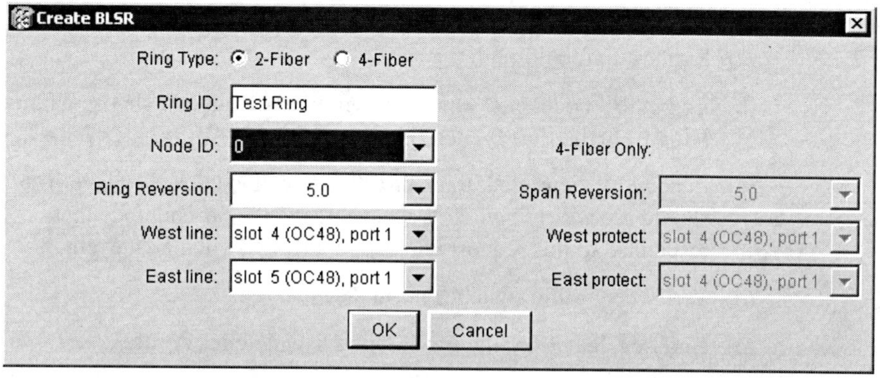

> **NOTE** Some or all of the following alarms display during BLSR setup: E-W MISMATCH, RING MISMATCH, APSCIMP, APSDFLTK, BLSROSYNC. The alarms will clear after you configure all the nodes in the BLSR.

Step 6 Complete Steps 2 through 5 at each node that you add to the BLSR.

Step 7 After you configure the last BLSR node, wait for the BLSR Ring Map Change dialog box to display.

> **NOTE** The dialog box can take 10 to 30 seconds to appear. The dialog box will not display if SDCC termination alarms, such as EOC or BLSR alarms (such as E-W MISMATCH and RING MISMATCH), are present. If an SDCC alarm is present, review the DCC provisioning at each node and follow the "Create the BLSR DCC Terminations" procedure. If BLSR alarms have not cleared, repeat Steps 1 through 6 at each node, making sure each node is provisioned correctly.

Step 8 In the BLSR Ring Map Change dialog box, click **Yes**.

Step 9 In the BLSR Ring Map dialog box, verify that the ring map contains all the nodes you provisioned in the expected order. If so, click **Accept**. If the nodes do not appear, or are not in the expected order, repeat Steps 1 through 8 accurately.

Step 10 Switch to network view and verify the following:

A green span line appears between all BLSR nodes. All E-W MISMATCH, RING MISMATCH, APSCIMP, DFLTK, and BLSROSYNC alarms are cleared.

Step 11 Test the BLSR using testing procedures normal for your site using the following steps:

(a) Run test traffic through the ring.

(b) Log in to a node, click the **Maintenance > Ring** tab, choose **Manual Ring** from the East Switch list, and then click **Apply**.

(c) In network view, click the **Conditions** tab and click **Retrieve**. You should see a Ring Switch West event, and the far-end node that responded to this request will report a Ring Switch East event.

(d) Verify that traffic switches normally.

(e) Choose **Clear** from the East Switch list and click **Apply**.

(f) Repeat Substeps A through D for the west switch.

(g) Disconnect the fibers at one node and verify that traffic switches normally.

Upgrading from Two-Fiber to Four-Fiber BLSR

Two-fiber OC-48 or OC-192 BLSRs can be upgraded to four-fiber BLSRs to provide added bandwidth and greater protection. The ONS 15454 does not support four-fiber OC-12 BLSR. To upgrade, you need to install two OC-48 or OC-192 cards at each two-fiber BLSR node, and then log in to CTC and upgrade each node from two-fiber to four-fiber. The fibers that were divided into working and protect bandwidths for the two-fiber BLSR are now fully allocated for working BLSR traffic.

Step 1 Log in to one of the two-fiber BLSR nodes. In network view, do the following:

(a) Verify that all spans between BLSR nodes on the network map are green.

(b) Click the **Alarms** tab. Verify that no critical or major alarms are present, nor any facility alarms, such as LOS, LOF, AIS-L, SF, and SD. In a BLSR, these facility conditions might be reported as minor alarms.

(c) Click the **Conditions** tab and click **Retrieve Conditions**. Verify that no ring switches are active. If trouble is indicated (for example, a major alarm exists), resolve the problem before proceeding to Step 2.

Step 2 Install two OC-48 or OC-192 cards at each BLSR node. You must install the same OC-N card rate as the two-fiber BLSR.

Step 3 Enable the ports for each new OC-N card:

(a) Display the card in card view.

(b) Click the **Provisioning > Line** tab.

(c) Click **Status** and choose **In Service**.

(d) Click **Apply**.

(e) Repeat Substeps A through D for each new OC-N card at each BLSR node.

Step 4 Connect the fiber to the new cards. Use the same east-west connection scheme that was used to create the two-fiber connections, as shown in Figure 9-23.

Step 5 Test the new fiber connections using procedures standard for your site. For example, pull a TX fiber for a protect card and verify that an LOS alarm displays for the appropriate remote RX card. Perform this fiber test for every span in the BLSR protect ring.

Step 6 Perform a span lockout at each BLSR node:

(a) At one of the BLSR nodes, switch to node view. Click the **Maintenance > Ring** tab.

(b) Under West Switch for the two-fiber BLSR you will convert, choose **Lockout Span**. Click **Apply**.

(c) Under East Switch, choose **Lockout Span**. Click **Apply**.

(d) Repeat Substeps A through C at each node in the two-fiber BLSR.

Step 7 Upgrade each node from two-fiber to four-fiber BLSR:

(a) At one of the BLSR nodes, switch to node view. Click the **Provisioning > Ring** tab.

(b) Choose the two-fiber BLSR. Click **Upgrade**.

(c) In the Upgrade BLSR dialog box, complete the following:

- **Span Reversion**—Set the amount of time that will pass before the traffic reverts to the original working path following a span reversion. The default is 5 minutes.

- **West Protect**—Assign the east BLSR port that will connect to the east protect fiber from the pull-down menu. This is slot 6 in Figure 9-23.

- **East Protect**—Assign the east BLSR port that will connect to the east protect fiber from the pull-down menu. This is slot 13 in Figure 9-23.

(d) Click **OK**.

(e) Complete Substeps A through D at each two-fiber BLSR node.

Step 8 Clear the span lockout:

(a) Display a BLSR node in node view. Click the **Maintenance > Ring** tab.

(b) Under West Switch, choose **Clear**. Click **Apply**.

(c) Under East Switch, choose **Clear**. Click **Apply**.

(d) Repeat Substeps A through C at each node in the new four-fiber BLSR.

(e) Switch to network view. Verify that no critical or major alarms are present. Also verify that facility alarms, such as LOS, LOF, AIS-L, SF, and SD, are not present.

Step 9 Test the four-fiber BLSR using procedures in Step 11 of the "Provision the BLSR" procedure.

Figure 9-26 shows the four-fiber BLSR after the upgrade from two-fiber BLSR has been performed.

Figure 9-26 *Four-Fiber BLSR After Upgrade*

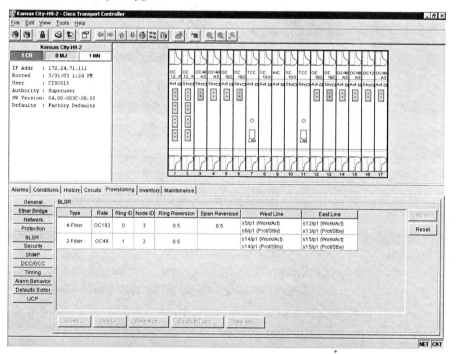

Adding and Removing BLSR Nodes

To add or remove a node, you force a protection switch to route traffic away from the span where you will add or remove the node. Figure 9-27 shows a three-node BLSR before the new node is added. You can add only one node at a time to an ONS 15454 BLSR. To add node 3, for example, follow these steps:

Step 1   Force a protection switch on the node 1 (slot 5, west) and node 4 (slot 12, east) span. The protection switch forces traffic away from the fibers that you will remove and reconnect to the added node.

Step 2   Remove fibers from node 1/slot 5 and node 4/slot 12, and then, using additional fibers, connect node 1 and node 4 to node 3.

Step 3   Remove the protection switch to route traffic through the added node.

Figure 9-27   *Three-Node BLSR*

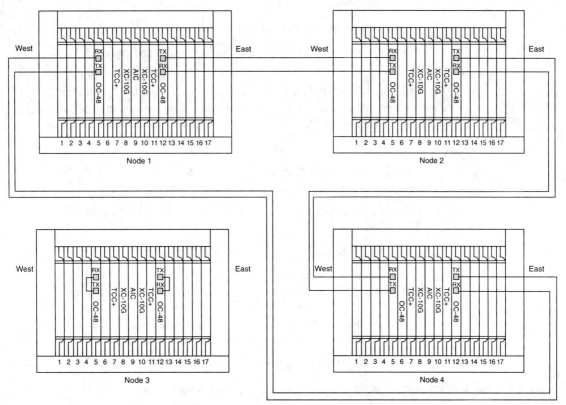

Add a BLSR Node

Perform these steps onsite and not from a remote location to add a BLSR node:

Step 1 Create a shelf schematic, similar to Figure 9-27, for the BLSR installation where you will add the node. In the diagram, identify the nodes, cards (slots), and spans (east or west) that will connect to the new node. This information is essential to complete this procedure without error. In Figure 9-27, you would identify slot 5 (west) on node 1, and slot 12 (east) on node 4.

Step 2 Log in to CTC and display the BLSR nodes in network view. Verify the following:

- All BLSR spans on the network map are green.
- On the Alarms tab, no critical or major alarms are present, nor any facility alarms, such as LOS, LOF, AIS-L, SF, and SD. In a BLSR, these facility conditions may be reported as minor alarms.
- On the Conditions tab, no ring switches are active.
- If a major alarm exists, resolve the problem before proceeding.

Step 3 Install the OC-N cards in the ONS 15454 that you will add to the BLSR using the "Install the BLSR Trunk Cards" procedure. Ensure that fiber cables are available to connect to the cards. Run test traffic through the node to ensure the cards function properly.

Step 4 Log in to the new node and complete the BLSR setup:

(a) Provision the SONET DCC using the "Create the BLSR DCC Terminations" procedure.

(b) Configure the BLSR timing using the "Provisioning ONS 15454 Timing" procedure.

(c) Enable the BLSR ports using the "Enable BLSR Ports" procedure.

(d) If the new node will connect to third-party equipment that cannot transport the K3 byte, use the "Remap the K3 Byte" procedure to remap the OC-48AS card's trunk card that connects to the third-party equipment. Make sure the trunk card at the other end of the span is mapped to the same byte set on the new node.

(e) Provision the BLSR using the "Provision the BLSR" procedure.

Step 5 Log in to the node that will connect to the new node through its east port (node 4 in Figure 9-27).

Step 6 Switch protection on the east port:

    (a) Click the **Maintenance > Ring** tab.

    (b) From the East Switch list, choose **Force Ring**. Click **Apply**.

Performing a force switch generates a manual switch request on an equipment (MANUAL-REQ) alarm. This is normal. Traffic is unprotected during a protection switch.

Step 7 Log in to the node that will connect to the new node through its west port (node 1 in Figure 9-27).

Step 8 Switch protection on the west port:

    (a) Click the **Maintenance > Ring** tab.

    (b) From the West Switch list, choose **Force Ring**. Click **Apply**.

Step 9 Following the schematic created in Step 1, remove the fiber connections from the two nodes that will connect directly to the new node:

    (a) Remove the east fiber from the node that will connect to the west port of the new node. In the Figure 9-27 example, this is node 4/slot 12.

    (b) Remove the west fiber from the node that will connect to the east port of the new node. In the Figure 9-27 example, this is node 1/slot 5.

Step 10 Replace the removed fibers with fibers that are connected to the new node. Connect the west port to the east port and the east port to the west port. Figure 9-28 shows the BLSR after the node is connected.

Step 11 Log out of CTC and then log back in to any node in the BLSR.

Step 12 In node view, choose the **Provisioning > Ring** tab and click **Ring Map**.

Step 13 In the BLSR Map Ring Change dialog box, click **Yes**.

Step 14 In the BLSR Ring Map dialog box, verify that the new node is added. If it is, click **Accept**. If it does not appear, log in to the new node. Verify that the BLSR is provisioned correctly according to the "Provision the BLSR" procedure, and then repeat Steps 12 through 13. If the node still does not appear, repeat the steps in the procedure making sure that no errors were made.

Step 15 From the **Go To** menu, choose **Network View**. Click the **Circuits** tab. Wait until all the circuits are discovered. The circuits that pass through the new node will be shown as incomplete.

Figure 9-28 *Four-Node BLSR*

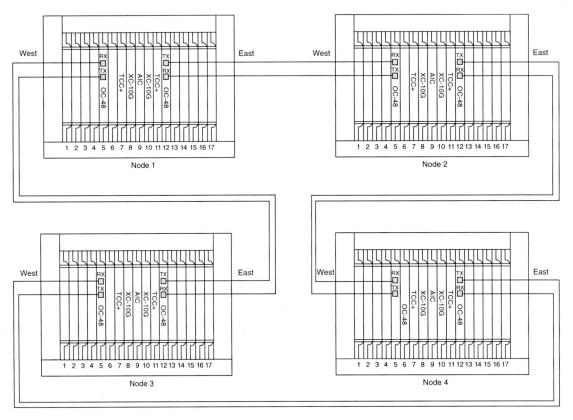

Step 16 In network view, right-click the new node and choose **Update Circuits with the New Node** from the shortcut menu. Verify that the number of updated circuits displayed in the dialog box is correct.

Step 17 Choose the **Circuits** tab and verify that no incomplete circuits are present.

Step 18 Clear the protection switch for the node that uses its east port to connect to the new node, and for the node that uses its west port to connect to the new node:

(a) To clear the protection switch from the east port, display the **Maintenance > Ring** tabs. From the East Switch list, choose **Clear**. Click **Apply**.

(b) To clear the protection switch from the west port, choose **Clear** from the West Switch list. Click **Apply**.

Remove a BLSR Node

The "Remove a BLSR" node procedure minimizes traffic outages during node deletions. You might need to delete and create circuits that pass through the node to be deleted if the circuit enters and exits the node on different STSs. This occurrence is rare and only applies to circuits created with Release 2.x software. Traffic will be lost when you delete and re-create circuits that passed through the deleted node. This section describes the steps used to remove a BLSR node:

Step 1 Before you commence this procedure, ensure that you have collected the following information:

- Which node is connected through its east port to the node that will be deleted. If you delete node 1 in Figure 9-28, for example, node 3 is the node connected through its east port to node 1.

- Which node is connected through its west port to the node that will be deleted. In Figure 9-28, node 2 is connected to node 1 through its west port.

Step 2 Log in to a node on the same BLSR as the node you will remove. Do not log in to the node that you will remove.

Step 3 Display the BLSR nodes in network view and verify the following:

- All BLSR spans on the network map are green.

- No critical or major alarms (LOF, LOS, ASP, or ASL) are displayed on the Alarms tab.

- On the Conditions tab, no ring switches are active.

- If trouble is indicated (for example, a critical or major alarm exists), resolve the problem before proceeding.

Step 4 Display the node that you will remove in node view.

Step 5 Delete all the circuits that originate or terminate in that node. If a circuit has multiple drops, delete only the drops that terminate on the node you want to delete:

(a) Click the **Circuits** tab. The circuits that use this node display.

(b) Choose circuits that originate or terminate on the node. Click **Delete**.

(c) Click **Yes** when prompted.

(d) If a multidrop circuit has drops at the node that will be removed, choose the circuit, click **Edit**, and remove the drops.

Step 6 Complete this step if circuits that were created using CTC Release 2.x. pass through the node that will be deleted:

(a) On the Circuits tab of the node that will be deleted, choose a circuit and click **Edit**.

(b) In the Edit Circuits window, check **Show Detailed Map**.

(c) Verify that the circuits enter and exit the node on the same STS. If a circuit enters on s5/p1/s1 (slot 5, port 1, sts1), verify that it exits on STS1. If a circuit enters/exits on different STSs, write down the name of the circuit. You will delete and then re-create these circuits in Substep E.

(d) From the View menu, choose **Go to Network View**, and then click the **Circuits** tab.

(e) Delete and then re-create each circuit recorded in Substep C that entered/exited the node to be deleted on different STSs. To delete the circuit, choose the circuit in the Circuits window, and then click the **Delete** button. To create the circuit, go to the "Create an Automatically Routed Circuit" procedure.

(f) Repeat Substeps A through E for each circuit displayed on the Circuits tab.

(g) Repeat Substeps A through C for each circuit displayed on the Circuits tab.

Step 7 Use information recorded in Step 1 to switch traffic away from the ports of neighboring nodes that will be disconnected when the node is removed:

(a) Open the neighboring node that is connected through its east port to the removed node.

(b) Click the **Maintenance > Ring** tab.

(c) From the East Switch list, choose **Force Ring**. Click **Apply**.

(d) Open the node that is connected through its west port to the removed node.

(e) Click the **Maintenance > Ring** tab.

(f) From the West Switch list, choose **Force Ring**. Click **Apply**.

Step 8 Remove all fiber connections between the node being removed and the two neighboring nodes.

Step 9 Reconnect the two neighboring nodes directly, west port to east port.

Step 10 If the removed node contained trunk OC-48AS cards with K3 bytes mapped to an alternate byte, use the "Remap the K3 Byte" procedure to verify and remap, if needed, the BLSR extended bytes on the newly connected neighboring nodes.

Step 11 Close CTC and then log in to a node on the reduced ring.

Step 12 Wait for the BLSR Map Ring Change dialog box to display. (If the dialog box does not display after 10 to 15 seconds, choose the **Provisioning > Ring** tabs and click **Ring Map**.) When the dialog box displays, click **Yes**.

Step 13 In the BLSR Ring Map dialog box, click **Accept**.

Step 14 Clear the protection switches on the neighboring nodes:

  (a) Open the node with the protection switch on its east port.

  (b) Click the **Maintenance > Ring** tab and choose **Clear** from the East Switch list. Click **Apply**.

  (c) Open the node with the protection switch on its west port.

  (d) Click the **Maintenance > Ring** tab and choose **Clear** from the West Switch list. Click **Apply**.

Step 15 If a BITS clock is not used at each node, check that the synchronization is set to one of the eastbound or westbound BLSR spans on the adjacent nodes. If the removed node was the BITS timing source, use a new node as the BITS source or choose internal synchronization at one node where all other nodes will derive their timing. For information about ONS 15454 timing, see the "Provisioning ONS 15454 Timing" section.

Moving BLSR Trunk Cards

The procedure of moving BLSR trunk cards is *service affecting* and needs to be performed with caution. It is recommended that the carrier or service provider announce a service outage window during the execution of this procedure, if performed in a production or operational environment. To change BLSR trunk cards, you drop one node at a time from the current BLSR. This applies to all BLSR nodes where cards will change slots. Review all the steps before you proceed.

Figure 9-29 shows a four-node OC-48 BLSR using trunk cards in slots 6 and 12 at all four nodes. In this example, trunk cards will be moved at node 4 from slots 6 and 12 to slots 5 and 6. To do this, node 4 is temporarily removed from the active BLSR ring while the trunk cards are switched.

Figure 9-29 *Four-Node BLSR Before the Trunk Card Switch*

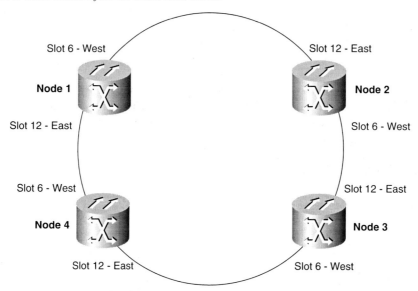

Figure 9-30 shows the four-node BLSR after the trunk card switch.

Figure 9-30 *Four-Node BLSR After the Trunk Card Switch*

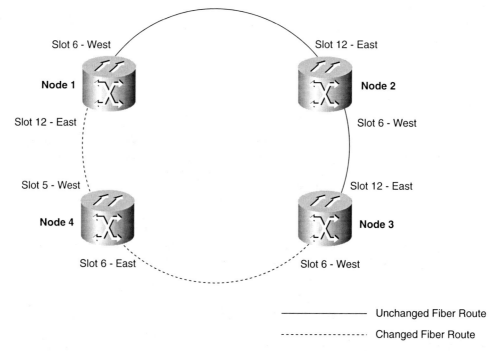

Move a BLSR Trunk Card

Use the following steps to move one BLSR trunk card to a different slot. Use this procedure for each card you want to move. Although the procedure is for OC-48 BLSR trunk cards, you can use the same procedure for OC-12, OC-48AS, and OC-192 cards. Ensure that the ONS 15454 nodes have CTC Release 2.0 or later and do not have any active alarms for the OC-48 or OC-12 cards or the BLSR configuration. This section describes the steps used to move a BLSR trunk card:

Step 1 Log in to CTC and display the BLSR nodes in network view. Verify the following:

- All BLSR spans on the network map are green.
- On the Alarms tab, no critical or major alarms are present, nor any facility alarms, such as LOS, LOF, AIS-L, SF, and SD. In a BLSR, these facility conditions might be reported as minor alarms.
- On the Conditions tab, no ring switches are active.
- If trouble is indicated (for example, a critical or major alarm exists), resolve the problem before proceeding.

Step 2 Switch traffic away from the node where the trunk card will be switched:

(a) Log in to the node that is connected through its east port to the node where the trunk card will be moved. (In the example shown in Figure 9-29, this is node 1.) Click the **Maintenance > Ring** tab.

(b) From the East Switch list, choose **Force Ring** and click **Apply**.

(c) When you perform a manual switch, a manual switch request equipment alarm (MANUAL-REA) is generated. This is normal.

(d) Log in to the node that is connected through its west port to the node where the trunk card will be moved. (In the example shown in Figure 9-29, this is node 3.) Click the **Maintenance > Ring** tab.

(e) From the West Switch list, choose **Force Ring** and click **Apply**.

Step 3 Log in to the node where the trunk card you will move is installed.

Step 4 Click the **Circuits** tab. Note down the circuit information or, from the File menu, choose **Print** or **Export** to print or export the information. This information will be needed to restore the circuits later.

Step 5 Delete the circuits on the card you remove:

(a) Highlight the circuit(s). To choose multiple circuits, press the **Shift** or **Ctrl** key.

(b) Click **Delete**.

(c) In the Delete Circuit dialog box, click **Yes**.

Step 6 Delete the SONET DCC termination on the card you remove:

   (a) Click the **Provisioning > SONET DCC** tab.

   (b) From the SDCC Terminations list, click the SONET DCC you need to delete and click **Delete**.

Step 7 Disable the ring on the current node:

   (a) Click the **Provisioning > Ring** tab.

   (b) Highlight the ring and click **Delete**.

   (c) On the confirmation message, confirm that this is the ring you want to delete. If so, click **Yes**.

Step 8 If an OC-N card is a timing source, click the **Provisioning > Timing** tab and set timing to **Internal**.

Step 9 Place the ports on the card out of service:

   (a) Double-click the card.

   (b) On the Provisioning > Line tab in the Status section, choose **Out of Service** for each port.

Step 10 Physically remove the card.

Step 11 Insert the card into its new slot and wait for the card to boot.

Step 12 To delete the card from its former slot, right-click the card in node view and choose **Delete** from the list of options.

Step 13 Place the port(s) back in service:

   (a) To open the card, double-click or right-click the card and choose **Open**.

   (b) Click the **Provisioning** tab.

   (c) From Status, choose **In Service**.

   (d) Click **Apply**.

Step 14 Follow the steps described in the "BLSR Configuration" section to re-enable the ring using the same cards (in their new slots) and ports for east and west. Use the same BLSR ring ID and node ID that was used before the trunk card was moved.

Step 15 Re-create the circuits that were deleted. See the "Create an Automatically Routed Circuit" procedure for instructions.

Step 16 If you use line timing and the card you move is a timing reference, re-enable the timing parameters on the card. See the "Provisioning ONS 15454 Timing" procedure for instructions.

Subtending Ring Configuration

The ONS 15454 supports up to 32 SONET DCCs with the TCC2 card in Release 4.x. The TCC2 hardware is ready to support up to 84 DCC terminations. Each ring requires two DCCs and a single ONS 15454 node can terminate and groom any one of the following ring combinations:

- 16 UPSRs
- 15 UPSRs + 1 BLSR (2-fiber or 4-fiber)
- 14 UPSRs + 2 BLSRs (one 4-fiber maximum)

Subtending rings offers the network designer versatility in layout of the fiber infrastructure. Subtending rings also reduces the number of nodes and cards required in a design and helps reduce external shelf-to-shelf cabling. Figure 9-31 shows an ONS 15454 with multiple subtending rings. It is possible to subtend a UPSR from a BLSR as well as subtend a BLSR from a UPSR.

NOTE The TCC2 card is based on a 400-MHz CPU with 256 MB flash and 256 MB RAM. The TCC2 card is *hardware-ready* to support up to 84 DCC terminations. However, the current Release 4.x of software supports only up to 32 DCC terminations.

Figure 9-31 *ONS Node with Multiple Subtending Rings*

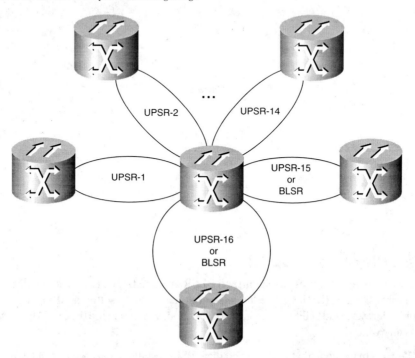

Figure 9-32 shows a UPSR subtending from a BLSR. In this example, node 3 is the only node serving both the BLSR and UPSR. OC-N cards in slots 5 and 12 serve the BLSR, and OC-N cards in slots 6 and 13 serve the UPSR.

Figure 9-32 *UPSR Subtending from BLSR*

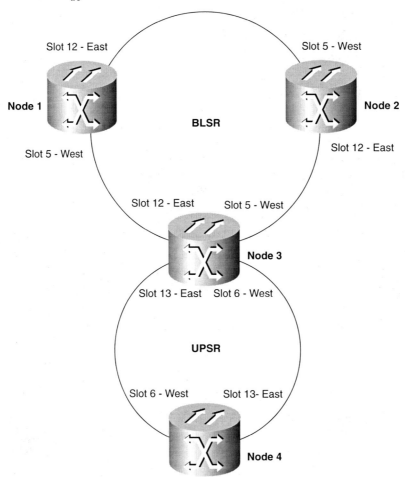

Subtend a UPSR from a BLSR

This procedure requires an established BLSR and one BLSR node with OC-N cards and fibers to carry the UPSR. The procedure also assumes you can set up a UPSR. For UPSR setup procedures, see the "UPSR Configuration" section. The procedure described here uses Figure 9-32 as an example:

Step 1 In the node that will subtend the UPSR, install the OC-N cards that will serve as the UPSR trunk cards (node 3, slots 6 and 13).

Step 2 Attach fibers from these cards to the UPSR trunk cards on the UPSR nodes. In Figure 9-32, slot 6 node 3 connects to slot 13/node 5, and slot 13 connects to slot 6/node 6.

Step 3 From node view, click the **Provisioning > SONET DCC** tab.

Step 4 Click **Create**.

Step 5 In the Create SDCC Terminations dialog box, click the slot and port that will carry the UPSR.

Step 6 Click **OK**. The selected slots/ports display in the SDCC Terminations section.

Step 7 Put the ports that you will use for the UPSR in service:

  (a) In the node view, double-click the UPSR trunk card.

  (b) Click the **Provisioning > Line** tab. Under Status, choose **In Service**.

  (c) Click **Apply**.

  (d) Repeat Substeps A through C for the second UPSR trunk card.

Step 8 Follow Steps 1 through 7 for the other nodes you will use for the UPSR.

Step 9 Go to network view to view the subtending ring.

Subtend a BLSR from a UPSR

This procedure requires an established UPSR and one UPSR node with OC-N cards and fibers to connect to the BLSR. The procedure also assumes you can set up a BLSR. For BLSR setup procedures, see the "BLSR Configuration" section.

The procedure described here uses Figure 9-33 as an example.

Step 1 In the node that will subtend the BLSR, install the OC-N cards that will serve as the BLSR trunk cards (node 3, slots 6 and 13).

Step 2 Attach fibers from these cards to the BLSR trunk cards on the BLSR nodes. In Figure 9-33, slot 6/node 3 connects to slot 13/node 5, and slot 13 connects to slot 6/node 6.

Step 3 From node view, click the **Provisioning > SONET DCC** tab.

Step 4 Click **Create**.

Step 5 In the Create SDCC Terminations dialog box, click the slot and port that will carry the BLSR.

Step 6 Click **OK**.

Step 7 The selected slots/ports display under SDCC Terminations.

Figure 9-33 *BLSR Subtending from UPSR*

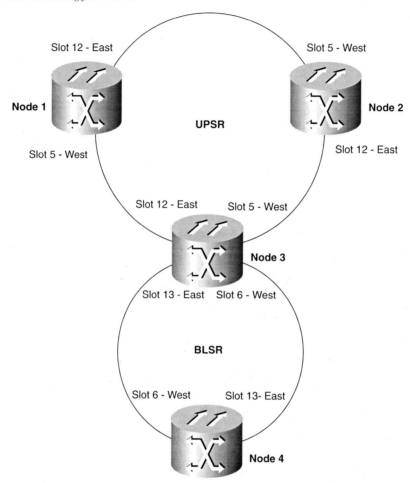

Step 8 Put the ports that you will use for the BLSR in service:

  (a) In the node view, double-click the BLSR trunk card.

  (b) Click the **Provisioning > Line** tabs. Under Status, choose **In Service**.

  (c) Click **Apply**.

  (d) Repeat Substeps A through C for the second BLSR trunk card.

Step 9 Use the "Provision the BLSR" procedure to configure the BLSR.

Step 10 Follow Steps 1 through 8 for the other nodes that will be in the BLSR.

Step 11 Go to the network view to see the subtending ring.

Subtend a BLSR from a BLSR

The ONS 15454 can support two BLSRs on the same node. This capability enables you to deploy an ONS 15454 in applications requiring SONET DCS (digital cross-connect system) or multiple SONET ADMs (add/drop multiplexers). After subtending two BLSRs, you can route circuits from nodes in one ring to nodes in the second ring. This procedure requires an established BLSR and one BLSR node with OC-N cards and fibers to carry the BLSR. The procedure also assumes you know how to set up a BLSR. For BLSR setup procedures, see the "BLSR Configuration" section.

The procedure described here uses Figure 9-34 as an example.

Figure 9-34 *BLSR Subtending from BLSR*

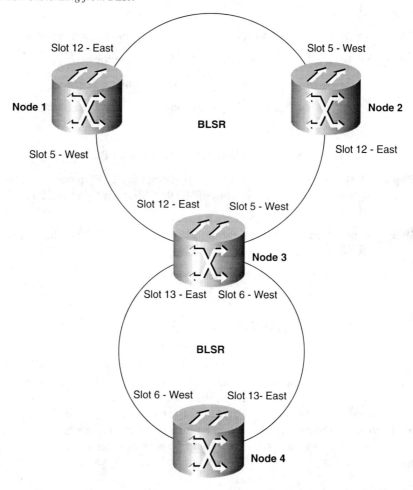

Step 1 In the node that will subtend the BLSR, install the OC-N cards that will serve as the BLSR trunk cards (node 3, slots 6 and 13).

Step 2 Attach fibers from these cards to the BLSR trunk cards on the BLSR nodes. In Figure 9-34, node 3/slot 6 connects to node 5/slot 13, and slot 13 connects to node 6/slot 6.

Step 3 From the node view, click the **Provisioning > SONET DCC** tab.

Step 4 Click **Create**.

Step 5 In the Create SDCC Terminations dialog box, click the slot and port that will carry the BLSR.

Step 6 Click **OK**.

Step 7 The selected slots/ports display in the SDCC Terminations section.

Step 8 Put the ports that you will use for the BLSR in service:

    (a) In node view, double-click the BLSR trunk card.

    (b) Click the **Provisioning > Line** tab. Under Status, choose **In Service**.

    (c) Click **Apply**.

    (d) Repeat Substeps A through C for the second BLSR trunk card.

Step 9 To configure the BLSR, use the "Provision the BLSR" procedure. The subtending BLSR must have a ring ID that differs from the ring ID of the first BLSR.

Step 10 Follow Steps 1 through 8 for the other nodes that will be in the subtending BLSR.

Step 11 Display the network view to see the subtending ring.

Linear ADM Configurations

ONS 15454s can be configured as linear ADMs by configuring one set of OC-N cards as the working path and a second set as the protect path. Unlike rings, linear (point-to-point) ADMs require that the OC-N cards at each node be in 1+1 protection to ensure that a break to the working line is automatically routed to the protect line. Figure 9-35 shows three ONS 15454s in a linear ADM configuration. Working traffic flows from slot 5/node A to slot 5/node B, and from slot 13/node B to slot 13/node C. The protect path is created by placing slot 5 in 1+1 protection with slot 4 at nodes 1 and 2, and slot 13 in 1+1 protection with slot 14 at nodes 2 and 3.

Figure 9-35 *Linear ADM*

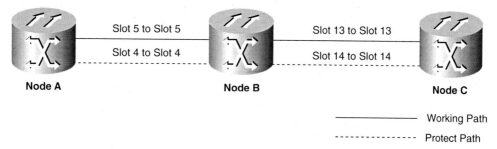

Create a Linear ADM

The following steps must be completed for each node that will be included in the linear ADM.

See Figure 9-35 for the procedure described here:

Step 1 Complete the general setup information for the node. For procedures, see the "Setting Up Basic Node Information" section.

Step 2 Set up the network information for the node. For procedures, see the "Setting Up Network Information" section.

Step 3 Set up 1+1 protection for the OC-N cards in the ADM. In Figure 9-35, slots 5 and 13 are the working ports and slots 4 and 14 are the protect ports. In this example, one protection group is set up for node A (slots 4 and 5), two for node B (slots 4 and 5, and 13 and 14), and one for node C (slots 13 and 14). To create protection groups, see the "Creating Protection Groups" section.

Step 4 For OC-N ports connecting ONS 15454s, set up the SONET DCC terminations:

(a) Log in to a linear ADM node and click the **Provisioning > SONET DCC** tabs.

(b) In the SDCC Terminations section, click **Create**.

(c) In the Create SDCC Terminations dialog box, choose the working port, and then click **OK**.

> **NOTE** Terminating nodes A and C in Figure 9-35 will have one SDCC, whereas intermediate nodes (node B) will have two SDCCs.

Step 5 Use the "Provisioning ONS 15454 Timing" section to set up the node timing. If a node uses line timing, set the working OC-N card as the timing source.

Step 6 Place the OC-N ports in service:

(a) Open an OC-N card that is connected to the linear ADM.

(b) On the Provisioning > Line tab under Status, choose **In Service**.

(c) Click **Apply**.

Step 7 Repeat Step 6 for each OC-N card connected to the linear ADM.

Convert a Linear ADM to UPSR

The following procedure describes how to convert a three-node linear ADM to a UPSR ring. You need a SONET test set to monitor traffic while you perform this procedure. This procedure will affect service and must be performed only during a scheduled outage.

See Figure 9-36 for the procedure described here:

Figure 9-36 *Linear ADM to UPSR Conversion*

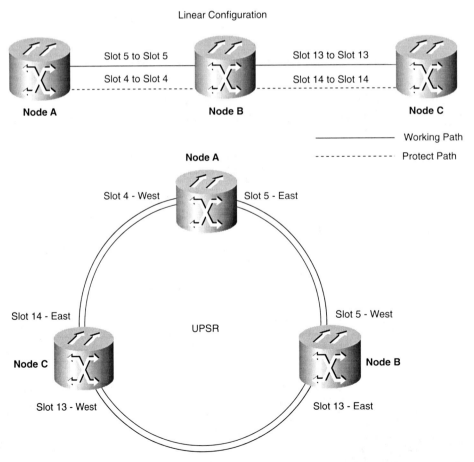

Step 1 Start CTC and log in to one of the nodes that you want to convert from linear to ring.

Step 2 Click the **Maintenance > Protection** tab.

Step 3 Under Protection Groups, choose the 1+1 protection group, which is the group that supports the 1+1 span cards.

Step 4 Under the selected group, verify that the working slot port is shown as Working-Active. If so, go to Step 5. If the working slot says Working-Standby and the protect slot says Protect-Active, switch traffic to the working slot:

   (a) Under Selected Group, choose the protect slot (that is, the slot that says Protect-Active).

   (b) From the Switch Commands, choose **Manual**.

   (c) Click **Yes** in the confirmation dialog box.

   (d) Under Selected Group, verify that the working slot port says Working-Active. If so, continue to Substep E. If not, clear the conditions that prevent the card from carrying working traffic before proceeding.

   (e) From the Switch Commands, choose **Clear**. A Confirm Clear Operation dialog displays.

   (f) Click **Yes** in the confirmation dialog box.

Step 5 Repeat Step 4 for each group in the 1+1 Protection Groups list at all nodes that will be converted.

Step 6 For each node, delete the 1+1 OC-N protection group that supports the linear ADM span:

NOTE   Deleting a 1+1 protection group may cause unequipped path (UNEQ-P) alarms to occur.

   (a) Click the **Provisioning > Protection** tab.

   (b) From the Protection Groups list, choose the 1+1 group you want to delete. Click **Delete**.

   (c) Click **Yes** in the confirmation dialog box.

   (d) Verify that no traffic disruptions are indicated on the test set. If disruptions occur, do not proceed. Re-create the protection group and isolate the cause of the disruption.

   (e) Continue deleting 1+1 protection groups while monitoring the existing traffic with the test set.

Step 7 Physically remove one of the protect fibers running between the middle and end nodes. In Figure 9-36, for example, the fiber from node B/slot 14 to node C/slot 14 is removed. The corresponding OC-48 card will go into an LOS condition for that fiber and port.

Step 8 Physically reroute the other protect fiber to connect the two end nodes. In the example, the fiber between node A/slot 4 and node B/slot 4 is rerouted to connect node A/slot 4 to node C/slot 14. If you leave the OC-N cards in place, go to Step 13. If you remove the cards, complete Steps 9 to 12. In this example, cards in node B/slots 4 and 14 are removed.

Step 9 Place the cards in slots 4 and 14 out of service for node B:

(a) Display the first card in card view and click the **Provisioning > Line** tab.

(b) Under Status, choose **Out of Service**. Click **Apply**.

(c) Repeat Substeps A and B for the second card.

Step 10 Delete the equipment records for the cards:

(a) Display the node view.

(b) Right-click the card you just took out of service (slot 4) and choose **Delete Card**.

(c) Click **Yes** in the confirmation dialog box.

(d) Repeat Substeps A through C for the second card (slot 14).

Step 11 Save all circuit information:

(a) In node view, click the **Provisioning > Circuits** tab.

(b) Record the circuit information using one of the following procedures:

- From the File menu, choose **Print** to print the Circuits table.

  Or

- From the File menu, choose **Export** to export the circuit data in HTML, CSV (comma-separated values), or TSV (tab-separated values). Click **OK** and save the file in a temporary directory.

Step 12 Remove the OC-N cards that are no longer connected to the end nodes (slots 4 and 14 in the example).

Step 13 Display one of the end nodes (node A or node C in the example).

Step 14 Click the **Provisioning > SONET DCC** tab.

Step 15 In the SDCC Terminations section, click **Create**.

Step 16 In the Create SDCC Terminations dialog box, choose the slot port that had been the protect slot in the linear ADM. (For example, for node A, this would be slot 4/port 1 [OC-48].)

Step 17 Click **OK**. An EOC SDCC alarm will occur until an SDCC termination is created on the adjacent node.

Step 18 Go to the node on the opposite end (node C in the example) and repeat Steps 14 through 17.

Step 19 Delete and re-enter the circuits one at a time. (See the "Creating Circuits and VT Tunnels" section.) Deleting circuits affects service.

You can create the circuits automatically or manually. However, circuits must be protected. When they were built in the linear ADM, they were protected by the protect path on node A/slot 4 to node B/slot 4 to node C/slot 14. With the new UPSR, circuits should also be created with protection. Deleting the first circuit and re-creating it to the same card port should restore the circuit immediately. Monitor your SONET test set to verify that the circuit was deleted and restored.

Step 20 You should also verify that the new circuit path for the clockwise (CW) fiber from node 1 to node 3 is working. To do this, switch to network view and move your cursor to the green span between nodes 1 and 3.

Step 21 Although the cursor shows only the first circuit created, do not panic if the other circuits are not present. Verify with the SONET test set that the original circuits and the new circuits are operational. The original circuits were created on the counterclockwise linear path.

Step 22 Go to the network map to view the newly created ring.

Convert a Linear ADM to a BLSR

The following procedure describes how to convert a three-node linear ADM to a BLSR. You need a SONET test set to monitor traffic while you perform this procedure. This procedure will affect service and must be performed only during a scheduled outage.

See Figure 9-37 for the procedure described here:

Step 1 Start CTC and log in to one of the nodes that you want to convert from linear to ring.

Step 2 Click the **Maintenance > Protection** tab.

Step 3 Under Protection Groups, choose the 1+1 protection group, which is the group that supports the 1+1 span cards.

Figure 9-37 *Linear ADM to BLSR Conversion*

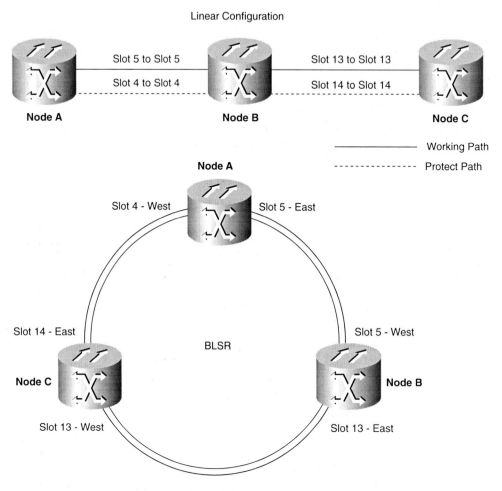

Step 4 Under Selected Group, verify that the working slot port is shown as Working-Active. If so, go to Step 5. If the working slot says Working-Standby and the protect slot says Protect-Active, switch traffic to the working slot:

(a) Under Selected Group, choose the protect slot (that is, the slot that says Protect-Active).

(b) From the Switch Commands, choose **Manual**.

(c) Click **Yes** in the confirmation dialog box.

(d) Verify that the working slot is carrying traffic. If it is, continue to Substep E. If not, clear the conditions that prevent the card from carrying working traffic before proceeding.

(e) From the Switch Commands, choose **Clear**. A Confirm Clear Operation dialog displays.

(f) Click **Yes** in the confirmation dialog box.

Step 5 Repeat Step 4 for each group in the 1+1 Protection Groups list at all nodes that will be converted.

Step 6 For each node, delete the 1+1 OC-N protection group that supports the linear ADM span:

(a) Click the **Provisioning > Protection** tab.

(b) From the Protection Groups list, choose the group you want to delete. Click **Delete**.

(c) Click **Yes** in the confirmation dialog box.

(d) Verify that no traffic disruptions are indicated on the SONET test set. If disruptions occur, do not proceed. Add the protection group and begin troubleshooting procedures to find out the cause of the disruption.

NOTE Deleting a 1+1 protection group can cause unequipped path (UNEQ-P) alarms to occur.

Step 7 Physically remove one of the protect fibers running between the middle and end nodes. In the example, the fiber running from node B/slot 14 to node C/slot 14 is removed. The corresponding end-node trunk card will display an LOS alarm.

Step 8 Physically reroute the other protect fiber so that it connects the two end nodes. In the example, the fiber between node A/slot 4 and node B/slot 4 is rerouted to connect node A/slot 4 to node C/slot 14. If you leave the OC-N cards in place, go to Step 13. If you remove the cards, complete Steps 9 through 12. (In this example, cards in node B/slots 4 and 14 are removed.)

Step 9 In the middle node B, place the cards in slots 4 and 14 out of service:

(a) Display the first card in card view, and then click the **Provisioning > Line** tab.

(b) Under Status, choose **Out of Service**. Click **Apply**.

(c) Repeat Substeps A and B for the second card.

Step 10 Delete the equipment records for the cards:

(a) From the View menu, choose **Node View**.

(b) Right-click the card you just took out of service (slot 4) and choose **Delete Card**. (You can also go to the Inventory tab, choose the card, and click **Delete**.)

(c) Click **Yes** in the confirmation dialog box.

(d) Repeat Substeps A through C for the second card (slot 13).

Step 11 Save all circuit information:

(a) In node view, click the **Provisioning > Circuits** tab.

(b) Record the circuit information using one of the following procedures:

- From the File menu, choose **Print** to print the circuits table.

  Or

- From the File menu, choose **Export** to export the circuit data in HTML, CSV, or TSV. Click **OK** and save the file in a temporary directory.

Step 12 Remove the OC-N cards that are no longer connected to the end nodes (slots 4 and 14 in the example).

Step 13 Log in to an end node. In node view, click the **Provisioning > SONET DCC** tab.

Step 14 In the SDCC Terminations section, click **Create**.

Step 15 Highlight the slot that is not already in the SDCC Terminations list (in this example, port 1 of slot 4 (OC-48) on node A.

Step 16 Click **OK**. An EOC SDCC alarm will occur until the DCC is created on the other node (in the example, node C/slot 14).

Step 17 Display the node on the opposite end (node C) and repeat Steps 13 through 16.

Step 18 For circuits running on a BLSR protect STS (STSs 7 through 12 for an OC-12 BLSR, STSs 25-48 for an OC-48 BLSR), delete and re-create the circuit:

(a) Delete the first circuit.

(b) Re-create the circuit on STSs 1 through 6 (for an OC-12 BLSR) or 1 through 24 (for an OC-48 BLSR) on the fiber that served as the protect fiber in the linear ADM. During circuit creation, deselect Route

Automatically and Fully Protected Path in the Circuit Creation dialog box so that you can manually route the circuit on the appropriate STSs. See the "Create a Unidirectional Circuit with Multiple Drops" procedure for more information.

(c) Repeat Substeps A and B for each circuit residing on a BLSR protect STS. Deleting circuits affects traffic.

Step 19 Follow all procedures in the "BLSR Configuration" section to configure the BLSR. The ring should have an east-west logical connection. Although it might not physically be possible to connect the OC-N cards in an east-west pattern, it is strongly recommended. If the network ring that is already passing traffic does not enable you to connect fiber in this manner, logical provisioning can be performed to satisfy this requirement. Be sure to assign the same ring ID and different node IDs to all nodes in the BLSR. Do not accept the BLSR ring map until all nodes are provisioned. An E-W Mismatch alarm will occur until all nodes are provisioned.

Step 20 Display the network map to view the newly created ring.

Pat-Protected Mesh Networking (PPMN)

PPMNs include multiple ONS 15454 SONET topologies, such as a combination of linear ADM and UPSR that extend the protection provided by a single UPSR to the meshed architecture of several interconnecting rings. In a PPMN, circuits travel diverse paths through a network of single or multiple meshed rings. When you create circuits, you can have CTC automatically route circuits across the PPMN, or you can manually route them. You can also choose levels of circuit protection. If you choose full protection, for example, CTC creates an alternate route for the circuit in addition to the main route. The second route follows a unique path through the network between the source and destination and sets up a second set of cross-connections.

In Figure 9-38, for example, a circuit is created from node A to node G. CTC determines that the shortest route between the two nodes passes through node B shown by the dotted line, and automatically creates cross-connections at nodes A, B, and G to provide the primary circuit path.

If full protection is selected, CTC creates a second unique route between nodes A and G, which, in this example, passes through nodes E and F. Cross-connections are automatically created at nodes A, E, F, and G, shown by the dashed line. If a failure occurs on the primary path, traffic switches to the second circuit path. In this example, node G switches from the traffic coming in from node B to the traffic coming in from node F and service resumes. The switch occurs within 50 ms.

Figure 9-38 *Path-Protected Mesh Network*

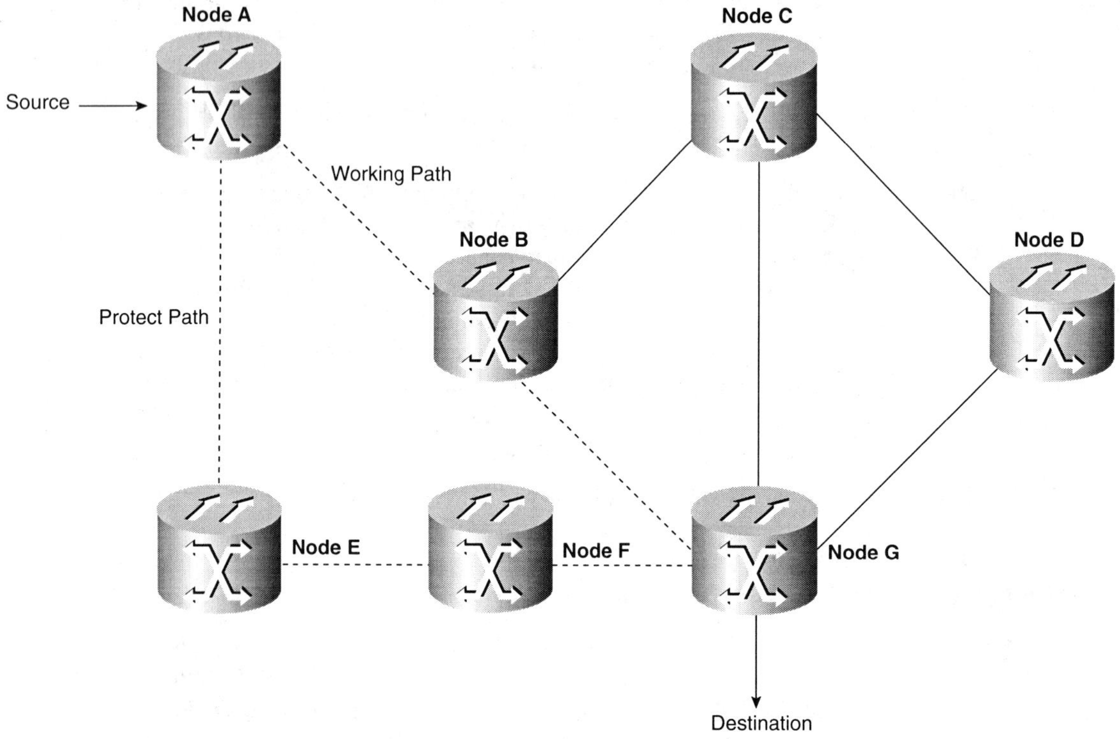

Circuit Provisioning

Provisioning a UPSR, BLSR, linear ADM, or PPMN is not complete without assigning circuits for transport of user traffic. This section explains how to create and administer various types of Cisco ONS 15454 circuits and tunnels. DCC tunnels can also be created to tunnel third-party equipment signaling through ONS 15454 networks. *Cross-connects* refers to the connections that occur within a single ONS 15454 that allow a circuit to enter on one port and exit on another, whereas the term *circuit* refers to the series of connections from a traffic source (where traffic ingresses the ONS network) to the drop or destination (where traffic egresses an ONS network). Various types of STS and VT1.5 circuits can be provisioned across and within ONS 15454 nodes with different attributes assigned to them. These attributes include the following:

- One-way, half-duplex circuits
- Two-way, full-duplex circuits
- Broadcast circuits
- Revertive or nonrevertive circuits
- Circuit size

- Automatic circuit routing
- Manual circuit routing
- Circuit path protection
- Protected source and destination cards and ports

Circuits can be routed automatically or can be manually routed. These circuits could be unidirectional or bidirectional, revertive or nonrevertive. The CTC Auto Range feature eliminates the need to individually build circuits of the same type, because CTC can create additional sequential circuits if you specify the number of circuits you need and build the first circuit. The additional circuits will inherit the attributes of the first circuit created. This is known as *A-to-Z provisioning* in CTC jargon. CTC enables you to provision circuits even before the cards can be installed. To provision an empty slot, right-click it and choose a card from the shortcut menu. However, circuits will not carry traffic until the cards are installed and their ports are placed in service. These circuits will carry traffic only after a signal is received. If you want to route circuits on protected drops, you need to create the card protection groups before creating circuits.

Create an Automatically Routed Circuit

This section describes the steps used to create an automatically routed circuit:

Step 1 Log in to an ONS 15454 and click the **Circuits** tab.

Step 2 Click **Create**.

Step 3 In the Circuit Creation dialog box, complete the following fields:

- **Name**—Assign a name to the circuit. The name can be alphanumeric and up to 32 characters (including spaces). If you leave the Name field blank, CTC assigns a default name to the circuit. If you work for a LEC or IXC, this field would be assigned with the customer circuit ID.

- **Type**—Choose the type of circuit you want to create: STS, VT (VT1.5), or VT tunnel. The circuit type determines the circuit-provisioning options that can be provisioned.

- **Size**—Choose the circuit size (STS circuits only). The "c" indicates concatenated STSs.

- **Bidirectional**—Check this box to create a two-way circuit. If this box is unchecked, CTC will create a unidirectional circuit. Only STS and VT circuits can be configured as unidirectional. VT tunnels are always bidirectional.

- **Number of Circuits**—Enter the number of circuits you want to create. If you enter more than one, you can use autoranging to create the additional circuits automatically. Otherwise, CTC returns to the Circuit Source page after you create each circuit until you finish creating the number of circuits specified here.

- **Auto Ranged**—This option enables you to choose the source and destination of one circuit. CTC automatically determines the source and destination for the remaining number of circuits and creates the circuits. To determine the source and destination, CTC increments the most specific part of the endpoints. An endpoint can be a port, an STS, or a VT/DS1. If CTC runs out of choices, or selects an endpoint that is already in use, CTC stops and allows you to either choose a valid endpoint or cancel. If you choose a valid endpoint and continue, autoranging begins after you click Finish for the current circuit.

- **Protected Drops**—If this box is checked, CTC displays only protected cards and ports (1:1, 1:N, 1+1, or BLSR protection) as choices for the circuit source and destination.

Step 4 Set the UPSR selector defaults for UPSR circuits:

- **Revertive**—Check this box if you want traffic to revert to the working path when the conditions that diverted it to the protect path are repaired. If Revertive is not chosen, traffic remains on the protect path after the switch.

- **Reversion Time**—If Revertive is checked, set the reversion time. This is the amount of time that will elapse before the traffic reverts to the working path. Traffic can revert when conditions causing the switch are cleared. (The default reversion time is 5 minutes.)

- **SF Threshold**—Set the UPSR path-level signal failure bit error rate (BER) thresholds (STS circuits only).

- **SD Threshold**—Set the UPSR path-level signal degrade BER thresholds (STS circuits only).

- **Switch on PDI-P**—Check this box if you want traffic to switch when an STS payload defect indicator is received (STS circuits only).

An example of provisioning the circuit attributes is shown in Figure 9-39.

Figure 9-39 *Circuit Creation Attributes*

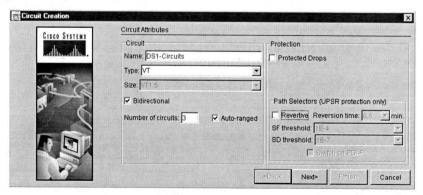

Step 5 Click **Next**.

Step 6 In the Circuit Source dialog box, set the circuit source. Options include Node, Slot, Port, STS, and VT/DS1. The options that display depend on the circuit type and circuit properties you selected in Step 3 and the cards installed in the node. Click **Use Secondary Source** if you need to create a UPSR bridge/selector circuit entry point in a multivendor UPSR.

Step 7 Click **Next**.

An example of provisioning the circuit source is shown in Figure 9-40.

Figure 9-40 *Provisioning the Circuit Source*

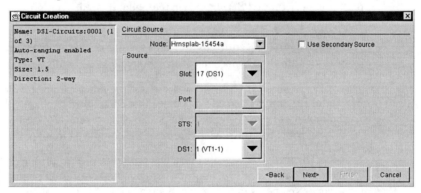

Step 8 In the Circuit Destination dialog box, enter the appropriate information for the circuit destination. If the circuit is bidirectional, you can click **Use Secondary Destination** if you need to create a UPSR bridge/selector circuit destination point in a multivendor UPSR. To add secondary destinations to unidirectional circuits, see the "Create a Unidirectional Circuit with Multiple Drops" procedure.

An example of provisioning the circuit destination is shown in Figure 9-41.

Figure 9-41 *Provisioning the Circuit Destination*

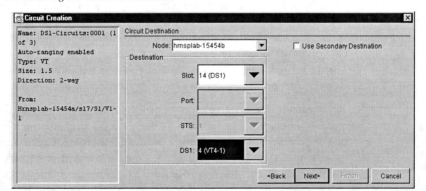

Step 9 Click **Next**.

Step 10 Under Circuit Routing Preferences, choose **Route Automatically**. The following options are available:

- **Using Required Nodes/Spans**—If selected, you can specify nodes and spans to include or exclude in the CTC-generated circuit route.

- **Review Route Before Creation**—If selected, you can review and edit the circuit route before the circuit is created.

Step 11 If you want the circuit routed on a protected path, choose **Fully Protected Path**. Otherwise, go to Step 12. CTC creates a primary and alternate circuit route (virtual UPSR) based on the nodal diversity option you choose:

- **Nodal Diversity Required**—Ensures that the primary and alternate paths within PPMN portions of the complete circuit path are node diverse.

- **Nodal Diversity Desired**—Specifies that node diversity should be attempted; if node diversity is not possible, however, CTC creates link-diverse paths for the PPMN portion of the complete circuit path.

- **Link Diversity Only**—Specifies that only link-diverse primary and alternate paths for PPMN portions of the complete circuit path are needed. The paths may be node diverse, but CTC does not check for node diversity.

Step 12 Click **Finish** or **Next** depending on whether you selected **Using Required Nodes/Spans** and/or **Review Route Before Creation**:

- **Using Required Nodes/Spans**—If selected, click **Next** to display the Circuit Route Constraints panel. On the circuit map, click a node or span and click **Include** (to include the node or span in the circuit) or **Exclude** (to exclude the node/span from the circuit). The order in which you choose included nodes and spans sets the circuit sequence. Click spans twice to change the circuit direction. After you add the spans and nodes, you can use the Up and Down buttons to change their order, or click Remove to remove a node or span. When you are finished, click **Finish** or **Next**, depending on whether you selected Review Route Before Creation.

- **Review Route Before Creation**—If selected, click **Next** to display the route for you to review. To add or delete a circuit span, choose a node on the circuit route. Blue arrows show the circuit route. Green arrows indicate spans that you can add. Click a span arrowhead, and then click **Include** to include the span or **Remove** to remove the span.

Circuit Provisioning

When you click **Finish**, CTC creates the circuit and returns to the Circuits window. If you entered more than one in Number of Circuits in the Circuit Attributes dialog box in Step 3, the Circuit Source dialog box displays so that you can create the remaining circuits. If Auto Ranged is checked, CTC automatically creates the number of sequential circuits that you entered in Number of Circuits. Otherwise, go on to Step 13.

An example of provisioning circuit routing preferences is shown in Figure 9-42.

Figure 9-42 *Circuit Routing Preferences*

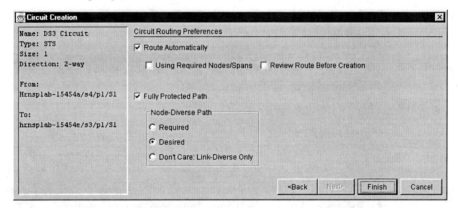

Step 13 If you provision circuits before installing the traffic cards and enabling their ports, you must install the cards and enable the ports before circuits will carry traffic.

Create a Manually Routed Circuit

This section describes the steps used to create a manually routed circuit:

Step 14 Log in to an ONS 15454 and click the Circuits tab.

Step 15 Click **Create**.

Step 16 In the Circuit Creation dialog box, complete the following fields:

- **Name**—Assign a name to the circuit. The name can be alphanumeric and up to 32 characters (including spaces). If you leave the Name field blank, CTC assigns a default name to the circuit. If you work for a LEC or IXC, this field would be assigned with the customer circuit ID.

- **Type**—Choose the type of circuit you want to create: STS, VT (VT1.5), or VT tunnel. The circuit type determines the circuit-provisioning options that can be provisioned.

- **Size**—Choose the circuit size (STS circuits only). The "c" indicates concatenated STSs.
- **Bidirectional**—Check this box to create a two-way circuit. If this box is unchecked, CTC will create a unidirectional circuit. Only STS and VT circuits can be configured as unidirectional. VT tunnels are always bidirectional.
- **Number of Circuits**—Type the number of circuits you want to create. CTC returns to the Circuit Source page after you create each circuit until you finish creating the number of circuits specified here.
- **Auto Ranged**—This option is not available with manual circuit routing.
- **Protected Drops**—If this box is checked, CTC displays only protected cards and ports (1:1, 1:N, 1+1, or BLSR protection) as choices for the circuit source and destination.

Step 17 Set the UPSR selector defaults for UPSR circuits:

- **Revertive**—Check this box if you want traffic to revert to the working path when the conditions that diverted it to the protect path are repaired. If Revertive is not chosen, traffic remains on the protect path after the switch.
- **Reversion Time**—If Revertive is checked, set the reversion time. This is the amount of time that will elapse before the traffic reverts to the working path. Traffic can revert when conditions that caused the switch are cleared. (The default reversion time is 5 minutes.)
- **SF Threshold**—Set the UPSR path-level signal failure BER thresholds (STS circuits only).
- **SD Threshold**—Set the UPSR path-level signal degrade BER thresholds (STS circuits only).
- **Switch on PDI-P**—Check this box if you want traffic to switch when an STS payload defect indicator is received (STS circuits only).

Step 18 Click **Next**.

Step 19 In the Circuit Source dialog box, set the circuit source. Options include Node, Slot, Port, STS, and VT/DS1. The options that display depend on the circuit type and circuit properties you selected in Step 3 and the cards installed in the node. Click **Use Secondary Source** if you need to create a UPSR bridge/selector circuit entry point in a multivendor UPSR.

Step 20 Click **Next**.

Step 21 In the Circuit Destination dialog box, enter the appropriate information for the circuit destination. If the circuit is bidirectional, you can click **Use Secondary Destination** if you need to create a UPSR bridge/selector circuit destination point in a multivendor UPSR.

Step 22 Click **Next**.

Step 23 Under Circuit Routing Preferences, deselect **Route Automatically**.

Step 24 If you want the circuit routed on a protected path, choose **Fully Protected Path**. Otherwise, go to Step 12. CTC creates a primary and alternate circuit route (virtual UPSR) based on the nodal-diversity option you choose:

- **Nodal Diversity Required**—Ensures that the primary and alternate paths within PPMN portions of the complete circuit path are nodally diverse.

- **Nodal Diversity Desired**—Specifies that node diversity should be attempted; if node diversity is not possible, however, CTC creates link-diverse paths for the PPMN portion of the complete circuit path.

- **Link Diversity Only**—Specifies that only link-diverse primary and alternate paths for PPMN portions of the complete circuit path are needed. The paths might be node diverse, but CTC does not check for node diversity.

An example of manually provisioning circuit routing preferences is shown in Figure 9-43.

Figure 9-43 *Manual Circuit Routing*

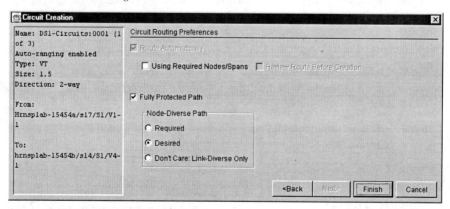

Step 25 Click **Next**. The Route Review and Edit panel is displayed for you to manually route the circuit. The green arrows pointing from the source node to other network nodes indicate spans that are available for routing the circuit.

Step 26 Set the circuit route:

    (a) Click the arrowhead of the span you want the circuit to travel.

    (b) If you want to change the source STS or VT, change it in the Source STS or Source VT fields.

    (c) Click **Add Span.** The span is added to the Included Spans list, and the span arrow turns blue.

Step 27 Repeat Step 13 until the circuit is provisioned from the source to the destination node. When provisioning a protected circuit, you need to choose only one path of BLSR or 1+1 spans from the source to the drop. If you choose unprotected spans as part of the path, choose two different paths for the unprotected segment of the path.

Step 28 After provisioning the circuit, click **Finish**. If you entered more than one in Number of Circuits in the Circuit Attributes dialog box in Step 3, the Circuit Source dialog box displays so that you can create the remaining circuits.

Step 29 If you are provisioning circuits before installing the traffic cards and enabling their ports, you must install the cards and enable the ports before circuits will carry traffic.

Creating Multiple Drops for Unidirectional Circuits

Unidirectional circuits can have multiple drops for use in broadcast circuit schemes. In broadcast scenarios, one source transmits traffic to multiple destinations, but traffic is not returned back to the source. When a unidirectional circuit is created, the card that does not have its backplane RX input terminated with a valid input signal generates an LOS alarm. To mask the alarm, an alarm profile suppressing the LOS alarm must be created and applied to the port that does not have its RX input terminated.

Step 1 Use the "Create an Automatically Routed Circuit" procedure to create a circuit. To make it unidirectional, clear the **Bidirectional** check box on the Circuit Creation dialog box.

Step 2 After the unidirectional circuit is created, in node or network view click the **Circuits** tab.

Step 3 Choose the unidirectional circuit and click **Edit**.

Step 4 On the Drops tab of the Edit Circuits dialog box, click **Create** or, if Show Detailed Map is selected, right-click a node on the circuit map and choose **Add Drop**.

Step 5 In the Define New Drop dialog box, complete the appropriate fields to define the new circuit drop: Node, Slot, Port, STS, VT (if applicable).

Step 6 Click **OK**.

Step 7 If you need to create additional drops, repeat Steps 4 through 6. If not, click **Close**.

Step 8 Verify the new drops on the Edit Circuit map:

If Show Detailed Map is selected: a "D" enclosed by circles appears on each side of the node graphic. If Show Detailed Map is not selected: "Drop #1, Drop #2" appears under the node graphic.

Creating Monitor Circuits

Secondary circuits can be set up to monitor traffic on primary bidirectional circuits. Figure 9-44 shows an example of a monitor circuit. At node 1, a VT1.5 is dropped from port 1 of an EC1-12 card. To monitor the VT1.5 traffic, test equipment is plugged into port 2 of the EC1-12 card and a monitor circuit to port 2 is provisioned in CTC. Circuit monitors are one-way. The monitor circuit in Figure 9-44 is used to monitor VT1.5 traffic received by port 1 of the EC1-12 card.

Figure 9-44 *Monitor Circuit Schematic*

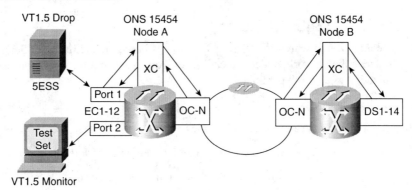

NOTE Monitor circuits cannot be used with EtherSwitch circuits. For unidirectional circuits, a drop must be created to the port where the test equipment is attached.

Step 1 Log in to CTC.

Step 2 In node view, click the **Circuits** tab.

Step 3 Choose the bidirectional circuit that you want to monitor. Click **Edit**.

Step 4 In the Edit Circuit dialog box, click the **Monitors** tab.

The Monitors tab displays ports that you can use to monitor the circuit selected in Step 3.

Step 5 On the Monitors tab, choose a port. The monitor circuit displays traffic coming into the node at the card/port you choose. In Figure 9-44, you would choose either the DS1-14 card (to test circuit traffic entering node 2 on the DS1-14) or the OC-N card at node 1 (to test circuit traffic entering node 1 on the OC-N card).

Step 6 Click **Create Monitor Circuit**.

Step 7 In the Circuit Creation dialog box, choose the destination node, slot, port, and STS for the monitored circuit. In the Figure 9-44 example, this is port 2 on the EC1-12 card. Click **Next**.

Step 8 In the Circuit Creation dialog box confirmation, review the monitor circuit information. Click **Finish**.

Step 9 In the Edit Circuit dialog box, click **Close**. The new monitor circuit displays on the Circuits tab.

Searching for ONS 15454 Circuits

CTC enables you to search for ONS 15454 circuits based on circuit name. Searches can be conducted at the network, node, and card level. You can search for whole words and include capitalization as a search parameter.

Step 1 Log in to CTC.

Step 2 Switch to the appropriate CTC view:

- **Network view** to conduct searches at the network level
- **Node view** to conduct searches at the network or node level
- **Card view** to conduct searches at the card, node, or network level

Step 3 Click the **Circuits** tab.

Step 4 If you are in node or card view, choose the scope for the search in the Scope field.

Step 5 Click **Search**.

Step 6 In the Circuit Name Search dialog box, complete the following:

- **Find What**—Enter the text of the circuit name you want to find.
- **Match Whole Word Only**—If checked, CTC selects circuits only if the entire word matches the text in the Find What field.

- **Match Case**—If checked, CTC selects circuits only when the capitalization matches the capitalization entered in the Find What field.
- **Direction**—Choose the direction for the search. Searches are conducted up or down from the currently selected circuit.

Step 7 Click **Find Next**.

Step 8 Repeat Steps 6 and 7 until you have finished, and then click **Cancel**.

Editing UPSR Circuits

Use the Edit Circuits window to change UPSR selectors and switch protection paths. This window enables you to do the following:

- View the UPSR circuit's working and protection paths
- Edit the reversion time
- Edit the signal fail/signal degrade thresholds
- Change PDI-P settings, perform maintenance switches on the circuit selector, and view switch counts for the selectors
- Display a map of the UPSR circuits to better see circuit flow between nodes

This section describes the steps used to edit UPSR circuits:

Step 1 Log in to the source or drop node of the UPSR circuit.

Step 2 Click the **Circuits** tab.

Step 3 Click the circuit you want to edit, and then click **Edit**.

Step 4 On the Edit Circuit window, click the **UPSR** tab.

Step 5 Edit the UPSR selectors:

- **Reversion Time**—Controls whether traffic reverts to the working path when conditions that diverted it to the protect path are repaired. If you choose Never, traffic does not revert. Selecting a time sets the amount of time that will elapse before traffic reverts to the working path.
- **SF Ber Level**—Sets the UPSR signal failure BER threshold (STS circuits only).
- **SD Ber Level**—Sets the UPSR signal degrade BER threshold (STS circuits only).
- **PDI-P**—When checked, traffic switches if an STS payload defect indication is received (STS circuits only).

- **Switch State**—Switches circuit traffic between the working and protect paths. The color of the Working Path and Protect Path fields indicates the active path. Normally, the working path is green and the protect path is purple. If the protect path is green, working traffic has switched to the protect path.
- **Clear**—Removes a previously set switch command.
- **Lockout of Protect**—Prevents traffic from switching to the protect circuit path.
- **Force to Working**—Forces traffic to switch to the working circuit path, regardless of whether the path is error free.
- **Force to Protect**—Forces traffic to switch to the protect circuit path, regardless of whether the path is error free.
- **Manual to Working**—Switches traffic to the working circuit path when the working path is error free.
- **Manual to Protect**—Switches traffic to the protect circuit path when the protect path is error free.

NOTE   The Force and Lockout commands override normal protection switching mechanisms. Applying these commands incorrectly can cause traffic outages.

Step 6   Click **Apply**.

Creating a Path Trace

The SONET J1 path trace is a repeated, fixed-length string comprised of 64 consecutive J1 bytes. This string can be used to monitor interruptions or changes to circuit traffic. Most ONS 15454 electrical cards can transmit and receive the J1 field, whereas optical cards can only receive it. Certain ONS 15454 cards do not support the J1 byte. Specific card documentation must be consulted to verify J1 byte support. The J1 path trace transmits a repeated, fixed-length string. If the string received at a circuit drop port does not match the string the port expects to receive, an alarm is raised.

Two path trace modes are available:

- **Automatic**—The receiving port assumes the first J1 string it receives is the baseline J1 string.
- **Manual**—The receiving port uses a string that you manually enter as the baseline J1 string.

To set up a path trace on an ONS 15454 circuit, follow the steps detailed here:

NOTE To perform this procedure for ONS 15454 nodes, you must have an STS circuit using a DS1, DS3E, DS3XM-6, or G1000-4 card at the circuit source and drop ports, or an STS circuit passing through an EC1, OC-3, OC-48AS, or OC-192 card.

Step 1 Log in to the circuit source node and click the **Circuits** tab.

Step 2 Choose the circuit you want to trace, and then click **Edit**.

Step 3 On the Edit Circuit window, click **Show Detailed Map** at the bottom of the window.

Step 4 On the detailed circuit map, right-click the source port for the circuit and choose **Edit Path Trace** from the shortcut menu.

Step 5 On the Circuit Path Trace window in the New Transmit String field (this field is available only on DS1, DS3E, DS3XM-6, and G1000-4 cards), enter the string that you want the source port to transmit. For example, you could enter the node IP address, node name, circuit name, or another string. If the New Transmit String field is left blank, the J1 transmits an empty string.

Step 6 Click **Apply** but do not close the window.

Step 7 Return to the Edit Circuit window.

Step 8 On the circuit map, right-click the drop port for the circuit and choose **Edit Path Trace** from the shortcut menu.

Step 9 In the Circuit Path Trace window in the New Transmit String field (this field is available only on DS1, DS3E, DS3XM-6, and G1000-4 cards), enter the string that you want the drop port to transmit. If the field is left blank, the J1 transmits an empty string.

Step 10 If you set Path Trace Mode to **Manual** in Step 11, enter the string that the drop port should expect to receive in the New Expected String field. This string must match the New Transmit String entered for the source port in Step 5. (When you click **Apply** in Step 12, this string becomes the Current Expected String.)

Step 11 In the Path Trace Mode field, choose one of the following options:

- **Auto**—Assumes the first string received from the source port is the baseline string. An alarm is raised when a string that differs from the baseline is received.

- **Manual**—Uses the Current Expected String field as the baseline string. An alarm is raised when a string that differs from the current expected string is received.

Step 12 Click **Apply**, and then click **Close**.

Step 13 Display the Circuit Path Trace window for the source port from Step 5.

Step 14 If you set the Path Trace Mode to **Manual** in Step 15, enter the string the source port should expect to receive in the New Expected String field. This string must match the New Transmit String entered for the source port in Step 9.

Step 15 In the Path Trace Mode field, choose one of the following options:

- **Auto**—Assumes that the first string received from the drop port is the baseline string. An alarm is raised when a string that differs from the baseline is received.

- **Manual**—Uses the Current Expected String field as the baseline string. An alarm is raised when a string that differs from the current expected string is received.

Step 16 Click **Apply** and click **Close**.

Step 17 After you set up the path trace, the received string is displayed in the Received box on the Path Trace Setup window. Click **Switch Mode** to toggle between ASCII and hexadecimal display. Click the **Reset** button to reread values from the port. Click **Default** to return to the path trace default settings. (Path Trace Mode is set to Off and the New Transmit and New Expected Strings are null.)

Provisioning SONET DCC Tunnels

SONET provides four DCCs for network element operations, administration, maintenance, and provisioning: one on the SONET section layer known as the *section DCC (SDCC)*, and three on the SONET line layer known as *line DCCs (LDCCs)*. The ONS 15454 uses the SDCC for management and provisioning. The LDCCs and the SDCC can be used to tunnel third-party SONET equipment across ONS 15454 networks when the SDCC is not used for ONS 15454 DCC terminations. A DCC tunnel endpoint is defined by slot, port, and DCC, where DCC can be either the SDCC or one of the LDCCs. You can link an SDCC to an LDCC, and an LDCC and link SDCCs to SDCCs. To create a DCC tunnel, you connect the tunnel endpoints from one ONS 15454 optical port to another. Each ONS 15454 can support up to 32 DCC tunnel connections.

Table 9-4 shows the DCC tunnels that you can create.

Table 9-4 *SONET DCC tunnels*

DCC	SONET Layer	SONET Bytes	OC-3 (All Ports)	OC-12, OC-48, OC-192
SDCC	Section	D1–D3	Yes	Yes
Tunnel 1	Line	D4–D6	No	Yes
Tunnel 2	Line	D7–D9	No	Yes
Tunnel 3	Line	D10–D12	No	Yes

Figure 9-45 shows a DCC tunnel example. Generic ADM equipment is connected to OC-3 cards at node A/slot 3/port 1 and node C/slot 3/port 1. Each ONS 15454 node is connected by OC-48 trunk cards. In the example, three tunnel connections are created: one at node A (OC-3 to OC-48), one at node B (OC-48 to OC-48), and one at node C (OC-48 to OC-3).

Figure 9-45 *SONET DCC Tunnel*

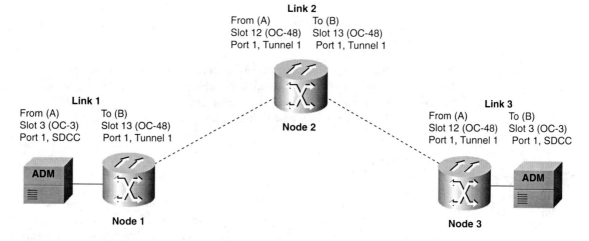

SONET DCC tunnel rules are as follows:

- Each ONS 15454 can have up to 32 DCC tunnel connections.
- Each ONS 15454 can have up to 32 SDCC terminations with the TCC2 card.
- An SDCC that is terminated cannot be used as a DCC tunnel endpoint.
- An SDCC that is used as an DCC tunnel endpoint cannot be terminated.
- All DCC tunnel connections are bidirectional.

This section describes the steps used to provision SONET DCC tunnels:

Step 1 Log in to an ONS 15454 that is connected to the non-ONS 15454 network.

Step 2 Click the **Provisioning > SONET DCC** tab.

Step 3 Beneath the DCC Tunnel Connections area (bottom right of the screen), click **Create**.

Step 4 In the Create DCC Tunnel Connection dialog box, choose the tunnel endpoints from the From (A) and To (B) lists.

> **NOTE** You cannot use the SDCC listed under SDCC Terminations (left side of the window) for tunnel connections. These are used for ONS 15454 optical connections.

Step 5 Click **OK**.

Step 6 Put the ports hosting the DCC tunnel in service:

(a) Double-click the card hosting the DCC in the shelf graphic or right-click the card on the shelf graphic and choose **Open**.

(b) Click the **Provisioning > Line** tab.

(c) Under Status, choose **In Service**.

(d) Click **Apply**.

Step 7 DCC provisioning is now complete for one node. Repeat these steps for all slots/ports that are part of the DCC tunnel, including any intermediate nodes that will pass traffic from third-party equipment. The procedure is confirmed when the third-party network elements successfully communicate over the newly established DCC tunnel.

Summary

It is important that you understand the theory behind SONET and TDM before jumping into provisioning mode. This chapter discusses the provisioning of the MSPP, SONET UPSR and BLSR, and TDM circuits. Prior to the provisioning of TDM circuits, one must perform the basic node provisioning for the SONET infrastructure. This chapter uses the ONS 15454 as an example because it exemplifies most of the SONET provisioning aspects. The ONS 15454 is also the most deployed MSPP device in the metro edge and core. This chapter covers initial node provisioning tasks and the provisioning of SONET circuits. Before you begin your ONS node setup, ensure that you have the prerequisite information including node name, contact, location, current date, and time. If the ONS 15454 will be connected to an OAM network, you will also need the IP address and subnet mask assigned to the node and the IP address of the default router.

The ONS 15454 provides several card protection methods. When you set up protection for ONS 15454 cards, you must choose between maximum protection and maximum slot availability. The highest protection reduces the number of available card slots, whereas the highest slot availability reduces the protection. SONET timing parameters must be set for each ONS 15454. In typical ONS 15454 networks, one node is always set to external timing. The externally timed node derives its timing from a BITS source wired to the BITS backplane pins. The BITS source, in turn, derives its timing from a PRS such as a Stratum 1 clock or GPS signal. The other nodes are set to line. The line nodes derive timing from the externally timed node through the OC-N trunk cards. You can set three timing references for each ONS 15454. The first two references are typically two BITS-level sources, or two line-level sources optically connected to a node with a BITS source. The third reference is the internal clock provided on every ONS 15454 TCC card. This clock is an ST3. If an ONS 15454 becomes isolated, timing is maintained at the ST3 level. The Inventory option displays information about cards installed in the ONS 15454 node including part numbers, serial numbers, hardware revisions, and equipment types. The tab provides a central location to obtain information about installed cards, software, and firmware versions.

UPSR provide duplicate fiber paths around the ring. Working traffic flows in one direction, and protection traffic flows in the opposite direction. If a problem occurs in the working traffic path, the receiving node switches to the path coming from the opposite direction. CTC automates UPSR ring configuration. UPSR traffic is defined within the ONS 15454 on a circuit-by-circuit basis. If a path-protected circuit is not defined within a 1+1 or BLSR line protection scheme and path protection is available and specified, CTC uses UPSR as the default. Because each traffic path is transported around the entire ring, UPSRs are best suited for networks where traffic concentrates at one or two locations and is not widely distributed. UPSR capacity is equal to its bit rate. Services can originate and terminate on the same UPSR, or they can be passed to an adjacent access or interoffice ring for transport to the service-terminating location.

The ONS 15454 can support two concurrent BLSRs. Each BLSR can have up to 32 ONS 15454s. Because the working and protect bandwidths must be equal, you can create only OC-12 (two-fiber only), OC-48, or OC-192 BLSRs. In two-fiber BLSRs, each fiber is divided into working and protect bandwidths. The CTC circuit routing routines calculate the "shortest path" for circuits based on many factors including requirements set by the circuit provisioner, traffic patterns, and distance.

Four-fiber BLSRs double the bandwidth of two-fiber BLSRs. Because they allow span switching as well as ring switching, four-fiber BLSRs increase the reliability and flexibility of traffic protection. Two fibers are allocated for working traffic and two fibers for protection. To implement a four-fiber BLSR, you must install four OC-48 or OC-48AS cards, or four OC-192 cards at each BLSR node. Four-fiber BLSRs provide span and ring switching. Span switching occurs when a working span fails and traffic switches to the protect fibers between the failed span and then returns to the working fibers. Multiple span switches can occur at the same time. Ring switching occurs when a span switch cannot recover traffic, such as when both the working and protect fibers fail on the same span. In a ring switch, traffic is routed to the protect fibers throughout the full ring.

PPMNs include multiple ONS 15454 SONET topologies, such as a combination of linear ADM and UPSR that extend the protection provided by a single UPSR to the meshed architecture of several interconnecting rings. In a PPMN, circuits travel diverse paths through a network of single or multiple meshed rings. When you create circuits, you can have CTC automatically route circuits across the PPMN, or you can manually route them. You can also choose levels of circuit protection. If you choose full protection, for example, CTC creates an alternate route for the circuit in addition to the main route. The second route follows a unique path through the network between the source and destination and sets up a second set of cross-connections.

Provisioning a UPSR, BLSR, linear ADM, or PPMN is not complete without assigning circuits for transport of user traffic. DCC tunnels can also be created to tunnel third-party equipment signaling through ONS 15454 networks. *Cross-connects* refers to the connections that occur within a single ONS 15454 that allow a circuit to enter on one port and exit on another, whereas the term *circuit* refers to the series of connections from a traffic source (where traffic ingresses the ONS network) to the drop or destination (where traffic egresses an ONS network).

This chapter includes the following sections:

- **Provisioning the SDH MSPP**—This section introduces the provisioning aspect of the Synchronous Digital Hierarchy (SDH) multiservice provisioning platform (MSPP), with a reminder that provisioning needs a solid understanding of SDH theory.
- **Initial Provisioning Tasks**—This section discusses the Cisco Transport Controller (CTC) craft interface as well as setup of basic node information including node name, contact, location, current date, and time. The configuration of users and security levels is also discussed in this section.
- **Provisioning of Protection Groups**—This section discusses the protection types that can be set up for ONS 15454 SDH cards as well as ways of provisioning protection groups. This section also discusses the steps involved in the editing and deletion of protection groups.
- **ONS 15454 Timing for SDH**—This section discusses SDH timing and the configuration of ONS 15454 SDH timing.
- **Node Inventory**—Node inventory provides a central location to obtain information about installed cards, software, and firmware versions.
- **IP Networking of ONS 15454 SDH Nodes for OAM&P**—This section explains how to set up Cisco ONS 15454 SDHs in an IP operations, administration, maintenance, and provisioning (OAM&P) network. ONS 15454 SDHs can be connected in many different ways within an IP environment.
- **SNCP Configuration**—This section discusses the installation of subnetwork connection protection (SNCP) trunk cards, DCC terminations, timing, and enabling of ports with respect to SNCP configuration.
- **MS-SPRing Configuration**—This section discusses ONS multiplex section–shared protection ring (MS-SPRing) configuration.
- **Subtending Ring Configurations**—This section discusses subtending rings and their benefits in terms of optical network design.
- **Linear ADM Configurations**—This section discusses the configuration of ONS 15454 SDH MSPPs as linear add/drop multiplexers (ADMs).
- **Extended SNCP Mesh Networks**—Extended SNCP (ESNCP) mesh networks include multiple ONS 15454 SDH topologies, such as a combination of linear ADM and SNCP, that extend the protection provided by a single SNCP to the meshed architecture of several interconnecting rings.
- **SDH Circuit Provisioning**—This section explains how to create and administer various types of Cisco ONS 15454 SDH circuits and tunnels. DCC tunnels can also be created to tunnel third-party equipment signaling through ONS 15454 SDH networks.

CHAPTER 10

Provisioning the Multiservice SDH MSPP

Provisioning the SDH MSPP

It is important that you understand the theory behind Synchronous Digital Hierarchy (SDH) and time-division multiplexing (TDM) before jumping into provisioning mode. This chapter discusses the provisioning of the SDH multiservice provisioning platform (MSPP), subnetwork connection protection (SNCP), multiplex section–shared protection ring (MS-SPRing), and TDM circuits. Prior to the provisioning of TDM circuits, one must perform the basic node provisioning for the SDH infrastructure. We use the ONS 15454 SDH as an example because it exemplifies most of the SDH provisioning aspects. Chapter 9, "Provisioning the Multiservice SONET MSPP," covered SONET configuration for the ONS 15454. This chapter covers SDH configuration for the ONS 15454 SDH MSPP. Notice that CTC has retained its look and feel for SDH. Many of the SDH parameters have SONET equivalents or analogies.

Initial Provisioning Tasks

For initial provisioning tasks, ensure that you have the prerequisite information including the node name, contact, location, current date, and time. If the ONS 15454 SDH will be connected to an operations, administration, maintenance, and provisioning (OAM&P) network, you also need the IP address and subnet mask assigned to the node and the IP address of the default router. If Dynamic Host Configuration Protocol (DHCP) is used, you need the IP address of the DHCP server. If you intend to create card protection groups, you need the card protection scheme that will be used and what cards will be included in it. Proper network design requires an analysis of the requirements and a determination of the SDH protection topology that will be used for the network.

The ONS Craft Interface

Cisco Transport Controller (CTC) is a graphical user interface (GUI)-based craft tool that is used to provision the ONS 15454 SDH. The CTC converts user input into corresponding Transaction Language 1 (TL-1) commands accepted by the MSPP. Alternatively, the TL-1 interface or Cisco Transport Manager (CTM) could be used. CTM is typically used in a network operations center (NOC) operations environment to configure and provision customer circuits. CTC is more suited for smaller networks and is used by engineers and technicians during deployment. CTC software is pre-installed on the TCC. If the CTC release is upgraded, new software needs to be installed on the TCC. CTC is downloaded from the Timing and Control Card (TCC) and installed on the PC or workstation automatically after the PC or workstation logs in to the ONS 15454 SDH. To connect to an ONS 15454 SDH, you enter the ONS 15454 SDH IP address in the URL field of a web browser. Figure 10-1 shows the CTC console.

Figure 10-1 *ONS 15454 SDH CTC Console*

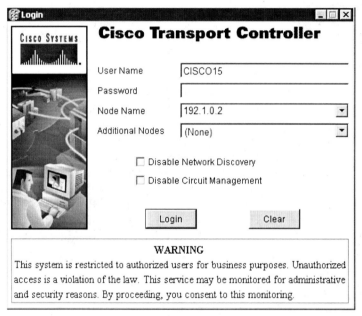

You can connect a PC to the ONS 15454 SDH using the RJ-45 LAN port on the TCC or the LAN 1 pins on the ONS 15454 SDH backplane. Each ONS 15454 SDH must have a unique IP address that you use to access the ONS 15454 SDH. The address is displayed on the front-panel LCD. The initial IP address, 192.1.0.2, is the default address for ONS 15454 SDH access and configuration. Each computer used to communicate with the ONS 15454 SDH should have only one IP address. Each ONS 15454 SDH comes preconfigured with a user account that has superuser privileges. The user ID for this account is CISCO15. The CISCO15 account is not assigned a password and cannot be deleted unless another account with superuser privileges is created. This account can be used to create other users. It is recommended that a password be assigned to the CISCO15 account. To create one, click the **Provisioning > Security** tab after you log in, and then change the CISCO15 password.

NOTE To use CTC with the ONS 15454 SDH Release 4.x system software, the PC or workstation must have a web browser with Java Runtime Environment (JRE) 1.3 or higher installed.

NOTE When you upgrade CTC software, the TCC stores the older CTC version as the protect CTC version, and the newer CTC release becomes the working version. You can view the software versions that are installed on an ONS 15454 SDH by selecting the **Maintenance** tab followed by the **Software** subtab. Select these tabs in *node view* to display the software installed on one node. Select the tabs in network view to display the software versions installed on all network nodes.

Set Up Basic Node Information

Setting basic information for each Cisco ONS 15454 SDH node is one of the first provisioning tasks you perform. This information includes node name, location, contact, and timing. Completing the information for each node facilitates ONS 15454 SDH management, particularly when the node is connected to a large ONS 15454 SDH network. Complete the following steps listed here for basic node setup. The procedure described here is for the SDH implementations of the MSPP. Figure 10-2 shows an example of basic node configuration.

Figure 10-2 *Basic Node Configuration*

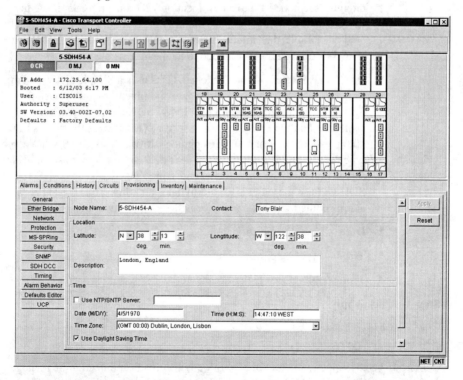

Step 1 Log in to the ONS 15454 SDH node. The CTC node view displays.

Step 2 Click the **Provisioning > General** tab.

Step 3 Enter the following information:

- **Node Name**—Type a name for the node. For TL-1 compliance, names must begin with an alpha character and have no more than 20 alphanumeric characters.

- **Contact**—Type the name of the node contact person and the phone number. (Optional)

- **Location**—Type the node location (for example, a city name or specific office location). (Optional)

- **Latitude**—Enter the node latitude: N (north) or S (south), degrees, and minutes. (Optional)

- **Longitude**—Enter the node longitude: E (east) or W (west), degrees, and minutes. (Optional)

  CTC uses the latitude and longitude to position node icons on the network view map. (You can also position nodes manually by pressing **Ctrl** and dragging the node icon to a new location.) To convert a coordinate in degrees to degrees and minutes, multiply the number after the decimal by 60. For example, the latitude 29.36958 converts to 29 degrees, 22 minutes (.36958 * 60 = 22.1748, rounded to the nearest whole number).

- **Use SNTP Server**—When checked, CTC uses a Simple Network Time Protocol (SNTP) server to set the date and time of the node. Using an SNTP server ensures that all ONS 15454 SDH network nodes use the same date and time reference. The server synchronizes the node's time after power outages or software upgrades. If you check Use SNTP Server, type the server's IP address in the next field. If you do not use an SNTP server, complete the Date, Time, and Time Zone fields. The ONS 15454 SDH will use these fields for alarm dates and times. (CTC displays all alarms in the login node's time zone for cross-network consistency.)

- **Date**—Type the current date.

- **Time**—Type the current time.

- **Time Zone**—Choose the time zone.

Step 4 Click **Apply**.

Set Up Network Information

ONS 15454 SDHs mostly operate in network environments and need to be managed by an element manager, such as CTM within an operations support system (OSS). The ONS ships with a default IP address of 192.1.0.2. Before you connect an ONS 15454 SDH to other ONS 15454 SDHs or to a LAN, the default IP address must be changed to reflect a valid address of the IP subnet that the node would reside on. Within IP networks, ONS 15454 SDHs often exist as hosts on management IP subnets. Complete the following steps for basic network setup. The procedure described here is for the SDH implementations of the MSPP. Figure 10-3 shows an example of IP network setup.

Figure 10-3 *IP Network Setup*

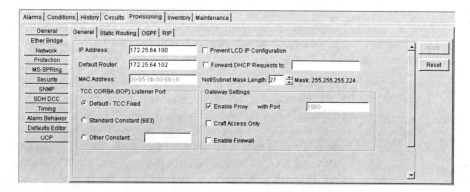

Step 1 From the CTC node view, click the **Provisioning > Network** tab.

Step 2 Enter the following information:

- **IP Address**—Type the IP address assigned to the ONS 15454 SDH node.

- **Prevent LCD IP Configuration**—If checked, this option prevents the ONS 15454 SDH IP address from being changed using the LCD front panel.

- **Default Router**—If the ONS 15454 SDH must communicate with a device on a network to which the ONS 15454 SDH is not connected, the ONS 15454 SDH forwards the packets to the default router. Type the IP address of the default router in this field. If the ONS 15454 SDH is not connected to a LAN, leave the field blank.

- **Subnet Mask Length**—If the ONS 15454 SDH is part of a subnet, type the subnet mask length (decimal number representing the subnet mask length in bits) or click the arrows to adjust the subnet mask length. The subnet mask length is the same for all ONS 15454 SDHs in the same subnet.

NOTE The MAC Address is *read only*. It displays the ONS 15454 SDH address as it is identified on the IEEE 802 Media Access Control (MAC) layer.

- **Forward DHCP Request To**—When checked, this option forwards Dynamic Host Configuration Protocol requests to the IP address entered in the Request To field. DHCP is a TCP/IP protocol that enables CTC computers to get temporary IP addresses from a server. If you enable DHCP, CTC computers that are directly connected to an ONS 15454 SDH node can obtain temporary IP addresses from the DHCP server.

- **TCC CORBA (IIOP) Listener Port**—Sets a listener port to allow communication with the ONS 15454 SDH through firewalls.
- **Gateway Settings**:
  - **Craft Access Only**—When this choice is enabled, the ONS 15454 SDH neither installs nor advertises default or static routes. CTC computers can communicate with the ONS 15454 SDH, but they cannot communicate directly with any other DCC-connected ONS 15454 SDH.
  - **Enable Proxy**—When this choice is enabled, the ONS 15454 SDH responds to CTC client requests with a list of DCC-connected ONS 15454 SDH nodes for which the node serves as a proxy. The CTC client establishes connections through the proxy server for any ONS 15454 SDH in the returned list. By using the proxy, the CTC client can connect to nodes that the PC on which the CTC client runs cannot access. If Enable Proxy is off, the node responds to CTC requests with an empty list, indicating that it is not willing to serve as a proxy.
  - **Enable Firewall**—If this choice is selected, the node prevents IP traffic from being routed between the DCC and the LAN port. The ONS 15454 SDH can communicate with machines connected to the LAN port or connected through the DCC. However, the DCC-connected machines cannot communicate with the LAN-connected machines, and the LAN-connected machines cannot communicate with the DCC-connected machines. A CTC node using the LAN to reach the node connected to the firewall can use the proxy capability to manage the unreachable, DCC-connected nodes. CTC connected to a DCC-connected node can only manage other DCC-connected nodes and the firewall itself.

Step 3 Click **Apply**.

Step 4 Click **Yes** on the confirmation dialog box.

Both ONS 15454 SDH TCC cards will reboot, one at a time.

NOTE You can also change the ONS 15454 SDH IP address, subnet mask, and default router address using the Slot, Status, and Port buttons on the front panel LCD.

User and Security Provisioning

Up to 500 users can be added to one ONS 15454 SDH. Each ONS 15454 SDH user can be assigned one of the following security levels:

- **Retrieve users**—Can retrieve and view CTC information but cannot set or modify parameters

- **Maintenance users**—Can access only the ONS 15454 SDH maintenance options
- **Provisioning users**—Can access provisioning and maintenance options
- **Superusers**—Can perform all the functions of the other security levels as well as set names, passwords, and security levels for other users

Table 10-1 shows the actions that each user can perform in node view.

Table 10-1 *ONS 15454 SDH Security Levels for Node View*

CTC Tab	Subtab	Actions	Retrieve	Maintenance	Provisioning	Superuser
Alarms	None	Synchronize alarms	Yes	Yes	Yes	Yes
		Filter alarms	Yes	Yes	Yes	Yes
		Delete cleared alarms	Yes	Yes	Yes	Yes
Conditions	None	Retrieve	Yes	Yes	Yes	Yes
		Filter	Yes	Yes	Yes	Yes
History	Session	Filter	Yes	Yes	Yes	Yes
	Node	Retrieve alarms/events	Yes	Yes	Yes	Yes
Circuits	None	Create/delete/edit/upgrade	No	No	Yes	Yes
		Search	Yes	Yes	Yes	Yes
Provisioning	General	Edit	No	No	Yes	Yes
	EtherBridge	Spanning trees: edit	No	No	Yes	Yes
		Thresholds: create/delete	No	No	Yes	Yes
	Network	General: edit	No	No	No	Yes
		Static routing: create, edit, or delete	No	No	Yes	Yes
		Open Shortest Path First (OSPF): edit	Edit area id	Edit area id	Yes	Yes
		Routing Information Protocol (RIP): edit	No	No	Yes	Yes

continues

Table 10-1 *ONS 15454 SDH Security Levels for Node View (Continued)*

CTC Tab	Subtab	Actions	Retrieve	Maintenance	Provisioning	Superuser
	Protection	Create/delete/edit	No	No	Yes	Yes
		Browse groups	Yes	Yes	Yes	Yes
	MS-SPRing	All (MS-SPRing), view ring map, and squelch table	No	Partial view	Partial edit	Yes
	Security	Create/delete	No	No	No	Yes
		Change password	Same user	Same user	Same user	All users
	SNMP	Create/delete/edit	No	No	No	Yes
		Browse trap destinations	Yes	Yes	Yes	Yes
	SDH DCC	Create, delete, or edit	No	No	No	Yes
	Timing	Edit	No	No	Yes	Yes
	Alarming	Edit	No	No	Yes	Yes
Inventory	None	Delete	No	No	Yes	Yes
		Reset	No	Yes	Yes	Yes
		View equipment information	No	Yes	Yes	Yes
Maintenance	Database	Backup/restore	No	No	No	Yes
	EtherBridge	Spanning tree: retrieve	Yes	Yes	Yes	Yes
		Spanning tree: clear/clear all	No	Yes	Yes	Yes
		MAC table: retrieve	Yes	Yes	Yes	Yes
		MAC table: clear/clear all		Yes	Yes	Yes
		Trunk utilization refresh	Yes	Yes	Yes	Yes

Table 10-1 *ONS 15454 SDH Security Levels for Node View (Continued)*

CTC Tab	Subtab	Actions	Retrieve	Maintenance	Provisioning	Superuser
	Protection	Switch/lock out operations	No	Yes	Yes	Yes
	MS-SPRing	MS-SPRing maintenance	No	Yes	Yes	Yes
	Software	Download/ upgrade/ activate/revert	No	No	No	Yes
	XC Cards	Protection switches	No	Yes	Yes	Yes
	Diagnostic	Retrieve diagnostics	No	Yes	Yes	Yes
		Lamp test	No	Yes	Yes	Yes
	Timing	Edit	No	Yes	Yes	Yes
	Audit	Retrieve	Yes	Yes	Yes	Yes
	Routing Table	Refresh	Yes	Yes	Yes	Yes
	Test Access	Refresh	Yes	Yes	Yes	Yes

Each ONS 15454 SDH user has a specified amount of time that he or she can leave the system idle before the CTC window is locked. The lockouts prevent unauthorized users from making changes. Higher-level users have shorter idle times, as shown in Table 10-2.

Table 10-2 *ONS 15454 SDH User Idle Times*

Security Level	Idle Time
Superuser	15 minutes
Provisioning	30 minutes
Maintenance	60 minutes
Retrieve	Unlimited

You can perform ONS 15454 SDH user management tasks from network or node view. In network view, you can add, edit, or delete users from multiple nodes at one time. If you perform user management tasks in node view, you can only add, edit, or delete users from that node. Figure 10-4 shows an example of user creation and setup.

NOTE You must add the same username and password to each node the user will access.

Figure 10-4 *User Creation and Setup*

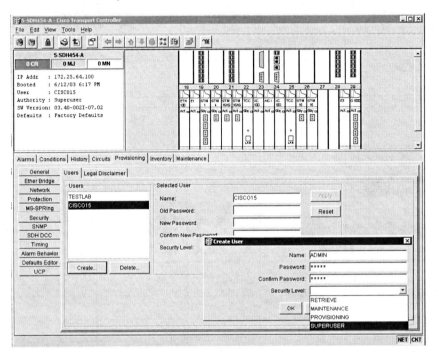

Create New Users

This section describes the steps used to create new users:

Step 1 In network view, choose the **Provisioning > Security** tab.

Step 2 On the Security pane, click **Create**.

Step 3 In the Create User dialog box, enter the following:

- **Name**—Type the user name.

- **Password**—Type the user password. The password must be a minimum of 6 and a maximum of 10 alphanumeric (a–z, A–Z, 0–9) and special characters (+, #, %), where at least 2 characters are nonalphabetic and at least 1 character is a special character.

- **Confirm Password**—Type the password again to confirm.

- **Security Level**—Choose the user's security level.

Step 4 Under Choose Applicable Nodes, deselect any nodes you do not want to add the user. (All network nodes are chosen by default.)

Step 5 Click **OK**.

Edit a User

This section describes the steps used to edit a user:

Step 1 In network view, choose the **Provisioning > Security** tab.

Step 2 Click **Change**.

Step 3 In the Change User dialog box, edit user information: name, password, password confirmation, and/or security level. (A superuser does not need to enter an old password. Other users must enter their old passwords when changing their password.)

Step 4 If you do not want the user changes to apply to all network nodes, deselect the unchanged nodes in the Change Users dialog box.

Step 5 Click **OK**.

Delete a User

This section describes the steps used to delete a user:

Step 1 In network view, choose the **Provisioning > Security** tab.

Step 2 Click **Delete**.

Step 3 In the Delete User dialog box, enter the name of the user you want to delete.

Step 4 If you do not want to delete the user from all network nodes, deselect the nodes.

Step 5 Click **OK**, and then click **Apply**.

Provisioning of Protection Groups

There are several card protection mechanisms for the ONS 15454 SDH. When you set up protection for ONS 15454 SDH cards, you must select between maximum protection and maximum slot availability. The highest protection reduces the number of available card slots, whereas the highest slot availability reduces the protection. Table 10-3 shows the protection types that can be set up for ONS 15454 SDH cards.

Table 10-3 *Protection Types*

Type	Cards	Description
1:1	E1-N-14 E1-42 E3-12 DS3i-N-12	Pairs 1 working card with 1 protect card. Install the protect card in an odd-numbered slot and the working card in an even-numbered slot. Use the slot that is next to the protect slot, toward the center of the shelf. For example: protect in slot 1, working in slot 2; protect in slot 3, working in slot 4; protect in slot 15, working in slot 14.

continues

Table 10-3 *Protection Types (Continued)*

Type	Cards	Description
1:N	E1-N-14 E1-42 DS3i-N-12	Assigns 1 protect card for several working cards. The maximum is 1:5. Protect cards (E1-N-14, DS3i-N-12) must be installed in slots 3 or 15, and the cards they protect must be on the same side of the shelf. Protect cards must match the cards they protect. For example, an E1-N-14 can protect only an E1-N-14 card. If a failure clears, traffic reverts to the working card after the reversion time has elapsed.
1+1	Any optical	Pairs a working optical port with a protect optical port. Protect ports must match the working ports. For example, port 1 of an STM-N card can be protected only by port 1 of another STM-N card. Cards do not need to be in adjoining slots.
Unprotected	Any	Unprotected cards can cause signal loss if a card fails or incurs a signal error. However, because no card slots are reserved for protection, unprotected schemes maximize the service available for use on the ONS 15454 SDH. Unprotected is the default protection type.

Figure 10-5 shows an example of protection group creation.

Figure 10-5 *Protection Group Creation*

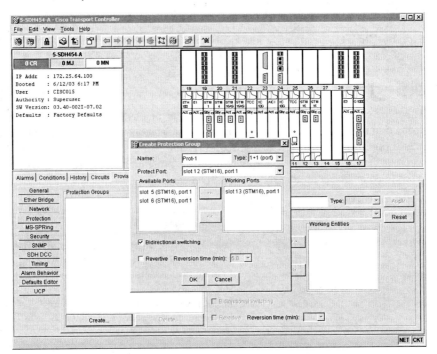

Create a Protection Group

This section describes the steps used to create a protection group:

Step 1 From the CTC node view, click the **Provisioning > Protection** tab.

Step 2 Under Protection Groups, click **Create**.

Step 3 In the Create Protection Group dialog box, enter the following:

- **Name**—Type a name for the protection group. The name can have up to 32 alphanumeric characters.

- **Type**—Choose the protection type (1:1, 1:N, or 1+1) from the drop-down menu. The protection selected determines the cards that are available to serve as protect and working cards. In the case of an SDH implementation, for example, only E1-N-14 and DS3i-N-12 cards display.

- **Protect Card or Port**—Choose the protect card (if using 1:1 or 1:N) or protect port (if using 1+1) from the drop-down menu.

  Based on these selections, a list of available working cards or ports displays under Available Cards or Available Ports.

Step 4 From the Available Cards or Available Ports list, choose the card or port that you want to be the working card or port (the card(s) or port(s) that will be protected by the card or port chosen in Protect Cards or Protect Ports). Click the top arrow button to move each card/port to the Working Cards or Working Ports list.

Step 5 Complete the remaining fields:

- **Bidirectional Switching**—(Optical cards only) Click if you want both transmit and receive channels to switch if a failure occurs to one.

- **Revertive**—If checked, the ONS 15454 SDH reverts traffic to the working card or port after failure conditions remain corrected for the amount of time entered in Reversion Time.

- **Reversion Time**—If Revertive is checked, enter the amount of time following failure condition correction that the ONS 15454 SDH should switch back to the working card or port.

Step 6 Click **OK**.

NOTE Before running traffic on a protected card within a protection group, enable the ports of all protection group cards.

Enable Ports

This section describes the steps used to enable ports:

Step 1 Log in to the node in CTC and display the card you want to enable in card view.

Step 2 Click the **Provisioning > Line** tab.

Step 3 Under the Status column, choose **In Service**.

Step 4 Click **Apply**.

Edit Protection Groups

This section describes the steps used to edit protection groups:

Step 1 From the CTC node view, click the **Provisioning > Protection** tab.

Step 2 In the Protection Groups section, choose a protection group.

Step 3 In the Selected Group section, edit the fields as appropriate.

Step 4 From the Available Cards or Available Ports list, choose the card or port that you want to be the working card or port that will be protected by the card or port chosen in Protect Cards or Protect Ports. Click the top arrow button to move each card or port to the Working Cards or Working Ports list. Complete the remaining fields as needed.

Step 5 Click **Apply**.

Delete Protection Groups

This section describes the steps used to delete protection groups:

Step 1 From the CTC node view, click the **Maintenance > Protection** tab.

Step 2 Verify that working traffic is not running on the protect card:

(a) In the Protection Groups section, choose the group you want to delete.

(b) In the Selected Group section, verify that the protect card is in standby mode. If it is in standby mode, continue with Step 3. If it is active mode, complete Substep C.

(c) If the working card is in standby mode, manually switch traffic back to the working card. In the Selected Group pane, click the working card and then click **Manual**. Verify that the protect card switches to standby mode and the working card is active. If it does, continue with Step 3. If the protect card is still active, do not continue. Begin troubleshooting procedures or call technical support.

Step 3 From the node view, click the **Provisioning > Protection** tab.

Step 4 In the Protection Groups section, click a protection group.

Step 5 Click **Delete**.

ONS 15454 Timing for SDH

SDH timing parameters must be set for each ONS 15454 SDH. Each ONS 15454 SDH independently accepts its timing reference from one of three sources:

- **External**—Building-integrated timing supply (BITS) pins on the ONS 15454 SDH backplane for SDH nodes or the Timing A and Timing B connectors on the MIC-C/T/P FMEC in slot 24 for SDH nodes.
- **Line**—An STM-N card installed in the ONS 15454 SDH that is connected to a node receiving timing from a BITS source.
- **Internal**—The internal ST3 clock on the TCC card.

In typical ONS 15454 SDH networks, one node is always set to external. In the case of SDH networks, the external node derives its timing from a MIC-C/T/P FMEC timing connector. The MIC-C/T/P FMEC, in turn, derives its timing from a primary reference source (PRS), such as a Stratum 1 clock or global positioning satellite (GPS) signal. The other nodes are set to line. The line nodes derive timing from the externally timed node through the STM-N trunk cards. You can set three timing references for each ONS 15454 SDH. In the case of SDH, the first two references are typically two FMEC-level sources, or two line-level sources optically connected to a node with a BITS source. The third reference is the internal oscillator provided on the ONS 15454 SDH TCC-I card. This clock is a Stratum 3 (ST3). If an ONS 15454 SDH becomes isolated, timing is maintained at the ST3 level.

Synchronization Status Messaging (SSM) is an SDH protocol that communicates information about the quality of the timing source. SSM messages are carried in the S1 byte of the SDH line layer. They enable SDH devices to automatically select the highest quality timing reference and to avoid timing loops. SSM messages are either generation 1 or generation 2. Generation 1 is the first and most widely deployed SSM message set. Generation 2 is a newer version. If you enable SSM for the ONS 15454 SDH, consult your timing reference documentation to determine which message set to use.

The SSM supported in SDH is G.811, STU, G812T, G812L, SETS, DUS (ordered from high quality to low quality). SSM messages are carried on bits 5 to 8 of SDH overhead byte S1. They enable SDH devices to automatically select the highest-quality timing reference and to avoid timing loops. The procedure described here is for SDH implementations of the MSPP. Figure 10-6 shows an example of provisioning of ONS 15454 SDH timing.

Figure 10-6 *Provisioning ONS Timing*

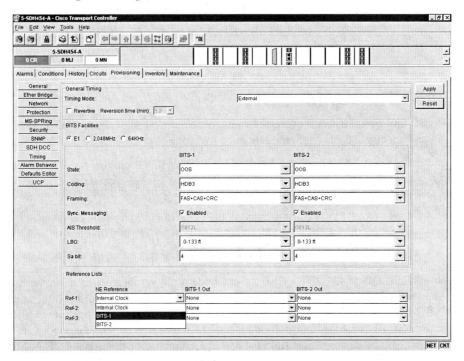

Provisioning ONS 15454 SDH Timing

This section describes the steps used to provision SDH timing:

Step 1 From the CTC node view, click the **Provisioning > Timing** tab.

Step 2 In the General Timing section, complete the following information:

- **Timing Mode**—Set to External if the ONS 15454 SDH derives its timing from a BITS source wired to the backplane pins in case of SDH or from MIC-C/T/P FMEC in case of SDH. Set the mode to Line if timing is derived from an STM-N card that is optically connected to the timing node. A third option, Mixed, enables you to set external and line timing references.

NOTE Mixed timing can cause timing loops and must be used with caution.

- **Revertive**—If checked, the ONS 15454 SDH reverts to a primary reference source after the conditions that caused it to switch to a secondary timing reference are corrected.

- **Revertive Time**—If Revertive is checked, indicate the amount of time the ONS 15454 SDH will wait before reverting back to its primary timing source.

Step 3 In the BITS Facilities section, complete the following information:

> **NOTE** The BITS Facilities section sets the parameters for BITS1 and BITS2 timing references. Many of these settings are determined by the timing source manufacturer. If equipment is timed through BITS Out, you can set timing parameters to meet the requirements of the equipment.

- **E1, 2.048 MHz, 64 KHz**—Choose **E1**, **2.048 MHz** or **64 KHz** depending on the signal supported in your network. For example, 64 kHz is used in Japan. E1, 2.048 MHz, and 64 kHz are physical signal modes used to transmit the external clock (from a GPS, for example) to BITS.

- **State**—Set the BITS reference to **IS** (in service) or **OOS** (out of service). For nodes set to line timing with no equipment timed through BITS Out, set State to OOS. For nodes using external timing or line timing with equipment timed through BITS Out, set State to IS.

- **Coding**—Set to the coding used by your BITS reference to either **HDB3** or **AMI** for SDH implementations. If you selected 2.048 MHz or 64 kHz, the coding option is disabled.

- **Framing**—Set to the framing used by your BITS reference to either unframed, FAS, FAS+CAS, FAS+CRC, or FAS+CAS+CRC. If you selected 2.048 MHz or 64 kHz, the framing option is disabled.

- **Sync Messaging**—Check to enable SSM. SSM is used to deliver clock quality. The SSM supported in SDH is G811, STU, G812T, G812L, SETS, DUS (ordered from high quality to low quality). If you selected 2.048 MHz or 64 kHz, the SSM option is disabled.

- **AIS Threshold**—Sets the quality level where a node sends an alarm indication signal (AIS) from the BITS1 Out and BITS2 Out backplane pins or FMEC connectors. When a node times at or below the AIS threshold quality, AIS is sent (used when SSM is disabled or framing is set to SF, unframed, FAS, or FAS+CAS).

Step 4 Under Reference Lists, complete the following information:

- **NE Reference**—Enables you to define three timing references (Ref 1, Ref 2, and Ref3). The node uses Reference 1 unless a failure occurs to that (in which case, the node uses Reference 2). If that fails, the node uses Reference 3, which is typically set to Internal Clock. This is the

Stratum 3 clock provided on the TCC card. The options displayed depend on the Timing Mode setting.

> **NOTE** Reference lists define up to three timing references for the node and up to six BITS Out references. BITS Out references define the timing references used by equipment that can be attached to the node's BITS Out pins on the backplane or MIC-C/T/P FMEC Timing A and Timing B Out connectors. If you attach equipment to BITS Out pins or Timing A Out/Timing B Out connectors, you normally attach it to a node with line mode because equipment near the external timing reference can be directly wired to the reference.

- **Timing Mode set to External**—Options are BITS1, BITS2, and Internal Clock.
- **Timing Mode set to Line**—Options are the node's working optical cards and Internal Clock. Select the cards/ports that are directly or indirectly connected to the node wired to the BITS source (that is, the node's trunk cards). Set Reference 1 to the trunk card that is closest to the BITS source. If slot 5 is connected to the node wired to the BITS source, for instance, select slot 5 as Reference 1.
- **Timing Mode set to Mixed**—Both BITS and optical cards are available, enabling you to set a mixture of external BITS and optical trunk cards as timing references.
- **BITS1 Out/BITS2 Out**—Define the timing references for equipment wired to the BITS Out backplane pins or Timing A Out/Timing B Out FMEC connectors. Normally, BITS Out or Timing Out is used with line nodes, so the options displayed are the working optical cards. BITS1 and BITS2 Out or Timing A Out/Timing B Out are enabled as soon as BITS1 and BITS2 facilities are placed in service.

Set Up of Internal Timing

If no BITS source is available, you can set up internal timing by timing all nodes in the ring from the internal clock of one node.

> **NOTE** Internal timing is Stratum 3 and not intended for permanent use. All ONS 15454 SDHs should be timed to a Stratum 2 or better primary reference source. In case of SDH, CTC refers to Timing A and Timing B as BITS1 and BITS2. The MIC-C/T/P FMEC connector is labeled as Timing A and Timing B.

This section describes the steps used to set up internal timing:

Step 1 Log in to the node that will serve as the timing source.

Step 2 In CTC node view, click the **Provisioning > Timing** tab.

Step 3 In the General Timing section, enter the following:

- **Timing Mode**—Set to External.
- **Revertive**—Is not relevant for internal timing. Leave at the default value.
- **Revertive Time**—Default setting of 5 minutes is sufficient.

Step 4 In the BITS Facilities section, enter the following information:

- **E1, 2.048 MHz, 64 kHz (SDH)**—Choose **E1, 2.048 MHz** or **64 kHz** depending on the signal supported in your network. For example, 64 kHz is used in Japan. E1, 2.048 MHz, and 64 kHz are physical signal modes used to transmit the external clock (from a GPS, for example) to BITS.
- **State**—Set BITS1 and BITS2 to **OOS** (out of service).
- **Coding**—Not relevant for internal timing. Leave the field at its default value of B8ZS or HDB3.
- **Framing**—Not relevant for internal timing. Leave the field at its default value of ESF or FAS+CAS+CRC.
- **Sync Messaging**—Checked.
- **AIS Threshold**—Not available.
- **LBO[SDH]**—Not relevant for internal timing; line build out (LBO) relates to the BITS cable length.
- **Sa Bit[SDH]**—Not relevant for internal timing; the Sa bit is used to transmit the SSM message.

Step 5 In the Reference Lists section, enter the following information.

NE Reference:

- **Ref1**—Set to **Internal Clock**.
- **Ref2**—Set to **Internal Clock**.
- **Ref3**—Set to **Internal Clock**.
- **BITS1 Out/BITS2 Out**—Set to None.

Step 6 Click **Apply**.

Step 7 Log in to a node that will be timed from the node set up in Steps 1 through 6.

Step 8 In CTC node view, click the **Provisioning > Timing** tab.

Step 9 In the General Timing section, enter the same information as entered in Step 3, except for the following:

- **Timing Mode** — Set to **Line**.
- **Revertive (SDH)** — Not relevant for internal timing; the default setting (checked) is sufficient.
- **Revertive Time (SDH)** — The default setting (5 minutes) is sufficient.

NE Reference/Reference Lists:

- **Ref1** — Set to the STM-N trunk card with the closest connection to the node serving as the timing source.
- **Ref2** — Set to the STM-N trunk card with the next closest connection to the node serving as the timing source.
- **Ref3** — Set to **Internal Clock**.

Step 10 Click **Apply**.

Step 11 Repeat Steps 7 through 10 at each node that will be timed by the node serving as the timing source.

Node Inventory

The Inventory option displays information about cards installed in the ONS 15454 SDH node including part numbers, serial numbers, hardware revisions, and equipment types. The tab provides a central location to obtain information about installed cards, software, and firmware versions. Using the ONS 15454 SDH export feature, you can export inventory data from ONS 15454 SDH nodes into spreadsheet and database programs to consolidate ONS 15454 SDH information for network inventory management and reporting. You can preprovision a slot before the card is installed by right-clicking the slot in node view and selecting a card type. The Inventory tab displays the following information about the cards installed in the ONS 15454 SDH:

- **Location** — The slot where the card is installed
- **Eqpt Type** — Equipment type the slot is provisioned for
- **Actual Eqpt Type** — The actual card that is installed in the slot
- **HW Part #** — Card part number
- **HW Rev** — Card revision number
- **Serial #** — Card serial number unique to each card

- **CLEI Code**—Common Language Equipment Identifier code
- **Firmware Rev**—Revision number of the software used by the ASIC chip installed on the card

Figure 10-7 shows an example of ONS 15454 SDH inventory.

Figure 10-7 *ONS 15454 SDH Inventory*

Alarms | Conditions | History | Circuits | Provisioning | Inventory | Maintenance

Location	Eqpt Type	Actual Eqpt Type	HW Part #	HW Rev	Serial #	CLEI Code	Firmware Rev
Chassis	BACKPLANE_454SDH	BACKPLANE	800-08708-01	A0	TBC06251472	WMM7V00ARA	
1	ETH100	E100T-12-G	800-19924-02	A0	SAG06412D53	SN4PFGXBAA	57-4504-01-A0
2	E1	E1N-14	800-08382-01	A1	SAG063700Q3	SOI4NW0GAA	57-5048-03
3	STM1	OC3-IR-4	800-18808-01	A0	SAG06310BR9	SOI4V2XGAA	76-99-00009-005a
4	STM4	OC12-LR-1	800-18317-01	A0	SAG06310AEG	SOI4W3YGAA	76-99-00011-004a
5	STM16	OC48AS-IR-1310	800-21571-01	A0	SAG06340CL1	WM2IW99DAA	57-4361-04
6	STM16	OC48AS-IR-1310	800-21571-01	A0	SAG06380C9A	WM2IW99DAA	57-4361-04
7	TCC	TCCI	800-09008-01	B2	SAG06340A6L	WM4C750BAA	57-5276-02
8	XC10G	XC10G	800-19977-02	A0	SAG06380E6Q	WM5ISXRCAA	85-3867-02_A0
9	AICI	AIC-I	800-17764-01	A0	SAG06483K99	SOC2AA0AAA	NOT APPLICABLE
10	XC10G	XC10G	800-19977-02	A0	SAG06380EE7	WM5ISXRCAA	85-3867-02_A0
11	TCC	TCCI	800-09008-01	B2	SAG06310ATE	WM4C750BAA	57-5276-02
12	STM16	OC48ELR-1533.47-100	800-18321-01	A0	SAG06320CAK	WMTRRG0DAA	57-4322-01-A0
13	STM16	OC48ELR-1533.47-100	800-18321-01	A0	SAG06320CAR	WMTRRG0DAA	57-4322-01-A0
16	E3	E3-12	800-08903-01	A0	SAG0636CDQL	SOI4U1WGAA	57-5255-03
17	G1000	G1000-4	800-08578-02	A1	SAG063924UN	SNP8KW0KAB	57-5187-03
19	FMEC_SMZ_E1	FMEC-E1	800-08437-02	A0	SAG06340DSW	SOI4NXPGAB	NOT APPLICABLE
21	FMEC_SMZ_E1	FMEC-E1	800-08437-02	A0	SAG06340DSV	SOI4NXPGAB	NOT APPLICABLE
23	ALM_PWR	ALARM	800-08433-01	A0	SAG06300K0M	WM4C860BAA	NOT APPLICABLE
24	CRFT_TMG	CRAFT	800-08432-01	A0	SAG06330G90	SOC2JJ0AAA	NOT APPLICABLE
28	FMEC_SMZ_E3	FMEC-E3	800-08431-02	A0	SAG06300J2Z	SOI4PZSGAB	NOT APPLICABLE
29	FMEC_SMZ_E3	FMEC-E3	800-08431-02	A0	SAG06300J40	SOI4PZSGAB	NOT APPLICABLE
Chassis	FAN_TRAY	FTA	800-14725-02	A0	TBC06365893	WMMYACTGAA	

Delete | Reset

IP Networking of ONS 15454 SDH Nodes for OAM&P

This section explains how to set up ONS 15454 SDHs in an IP OAM&P network. ONS 15454 SDHs can be connected in many different ways within an IP environment:

- They can be connected to LANs through direct connections or a router.
- IP subnetting can create ONS 15454 SDH node groups that enable you to provision non-DCC-connected nodes in a network.
- Different IP functions and protocols can be used to achieve specific network goals. For example, Proxy Address Resolution Protocol (ARP) enables one LAN-connected ONS 15454 SDH to serve as a gateway for ONS 15454 SDHs that are not connected to the LAN.
- You can create static routes to enable connections among multiple CTC sessions with ONS 15454 SDHs that reside on the same subnet but have different destination IP addresses.
- If ONS 15454 SDHs are connected to OSPF networks, ONS 15454 SDH network information is automatically communicated across multiple LANs and WANs.

CTC and ONS Nodes on the Same IP Subnet

Figure 10-8 shows a basic ONS 15454 SDH LAN configuration with the ONS 15454 SDHs and CTC computer residing on the same 192.168.1.0 subnet. All ONS 15454 SDHs connect to LAN A, and all ONS 15454 SDHs have DCC connections configured.

Figure 10-8 *CTC and ONS Nodes on the Same IP Subnet*

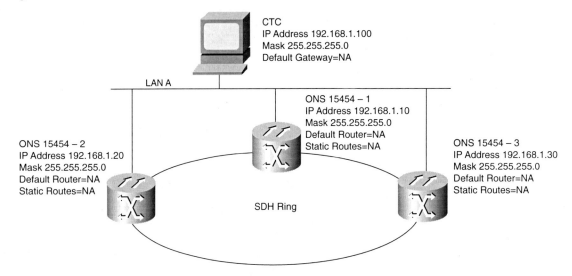

CTC and ONS Nodes on Separate IP Subnets

In this scenario, the CTC computer resides on subnet 192.168.1.0 and attaches to LAN A, as shown in Figure 10-9. The ONS 15454 SDHs reside on a different subnet (192.168.2.0) and attach to LAN B. A router connects LAN A to LAN B. The IP address of router interface A is set to 192.168.1.1, and the IP address of router interface B is set to 192.168.2.1.

On the CTC computer, the default gateway is set to router interface A. If the LAN uses DHCP, the default gateway and IP address are assigned automatically. In this example, a DHCP server is not available.

Using Proxy ARP to Enable an ONS 15454 SDH Gateway

In this scenario, as shown in Figure 10-10, ONS nodes 2 and 3 connect to ONS 15454 SDH-1 through the SDH DCC. Because all three ONS 15454 SDHs are on the same IP subnet, Proxy ARP enables ONS 15454 SDH-1 to serve as a gateway for ONS 15454 SDHs 2 and 3.

Figure 10-9 *CTC and ONS Nodes on Separate IP Subnets*

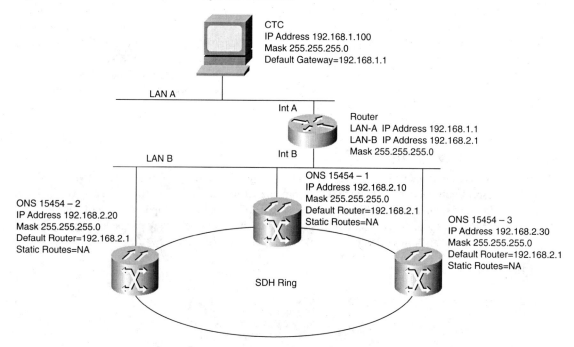

Figure 10-10 *Using Proxy ARP to Enable an ONS Gateway*

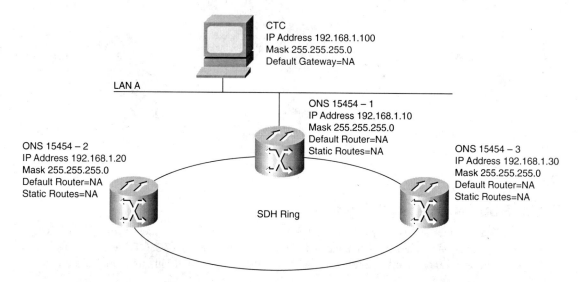

470 Chapter 10: Provisioning the Multiservice SDH MSPP

Proxy ARP enables one LAN-connected ONS 15454 SDH to respond to the ARP request for ONS 15454 SDHs not connected to the LAN. ONS 15454 SDH Proxy ARP requires no user configuration. For this to occur, the DCC-connected ONS 15454 SDHs must reside on the same IP subnet. When a LAN device, such as CTC, sends an ARP request to an ONS 15454 SDH that is not connected to the LAN, the gateway ONS 15454 SDH returns its MAC address to the LAN device. The LAN device then sends the datagram for the remote ONS 15454 SDH to the MAC address of the proxy ONS 15454 SDH. The proxy ONS uses the DCC channel to communicate with the remote ONS devices.

Default Gateway on the CTC

In this scenario, ONS nodes 2 and 3 reside on separate subnets, 192.168.2.0 and 192.168.3.0, respectively. ONS node 1 and the CTC computer are on subnet 192.168.1.0. The network includes different subnets because Proxy ARP is not used. For the CTC computer to communicate with ONS nodes 2 and 3, ONS node 1 is entered as the default gateway on the CTC computer. Figure 10-11 shows a default gateway on the CTC workstation.

Figure 10-11 *Default Gateway on the CTC*

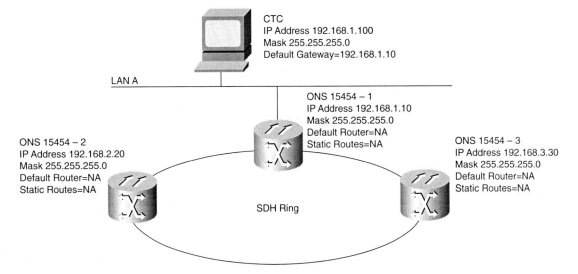

Static Routes on the ONS Node

Static routes are used by ONS 15454 SDHs residing on one IP subnet to connect to CTC sessions on another subnet connected via a router. In Figure 10-12, CTC residing on subnet 192.168.1.0 connects to a router through interface A. The router in this example is not configured for OSPF. The ONS nodes residing on subnet 192.168.2.0 are connected through ONS node 1 to the router through interface B. Proxy ARP enables ONS 15454 SDH node 1 as a gateway for ONS nodes 2 and 3. To connect to CTC computers on LAN A, a static route is created on ONS node 1.

Figure 10-12 *Static Route with a Single CTC Destination*

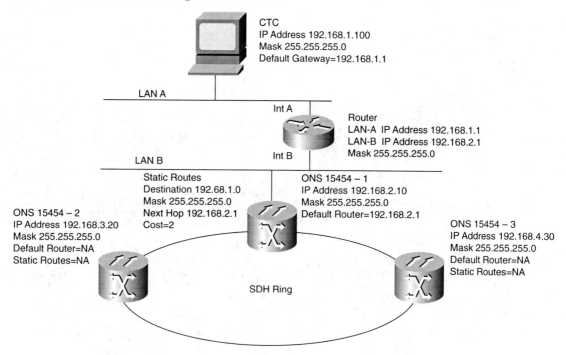

The destination and subnet mask entries control access to the ONS 15454 SDHs:

- If a single CTC computer is connected to router, enter the complete CTC host route IP address as the destination with a subnet mask of 255.255.255.255.
- If CTC computers on a subnet are connected to router, enter the destination subnet (in this example, 192.168.1.0) and a subnet mask of 255.255.255.0.
- If all CTC computers are connected to router, enter a destination of 0.0.0.0 and a subnet mask of 0.0.0.0. Figure 10-13 shows an example of this configuration.
- The IP address of router interface B is entered as the next hop, and the cost in number of hops from source to destination is 2.

Figure 10-13 *Static Route with Multiple LAN Destinations*

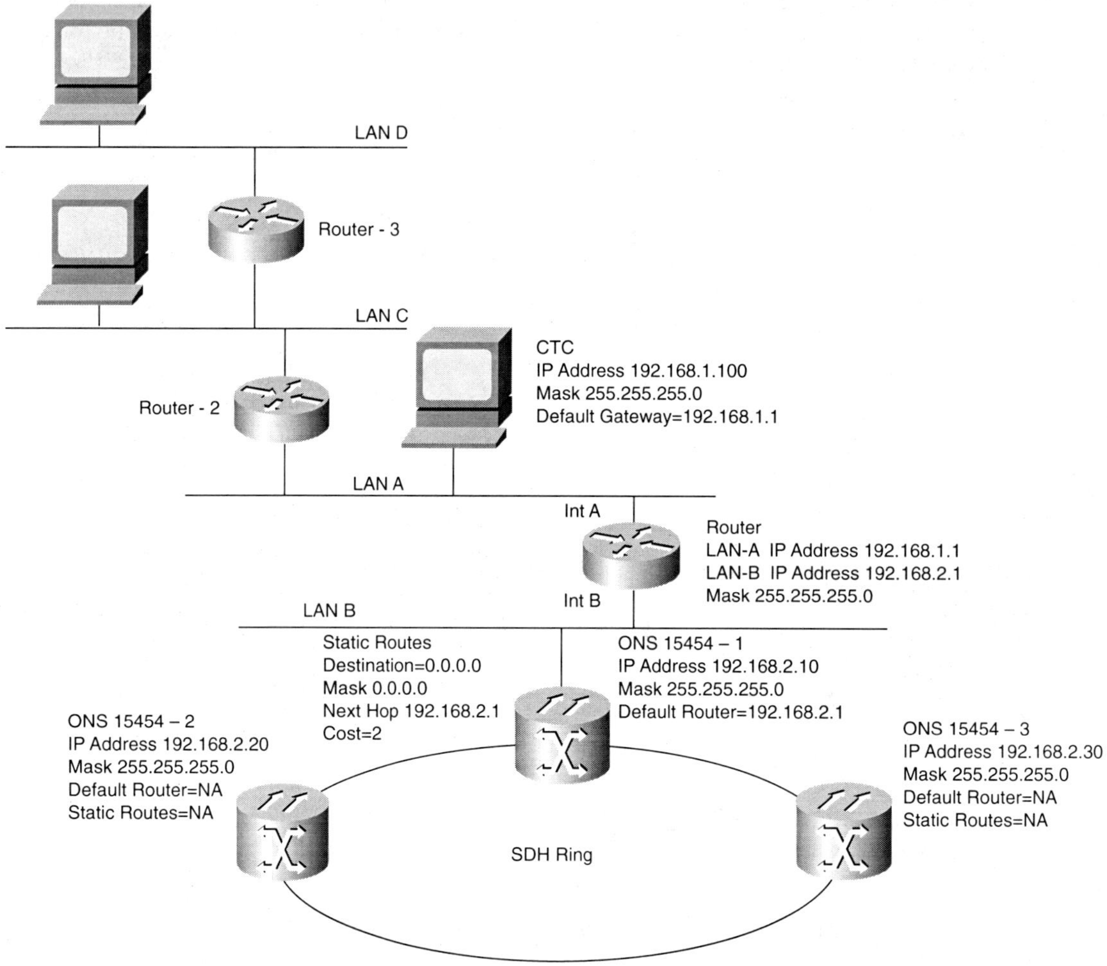

Use the following steps to create a static route on the ONS node:

Step 1 Log in to the ONS 15454 SDH and choose the **Provisioning > Network** tab.

Step 2 Click the **Static Routing** tab and then click **Create**.

Step 3 In the Create Static Route dialog box, enter the following:

- **Destination**—Enter the IP address of the computer running CTC. To limit access to one computer, enter the full IP address (in the example, 192.168.1.100). To allow access to all computers on the 192.168.1.0 subnet, enter **192.168.1.0** and a subnet mask of **255.255.255.0**. You can enter a destination of 0.0.0.0 to allow access to all CTC computers that connect to the router.

- **Mask**—Enter a subnet mask. If the destination is a host route (single CTC computer), enter a 32-bit subnet mask (255.255.255.255). If the destination is a subnet, adjust the subnet mask accordingly (for example, 255.255.255.0). If the destination is 0.0.0.0, enter a subnet mask of 0.0.0.0 to provide access to all CTC computers.
- **Next Hop**—Enter the IP address of the router port (in this example, 192.168.90.1) or the node IP address if the CTC computer is connected to the node directly.
- **Cost**—Enter the number of hops between the ONS 15454 SDH and the computer. In this example, the cost is two, one hop from the ONS 15454 SDH to the router and a second hop from the router to the CTC workstation.

Step 4 Click **OK**. Verify that the static route displays in the Static Route window, or ping the node.

Static Routes for Multiple CTCs

This scenario illustrates the static routes used when multiple CTC computers need to access ONS nodes residing on the same subnet, as shown in Figure 10-14. In this scenario, CTC nodes 1 and 2, as well as all ONS nodes, are on the same IP subnet 192.168.1.0. ONS node 1 and CTC-1 are attached to LAN A. ONS node 2 and CTC-2 are attached to LAN B. Static routes are added to ONS node 1 pointing to CTC-1, and to ONS node 2 pointing to CTC-2. The static routes are entered from the node's perspective.

Figure 10-14 *Static Routes on the ONS Nodes*

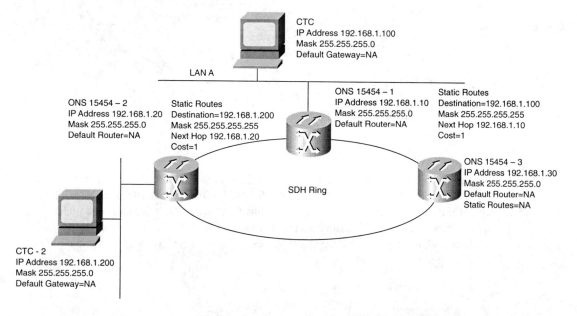

ONS OSPF Configuration

ONS 15454 SDHs use the OSPF protocol in internal ONS networks for node discovery, circuit routing, and node management. You can enable OSPF on the ONS 15454 SDHs so that the ONS node topology is sent to OSPF routers on a LAN. Advertising the ONS 15454 SDH network topology to LAN routers eliminates the need to manually enter static routes for ONS 15454 SDH subnets.

Figure 10-15 shows the same network enabled for OSPF, and Figure 10-16 shows the same network without OSPF. Static routes must be manually added to the router for CTC computers on LAN A to communicate with ONS nodes 2 and 3 because these nodes reside on different subnets.

Figure 10-15 *ONS Configuration with OSPF Enabled*

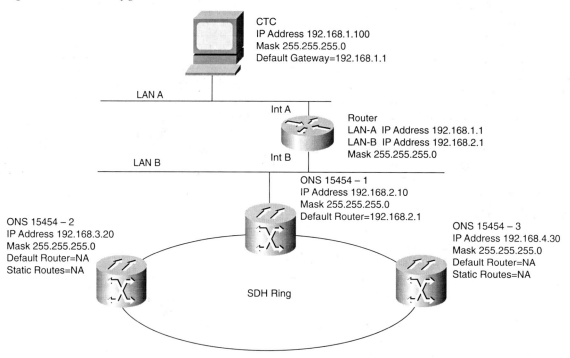

Use the following procedure to enable OSPF on each ONS 15454 SDH node that you want included in the OSPF network topology. ONS 15454 SDH OSPF settings must match the router OSPF settings, so you will need to get the OSPF area ID, hello and dead intervals, and authentication key (if OSPF authentication is enabled) from the router to which the ONS 15454 SDH network is connected before enabling OSPF. Figure 10-17 shows an example of OSPF provisioning.

IP Networking of ONS 15454 SDH Nodes for OAM&P   475

Figure 10-16 *ONS Configuration with OSPF Disabled*

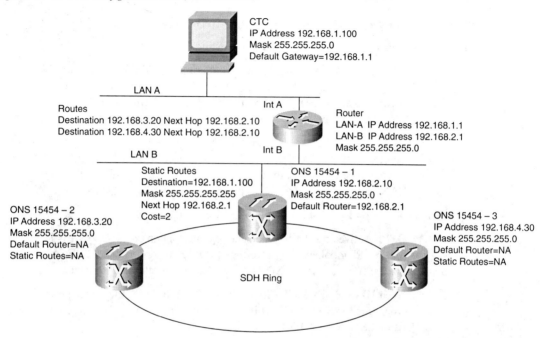

Figure 10-17 *OSPF Configuration for the ONS 15454 SDH*

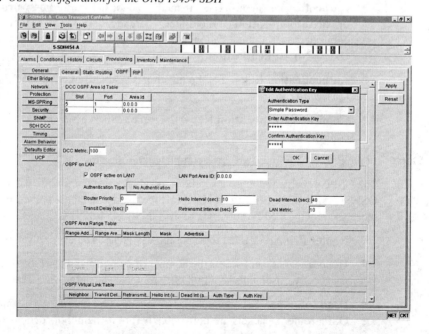

Step 1 Log in to the ONS 15454 SDH node.

Step 2 In node view, choose the **Provisioning > Network > OSPF** tab.

Step 3 On the upper-left side, complete the following:

- **DCC OSPF Area ID**—Enter the number that identifies the ONS 15454 SDHs as a unique OSPF area. The OSPF area number can be an integer between 0 and 4,294,967,295, and it can take a form similar to an IP address. The number must be unique to the LAN OSPF area.

- **DCC Metric**—This value is normally unchanged. It sets a cost for sending packets across the DCC, which is used by OSPF routers to calculate the shortest path. This value should always be higher than the LAN metric. The default DCC metric is 100.

Step 4 In the OSPF on LAN area, complete the following:

- **OSPF Active on LAN**—When checked, this option enables ONS 15454 SDH OSPF topology to be advertised to OSPF routers on the LAN. Enable this field on ONS 15454 SDHs that directly connect to OSPF routers.

- **Area ID for LAN Port**—Enter the OSPF area ID for the router port where the ONS 15454 SDH is connected. (This number is different from the DCC area ID.)

Step 5 In the Authentication area, complete the following:

- **Authentication Type**—Use the menu to choose **Simple Password** or **No Authentication**. If the router where the ONS 15454 SDH is connected uses authentication, choose **Simple Password**. Otherwise, choose **No Authentication**.

- **Authentication Key**—If authentication is enabled, enter the OSPF key (password).

- **Confirm Authentication Key**—Enter the OSPF key again for confirmation purposes.

Step 6 In the Priority and Intervals area, complete the following:

The OSPF priority and intervals default to values most commonly used by OSPF routers. In the Priority and Interval area, verify that these values match those used by the OSPF router where the ONS 15454 SDH is connected.

- **Router Priority**—Used to select the designated router for a subnet.

- **Hello Interval (sec)**—Sets the number of seconds between OSPF hello packet advertisements sent by OSPF routers. The default is 10 seconds.

- **Dead Interval**—Sets the number of seconds that will pass while an OSPF router's packets are not visible before its neighbors declare the router down. The default is 40 seconds.

- **Transit Delay (sec)**—Indicates the service speed. The default is 1 second.
- **Retransmit Interval (sec)**—Sets the time that will elapse before a packet is resent. The default is 5 seconds.
- **LAN Metric**—Sets a cost for sending packets across the LAN. This value should always be lower than the DCC metric. The default is 10.

Step 7 In the OSPF Area Range Table area, complete the following:

Area range tables consolidate the information that is propagated outside an OSPF area border. One ONS 15454 SDH in the ONS 15454 SDH OSPF area is connected to the OSPF router. An area range table on this node points the router to the other nodes that reside within the ONS 15454 SDH OSPF area.

To create an area range table, follow these substeps:

(a) Under OSPF Area Range Table, click **Create**.

(b) In the Create Area Range dialog box, enter the following:

- **Range Address**—Enter the area IP address for the ONS 15454 SDHs that reside within the OSPF area. For example, if the ONS 15454 SDH OSPF area includes nodes with IP addresses 10.10.20.100, 10.10.30.150, 10.10.40.200, and 10.10.50.250, the range address would be 10.10.0.0.
- **Range Area ID**—Enter the OSPF area ID for the ONS 15454 SDHs. This is either the ID in the DCC OSPF Area ID field or the ID in the area ID for LAN Port field.
- **Mask Length**—Enter the subnet mask length. In the Range Address example, this is 16.
- **Mask**—Displays the subnet mask used to reach the destination host or network.
- **Advertise**—Check this option if you want to advertise the OSPF range table.

(c) Click **OK**.

Step 8 All OSPF areas must be connected to area 0. If the ONS 15454 SDH OSPF area is not physically connected to area 0, use the following steps to create a virtual link table that will provide the disconnected area with a logical path to area 0:

(a) Under OSPF Virtual Link Table, click **Create**.

(b) In the Create Virtual Link dialog box, complete the following fields. (The OSPF settings must match OSPF settings for the ONS 15454 SDH OSPF area.)

- **Neighbor**—Enter the router ID of the area 0 router.

- **Transit Delay (sec)**—The service speed. The default is 1 second.
- **Hello Interval (sec)**—The number of seconds between OSPF hello packet advertisements sent by OSPF routers. The default is 10 seconds.
- **Auth Type**—If the router where the ONS 15454 SDH is connected uses authentication, choose **Simple Password**. Otherwise, set it to **No Authentication**.
- **Retransmit Interval (sec)**—Sets the time that will elapse before a packet is resent. The default is 5 seconds.
- **Dead Interval (sec)**—Sets the number of seconds that will pass while an OSPF router's packets are not visible before its neighbors declare the router down. The default is 40 seconds.

(c) Click **OK**.

Step 9 After entering ONS 15454 SDH OSPF area data, click **Apply**.

NOTE If you changed the area ID, the TCC cards will reset, one at a time.

SNCP Configuration

Subnetwork connection protection (SNCP) rings can be provisioned using CTC. SNCP rings provide duplicate fiber paths in the network. Working traffic flows in one direction, and protection traffic flows in the opposite direction. If a fiber break occurs in the working path, the receiving node switches to the protect path coming from the opposite direction. Switching in SNCP networks normally occurs at the end of the path and is triggered by defects or alarms along the path.

The SNCP network can be divided into a number of interconnected subnetworks. Within each subnetwork, protection is provided at the path level, and the automatic protection switching between two paths is provided at the subnetwork boundaries. The node at the end of the path and the intermediate nodes in the path select the best traffic signal. The virtual container is not terminated at the intermediate node; instead, it compares the quality of the signal on the two incoming ports and selects the better signal. CTC uses an SNCP ring as the default protection mechanism. The ONS 15454 SDH uses the SDH regenerator section DCC (SDCC) for data communications. Because the multiplex section DCCs are not used, they are available to SDH tunnel DCCs from third-party equipment across ONS 15454 SDH networks.

The following steps are involved with setting up an SNCP Ring:

Step 1 Install the SNCP ring trunk cards.

Step 2 Configure the SNCP ring DCC terminations and place ports in service.

Step 3 Configure the external, line, or mixed timing for the ONS 15454 SDH.

Step 4 Set up the SNCP circuits.

Install the SNCP Ring Trunk Cards

This section describes the steps used to install SNCP trunk cards:

Step 1 Install the SNCP trunk cards. These are STM-N cards that you will use as SNCP trunk cards. You can install the STM-1, STM-4, and STM-16 cards in slots 1 to 6 and 12 to 17. The STM-64 card can be installed only in slots 5, 6, 12, or 13.

Step 2 The installed STM-N cards will boot up by themselves. Let the trunk cards boot up to steady state.

Step 3 Install the fiber for east and west STM-N card ports, as shown in Figure 10-18. It's always a good idea to install the west card port in the farthest slot to the left and the east card port in the farthest slot to the right. Figure 10-18 shows fiber connections for a four-node SNCP ring with trunk cards in slot 5 (west) and slot 12 (east).

Figure 10-18 *4 Node SNCP Ring*

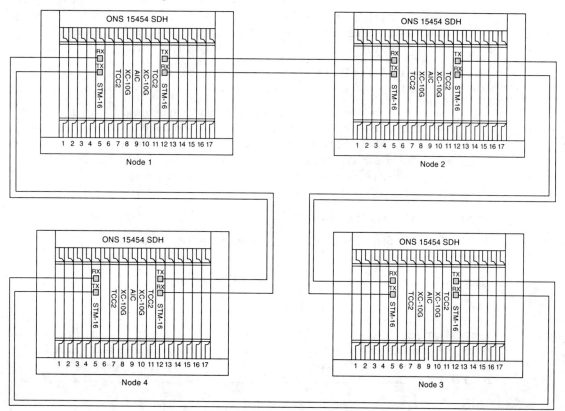

Step 4 Plug fiber from the transmit (TX) connector of an STM-N card at one node into the receive (RX) connector of an STM-N card at the adjacent node. The card displays a signal failure (SF) LED if TX and RX fibers are mismatched, after the DCCs are on and the ports are in service. Do not attempt to troubleshoot high-powered lasers by looking into the fiber for visible light.

Configure the SNCP Ring DCC Terminations and Place Ports in Service

This section describes the steps used to configure SNCP DCC terminations and place ports in service:

Step 1 Connect CTC to the first node that you will provision for the SNCP ring.

Step 2 Click the **Provisioning > SDH DCC** tab.

Step 3 In the SDCC Terminations section, click **Create**.

Step 4 In the Create SDCC Terminations dialog box, press the **Ctrl** key and click the two slots/ports that will serve as the SNCP ports at the node. For example, slot 5 (STM-16)/port 1 and slot 14 (STM-16)/port 1.

Step 5 Choose the **Set to IS, if allowed** option that places the trunk card ports in service.

Select the port state option that fits your network requirements. You can select the OOS-MT option if SNCP is provisioned prior to the physical installation of the cards and fiber. This option enables you to avoid DCC termination alarms until all DCCs are configured on the SNCP ring. You can also enable or disable OSPF on the DCC according to your network requirements.

Step 6 Click **OK**. The slots and ports appear in the SDCC Terminations list.

Step 7 Complete Steps 3 to 6 for each SNCP node that will participate in the SNCP ring.

Figure 10-19 shows an example of an SNCP SDH DCC configuration.

NOTE The next two steps involve configuration of ONS 15454 SDH timing (external, line, or mixed timing) and the setting up of SNCP circuits. See the "Provisioning ONS 15454 SDH Timing" and "SNCP Configuration" sections for these procedures.

Figure 10-19 *DCC Provisioning for SNCP*

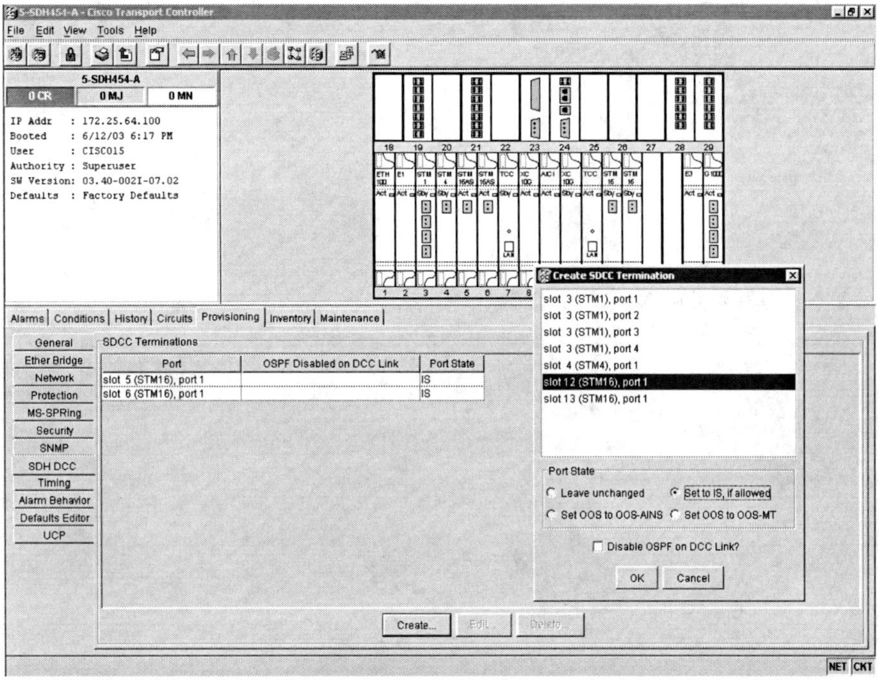

Enable the SNCP Ports

This section describes the steps used to enable SNCP ports:

Step 1 Log in to the first SNCP node.

Step 2 Double-click one of the cards that you configured as an SDCC termination.

Step 3 Click the **Provisioning > Line** tab.

Step 4 Under State, choose **In Service** for each port that you want enabled.

Step 5 Repeat Steps 2 through 4 for the second card.

Step 6 Click **Apply**.

The preceding steps describe SNCP configuration for a single node. Figure 10-20 shows an example. Apply the same procedures to configure the additional nodes. To create circuits, see the "SDH Circuit Provisioning" section.

Figure 10-20 *Enabling SNCP Ports for the ONS 15454 SDH*

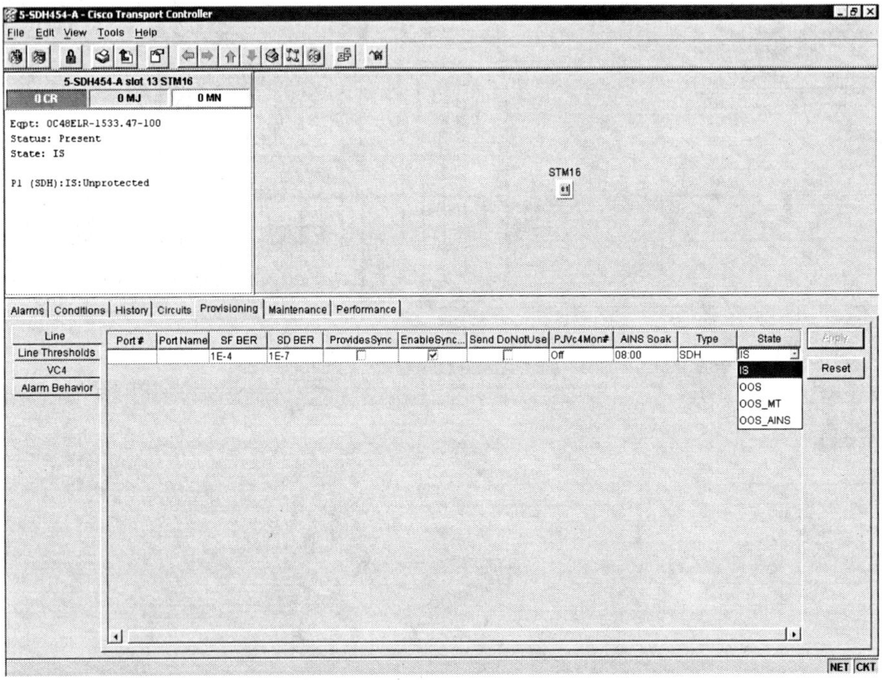

Adding and Removing Nodes from an SNCP Ring

This section explains how to add and remove nodes in an ONS 15454 SDH SNCP ring configuration. To add or remove a node in an SNCP ring, you need to perform two basic procedures:

Step 1 Switch SNCP traffic away from the area of the ring that will be affected by the node insertion or removal.

Step 2 Add or remove the node and reconfigure.

Switch SNCP Ring Traffic

This section describes the steps used to switch SNCP traffic:

Step 1 From CTC, display the network view.

Step 2 Right-click the span that will be cut to add or delete a node and choose **Circuits** from the shortcut menu.

Step 3 In the Circuits on Span dialog box, choose a protection option from the Perform SNCP span switching menu:

- **CLEAR**—Removes a previously set switch command.

- **MANUAL_SWITCH AWAY**—Switches the span if the new span is error free.
- **FORCE_SWITCH AWAY**—Forces the span to switch, even if the path has signal degrade (SD) or signal failure (SF) conditions. Force switch states have a higher priority than manual switches.
- **LOCKOUT OF PROTECTION**—Prevents traffic from switching to the protect circuit path under any circumstances. Of all switch states, lockout has the highest priority.

Step 4 Click **Apply**.

Step 5 When the confirmation dialog box appears, click **Yes** to confirm the protection switching. The column under Switch State changes to your chosen level of protection.

Step 6 Click **Close** after switch state changes.

Add an SNCP Node

This section describes the steps used to add an SNCP node:

Step 1 Start CTC for one of the SNCP ring nodes and display the network view.

Step 2 Clear any alarms or conditions on the ring nodes.

Step 3 At the node that you will add to the SNCP, complete the following steps:

(a) Verify that the STM-N cards are installed and fiber is available to connect to the other nodes.

(b) Run test traffic through the cards that will connect to the SNCP.

(c) Complete the procedure for setting up an SNCP ring to provision the new node.

Step 4 Start CTC for a node that will physically connect to the new node.

Step 5 See the "Switch SNCP Ring Traffic" procedure to initiate a FORCE switch to move traffic away from the span that will connect to the new node. Traffic is not protected during a protection switch.

Step 6 Two nodes will connect directly to the new node. Remove their fiber connections as follows:

(a) Remove the east fiber connection from the node that will connect to the west port of the new node.

(b) Remove the west fiber connection from the node that will connect to the east port of the new node.

Step 7 Replace the removed fiber connections with connections from the new node. This step must be performed onsite at the new node.

Step 8 Log out of CTC, and then log back in to the new node in the ring.

Step 9 Display the network view. The new node should appear in the network view. Wait for a few minutes to allow all the nodes to appear.

Step 10 Click the **Circuits** tab and wait for all the circuits to appear including spans. Circuits that will pass through the new node display as incomplete.

Step 11 In the network view, right-click the new node and choose **Update Circuits with New Node** from the list of options.

Step 12 Click the **Circuits** tab and verify that no incomplete circuits display. If incomplete circuits display, repeat Step 10.

Step 13 Use the "Switch SNCP Ring Traffic" procedure to clear the protection switch.

Remove an SNCP Node

This section describes the steps used to remove an SNCP node:

Step 1 Start CTC for one of the SNCP ring nodes and display the network view. Clear any alarms or conditions on the ring nodes.

Step 2 Complete the "Switch SNCP Ring Traffic" procedure to initiate a FORCE switch to move traffic away from the node you will remove. Initiate a FORCE switch on all spans connected to the node you remove. Traffic is not protected during a forced protection switch.

Step 3 Log in to the node that you will remove and display the node view.

Step 4 Delete circuits that originate or terminate in that node. (If a circuit has multiple drops, delete only the drops that terminate on the node you delete.)

    (a) Click the **Circuits** tab.

    (b) Choose the circuit(s) to delete. To choose multiple circuits, press the **Shift** or **Ctrl** key while selecting circuits.

    (c) Click **Delete**.

    (d) Click **Yes** when prompted.

Step 5 From the node that will be deleted, remove the east and west span fibers. At this point, the node is no longer a part of the ring.

Step 6 Reconnect the span fibers of the nodes that remain in the ring.

Step 7 Log out of CTC, and then log back into a node in the ring.

Step 8 Click the **Alarms** tab of each newly connected node and verify that the span cards are free of alarms. Resolve any alarms before proceeding.

Step 9 If the removed node was the BITS, choose a new node as the BITS source or choose another node as the master timing node.

Step 10 See the "Switch SNCP Ring Traffic" procedure to clear the protection switch.

Figure 10-21 shows an example of SNCP circuit deletion.

Figure 10-21 *SNCP Circuit Deletion*

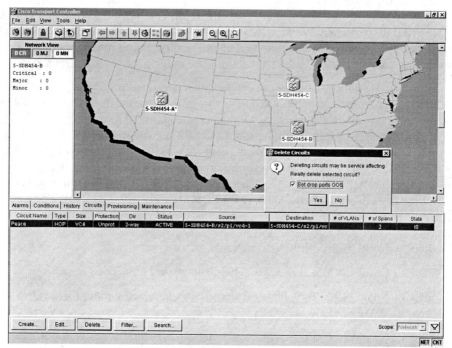

MS-SPRing Configuration

There are two types of MS-SPRings: two-fiber and four-fiber. Two-fiber MS-SPRings share service and protection equally, but only two physical fibers are required. With four-fiber MS-SPRings, the nodes on both sides of the failed span perform a span switch and use the second pair of fibers as the new working route.

MS-SPRings share the ring bandwidth equally between working and protection traffic. Half of the payload bandwidth is reserved for protection in each direction. An MS-SPRing node can terminate traffic it receives from either side of the ring and is suited for distributed node-to-node traffic applications, such as interoffice networks and access networks. MS-SPRings allow bandwidth to be reused around the ring and can carry more traffic than a network with traffic flowing through one central hub. When properly configured, MS-SPRings can also carry more traffic than an SNCP operating at the same STM-N rate.

The ONS 15454 SDH can support a number of ring combinations with Release 4.x of system software if the total DCC usage is equal to or less than 32 DCCs. Each MS-SPRing can have up to 32 ONS 15454 SDH nodes. Because the working and protect bandwidths must be equal, you can create only STM-4 (two-fiber only), STM-16, or STM-64 MS-SPRings. MS-SPRings with 16 or fewer nodes meet the ITU-T G.841 switch time requirement.

In two-fiber MS-SPRings, each fiber is divided into working and protect bandwidths. In an STM-16 MS-SPRing, for example, VC4s 1 to 8 carry the working traffic, and VC4s 9 to 16 are reserved for protection. Working traffic (VC4s 1 to 8) travels in one direction on one fiber and in the opposite direction on the second fiber. The CTC circuit routing routines calculate the shortest path for circuits based on requirements, traffic patterns, and distance.

Four-fiber MS-SPRings double the bandwidth of two-fiber MS-SPRings. Four-fiber MS-SPRings increase the reliability and flexibility of traffic protection because they allow span switching as well as ring switching. Two fibers are allocated for working traffic and two fibers for protection. To implement a four-fiber MS-SPRing, you must install four STM-16 cards or four STM-64 cards at each MS-SPRing node. MS-SPRing is not supported on STM-4. The ONS 15454 SDH uses the SDH regenerator section DCC (SDCC) for data communications. Because the multiplex section DCCs are not used, they are available to tunnel DCCs from third-party equipment across ONS 15454 SDH networks.

The following steps are involved with setting up an MS-SPRing ring:

Step 1 Install the MS-SPRing trunk cards.

Step 2 Create the MS-SPRing DCC terminations and place ports in service.

Step 3 Set up MS-SPRing timing (external, line, or mixed timing) for the ONS 15454 SDH.

Step 4 If an MS-SPRing span passes through equipment that cannot transparently transport the K3 byte, remap the MS-SPRing extension byte on the trunk cards at each end of the span.

Step 5 Provision the MS-SPRing.

Install the MS-SPRing Trunk Cards

This section describes the steps used to install MS-SPRing trunk cards:

Step 1 Install the STM-4, STM-16, or STM-64 cards that will serve as the MS-SPRing trunk cards. You can install the STM-4 and STM-16 cards in slots 1 to 6 and 12 to 17. The STM-64 card can be installed only in slots 5, 6, 12, or 13.

Step 2 The installed STM-N cards will boot up by themselves. Let the trunk cards boot up to steady state.

Step 3 Install the fiber for east and west STM-N card ports, as shown in Figure 10-22. It's always a good idea to install the west card port in the farthest slot to the left and the east card port in the farthest slot to the right. Plug fiber from the TX connector of an STM-N card at one node into the RX connector of an STM-N card at the adjacent node. The card displays an SF LED if TX and RX fibers are mismatched, after the DCCs are on and the ports are in service. Do not attempt to troubleshoot high-powered lasers by looking into the fiber for visible light.

Figure 10-22 *Two-Fiber, Four-Node MS-SPRing Ring*

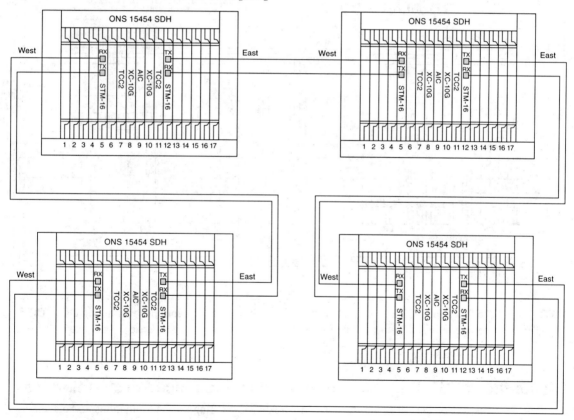

Step 4 For four-fiber MS-SPRings, use the same east-west connection pattern for the working and protect fibers. Do not mix working and protect card connections. The MS-SPRing will not function if working and protect cards are interconnected. Figure 10-23 shows fiber connections for a four-fiber

488 Chapter 10: Provisioning the Multiservice SDH MSPP

MS-SPRing. Slot 5 (west) and slot 12 (east) carry the working traffic. Slot 6 (west) and slot 13 (east) carry the protect traffic.

Figure 10-23 *Four-Fiber, Four-Node MS-SPRing Ring*

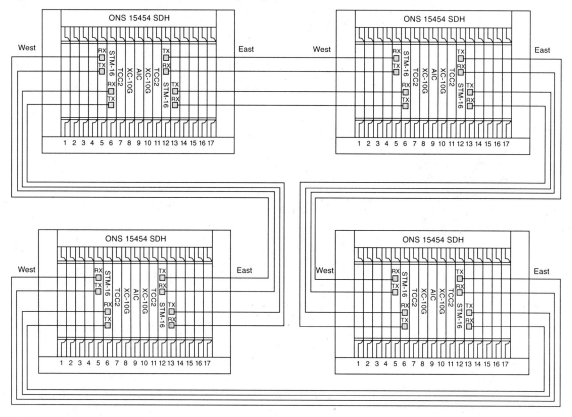

NOTE The SDH and SONET versions of the Cisco ONS 15454 do not interoperate via DCC. DCC interoperability is not available for ONS 15454 SDH Software Release 3.4 and earlier.

Create the MS-SPRing DCC Terminations and Place Ports in Service

This section describes the steps used to create MS-SPRing DCC terminations and place ports in service:

Step 1 Start CTC for the first node that you will provision for the MS-SPRing.

Step 2 Click the **Provisioning > SDH DCC** tab.

Step 3 In the SDCC Terminations section, click **Create**.

MS-SPRing Configuration   489

Step 4 In the Create SDCC Terminations dialog box, press **Ctrl** and click the two slots/ports that will serve as the MS-SPRing ports at the node. For example, slot 5 (STM-16)/port 1 and slot 12 (STM-16)/port 1. For four-fiber MS-SPRings, provision the working cards, but not the protect cards, as DCC terminations.

Step 5 Choose the **Set to IS, if allowed** option that places the trunk card ports in service. Choose the port state option that fits your network requirements. You may choose the **OOS-MT** option if SNCP is provisioned prior to the physical installation of the cards and fiber. This option enables you to avoid DCC termination alarms until all DCCs are configured on the SNCP ring.

Step 6 Enable or disable OSPF on the DCC according to your network requirements.

Step 7 Click **OK**.

The slots/ports appear in the SDCC Terminations list. Figure 10-24 shows an example of SDH DCC provisioning for MS-SPRing.

Figure 10-24 *SDH DCC Provisioning for MS-SPRing*

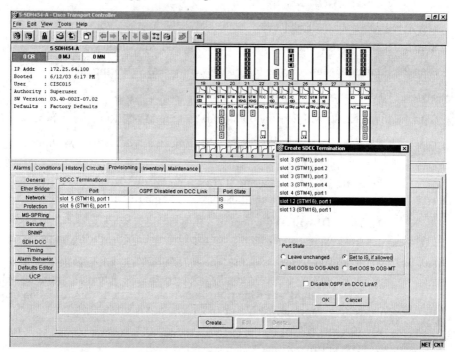

Step 8 Complete Steps 3 to 7 at each node that will be in the MS-SPRing.

Step 9 After configuring the SDH DCC, set the timing for the nodes as described in the "Provisioning ONS 15454 SDH Timing" section.

Remap the K3 Byte

This section describes the steps used to remap the K3 byte:

Step 1 Start CTC for one of the nodes that connects to the third-party equipment.

Step 2 Double-click the STM-16 card that connects to the third-party equipment. The card view displays.

Step 3 Click the **Provisioning > Line** tab.

Step 4 Click MS-**SPRing Ext Byte** and choose the alternate byte: **Z2**, **E2**, or **F1**.

Step 5 Click **Apply**.

Step 6 (Four-fiber MS-SPRing only) Repeat Steps 2 to 5 for each protect card.

Step 7 (Two-fiber MS-SPRing only) Repeat Steps 2 to 5 at the node and card on the other end of the MS-SPRing span.

Provision the MS-SPRing

This section describes the steps used to provision the MS-SPRing:

Step 1 Start CTC for a node in the MS-SPRing and go to node view.

Step 2 From node view, choose the **Provisioning > MS-SPRing** tab.

Step 3 Click the **Create MS-SPRing** button.

Step 4 In the Create MS-SPRing dialog box, set the MS-SPRing properties:

- **Ring Type**—Select the MS-SPRing ring type, either two-fiber or four-fiber.

- **Ring ID**—Assign a ring ID (a number between 0 and 9999). Nodes in the same MS-SPRing must have the same ring ID.

- **Node ID**—Enter the node ID. If the node is being added to an MS-SPRing, use an ID that is not used by other MS-SPRing nodes.

- **Ring Reversion**—Set the amount of time that will pass before the traffic reverts to the original working path. The default is 5 minutes. All nodes in an MS-SPRing ring should have the same ring reversion setting, especially if Never (nonrevertive) is selected.

- **West Line**—Enter the slot port on the node that will connect to the MS-SPRing via the node's west line.

- **East Line**—Enter the slot port on the node that will connect to the MS-SPRing via the node's east line.

For four-fiber MS-SPRings, complete the following:

- **Span Reversion**—Complete the Span Reversion field for four-fiber MS-SPRings. Choose the amount of time that will elapse before the traffic reverts to the original working path following a traffic failure. The default is 5 minutes. Span reversions can be set to Never. If you set a ring reversion time, the times must be the same for both ends of the span. It is recommended that you use the same span reversion time throughout the ring to avoid reversion time mismatches.

- **West Protect**—Enter the slot port on the node that will connect to the MS-SPRing via the node's west protect fiber(s).

- **East Protect**—Enter the slot port on the node that will connect to the MS-SPRing via the node's east protect fiber(s).

Step 5 Click **OK**. Figure 10-25 shows an example of MS-SPRing provisioning. Now, go to network view.

Figure 10-25 *MS-SPRing Creation*

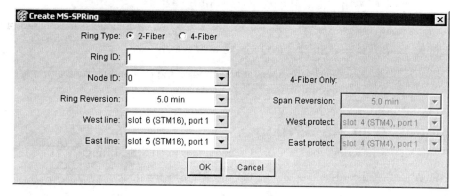

Step 6 Click the span you want to include in the MS-SPRing, and then click the **Add Span** button. Perform this step for each span you add to the MS-SPRing.

Step 7 Click the **Finish** button after you have chosen enough spans to create a two-fiber or four-fiber MS-SPRing.

> **NOTE** Some or all of the following alarms display during MS-SPRing setup: E-W MISMATCH, RING MISMATCH, APSCIMP, APSDFLTK, and MSSP-OOSYNC. These alarms should clear automatically.

Step 8 From network view, verify the following:

- A green span line appears between all MS-SPRing nodes.
- All E-W MISMATCH, RING MISMATCH, APSCIMP, DFLTK, and MSSP-OOSYNC alarms are cleared.

Step 9 Test the MS-SPRing using the following test procedure:

(a) Run test traffic through the ring.

(b) From network view, click the **Provisioning > MS-SPRing** tab.

(c) Click the ring, and then click **Edit**.

(d) Right-click an east port and choose **MANUAL RING** from the Set East Protection Operation list. Click **Apply**.

(e) Click the **Conditions** tab, and then click **Retrieve**. You should see a Ring Switch West event, and the far-end node that responded to this request should report a Ring Switch East event.

(f) Verify that traffic switches normally.

(g) Choose **Clear** from the Set East Protection Operation list and click **Apply**.

(h) Repeat Steps A to G for the west switch.

(i) Disconnect the fibers at any node on the ring and verify that traffic switches normally.

Adding Nodes to an MS-SPRing

This section describes the steps used to add nodes to an MS-SPRing:

Step 1 Install cards and configure the new MS-SPRing node.

Step 2 Switch MS-SPRing traffic before connecting a new node.

Step 3 Connect fiber to the new node.

Step 4 Provision the ring for the new node.

Install Cards and Configure the New MS-SPRing Node

This section describes the steps used to install cards and configure the new MS-SPRing node:

Step 1 Install the STM-4, STM-16, or STM-64 cards that you will add to the MS-SPRing. You can install the STM-4 and STM-16 cards in slots 1 to 6 and slots 12 to 17. The STM-64 card can be installed only in slots 5, 6, 12, or 13.

Step 2 Allow the cards to boot.

Step 3 Log in to the new node and complete the initial configuration.

Step 4 Provision the SDH DCC and place ports in service for the new node's cards.

Step 5 Configure the MS-SPRing timing (external, line, or mixed timing).

Step 6 If the new node will connect to third-party equipment that cannot transport the K3 byte, remap trunk cards that connect to the third-party equipment. Make sure the trunk card at the other end of the span is mapped to the same byte that is set on the new node.

Step 7 Complete the "Provision the MS-SPRing" procedure.

Figure 10-26 shows a three-node MS-SPRing before the new node is added.

Figure 10-26 *Two-Fiber, Three-Node MS-SPRing*

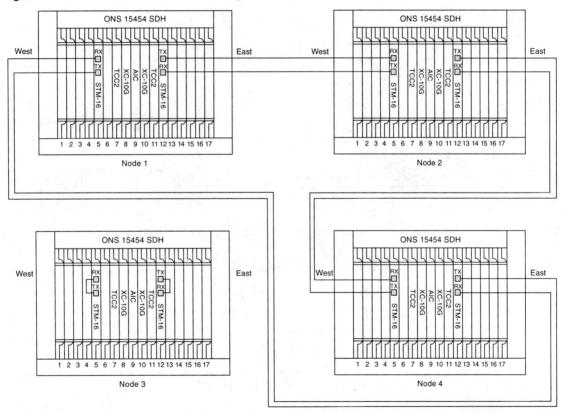

Switch MS-SPRing Traffic Before Connecting a New Node

This section describes the steps used to switch MS-SPRing traffic before connecting a new node:

Step 1 Log in to the existing node that will connect to the new node through its east port. This would be node 4 in Figure 10-27. Remember that traffic is unprotected during a protection switch.

Figure 10-27 *Converted Four-Node MS-SPRing*

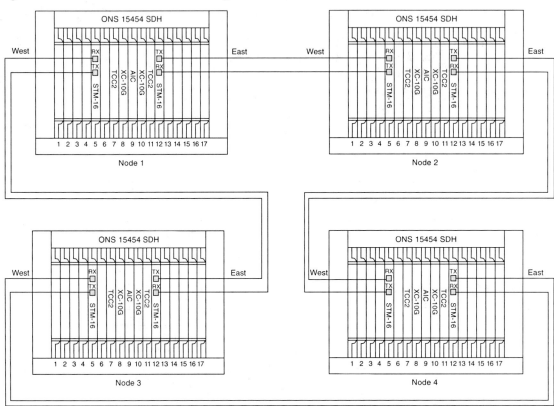

- **Step 2** Switch protection on the node's east port:

    (a) Click the **Maintenance > MS-SPRing** tab.

    (b) From the East Switch list, choose **FORCE RING**. Click **Apply**. Performing a FORCE switch generates a manual switch request on an equipment (MANUAL-REQ) alarm. This is normal.

- **Step 3** Log in to the existing node that will connect to the new node through its west port. This would be node 4 in Figure 10-26.

- **Step 4** Switch protection on the node's west port:

    (a) Click the **Maintenance > MS-SPRing** tab.

    (b) From the West Switch list, choose **FORCE RING**. Click **Apply**.

Connect Fiber to the New Node

This section describes the steps used to connect fiber to the new node:

Step 1 Remove the fiber connections from the two nodes that will connect directly to the new node.

   (a) Remove the east fiber from the node that will connect to the west port of the new node. In Figure 10-26, this is node 4/slot 12.

   (b) Remove the west fiber from the node that will connect to the east port of the new node. In Figure 10-26, this is node 1/slot 5.

Step 2 Replace the removed fibers with fibers connected from the new node. Connect the west port to the east port and the east port to the west port. Figure 10-27 shows the MS-SPRing example after the node is connected.

NOTE The new node will not appear in the ring until you exit CTC, restart, and provision the ring to accept the new node.

Provision the Ring for the New Node

This section describes the steps used to provision the ring for the new node:

Step 1 Start CTC again from any node in the MS-SPRing.

Step 2 In node view, choose the **Provisioning > MS-SPRing** tab.

Step 3 Click a ring, and then click **Ring Map**.

Step 4 In the MS-SPRing Map Ring Change dialog box, click **Yes**.

Step 5 In the MS-SPRing Map dialog box, verify that the new node is added. If it is, click **Accept**. If it does not appear, start CTC for the new node. Verify that the MS-SPRing is provisioned correctly and repeat Steps 1 to 4 in this procedure. If the node still does not appear, repeat all procedures for adding a node, making sure that no errors were made.

Step 6 Display the network view and click the **Circuits** tab. Wait until all the circuits are discovered. The circuits that pass through the new node will be shown as incomplete.

Step 7 Right-click the new node and choose **Update Circuits with New Node** from the shortcut menu. Verify that the number of updated circuits displayed in the dialog box is correct.

Step 8 Choose the **Circuits** tab and verify that no incomplete circuits are present.

Step 9 Clear the protection switch on the existing node using its east port to connect to the new node. Then, clear the protection switch on the existing node using its west port to connect to the new node.

(a) To clear the protection switch from the east port, display the node view and then display the **Maintenance > MS-SPRing** tab. From the East Switch list, select **CLEAR**. Click **Apply**.

(b) To clear the protection switch from the west port, display the node view and then display the **Maintenance > MS-SPRing** tab. From the West Switch list, select **CLEAR**. Click **Apply**.

Removing Nodes from an MS-SPRing

This section explains how to remove nodes in an ONS 15454 SDH MS-SPRing configuration:

Step 1 Start CTC for a node on the same MS-SPRing as the node you will remove, and display the network view. Clear any alarms or conditions on the network.

Step 2 Use the following subprocedure to delete all the circuits that originate or terminate in that node. If a circuit has multiple drops, delete only the drops that terminate on the node you want to delete.

(a) Click the **Circuits** tab. The circuits that use this node display.

(b) Select circuits that originate or terminate on the node. Click **Delete**.

(c) Click **Yes** when prompted.

(d) If a multidrop circuit has drops at the node that will be removed, choose the circuit, click **Edit**, and remove the drops.

Step 3 Switch traffic away from the ports of neighboring nodes that will be disconnected when the node is removed. Remember that traffic is unprotected during the protection switch.

(a) Start CTC for the neighboring node that is connected through its east port to the removed node.

(b) Click the **Maintenance > MS-SPRing** tab.

(c) From the East Switch list, select **FORCE RING**, and then click **Apply**.

(d) Start CTC for the node that is connected through its west port to the removed node.

(e) Click the **Maintenance > MS-SPRing** tab.

(f) From the West Switch list, select **FORCE RING**, and then click **Apply**.

Step 4 Remove all fiber connections between the node being removed and the two neighboring nodes.

Step 5 Reconnect the two neighboring nodes directly, west port to east port.

Step 6 If the removed node contained trunk STM-16 cards with K3 bytes mapped to an alternate byte, use the "Remap the K3 Byte" procedure to verify and remap the MS-SPRing extended bytes on the newly connected neighboring nodes.

Step 7 Exit CTC, and then start CTC for a node on the reduced ring.

Step 8 Wait for the MS-SPRing Map Ring Change dialog box to display. When the dialog box displays, click **Yes**. If the dialog box does not display after 10 to 15 seconds, choose the **Provisioning > MS-SPRing** tab and click **Ring Map**.

Step 9 In the MS-SPRing Ring Map dialog box, click **Accept**.

Step 10 Display the network view and choose the **Circuits** tab.

Step 11 Delete and recreate any incomplete circuits. Recreate the incomplete circuits one at a time.

Step 12 Clear the protection switches on the neighboring nodes.

  (a) Display the node with the protection switch on its east port.

  (b) Click the **Maintenance > MS-SPRing** tab and select **CLEAR** from the East Switch list. Click **Apply**.

  (c) Start CTC for the node with the protection switch on its west port.

  (d) Click the **Maintenance > MS-SPRing** tab and select **CLEAR** from the West Switch list. Click **Apply**.

Step 13 If a BITS clock is not used at each node, check that the synchronization is set to one of the eastbound or westbound MS-SPRing spans on the adjacent nodes. If the removed node was the BITS timing source, use a new node as the BITS source or choose internal synchronization at a designated node, from where all other nodes will derive their timing.

Upgrading a Two-Fiber MS-SPRing to a Four-Fiber MS-SPRing

Two-fiber STM-16 or STM-64 MS-SPRings can be upgraded to four-fiber MS-SPRings. To upgrade your network, you need to install two additional STM-16 or STM-64 cards at each two-fiber MS-SPRing node, run CTC, and upgrade the MS-SPRing from two-fiber to four-fiber. The fibers that were divided into working and protect bandwidths for the two-fiber MS-SPRing would be fully allocated for working MS-SPRing traffic. Ensure that the optical transmit and receive levels are in their acceptable range. This section describes the steps used to upgrade a two-fiber MS-SPRing to a four-fiber MS-SPRing:

Step 1 Start CTC for one of the two-fiber MS-SPRing nodes and display the network view. Clear any alarms or conditions.

Step 2 Install two additional STM-16 or STM-64 cards at each MS-SPRing node. You must install the same STM-N card rate as the two-fiber ring.

Step 3 Connect the fiber to the new cards using the same east-west connection scheme that connected the two-fiber connections.

Step 4 Set the card ports in service for each new STM-N card.

Step 5 Test the new fiber connections using the following procedure: Pull a TX fiber for a protect card and verify that an LOS alarm displays for the appropriate RX card. Perform this test for every span in the MS-SPRing protect ring.

Step 6 Upgrade the MS-SPRing.

> **NOTE** The Upgrade button will be disabled until the ring is upgradable, which means that there would have to be enough protect cards in each shelf with ports in service, no SDCC terminations, and no other protection groups. If all of your nodes meet these conditions, the Upgrade to Four-Fiber button is enabled.

(a) Display the network view and click the **Provisioning > MS-SPRing** tab.

(b) Click the two-fiber MS-SPRing you want to upgrade and click the **Upgrade to Four-Fiber** button.

(c) In the Upgrade MS-SPRing dialog box, click a span reversion time from the pull-down menu and click **Next**. The span reversion time sets the amount of time that will elapse before the traffic reverts to the original working path following a traffic failure. The default is 5 minutes.

(d) Complete the following from the protection pull-down menus:

- **West Protect** — Assign the east MS-SPRing port that will connect to the east protect fiber.

- **East Protect** — Assign the east MS-SPRing port that will connect to the east protect fiber.

(e) Click **Finish**.

Step 7 Test the MS-SPRing using the following procedure:

(a) Run test traffic through the ring.

(b) Log in to a node on the ring, click the **Maintenance > MS-SPRing** tab, and select **MANUAL RING** from the East Switch list. Click **Apply**.

(c) In network view, click the **Conditions** tab and click **Retrieve**. You should see a Ring Switch West event, and the far-end node that responded to this request should report a Ring Switch East event.

(d) Verify that traffic switches normally.

(e) Select **Clear** from the East Switch list and click **Apply**.

(f) Repeat Steps A to E for the west switch.

(g) Disconnect the fibers at any node on the ring and verify that traffic switches normally.

Figure 10-28 shows the four-fiber MS-SPRing after the upgrade from the two-fiber MS-SPRing has been performed.

Figure 10-28 *Four-Fiber MS-SPRING After Upgrade*

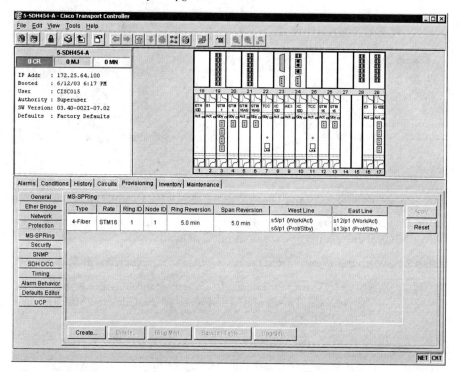

Moving MS-SPRing Trunk Cards

To move MS-SPRing trunk cards, you drop one node at a time from the current MS-SPRing. This procedure disrupts service during the time needed to complete the steps in the following procedure. Service disruption applies to all MS-SPRing nodes where cards will change slots. It is prudent to announce a service outage while implementing this procedure.

Figure 10-29 shows a four-node STM-16 MS-SPRing using trunk cards in slots 6 and 12 at all four nodes. In this example, the user moves trunk cards at node 4 in slots 6 and 12 to slots 5 and 6. Node 4 must be temporarily removed from the active MS-SPRing while the trunk cards are moved.

Figure 10-29 *Four-Node MS-SPRing Ring Before Trunk Card Switch*

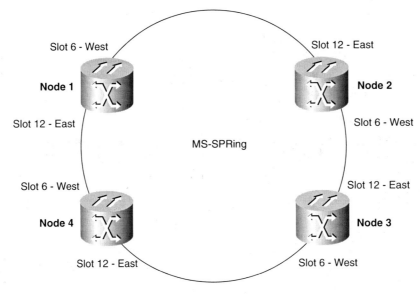

This section describes the steps used to move MS-SPRing trunk cards:

Step 1 Start CTC for one of the MS-SPRing nodes and display the network view. Clear any alarms or conditions in the network.

Step 2 Switch traffic away from the node where the trunk card will be moved:

(a) Start CTC for the node that is connected through its east port to the target node. This is node 1 in Figure 10-29. Click the **Maintenance > MS-SPRing** tab.

(b) From the East Switch list, select **FORCE RING** and click **Apply**.

When you perform a manual switch, a manual switch request equipment alarm (MANUAL-REQ) is generated. This is normal. Remember that traffic is unprotected during a protection switch.

    (c) Start CTC for the node that is connected through its west port to the target node. This is node 3 in Figure 10-29. Click the **Maintenance > MS-SPRing** tab.

    (d) From the West Switch list, select **FORCE RING** and click **Apply**.

Step 3 Start CTC on the target node.

Step 4 Click the **Circuits** tab. Write down the circuit information or, from the File menu, choose **Print** or **Export** to print or export the information. You will need this information to restore the circuits later.

Step 5 Delete the circuits on the card you remove:

    (a) Highlight the circuit(s). To choose multiple circuits, press the **Shift** or **Ctrl** key while choosing circuits.

    (b) Click **Delete**.

    (c) In the Delete Circuit dialog box, click **Yes**.

Step 6 Delete the SDH DCC termination on the card you remove:

    (a) Click the **Provisioning > SDH DCC** tab.

    (b) From the SDCC Terminations list, click the SDH DCC you need to delete and click **Delete**.

    (c) Click the **Set Unused Port out of Service** check box.

Step 7 Disable the ring on the target node:

    (a) Click the **Provisioning > MS-SPRing** tab.

    (b) Highlight the ring and click **Delete**.

    (c) In the confirmation message, confirm that this is the ring you want to delete. If so, click **Yes**.

Step 8 If an STM-N card is a timing source, choose the **Provisioning > Timing** tab and set timing to **Internal**.

Step 9 Physically remove the card.

Step 10 Insert the card into its new slot and wait for the card to boot.

Step 11 To delete the card in CTC from its former slot, right-click the card in node view and choose **Delete Card** from the list of options.

Step 12 Place the port(s) back in service.

Step 13 Re-enable the ring using the same cards (in their new slots) and ports for east and west. Use the same MS-SPRing ring ID and node ID that was used before the trunk card was moved.

Step 14 Recreate the circuits that were deleted. See the "Create an Automatically Routed High-Order Path Circuit" procedure.

Step 15 If you use line timing and the card you move is a timing reference, re-enable the timing parameters on the card. Figure 10-30 shows the ring after the change.

Figure 10-30 *Four-Node MS-SPRing Ring After Trunk Card Switch*

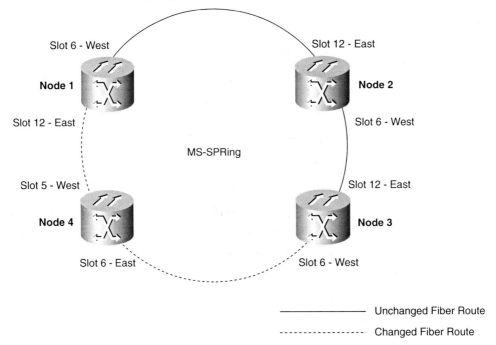

Subtending Ring Configurations

The ONS 15454 SDH supports up to 32 SDH DCCs with Release 4.x of the system software. This means that a single ONS 15454 SDH node can terminate and groom any one of the following ring combinations:

- 16 SNCP rings, or
- 15 SNCP rings + 1 MS-SPRing (2-fiber or 4-fiber), or
- 14 SNCP rings + 2 MS-SPRings (1 4-fiber maximum)

Subtending rings offers the network designer versatility in layout of the fiber infrastructure. Subtending rings also reduces the number of nodes and cards required in a design and helps reduce external shelf-to-shelf cabling. Figure 10-31 shows an ONS 15454 SDH with multiple subtending rings. It is possible to subtend an SNCP ring from an MS-SPRing as well as subtend an MS-SPRing from an SNCP ring.

NOTE The TCC2 card is based on a 400-MHz CPU with 256 MB Flash and 256 MB RAM. The TCC2 card is *hardware-ready* to support up to 84 DCC terminations. However, the current Release 4.x of the software supports only up to 32 DCC terminations.

Figure 10-31 *ONS 15454 SDH Node with Multiple Subtending Rings*

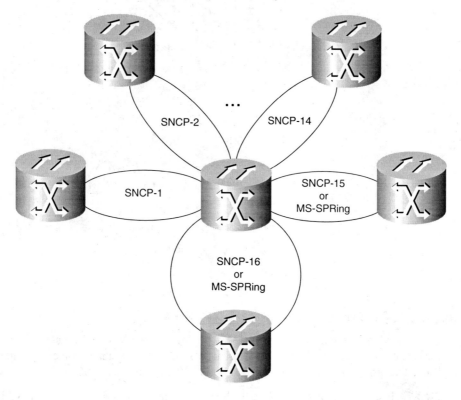

Figure 10-32 shows an SNCP ring subtending from a MS-SPRing. In this example, node 3 is the only node serving both the MS-SPRing and SNCP ring. STM-N cards in slots 5 and 12 serve the MS-SPRing, and STM-N cards in slots 6 and 13 serve the SNCP ring.

Figure 10-32 *SNCP Ring Subtending from MS-SPRing*

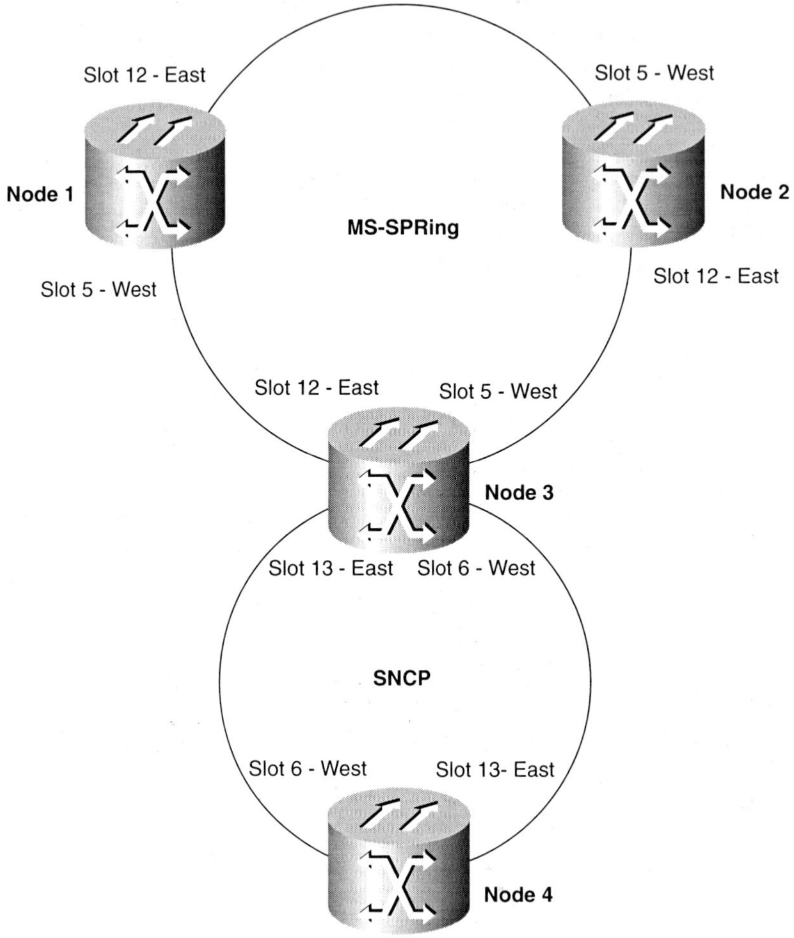

Subtend an SNCP Ring from an MS-SPRing

This procedure requires an established MS-SPRing and one MS-SPRing node with STM-N cards and fibers to carry the SNCP ring. The procedure also assumes you can set up an SNCP Ring. The procedure described here uses Figure 10-32 as an example.

Step 1 In the node that will subtend the SNCP ring, install the STM-N cards that will serve as the SNCP trunk cards (node 3, slots 6 and 13).

Step 2 Attach fibers from these cards to the SNCP trunk cards on the SNCP. In Figure 10-32, slot 6 node 3 connects to slot 13/node 5, and slot 13 connects to slot 6/node 6.

Step 3 From node view, click the **Provisioning > SDH DCC** tab.

Step 4 Click **Create**.

Step 5 In the Create SDCC Terminations dialog box, click the slot and port that will carry the SNCP ring.

Step 6 Choose the **Set to IS, if allowed** option. This places the trunk card ports in service.

> **NOTE** Choose the port state option that fits your network requirements. You can choose the **OOS-MT** option if SNCP is provisioned prior to the physical installation of the cards and fiber. This option enables you to avoid DCC termination alarms until all DCCs are configured on the SNCP ring.

Step 7 (Optional) Under the port state options, you can also enable or disable OSPF on the DCC according to your requirements.

Step 8 Click **OK**. The chosen slots/ports display in the SDCC Terminations section.

Step 9 Follow Steps 1 to 8 for other nodes on the SNCP ring.

Step 10 Go to network view to view the subtending ring.

Subtend an MS-SPRing from an SNCP Ring

This procedure requires an established SNCP ring and one SNCP node with STM-N cards and fibers to connect to the MS-SPRing. The procedure also assumes you can set up an MS-SPRing. For MS-SPRing setup procedures, see the "Provision the MS-SPRing" section.

The procedure described here uses Figure 10-33 as an example.

Step 1 In the node that will subtend the MS-SPRing, install the STM-N cards that will serve as the MS-SPRing trunk cards (node 3, slots 6 and 13).

Step 2 Attach fibers from these cards to the MS-SPRing trunk cards on the MS-SPRing nodes. In Figure 10-33, slot 6/node 3 connects to slot 13/node 5, and slot 13 connects to slot 6/node 6.

Step 3 From node view, click the **Provisioning > SDH DCC** tab.

Step 4 Click **Create**.

Step 5 In the Create SDCC Terminations dialog box, click the slot and port that will carry the MS-SPRing.

Step 6 Choose the **Set to IS, if allowed** option. This places the trunk card ports in service.

Figure 10-33 *MS-SPRing Subtending from SNCP Ring*

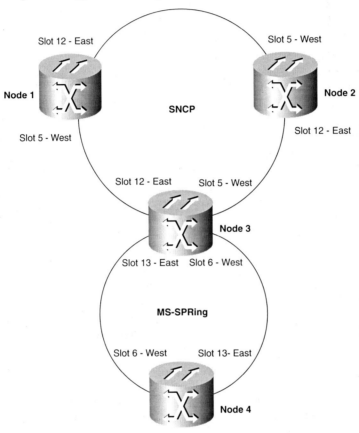

NOTE	Choose the port state option that fits your network requirements. You can choose the **OOS-MT** option if the MS-SPRing is provisioned prior to the physical installation of the cards and fiber. This option enables you to avoid DCC termination alarms until all DCCs are configured on the MS-SPRing.

Step 7 (Optional) Under the port state options, you can also enable or disable OSPF on the DCC according to your requirements.

Step 8 Click **OK**.

Step 9 The chosen slots/ports display under SDCC Terminations.

Step 10 Configure the MS-SPRing. See the "Provision the MS-SPRing" procedure.

Step 11 Follow Steps 1 to 10 for the other nodes that will be in the MS-SPRing.

Step 12 Display the network view to see the subtending ring.

Subtend an MS-SPRing from an MS-SPRing

The ONS 15454 SDH can support two MS-SPRings on the same node. This capability enables you to deploy an ONS 15454 SDH in applications requiring SDH DCS (digital cross-connect system) or multiple SDH ADMs. After subtending two MS-SPRings, you can route circuits from nodes in one ring to nodes in the second ring. This procedure requires an established MS-SPRing and one MS-SPRing node with STM-N cards and fibers to carry the MS-SPRing. The procedure also assumes you know how to set up an MS-SPRing. For MS-SPRing setup procedures, see the "Provision the MS-SPRing" section.

The procedure described here uses Figure 10-34 as an example.

Figure 10-34 *MS-SPRing Subtending from MS-SPRing*

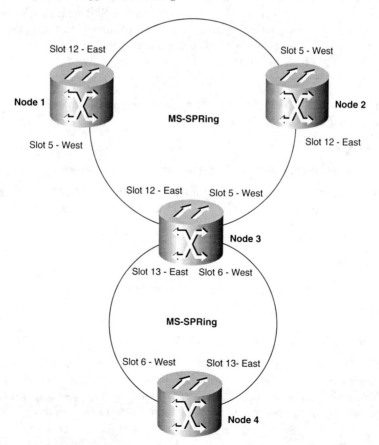

Step 1	In the node that will subtend the MS-SPRing, install the STM-N cards that will serve as the MS-SPRing trunk cards (node 3, slots 6 and 13).
Step 2	Attach fibers from these cards to the MS-SPRing trunk cards on the MS-SPRing nodes. In Figure 10-34, node 3/slot 6 connects to node 5/slot 13, and slot 13 connects to node 6/slot 6.
Step 3	From the node view, click the **Provisioning > SDH DCC** tab.
Step 4	Click **Create**.
Step 5	In the Create SDCC Terminations dialog box, click the slot and port that will carry the MS-SPRing.
Step 6	Choose the **Set to IS, if allowed** option. This places the trunk card ports in service.

> **NOTE** Choose the port state option that fits your network requirements. You can choose the **OOS-MT** option if MS-SPRing is provisioned prior to the physical installation of the cards and fiber. This option enables you to avoid DCC termination alarms until all DCCs are configured on the MS-SPRing.

Step 7	(Optional) Under the port state options, you can also enable or disable OSPF on the DCC according to your requirements.
Step 8	Click **OK**.
Step 9	The chosen slots/ports display under SDCC Terminations.
Step 10	To configure the MS-SPRing, use the "Provision the MS-SPRing" procedure. The subtending MS-SPRing must have a ring ID that differs from the ring ID of the first MS-SPRing.
Step 11	Follow Steps 1 to 10 for the other nodes that will be in the subtending MS-SPRing.
Step 12	Display the network view to see the subtending ring.

Linear ADM Configurations

ONS 15454 SDHs can be configured as linear ADMs by configuring one set of STM-N cards as the working path and a second set as the protect path. Unlike rings, linear (point-to-point) ADMs require that the STM-N cards at each node be in 1+1 protection to ensure that a break to the working line is automatically routed to the protect line. Figure 10-35 shows three ONS

15454 SDHs in a linear ADM configuration. Working traffic flows from slot 5/node A to slot 5/node B, and from slot 13/node B to slot 13/node C. The protect path is created by placing slot 5 in 1+1 protection with slot 4 at nodes 1 and 2, and slot 13 in 1+1 protection with slot 14 at nodes 2 and 3.

Figure 10-35 *Linear ADM*

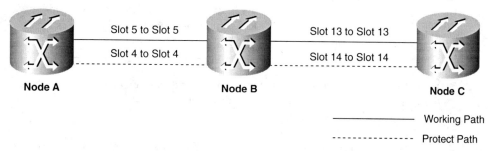

Create a Linear ADM

The following steps must be completed for each node that will be included in the linear ADM.

Refer to the illustration in Figure 10-35 for the procedure described here.

Step 1 Complete the general setup information for the node. For procedures, see the "Set Up Basic Node Information" section.

Step 2 Set up the network information for the node. For procedures, see the "Set Up Network Information" section.

Step 3 Set up 1+1 protection for the STM-N cards in the Linear ADM. In Figure 10-35, slots 5 and 13 are the working ports and slots 4 and 14 are the protect ports. In this example, one protection group is setup for node A (slots 4 and 5), two for node B (slots 4 and 5, and 13 and 14), and one for node C (slots 13 and 14). To create protection groups, see the "Creating a Protection Group" section.

Step 4 For STM-N ports connecting ONS 15454 SDHs, set up the SDH DCC terminations:

(a) Log in to a linear ADM node and choose the **Provisioning > SDH DCC** tab.

(b) In the SDCC Terminations section, click **Create**.

> **NOTE** Terminating nodes A and C in Figure 10-35 will have one SDCC, whereas intermediate nodes (node B) will have two SDCCs.

Step 5 Choose the **Set to IS, if allowed** option. This places the trunk card ports in service.

> **NOTE** Choose the port state option that fits your network requirements. You can choose the **OOS-MT** option if the linear ADM is provisioned prior to the physical installation of the cards and fiber. This option enables you to avoid DCC termination alarms until all DCCs are configured on the Linear ADM.

Step 6 (Optional) Under the port state options, you can also enable or disable OSPF on the DCC according to your requirements.

Step 7 Click **OK**.

The slots/ports appear in the SDCC Terminations list.

Step 8 Complete Step 1 to 7 at each node.

Step 9 Use the "Provisioning ONS 15454 SDH Timing" section to set up the node timing. If a node uses line timing, set the working STM-N card as the timing source.

Convert a Linear ADM to an SNCP Ring

The following procedures describe how to convert a three-node linear ADM to an SNCP ring. You need an SDH test set to monitor traffic while you perform these procedures. This procedure affects service and must be performed only during a scheduled outage.

Refer to the illustration in Figure 10-36 for the procedure described here:

Step 1 Start CTC and log in to one of the nodes that you want to convert from linear to ring.

Step 2 Click the **Maintenance > Protection** tab.

Step 3 Under Protection Groups, choose the 1+1 protection group, which is the group that supports the 1+1 span cards.

Step 4 Under the chosen group, verify that the working slot port is shown as Working-Active. If so, go to Step 5. If the working slot says Working-Standby and the protect slot says Protect-Active, switch traffic to the working slot:

  (a) Under Selected Group, choose the protect slot (that is, the slot that says Protect-Active).

  (b) From the Switch Commands, choose **Manual**.

(c) Click **Yes** in the confirmation dialog box.

(d) Under Selected Group, verify that the working slot port says Working-Active. If so, continue to Substep E. If not, clear the conditions that prevent the card from carrying working traffic before proceeding.

(e) From the Switch Commands, choose **Clear**. A Confirm Clear Operation dialog box displays.

(f) Click **Yes** in the confirmation dialog box.

Figure 10-36 *Linear ADM to SNCP Conversion*

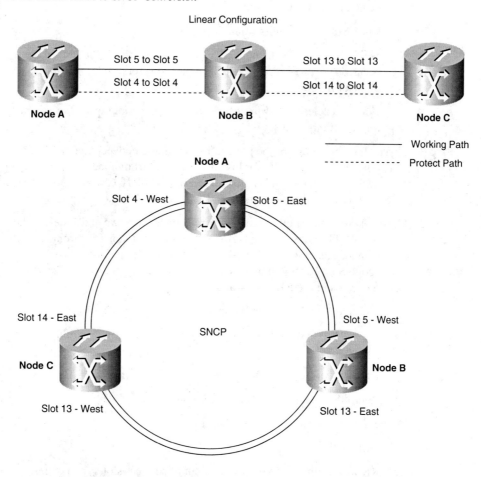

Step 5 Repeat Step 4 for each group in the 1+1 Protection Groups list at all nodes that will be converted.

Step 6 For each node, delete the 1+1 STM-N protection group that supports the linear ADM span:

> **NOTE** Deleting a 1+1 protection group can cause unequipped path (UNEQ-P) alarms to occur.

  (a) Click the **Provisioning > Protection** tab.

  (b) From the Protection Groups list, select the 1+1 group you want to delete. Click **Delete**.

  (c) Click **Yes** in the confirmation dialog box.

  (d) Verify that no traffic disruptions are indicated on the test set. If disruptions occur, do not proceed. Recreate the protection group and isolate the cause of the disruption.

  (e) Continue deleting 1+1 protection groups while monitoring the existing traffic with the test set.

Step 7 Physically remove one of the protect fibers running between the middle and end nodes. In Figure 10-36, for instance, the fiber from node B/slot 14 to node C/slot 14 is removed. The corresponding STM-16 card will go into an LOS condition for that fiber and port.

Step 8 Physically reroute the other protect fiber to connect the two end nodes. In the example, the fiber between node A/slot 4 and node B/slot 4 is rerouted to connect node A/slot 4 to node C/slot 14. If you leave the STM-N cards in place, go to Step 13. If you remove the cards, complete Steps 9 to 12. In this example, cards in node B/slots 4 and 14 are removed.

Step 9 Place the cards in slots 4 and 14 out of service for node B:

  (a) Display the first card in card view and choose the **Provisioning > Line** tab.

  (b) Under Status, choose **Out of Service**. Click **Apply**.

  (c) Repeat Substeps A and B for the second card.

Step 10 Delete the equipment records for the cards:

  (a) Display the node view.

  (b) Right-click the card you just took out of service (slot 4) and choose **Delete Card**.

  (c) Click **Yes** in the confirmation dialog box.

  (d) Repeat Substeps A through C for the second card (slot 14).

Step 11 Save all circuit information.

    (a) In node view, choose the **Provisioning > Circuits** tab.

    (b) Record the circuit information using one of the following procedures:

- From the File menu, choose **Print** to print the circuits table.
- From the File menu, choose **Export** to export the circuit data in HTML, CSV (comma-separated values), or TSV (tab-separated values). Click **OK** and save the file in a temporary directory.

Step 12 Remove the STM-N cards that are no longer connected to the end nodes (slots 4 and 14 in the example).

Step 13 Display one of the end nodes (node A or node C in the example).

Step 14 Click the **Provisioning > SDH DCC** tab.

Step 15 In the SDCC Terminations section, click **Create**.

Step 16 In the Create SDCC Terminations dialog box, choose the slot/port that had been the protect slot in the linear ADM—for example, for node A, this would be slot 4/port 1 (STM-16).

Step 17 Click **OK**. An EOC SDCC alarm will occur until an SDCC termination is created on the adjacent node.

Step 18 Go to the node on the opposite end (node C in the example) and repeat Steps 14 through 17.

Step 19 Delete and re-enter the circuits one at a time. Deleting circuits affects service.

You can create the circuits automatically or manually. However, circuits must be protected. When they were built in the linear ADM, they were protected by the protect path on node A/slot 4 to node B/slot 4 to node C/slot 14. With the new SNCP ring, circuits should also be created with protection. Deleting the first circuit and re-creating it to the same card port should restore the circuit immediately.

Step 20 Monitor your SDH test set to verify that the circuit was deleted and restored.

Step 21 You should also verify that the new circuit path for the clockwise (CW) fiber from node 1 to node 3 works. To do this, switch to network view and move your cursor to the green span between nodes 1 and 3. Although the cursor shows only the first circuit created, do not panic if the other circuits are not present. Verify with the SDH test set that the original circuits and the new circuits are operational. The original circuits were created on the counterclockwise linear path.

Step 22 Go to the network map to view the newly created ring.

Convert a Linear ADM to an MS-SPRing

The following procedures describe how to convert a three-node linear ADM to an MS-SPRing. You need an SDH test set to monitor traffic while you perform these procedures. This procedure affects service and must be performed only during a scheduled outage.

Refer to the illustration in Figure 10-37 for the procedure described here.

Figure 10-37 *Linear ADM to MS-SPRing Conversion*

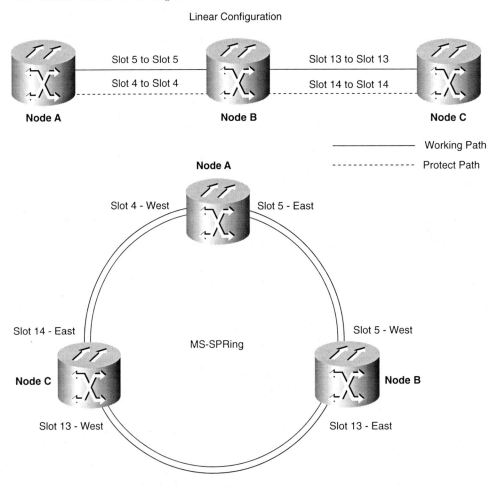

Step 1 Start CTC and log in to one of the nodes that you want to convert from linear to ring.

Step 2 Click the **Maintenance > Protection** tab.

Step 3 Under Protection Groups, choose the 1+1 protection group, which is the group that supports the 1+1 span cards.

Step 4 Under Selected Group, verify that the working slot/port is shown as Working-Active. If so, go to Step 5. If the working slot says Working-Standby and the protect slot says Protect-Active, switch traffic to the working slot:

(a) Under Selected Group, select the protect slot (that is, the slot that says Protect-Active).

(b) From the Switch Commands, choose **Manual**.

(c) Click **Yes** in the confirmation dialog box.

(d) Verify that the working slot is carrying traffic. If it is, continue to Substep E. If not, clear the conditions that prevent the card from carrying working traffic before proceeding.

(e) From the Switch Commands, choose **Clear**. A Confirm Clear Operation dialog box displays.

(f) Click **Yes** in the confirmation dialog box.

Step 5 Repeat Step 4 for each group in the 1+1 Protection Groups list at all nodes that will be converted.

Step 6 For each node, delete the 1+1 STM-N protection group that supports the linear ADM span:

(a) Click the **Provisioning > Protection** tab.

(b) From the Protection Groups list, choose the group you want to delete. Click **Delete**.

(c) Click **Yes** in the confirmation dialog box.

(d) Verify that no traffic disruptions are indicated on the SDH test set. If disruptions occur, do not proceed. Add the protection group and begin troubleshooting procedures to find out the cause of the disruption.

NOTE Deleting a 1+1 protection group can cause unequipped path (UNEQ-P) alarms to occur.

Step 7 Physically remove one of the protect fibers running between the middle and end nodes. In the example, the fiber running from node B/slot 14 to node C/slot 14 is removed. The corresponding end-node trunk card will display an LOS alarm.

Step 8 Physically reroute the other protect fiber so it connects the two end nodes. In the example, the fiber between node A/slot 4 and node B/slot 4 is rerouted to connect node A/slot 4 to node C/slot 14. If you leave the STM-N cards in place, go to Step 13. If you remove the cards, complete Steps 9 to 12. (In this example, cards in node B/slots 4 and 14 are removed.)

Step 9 In the middle node B, place the cards in slots 4 and 14 out of service:

(a) Display the first card in card view, and then choose the **Provisioning > Line** tab.

(b) Under Status, select **Out of Service**. Click **Apply**.

(c) Repeat Substeps A and B for the second card.

Step 10 Delete the equipment records for the cards:

(a) From the view menu, choose **Node View**.

(b) Right-click the card you just took out of service (slot 4) and select **Delete Card**. (You can also go to the Inventory tab, choose the card, and click **Delete**.)

(c) Click **Yes** in the confirmation dialog box.

(d) Repeat Substeps A through C for the second card (slot 13).

Step 11 Save all circuit information:

(a) In node view, choose the **Provisioning > Circuits** tab.

(b) Record the circuit information using one of the following procedures:

- From the File menu, choose **Print** to print the circuits table.

- From the File menu, choose **Export** to export the circuit data in HTML, CSV, or TSV. Click **OK** and save the file in a temporary directory.

Step 12 Remove the STM-N cards that are no longer connected to the end nodes (slots 4 and 14 in the example).

Step 13 Log in to an end node. In node view, click the **Provisioning > SDH DCC** tab.

Step 14 In the SDCC Terminations section, click **Create**.

Step 15 Highlight the slot that is not already in the SDCC Terminations list (in this example, port 1 of slot 4 [STM-16] on node A).

Step 16 Click **OK**. An EOC SDCC alarm will occur until the DCC is created on the other node; in the example, node C/slot 14.

Step 17 Display the node on the opposite end (node C) and repeat Steps 13 through 16.

Step 18 For circuits running on a MS-SPRing protect VC4 (VC4 3 to 4 for an STM-4 MS-SPRing, VC4s 9 to 16 for an STM-16 MS-SPRing, and VC4s 33 to 64 for an STM-64), delete and recreate the circuit:

(a) Delete the first circuit by clicking the **Circuits** tab. Choose the circuit, click **Delete**, and click **Yes** when prompted.

(b) Recreate the circuit on VC4s 3 to 4 (for an STM-4 MS-SPRing), VC4s 9 to 16 (for an STM-16 MS-SPRing), or VC4s 33 to 64 (for an STM-64 MS-SPRing) on the fiber that served as the protect fiber in the linear ADM. During circuit creation, uncheck **Route Automatically** and **Fully Protected Path** in the Circuit Creation dialog box so that you can manually route the circuit on the appropriate VC4s. See the "Create a Unidirectional Circuit with Multiple Drops" procedure for more information.

(c) Repeat Substeps A and B for each circuit residing on an MS-SPRing protect VC4. Deleting circuits affects traffic.

Step 19 Follow all procedures in the "Provision the MS-SPRing" section to configure the MS-SPRing. The ring should have an east-west logical connection. Although it may not physically be possible to connect the STM-N cards in an east-west pattern, it is strongly recommended. If the network ring that is already passing traffic does not provide the opportunity to connect fiber in this manner, logical provisioning can be performed to satisfy this requirement. Be sure to assign the same ring ID and different node IDs to all nodes in the MS-SPRing. Do not accept the MS-SPRing ring map until all nodes are provisioned. An E-W Mismatch alarm will occur until all nodes are provisioned.

Step 20 Display the network map to view the newly created ring.

Extended SNCP Mesh Networks

ESCNP mesh networks include multiple ONS 15454 SDH topologies, such as a combination of linear ADM and SNCP rings that extend the protection provided by a single SNCP ring to the meshed architecture of several interconnecting rings. In an ESCNP mesh, circuits travel diverse paths through a network of single or multiple meshed rings. When you create circuits, you can have CTC automatically route circuits across the ESCNP mesh, or you can manually route them. You can also choose levels of circuit protection. If you choose full protection, however, CTC creates an alternate route for the circuit in addition to the main route. The second route follows a unique path through the network between the source and destination and sets up a second set of cross-connections.

In Figure 10-38, for example, a circuit is created from node A to node G. CTC determines that the shortest route between the two nodes passes through node B shown by the dotted line, and automatically creates cross-connections at nodes A, B, and G to provide the primary circuit path.

Figure 10-38 *Extended SNCP Mesh*

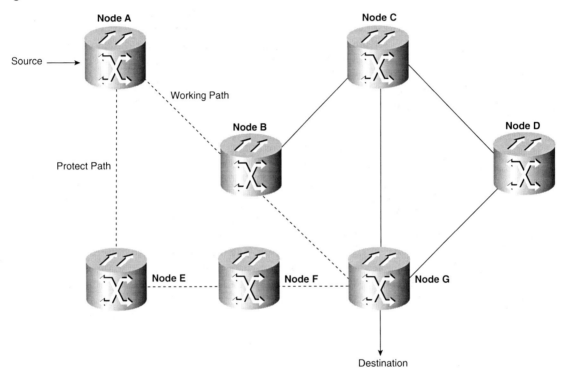

If full protection is selected, CTC creates a second unique route between nodes A and G, which, in this example, passes through nodes E and F. Cross-connections are automatically created at nodes A, E, F, and G, shown by the dashed line. If a failure occurs on the primary path, traffic switches to the second circuit path. In this example, node G switches from the traffic coming in from node B to the traffic coming in from node F and service resumes. The switch occurs within 50 ms.

SDH Circuit Provisioning

Provisioning an SNCP, MS-SPRing, linear ADM, or ESNCP mesh network is not complete without assigning circuits for transport of user traffic. This section explains how to create and administer various types of Cisco ONS 15454 SDH circuits and tunnels. DCC tunnels can also be created to tunnel third-party equipment signaling through ONS 15454 SDH networks. Cross-connects are the connections that occur within a single ONS 15454 SDH that allow a circuit to enter on one port and exit on another, whereas the term *circuit* refers to the series of connections from a traffic source (where traffic ingresses the ONS network) to the drop or destination (where traffic egresses an ONS network). Various types of SDH circuits can be created using the ONS 15454 SDH including the following:

- VC high-order path circuits
- VC low-order path tunnels for port grouping
- Multiple drops for unidirectional circuits
- Monitor circuits
- DCC tunnels

Various types of VC high-order path circuits and VC low-order path tunnels can be provisioned across and within ONS 15454 SDH nodes with different attributes assigned to them. These attributes include the following:

- One-way, half-duplex circuits
- Two-way, full-duplex circuits
- Broadcast circuits
- Circuit size
- Revertive or nonrevertive
- Automatic circuit routing
- Manual circuit routing
- Circuit path protection
- Protected sources and destinations

Circuits can be routed automatically or can be manually routed. These circuits could be unidirectional or bidirectional.

Protection channel access (PCA) circuits can be provisioned to carry traffic on MS-SPRing protection channels when conditions are fault-free. Traffic routed on MS-SPRing protection channels is also known as *extra traffic*. Extra traffic has lower priority than the traffic on the working channels and has no means for protection. During ring or span switches, protection channel circuits are preempted and squelched. In a two-fiber STM-16 MS-SPRing, for example, VC4s 9 through 16 can carry extra traffic when no ring switches are active, but protection channel circuits on these VC4s are preempted when a ring switch occurs. When the conditions that caused the ring switch are corrected and the ring switch removed, protection channel circuits are restored (assuming the MS-SPRing is provisioned as revertive).

Provisioning traffic on MS-SPRing protection channels is performed during circuit provisioning. If MS-SPRings are provisioned as nonrevertive, protection channel circuits will not be restored following a ring or span switch until the MS-SPRing is manually switched. It is important to remember that protection channel circuits will be routed on working channels when you upgrade an MS-SPRing, either from a two-fiber to a four-fiber, or from one optical speed to a higher one. If you upgrade a two-fiber STM-16 MS-SPRing to an STM-64, for example, VC4s 9 to 16 on the STM-16 MS-SPRing become working channels on the STM-64 MS-SPRing.

The CTC autorange feature eliminates the need to individually build circuits of the same type, because CTC can create additional sequential circuits if you specify the number of circuits you need and build the first circuit. The additional circuits will inherit the attributes of the first circuit created. This is known as *A-to-Z provisioning* in CTC jargon. CTC enables you to provision circuits even before the cards can be installed. To provision an empty slot, right-click it and choose a card from the shortcut menu. However, circuits will not carry traffic until the cards are installed and their ports are placed in service. These circuits will carry traffic only after a signal is received. If you want to route circuits on protected drops, you need to create the card protection groups before creating circuits.

Create an Automatically Routed Circuit

This section describes the steps used to create an automatically routed circuit:

Step 1   Log in to an ONS 15454 SDH on the network.

Step 2   Choose **Go to Network View** from the View menu.

Step 3   Choose the **Circuits** tab and click **Create**.

Step 4   In the Circuit Creation dialog box, complete the following fields:

- **Name**—Assign a name to the circuit. The name can be alphanumeric and up to 48 characters (including spaces). If you leave the Name field blank, CTC assigns a default name to the circuit. If you work for a LEC or IXC, this field is assigned with the customer circuit ID. If you create a monitor circuit, remember to keep the name within 44 characters, because CTC will add "_MON" (4 characters) to the end of the user assigned name.

- **Type**—Select the type of circuit you want to create: **VC_HO_Path _Circuit (HOP)**. The circuit type determines the circuit-provisioning options that can be provisioned.

- **Size**—Select the circuit size. VC high-order path circuits can be VC4, VC4-2c, VC4-3c, VC4-4c, VC4-8c, VC4-16c, or VC4-64c for optical cards and some Ethernet cards depending on the card type. Of the Ethernet cards, only the G-1000 can use VC4-3c and VC4-8c. The "c" indicates concatenated VC4s.

- **Bidirectional**—Check this box to create a two-way circuit. If this box is unchecked, CTC will create a unidirectional circuit.

- **Number of Circuits**—Enter the number of circuits you want to create. If you enter more than one, you can use autoranging to create the additional circuits automatically. Otherwise, CTC returns to the Circuit Source page after you create each circuit until you finish creating the number of circuits specified here.

- **Auto Ranged**—This option enables you to select the source and destination of one circuit. CTC automatically determines the source and destination for the remaining number of circuits and creates the circuits.

- **State**—Choose one of the following service states to apply to the circuit:

  - **OOS**—The circuit is out of service. Traffic is not passed on the circuit until it is in service.
  - **IS**—The circuit is in service.
  - **OOS-AINS**—(Default) The circuit is in service when it receives a valid signal. Until then, the circuit is out of service.
  - **OOS-MT**—The circuit is in maintenance state. The maintenance state does not interrupt traffic flow. This state suppresses alarms and conditions and allows loopbacks to be performed on the circuit. Use OOS-MT for circuit testing or to suppress circuit alarms temporarily. Change the state to IS, OOS, or OOS-AINS when testing is complete.

- **Apply to Drop Ports**—Check this box to apply the state chosen in the State field to the circuit source and destination ports. CTC applies the circuit state to the ports if the circuit is in full control of the port. If not, a Warning dialog box displays the ports where the circuit state could not be applied. If not checked, CTC will not change the state of the source and destination ports. LOS alarms display if in service (IS) ports are not receiving signals.

- **Inter-domain (UCP) SLA**—If the circuit will travel on a unified control plane (UCP) channel, enter the service level agreement (SLA) number. Otherwise, leave the field set to 0.

- **Protected Drops**—If this box is checked, CTC displays protected cards and ports only (1:1, 1:N, 1+1, or MS-SPRing protection) as choices for the circuit source and destination.

Figure 10-39 shows an example of provisioning the circuit attributes.

Figure 10-39 *Circuit Creation Attributes*

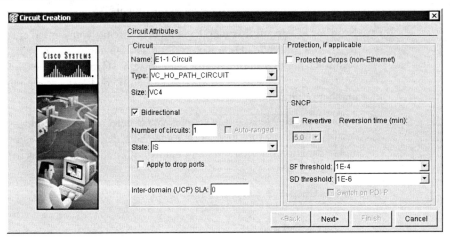

Step 5 If the circuit will be routed on an SNCP, set the following SNCP path selector defaults as follows:

- **Revertive**—Check this box if you want traffic to revert to the working path when the conditions that diverted it to the protect path are repaired. If Revertive is not chosen, traffic remains on the protect path after the switch.

- **Reversion Time**—If Revertive is checked, set the reversion time. This is the amount of time that will elapse before the traffic reverts to the working path. Traffic can revert when conditions causing the switch are cleared. (The default reversion time is 5 minutes.)

- **SF Threshold**—Choose from one E3, one E4, or one E5.

- **SD Threshold**—Choose from one E5, one E6, one E7, one E8, or one E9.

- **Switch on PDI-P**—Check this box if you want traffic to switch when a VC4 payload defect indicator is received (VC4 circuits only).

Step 6 Click **Next**.

Step 7 In the Circuit Source dialog box, set the circuit source. Options include node, slot, port, and VC4. The options that display depend on the circuit type and circuit properties you chose in Step 4 and the cards installed in the node. Click **Use Secondary Source** if you need to create an SNCP bridge-selector circuit entry point in a multivendor SNCP. Figure 10-40 shows an example of a Circuit Source dialog box.

Figure 10-40 *Circuit Source Creation*

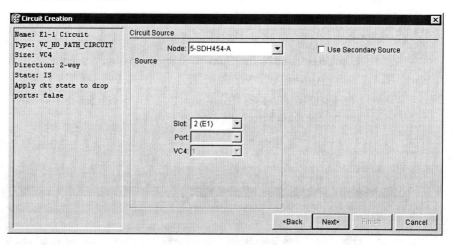

Step 8 Click **Next**.

Step 9 In the Circuit Destination dialog box, enter the appropriate information for the circuit destination. If the circuit is bidirectional, you can click **Use Secondary Destination** if you need to create an SNCP bridge-selector circuit destination point in a multivendor SNCP. To add secondary destinations to unidirectional circuits.

Figure 10-41 shows an example of a Circuit Destination dialog box.

Figure 10-41 *Circuit Destination Creation*

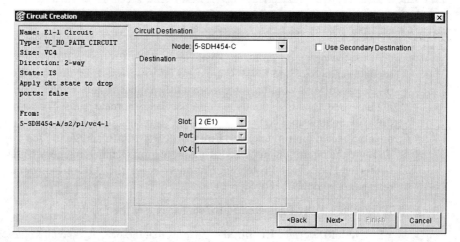

Step 10 Click **Next**.

Step 11 Under Circuit Routing Preferences, choose **Route Automatically**. The following options are available:

- **Using Required Nodes/Spans**—If selected, you can specify nodes and spans to include or exclude in the CTC-generated circuit route.

- **Review Route Before Creation**—If selected, you can review and edit the circuit route before the circuit is created.

Figure 10-42 shows an example of a Circuit Routing Preferences dialog box.

Figure 10-42 *Automatic Circuit Routing Preferences*

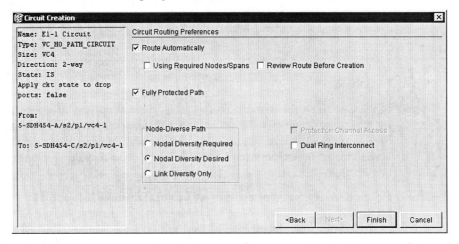

Step 12 Set the circuit path protection as follows:

To route the circuit on a protected path, leave **Fully Protected Path** checked (default) and go to Step 13. CTC creates a fully protected circuit route based on the path diversity option you choose. Fully protected paths might or might not have SNCP path segments with primary and alternate paths. The path diversity options apply only to SNCP path segments, if any exist. To create an unprotected circuit, uncheck **Fully Protected Path** and go to Step 14. To route the circuit on an MS-SPRing protection channel, uncheck **Fully Protected Path**, check **Protection Channel Access**, click **OK** on the Warning dialog box, and then go to Step 14.

Step 13 If you chose **Fully Protected Path**, choose one of the following:

- **Nodal Diversity Required**—Ensures that the primary and alternate paths within the extended SNCP mesh network portions of the complete circuit path are node diverse.

- **Nodal Diversity Desired**—Specifies that node diversity should be attempted; if node diversity is not possible, however, CTC creates link diverse paths for the extended SNCP mesh network portion of the complete circuit path.

- **Link Diversity Only**—Specifies that only link-diverse primary and alternate paths for extended SNCP mesh network portions of the complete circuit path are needed. The paths can be node diverse, but CTC does not check for node diversity.

Step 14 If you chose Using **Required Nodes/Spans** in Step 11, complete the following substeps. If you did not choose this check box, proceed to Step 15.

(a) Click **Next** to display the Circuit Route Constraints screen.

(b) On the circuit map, click a node or span (link).

(c) Click **Include** to include the node or span in the circuit. Click **Exclude** to exclude the node or span from the circuit. The order in which you select included nodes and spans sets the circuit sequence.

(d) Click **Spans** twice to change the circuit direction.

(e) Repeat Step C for each node or span you want to include or exclude.

(f) Review the circuit route. After you add the spans and nodes, you can use the **Up** and **Down** buttons to change the circuit routing order.

(g) Click **Remove** to remove a node or span.

(h) If you selected **Review Route Before Creation** in Step 11, click **Next** and follow Step 15. When you finish, click **Finish** and go to Step 16.

Step 15 If you chose **Review Route Before Creation** in Step 11, click **Next** to display the route. If you did not choose this check box, proceed to Step 16. The following options are available:

- To add or delete a circuit span, select a node on the circuit route.

- Blue arrows show the circuit route.

- Green arrows indicate spans that you can add.

- Click a span arrowhead, and then click **Include** to include the span or **Remove** to remove the span.

Step 16 After you click **Finish**, CTC creates the circuit and returns to the Circuits window. If you entered more than 1 in the Number of Circuits field in Step 4, the Circuit Source dialog box displays so that you can create the remaining circuits. If Auto Ranged is checked, CTC automatically creates the number of sequential circuits that you entered in Number of Circuits. Otherwise, proceed to Step 17.

Step 17 If you left the ports out of service in Step 4, the cards must be installed and the ports placed in service before circuits will carry traffic.

Create a Manually Routed Circuit

This section describes the steps used to create a manually routed circuit:

Step 1 Log in to an ONS 15454 SDH on the network.

Step 2 Choose **Go to Network View** from the view menu.

Step 3 Choose the **Circuits** tab and click **Create**.

Step 4 In the Circuit Creation dialog box, complete the following fields:

- **Name**—Assign a name to the circuit. The name can be alphanumeric and up to 48 characters (including spaces). If you leave the Name field blank, CTC assigns a default name to the circuit. If you work for an LEC or IXC, this field is assigned with the customer circuit ID. If you create a monitor circuit, remember to keep the name within 44 characters because CTC will add "_MON" (4 characters) to the end of the user assigned name.

- **Type**—Select the type of circuit you want to create: **VC_HO_Path_Circuit (HOP)**. The circuit type determines the circuit-provisioning options that can be provisioned.

- **Size**—Select the circuit size. VC high-order path circuits can be VC4, VC4-2c, VC4-3c, VC4-4c, VC4-8c, VC4-16c, or VC4-64c for optical cards and some Ethernet cards depending on the card type. Of the Ethernet cards, only the G-1000 can use VC4-3c and VC4-8c. The "c" indicates concatenated VC4s.

- **Bidirectional**—Check this box to create a two-way circuit. If this box is unchecked, CTC will create an unidirectional circuit.

- **Number of Circuits**—Enter the number of circuits you want to create. If you enter more than 1, you can use autoranging to create the additional circuits automatically. Otherwise, CTC returns to the Circuit Source page after you create each circuit until you finish creating the number of circuits specified here.

- **Auto Ranged**—Deselect the box if you do not want to create multiple circuits automatically. This check box is automatically selected when you enter more than 1 in the Number of Circuits field. Leave the box selected if you want to create multiple optical circuits with the same source and destination and you want CTC to create the circuits automatically.

- **State** — Select one of the following service states to apply to the circuit:
    - **OOS** — The circuit is out of service. Traffic is not passed on the circuit until it is in service.
    - **IS** — The circuit is in service.
    - **OOS-AINS** — (Default) The circuit is in service when it receives a valid signal. Until then, the circuit is out of service.
    - **OOS-MT** — The circuit is in maintenance state. The maintenance state does not interrupt traffic flow. This state suppresses alarms and conditions and allows loopbacks to be performed on the circuit. Use OOS-MT for circuit testing or to suppress circuit alarms temporarily. Change the state to IS, OOS, or OOS-AINS when testing is complete.

- **Apply to Drop Ports** — Check this box to apply the state chosen in the State field to the circuit source and destination ports. CTC applies the circuit state to the ports if the circuit is in full control of the port. If not, a Warning dialog box displays the ports where the circuit state could not be applied. If not checked, CTC will not change the state of the source and destination ports. LOS alarms display if in service (IS) ports are not receiving signals.

- **Inter-domain (UCP) SLA** — If the circuit will travel on a UCP channel, enter the SLA number. Otherwise, leave the field set to 0.

- **Protected Drops** — If this box is checked, CTC displays protected cards and ports only (1:1, 1:N, 1+1, or MS-SPRing protection) as choices for the circuit source and destination.

Step 5 If the circuit will be routed on an SNCP, set the following SNCP path selector defaults as follows:

- **Revertive** — Check this box if you want traffic to revert to the working path when the conditions that diverted it to the protect path are repaired. If Revertive is not chosen, traffic remains on the protect path after the switch.

- **Reversion Time** — If Revertive is checked, set the reversion time. This is the amount of time that will elapse before the traffic reverts to the working path. Traffic can revert when conditions causing the switch are cleared. (The default reversion time is 5 minutes.)

- **SF Threshold** — Choose from one E3, one E4, or one E5.

- **SD Threshold** — Choose from one E5, one E6, one E7, one E8, or one E9.

- **Switch on PDI-P**—Check this box if you want traffic to switch when a VC4 payload defect indicator is received (VC4 circuits only).

Step 6 Click **Next**.

Step 7 In the Circuit Source dialog box, set the circuit source. Options include node, slot, port, and VC4. The options that display depend on the circuit type and circuit properties you chose in Step 4 and the cards installed in the node.

> **NOTE** E1 cards use VC4 circuits. All 12 of the E1 ports use VC4 bandwidth.

Step 8 Click **Use Secondary Source** if you need to create an SNCP bridge-selector circuit entry point in a multivendor SNCP.

Step 9 Click **Next**.

Step 10 In the Circuit Destination dialog box, enter the appropriate information for the circuit destination. If the circuit is bidirectional, you can click **Use Secondary Destination** if you need to create an SNCP bridge-selector circuit destination point in a multivendor SNCP.

Step 11 Click **Next**.

Step 12 Under Circuit Routing Preferences, deselect **Route Automatically**. Figure 10-43 shows an example of a Circuit Routing Preferences dialog box.

Figure 10-43 *Manual Circuit Routing Preferences*

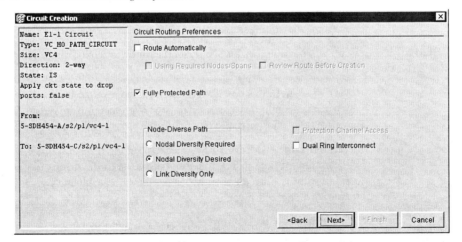

Step 13 Set the circuit path protection as follows:

To route the circuit on a protected path, choose **Fully Protected Path** and go to Step 14. CTC creates a fully protected circuit route based on the path diversity option you choose. Fully protected paths might or might not have SNCP path segments with primary and alternate paths. The path diversity options apply only to SNCP path segments, if any exist. To create an unprotected circuit, uncheck **Fully Protected Path** and go to Step 15. If you intend to route the circuit on an MS-SPRing protection channel, uncheck **Fully Protected Path**, check **Protection Channel Access**, click **OK** in the Warning dialog box, and go to Step 15. You must remember that circuits routed on PCA are not protected and are preempted during MS-SPRing and span switches.

Step 14 If you chose **Fully Protected Path** in Step 13, choose one of the following. If not, go to Step 15.

- **Nodal Diversity Required**—Ensures that the primary and alternate paths within the extended SNCP mesh network portions of the complete circuit path are node diverse.

- **Nodal Diversity Desired**—Specifies that node diversity is preferred, but if node diversity is not possible, CTC creates link-diverse paths for the extended SNCP mesh network portion of the complete circuit path.

- **Link Diversity Only**—Specifies that only link-diverse primary and alternate paths for extended SNCP mesh network portions of the complete circuit path are needed. The paths might be node diverse, but CTC does not check for node diversity.

Step 15 Click **Next**. The Route Review and Edit screen is displayed for you to manually route the circuit. The green arrows pointing from the source node to other network nodes indicate spans that are available for routing the circuit.

Step 16 Set the circuit route:

(a) Click the arrowhead of the span you want the circuit to travel.

(b) If you want to change the source VC4, change it in the Source VC4 fields.

(c) Click **Add Span**. The span is added to the Included Spans list, and the span arrow turns blue.

Step 17 Repeat Step 16 until the circuit is provisioned from the source to the destination node. When provisioning a protected circuit, you need to choose only one path of MS-SPRing or 1+1 spans from the source to the drop. In SNCP, you must choose both paths around the ring for the circuit to be protected.

Step 18 When the circuit is provisioned, click **Finish**. If you entered more than 1 in the Number of Circuits field in the Circuit Attributes dialog box in Step 4, the Circuit Source dialog box displays so that you can create the remaining circuits.

Step 19 If you left the ports out of service in Step 4, the cards must be installed and the ports placed in service before circuits will carry traffic.

Creating VC Low-Order Path Tunnels for Port Grouping

This section explains how to create VC low-order path tunnels for the E3 and DS3I cards. VC low-order path tunnels (VC_LO_PATH_TUNNEL) are automatically set to bidirectional with port grouping enabled. Three ports form a port group. In one E3 or one DS3I card, for example, there are four port groups: ports 1 to 3 = PG1, ports 4 to 6 = PG2, ports 7 to 9 = PG3, and ports 10 to 12 = PG4. CTC shows VC3-level port groups, but the XC10G creates only VC4-level port groups. VC4 tunnels must be used to transport VC3 signal rates. Circuits can be provisioned before or after the cards are installed. Tunnels are routed automatically. The following rules apply to port-grouped circuits:

- A port group goes through a VC_LO_PATH_TUNNEL circuit, with a set size of VC4.
- The circuit must be bidirectional and cannot use multiple drops.
- The circuit number must be set to 1.
- The Auto Ranged field must be set to Yes.
- The Use Secondary Destination field must be set to No.
- The Route Automatically field must be set to Yes.
- Monitor circuits cannot be created on a VC3 circuit in a port group.
- Circuits assigned to a state other than OOS using an E1, E3, and DS3I card for the source and destination ports can change to IS even though a signal is not present on the ports. Some cross-connects transition to IS, whereas others are OOS_AINS. A circuit state called *partial* might appear during a manual transition for some abnormal reason, such as a CTC crash, communication error, or one of the connections could not be changed.

The following procedure describes the steps for creating a low-order path tunnel for port grouping:

Step 1 Log in to a node on the network where you will create the circuit. Circuits can be created from the network view, node view, or card view.

Step 2 From the View menu, choose **Go to Network View**.

Step 3 Click the **Circuits** tab, and then click **Create**.

Step 4 In the Circuit Creation dialog box, complete the following fields:

- **Name**—Assign a name to the circuit. The name can be alphanumeric and up to 48 characters (including spaces). If you leave the Name field blank, CTC assigns a default name to the circuit. If you work for a LEC or IXC, this field would be assigned with the customer circuit ID. If you create a monitor circuit, remember to keep the name within 44 characters because CTC will add _MON (4 characters) to the end of the user-assigned name.

- **Type**—Choose the type of circuit you want to create: **VC_LO_Path_Tunnel**. The circuit type determines the circuit-provisioning options that can be provisioned.

- **Size**—Automatically set to VC4.

- **Bidirectional**—Automatically set to Bidirectional.

- **Number of Circuits**—This field automatically lists one port group. Three ports form one port group. In one E3 or one DS3I card, for example, there are four port groups: ports 1 to 3 = PG1, ports 4 to 6 = PG2, ports 7 to 9 = PG3, and ports 10 to 12 = PG4. Low-order path tunneling is performed at the VC3 level.

- **Auto Ranged**—The check box is automatically selected. If you choose the source and destination of one circuit, CTC automatically determines the source and destination for the remaining circuits in the Number of Circuits field and creates them. To determine the source and destination, CTC increments the most specific part of the endpoint. An endpoint can be a port or a VC4. If CTC cannot find a valid destination, or selects an endpoint that is already in use, CTC stops and allows you to either select a valid endpoint or cancel. If you select a valid endpoint and continue, autoranging begins after you click **Finish** for the current circuit.

- **State**—Choose one of the following service states to apply to the circuit:

  - **OOS**—The circuit is out of service. Traffic is not passed on the circuit until it is in service.
  - **IS**—The circuit is in service.
  - **OOS-AINS**—(Default) The circuit is in service when it receives a valid signal. Until then, the circuit is out of service.
  - **OOS-MT**—The circuit is in maintenance state. The maintenance state does not interrupt traffic flow. This state suppresses alarms and conditions and allows loopbacks to be performed on the circuit. Use OOS-MT for circuit testing or to suppress circuit alarms temporarily. Change the state to IS, OOS, or OOS-AINS when testing is complete.

- **Apply to Drop Ports**—Check this box to apply the state chosen in the State field to the circuit source and destination ports. CTC applies the circuit state to the ports if the circuit is in full control of the port. If not, a Warning dialog box identifies the ports where the circuit state could not be applied. If not checked, CTC will not change the state of the source and destination ports. LOS alarms display if IS ports are not receiving signals.

- **Inter-domain (UCP) SLA**—If the circuit will travel on a unified control plane (UCP) channel, enter the SLA number. Otherwise, leave the field set to 0.

- **Protected Drops**—If this box is checked, CTC displays protected cards and ports only (1:1, 1:N, 1+1, or MS-SPRing protection) as choices for the circuit source and destination.

Step 5 If the circuit will be routed on an SNCP, set the SNCP path selector defaults as follows:

- **Revertive**—Check this box if you want traffic to revert to the working path when the conditions that diverted the traffic to the protect path are repaired. If Revertive is not chosen, traffic remains on the protect path after the switch.

- **Reversion Time**—If Revertive is checked, set the reversion time. This is the amount of time that will elapse before the traffic reverts to the working path. Traffic can revert when conditions causing the switch are cleared. (The default reversion time is 5 minutes.)

- **SF Threshold**—Choose from one E3, one E4, or one E5.

- **SD Threshold**—Choose from one E5, one E6, one E7, one E8, or one E9.

- **Switch on PDI-P**—The check box is automatically deselected.

Step 6 Click **Next**.

Step 7 In the Circuit Source dialog box, set the circuit source. Options include node, slot, and VC4. The options displayed depend on the circuit type, the circuit properties chosen in Step 4, and the cards installed in the node.

Step 8 Click **Next**.

Step 9 In the Circuit Destination dialog box, enter the appropriate information for the circuit destination. Options include node, slot, and VC4. The options displayed depend on the circuit type, the circuit properties chosen in Step 4, and the cards installed in the node.

Step 10 Click **Next**. Under Circuit Routing Preferences, **Route Automatically** is chosen.

Step 11 Set the circuit path protection as follows:

To route the circuit on a protected path, leave Fully Protected Path checked (default) and go to Step 12. CTC creates a fully protected circuit route based on the path diversity option you choose. Fully protected paths might or might not have SNCP path segments with primary and alternate paths. The path diversity options apply only to SNCP path segments, if any exist.

To create an unprotected circuit, uncheck Fully Protected Path and go to Step 13.

To route the circuit on an MS-SPRing protection channel, uncheck Fully Protected Path, check Protection Channel Access, click **OK** on the Warning dialog box, and then go to Step 13.

> **NOTE** Circuits routed on PCA are not protected and are preempted during MS-SPRing and span switches.

Step 12 If you chose Fully Protected Path, choose one of the following. If you did not choose Fully Protected Path, go to Step 13.

- **Nodal Diversity Required**—Ensures that the primary and alternate paths within the extended SNCP mesh network portions of the complete circuit path are node diverse.

- **Nodal Diversity Desired**—Specifies that node diversity is preferred, but if node diversity is not possible, CTC creates link-diverse paths for the extended SNCP mesh network portion of the complete circuit path.

- **Link Diversity Only**—Specifies that only link-diverse primary and alternate paths for extended SNCP mesh network portions of the complete circuit path are needed. The paths might be node diverse, but CTC does not check for node diversity.

Step 13 Click **Finish**. CTC creates the circuit and returns to the Circuits window.

Step 14 If you left the ports out of service in Step 4, the cards must be installed and the ports placed in service before circuits will carry traffic.

Creating Multiple Drops for Unidirectional Circuits

Broadcast schemes incorporate unidirectional circuits with multiple drops. In broadcast scenarios, one source transmits traffic to multiple destinations, but traffic is not returned back to the source. When a unidirectional circuit is created, the card that does not have its backplane RX input terminated with a valid input signal generates a loss of service (LOS) alarm. To mask the alarm, an alarm profile suppressing the LOS alarm must be created and applied to the port that does not have its RX input terminated. This section describes the steps used to create multiple drops for unidirectional circuits:

Step 1 Create a circuit. To make it unidirectional, clear the **Bidirectional** check box on the Circuit Creation dialog box.

Step 2 After the unidirectional circuit is created, in node or network view choose the **Circuits** tab.

Step 3 Choose the unidirectional circuit and click **Edit**.

Step 4 On the Drops tab of the Edit Circuits dialog box, click **Create** or, if Show Detailed Map is chosen, right-click a node on the circuit map and choose **Add Drop**.

Step 5 On the Define New Drop dialog box, complete the appropriate fields to define the new circuit drop: node, slot, port, and VC4.

Step 6 Under Target Circuit State, choose one of the following service states to apply to the circuit:

- **OOS**—The circuit is out of service. Traffic is not passed on the circuit until it is in service.
- **IS**—The circuit is in service.
- **OOS-AINS**—The circuit is in service when it receives a valid signal; until then, the circuit is out of service.
- **OOS-MT**—The circuit is in a maintenance state. The maintenance state does not interrupt traffic flow; it suppresses alarms and conditions and allows loopbacks to be performed on the circuit. Use OOS-MT for circuit testing or to suppress circuit alarms temporarily. Change the state to IS, OOS, or OOS-AINS when testing is complete.
- **No Change**—(default) The circuit remains in the current state.

Step 7 Set the circuit path protection as follows:

To route the circuit on a protected path, leave **Fully Protected Path** checked (default) and go to Step 8. CTC creates a fully protected circuit route based on the path diversity option you choose. Fully protected paths might or might not have SNCP path segments with primary and alternate paths. The path diversity options apply only to SNCP path segments, if any exist.

To create an unprotected circuit, uncheck **Fully Protected Path** and go to Step 9.

To route the circuit on an MS-SPRing protection channel, uncheck **Fully Protected Path**, check **Protection Channel Access**, click **OK** on the Warning dialog box, and then go to Step 9. Circuits routed on PCA are not protected and are preempted during MS-SPRing and span switches.

Step 8 If you chose Fully Protected Path, choose one of the following. If you did not choose Fully Protected Path, go to Step 9.

- **Nodal Diversity Required**—Ensures that the primary and alternate paths within the extended SNCP mesh network portions of the complete circuit path are node diverse.

- **Nodal Diversity Desired**—Specifies that node diversity is preferred, but if node diversity is not possible, CTC creates link-diverse paths for the extended SNCP mesh network portion of the complete circuit path.

- **Link Diversity Only**—Specifies that only link-diverse primary and alternate paths for extended SNCP mesh network portions of the complete circuit path are needed. The paths might be node diverse, but CTC does not check for node diversity.

Step 9 Click **OK**.

Step 10 If you need to create additional drops, repeat Steps 4 through 9. If not, click **Close**.

Step 11 Verify the new drops on the Edit Circuit map.

If Show Detailed Map is selected, a "D" enclosed by circles appears on each side of the node graphic. If Show Detailed Map is not selected, "Drop #1, Drop #2" appear under the node graphic.

Creating Monitor Circuits

Secondary circuits can be set up to monitor traffic on primary bidirectional circuits. Figure 10-44 shows an example of a monitor circuit. At node 1, a VC4 is dropped from port 1 of an STM-1 card. To monitor the VC4 traffic, test equipment is plugged into port 2 of the STM-1 card and a monitor circuit to port 2 is provisioned in CTC. Circuit monitors are one-way. The monitor circuit in Figure 10-44 is used to monitor VC4 traffic received by port 1 of the STM-1 card.

Figure 10-44 *Monitor Circuit Schematic*

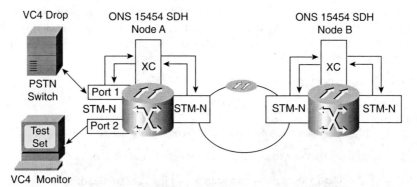

NOTE Monitor circuits cannot be used with EtherSwitch circuits. For unidirectional circuits, a drop must be created to the port where the test equipment is attached.

This section describes the steps used to create monitor circuits:

Step 1 Log in to CTC and choose the **Circuits** tab in node view.

Step 2 Choose the bidirectional circuit that you want to monitor. Click **Edit**.

Step 3 On the Edit Circuit dialog box, click the **Monitors** tab.

Step 4 The Monitors tab displays ports that you can use to monitor the circuit chosen in Step 2.

Step 5 On the Monitors tab, choose a port. The monitor circuit displays traffic coming into the node at the card/port you choose.

Step 6 Click **Create Monitor Circuit**.

Step 7 On the Circuit Creation dialog box, choose the destination node, slot, port, and VC4 for the monitored circuit. In the Figure 10-44 example, this is port 2 on the STM-1 card.

Step 8 If Use Secondary Destination is chosen, enter the slot, port, and VC4. Click **Next**.

Step 9 On the Circuit Creation dialog box confirmation, review the monitor circuit information. To route the monitor circuit on an MS-SPRing protection channel, check **Protection Channel Access** and click **OK** on the Warning dialog box. Click **Finish**.

Step 10 On the Edit Circuit dialog box, click **Close**. The new monitor circuit displays on the Circuits tab.

Searching for ONS 15454 SDH Circuits

CTC enables you to search for ONS 15454 SDH circuits based on circuit name. Searches can be conducted at the network, node, and card level. You can search for whole words and include capitalization as a search parameter. This section describes the steps used to search for ONS 15454 SDH circuits:

Step 1 Log in to CTC.

Step 2 Switch to the appropriate CTC view:

- Network view to conduct searches at the network level
- Node view to conduct searches at the network or node level
- Card view to conduct searches at the card, node, or network level

Step 3 Click the **Circuits** tab.

Step 4 If you are in node or card view, choose the scope for the search in the Scope field.

Step 5 Click **Search**.

Step 6 In the Circuit Name Search dialog box, complete the following:

- **Find What**—Enter the text of the circuit name you want to find.
- **Match Whole Word Only**—If checked, CTC selects circuits only if the entire word matches the text in the Find What field.
- **Match Case**—If checked, CTC selects circuits only when the capitalization matches the capitalization entered in the Find What field.
- **Direction**—Select the direction for the search. Searches are conducted up or down from the currently selected circuit.

Step 7 Click **Find Next**.

Step 8 Repeat Steps 6 and 7 until you are finished, and then click **Cancel**.

Editing SNCP Circuits

Use the Edit Circuits window to change SNCP selectors and switch protection paths. This window enables you to do the following:

- View the SNCP circuit's working and protection paths.
- Edit the reversion time.
- Edit the Signal Fail/Signal Degrade thresholds.
- Change PDI-P settings, perform maintenance switches on the circuit selector, and view switch counts for the selectors.
- Display a map of the SNCP circuits to better see circuit flow between nodes.

This section describes the steps used to edit SNCP circuits:

Step 1 Log in to the source or drop node of the SNCP circuit.

Step 2 Click the **Circuits** tab.

Step 3 Click the circuit you want to edit, and then click **Edit**.

Step 4 On the Edit Circuit window, click the **SNCP Selectors** tab.

Step 5 Edit the SNCP selectors:

- **Reversion Time**—Controls whether traffic reverts to the working path when conditions that diverted it to the protect path are repaired. If you select Never, traffic does not revert. Selecting a time sets the amount of time that will elapse before traffic reverts to the working path.

- **SF BER Level**—Sets the SNCP signal failure BER threshold (VC4 circuits only).
- **SD BER Level**—Sets the SNCP signal degrade BER threshold (VC4 circuits only).
- **PDI-P**—When checked, traffic switches if a VC4 payload defect indication is received (VC4 circuits only).
- **Switch State**—Switches circuit traffic between the working and protect paths. The color of the Working Path and Protect Path fields indicates the active path. Normally, the working path is green and the protect path is purple. If the protect path is green, working traffic has switched to the protect path.
  - **CLEAR**—Removes a previously set switch command.
  - **LOCKOUT OF PROTECT**—Prevents traffic from switching to the protect circuit path.
  - **FORCE TO WORKING**—Forces traffic to switch to the working circuit path, regardless of whether the path is error free.
  - **FORCE TO PROTECT**—Forces traffic to switch to the protect circuit path, regardless of whether the path is error free.
  - **MANUAL TO WORKING**—Switches traffic to the working circuit path when the working path is error free.
  - **MANUAL TO PROTECT**—Switches traffic to the protect circuit path when the protect path is error free.

NOTE The **FORCE** and **LOCKOUT** commands override normal protection switching mechanisms. Applying these commands incorrectly can cause traffic outages.

Step 6 Click **Apply**.

Creating a Path Trace

The SDH J1 path trace is a repeated, fixed-length string comprised of consecutive J1 bytes. This string can be used to monitor interruptions or changes to circuit traffic. Most ONS electrical cards can transmit and receive the J1 field, whereas optical cards can only receive it. Certain ONS cards do not support the J1 byte. Specific card documentation must be consulted to verify J1 byte support. The J1 path trace transmits a repeated, fixed-length string. If the string received at a circuit drop port does not match the string the port expects to receive, an alarm is raised.

There are two types of J1 bytes:

- **Low order (LO-J1)**—The electrical cards support LO-J1.
- **High order (HO-J1)**—The optical cards support HO-J1 (VC4) and cannot monitor the LO-J1 (VC3).

To set up a path trace on an ONS 15454 SDH circuit, follow the steps detailed here:

Step 1 Log in to a node on the network where you will create the path trace.

Step 2 From node view, click the **Circuits** tab.

Step 3 For the circuit you want to monitor, verify that the source and destination ports are on a card that can transmit and receive the path trace string. If neither port is on a transmit/receive card, you will not be able to complete this procedure. If one port is on a transmit/receive card and the other is on a receive-only card, you can set up the transmit string at the transmit/receive port and the receive string at the receive-only port, but you will not be able to transmit in both directions.

Step 4 Click the circuit you want to trace, and then click **Edit**.

Step 5 In the Edit Circuit window, check the **Show Detailed Map** box at the bottom of the window. A detailed map of the source and destination ports displays.

Step 6 Provision the circuit source transmit string:

(a) On the detailed circuit map, right-click the circuit source port and select **Edit Path Trace** from the shortcut menu.

(b) Select the format of the transmit string by clicking either the **16 byte** or the **64 byte** selection button.

(c) In the New Transmit String field, enter the circuit source transmit string. Enter a string that makes the source port easy to identify, such as the node IP address, node name, circuit name, or another string. If the New Transmit String field is left blank, the J1 transmits a string of null characters.

(d) Click **Apply**, and then click **Close**.

Step 7 Provision the circuit destination transmit string:

(a) In the Edit Circuit window, right-click the circuit destination port and select **Edit Path Trace** from the shortcut menu.

(b) In the New Transmit String field, enter the string that you want the circuit destination to transmit. Enter a string that makes the destination port easy to identify, such as the node IP address, node name, circuit name, or another string. If the New Transmit String field is left blank, the J1 transmits a string of null characters.

(c) Click **Apply**.

Step 8 Provision the circuit destination expected string:

(a) In the Circuit Path Trace window, enable the path trace expected string by choosing **Auto** or **Manual** from the Path Trace Mode drop-down menu.

- **Auto**—The first string received from the source port is the baseline. An alarm is raised when a string that differs from the baseline is received. Continue with Substep B.

- **Manual**—The string entered in the Current Expected String field is the baseline. An alarm is raised when a string that differs from the current expected string is received. Enter the string that the circuit destination should receive from the circuit source in the New Expected String field.

(b) Click the **Disable AIS on TIM-P** check box if you want to suppress the alarm indication signal when the path trace identifier mismatch path (TIM-P) alarm displays.

(c) Click **Apply**, and then click **Close**.

Step 9 Provision the circuit source expected string.

(a) In the Edit Circuit window, right-click the circuit source port and select **Edit Path Trace** from the shortcut menu.

(b) In the Circuit Path Trace window, enable the path trace expected string by selecting **Auto** or **Manual** from the Path Trace Mode drop-down menu.

- **Auto**—Uses the first string received from the port at the other end as the baseline string. An alarm is raised when a string that differs from the baseline is received. Continue with Substep C.

- **Manual**—Uses the Current Expected String field as the baseline string. An alarm is raised when a string that differs from the current expected string is received. Enter the string that the circuit source should receive from the circuit destination in the New Expected String field.

(c) Click the **Disable AIS on TIM-P** check box if you want to suppress the alarm indication signal when the TIM-P alarm displays.

(d) Click **Apply**, and then click **Close**.

Step 10 After you set up the path trace, the received string displays in the Received box on the path trace setup window. The following options are available:

- Click **Switch Mode** to toggle between ASCII and hexadecimal display.
- Click the **Reset** button to reread values from the port.

- Click **Default** to return to the path trace default settings. (Path trace mode is set to off and the new transmit and new expected strings are null.) Clicking Default will generate alarms if the port on the other end is provisioned with a different string. The expect and receive strings are updated every few seconds only if path trace mode is set to Auto or Manual.

When you display the detailed circuit window, path trace is indicated by an "M" (manual path trace) or an "A" (automatic path trace) at the circuit source and destination ports.

Monitor a Path Trace on STM-N Ports

This section describes the steps used to monitor a path trace on STM-N ports:

Step 1 Start CTC on a node in the network where path trace was provisioned on the circuit source and destination ports.

Step 2 Click **Circuits**.

Step 3 Choose the VC4 circuit that has path trace provisioned on the source and destination ports, and then click **Edit**.

Step 4 In the Edit Circuit window, click the **Show Detailed Map** box at the bottom of the window. A detailed circuit graphic showing source and destination ports is displays.

Step 5 On the detailed circuit map, right-click the circuit STM-N port and choose **Edit Path Trace** from the shortcut menu. The STM-N port must be on a receive-only card. If not, the Edit Path Trace menu item will not display.

Step 6 In the Circuit Path Trace window, enable the path trace expected string by choosing **Auto** or **Manual** from the Path Trace Mode drop-down menu:

- **Auto**—Uses the first string received from the port at the other end as the baseline string. An alarm is raised when a string that differs from the baseline is received. For STM-N ports, Auto is recommended, because Manual mode requires you to trace the circuit on the Edit Circuit window to determine whether the port is the source or destination path.

- **Manual**—Uses the Current Expected String field as the baseline string. An alarm is raised when a string that differs from the current expected string is received.

Step 7 If you set Path Trace Mode to Manual, enter the string that the STM-N port should receive in the New Expected String field. To do this, trace the circuit path on the detailed circuit window to determine whether the port is in the circuit source or destination path, and then set the New Expected String field to the string transmitted by the circuit source or destination. If you set Path Trace Mode to Auto, ignore the New Expected String field.

Step 8 The Disable AIS on TIM-P check box cannot be chosen.

> **NOTE** SDH Software Release 3.4 and earlier does not support changes to the disable AIS on TIM-P field. The STM-N path trace monitoring does not generate AIS on TIM-P.

Step 9 Click **Apply**, and then click **Close**.

Create a Half Circuit Using an STM-N Card as a Destination in an MS-SPRing or 1+1 Topology

The following procedure describes how to create STM-N circuits from a drop to an STM-N card on the same node in an MS-SPRing or 1+1 topology:

Step 1 From node view, go to the View menu and select **Go to Network View**.

Step 2 Click the **Circuits** tab, and then click **Create**.

Step 3 In the Circuit Creation dialog box, complete the following fields:

- **Name**—Assign a name to the circuit. The name can be alphanumeric and up to 48 characters (including spaces). If you leave the Name field blank, CTC assigns a default name to the circuit.

- **Type**—Choose **VC_HO_Path_Circuit (HOP)**. The circuit type determines the circuit-provisioning options that display. The E3 and DS3I cards must use VC low-order path tunnels.

- **Size**—Select the circuit size (VC_HO_Path_Circuits only). VC high-order path circuits can be VC4, VC4-2c, VC4-3c, VC4-4c, VC4-8c, VC4-16c, or VC4-64c for optical cards and some Ethernet cards depending on the card type. Of the Ethernet cards, only the G-1000 can use VC4-3c and VC4-8c. The "c" indicates concatenated VC4s.

- **Bidirectional**—(Default) Leave this box checked.

- **Number of Circuits**—Type the number of circuits you want to create. The default is 1.

- **Auto Ranged**—Deselect this box. This check box is automatically selected when you enter more than 1 in the Number of Circuits field.

- **State** — Choose one of the following service states to apply to the circuit:
    - **IS** — The circuit is in service.
    - **OOS** — The circuit is out of service. Traffic is not passed on the circuit until it is in service.
    - **OOS-AINS** — (Default) The circuit is in service when it receives a valid signal; until then, the circuit is out of service.
    - **OOS-MT** — The circuit is in a maintenance state. The maintenance state does not interrupt traffic flow; it suppresses alarms and conditions and allows loopbacks to be performed on the circuit. Use OOS-MT for circuit testing or to suppress circuit alarms temporarily. Change the state to IS, OOS, or OOS-AINS when testing is complete.
- **Apply to Drop Ports** — Check this box if you want to apply the state chosen in the State field to the circuit source and destination ports. CTC will apply the circuit state to the ports if the circuit is in full control of the port. If not, a Warning dialog box displays the ports where the circuit state could not be applied. If not checked, CTC will not change the state of the source and destination ports. LOS alarms display if IS ports do not receive signals.
- **Inter-domain (UCP) SLA** — If the circuit will travel on a UCP channel, enter the SLA number. Otherwise, leave the field set to 0.
- **Protected Drops** — Uncheck this box.

Step 4 Click **Next**.

Step 5 Provision the circuit source:

(a) From the Node pull-down menu, choose the node that will contain the circuit.

(b) From the Slot pull-down menu, choose the slot containing the card where the circuit will originate.

(c) From the Port pull-down menu, choose the port where the circuit will originate.

Step 6 Click **Next**.

Step 7 Provision the circuit destination:

(a) From the Node pull-down menu, select the node chosen in Step 5a.

(b) From the Slot pull-down menu, choose the STM-N or E1 card to map the high-order path circuit to a VC4.

(c) Select the destination VC4 from the submenus that display.

Step 8 Click **Finish**. If you entered more than 1 in the Number of Circuits field and did not choose Auto Ranged, the Circuit Creation dialog box is displayed so that you can create the remaining circuits. Repeat this procedure for each additional circuit. After completing the circuit(s), CTC displays the Circuits window.

Step 9 In the Circuits window, verify that the newly created circuits appear in the Circuits list.

Create a Half Circuit Using an STM-N as a Destination in an SNCP

The following procedure describes how to create an STM-N circuit from a drop to an STM-N card on the same node in an SNCP:

Step 1 From node view, go to the View menu and choose **Go to Network View**.

Step 2 Click the **Circuits** tab, and then click **Create**.

Step 3 In the Circuit Creation dialog box, complete the following fields:

- **Name**—Assign a name to the circuit. The name can be alphanumeric and up to 48 characters (including spaces). If you leave the field blank, CTC assigns a default name to the circuit.

- **Type**—Choose **VC_HO_Path_Circuit (HOP)**. The circuit type determines the circuit-provisioning options that display. The E3 and DS3I cards must use VC low-order path tunnels.

- **Size**—Choose the circuit size (**VC_HO_Path_Circuits only**). VC high-order path circuits can be VC4, VC4-2c, VC4-3c, VC4-4c, VC4-8c, VC4-16c, or VC4-64c for optical cards and some Ethernet cards depending on the card type. Of the Ethernet cards, only the G-1000 can use VC4-3c and VC4-8c. The "c" indicates concatenated VC4s.

- **Bidirectional**—(Default) Leave checked for this circuit.

- **Number of Circuits**—Type the number of circuits you want to create. The default is 1.

- **Auto Ranged**—Deselect this check box. This check box is automatically selected when you enter more than 1 in the Number of Circuits field.

- **State**—Choose one of the following service states to apply to the circuit:
    — **IS**—The circuit is in service.
    — **OOS**—The circuit is out of service. Traffic is not passed on the circuit until it is in service.

- **OOS-AINS**—(Default) The circuit is in service when it receives a valid signal; until then, the circuit is out of service.
- **OOS-MT**—The circuit is in a maintenance state. The maintenance state does not interrupt traffic flow; it suppresses alarms and conditions and allows loopbacks to be performed on the circuit. Use OOS-MT for circuit testing or to suppress circuit alarms temporarily. Change the state to IS, OOS, or OOS-AINS when testing is complete.

- **Apply to Drop Ports**—Check this box if you want to apply the state chosen in the State field to the circuit source and destination ports. CTC will apply the circuit state to the ports if the circuit is in full control of the port. If not, a Warning dialog box displays the ports where the circuit state could not be applied. If not checked, CTC will not change the state of the source and destination ports. LOS alarms display if IS ports do not receive signals.

- **Inter-domain (UCP) SLA**—If the circuit will travel on a UCP channel, enter the SLA number. Otherwise, leave the field set to 0.

- **Protected Drops**—Leave this box unchecked.

Step 4 If the circuit will be routed on an SNCP, set the SNCP path selector defaults as follows:

- **Revertive**—Check this box if you want traffic to revert to the working path when the conditions that diverted traffic to the protect path are repaired. If Revertive is not chosen, traffic remains on the protect path after the switch.

- **Reversion Time**—If Revertive is checked, set the reversion time. This is the amount of time that will elapse before the traffic reverts to the working path. Traffic can revert when conditions causing the switch are cleared. (The default reversion time is 5 minutes.)

- **SF Threshold**—Choose from one E3, one E4, or one E5.

- **SD Threshold**—Choose from one E5, one E6, one E7, one E8, or one E9.

- **Switch on PDI-P**—(VC4 circuits only) Check this box if you want traffic to switch when a VC4 payload defect indicator is received.

Step 5 Click **Next**.

Step 6 Provision the circuit source:

(a) From the Node pull-down menu, choose the node that will contain the circuit.

(b) From the Slot pull-down menu, choose the slot containing the card where the circuit will terminate.

(c) From the Port pull-down menu, choose the port where the circuit will terminate.

Step 7 Click **Next**.

Step 8 Provision the circuit destination:

(a) From the Node pull-down menu, select the node that will contain the circuit. This will be the same as the node chosen in Step 6 Substep A.

(b) Choose the choose the STM-N or E1 card to map the high-order path circuit to a VC4.

(c) Choose the destination VC4 from the submenus that display.

Step 9 Click **Use Secondary Destination** and repeat Steps 6 to 9 to define the secondary destination.

Step 10 Click **Finish**. If you entered more than one in the Number of Circuits field and did not check the Auto Ranged check box, the Circuit Creation dialog box is displayed so that you can create the remaining circuits. Repeat this procedure for each additional circuit. After completing the circuit(s), CTC displays the Circuits window.

In the Circuits window, verify that the newly created circuits appear in the Circuits list.

Filtering, Viewing, and Changing Circuit Options

Use the following procedures to set up circuit filters, view circuits on an ONS 15454 SDH span, change the state of a circuit, edit a circuit name, or change the color of active (working) and standby (protect) circuit spans.

Filter the Display of Circuits

This section describes the steps used to filter the display of circuits:

Step 1 Switch to the appropriate CTC view, as follows:

- To filter network circuits, from the View menu, and choose **Go to Network View**. To filter circuits that originate, terminate, or pass through a specific node, from the View menu, choose **Go to Other Node**, and then choose the node you want to search and click **OK**.

- To filter circuits that originate, terminate, or pass through a specific card, switch to node view, and then double-click the card on the shelf graphic to display the card in card view.

Step 2 Click the **Circuits** tab.

Step 3 Set the attributes for filtering the circuit display:

(a) Click the **Filter** button.

(b) In the Filter dialog box, set the filter attributes:

- **Name**—Enter a complete or partial circuit name to filter circuits based on circuit name; otherwise, leave the field blank.
- **Direction**—Choose one of the following options: **Any** (direction not used to filter circuits), **1-way** (display only one-way circuits), or **2-way** (display only two-way circuits).
- **Status**—Choose one of the following options: **Any** (status not used to filter circuits), **Active** (display only active circuits), **Incomplete** (display only incomplete circuits—that is, circuits missing a connection or span to form a complete path), or **Upgradable** (display only upgradable circuits).
- **Slot**—Enter a slot number to filter circuits based on source or destination slot; otherwise, leave the field blank.
- **Port**—Enter a port number to filter circuits based on source or destination port; otherwise, leave the field blank.
- **Type**—Choose one of the following options: **Any** (type not used to filter circuits), **VC_HO_PATH_CIRCUIT** (displays only high-order path circuits), **VC_LO_PATH_CIRCUIT** (displays only low-order path circuits), or **VC_LO_PATH_TUNNEL** (displays only low-order path tunnels).
- **Size**—Click the appropriate check boxes to filter circuits based on size: **VC3, VC4, VC4-2c, VC4-3c, VC4-4c, VC4-8c, VC4-16c,** or **VC4-64c.** The check boxes displayed depend on the entry selected in the Type field.

Step 4 Click **OK**. Circuits matching the attributes in the Filter Circuits dialog box displays in the Circuits window.

Step 5 To turn filtering off, click the **Filter** icon in the lower-right corner of the Circuits window. Click the icon again to turn filtering on, and click the **Filter** button to change the filter attributes.

View Circuits on a Span

This section describes the steps used to view circuits on a span:

Step 1 From the View menu on the node view, choose **Go to Network View**.

Step 2 Place your mouse cursor directly over the span (green line) containing the circuits you want to view, right-click, and choose one of the following:

- **Circuits**—To view MS-SPRing, SNCP, 1+1, or unprotected circuits on the span.
- **PCA Circuits**—To view circuits routed on a MS-SPRing protected channel. (This option does not display if the span you right-clicked is not an MS-SPRing span.)

Step 3 In the Circuits on Span dialog box, view the following information for circuits provisioned on the span:

- **VC4**—The number of VC4s used by the circuits.
- **VC3**—The number of VC3s used by the circuits.
- **TUG2**—The number of TUG2s used by the circuits.
- **VC12**—The number of VC12s used by the circuits.
- **SNCP**—(SNCP span only) Checked for SNCP circuits.
- **Circuit**—Displays the circuit name.
- **Switch State**—(SNCP span only) Displays the switch state of the circuit (that is, whether any span switches are active). For SNCP spans, switch types include CLEAR (no spans are switched), MANUAL SWITCH AWAY (a manual switch is active), FORCE SWITCH AWAY (a force switch is active), and LOCKOUT OF PROTECTION (a span lockout is active).

Change a Circuit State

This section describes the steps used to change circuit state:

Step 1 Click the **Circuits** tab.

Step 2 Click the circuit with the state you want to change.

Step 3 From the Tools menu, choose **Circuits > Set Circuit State**. Alternatively, you can click the **Edit** button, and then click the **State** tab on the Edit Circuits window.

Step 4 In the Set Circuit State dialog box, change the circuit state by choosing one of the following from the Target Circuit State drop-down menu:

- **IS**—Places the circuit in service
- **OOS**—Places the circuit out of service

SDH Circuit Provisioning

- **OOS-AINS**—Places the circuit in an out of service, auto in service state
- **OOS-MT**—Places the circuit in an out of service, maintenance state

Step 5 If you want to apply the state to the circuit source and destination ports, check the **Apply to Drop Ports** check box.

Step 6 Click **OK**.

NOTE CTC will not change the state of the circuit source and destination port in certain circumstances (for example, if the circuit size is smaller than the port and you change the state from IS to OOS). If this occurs, a message is displayed and you will need to change the port state manually.

Edit a Circuit Name

This section describes the steps used to edit the circuit name:

Step 1 Click the **Circuits** tab.

Step 2 Click the circuit you want to rename, and then click **Edit**.

Step 3 In the General tab, click the **Name** field and edit or rename the circuit. Names can be up to 48 alphanumeric and/or special characters. This is true except for monitor circuits. If you edit a monitor circuit on this circuit, do not make the name longer than 44 characters because monitor circuits will add _MON (4 characters) to the circuit name.

Step 4 Click the **Apply** button.

Step 5 From File menu, choose **Close**.

Step 6 In the Circuits window, verify that the circuit was correctly renamed.

Change Active and Standby Span Color

This section describes the steps used to change active and standby span color:

Step 1 From the Edit menu, choose **Preferences**.

Step 2 In the Preferences dialog box, click the **Circuits** tab.

Step 3 Complete one or more of the following steps, as required:

- To change the color of the active (working) span, go to Step 4.
- To change the color of the standby (protect) span, go to Step 5.
- To return active and standby spans to the default colors, go to Step 6.

Step 4 To change the color of the active span, follow these substeps:

(a) Next to Active Span Color, click the **Color** button.

(b) In the Pick a Color dialog box, click the color for the active span, or click the **Reset** button if you want the active span to display the last applied (saved) color.

(c) Click **OK** to close the Pick a Color dialog box. If you want to change the standby span color, go to Step 5. If not, click **OK** to save the change and close the Preferences dialog box, or click **Apply** to save the change and keep the Preferences dialog box displayed.

Step 5 To change the color of the standby span, follow these substeps:

(a) Next to Standby Span Color, click the **Color** button.

(b) In the Pick a Color dialog box, choose the color for the standby span, or click the **Reset** button if you want the standby span to display the last applied (saved) color.

(c) Click **OK** to save the change and close the Preferences dialog box, or click **Apply** to save the change and keep the Preferences dialog box displayed.

Step 6 If you want to return the active and standby spans to their default colors, follow these substeps:

(a) From the Edit menu, choose **Preferences**.

(b) In the Preferences dialog box, click the **Circuits** tab.

(c) Choose **Reset to Defaults**.

(d) Click **OK** to save the change and close the Preferences dialog box, or click **Apply** to save the change and keep the Preferences dialog box displayed.

Creating SDH DCC Tunnels

There are four data communications channels (DCCs) for network element operations, administration, maintenance, and provisioning within SDH. There is one DCC on the SDH regenerator section layer and three on the SDH multiplex section layer. The regeneration section DCC is also known as the section DCC or SDCC. The SDCC uses three SDH bytes, D1 through D3. The three multiplex section DCCs are known as Tunnel1, Tunnel2, and Tunnel3. Tunnel1 uses SDH bytes D4 through D6, Tunnel2 uses SDH bytes D7 through D9, and Tunnel3 uses SDH bytes D10 through D12.

The ONS 15454 SDH uses the SDCC for internodal data communications. The multiplex section DCCs are available to tunnel DCCs from third-party equipment across ONS 15454

SDH networks, because the ONS does not use them. If D4 through D12 are used as data DCCs, they cannot be used for DCC tunneling. Table 10-4 DCC1 shows the various kinds of SDH DCC tunnels.

Table 10-4 *DCC1 SDH DCC Tunnels*

DCC	SDH Layer	SDH Bytes	STM-1 (All Ports)	STM-4, STM-16, STM-64
SDCC	Regenerator section	D1–D3	Yes	Yes
Tunnel1	Multiplex section	D4–D6	No	Yes
Tunnel2	Multiplex section	D7–D9	No	Yes
Tunnel3	Multiplex section	D10–D12	No	Yes

An SDH DCC tunnel endpoint is defined by slot, port, and DCC, where DCC can be SDCC, Tunnel1, Tunnel2, or Tunnel3. You can link a SDCC to a multiplex section DCC (Tunnel1, Tunnel2, or Tunnel3) and a multiplex section DCC to a regenerator section DCC. You can also link multiplex section DCCs to multiplex section DCCs and link regenerator section DCCs to regenerator section DCCs. To create a DCC tunnel, you connect the tunnel endpoints from one ONS 15454 SDH optical port to another. Each ONS 15454 SDH can support up to 32 DCC tunnel connections. The STM-1 card can support SDCC tunnels only, whereas the STM-4, STM-16, and STM-64 cards can support SDCC as well as multiplex section DCC tunnels. Figure 10-45 shows an SDH DCC tunnel example. SDH ADMs are connected to STM-1 cards at node 1/slot 3/port 1 and node 3/slot 3/port 1. Each ONS 15454 SDH node is connected by STM-16 trunk cards. In the example, three tunnel connections are created, one at node 1 (STM-1 to STM-16), one at node 2 (STM-16 to STM-16), and one at node 3 (STM-16 to STM-1).

Figure 10-45 *SDH DCC Tunnel*

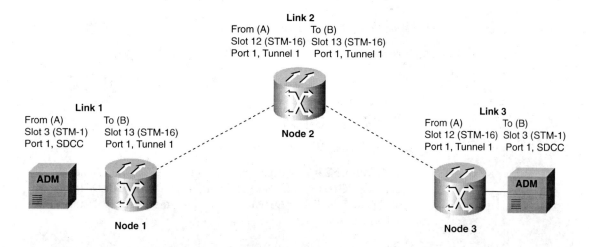

NOTE DCC tunneling is required for ONS 15454 SDH nodes transporting data through ONS 15454 SDH nodes, because a DCC will not function on a mixed network of ONS 15454 SDH nodes and ONS 15454 SONET nodes.

The following criteria applies to DCC tunnels:

- Each ONS 15454 SDH can have a maximum of 32 DCC tunnel connections.
- Each ONS 15454 SDH can have a maximum of 10 regenerator section DCC terminations.
- A regenerator section DCC that is terminated cannot be used as a DCC tunnel endpoint.
- A regenerator section DCC that is used as a DCC tunnel endpoint cannot be terminated.
- All DCC tunnel connections are bidirectional.

NOTE A multiplex section DCC cannot be used for tunneling if a data DCC is assigned. Each ONS 15454 SDH can have up to 32 DCC tunnel connections. Terminated regenerator section DCCs used by the ONS 15454 SDH cannot be used as a DCC tunnel endpoint, and a regenerator section DCC that is used as a DCC tunnel endpoint cannot be terminated. All DCC tunnel connections are bidirectional.

This section describes the steps used to create SDH DCC tunnels:

Step 1 Log in to an ONS 15454 SDH that is connected to the non-ONS 15454 SDH network.

Step 2 From the node view, choose **Go to Network View**.

Step 3 Click the **Provisioning > Overhead Circuits** tab.

Step 4 Click **Create**.

Step 5 In the Circuit Creation dialog box, provision the DCC tunnel as follows:

- **Name**—Type the tunnel name.
- **Type**—Choose one of the following options:
    - **DCC Tunnel D1–D3**—Enables you to choose either the regenerator section DCC (D1 through D3) or a multiplex section DCC (D4 through D6, D7 through D9, or D10 through D12) as the source or destination endpoints.

- **DCC Tunnel D4–D12**—Provisions the full multiplex section DCC as a tunnel:
    - **Source Node**—Choose the source node.
    - **Slot**—Choose the source slot.
    - **Port**—If displayed, select the source port.
    - **Channel**—Displayed if you selected DCC Tunnel D1-D3 as the tunnel type. Select one of the following:
        - **DCC1 (D1–D3)**—The regenerator section DCC
        - **DCC2 (D4–D6)**—Multiplex section DCC 1
        - **DCC3 (D7–D9)**—Multiplex section DCC 2
        - **DCC4 (D10–D12)**—Multiplex section DCC 3

Step 6 Click **OK**.

Step 7 Assign the ports hosting the DCC tunnel in service.

Step 8 Repeat these steps for all slots ports that are part of the DCC tunnel, including any intermediate nodes that will pass traffic from third-party equipment. The procedure is confirmed when the third-party network elements successfully communicate over the DCC tunnel.

Summary

It is important that you understand the theory behind SDH and TDM before jumping into provisioning mode. This chapter discussed the provisioning of the SDH MSPP, SNCP, MS-SPRing, and TDM. Prior to the provisioning of TDM or Ethernet circuits, one must configure the basic node provisioning for the SDH infrastructure. This chapter uses ONS 15454 SDH as an example because it exemplifies most of the SDH provisioning aspects. The ONS 15454 SDH is also the most deployed MSPP device in the metro edge and core. This chapter covered initial node provisioning tasks and the provisioning of SDH circuits. Before you begin your ONS node setup, ensure that you have the prerequisite information that includes node name, contact, location, current date, and time. If the ONS 15454 SDH will be connected to an OAM&P network, you also need the IP address and subnet mask assigned to the node and the IP address of the default router. If DHCP is used, you need the IP address of the DHCP server. If you intend to create card protection groups, you need the card protection scheme that will be used and what cards will be included in it. Proper network design requires an analysis of the requirements and a determination of the SDH protection topology that will be used for the network.

The ONS 15454 SDH provides several card protection methods. When you set up protection for ONS 15454 SDH cards, you must select between maximum protection and maximum slot availability. The highest protection reduces the number of available card slots, and the highest slot availability reduces the protection. SDH timing parameters must be set for each ONS 15454

SDH. Each ONS 15454 SDH independently accepts its timing reference from one of three sources: external, line, or internal.

In typical ONS 15454 SDH networks, one node is always set to external. The external node derives its timing from a BITS source wired to the BITS backplane pins. The BITS source, in turn, derives its timing from a PRS, such as a Stratum 1 clock or GPS signal. The other nodes are set to line. The line nodes derive timing from the externally timed node through the STM-N trunk cards. You can set three timing references for each ONS 15454 SDH. The first two references are typically two BITS-level sources, or two line-level sources optically connected to a node with a BITS source. The third reference is the internal oscillator of the ONS 15454 SDH TCC card. This clock is an ST3. If an ONS 15454 SDH becomes isolated, timing is maintained at the ST3 level.

The Inventory option displays information about cards installed in the ONS 15454 SDH node including part numbers, serial numbers, hardware revisions, and equipment types. The option provides a central location to obtain information about installed cards, software, and firmware versions.

SNCP provides duplicate fiber paths around the ring. Working traffic flows in one direction, and protection traffic flows in the opposite direction. If a problem occurs in the working traffic path, the receiving node switches to the path coming from the opposite direction. CTC automates SNCP ring configuration. SNCP traffic is defined within the ONS 15454 SDH on a circuit-by-circuit basis. If a path-protected circuit is not defined within a 1+1 or MS-SPRing line protection scheme and path protection is available and specified, CTC uses SNCP as the default. Because each traffic path is transported around the entire ring, SNCPs are best suited for networks where traffic concentrates at one or two locations and is not widely distributed. SNCP capacity is equal to its bit rate. Services can originate and terminate on the same SNCP, or they can be passed to an adjacent access or interoffice ring for transport to the service-terminating location.

The ONS 15454 SDH can support two concurrent MS-SPRings. Each MS-SPRing can have up to 32 ONS 15454 SDHs. Because the working and protect bandwidths must be equal, you can create only STM-4 (two-fiber only), STM-16, or STM-64 MS-SPRings. In two-fiber MS-SPRings, each fiber is divided into working and protect bandwidths. The CTC circuit routing routines calculate the "shortest path" for circuits based on many factors including requirements set by the circuit provisioner, traffic patterns, and distance.

Four-fiber MS-SPRings double the bandwidth of two-fiber MS-SPRings. Because they allow span switching as well as ring switching, four-fiber MS-SPRings increase the reliability and flexibility of traffic protection. Two fibers are allocated for working traffic, and two fibers for protection. To implement a four-fiber MS-SPRing, you must install four STM-16 or four STM-64 cards at each MS-SPRing node. Four-fiber MS-SPRings provide span and ring switching. Span switching occurs when a working span fails and traffic switches to the protect fibers between the failed span and then returns to the working fibers. Multiple span switches can occur at the same time. Ring switching occurs when a span switch cannot recover traffic, such as when both the working and protect fibers fail on the same span. In a ring switch, traffic is routed to the protect fibers throughout the full ring.

Extended SNCP meshed networks include multiple ONS 15454 SDH topologies, such as a combination of linear ADM and SNCP, that extend the protection provided by a single SNCP to the meshed architecture of several interconnecting rings. In an ESNCP mesh, circuits travel diverse paths through a network of single or multiple meshed rings. When you create circuits, you can have CTC automatically route circuits across the ESNCP mesh, or you can manually route them. You can also select levels of circuit protection. If you select full protection, for example, CTC creates an alternate route for the circuit in addition to the main route. The second route follows a unique path through the network between the source and destination and sets up a second set of cross-connections.

Provisioning an SNCP, MS-SPRing, linear ADM, or ESNCP mesh is not complete without assigning circuits for transport of user traffic. Circuits and tunnels of various sizes can be created to carry user traffic. Various DCC tunnels can also be created to tunnel third-party equipment signaling through ONS 15454 SDH networks. Cross-connects refer to the connections that occur within a single ONS 15454 SDH that allow a circuit to enter on one port and exit on another, and the term *circuit* refers to the series of connections from a traffic source (where traffic ingresses the ONS network) to the drop or destination (where traffic egresses an ONS network). Various types of VC circuits can be provisioned across and within ONS 15454 SDH nodes with different attributes assigned to them.

This chapter includes the following sections:

- **Ethernet and IP Services over SONET/SDH**—Ethernet services can be transported over Synchronous Optical Network (SONET) and Synchronous Digital Hierarchy (SDH) by integrating Ethernet modules in the ONS chassis. The ONS 15454/15454 SDH supports various Ethernet cards known as the G-Series, E-Series, ML-Series, and SL-Series cards.

- **G-Series Provisioning of Ethernet over SONET**—This section discusses the G-Series card and implementation of the card in various applications over SONET. G-Series card in production is known as the G1K-4 card. This card is a four-port Gigabit Ethernet card. The G1K-4 card is not a Gigabit Ethernet switch. It maps up to four Gigabit Ethernet interfaces onto a SONET transport network with bandwidth at the signal levels up to STS-48c per card.

- **E-Series Provisioning of Ethernet over SONET**—The E-Series cards incorporate Layer 2 switching within the card. This section discusses the various kinds of E-Series cards and implementation of the cards in various applications over SONET. E-Series cards support VLAN, IEEE 802.1Q, spanning tree, and IEEE 802.1D.

- **G-Series Provisioning of Ethernet over SDH**—This section discusses the G-Series card and implementation of the card in various applications over SDH. G-Series card in production is known as the G1K-4 card. This card is a four-port Gigabit Ethernet card. The G1K-4 card is not a Gigabit Ethernet switch. It maps up to four Gigabit Ethernet interfaces onto a SONET transport network with bandwidth at the signal levels up to VC4-16c per card.

- **E-Series Provisioning of Ethernet over SDH**—The E-Series cards incorporate Layer 2 switching within the card. This section discusses the various kinds of E-Series cards and implementation of the cards in various applications over SDH. E-Series cards support VLAN, IEEE 802.1Q, spanning tree, and IEEE 802.1D. Multicard EtherSwitch provisions two or more Ethernet cards to act as a single Layer 2 switch. It supports one VC4-2c or two VC4 circuits.

- **ML-Series Provisioning of Ethernet and IP over SONET and SDH**—The ML-Series cards are independent multilayer switches that can be integrated into the ONS 15454 SONET and SDH chassis. Addition of these cards makes the ONS 15454/15454 SDH Layer 2 and 3 aware, and endows the 15454/15454 SDH with full IP functionality. There are two versions of the ML-Series card in production: the 2-port Gigabit Ethernet card (ML1000-2) and 12-port Fast Ethernet card (ML100T-12).

CHAPTER 11

Ethernet, IP, and RPR over SONET and SDH

Ethernet and IP Services over SONET/SDH

The addition of Ethernet and IP functionality to the 15454 truly makes it a multiservice platform. The ONS 15454 and 15327 multiservice provisioning platforms (MSPPs), as well as the ONS 15600 MSSP, combine carrier-class TDM transport along with Ethernet and IP over SONET/SDH. The resiliency of the underlying SONET/SDH fabric leads to a solid deterministic infrastructure for transport of Ethernet and IP services. The MSPPs are of value to carriers that offer plain old telephone service (POTS), VoIP, data services and Greenfield pure IP service providers.

The Cisco ONS 15454 integrates Ethernet into the SONET and SDH time-division multiplexing (TDM) platforms. The ONS 15454 supports various Ethernet cards known as the G-Series, E-Series, and ML-Series. The G-Series card supports Gigabit Ethernet and is mainly used for transparent LAN services. The G-Series card does not support an Ethernet switch fabric and normally maps an external Ethernet source to the SONET or SDH STS-N or VC4-N. The E-Series card on the other hand, supports Ethernet switching, VLANs, IEEE 802.1Q, spanning tree, and IEEE 802.1D. The E-Series card is mainly used to implement Layer 2 customer virtual private networks (VPNs) using 802.1Q virtual LANs (VLANs) over the SONET/SDH infrastructure.

Scalability associated with the use of the Spanning Tree Protocol (STP) and 802.1Q tagging have been overcome by the ML-Series cards. The ML-Series cards support full multilayer switching and virtual routing and forwarding (VRF) lite, which enables service providers to create Layer 3 VPNs without having to implement MPLS or IPSec.

G-Series Provisioning of Ethernet over SONET

The G-Series card in production is known as the G1K-4 card. This card is a four-port Gigabit Ethernet card. The G1K-4 card is not a Gigabit Ethernet switch. The G1K-4 card maps up to four Gigabit Ethernet interfaces onto a SONET transport network. A single card provides scalable and provisionable transport bandwidth at the signal levels up to STS-48c per card. The card provides line rate forwarding for unicast, multicast, and broadcast

Ethernet frames. The G-Series card can also be configured to support jumbo frames defined as a maximum of 10,000 bytes. The G-Series card incorporates features optimized for carrier-class applications, such as high availability performance under software upgrades, all types of SONET equipment protection switches, and reprovisioning and support of Gigabit Ethernet traffic at full line rate. The G-Series card also supports full Transaction Language 1 (TL1)-based provisioning capability.

The G1K-4 card allows an Ethernet private line service to be provisioned and managed very much like a traditional SONET or SDH line. G1K-4 card applications include providing carrier-grade transparent LAN services (TLS), 100-Mbps Ethernet private line services (when combined with an external 100-Mbps Ethernet switch with Gigabit uplinks), and high availability transport for applications, such as storage over metropolitan-area networks (MANs). You can map the four ports on the G1K-4 independently to any combination of STS-1, STS-3c, STS-6c, STS-9c, STS-12c, STS-24c, and STS-48c circuit sizes, provided the sum of the circuit sizes that terminate on a card do not exceed STS-48c. The G1K-4 cards use the LAN Extension (LEX) protocol, which is a derivative of PPP over HDLC for SONET.

To support a Gigabit Ethernet port at full line rate, an STS circuit with a capacity greater or equal to 1 Gbps (bidirectional 2 Gbps) is needed. An STS-24c is the minimum circuit size that can support a Gigabit Ethernet port at full line rate. The G1K-4 supports a maximum of two ports at full line rate. Ethernet cards can be placed in any of the 12 multipurpose card slots. In most configurations, at least two of the 12 slots need to be reserved for optical trunk cards, such as the OC-192 card. The reserved OC-N slots give the ONS 15454 a practical maximum of 10 G1K-4 cards. The G1K-4 card requires the XC10G card to operate. The G1K-4 card is not compatible with XC or XCVT cards. The G1K-4 transmits and monitors the SONET J1 Path Trace byte in the same manner as ONS 15454 DS-N cards. The G1K-4 uses STS cross-connects only. Virtual tributary (VT)-level cross-connects are not used with the G1K-4. All SONET-side STS circuits used for the circuit must be contiguous in numbering. The G1K-4 circuits connect with OC-N cards or other G1K-4 cards and are not compatible with E-Series Ethernet cards. However, the G-Series cards can communicate with the ML-Series cards using the LEX protocol.

The G1K-4 card transports any Layer 3 protocol that can be encapsulated and transported over Gigabit Ethernet, such as IP or IPX, over a SONET network. The data is transmitted on the Gigabit Ethernet fiber into the standard Cisco gigabit interface converter (GBIC) on a G1K-4 card. The G1K-4 card transparently maps Ethernet frames into the SONET payload by multiplexing the payload onto a SONET OC-N card. When the SONET payload reaches the destination node, the process is reversed and the data is transmitted from the standard Cisco GBIC in the destination G1K-4 card onto the Gigabit Ethernet fiber.

The G1K-4 card discards certain types of erroneous Ethernet frames rather than transport them over SONET. Erroneous Ethernet frames include corrupted frames with cyclic redundancy check (CRC) errors and undersized frames that do not conform to the minimum 60-byte-length Ethernet standard. The G1K-4 card forwards valid frames unmodified over the SONET network. Information in the headers is not affected by the encapsulation and transport. For example, packets with formats that include IEEE 802.1Q information will travel through the process unaffected.

The G1K-4 card supports IEEE 802.3x flow control and frame buffering to reduce data traffic congestion. Each port has 512 KB of buffer memory available for receive and transmit channels. When the buffer memory on the Ethernet port nears capacity, the ONS 15454 uses IEEE 802.3x flow control to send back a pause frame to the source at the opposite end of the Gigabit Ethernet connection. The *pause* frame instructs that source to stop sending packets for a specific period of time. The sending station waits the requested time before sending more data. Figure 11-1 shows pause frames being sent from the ONS 15454 to the sources of the data. The G1K-4 card does not respond to pause frames received from client devices.

Figure 11-1 *G-Series Point-to-Point Circuit*

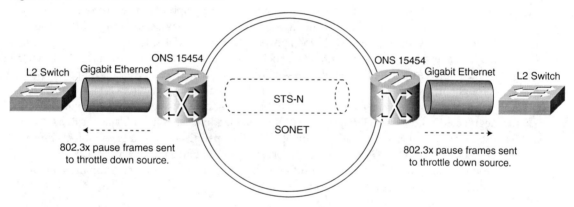

This flow-control mechanism matches the sending and receiving device throughput to that of the bandwidth of the Synchronous Transport Module (STM) circuit. For example, a router can transmit to the Gigabit Ethernet port on the G1K-4 card. This particular data rate can occasionally exceed 622 Mbps, but the ONS 15454 SONET circuit assigned to the G1K-4 card port can only be STS-12 (622.08 Mbps). In this example, the ONS 15454 sends out a pause frame and requests that the router delay its transmission for a certain period of time. With a flow-control capability combined with the substantial per-port buffering capability, a private line service provisioned at less than full line rate capacity (STS-24c) is nevertheless very efficient because frame loss can be controlled to a large extent. Because of these characteristics, the link autonegotiation and flow-control capability on the attached Ethernet device must be correctly provisioned for successful link autonegotiation and flow control on the G1K-4. If link autonegotiation fails, the G1K-4 does not use flow control (default).

The G1K-4 supports end-to-end Ethernet link integrity. This capability is integral to providing an Ethernet private line service and correct operation of Layer 2 and Layer 3 protocols on the attached Ethernet devices at each end. End-to-end Ethernet link integrity essentially means that if any part of the end-to-end path fails, the entire path fails. Failure of the entire path is ensured by turning off the transmit lasers at each end of the path. The attached Ethernet devices recognize the disabled transmit laser as a loss of carrier and consequently an inactive link. Some customer premises equipment (CPE) network devices can be configured to ignore a loss-of-carrier condition. If such a device attaches to a G1K-4 card at one end, alternative techniques (such as use of Layer 2 or Layer 3 protocol keep alive messages) are required to route traffic around failures. The response time of such alternate techniques is typically much longer than techniques that use link state as an indication of an error condition.

The end-to-end Ethernet link integrity feature of the G1K-4 can be used in combination with Gigabit EtherChannel (GEC) capability on attached devices. The combination provides an Ethernet traffic restoration scheme that has a faster response time than techniques, such as spanning tree rerouting, yet is more bandwidth efficient because spare bandwidth does not need to be reserved. The G1K-4 supports GEC, which is a Cisco proprietary standard similar to the IEEE link-aggregation standard (IEEE 802.3ad).

Although the G1K-4 card does not actively run GEC, it supports the end-to-end GEC functionality of attached Ethernet devices. If two Ethernet devices running GEC connect through G1K-4 cards to an ONS 15454 network, the ONS 15454-side network is transparent to the EtherChannel devices. The EtherChannel devices operate as if they are directly connected to each other. Any combination of G1K-4 parallel circuit sizes can be used to support GEC throughput. GEC provides line-level active redundancy and protection (1:1) for attached Ethernet equipment. It can also bundle parallel G1K-4 data links together to provide more aggregated bandwidth. STP operates as if the bundled links are one link and permits GEC to use these multiple parallel paths. Without GEC, STP permits only a single nonblocked path. GEC can also provide G1K-4 card–level protection or redundancy because it can support a group of ports on different cards (or different nodes) so that if one port or card has a failure, traffic is rerouted over the other port or card.

G1K-4 Port Provisioning

This section explains how to provision Ethernet ports on a G1K-4 card. Most provisioning requires filling in two fields: Enabled and Flow Control Negotiation. You can also configure the maximum frame size permitted, either jumbo or 1548 bytes.

Step 1   Click the **CTC node view** and double-click the **G1K-4 card** graphic to open the card.

Step 2   Click the **Provisioning > Port** tab. Figure 11-2 shows the Provisioning tab with the Port subtab selected.

Step 3   If you want to label the port, double-click the **Port Name** heading. Click anywhere else on the screen to save the change.

Step 4   Click the Enabled check box(es) to activate the corresponding Ethernet port(s).

Step 5   To disable/enable flow-control negotiation, click the **Flow Control Neg** check box. Flow-control negotiation is enabled by default.

Figure 11-2 *G1K-4 Port Provisioning*

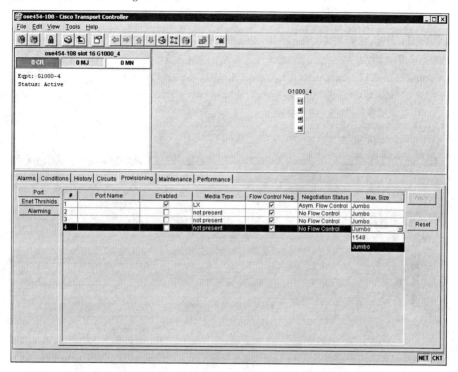

> **NOTE** Flow control is enabled only when the attached device is set for autonegotiation. If autonegotiation has been provisioned on the attached device, but the negotiation status indicates no flow control, check the autonegotiation settings on the attached device for interoperation with the asymmetric flow-control capability of the G1K-4.

Step 6 To permit the acceptance of jumbo-size Ethernet frames, click the **Max Size** column to reveal the pull-down menu and choose **Jumbo**. The maximum accepted frame size is set to Jumbo by default.

Step 7 Click **Apply**.

G1K-4 Point-to-Point Circuit Provisioning

G1K-4 cards support point-to-point circuit configuration. Provisionable circuit sizes are STS-1, STS-3c, STS-6c, STS-9c, STS-12c, STS-24c, and STS-48c. Each Ethernet port maps to a unique Synchronous Transport Signal (STS) circuit on the SONET side of the G1K-4. The two ends of the circuit must terminate at GE switches. The G1K-4 supports any combination of up to four circuits from the list of valid circuit sizes; however, the circuit sizes can add up to no more than 48 STSs.

Step 1 Log in to an ONS 15454 that you will use as one of the Ethernet circuit endpoints.

Step 2 In CTC node view, click the **Circuits** tab and click **Create**. The Circuit Creation (Circuit Attributes) dialog box opens. An example is shown in Figure 11-3.

Figure 11-3 *G1K-4 Circuit Attributes*

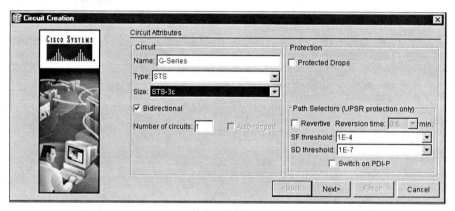

Step 3 In the Name field, type a name for the circuit.

Step 4 From the **Type** pull-down menu, choose **STS**. The VT and VT tunnel types do not apply to Ethernet circuits.

Step 5 Choose the size of the circuit from the Size pull-down menu. The valid circuit sizes for a G1K-4 circuit are STS-1, STS-3c, STS-6c, STS-9c, STS-12c, STS-24c, and STS-48c.

Step 6 Verify that the **Bidirectional** check box is checked and click **Next**. The Circuit Creation (Circuit Source) dialog box opens. An example is shown in Figure 11-4.

Step 7 Choose the circuit source node from the Node menu. Either end node can be the circuit source.

Step 8 From the Slot menu, choose the slot containing the G1K-4 card that you will use for one end of the point-to-point circuit.

Step 9 From the **Port** menu, choose a port.

Step 10 Click **Next**. The Circuit Creation (Destination) dialog box opens.

Step 11 Choose the circuit destination from the **Node** menu.

Figure 11-4 *G1K-4 Circuit Source*

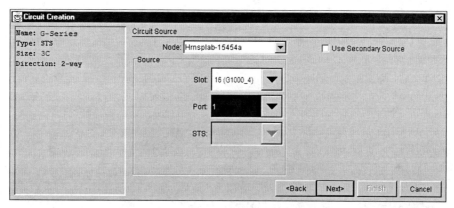

Step 12 From the **Slot** menu, choose the slot that holds the G1K-4 card that you will use for the other end of the point-to-point circuit.

Step 13 From the **Port** menu, choose a port.

Step 14 Click **Next**. The Circuit Creation (Circuit Routing Preferences) dialog box opens.

Step 15 Confirm that the following information about the point-to-point circuit is correct:

- Circuit name
- Circuit type
- Circuit size
- ONS 15454 nodes included in the circuit

Step 16 Click **Finish**.

Step 17 Provision the Ethernet ports as needed.

G1K-4 Manual Cross-Connect Provisioning

ONS 15454s require end-to-end DCC connectivity between nodes for normal provisioning of Ethernet circuits using Cisco Transport Controller (CTC). When attempting to connect via third-party equipment, OSI/TARP-based equipment does not allow tunneling of the ONS 15454 TCP/IP-based data communications channel (DCC). To circumvent a lack of a continuous DCC, the Ethernet circuit must be manually cross-connected to an STS-N channel riding through the non-ONS SONET network. This allows an Ethernet circuit to run from ONS node to ONS node while using the third-party non-ONS SONET network. The connectivity is shown in Figure 11-5. This section describes the procedure to configure G1K-4 manual cross-connects.

Figure 11-5 *G1K-4 Manual Cross-Connects*

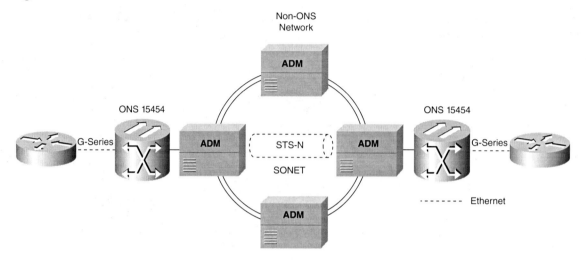

Step 1 Display CTC for one of the ONS 15454 Ethernet circuit endpoint nodes.

Step 2 Click the **Circuits** tab, and then click **Create**.

Step 3 The Circuit Creation (Circuit Attributes) dialog box opens.

Step 4 In the Name field, type a name for the circuit.

Step 5 From the Type pull-down menu, choose **STS**. The VT and VT tunnel types do not apply to Ethernet circuits.

Step 6 Choose the size of the circuit from the Size pull-down menu. The valid circuit sizes for a G1K-4 circuit are STS-1, STS-3c, STS-6c, STS-9c, STS-12c, STS-24c, and STS-48c.

Step 7 Verify that the **Bidirectional** check box is checked and click **Next**. The Circuit Creation (Circuit Source) dialog box opens.

Step 8 Choose the circuit source node from the Node menu.

Step 9 From the Slot menu, choose the slot containing the Ethernet card.

Step 10 From the Port menu, choose a port.

Step 11 Click **Next**. The Circuit Creation (Destination) dialog box opens.

Step 12 From the Node menu, choose the current node as the circuit destination.

Step 13 From the Slot menu, choose the optical card that will carry the circuit.

Step 14 Choose the STS that will carry the circuit from the STS menu, and then click **Next**.

Step 15 Confirm that the following information is correct:

- Circuit name
- Circuit type
- Circuit size
- ONS 15454 nodes included in this circuit

Step 16 Click **Finish**.

Step 17 You now need to provision the Ethernet ports.

Step 18 To complete the procedure, repeat Steps 1 through 16 at the second ONS 15454.

E-Series Provisioning of Ethernet over SONET

The E-Series cards incorporate Layer 2 switching within the card. The ONS 15454 E-Series cards include the E100T-12/E100T-G and E1000-2/E1000-2-G cards. E-Series cards support VLAN, IEEE 802.1Q, spanning tree, and IEEE 802.1D. An ONS 15454 holds a maximum of 10 Ethernet cards, and you can insert Ethernet cards in any multipurpose slot. The E100T-12/E100T-G cards provide 12 switched, IEEE 802.3-compliant 10/100 BASE-T Ethernet ports. The ports detect the speed of an attached device via autonegotiation and automatically connect at the appropriate speed and duplex mode, either half or full duplex, and determine whether to enable or disable flow control. The E100T-G is the functional equivalent of the E100T-12.

The E1000-2/E1000-2-G cards provides two switched, IEEE 802.3-compliant GE (1000-Mbps) ports that support full-duplex operation. The E1000-2 is the functional equivalent of the E1000-2-G. An ONS 15454 using XC10G cards requires the *G* versions of the E-Series Ethernet cards. Ethernet circuits can link ONS nodes through point-to-point, SPR (SPR), or hub-and-spoke configurations. Two nodes usually connect with a point-to-point configuration. More than two nodes usually connect with an SPR configuration or a hub-and-spoke configuration.

Multicard EtherSwitch provisions two or more Ethernet cards to act as a single Layer 2 switch. It supports one STS-6c SPR, two STS-3c SPRs, or six STS-1 SPRs. The bandwidth of the single switch formed by the Ethernet cards matches the bandwidth of the provisioned Ethernet circuit up to STS-6c worth of bandwidth.

Single-card EtherSwitch allows each Ethernet card to remain a single switching entity within the ONS 15454 shelf. This option allows a full STS-12c worth of bandwidth between two Ethernet circuit points. Single-card bandwidth options are as follows:

- STS 12c
- STS 6c + STS 6c
- STS 6c + STS 3c + STS 3c
- STS 6c + 6 STS-1s

- STS 3c + STS 3c + STS 3c + STS 3c
- STS 3c + STS 3c + 6 STS-1s
- 12 STS-1s

VLANs can be created after the Ethernet circuits have been provisioned. The ONS 15454 network supports a maximum of 509 user-provisionable VLANs using the E-Series cards. The Ethernet ports can then be provisioned and assigned to VLANs. Specific sets of ports define the broadcast domain for the ONS 15454. The definition of VLAN ports includes all Ethernet and packet-switched SONET port types. All VLAN IP address discovery, flooding, and forwarding are limited to these ports.

The ONS 15454 802.1Q-based VLAN mechanism provides logical isolation of subscriber LAN traffic over a common SONET transport infrastructure. Each subscriber has an Ethernet port at each site, and each subscriber is assigned to a VLAN. Although the subscriber's VLAN data flows over shared circuits, the service appears to the subscriber as a private data transport infrastructure. It is precisely this logical closed user-group mechanism that enables service providers to offer virtual private networking (VPN) services based on VLAN membership.

Provision E-Series Ethernet Ports

E-Series port provisioning requires filling in two fields: Enabled and Mode. This is shown in Figure 11-6. However, you can also map incoming traffic to a low-priority or a high-priority queue using the Priority column, and you can enable spanning tree with the STP Enabled column. The Status column displays information about the port's current operating mode, and the STP State column provides the current spanning tree status.

Step 1 Display CTC and double-click the card graphic to open the Ethernet card.

Step 2 Click the **Provisioning > Port** tab. Figure 11-6 shows the Provisioning tab with the Port function subtab selected.

Step 3 From the Port screen, choose the appropriate mode for each Ethernet port. Valid choices for the E100T-12/E100T-G card are Auto, 10 Half, 10 Full, 100 Half, or 100 Full. Valid choices for the E1000-2/E1000-2-G card are 1000 Full or Auto.

Both 1000 Full and Auto mode set the E1000-2 port to the 1000-Mbps and full-duplex operating mode; however, flow control is disabled when 1000 Full is selected. Choosing Auto mode enables the E1000-2 card to autonegotiate flow control. Flow control is a mechanism that prevents network congestion by ensuring that transmitting devices do not overwhelm receiving devices with data. The E1000-2 port handshakes with the connected network device to determine whether that device supports flow control.

Figure 11-6 *E-100 Port Provisioning*

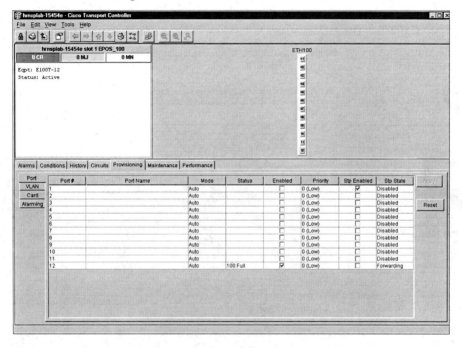

Step 4 Click the Enabled check box(es) to activate the corresponding Ethernet port(s).

Step 5 Click **Apply**.

Step 6 Your Ethernet ports are now provisioned and ready to be configured for VLAN membership.

Step 7 Repeat this procedure for all other cards that will be in the VLAN.

E-Series EtherSwitch Point-to-Point Circuit Provisioning

The ONS 15454 can set up a straight point-to-point Ethernet circuit as single card or multicard. A multicard EtherSwitch limits bandwidth to STS-6c of bandwidth between two Ethernet circuit points, but allows adding nodes and cards and creating an SPR. Figure 11-7 shows an E-Series multicard EtherSwitch point-to-point implementation.

A single-card EtherSwitch allows each Ethernet card to remain a single switching entity within the ONS chassis. The single-card EtherSwitch allows a full STS-12c of bandwidth between two Ethernet circuit points. Figure 11-8 shows an E-Series single-card EtherSwitch point-to-point implementation.

Figure 11-7 *Multicard EtherSwitch Point-to-Point Circuit*

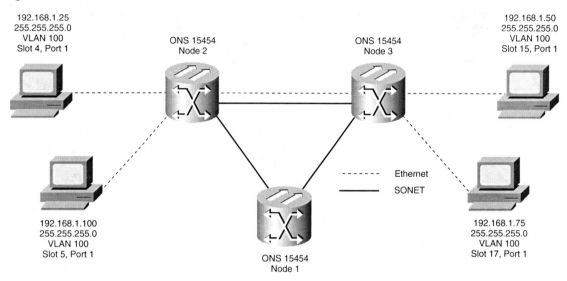

Figure 11-8 *Single-Card EtherSwitch Point-to-Point Circuit*

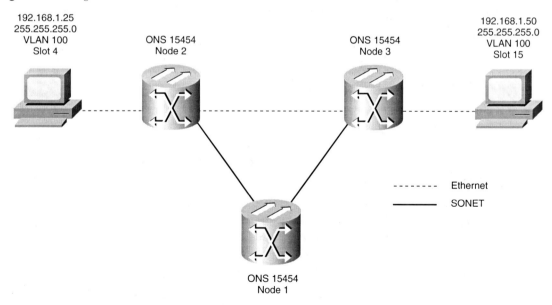

Step 1 Display CTC for one of the ONS 15454 Ethernet circuit endpoint nodes.

Step 2 Double-click one of the Ethernet cards that will carry the circuit.

Step 3 Click the **Provisioning > Card** tab.

Step 4 If you are building a multicard EtherSwitch point-to-point circuit, follow these substeps:

(a) Under Card Mode, verify that **Multicard EtherSwitch Group** is checked.

(b) If Multicard EtherSwitch Group is not checked, check it and click **Apply**.

(c) Repeat Steps 2 through 4 for all other Ethernet cards in the ONS 15454 that will carry the circuit.

If you are building a single-card EtherSwitch circuit, follow these substeps:

(a) Under Card Mode, verify that Single-Card EtherSwitch is checked.

(b) If Single-Card EtherSwitch is not checked, check it and click **Apply**.

Step 5 Navigate to the other ONS 15454 Ethernet circuit endpoint.

Step 6 Repeat Steps 2 through 5.

Step 7 Click the **Circuits** tab and click **Create**. The Circuit Creation (Circuit Attributes) dialog box opens.

Step 8 In the Name field, type a name for the circuit.

Step 9 From the Type pull-down menu, choose **STS**.

Step 10 Choose the size of the circuit from the Size pull-down menu.

The valid circuit sizes for an Ethernet multicard circuit are STS-1, STS-3c, and STS-6c. The valid circuit sizes for an Ethernet single-card circuit are STS-1, STS-3c, STS-6c, and STS-12c.

Step 11 Verify that the **Bidirectional** check box is checked and click **Next**. The Circuit Creation (Circuit Source) dialog box opens. An example is shown in Figure 11-9.

Step 12 Choose the circuit source from the Node menu. Either end node can be the circuit source.

Step 13 If you are building a multicard EtherSwitch circuit, choose **Ethergroup** from the Slot menu and click **Next**.

Step 14 If you are building a single-card EtherSwitch circuit, from the Slot menu, choose the Ethernet card for which you enabled the single-card EtherSwitch and click **Next**.

Step 15 The Circuit Creation (Destination) dialog box opens.

Step 16 Choose the circuit destination from the Node menu. Choose the node that is not the source.

Figure 11-9 *E-Series Circuit Creation*

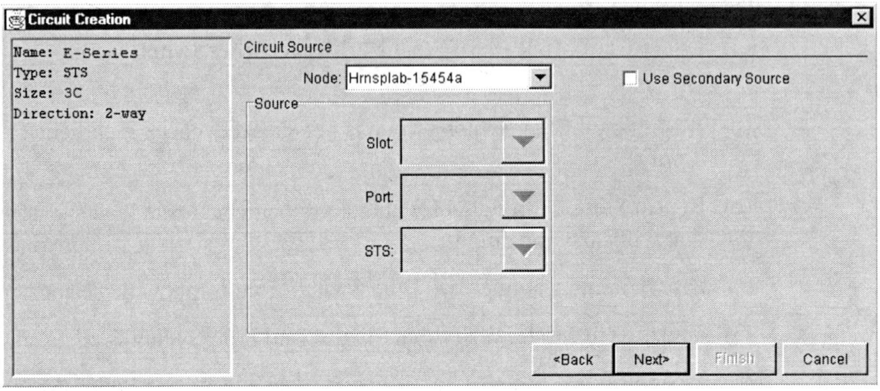

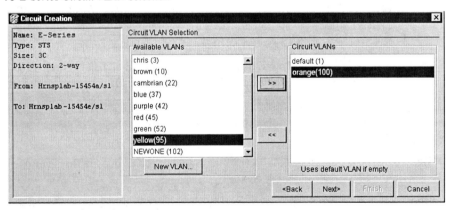

Step 17 If you are building multicard EtherSwitch circuits, choose **Ethergroup** from the Slot menu and click **Next**.

If you are building a single-card EtherSwitch circuit, from the Slot menu, choose the Ethernet card for which you enabled the single-card EtherSwitch and click **Next**.

The Circuit Creation (Circuit VLAN Selection) dialog box opens. An example is shown in Figure 11-10.

Figure 11-10 *E-Series Circuit VLAN Selection*

Step 18 Create the VLAN:

  (a) Click the **New VLAN** tab.

  (b) Assign an easily identifiable name to your VLAN.

  (c) Assign a VLAN ID.

(d) Click **OK**.

(e) Highlight the VLAN name and click the **>>** tab to move the available VLAN(s) to the Circuit VLANs column.

Step 19 Click **Next**. The Circuit Creation (Circuit Routing Preferences) dialog box opens.

Step 20 Confirm that the following information about the point-to-point circuit is correct:

- Circuit name
- Circuit type
- Circuit size
- VLANs on the circuit
- ONS 15454 nodes included in the circuit

Step 21 Click **Finish**.

Step 22 You now need to provision the Ethernet ports and assign ports to VLANs.

E-Series Shared Packet Ring Provisioning

The SPR provisioning steps for the multicard EtherSwitch is described in this section. The steps are generic and can be adapted to any SPR architecture. An example is shown in Figure 11-11.

Step 1 Display CTC for one of the ONS 15454 Ethernet circuit endpoints.

Step 2 Double-click one of the Ethernet cards that will carry the circuit.

Step 3 Click the **Provisioning > Card** tab.

Step 4 Under Card Mode, verify that **Multicard EtherSwitch Group** is checked.

Step 5 If Multicard EtherSwitch Group is not checked, check it and click **Apply**.

Step 6 Display the node view.

Step 7 Repeat Steps 2 through 6 for all other Ethernet cards in the ONS 15454 that will carry the SPR.

Step 8 Navigate to the other ONS 15454 endpoint.

Step 9 Repeat Steps 2 through 7.

Step 10 Click the **Circuits** tab and click **Create**. The Circuit Creation (Circuit Attributes) dialog box opens.

Step 11 In the Name field, type a name for the circuit.

Step 12 From the Type pull-down menu, choose **STS**.

Figure 11-11 *Multicard EtherSwitch SPR*

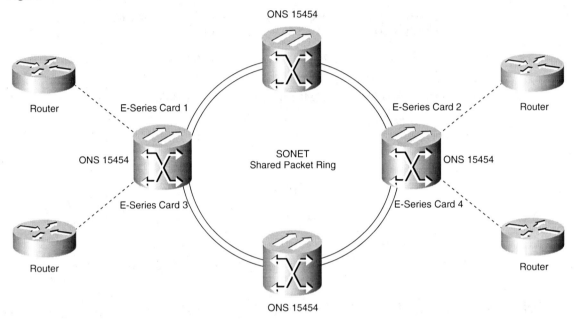

Step 13 From the Size pull-down menu, choose the size of the circuit. For SPR Ethernet, valid circuit sizes are STS-1, STS-3c, and STS-6c.

Step 14 Verify that the **Bidirectional** check box is checked.

Step 15 Click **Next**.

Step 16 The Circuit Creation (Circuit Source) dialog box opens. From the Node menu, choose the circuit source. Any SPR node can serve as the circuit source.

Step 17 Choose **Ethergroup** from the Slot menu and click **Next**.

The Circuit Creation (Circuit Destination) dialog box opens.

Step 18 Choose the circuit destination from the Node menu.

Except for the source node, any SPR node can serve as the circuit destination.

Step 19 Choose **Ethergroup** from the Slot menu and click **Next**. The Circuit Creation (Circuit VLAN Selection) dialog box opens.

Step 20 Create the VLAN:

(a) Click the **New VLAN** tab. The Circuit Creation (Define New VLAN) dialog box opens.

(b) Assign an easily identifiable name to your VLAN.

(c) Assign a VLAN ID. This VLAN ID number must be unique. It is usually the next-available number not already assigned to an existing VLAN (between 2 and 4093). Each ONS 15454 network supports a maximum of 509 user-provisionable VLANs.

(d) Click **OK**.

(e) Highlight the VLAN name and click the **>>** tab to move the VLAN(s) from the Available VLANs column to the Circuit VLANs column. By moving the VLAN from the Available VLANs column to the Circuit VLANs column, all the VLAN traffic is forced to use the SPR circuit you created.

Step 21 Click **Next**.

Step 22 Uncheck the **Route Automatically** check box and click **Next**.

Step 23 Click either span (green arrow) leading from the source node. The span turns white. An example is shown in Figure 11-12.

Figure 11-12 *Adding a Span for E-Series SPR*

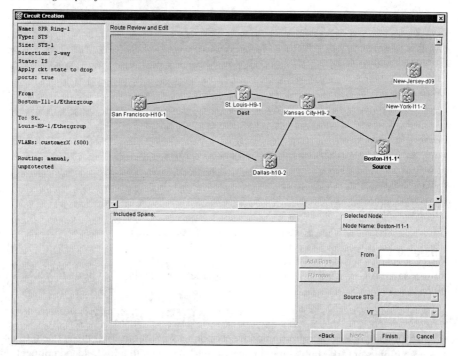

Step 24 Click **Add Span**.

The span turns blue and adds the span to the Included Spans field.

574 Chapter 11: Ethernet, IP, and RPR over SONET and SDH

Step 25 Click the node at the end of the blue span.

Step 26 Click the green span leading to the next node.

The span turns white.

Step 27 Click **Add Span**.

The span turns blue.

Step 28 Repeat Steps 24 through 27 for every node remaining in the ring.

Step 29 Verify that the new circuit is correctly configured. An example is shown in Figure 11-13.

Figure 11-13 *Viewing the Span for E-Series SPR*

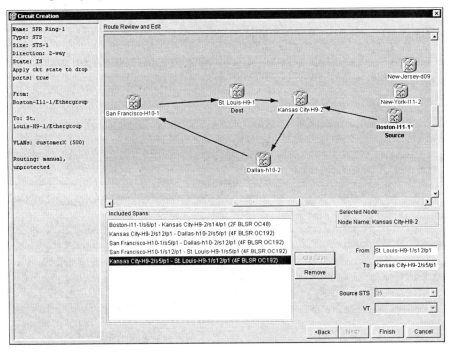

Step 30 Click **Finish**.

Step 31 You now need to provision the Ethernet ports and assign ports to VLANs.

E-Series Hub-and-Spoke Ethernet Circuit Provisioning

The hub-and-spoke configuration connects point-to-point circuits (the spokes) to an aggregation point (the hub). In many cases, the hub links to a high-speed connection and the spokes are Ethernet cards. Figure 11-14 shows a sample hub-and-spoke ring. This section provides generic steps for creating a hub-and-spoke Ethernet circuit configuration.

Figure 11-14 *Hub-and-Spoke Ethernet Circuit*

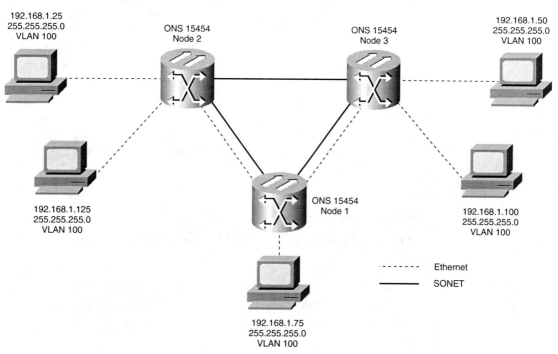

Step 1 Display CTC for one of the ONS 15454 Ethernet circuit endpoints.

Step 2 Double-click the Ethernet card that will create the circuit.

Step 3 Click the **Provisioning > Card** tab.

Step 4 Under Card Mode, check the **Single-Card EtherSwitch** check box. If Single-Card EtherSwitch is not checked, check it and click **Apply**.

Step 5 Navigate to the other ONS 15454 endpoint and repeat Steps 2 through 4.

Step 6 Display the node view or network view.

Step 7 Click the **Circuits** tab and click **Create**. The Circuit Creation (Circuit Attributes) dialog box opens.

Step 8 In the Name field, type a name for the circuit.

Step 9 From the **Type** pull-down menu, choose **STS**.

Step 10 Choose the size of the circuit from the Size pull-down menu.

Step 11 Verify that the **Bidirectional** check box is checked and click **Next**.

The Circuit Creation (Circuit Source) dialog box opens.

Step 12 From the Node menu, choose the circuit source.

Either end node can be the circuit source.

From the Slot menu, choose the Ethernet card where you enabled the single-card EtherSwitch and click **Next**. The Circuit Creation (Circuit Destination) dialog box opens.

Step 13 Choose the circuit destination from the Node menu.

Choose the node that is not the source.

Step 14 From the Slot menu, choose the Ethernet card where you enabled the single-card EtherSwitch and click **Next**. The Circuit Creation (Circuit VLAN Selection) dialog box opens.

Step 15 Create the VLAN:

(a) Click the **New VLAN** tab.

The Circuit Creation (Define New VLAN) dialog box opens.

(b) Assign an easily identifiable name to your VLAN.

(c) Assign a VLAN ID. This should be the next available number (between 2 and 4093) not already assigned to an existing VLAN. Each ONS 15454 network supports a maximum of 509 user-provisionable VLANs.

(d) Click **OK**.

(e) Highlight the VLAN name and click the >> tab to move the VLAN(s) from the Available VLANs column to the Circuit VLANs column.

Step 16 Click **Next**. The Circuit Creation (Circuit Routing Preferences) dialog box opens.

Step 17 Confirm that the following information about the point-to-point circuit is correct:

- Circuit name
- Circuit type
- Circuit size
- VLANs that will be transported across this circuit
- ONS 15454 nodes included in this circuit

Step 18 Click **Finish**. You must now provision the second circuit and attach it to the already created VLAN.

Step 19 Log in to the ONS 15454 Ethernet circuit endpoint for the second circuit.

Step 20 Double-click the Ethernet card that will create the circuit. The CTC card view displays.

Step 21 Click the **Provisioning > Card** tab.

Step 22 Under Card Mode, check **Single-Card EtherSwitch**. If the **Single-Card EtherSwitch** check box is not checked, check it and click **Apply**.

Step 23 Log in to the other ONS 15454 endpoint for the second circuit and repeat Steps 21 through 23.

Step 24 Display the CTC node view.

Step 25 Click the **Circuits** tab and click **Create**.

Step 26 Choose **STS** from the Type pull-down menu.

Step 27 Choose the size of the circuit from the Size pull-down menu.

Step 28 Verify that the **Bidirectional** check box is checked and click **Next**.

Step 29 Choose the circuit source from the Node menu and click **Next**. Either end node can be the circuit source.

Step 30 Choose the circuit destination from the Node menu. Choose the node that is not the source.

Step 31 From the Slot menu, choose the Ethernet card where you enabled the single-card EtherSwitch and click **Next**. The Circuit Creation (Circuit VLAN Selection) dialog box is displayed.

Step 32 Highlight the VLAN that you created for the first circuit and click the >> tab to move the VLAN(s) from the Available VLANs column to the Selected VLANs column.

Step 33 Click **Next** and click **Finish**.

Step 34 You now need to provision the Ethernet ports and assign these ports to the VLANs.

Provision an E-Series Single-Card EtherSwitch Manual Cross-Connect

ONS 15454s require end-to-end DCC connectivity between nodes for normal provisioning of Ethernet circuits using CTC. When attempting to connect via third-party equipment, OSI/TARP-based equipment does not allow tunneling of the ONS 15454 TCP/IP-based DCC. To circumvent a lack of a continuous DCC, the Ethernet circuit must be manually cross-connected to an STS-N channel riding through the non-ONS SONET network. This allows an Ethernet circuit to run from ONS node to ONS node while using the third-party non-ONS SONET network. The connectivity is shown in Figure 11-15. This section describes the procedure to configure E-Series single-card EtherSwitch manual cross-connects:

Step 1 Display CTC for one of the ONS 15454 Ethernet circuit endpoints.

Step 2 Double-click one of the Ethernet cards that will carry the circuit.

Figure 11-15 *Ethernet Manual Cross-Connects*

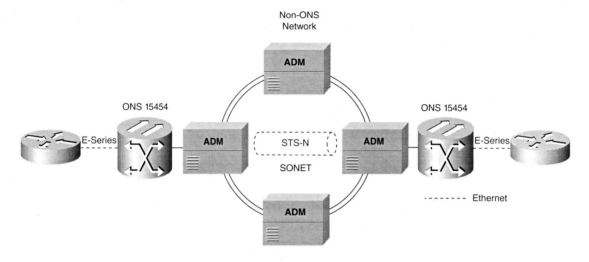

Step 3 Click the **Provisioning > Card** tab.

Step 4 Under Card Mode, verify that **Single-Card EtherSwitch** is checked.

Step 5 If the Single-Card EtherSwitch is not checked, check it and click **Apply**.

Step 6 Display the node view.

Step 7 Click the **Circuits** tab and click **Create**. The Circuit Creation (Circuit Attributes) dialog box opens.

Step 8 In the Name field, type a name for the circuit.

Step 9 From the Type pull-down menu, choose **STS**.

Step 10 Choose the size of the circuit from the Size pull-down menu. The valid circuit sizes for an Ethernet multicard circuit are STS-1, STS-3c, and STS-6c.

Step 11 Verify that the **Bidirectional** check box is checked and click **Next**. The Circuit Creation (Circuit Source) dialog box opens.

Step 12 From the Node menu, choose the current node as the circuit source.

Step 13 From the Slot menu, choose the Ethernet card that will carry the circuit and click **Next**. The Circuit Creation (Circuit Destination) dialog box opens.

Step 14 From the Node menu, choose the current node as the circuit destination.

Step 15 From the Slot menu, choose the optical card that will carry the circuit.

Step 16 Choose the STS that will carry the circuit from the STS menu and click **Next**. The Circuit Creation (Circuit VLAN Selection) dialog box opens.

Step 17 Create the VLAN:

    (a) Click the **New VLAN** tab. The Circuit Creation (Define New VLAN) dialog box opens.

    (b) Assign an easily identifiable name to your VLAN.

    (c) Assign a VLAN ID.

    The VLAN ID should be the next available number (between 2 and 4093) that is not already assigned to an existing VLAN. Each ONS 15454 network supports a maximum of 509 user-provisionable VLANs.

    (d) Click **OK**.

    (e) Highlight the VLAN name and click the arrow >> tab to move the VLAN(s) from the Available VLANs column to the Circuit VLANs column.

Step 18 Click **Next**.

The Circuit Creation (Circuit Routing Preferences) dialog box opens.

Step 19 Confirm that the following information is correct:

- Circuit name
- Circuit type
- Circuit size
- VLANs on this circuit
- ONS 15454 nodes included in this circuit

Step 20 Click **Finish**.

Step 21 You now need to provision the Ethernet ports and assign ports to VLANs.

Step 22 After assigning the ports to the VLANs, repeat Steps 1 through 19 at the second ONS 15454 Ethernet manual cross-connect endpoint.

Provision an E-Series Multicard EtherSwitch Manual Cross-Connect

Step 1 Display CTC for one of the ONS 15454 Ethernet circuit endpoints.

Step 2 Double-click one of the Ethernet cards that will carry the circuit.

Step 3 Click the **Provisioning > Card** tab.

Step 4 Under Card Mode, verify that **Multicard EtherSwitch Group** is checked. If the Multicard EtherSwitch Group is not checked, check it and click **Apply**.

Step 5 Display the node view.

Step 6 Repeat Steps 2 through 5 for any other Ethernet cards in the ONS 15454 that will carry the circuit.

Step 7   Click the **Circuits** tab and click **Create**. The Circuit Creation (Circuit Attributes) dialog box opens.

Step 8   In the Name field, type a name for the circuit.

Step 9   From the Type pull-down menu, choose **STS**.

Step 10 Choose the size of the circuit from the Size pull-down menu. The valid circuit sizes for an Ethernet multicard circuit are STS-1, STS-3c, and STS-6c.

Step 11 Verify that the **Bidirectional** check box is checked and click **Next**. The Circuit Creation (Circuit Source) dialog box opens.

Step 12 From the Node menu, choose the current node as the circuit source.

Step 13 Choose **Ethergroup** from the Slot menu and click **Next**. The Circuit Creation (Circuit Destination) dialog box opens.

Step 14 From the Node menu, choose the current node as the circuit destination.

Step 15 Choose **Ethergroup** from the Slot menu and click **Next**.

Step 16 Create the VLAN:

  (a) Click the **New VLAN** tab. The Circuit Creation (Define New VLAN) dialog box opens.

  (b) Assign an easily identifiable name to your VLAN.

  (c) Assign a VLAN ID.

  (d) The VLAN ID should be the next available number (between 2 and 4093) that is not already assigned to an existing VLAN. Each ONS 15454 network supports a maximum of 509 user-provisionable VLANs.

  (e) Click **OK**.

  (f) Highlight the VLAN name and click the arrow >> tab to move the VLAN(s) from the Available VLANs column to the Circuit VLANs column.

Step 17 Click **Next**. The Circuit Creation (Circuit Routing Preferences) dialog box opens.

Step 18 Confirm that the following information is correct:

- Circuit name
- Circuit type
- Circuit size
- VLANs on this circuit
- ONS 15454 nodes included in this circuit

Step 19 Click **Finish**. You now need to provision the Ethernet ports and assign ports to VLANs.

Step 20 Highlight the circuit and click **Edit**. The Edit Circuit dialog box opens.

Step 21 Click **Drops** and click **Create**. The Define New Drop dialog box opens.

Step 22 From the Slot menu, choose the optical card that links the ONS 15454 to the non-ONS 15454 equipment.

Step 23 From the Port menu, choose the appropriate port.

Step 24 From the STS menu, choose the STS that matches the STS of the connecting non-ONS 15454 equipment.

Step 25 Click **OK**. The Edit Circuit dialog box opens.

Step 26 Confirm the circuit information that displays in the Circuit Information dialog box and click **Close**.

Step 27 Repeat Steps 1 through 26 at the second ONS 15454 Ethernet manual cross-connect endpoint.

Provision Ethernet Ports for VLAN Membership

The ONS 15454 enables you to configure the VLAN membership and IEEE 802.1Q tag handling of individual Ethernet ports:

Step 1 Display the CTC card view for the Ethernet card.

Step 2 Click the **Provisioning > VLAN** tab. An example is shown in Figure 11-16.

Step 3 To add a port in a VLAN, click the port and choose either **Tagged** or **Untag**. If a port is marked Untag, the ONS 15454 will tag ingress frames and strip tags from egress frames. If a port is marked **Tagged**, the ONS 15454 will handle ingress frames according to VLAN ID, but egress frames will not have their tags removed. A port marked with the -- symbol does not belong to any VLAN.

If a port is a member of only one VLAN, go to that VLANs row and choose **Untag** from the Port column. Choose -- for all the other VLAN rows in that Port column. The VLAN with Untag selected can connect to the port, but other VLANs cannot access that port.

If a port is a trunk port that connects multiple VLANs to an external trunking device, it must have tagging (802.1Q) enabled for all the VLANs that connect to that external device. Choose **Tagged** at all VLAN rows that need to be trunked. Choose **Untag** at one or more VLAN rows in the trunk port's column that do not need to be trunked (for example, the default VLAN). Each Ethernet port must be attached to at least one untagged VLAN.

Step 4 After each port is assigned to its appropriate VLAN, click **Apply**.

Figure 11-16 *Assigning Ports to VLANs*

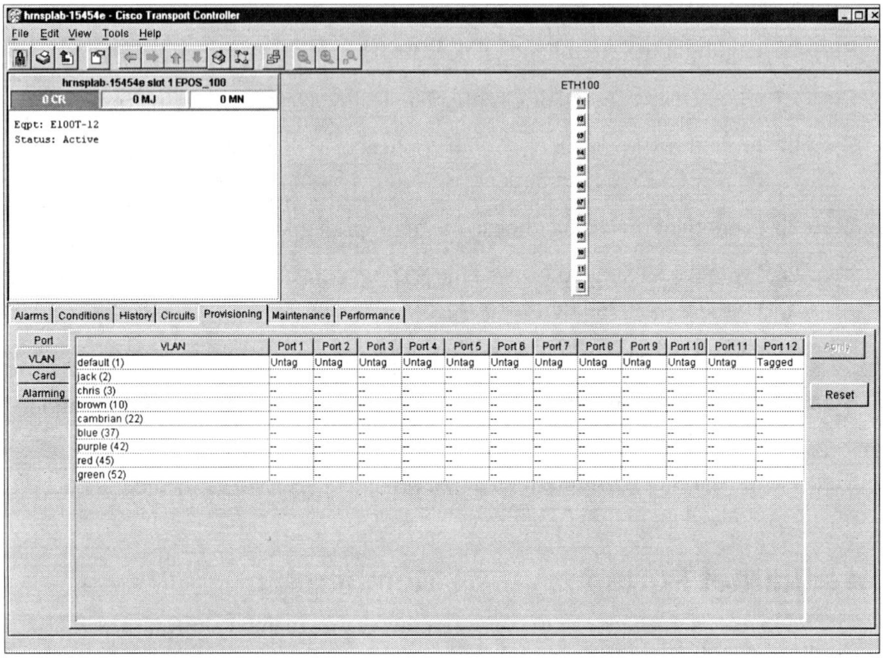

Enable E-Series Spanning Tree on Ethernet Ports

The IEEE 802.1D Spanning Tree Protocol (STP) is supported by the E-Series card. When an Ethernet card is installed, STP operates over all packet-switched ports including Ethernet and SONET ports. STP is disabled by default on Ethernet ports, but can be enabled if necessary. On SONET interface ports, STP is activated by default and cannot be disabled.

The Ethernet card can enable STP on the Ethernet ports to allow redundant paths to the attached Ethernet equipment. STP spans cards so that both equipment and facilities are protected against failure. STP detects and eliminates network loops. When STP detects multiple paths between any two network hosts, STP blocks ports until only one path exists between any two network hosts. The single path eliminates possible bridge loops. This is crucial for SPRs, which include inherent looped links. To remove loops, STP defines a tree that spans all the switches in an extended network. STP forces certain redundant data paths into a standby (blocked) state. If one network segment in the STP becomes unreachable, the spanning-tree algorithm reconfigures the spanning-tree topology and reactivates the blocked path to re-establish the link.

STP operation is transparent to end stations that do not discriminate between connections to a single LAN segment or to a switched LAN with multiple segments. The E-Series card supports one STP instance per circuit and a maximum of eight STP instances per ONS 15454. The ONS 15454 can operate multiple instances of STP to support VLANs in a looped topology. You can dedicate separate circuits across the SONET ring for different VLAN groups.

For example, one circuit for private TLS services and one circuit for Internet access. Each circuit would run its own STP to maintain VLAN connectivity in a multiring environment. Table 11-1 indicates the E-Series card STP default values. These preconfigured values work well for most implementations. These values can be modified if necessary.

Table 11-1 *E-Series Card STP Default Values*

STP Variable	Default Value	Value Range
Priority	32,768	0–65,535
Bridge Max Age	20 seconds	6–40 seconds
Bridge Hello Time	2 seconds	1–10 seconds
Bridge Forward Delay	15 seconds	4–30 seconds

Step 1 Display the CTC card view.

Step 2 Click the **Provisioning > Port** tab.

Step 3 In the left column, find the applicable port number and check the **STP Enabled** check box to enable STP for that port.

Step 4 Click **Apply**.

Step 5 At the node view, click the **Maintenance > Etherbridge > Spanning Trees** tab to view spanning-tree parameters. An example is shown in Figure 11-17.

Figure 11-17 *Assigning Ports to VLANs*

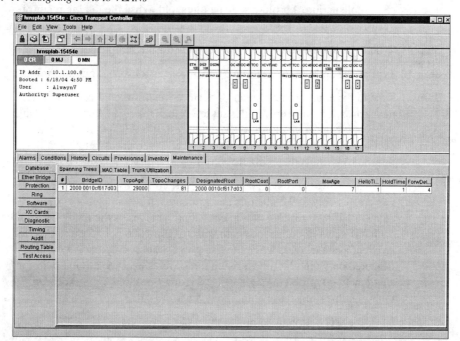

Retrieve the MAC Table Information

Step 1 Click the **Maintenance > EtherBridge > MAC Table** tab.

Step 2 Choose the appropriate Ethernet card or Ethergroup from the Layer 2 Domain pull-down menu.

Step 3 Click **Retrieve** for the ONS 15454 to retrieve and display the current MAC IDs.

Creating Ethernet RMON Alarm Thresholds

Step 1 Display the CTC node view.

Step 2 Click the **Provisioning > Etherbridge > Thresholds** tab.

Step 3 Click **Create**. The Create Ethernet Threshold dialog box opens.

Step 4 From the Slot menu, choose the appropriate Ethernet card.

Step 5 From the Port menu, choose the Port on the Ethernet card.

Step 6 From the Variable menu, choose the Ethernet threshold variable.

Step 7 From Alarm Type, indicate whether the rising threshold, falling threshold, or both the rising and falling thresholds will trigger the event.

Step 8 From the Sample Type pull-down menu, choose either **Relative** or **Absolute**. Relative restricts the threshold to use the number of occurrences in the user-set sample period. Absolute sets the threshold to use the total number of occurrences, regardless of any time period.

Step 9 Type in an appropriate number of seconds for the sample period.

Step 10 Type in the appropriate number of occurrences for the rising threshold.

Step 11 Type in the appropriate number of occurrences for the falling threshold. In most cases, a falling threshold is set lower than the rising threshold. A falling threshold is the counterpart to a rising threshold. When the number of occurrences is above the rising threshold and then drops below a falling threshold, it resets the rising threshold.

When the network problem that caused 1001 collisions in 15 minutes subsides and creates only 799 collisions in 15 minutes, for example, occurrences fall below a falling threshold of 800 collisions. This resets the rising threshold so that if network collisions again spike over a 1000 per 15-minute period, an event again triggers when the rising threshold is crossed. An event is triggered only the first time a rising threshold is exceeded. (Otherwise, a single network problem might cause a rising threshold to be exceeded multiple times and cause a flood of events.)

Step 12 Click the **OK** button to complete the procedure.

G-Series Provisioning of Ethernet over SDH

The G-Series card in production is known as the G1K-4 card. The Cisco ONS 15454 SDH integrates Ethernet into an SDH TDM platform. The ONS 15454 SDH supports the G-Series G1K-4 Ethernet card. The G1K-4 card maps up to four GE interfaces onto an SDH transport network. A single card provides scalable and provisionable transport bandwidth at the signal levels up to VC4-16C per card. The card provides line rate forwarding for unicast, multicast, and broadcast Ethernet frames and can be configured to support up to 10,000 byte jumbo frames. The G-Series card incorporates features optimized for carrier-class applications, such as high-availability performance under software upgrades and all types of SDH equipment protection switches. Hitless reprovisioning and support of Gigabit Ethernet traffic at full line rate.

The G1K-4 card allows an Ethernet private line service to be provisioned and managed very much like a traditional SDH line. G1K-4 card applications include providing carrier-grade TLS, 100-Mbps Ethernet private line services (when combined with an external 100-Mbps Ethernet switch with Gigabit uplinks), and high availability transport for applications, such as storage over MANs. You can map the four ports on the G1K-4 independently to any combination of VC4, VC4-2c, VC4-3c, VC4-8c, and VC4-16c circuit sizes, provided the sum of the circuit sizes that terminate on a card do not exceed VC4-16c. The G1K-4 cards use the LEX protocol, which is a derivative of PPP over HDLC for SDH.

To support a GE port at full line rate, an STM circuit with a capacity greater or equal to 1 Gbps (bidirectional 2 Gbps) is needed. A VC4-8c is the minimum circuit size that can support a GE port at full line rate. The G1K-4 supports a maximum of two ports at full line rate.

Ethernet cards can be placed in any of the 12 multipurpose card slots. In most configurations, at least two of the 12 slots need to be reserved for optical trunk cards, such as the STM-64 card. The reserved slots give the ONS 15454 SDH a practical maximum of 10 G1K-4 cards. The G1K-4 card requires the XC10G card to operate. The G1K-4 transmits and monitors the SDH J1 Path Trace byte in the same manner as ONS 15454 SDH cards. The G1K-4 uses VC3-level cross-connects only. VC12-level cross-connects are not supported by the G1K-4. All SDH-side VC3 circuits used for the circuit must be contiguous in numbering. The G1K-4 circuits connect with STM-N cards or other G1K-4 cards and are not compatible E-Series Ethernet cards. However, the G-Series cards can communicate with the ML-Series cards using the LEX protocol.

The G1K-4 card transports any Layer 3 protocol that can be encapsulated and transported over Gigabit Ethernet, such as IP or IPX, over an SDH network. The data is transmitted on the GE fiber into the standard Cisco GBIC on a G1K-4 card. The G1K-4 card transparently maps Ethernet frames into the SDH payload by multiplexing the payload onto an SDH STM-N card. When the SDH payload reaches the destination node, the process is reversed and the data is transmitted from the standard Cisco GBIC in the destination G1K-4 card onto the GE fiber.

The G1K-4 card discards certain types of erroneous Ethernet frames rather than transport them over SDH. Erroneous Ethernet frames include corrupted frames with CRC errors and undersized frames that do not conform to the minimum 60-byte-length Ethernet standard.

The G1K-4 card forwards valid frames unmodified over the SDH network. Information in the headers is not affected by the encapsulation and transport. For example, packets with formats that include IEEE 802.1Q information will travel through the process unaffected.

The G1K-4 card supports IEEE 802.3x flow control and frame buffering to reduce data traffic congestion. Each port has 512 KB of buffer memory available for *receive* and *transmit* channels. When the buffer memory on the Ethernet port nears capacity, the ONS 15454 SDH uses IEEE 802.3x flow control to send back a pause frame to the source at the opposite end of the GE connection. The pause frame instructs that source to stop sending packets for a specific period of time. The sending station waits the requested time before sending more data. Figure 11-18 shows pause frames being sent from the ONS 15454 SDH to the sources of the data. The G1K-4 card does not respond to pause frames received from client devices.

Figure 11-18 *G-Series Point-to-Point Circuit*

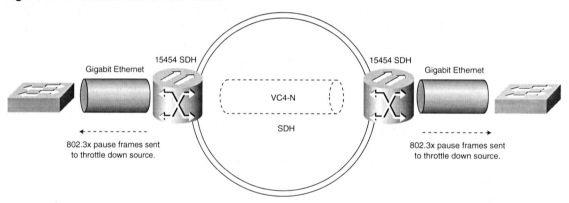

This flow-control mechanism matches the sending and receiving device throughput to that of the bandwidth of the STM circuit. For example, a router can transmit to the GE port on the G1K-4 card. This particular data rate can occasionally exceed 622 Mbps, but the ONS 15454 SDH circuit assigned to the G1K-4 card port can only be VC4-4c (622.08 Mbps). In this example, the ONS 15454 SDH sends out a pause frame and requests that the router delay its transmission for a certain period of time. With a flow-control capability combined with the substantial per-port buffering capability, a private line service provisioned at less than full line rate capacity (VC4-8c) is nevertheless very efficient because frame loss can be controlled to a large extent. Because of these characteristics, the link autonegotiation and flow-control capability on the attached Ethernet device must be correctly provisioned for successful link autonegotiation and flow control on the G1K-4. If link autonegotiation fails, the G1K-4 does not use flow control (default).

The G1K-4 supports end-to-end Ethernet link integrity. This capability is integral to providing an Ethernet private line service and correct operation of Layer 2 and Layer 3 protocols on the attached Ethernet devices at each end. End-to-end Ethernet link integrity essentially means that if any part of the end-to-end path fails, the entire path fails. Failure of the entire path is ensured by turning off the transmit lasers at each end of the path. The attached Ethernet devices recognize the disabled transmit laser as a loss of carrier and consequently an inactive link. Some CPE network devices can be configured to ignore a loss-of-carrier condition. If such a device attaches to a G1K-4 card at one end, alternative techniques (such as use of Layer 2 or Layer 3 protocol keepalive messages) are required to route traffic around failures. The response time of such alternative techniques is typically much longer than techniques that use link state as an indication of an error condition.

The end-to-end Ethernet link-integrity feature of the G1K-4 can be used in combination with GEC capability on attached devices. The combination provides an Ethernet traffic-restoration scheme that has a faster response time than techniques, such as spanning-tree rerouting, yet is more bandwidth efficient because spare bandwidth does not need to be reserved. The G1K-4 supports GEC, which is a Cisco proprietary standard similar to the IEEE link-aggregation standard (IEEE 802.3ad).

Although the G1K-4 card does not actively run GEC, it supports the end-to-end GEC functionality of attached Ethernet devices. If two Ethernet devices running GEC connect through G1K-4 cards to an ONS 15454 SDH network, the ONS 15454 SDH-side network is transparent to the EtherChannel devices. The EtherChannel devices operate as if they are directly connected to each other. Any combination of G1K-4 parallel circuit sizes can be used to support GEC throughput. GEC provides line-level active redundancy and protection (1:1) for attached Ethernet equipment. It can also bundle parallel G1K-4 data links together to provide more aggregated bandwidth. STP operates as if the bundled links are one link and permits GEC to use these multiple parallel paths. Without GEC, STP permits only a single nonblocked path. GEC can also provide G1K-4 card-level protection or redundancy because it can support a group of ports on different cards (or different nodes) so that if one port or card has a failure, traffic is rerouted over the other port or card.

Provision G1K-4 Ethernet Ports

Step 1 From the node view, double-click the **G1K-4 card** graphic to open the card view.

Step 2 Click the **Provisioning > Port** tab. An example is shown in Figure 11-19.

Step 3 For each G1K-4 port, provision the following parameters:

- **Port Name**—If you want to label the port, type the port name.
- **State**—Choose **IS** or **OOS-AINS** to activate or prepare a port for service. The following port states are available:
  - **IS**—The circuit is in service.
  - **OOS**—The circuit is out of service. Traffic is not passed on the circuit until it is in service.
  - **OOS-AINS**—(Default) The circuit is in service when it receives a valid signal; until then, the circuit is out of service.

— **OOS-MT**—The circuit is in a maintenance state. The maintenance state does not interrupt traffic flow; it suppresses alarms and conditions and allows loopbacks to be performed on the circuit. Use OOS-MT for circuit testing or to suppress circuit alarms temporarily. Change the state to IS, OOS, or OOS-AINS when testing is complete.

Figure 11-19 *Provisioning G1K-4 ports for SDH*

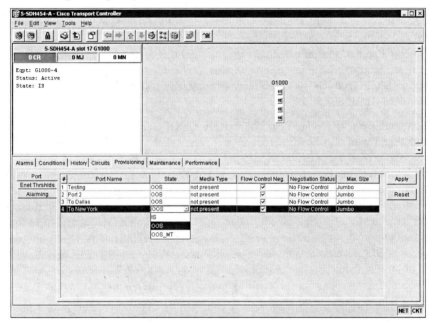

- **Flow Control Neg**—Click this check box to enable flow-control negotiation on the port (default). If you do not want to enable flow control, uncheck the box.

- **Max Size**—To permit the acceptance of jumbo-size Ethernet frames, choose **Jumbo** (default). If you do not want to permit jumbo-size Ethernet frames, choose **1548**.

- To activate flow control, the Ethernet device attached to the G1K-4 card must be set to autonegotiation. If flow control is enabled but the negotiation status indicates no flow control, check the autonegotiation settings on the attached Ethernet device.

- The maximum frame size of 1548 bytes, rather than the common maximum frame size of 1518 bytes, enables the port to accept valid Ethernet frames that use protocols such as Multiprotocol Label Switching (MPLS). Protocols, such as MPLS, add bytes in the shim header that can cause the frame size to exceed the common 1518-byte maximum.

Step 4 Click **Apply**.

Step 5 Refresh the Ethernet statistics:

   (a) Click the **Performance > Statistics** tab.

   (b) Click the **Refresh** button.

Provision a G1K-4 EtherSwitch Circuit

G1K-4 cards support point-to-point circuit configuration. Provisionable circuit sizes are VC4, VC4-2c, VC4-3c, VC4-4c, VC4-8c, and VC4-16c. Each Ethernet port maps to a unique VC4 circuit on the SDH side of the G1K-4. The two ends of the circuit must terminate at GE switches. The G1K-4 supports any combination of up to four circuits from the list of valid circuit sizes; however, the circuit sizes can add up to no more than VC4-16c.

Step 1 From the node view, click the **Circuits** tab and click **Create**. An example is shown in Figure 11-20.

Figure 11-20 *Provisioning G1K-4 EtherSwitch Circuit*

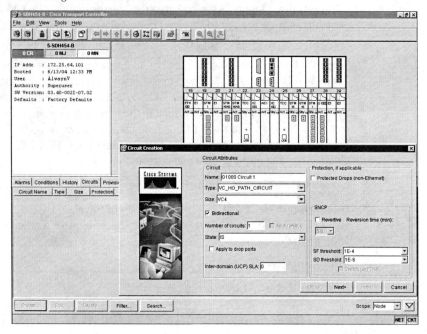

Step 2 In the Create Circuits dialog box, complete the following fields:

- **Name**—Assign a name to the circuit. The name can be alphanumeric and up to 48 characters (including spaces). If you leave the field blank, CTC assigns a default name to the circuit.

- **Type**—Choose **VC_HO_PATH_CIRCUIT**.
- **Size**—Choose the circuit size. Valid circuit sizes for a G1K-4 circuit are VC4, VC4-2c, VC4-3c, VC4-4c, VC4-8c, and VC4-16c.
- **Bidirectional**—Leave checked for this circuit (default).
- **Number of Circuits**—Leave set to 1 (default).
- **State**—Choose a service state to apply to the circuit:
  - **IS**—The circuit is in service.
  - **OOS**—The circuit is out of service. Traffic is not passed on the circuit until it is in service.
  - **OOS-AINS**—(Default) The circuit is in service when it receives a valid signal; until then, the circuit is out of service.
  - **OOS-MT**—The circuit is in a maintenance state. The maintenance state does not interrupt traffic flow; it suppresses alarms and conditions and allows loopbacks to be performed on the circuit. Use OOS-MT for circuit testing or to suppress circuit alarms temporarily. Change the state to IS, OOS, or OOS-AINS when testing is complete.
- **Apply to Drop Ports**—Check this box to apply the state chosen in the **State** field (IS or OOS-MT only) to the Ethernet circuit source and destination ports. You cannot apply OOS-AINS to G1K-4 Ethernet card ports. CTC will apply the circuit state to the ports if the circuit is in full control of the port. If not, a Warning dialog box displays the ports where the circuit state could not be applied. If not checked, CTC will not change the state of the source and destination ports.
- **Create Cross-Connects Only (TL1-Like)**—Uncheck this box.
- **Inter-domain (UCP) SLA**—If the circuit will travel on a UCP channel, enter the service level agreement (SLA) number. Otherwise, leave the field set to 0.
- **Auto Ranged**—Not available.
- **Protected Drops**—Leave unchecked.

Step 3 If the circuit will be routed on an SNCP, set the SNCP path selectors.

Step 4 Click **Next**.

Step 5 Provision the circuit source. An example is shown in Figure 11-21.

(a) From the Node pull-down menu, choose the circuit source node. Either end node can be the point-to-point circuit source.

(b) From the Slot pull-down menu, choose the slot containing the G1K-4 card that you will use for one end of the point-to-point circuit.

(c) From the Port pull-down menu, choose a port.

Figure 11-21 *EtherSwitch Circuit Source*

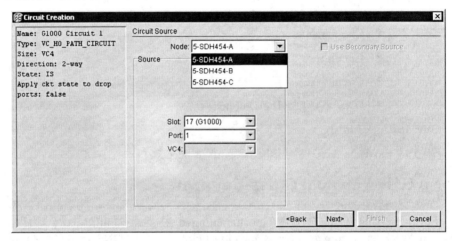

Step 6 Click **Next**.

Step 7 Provision the circuit destination. An example is shown in Figure 11-22.

(a) From the Node pull-down menu, choose the circuit destination node.

Figure 11-22 *EtherSwitch Circuit Destination*

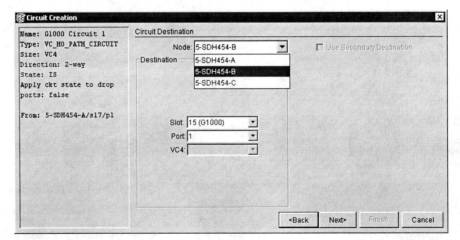

(b) From the Slot pull-down menu, choose the slot containing the G1K-4 card that you will use for other end of the point-to-point circuit.

(c) From the Port pull-down menu, choose a port.

Step 8 Click **Next**. The Circuits window appears.

Step 9 Confirm that the following information about the point-to-point circuit is correct:

- Circuit name
- Circuit type
- Circuit size
- ONS 15454 SDH circuit nodes

Step 10 Click **Finish**.

Step 11 Provision the G1K-4 Ethernet ports as needed.

Provision a G1K-4 Manual Cross-Connect

ONS 15454s require end-to-end DCC connectivity between nodes for normal provisioning of Ethernet circuits using CTC. When attempting to connect via third-party SDH equipment, OSI/TARP-based equipment does not allow tunneling of the ONS 15454 TCP/IP-based DCC. To circumvent a lack of continuous DCC, the Ethernet circuit must be manually cross-connected to a VC4-N channel riding through the non-ONS SDH network. This allows an Ethernet circuit to run from ONS node to ONS node while using the third-party non-ONS SDH network. The connectivity is shown in Figure 11-23. This section describes the procedure to configure G1K-4 manual cross-connects.

Figure 11-23 *G1K-4 Manual Cross-Connects*

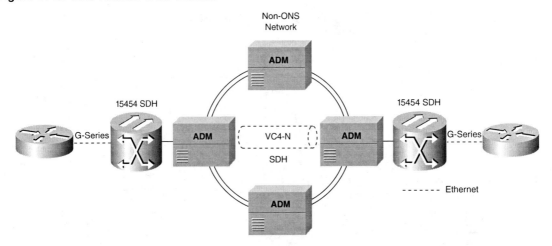

Step 1 From the node view, click the **Circuits** tab and click **Create**.

Step 2 In the Create Circuits dialog box, complete the following fields:

- **Name**—Assign a name to the source cross-connect. The name can be alphanumeric and up to 48 characters (including spaces). If you leave the field blank, CTC assigns a default name to the source cross-connect.
- **Type**—Choose **VC_HO_PATH_CIRCUIT**.
- **Size**—Choose the size of the circuit that will be carried by the cross-connect. The valid circuit sizes for a G1K-4 circuit are VC4, VC4-2c, VC4-3c, VC4-4c, VC4-8c, and VC4-16c.
- **Bidirectional**—Leave checked for this cross-connect (default).
- **Number of Circuits**—Leave set to 1 (default).
- **Auto Ranged**—Not available.
- **State**—Choose a service state to apply to the circuit after it is created:
  - **IS**—The circuit is in service.
  - **OOS**—The circuit is out of service. Traffic is not passed on the circuit until it is in service.
  - **OOS-AINS**—(Default) The circuit is in service when it receives a valid signal; until then, the circuit is out of service.
  - **OOS-MT**—The circuit is in a maintenance state. The maintenance state does not interrupt traffic flow; it suppresses alarms and conditions and allows loopbacks to be performed on the circuit. Use OOS-MT for circuit testing or to suppress circuit alarms temporarily. Change the state to IS, OOS, or OOS-AINS when testing is complete.
- **Apply to Drop Ports**—Uncheck this box.
- **Create Cross-Connects Only (TL1-Like)**—Uncheck this box.
- **Inter-domain (UCP) SLA**—If the circuit will travel on a UCP channel, enter the SLA number. Otherwise, leave the field set to 0.
- **Protected Drops**—Leave unchecked.

Step 3 If the circuit carried by the cross-connect will be routed on an SNCP, set the SNCP path selectors:

- **Revertive**—Check this box if you want traffic to revert to the working path when the conditions that diverted it to the protect path are repaired. If you do not choose Revertive, traffic remains on the protect path after the switch.

- **Reversion Time**—If Revertive is checked, choose the reversion time. Click the **Reversion Time** field and choose a reversion time from the pull-down menu. The range is 0.5 to 12.0 minutes. The default is 5.0 minutes. This is the amount of time that will elapse before the traffic reverts to the working path. Traffic can revert when conditions causing the switch are cleared.
- **SF Threshold**—Choose from one E3, one E4, or one E5.
- **SD Threshold**—Choose from one E5, one E6, one E7, one E8, or one E9.
- **Switch on PDI-P**—Check this box if you want traffic to switch when a VC4 payload defect indicator is received (VC4 circuits only).

Step 4 Click **Next**.

Step 5 Provision the circuit source.

(a) From the Node pull-down menu, choose the circuit source node.

(b) From the Slot pull-down menu, choose the G1K-4 that will be the cross-connect source.

(c) From the Port pull-down menu, choose the cross-connect source port.

Step 6 Click **Next**.

Step 7 Provision the circuit destination.

(a) From the Node pull-down menu, choose the cross-connect source node selected in Step 5. (For Ethernet cross-connects, the source and destination nodes are the same.)

(b) From the Slot pull-down menu, choose the STM-N card that is connected to the non-ONS equipment.

(c) Depending on the STM-N card, choose the port and/or VC4 from the Port and VC4 pull-down menus.

Step 8 Click **Next**.

Step 9 Verify the cross-connection information:

- Circuit name
- Circuit type
- Circuit size
- ONS 15454 SDH circuit nodes

If the information is not correct, click the **Back** button and repeat the procedure with the correct information.

Step 10 Click **Finish**.

Step 11 Provision the G1K-4 Ethernet ports as needed.

Step 12 To complete the procedure, repeat Steps 1 to 10 at the second ONS 15454 SDH.

E-Series Provisioning of Ethernet over SDH

The E-Series cards incorporate Layer 2 switching within the card. The ONS 15454 SDH E-Series cards include the E100T-G and the E1000-2-G cards. The E100T-G card provides 12 switched, IEEE 802.3-compliant 10/100 BASE-T Ethernet ports. The ports detect the speed of an attached device via autonegotiation and automatically connect at the appropriate speed and duplex mode—either half or full duplex—and determine whether to enable or disable flow control.

The E1000-2-G cards provide two switched, IEEE 802.3-compliant GE (1000-Mbps) ports that support full-duplex operation. An ONS 15454 SDH requires the *G* versions of the E-Series Ethernet cards. Ethernet circuits can link ONS nodes through point-to-point, SPR, or hub-and-spoke configurations. Two nodes usually connect with a point-to-point configuration. More than two nodes usually connect with an SPR configuration or a hub-and-spoke configuration.

Multicard EtherSwitch provisions two or more Ethernet cards to act as a single Layer 2 switch. It supports one VC4-2c or two VC4 circuits. The bandwidth of the single switch formed by the Ethernet cards matches the bandwidth of the provisioned Ethernet circuit up to VC4-2c worth of bandwidth.

Single-card EtherSwitch allows each Ethernet card to remain a single switching entity within the ONS 15454 SDH shelf. This option allows a full VC4-4c worth of bandwidth between two Ethernet circuit points. Single-card bandwidth options are as follows:

- VC4-4c
- VC4-2c + VC4-2c
- VC4-2c + VC4 + VC4
- VC4 + VC4 + VC4 + VC4

VLANs can be created after the Ethernet circuits have been provisioned. The ONS 15454 SDH network supports a maximum of 509 user-provisionable VLANs using the E-Series cards. The Ethernet ports can then be provisioned and assigned to VLANs. Specific sets of ports define the broadcast domain for the ONS 15454 SDH. The definition of VLAN ports includes all Ethernet and packet-switched SDH port types. All VLAN IP address discovery, flooding, and forwarding are limited to these ports.

The ONS 15454 SDH 802.1Q-based VLAN mechanism provides logical isolation of subscriber LAN traffic over a common SDH transport infrastructure. Each subscriber has an Ethernet port at each site, and each subscriber is assigned to a VLAN. Although the subscriber's VLAN data flows over shared circuits, the service appears to the subscriber as a private data transport infrastructure. It is precisely this logical closed user-group mechanism that enables service providers to offer VPN services based on VLAN membership.

Provision E-Series Ethernet Ports

Step 1 Display CTC and double-click the card graphic to open the Ethernet card.

Step 2 Click the **Provisioning > Ether Port** tab. An example is shown in Figure 11-24.

Figure 11-24 *Provisioning E-Series Ports*

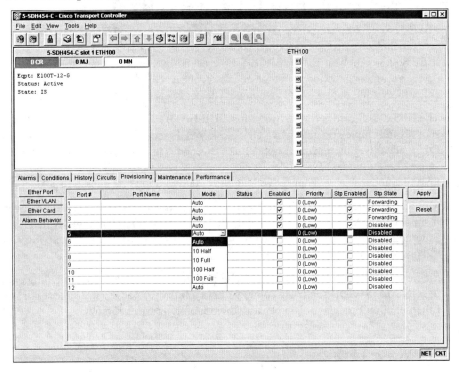

Step 3 From the Port window, choose the appropriate mode for each Ethernet port. The following are valid choices for the E100T-G card:

- Auto
- 10 Half
- 10 Full
- 100 Half
- 100 Full

The following are valid choices for the E1000-2-G card:

- Auto
- 1000 Full

E-Series Provisioning of Ethernet over SDH

Both 1000 Full and Auto mode set the E1000-2-G port to the 1000-Mbps and full-duplex operating mode. However, flow control is disabled when 1000 Full is selected. Choosing Auto mode enables the E1000-2-G card to autonegotiate flow control. Flow control is a mechanism that prevents network congestion by ensuring that transmitting devices do not overwhelm receiving devices with data. The E1000-2-G port handshakes with the connected network device to determine whether that device supports flow control.

Step 4 Click the **Enabled** check box(es) to activate the corresponding Ethernet port(s).

Step 5 Click **Apply**. Your Ethernet ports are now provisioned and ready to be configured for VLAN membership.

Step 6 Repeat this procedure for all other cards that will be in the VLAN.

Provision an E-Series EtherSwitch Point-to-Point Circuit (Multicard or Single Card)

The ONS 15454 SDH can set up a point-to-point (straight) Ethernet circuit as single card or multicard. Multicard EtherSwitch limits bandwidth to VC4-2c of bandwidth between two Ethernet circuit points, but allows adding nodes and cards and creating an SPR. Figure 11-25 shows an E-Series multicard EtherSwitch point-to-point implementation.

Figure 11-25 *Multicard EtherSwitch Point-to-Point Circuit*

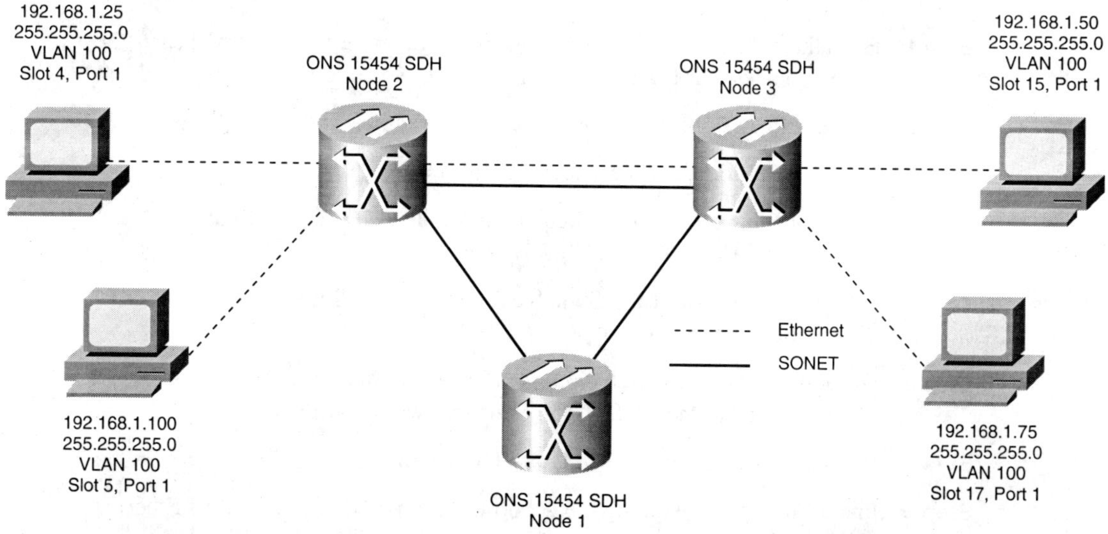

Single-card EtherSwitch allows each Ethernet card to remain a single switching entity within the ONS chassis. Single-card EtherSwitch allows a full VC4-4c of bandwidth between two Ethernet circuit points. Figure 11-26 shows an E-Series single-card EtherSwitch point-to-point implementation.

Figure 11-26 *Single-Card EtherSwitch Point-to-Point Circuit*

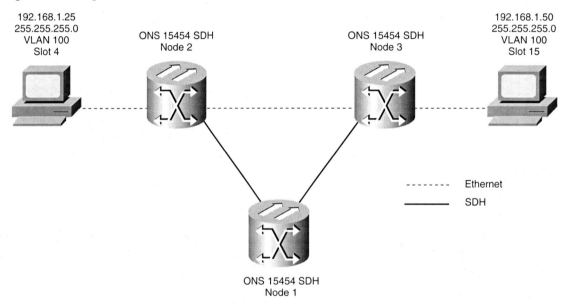

Step 1 From the node view, double-click one of the Ethernet cards that will carry the circuit. Change card settings only if there are no circuits using this card.

Step 2 Click the **Provisioning > Ether Card** tab. An example is shown in Figure 11-27.

Step 3 Under Card Mode, choose one of the following:

- For Multicard EtherSwitch circuit groups, choose **Multicard EtherSwitch Group**. Click **Apply**.

- For single-card EtherSwitch circuits, choose **Single-card EtherSwitch**. Click **Apply**.

Step 4 For multicard EtherSwitch circuits only, repeat Steps 1 to 3 for all other Ethernet cards in the ONS 15454 SDH that will carry the circuit.

Step 5 From the View menu, choose **Go to Other Node**.

Step 6 In the Choose Node dialog box, choose the other ONS 15454 SDH Ethernet circuit endpoint node and repeat Steps 1 to 5.

Figure 11-27 *Provisioning E-Series EtherSwitch Circuits*

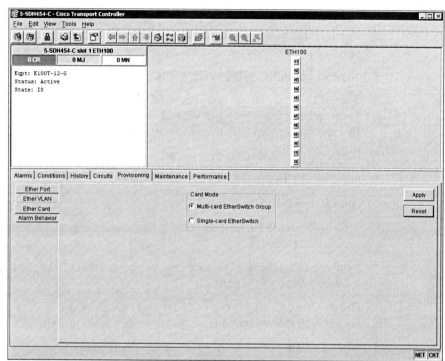

Step 7 Click the **Circuits** tab and click **Create**. An example is shown in Figure 11-28.

Figure 11-28 *E-Series Circuit Attributes*

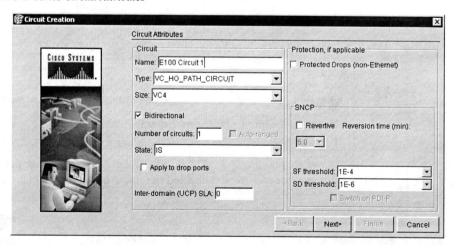

Step 8 In the Create Circuits dialog box, complete the following fields:

- **Name**—Assign a name to the circuit. The name can be alphanumeric and up to 48 characters (including spaces). If you leave the field blank, CTC assigns a default name to the circuit.
- **Type**—Choose **VC_HO_PATH_CIRCUIT**.
- **Size**—Choose the circuit size. The valid circuit sizes for an Ethernet Multicard circuit are VC4 and VC4-2c. The valid circuit sizes for an Ethernet Single-card circuit are VC4, VC4-2c, and VC4-4c.
- **Bidirectional**—Leave the default unchanged (checked).
- **Number of Circuits**—Leave the default unchanged (one).
- **Auto Ranged**—Not available.
- **State**—Choose **IS** (in service). Ethergroup circuits are stateless, and always in service.
- **Apply to Drop Ports**—Uncheck this box; states cannot be applied to E-Series Ethernet card ports.
- **Create Cross-Connects Only (TL1-Like)**—Uncheck this box; it does not apply to Ethernet circuits.
- **Inter-domain (UCP) SLA**—If the circuit will travel on a UCP channel, enter the SLA number. Otherwise, leave the field set to 0.
- **Protected Drops**—Leave the default unchanged (unchecked).

Step 9 If the circuit will be routed on an SNCP, set the SNCP path selectors.

Step 10 Click **Next**.

Step 11 Provision the circuit source. An example is shown in Figure 11-29.

    (a) From the Node pull-down menu, choose one of the EtherSwitch circuit endpoint nodes. (Either end node can be the EtherSwitch circuit source.)

    (b) From the Slot pull-down menu, choose one of the following:

- If you are building a multicard EtherSwitch circuit, choose **Ethergroup**.
- If you are building a single-card EtherSwitch circuit, choose the Ethernet card where you enabled the single-card EtherSwitch.

Step 12 Click **Next**.

Figure 11-29 *E-Series Circuit Source*

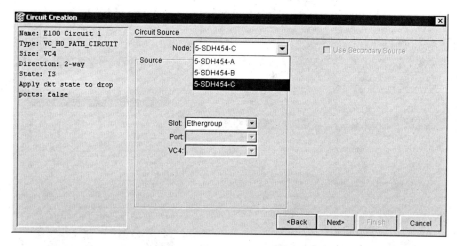

Step 13 Provision the circuit destination. An example is shown in Figure 11-30.

    (a) From the Node pull-down menu, choose the second EtherSwitch circuit endpoint node.

    (b) From the Slot pull-down menu, choose one of the following:

        — If you are building a multicard EtherSwitch circuit, choose **Ethergroup**.

        — If you are building a single-card EtherSwitch circuit, choose the Ethernet card where you enabled the single-card EtherSwitch.

Figure 11-30 *E-Series Circuit Destination*

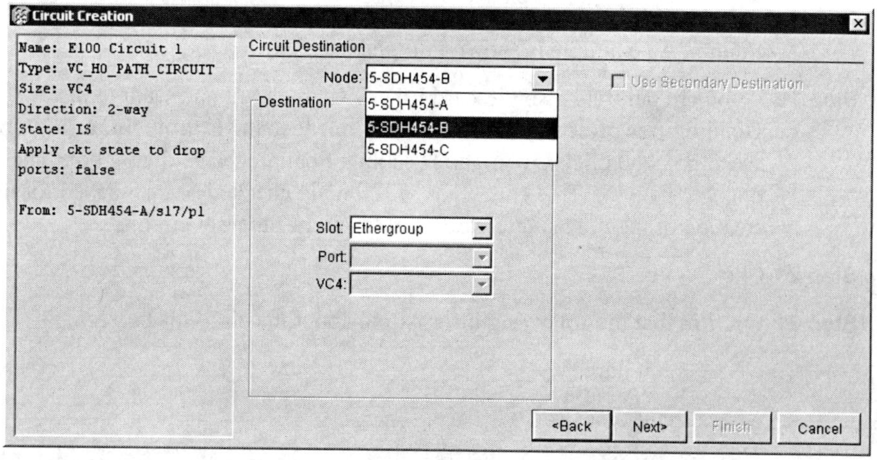

Step 14 Click **Next**.

Step 15 If the desired VLAN already exists, go to Step 18. Under Circuit VLAN Selection, click **New VLAN**. An example is shown in Figure 11-31.

Figure 11-31 *E-Series VLAN Provisioning*

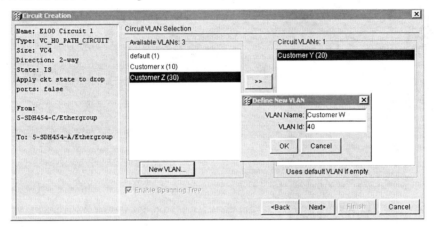

Step 16 In the New VLAN dialog box, complete the following:

- **VLAN Name**—Assign an easily identifiable name to your VLAN.

- **VLAN ID**—Assign a VLAN ID. The VLAN ID should be the next available number between 2 and 4093 that is not already assigned to an existing VLAN. Each ONS 15454 SDH network supports a maximum of 509 user-provisionable VLANs.

Step 17 Click **OK**.

Step 18 Under Circuit VLAN Selection, highlight the VLAN name and click the arrow (>>) button to move the available VLAN(s) to the Circuit VLANs column. An example is shown in Figure 11-32.

Step 19 If you are building a single-card EtherSwitch circuit and want to disable spanning-tree protection on this circuit, uncheck the **Enable Spanning Tree** check box and click **OK** in the Disabling Spanning Tree dialog box. The Enable Spanning Tree check box will remain checked or unchecked for the creation of the next single-card point-to-point Ethernet circuit.

Step 20 Click **Next**.

Step 21 Confirm that the following information about the circuit is correct:

- Circuit name
- Circuit type
- Circuit size
- ONS 15454 SDH circuit nodes

Figure 11-32 *Assigning the Available VLANs*

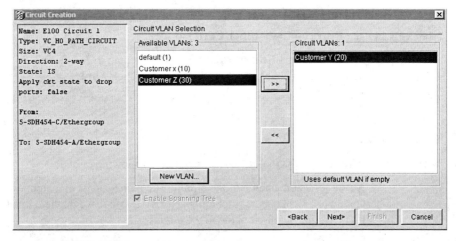

Step 22 Click **Finish**.

Step 23 Provision Ethernet ports and assign them for VLAN Membership.

Provision an E-Series Shared Packet Ring Circuit

The SPR provisioning steps for the multicard EtherSwitch is described in this section. The steps are generic and can be adapted to any SPR architecture. An example is shown in Figure 11-33.

Figure 11-33 *EtherSwitch SPR*

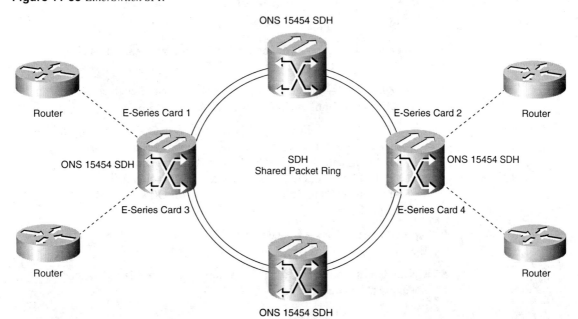

Step 1 From the node view, double-click one of the Ethernet cards that will carry the circuit. Change card settings only if there are no circuits using this card.

Step 2 Click the **Provisioning > Ether Card** tab.

Step 3 Verify that **Multicard EtherSwitch Group** is selected. If Multicard EtherSwitch Group is not selected, choose it and click **Apply**.

Step 4 Repeat Steps 1 to 3 for all other Ethernet cards in the ONS 15454 SDH that will carry the SPR.

Step 5 Click the **Circuits** tab, and then click **Create**.

Step 6 In the Create Circuits dialog box, complete the following fields:

- **Name**—Assign a name to the circuit. The name can be alphanumeric and up to 48 characters (including spaces). If you leave the field blank, CTC assigns a default name to the circuit.

- **Type**—Choose **VC_HO_PATH_CIRCUIT**.

- **Size**—Choose the circuit size. For SPR Ethernet, valid circuit sizes are VC4 or VC4-2c.

- **Bidirectional**—Leave checked for this circuit (default).

- **Number of Circuits**—Leave set to 1 (default).

- **Auto Ranged**—Unavailable.

- **State**—Choose a service state to apply to the circuit:

    — **IS**—The circuit is in service.

    — **OOS**—The circuit is out of service. Traffic is not passed on the circuit until it is in service.

    — **OOS-AINS**—(Default) The circuit is in service when it receives a valid signal; until then, the circuit is out of service.

    — **OOS-MT**—The circuit is in a maintenance state. The maintenance state does not interrupt traffic flow; it suppresses alarms and conditions and allows loopbacks to be performed on the circuit. Use OOS-MT for circuit testing or to suppress circuit alarms temporarily. Change the state to IS, OOS, or OOS-AINS when testing is complete.

- **Apply to Drop Ports**—Uncheck this box; states cannot be applied to E-Series Ethernet card ports.

- **Create Cross-Connects Only (TL1-Like)**—Uncheck this box; it does not apply to Ethernet circuits.

- **Inter-domain (UCP) SLA**—If the circuit will travel on a UCP channel, enter the SLA number. Otherwise, leave the field set to 0.
- **Protected Drops**—Leave unchecked.

Step 7 If the circuit will be routed on an SNCP, set the SNCP path selectors.

Step 8 Click **Next**.

Step 9 Provision the circuit source:

(a) From the Node pull-down menu, choose one of the SPR circuit endpoint nodes. (Either end node can be the SPR circuit source.)

(b) From the Slot pull-down menu, choose **Ethergroup**.

Step 10 Click **Next**.

Step 11 Provision the circuit destination.

(a) From the Node pull-down menu, choose the second SPR circuit endpoint node.

(b) From the Slot pull-down menu, choose **Ethergroup**.

Step 12 Click **Next**.

Step 13 Review the VLANs listed under Available VLANs. If the VLAN you want to use is displayed, go to Step 15. If you need to create a new VLAN, complete the following substeps:

(a) Click the **New VLAN** button.

(b) In the New VLAN dialog box, complete the following:

- **VLAN Name**—Assign an easily identifiable name to your VLAN.
- **VLAN ID**—Assign a VLAN ID. The VLAN ID should be the next available number between 2 and 4093 that is not already assigned to an existing VLAN. Each ONS 15454 SDH network supports a maximum of 509 user-provisionable VLANs.

(c) Click **OK**.

Step 14 Click the VLAN you want to use in the Available VLANs column, and then click the Arrow (>>) button to move the VLAN to the Circuit VLANs column. Moving the VLAN from Available VLANs to Circuit VLANs forces all the VLAN traffic to use the SPR you are creating.

Step 15 Click **Next**.

Step 16 Under Circuit Routing Preferences, uncheck the **Route Automatically** check box and click **Next**.

Step 17 Under Route Review and Edit panel, click the source node, and then click either span (green arrow) leading from the source node. The span turns white.

Step 18 Click **Add Span**. The span turns blue. CTC adds the span to the Included Spans list.

Step 19 Click the node at the end of the blue span.

Step 20 Click the green span with the source node from Step 17. The span turns white.

Step 21 Click **Add Span**. The span turns blue. An example is shown in Figure 11-34.

Figure 11-34 *E-Series Circuit Routing*

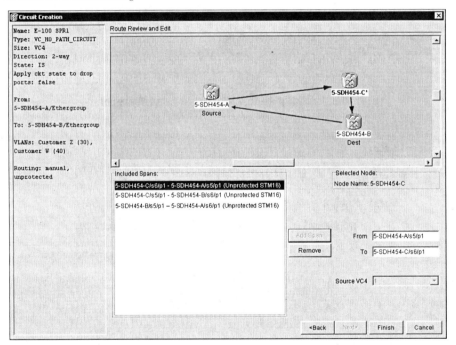

Step 22 Repeat Steps 18 to 21 for every node in the ring.

Step 23 Verify that the new circuit is correctly configured. If the circuit information is not correct, click the **Back** button and repeat the procedure with the correct information. If the circuit is incorrect, you can also click **Finish**, delete the completed circuit, and begin the procedure again.

Step 24 Click **Finish**.

Step 25 Provision E-Series Ethernet ports and assign VLAN membership.

Provision an E-Series Hub-and-Spoke Ethernet Circuit

The hub-and-spoke configuration connects point-to-point circuits (the spokes) to an aggregation point (the hub). In many cases, the hub links to a high-speed connection and the spokes are Ethernet cards. Figure 11-35 shows a sample hub-and-spoke ring. This section provides generic steps for creating a hub-and-spoke Ethernet circuit configuration.

Figure 11-35 *Hub-and-Spoke Ethernet Circuit*

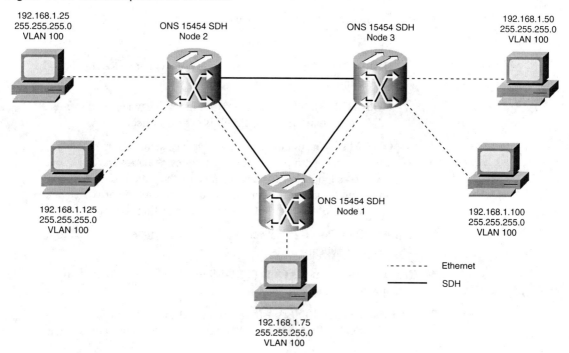

Step 1 From the node view, double-click the Ethernet card that will carry the circuit. Change card settings only if there are no circuits using this card.

Step 2 Click the **Provisioning > Ether Card** tab.

Step 3 Under Card Mode, choose **Single-card EtherSwitch** and click **Apply**.

Step 4 Navigate to the other ONS 15454 SDH endpoint node of the hub-and-spoke circuit and repeat Steps 1 to 3.

Step 5 Click the **Circuits** tab, and then click **Create**.

Step 6 In the Create Circuits dialog box, complete the following fields:

- **Name**—Assign a name to the circuit. The name can be alphanumeric and up to 48 characters (including spaces). If you leave the field blank, CTC assigns a default name to the circuit.

- **Type**—Choose **VC_HO_PATH_CIRCUIT**.
- **Size**—Choose the circuit size.
- **Bidirectional**—Leave checked for this circuit (default).
- **Number of Circuits**—Leave set to 1 (default).
- **Auto Ranged**—Not available.
- **State**—Choose a service state to apply to the circuit:
    - **IS**—The circuit is in service.
    - **OOS**—The circuit is out of service. Traffic is not passed on the circuit until it is in service.
    - **OOS-AINS**—(Default) The circuit is in service when it receives a valid signal; until then, the circuit is out of service.
    - **OOS-MT**—The circuit is in a maintenance state. The maintenance state does not interrupt traffic flow; it suppresses alarms and conditions and allows loopbacks to be performed on the circuit. Use OOS-MT for circuit testing or to suppress circuit alarms temporarily. Change the state to IS, OOS, or OOS-AINS when testing is complete.
- **Apply to Drop Ports**—Uncheck this box; states cannot be applied to E-Series Ethernet card ports.
- **Create Cross-Connects Only (TL1-Like)**—Uncheck this box; it does not apply to Ethernet circuits.
- **Inter-domain (UCP) SLA**—If the circuit will travel on a UCP channel, enter the SLA number. Otherwise, leave the field set to 0.
- **Protected Drops**—Leave unchecked.

Step 7 If the circuit will be routed on an SNCP, set the SNCP path selectors.

Step 8 Click **Next**.

Step 9 Provision the circuit source:

(a) From the Node pull-down menu, choose one of the hub-and-spoke circuit endpoint nodes. (Either end node can be the circuit source.)

(b) From the Slot pull-down menu, choose the Ethernet card where you enabled the Single-card EtherSwitch in Step 3.

Step 10 Click **Next**.

Step 11 Provision the circuit destination.

(a) From the Node pull-down menu, choose the second EtherSwitch circuit endpoint node.

(b) From the Slot pull-down menu, choose the Ethernet card where you enabled the Single-card EtherSwitch in Step 3.

Step 12 Click **Next**.

Step 13 Review the VLANs listed under Available VLANs. If the VLAN you want to use is displayed, go to Step 15. If you need to create a new VLAN, complete the following steps:

(a) Click the **New VLAN** button.

(b) In the New VLAN dialog box, complete the following:

- **VLAN Name**—Assign an easily identifiable name to your VLAN.
- **VLAN ID**—Assign a VLAN ID. The VLAN ID should be the next available number between 2 and 4093 that is not already assigned to an existing VLAN. Each ONS 15454 SDH network supports a maximum of 509 user-provisionable VLANs.

(c) Click **OK**.

Step 14 Click the **VLAN** you want to use in the Available VLANs column, then click the arrow (>>) button to move the VLAN to the Circuit VLANs column. Moving the VLAN from Available VLANs to Circuit VLANs forces all the VLAN traffic to use the SPR you create.

Step 15 Click **Next**.

Step 16 Confirm that the following information about the hub-and-spoke circuit is correct:

- Circuit name
- Circuit type
- Circuit size
- VLAN names
- ONS 15454 SDH circuit nodes

If the circuit information is not correct, click the **Back** button and repeat the procedure with the correct information. You can also click **Finish**, delete the completed circuit, and start the procedure from the beginning.

Step 17 Click **Finish**.

Step 18 Navigate to an ONS 15454 SDH that will be an endpoint for the second Ethernet circuit.

Step 19 Double-click the Ethernet card that will carry the circuit.

Step 20 Click the **Provisioning > Ether Card** tab.

Step 21 Under Card Mode, choose **Single-Card EtherSwitch** and click **Apply**.

Step 22 From the View menu, choose **Go to Other Node**.

Step 23 In the Select Node dialog box, choose the other endpoint node for the second circuit and repeat Steps 19 to 21 at that node.

Step 24 Click the **Circuits** tab, and then click **Create**.

Step 25 In the Create Circuits dialog box, complete the following fields:

- **Name**—Assign a name to the circuit. The name can be alphanumeric and up to 48 characters (including spaces). If you leave the field blank, CTC assigns a default name to the circuit.

- **Type**—Choose **STS**.

- **Size**—Choose the circuit size.

- **Bidirectional**—Leave checked for this circuit.

- **Number of Circuits**—Leave set to one (default).

- **Auto Ranged**—Not available.

- **State**—Choose a service state to apply to the circuit:

    — **IS**—The circuit is in service.

    — **OOS**—The circuit is out of service. Traffic is not passed on the circuit until it is in service.

    — **OOS-AINS**—(Default) The circuit is in service when it receives a valid signal; until then, the circuit is out of service.

    — **OOS-MT**—The circuit is in a maintenance state. The maintenance state does not interrupt traffic flow; it suppresses alarms and conditions and allows loopbacks to be performed on the circuit. Use OOS-MT for circuit testing or to suppress circuit alarms temporarily. Change the state to IS, OOS, or OOS-AINS when testing is complete.

- **Apply to Drop Ports**—Uncheck this box; states cannot be applied to E-Series Ethernet card ports.

- **Create Cross-Connects Only (TL1-Like)**—Uncheck this box because it does not apply to Ethernet circuits.

- **Inter-domain (UCP) SLA**—If the circuit will travel on a UCP channel, enter the SLA number. Otherwise, leave the field set to 0.

- **Protected Drops**—Leave unchecked.

Step 26 If the circuit will be routed on an SNCP, set the SNCP path selectors.

E-Series Provisioning of Ethernet over SDH   611

Step 27 Click **Next**.

Step 28 Provision the circuit source.

    (a) From the Node pull-down menu, choose one of the hub-and-spoke circuit endpoint nodes. (Either end node can be the circuit source.)

    (b) From the Slot pull-down menu, choose the Ethernet card where you enabled the single-card EtherSwitch in Step 21.

Step 29 Click **Next**.

Step 30 Provision the circuit destination.

    (a) From the Node pull-down menu, choose the second EtherSwitch circuit endpoint node.

    (b) From the Slot pull-down menu, choose the Ethernet card where you enabled the single-card EtherSwitch.

Step 31 Click **Next**.

Step 32 Highlight the VLAN that you created for the first circuit and click the arrow (>>) button to move the VLAN(s) from the Available VLANs column to the Selected VLANs column.

Step 33 Click **Next**.

Step 34 Confirm that the following information about the second hub-and-spoke circuit is correct:

- Circuit name
- Circuit type
- Circuit size
- VLAN names
- ONS 15454 SDH circuit nodes

If the circuit information is not correct, click the **Back** button and repeat the procedure with the correct information. You can also click **Finish**, delete the completed circuit, and start the procedure from the beginning.

Step 35 Click **Finish**.

Step 36 Provision the E-Series Ethernet ports and assign VLAN membership.

Provision an E-Series Single-Card EtherSwitch Manual Cross-Connect

ONS 15454s require end-to-end DCC connectivity between nodes for normal provisioning of Ethernet circuits using CTC. When attempting to connect via third-party equipment,

OSI/TARP-based equipment does not allow tunneling of the ONS 15454 SDH TCP/IP-based DCC. To circumvent a lack of continuous DCC, the Ethernet circuit must be manually cross-connected to a VC4-N channel riding through the non-ONS SDH network. This allows an Ethernet circuit to run from ONS node to ONS node while using the third-party non-ONS SDH network. The connectivity is shown in Figure 11-36. This section describes the procedure to configure E-Series single-card EtherSwitch manual cross-connects.

Figure 11-36 *Ethernet Manual Cross-Connects*

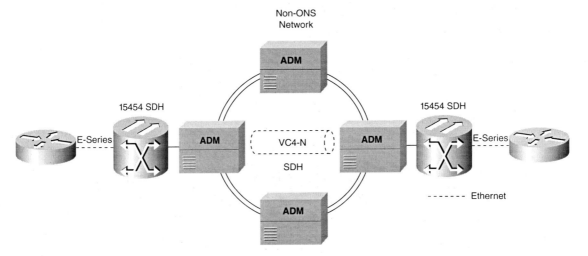

Step 1 From the node view, double-click the Ethernet card that will carry the cross-connect. In this procedure, cross-connect refers to a circuit connection created within the same node between the Ethernet card and an STM-N card connected to third-party equipment. You create cross-connects at the source and destination nodes so an Ethernet circuit can be routed from source to destination across third-party equipment.

Step 2 Click the **Provisioning > Ether Card** tab. Change card settings only if there are no circuits using this card.

Step 3 Under Card Mode, choose **Single-Card EtherSwitch** and click **Apply**.

Step 4 Click the **Circuits** tab, and then click **Create**.

Step 5 In the Create Circuits dialog box, complete the following fields:

- **Name**—Assign a name to the cross-connect. The name can be alphanumeric and up to 48 characters (including spaces). If you leave the field blank, CTC assigns a default name to the cross-connect.

- **Type**—Choose **VC_HO_PATH_CIRCUIT**.

- **Size**—Choose the cross-connect size. For single-card EtherSwitch, the available sizes are VC4, VC4-2c, and VC4-4c.
- **Bidirectional**—Leave checked for this cross-connect (default).
- **Number of Circuits**—Leave set to 1 (default).
- **Auto Ranged**—Not available.
- **State**—Choose a service state to apply to the circuit:
  - **IS**—The circuit is in service.
  - **OOS**—The circuit is out of service. Traffic is not passed on the circuit until it is in service.
  - **OOS-AINS**—(Default) The circuit is in service when it receives a valid signal; until then, the circuit is out of service.
  - **OOS-MT**—The circuit is in a maintenance state. The maintenance state does not interrupt traffic flow; it suppresses alarms and conditions and allows loopbacks to be performed on the circuit. Use OOS-MT for circuit testing or to suppress circuit alarms temporarily. Change the state to IS, OOS, or OOS-AINS when testing is complete.
- **Apply to Drop Ports**—Uncheck this box.
- **Create Cross-Connects Only (TL1-Like)**—Uncheck this box.
- **Inter-domain (UCP) SLA**—If the circuit will travel on a UCP channel, enter the SLA number. Otherwise, leave the field set to 0.
- **Protected Drops**—Leave unchecked.

Step 6 If the circuit carried by the cross-connect will be routed on an SNCP, set the SNCP path selectors:

- **Revertive**—Check this box if you want traffic to revert to the working path when the conditions that diverted it to the protect path are repaired. If you do not choose Revertive, traffic remains on the protect path after the switch.
- **Reversion Time**—If Revertive is checked, choose the reversion time. Click the **Reversion Time** field and choose a reversion time from the pull-down menu. The range is 0.5 to 12.0 minutes. The default is 5.0 minutes. This is the amount of time that will elapse before the traffic reverts to the working path. Traffic can revert when conditions causing the switch are cleared.
- **SF Threshold**—Choose from one E3, one E4, or one E5.

- **SD Threshold**—Choose from one E5, one E6, one E7, one E8, or one E9.
- **Switch on PDI-P**—Check this box if you want traffic to switch when a VC4 payload defect indicator is received (VC4 circuits only).

Step 7 Click **Next**.

Step 8 Provision the circuit source.

(a) From the Node pull-down menu, choose the cross-connect source node.

(b) From the Slot pull-down menu, choose the Ethernet card where you enabled the single-card EtherSwitch in Step 3.

Step 9 Click **Next**.

Step 10 Provision the circuit destination.

(a) From the Node pull-down menu, choose the cross-connect circuit source node selected in Step 8. (For Ethernet cross-connects, the source and destination nodes are the same.)

(b) From the Slot pull-down menu, choose the STM-N card that is connected to the non-ONS equipment.

(c) Depending on the STM-N card, choose the port and/or VC4 from the Port and VC4 pull-down menus.

Step 11 Click **Next**.

Step 12 Review the VLANs listed under Available VLANs. If the VLAN you want to use is displayed, go to Step 14. If you need to create a new VLAN, complete the following steps:

(a) Click the **New VLAN** button.

(b) In the New VLAN dialog box, complete the following:

- **VLAN Name**—Assign an easily identifiable name to your VLAN.
- **VLAN ID**—Assign a VLAN ID. The VLAN ID should be the next available number between 2 and 4093 that is not already assigned to an existing VLAN. Each ONS 15454 SDH network supports a maximum of 509 user-provisionable VLANs.

(c) Click **OK**.

Step 13 Click the VLAN you want to use in the Available VLANs column, then click the arrow (>>) button to move the VLAN to the Circuit VLANs column.

Step 14 Click **Next**. The Circuit Creation (Circuit Routing Preferences) dialog box opens.

Step 15 Confirm that the following information about the Single-card EtherSwitch manual cross-connect is correct:

- Circuit name
- Circuit type
- Circuit size
- VLAN names
- ONS 15454 SDH nodes

If the information is not correct, click the **Back** button and repeat the procedure with the correct information.

Step 16 Click **Finish**.

Step 17 Provision E-Series Ethernet ports and assign VLAN membership.

NOTE   The appropriate VC4 circuit must exist in the non-ONS equipment to connect the two VC4 circuits from the ONS 15454 SDH Ethernet manual cross-connect endpoints.

Provision an E-Series Multicard EtherSwitch Manual Cross-Connect

Step 1 From the node view, double-click the Ethernet card where you want to create the cross-connect. In this procedure, cross-connect refers to a circuit connection created within the same node between the Ethernet card and an STM-N card connected to third-party equipment. You create cross-connects at the source and destination nodes, so an Ethernet circuit can be routed from source to destination across third-party equipment.

Step 2 Click the **Provisioning > Ether Card** tab.

Step 3 Under Card Mode, choose **Multicard EtherSwitch Group** and click **Apply**.

Step 4 From the View menu, choose **Go to Parent View**.

Step 5 Repeat Steps 1 to 4 for any other Ethernet cards in the ONS 15454 SDH that will carry the circuit.

Step 6 Click the **Circuits** tab, and then click **Create**.

Step 7 In the Create Circuits dialog box, complete the following fields:

- **Name**—Assign a name to the source cross-connect. The name can be alphanumeric and up to 48 characters (including spaces). If you leave the field blank, CTC assigns a default name to the source cross-connect.

- **Type**—Choose **VC_HO_PATH_CIRCUIT**.
- **Size**—Choose the size of the circuit that will be carried by the cross-connect. For multicard EtherSwitch circuits, the available sizes are VC4 and VC4-2c.
- **Bidirectional**—Leave checked (default).
- **Number of Circuits**—Leave set to one (default).
- **Auto Ranged**—Not available.
- **State**—Choose a service state to apply to the circuit:
    - **IS**—The circuit is in service.
    - **OOS**—The circuit is out of service. Traffic is not passed on the circuit until it is in service.
    - **OOS-AINS**—(Default) The circuit is in service when it receives a valid signal; until then, the circuit is out of service.
    - **OOS-MT**—The circuit is in a maintenance state. The maintenance state does not interrupt traffic flow; it suppresses alarms and conditions and allows loopbacks to be performed on the circuit. Use OOS-MT for circuit testing or to suppress circuit alarms temporarily. Change the state to IS, OOS, or OOS-AINS when testing is complete.
- **Apply to Drop Ports**—Uncheck this box.
- **Create Cross-Connects Only (TL1-Like)**—Uncheck this box.
- **Inter-domain (UCP) SLA**—If the circuit will travel on a UCP channel, enter the SLA number. Otherwise, leave the field set to 0.
- **Protected Drops**—Leave unchecked.

Step 8 If the circuit carried by the cross-connect will be routed on an SNCP, set the SNCP path selectors:

- **Revertive**—Check this box if you want traffic to revert to the working path when the conditions that diverted it to the protect path are repaired. If you do not choose **Revertive**, traffic remains on the protect path after the switch.
- **Reversion Time**—If Revertive is checked, choose the reversion time. Click the **Reversion Time** field and choose a reversion time from the pull-down menu. The range is 0.5 to 12.0 minutes. The default is 5.0 minutes. This is the amount of time that will elapse before the traffic reverts to the working path. Traffic can revert when conditions causing the switch are cleared.
- **SF Threshold**—Choose from one E3, one E4, or one E5.

- **SD Threshold**—Choose from one E5, one E6, one E7, one E8, or one E9.
- **Switch on PDI-P**—Check this box if you want traffic to switch when a VC4 payload defect indicator is received (VC4 circuits only).

Step 9 Click **Next**.

Step 10 Provision the cross-connect source.

(a) From the Node pull-down menu, choose the cross-connect source node.

(b) From the Slot pull-down menu, choose **Ethergroup**.

Step 11 Click **Next**.

Step 12 From the Node pull-down menu under Destination, choose the circuit source node selected in Step 10. For Ethernet cross-connects, the source and destination nodes are the same. The Slot field automatically is provisioned for Ethergroup.

Step 13 Click **Next**.

Step 14 Review the VLANs listed under Available VLANs. If the VLAN you want to use is displayed, go to Step 16. If you need to create a new VLAN, complete the following steps:

(a) Click the **New VLAN** button.

(b) In the New VLAN dialog box, complete the following:

- **VLAN Name**—Assign an easily identifiable name to your VLAN.
- **VLAN ID**—Assign a VLAN ID. The VLAN ID should be the next available number between 2 and 4093 that is not already assigned to an existing VLAN. Each ONS 15454 SDH network supports a maximum of 509 user-provisionable VLANs.

(c) Click **OK**.

Step 15 Click the VLAN you want to use in the Available VLANs column, then click the arrow (>>) button to move the VLAN to the Circuit VLANs column.

Step 16 Click **Next**. The Circuit Creation (Circuit Routing Preferences) dialog box opens.

Step 17 In the left pane, verify the cross-connect information. In this task, *circuit* refers to the Ethernet cross-connect.

- Circuit name
- Circuit type

- Circuit size
- VLANs
- ONS 15454 SDH nodes

If the information is not correct, click the **Back** button and repeat the procedure with the correct information.

Step 18 Click **Finish**.

Step 19 Provision the E-Series Ethernet ports.

Step 20 Provision Ethernet ports for VLAN membership.

Step 21 From the View menu, choose **Go to Home View**.

Step 22 Click the **Circuits** tab.

Step 23 Highlight the circuit and click **Edit**. The Edit Circuit dialog box opens.

Step 24 Click **Drops**, and then click **Create**. The Define New Drop dialog box opens.

Step 25 From the Slot menu, choose the STM-N card that links the ONS 15454 SDH to the non-ONS 15454 SDH equipment.

Step 26 From the Port menu, choose the appropriate port.

Step 27 From the VC4 menu, choose the VC4 that matches the VC4 of the connecting non-ONS 15454 SDH equipment.

Step 28 Click **OK**.

Step 29 Confirm the circuit information that displays in the Edit Circuit dialog box and click **Close**.

Step 30 Repeat Steps 2 to 29 at the second ONS 15454 SDH Ethernet manual cross-connect endpoint. The appropriate VC4 circuit must exist in the non-ONS 15454 SDH equipment to connect the two ONS 15454 SDH Ethernet manual cross-connect endpoints.

Provision Ethernet Ports for VLAN Membership

The ONS 15454 SDH enables you to configure the VLAN membership and IEEE 802.1Q tag handling of individual Ethernet ports.

Step 1 Display the CTC card view for the Ethernet card.

Step 2 Click the **Provisioning > Ether VLAN** tab. An example is shown in Figure 11-37.

Figure 11-37 *Provisioning VLANs*

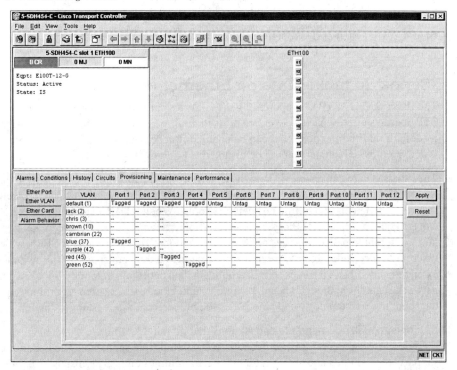

Step 3 To put a port in a VLAN, click the port and choose either **Tagged** or **Untag**. If a port is marked Untag, the ONS 15454 SDH will tag ingress frames and strip tags from egress frames. If a port is marked Tagged, the 15454 SDH will handle ingress frames according to VLAN ID, but egress frames will not have their tags removed. A port marked with the -- symbol does not belong to any VLAN.

If a port is a member of only one VLAN, go to that VLANs row and choose **Untag** from the Port column. Choose -- for all the other VLAN rows in that Port column. The VLAN with Untag selected can connect to the port, but other VLANs cannot access that port.

If a port is a trunk port that connects multiple VLANs to an external trunking device, you must have tagging (802.1Q) enabled for all the VLANs that connect to that external device. Choose **Tagged** at all VLAN rows that need to be trunked. Choose **Untag** at one or more VLAN rows in the trunk port's column that do not need to be trunked (for example, the default VLAN). Each Ethernet port must be attached to at least one untagged VLAN.

Step 4 After each port is assigned to its appropriate VLAN, click **Apply**.

Enable E-Series Spanning Tree on Ethernet Ports

The IEEE 802.1D STP is supported by the E-Series card. When an Ethernet card is installed, STP operates over all packet-switched ports including Ethernet and SDH ports. STP is disabled by default on Ethernet ports, but can be enabled if necessary. On SDH interface ports, STP is activated by default and cannot be disabled.

The Ethernet card can enable STP on the Ethernet ports to allow redundant paths to the attached Ethernet equipment. STP spans cards so that both equipment and facilities are protected against failure. STP detects and eliminates network loops. When STP detects multiple paths between any two network hosts, STP blocks ports until only one path exists between any two network hosts. The single path eliminates possible bridge loops. This is crucial for SPRs, which include inherent looped links. To remove loops, STP defines a tree that spans all the switches in an extended network. STP forces certain redundant data paths into a standby (blocked) state. If one network segment in the STP becomes unreachable, the spanning-tree algorithm reconfigures the spanning-tree topology and reactivates the blocked path to reestablish the link.

STP operation is transparent to end stations that do not discriminate between connections to a single LAN segment or to a switched LAN with multiple segments. The E-Series card supports one STP instance per circuit and a maximum of eight STP instances per ONS 15454 SDH. The ONS 15454 SDH can operate multiple instances of STP to support VLANs in a looped topology. You can dedicate separate circuits across the SDH ring for different VLAN groups—for example, one circuit for private TLS services and one circuit for Internet access. Each circuit would run its own STP to maintain VLAN connectivity in a multiring environment. Table 11-2 indicates the E-Series card STP default values. These preconfigured values work well for most implementations. Theses values can be modified if necessary.

Table 11-2 *E-Series Card STP Default Values*

STP Variable	Default Value	Value Range
Priority	32,768	0–65,535
Bridge Max Age	20 seconds	6–40 seconds
Bridge Hello Time	2 seconds	1–10 seconds
Bridge Forward Delay	15 seconds	4–30 seconds

Step 1 Display the CTC card view.

Step 2 Click the **Provisioning > Port** tab.

Step 3 In the left column, find the applicable port number and check the **STP Enabled** check box to enable STP for that port.

Step 4 Click **Apply**.

Step 5 At the node view, click the **Maintenance > Etherbridge > Spanning Trees** tab to view spanning-tree parameters. An example is shown in Figure 11-38.

Figure 11-38 *Spanning-Tree Parameters*

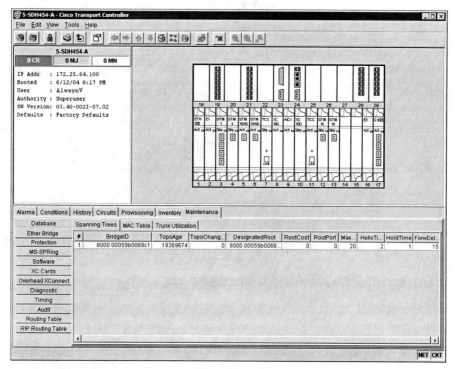

Retrieve the MAC Table Information

Step 1 Click the **Maintenance > EtherBridge > MAC Table** tab.

Step 2 Choose the appropriate Ethernet card or Ethergroup from the Layer 2 Domain pull-down menu.

Step 3 Click **Retrieve** for the ONS 15454 SDH to retrieve and display the current MAC IDs. An example is shown in Figure 11-39.

Figure 11-39 *MAC Table Information*

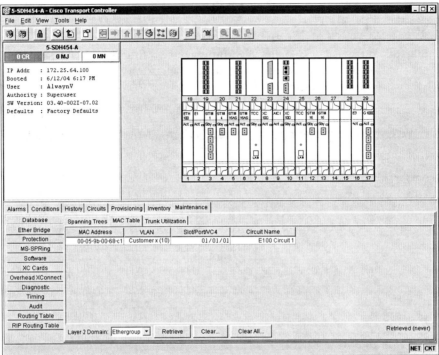

Creating Ethernet RMON Alarm Thresholds

Step 1 Display the CTC node view.

Step 2 Click the **Provisioning > Etherbridge > Thresholds** tabs.

Step 3 Click **Create**. The Create Ethernet Threshold dialog box opens. An example is shown in Figure 11-40.

Step 4 From the Slot menu, choose the appropriate Ethernet card.

Step 5 From the Port menu, choose the port on the Ethernet card.

Step 6 From the Variable menu, choose the Ethernet threshold variable.

Step 7 From Alarm Type, indicate whether the rising threshold, falling threshold, or both the rising and falling thresholds will trigger the event.

Step 8 From the Sample Type pull-down menu, choose either **Relative** or **Absolute**. Relative restricts the threshold to use the number of occurrences in the user-set sample period. Absolute sets the threshold to use the total number of occurrences, regardless of any time period.

Figure 11-40 *RMON Alarm Thresholds*

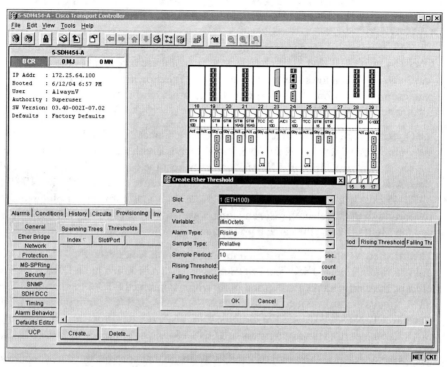

Step 9 Type in an appropriate number of seconds for the Sample Period.

Step 10 Type in the appropriate number of occurrences for the Rising Threshold.

Step 11 Type in the appropriate number of occurrences for the Falling Threshold. In most cases, a falling threshold is set lower than the rising threshold. A falling threshold is the counterpart to a rising threshold. When the number of occurrences is above the rising threshold and then drops below a falling threshold, it resets the rising threshold.

When the network problem that caused 1001 collisions in 15 minutes subsides and creates only 799 collisions in 15 minutes; for example, occurrences fall below a falling threshold of 800 collisions. This resets the rising threshold so that if network collisions again spike over a 1000 per 15 minute period, an event again triggers when the rising threshold is crossed. An event is triggered only the first time a rising threshold is exceeded (otherwise a single network problem might cause a rising threshold to be exceeded multiple times and cause a flood of events).

Step 12 Click the **OK** button to complete the procedure.

ML-Series Provisioning of Ethernet, IP, and RPR over SONET/SDH

The ML-Series cards are independent multilayer switches that can be integrated into the ONS 15454 SONET and SDH chassis. Addition of these cards makes the ONS 15454 Layer 2 and 3 aware, and endows the 15454 with full IP functionality. Two versions of the ML-Series card are in production; the 2-port Gigabit Ethernet card (ML1000-2) and 12-port Fast Ethernet card (ML100T-12).

An ONS 15454 SONET with a 10 Gigabit cross-connect card (XC10G) can host the card in any traffic card slot, whereas an ONS 15454 SONET with a cross-connect card (XC) or cross-connect virtual tributary card (XCVT) can only host the card in the four high-speed traffic slots. An ONS 15454 SDH can host the card in any traffic card slot. The ML100T-12 features 12 RJ-45 interfaces, and the ML1000-2 features two small form-factor pluggable (SFP) slots supporting short-wavelength (SX) and long-wavelength (LX) optical modules. The ML100T-12 and the ML1000-2 use the same hardware and software base and offer the same feature sets.

The ML-Series cards can be installed in either a North American ONS 15454 SONET chassis or in an International ONS 15454 SDH chassis. When installed in an ONS 15454 SONET chassis, the card features two virtual ports with a combined STS-48 maximum. The STS or STM circuits are provisioned through the ONS 15454 GUI (CTC) in the same manner as standard OC-N card STS or STM circuits. CTC also provides provisioning, inventory, SONET/SDH alarm reporting, and other standard ONS 15454 card functions for the ML-Series.

The ML-Series cards ship with Cisco IOS Software that controls the data functions of the card. The Cisco IOS image used by the ML-Series card is not permanently stored on the ML-Series card but in the Flash memory of the TCC+ or TCC2 card. During a hard reset, when a card is physically removed and reinserted, the Cisco IOS Software image is downloaded from the Flash memory of the TCC+/TCC2 to the memory cache of the ML-Series card. The cached image is then decompressed and initialized for use by the ML-Series card. During a soft reset, when the ML-Series card is reset through CTC or an IOS command, the ML-Series card checks its cache for an IOS image. If a valid and current IOS image exists, the ML-Series card decompresses and initializes the image. If the image does not exist, the ML-Series requests a new copy of the IOS image from the TCC. Caching of the IOS image provides a significant time savings when a warm reset is performed.

NOTE An ML-Series Cisco IOS image upgrade is accomplished only through CTC. Cisco IOS images for the ML-Series card are available only as part of an ONS 15454 software release.

CTC provides limited GUI-based configuration options for the ML-Series card. CTC offers ML-Series status information, SONET/SDH alarm management, Cisco IOS Telnet session initialization, Cisco IOS configuration file management, and SONET/SDH circuit provisioning.

The Cisco IOS command-line interface (CLI) is the primary user interface for the ML-Series card. Most Ethernet and Layer 3 configuration for the card can be done only via the Cisco IOS CLI. SONET/SDH circuits cannot be provisioned through IOS, but must be configured through CTC. SONET circuits can also be provisioned with TL1 on the ONS 15454. Users can access Cisco IOS in three ways:

- Card console port on the faceplate of the card
- SNMP management via an Ethernet port
- Telnet session via an Ethernet port

The Telnet sessions can be initiated through a terminal program on a workstation or through CTC. Table 11-3 shows the feature richness of the ML-Series card. The integration of Layer 2/3 switching with SONET/SDH in the ML-Series card opens up a whole new world of design options for the Optical network architect.

Table 11-3 *ML-Series Features*

Feature Set	Description
CTC features	Standard STS circuit provisioning for SONET virtual ports
	Standard STM circuit provisioning for SDH virtual ports
	SONET alarm reporting for path alarms and other ML-Series specific alarms on ONS 15454 SONET
	SDH alarm reporting for path alarms and other ML-Series specific alarms on ONS 15454 SDH
	Raw port statistics
	Standard inventory and card management functions
	Cisco IOS CLI Telnet sessions from CTC
	IOS startup configuration file management
General features	Cisco Discovery Protocol (CDP) support on Ethernet ports
	Dynamic Host Configuration Protocol (DHCP) relay
	Hot Standby Router Protocol (HSRP) over 10/100 Ethernet, Gigabit Ethernet, FEC, GEC, and Bridge Group Virtual Interface (BVI)
	Internet Control Message Protocol (ICMP)
	IRB routing mode support
	Simple Network Management Protocol (SNMP)
	Transaction Language 1 (TL1)
	Cisco IOS
	NEBS3 compliant

continues

Table 11-3 *ML-Series Features (Continued)*

Feature Set	Description
Layer 1.5 features	10/100BASE-TX half-duplex and full-duplex data transmission
	1000BASE-SX, 1000BASE-LX full-duplex data transmission
	Two SONET virtual ports with maximum bandwidth of STS-48c per card on ONS 15454 SONET
	Two SDH virtual ports with maximum bandwidth of VC4-16c per card on ONS 15454 SDH
Layer 2 features	Cisco HDLC SONET/SDH port encapsulation
	(Note: HDLC does not provide VLAN trunking support.)
	PPP/Bridge Control Protocol (PPP/BCP) SONET/SDH port encapsulation (VLAN trunking supported via BCP)
	LEX (PPP over HDLC) SONET/SDH port encapsulation (VLAN trunking supported via LEX)
	Packet over SONET/SDH (POS)
	POS channel (with LEX encapsulation only) support
	G-Series card compatible
	Interoperability with POS routers
Layer 2 bridging features	Layer 2 transparent bridging
	Layer 2 MAC learning, aging, and switching by hardware
	Spanning Tree Protocol (IEEE 802.1D) per bridge group
	Protocol tunneling
	A maximum of 255 active bridge groups
	Up to 60,000 MAC addresses per card, with a supported limit of 8000 per bridge group
	Integrated routing and bridging (IRB)
	VLAN features
	802.1P/Q-based VLAN trunking
	802.1Q VLAN tunneling
	802.1D Spanning Tree Protocol and 802.1W Rapid Spanning Tree Protocol
	IEEE 802.1Q-based VLAN routing and bridging
ML100T-12 Fast EtherChannel features	Aggregation of up to four Fast Ethernet ports
	Load sharing based on source and destination IP addresses of unicast packets
	Load sharing for bridge traffic based on MAC addresses
	IRB on the Fast EtherChannel
	IEEE 802.1Q trunking on the Fast EtherChannel
	Up to six active FEC port channels

Table 11-3 *ML-Series Features (Continued)*

Feature Set	Description
ML1000-2 Gigabit EtherChannel features	Aggregation of the two Gigabit Ethernet ports
	Load sharing for bridge traffic based on MAC addresses
	IRB on the Gigabit EtherChannel
	IEEE 802.1Q trunking on the Gigabit EtherChannel
QoS features	SLAs with 1-Mbps granularity
	Input policing
	Guaranteed bandwidth (weighted round-robin [WDRR] plus strict-priority scheduling)
	Classification based on Layer 2 priority, VLAN ID, Layer 3 ToS/DSCP, and port
	Low Latency Queuing support for unicast VoIP
Layer 3 features	Default routes
	IP unicast and multicast forwarding support
	Simple IP access control lists (ACLs) (both Layer 2 and Layer 3 forwarding path)
	Extended IP ACLs in software (control plane only)
	IP and IP multicast routing and switching between Ethernet ports
	Load balancing among equal cost paths based on source and destination IP addresses
	Up to 18,000 IP routes
	Up to 20,000 IP host entries
	Up to 40 IP multicast groups
Layer 3 routing protocol support	VPN routing and forwarding lite (VRF Lite)
	Intermediate System-to-Intermediate System (IS-IS) Protocol
	Routing Information Protocol (RIP and RIP II)
	Enhanced Interior Gateway Routing Protocol (EIGRP)
	Open Shortest Path First (OSPF) Protocol
	Protocol Independent Multicast (PIM)—Sparse, sparse-dense, and dense modes
	Secondary addressing
	Static routes
	Local proxy ARP
	Border Gateway Protocol (BGP)
	Classless interdomain routing (CIDR)

For the ONS 15454 SONET, ML-Series cards feature two SONET virtual ports with a maximum combined bandwidth of STS-48. Each port carries an STS circuit with a size of STS-1, STS-3c, STS-6c, STS-9c, STS-12c, or STS-24c. On the ONS 15454 SDH, ML-Series cards feature two SDH virtual ports with a maximum combined bandwidth of VC4-16c. Each port carries an STM circuit with a size of VC4, VC4-2C, VC4-3C, or VC4-8C.

The ML-Series supports three forms of SONET/SDH port encapsulation: Cisco HDLC, PPP-BCP, and LEX. Cisco HDLC is standard on most Cisco data devices. However, it does not offer VLAN trunking support. PPP-BCP is a popular standard linked to RFC 2878 that supports VLAN trunking via BCP. LEX is a protocol based on PPP over HDLC used by the G-Series cards. The LEX protocol also supports VLAN trunking. This allows interoperability between the ML-Series and POS routers, as well as interoperability with the G-Series Ethernet cards on the Cisco ONS 15454 SONET, ONS 15454 SDH, and ONS 15327. All three formats support bridging and routing, standard SONET/SDH payload scrambling, and HDLC frame check sequence.

The ML-Series offers Fast EtherChannel, Gigabit EtherChannel, and POS channel link aggregation. Link-aggregation groups multiple ports into a larger logical port and provides resiliency during the failure of any individual ports. The ML-Series supports a maximum of four Ethernet ports in Fast EtherChannel, two Ethernet ports in Gigabit EtherChannel, and two SONET/SDH virtual ports in the POS channel. The POS channel is supported only with LEX encapsulation. *Traffic flows* map to individual ports based on MAC SA/DA for bridged packets and IP SA/DA for routed packets. There is no support for policing or class-based packet priorities when link aggregation is configured.

ML-Series cards provide optical network designers with two options for creating customer VPNs: Ethernet VLANs and VRF lite VPNs. Standard VRF is an extension of IP routing that provides multiple routing instances and separate IP routing and forwarding tables for each VPN. VRF is used with internal MP-iBGP. MP-iBGP distributes the VRF information between routers to provide Layer 3 MPLS VPN. However, VRF lite stores VRF information locally and does not distribute the VRF information to connected equipment. VRF information directs traffic to the correct interfaces and subinterfaces when the traffic is received from customer routers or from service provider routers. VRF lite allows an ML-Series card, acting as customer equipment, to have multiple interfaces and subinterfaces with service provider equipment. The customer ML-Series card can then service multiple customers. Normal customer equipment serves a single customer.

Accessing the ML-Series Card

An IOS startup configuration file must be loaded and the ML-Series card must be installed and initialized prior to opening a Cisco IOS CLI session on CTC. The ML-Series card needs a startup configuration file to configure itself beyond the default configuration when the card is reset. If no startup configuration file exists in the TCC flash memory, the card will boot up to a default configuration. It is not possible to establish a Telnet connection to the card until a startup configuration file is loaded onto the ML-Series card. Until then, access to the card can be achieved only via the console port. Users can manually set up the startup configuration file through the serial console port and the Cisco IOS CLI configuration mode or load a Cisco IOS-supplied sample startup configuration file through CTC. Due to space limitations on the ML-Series card faceplate, the console port is an RJ-11 modular jack rather than the more common RJ-45 modular jack. Cisco supplies an RJ-11 to RJ-45 console cable adapter with each ML-Series card. After connecting the adapter, the console port functions like the standard Cisco RJ-45 console port. The data rate and character format of the console PC or terminal must match the console port default settings of 9600 baud, 8 data bits, 1 stop bit, and no parity.

NOTE The ML-Series card does not allow users to access the read-only memory monitor mode (ROMMON). The ML-Series card ROMMON is preconfigured to boot the correct Cisco IOS Software image for the ML-Series card. When the running configuration file is altered, a RUNCFG-SAVENEED condition appears in the CTC. This condition is a reminder to enter a **copy running-config startup-config** command in the Cisco IOS CLI, or the changes will be lost, when the ML-Series card reboots.

CTC enables a user to load the startup configuration file required by the ML-Series. A Cisco-supplied sample IOS startup configuration file, named Basic-IOS-startup-config.txt, is available on the Cisco ONS 15454 SONET/SDH software CD-ROM. CISCO15 is the IOS CLI default line password and the enable password for this configuration. Users can also create their own startup configuration file through the serial console port on the ML-Series card. During initialization, the ML-Series card first checks for a locally, valid cached copy of IOS. It will then either download the Cisco IOS Software image from the TCC or proceed directly to decompressing and initializing the image. If you are connected to the console port, you will notice that the Router> Prompt appears after IOS initialization. You will be able to enter the IOS configuration mode and set up the basic ML-Series configuration at this time.

CTC can load a Cisco IOS startup configuration file into the TCC card Flash before the ML-Series card is physically installed in the slot. When installed, the ML-Series card downloads and applies the Cisco IOS Software image and the preloaded Cisco IOS startup configuration file. Preloading the startup configuration file allows an ML-Series card to immediately operate as a fully configured card when inserted into the ONS 15454 SONET/SDH chassis.

If the ML-Series card is booted up prior to the loading of the Cisco IOS startup configuration file into TCC card Flash, the ML-Series card must be reset to use the Cisco IOS startup configuration file or the user can issue the **copy start run** IOS command at the CLI to configure the ML-Series card to use the Cisco IOS startup configuration file.

The following procedure details the initial loading of a Cisco IOS startup configuration file through CTC.

Step 1 At the card-level view of the ML-Series card, click the **IOS** tab. The CTC IOS window appears. An example is shown in Figure 11-41.

Figure 11-41 *ML-Series IOS CLI*

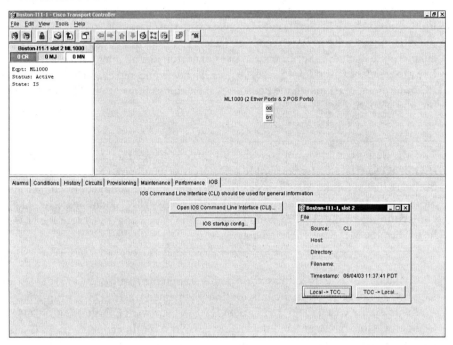

- **Step 2** Click the **IOS Startup Config** button. The Config File dialog box appears.
- **Step 3** Click the **Local -> TCC** button, as shown in Figure 11-41.
- **Step 4** The sample IOS startup configuration file can be installed from either the ONS 15454 SONET/SDH software CD or from a PC or network folder:

  To install the Cisco supplied startup config file from the ONS 15454 SONET/SDH software CD-ROM, insert the CD-ROM into the CD-ROM drive of the PC or workstation. Using the CTC Config File dialog box, navigate to the CD-ROM drive of the PC or workstation and double-click the **Basic-IOS-startup-config.txt** file

  To install a user config file from a PC or network folder, navigate to the folder containing the desired IOS startup config file and double-click the desired IOS startup config file.
- **Step 5** At the Are You Sure? dialog box, click the **Yes** button. The Directory and Filename fields on the configuration file dialog update to reflect that the IOS startup config file is loaded onto the TCC.

Step 6 To load the IOS startup config file from the TCC to the ML-Series card, follow these substeps:

  (a) If the ML-Series card has already been installed, right-click the ML-Series card at the node level CTC view and choose **Reset Card**. After the reset, the ML-Series card runs under the newly loaded IOS startup config.

  (b) If the ML-Series card is not yet installed, installing the ML-Series card into the slot will load and run the newly loaded IOS startup configuration on the ML-Series card.

NOTE When the Cisco IOS startup configuration file is downloaded and parsed at initialization, if there is an error in the parsing of this file, an ERROR-CONFIG alarm is reported and appears under the CTC alarms pane or in TL1. This typically occurs if there is an error in the startup config file and can be corrected.

After loading the startup config file, the ML-Series card must be reset to use the Cisco IOS startup configuration file. At this point, the user can access the IOS command line via CTC (as shown in Figure 11-42), the console port, or by telnetting to the node IP address and slot number.

Figure 11-42 *CTC IOS CLI*

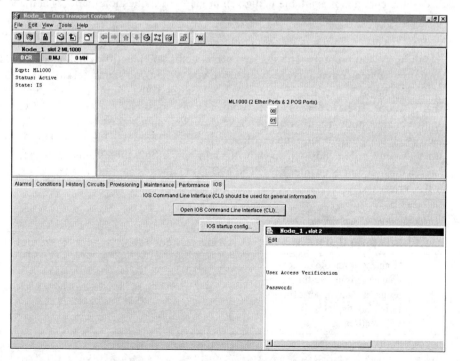

The virtual type terminal (vty) lines used for Telnet access are initially not fully configured. To gain Telnet access to the ML-Series card, one must configure the vty lines via the serial console connection or preload a startup configuration file that configures the vty lines. To Telnet into the ML-Series card, you must use the IP address and the total of the slot number plus 2000 as the Telnet port address. For an IP address of 150.100.1.5 and slot 2, for example, you would enter **or telnet 150.100.1.5:2002**.

You can also Telnet into the ML-Series card using the management port. Because there is no separate management port on ML-Series cards, any Fast Ethernet interface on the ML100T-12 card, any Gigabit Ethernet interface on the ML1000-2 card, or any POS interface on either ML-Series card can be configured as a management port. For the POS interface to exist, an STS or STM circuit must first be created through CTC or TL1. You can remotely Telnet into and configure the ML-Series card through the management port, but first you must configure an IP address so that the ML-Series card is reachable.

ML-Series IOS Command Modes

The ML-Series cards use a subset of the Cisco IOS command set. Appendix A, "ML-Series Command Reference," describes the commands that are unique to the ML-Series card, as well as IOS commands not supported by the ML-Series. The IOS user interface has several different modes. The commands available to you depend on which mode you are in. To get a list of the commands available in a given mode, type a question mark (**?**) at the system prompt.

Table 11-4 describes the most commonly used modes, how to enter the modes, and the resulting system prompts. The system prompt helps you identify which mode you are in and, therefore, which commands are available to you. When you start a session on the ML-Series card, you begin in user EXEC mode. A small subset of the commands is available in user EXEC mode. To have access to all commands, you must enter privileged EXEC mode, also called enable mode. From privileged EXEC mode, you can type in any EXEC command or access global configuration mode. Most of the EXEC commands are single-use commands, such as **show** commands, which show the current configuration status, and **clear** commands, which clear counters or interfaces. The EXEC commands are not saved across reboots of the ML-Series card.

Table 11-4 *ML-Series IOS Command Modes*

Mode	What You Use It For	How to Access	Prompt
User EXEC	Connect to remote devices, change terminal settings on a temporary basis, perform basic tests, and display system information.	Log in.	Router>

Table 11-4 *ML-Series IOS Command Modes (Continued)*

Mode	What You Use It For	How to Access	Prompt
Privileged EXEC (also called enable mode)	Set operating parameters. The privileged command set includes the commands in user EXEC mode, as well as the **configure** command. Use this command mode to access the other command modes.	From user EXEC mode, enter the **enable** command and the enable password.	Router#
Global configuration	Configure features that affect the system as a whole.	From privileged EXEC mode, enter the **configure terminal** command.	Router(config)#
Interface configuration	Enable features for a particular interface. Interface commands enable or modify the operation of a Fast Ethernet, Gigabit Ethernet, or POS port.	From global configuration mode, enter the **interface** *type number* command. For example, enter **interface fastethernet 0** for Fast Ethernet or **interface gigabitethernet 0** for Gigabit Ethernet interfaces or **interface pos 0** for POS interfaces.	Router(config-if)#
Line configuration	Configure the console port or vty line from the directly connected console or the virtual terminal used with Telnet.	From global configuration mode, enter the **line console 0** command to configure the console port or the **line vty** *line-number* command to configure a VTY line.	Router(config-line)#

The configuration modes enable you to make changes to the running configuration. If you later save the configuration, these commands are stored across ML-Series card reboots. You must start in global configuration mode. From global configuration mode, you can enter interface configuration mode, subinterface configuration mode, and a variety of protocol-specific modes. ROMMON is a separate mode used when the ML-Series card cannot boot properly. For example, your ML-Series card might enter ROM monitor mode if it does not find a valid system image when it is booting, or if its configuration file is corrupted at startup.

ML-Series interface characteristics include, but are not limited to, IP address, address of the port, data encapsulation method, and media type. These characteristics must be configured on the Fast Ethernet, Gigabit Ethernet, and POS interfaces via the IOS CLI. A fair knowledge of routing and switching is required to leverage the Layer 2/3 functionality of the ML-Series card. For details on IOS configuration, refer to the appropriate IOS configuration manual and command reference of the version of IOS you use. The following sections present ML-Series IOS configurations, based on the premise that the reader has a fairly detailed knowledge of routing, switching, and the IOS command set.

ML-Series Fast Ethernet Interface Configuration

Follow the steps in Table 11-5 to configure the IP address or bridge group number, autonegotiation, and flow control on a Fast Ethernet interface, beginning in global configuration mode.

Table 11-5 *ML-Series Fast Ethernet Interface Configuration Steps*

Step	Command Syntax	Description
Step 1	Router(config)# **interface fastethernet** *number*	Activates interface configuration mode to configure the Fast Ethernet interface.
Step 2	Router(config-if)# **ip address** *ip-address {subnet-mask* | **bridge-group** *bridge-group-number}*	Sets the IP address and IP subnet mask to be assigned to the interface. or Assigns a network interface to a bridge group.
Step 3	Router(config-if)# **[no] speed {10 | 100 | auto}**	Configures the transmission speed for 10 or 100 Mbps. If you set the speed or duplex for **auto**, you enable autonegotiation on the system—the ML-Series card matches the speed and duplex mode of the partner node.
Step 4	Router(config-if)# **[no] duplex {full | half | auto}**	Full duplex, half duplex, or autonegotiate.
Step 5	Router(config-if)# **flowcontrol send {on | off | desired}**	(Optional) Sets the send flow-control value for an interface. Flow control works only with port-level policing.
Step 6	Router(config-if)# **no shutdown**	Enables the interface by preventing it from shutting down.
Step 7	Router(config)# **end**	Returns to privileged EXEC mode.
Step 8	Router# **copy running-config startup-config**	(Optional) Saves your configuration changes to the TCC Flash.

The default setting for the negotiation mode is auto for the Fast Ethernet interface. Example 11-1 shows an initial configuration of a Fast Ethernet interface with an IP address, autonegotiated speed, and autonegotiated duplex.

Example 11-1 *Fast Ethernet Interface Configuration*

```
Router(config)# interface fastethernet 1
Router(config-if)# ip address 172.16.1.1 255.255.255.0
Router(config-if)# speed auto
Router(config-if)# duplex auto
Router(config-if)# no shutdown
Router(config-if)# end
Router# copy running-config startup-config
```

Gigabit Ethernet Interface Configuration

Follow the steps in Table 11-6 to configure the IP address or bridge group number, autonegotiation, and flow control on a Gigabit Ethernet interface, beginning in global configuration mode.

Table 11-6 *Gigabit Ethernet Interface Configuration Steps*

Step	Command	Description
Step 1	Router(config)# **interface gigabitethernet** *number*	Activates interface configuration mode to configure the Gigabit Ethernet interface.
Step 2	Router(config-if)# **ip address** *ip-address* {*subnet-mask*	**bridge-group** *bridge-group-number*}	Sets the IP address and subnet mask. or Assigns a network interface to a bridge group.
Step 3	Router(config-if)# [**no**] **negotiation auto**	Sets negotiation mode to **auto**. The Gigabit Ethernet port attempts to negotiate the link with the partner port. If you want the port to force the link up no matter what the partner port setting is, set the Gigabit Ethernet interface to **no negotiation auto**.
Step 4	Router(config-if)# **flowcontrol** {**send**	**receive**} {**on**	**off**	**desired**}	(Optional) Sets the send or receive flow-control value for an interface. Flow control works only with port-level policing.
Step 5	Router#(config-if)# **no shutdown**	Enables the interface by preventing it from shutting down.
Step 6	Router#(config)# **end**	Returns to privileged EXEC mode.
Step 7	Router# **copy running-config startup-config**	(Optional) Saves configuration changes to TCC Flash.

The default setting for the negotiation mode is auto for the Gigabit Ethernet interfaces. The Gigabit Ethernet port always operates at 1000 Mbps in full-duplex mode. Example 11-2 shows an initial configuration of a Gigabit Ethernet interface with autonegotiation and an IP address.

Example 11-2 *Gigabit Ethernet Interface Configuration*

```
Router(config)# interface gigabitethernet 0
Router(config-if)# ip address 172.16.2.1 255.255.255.0
Router(config-if)# negotiation auto
Router(config-if)# no shutdown
Router(config-if)# end
Router# copy running-config startup-config
```

POS Interface Configuration

POS combines PPP with SONET and SDH interfaces. The transmission rates supported by the ML-Series Cards are indicated in Table 11-7. Notice that the higher transmission rates are only supported between ML-Series terminations.

Table 11-7 *Transmission Rates Supported by ML-Series Cards*

Topology	Supported Sizes
Circuits terminated by two ML-Series cards	STS-1, STS-3c, STS-6c, STS-9c, STS-12c, and STS-24c (SONET) or VC4, VC4-2c, VC4-3c, VC4-4c, and VC4-8c (SDH)
Circuits terminated by G-Series card and ML-Series card	STS-1, STS-3c, STS-6c, STS-9c, STS-12c (SONET) or VC4, VC4-2c, VC4-3c, VC4-4c, and VC4-8c (SDH)
Circuits terminated by ML-Series card and External POS device	STS-3c and STS-12c (SONET) or VC4-2c and VC4-3c (SDH)
VCAT (virtual Concatenation) circuits	STS-1 or VC4 increments up to OC-N/STM-N speed

The section overhead (SOH) and line overhead (LOH) form the transport overhead (TOH), whereas the path overhead (POH) and actual payload (referred to as payload capacity) form the synchronous payload envelope (SPE). SONET framing has been explained in great detail in Chapter 5, "SONET Architectures." Each layer adds a number of overhead bytes to the SONET frame.

One of the overhead bytes in the SONET frame is the C2 byte. The SONET standard defines the C2 byte as the path signal label. The purpose of this byte is to communicate the payload type being encapsulated by the SONET framing overhead (FOH). The C2 byte allows a single interface to transport multiple payload types simultaneously. Table 11-8 lists C2 byte hex values.

Table 11-8 *C2 Byte Values*

Hex Value	SONET Payload Contents
00	Unequipped
01	Equipped non specific payload
02	VTs inside (default)
03	VTs in locked mode (no longer supported)
04	Asynchronous DS3 mapping
12	Asynchronous DS4NA mapping
13	Asynchronous Transfer Mode (ATM) cell mapping

Table 11-8 *C2 Byte Values (Continued)*

Hex Value	SONET Payload Contents
14	Distributed queue, dual-bus (DQDB) protocol cell mapping
15	Asynchronous FDDI mapping
16	IP inside PPP with scrambling
CF	IP inside PPP without scrambling
FE	Test signal mapping (see ITU-T G.707)

POS interfaces use a value of 0x16 or 0xCF in the C2 byte depending on whether ATM-style scrambling is enabled or not. RFC 2615, which defines PPP over SONET, mandates the use of these values based on the scrambling setting. If scrambling is enabled, POS interfaces use a C2 value of 0x16 (PPP and HDLC encapsulation). If scrambling is disabled, POS interfaces use a C2 value of 0xCF (PPP and HDLC encapsulation). LEX encapsulation uses a C2 value of 0x01, regardless of the scrambling setting.

Most POS interfaces that use a default C2 value of 0x16 (22 decimal) insert the **pos flag c2 22** command in the configuration, although this line does not appear in the running configuration because it is the default. Changing the C2 value from the default value does not affect POS scrambling settings. If a Cisco POS interface fails to come up when connected to a third-party device, confirm the scrambling and CRC settings as well as the advertised value in the C2 byte. Follow the steps in Table 11-9 to configure the POS interface on the ML-Series card beginning in global configuration mode.

Table 11-9 *ML-Series POS Interface Configuration Steps*

Step	Command	Description
Step 1	Router(config)# **interface pos** *number*	Activates interface configuration mode to configure the POS interface. The POS interface is created upon the creation of a SONET/SDH circuit.
Step 2	Router#(config-if)# **ip address** *ip-address* {*subnet-mask* | **bridge-group** *bridge-group-number*}	Sets the IP address and subnet mask. or Assigns a network interface to a bridge group.
Step 3	Router#(config-if)#**shutdown**	Manually shuts down the interface. Encapsulation changes on POS ports are allowed only when the interface is shut down (ADMIN_DOWN).
Step 4	Router#(config-if)# **encapsulation** *type*	Sets the encapsulation type. Valid values are as follows: **hdlc**—Cisco HDLC **lex**—LAN extension, special encapsulation for use with Cisco ONS G-Series Ethernet line cards **ppp**—Point-to-Point Protocol

continues

Table 11-9 *ML-Series POS Interface Configuration Steps (Continued)*

Step	Command	Description
Step 5	Router#(config-if)# **pos flag c2** *byte value*	(Optional) Sets the C2 byte value. Valid choices are 0 to 255 (decimal). The default value is 0x01 (hex) for LEX.
Step 6	Router(config-if)# **no keepalive**	(Optional) Turns off keepalive messages. Keepalive messages, although not required, are recommended. Default MTU values are as follows: **lex**—1500 bytes **hdlc**—4470 bytes **ppp**—4470 bytes
Step 7	Router(config-if)# **crc {16	32}**	(Optional) Sets the CRC value. If the device to which the POS module is connected does not support the default CRC value of 32, set both devices to use a value of 16.
Step 8	Router(config-if)# **mtu** *bytes*	(Optional) Configures the MTU size up to a maximum of 9000 bytes.
Step 9	Router(config-if)# **no pos scramble-spe**	(Optional) Disables payload scrambling on the interface. Payload scrambling is on by default.
Step 10	Router(config-if)# **pos report** **{pais	plop	prdi	plm	ptim	sd-ber-b3	sf-ber-b3	uneq-p}**	(Optional) Permits console logging of selected SONET/SDH alarms. The alarms are as follows: **pais** (path alarm indication signal) **plop** (path loss of pointer) **prdi** (path remote defect indication) **plm** (payload label, C2 mismatch) **ptim** (path trace identifier mismatch) **sd-ber-b3** (PBIP BER in excess of SD threshold) **sf-ber-b3** (PBIP BER in excess of SF threshold) **uneq-p** (Path Label Equivalent to Zero failure)
Step 11	Router(config-if)# **pos trigger delay** *millisecond*	(Optional) Delays triggering the line protocol of the interface from going down. Delay can be set from 200 to 2000 ms. If no time intervals are specified, the default delay is set to 200 ms.
Step 12	Router#(config-if)#**no shutdown**	Restarts the shutdown interface.

Table 11-9 *ML-Series POS Interface Configuration Steps (Continued)*

Step	Command	Description
Step 13	Router#(config)# **end**	Returns to privileged EXEC mode.
Step 14	Router# **copy running-config startup-config**	(Optional) Saves configuration changes to NVRAM.

Example 11-3 shows an initial configuration of a POS interface.

Example 11-3 *POS Interface Configuration*

```
Router(config)# interface pos number
Router(config-if)# ip address 172.16.3.1 255.255.255.0
Router(config-if)# negotiation auto
Router(config-if)# encapsulation ppp
Router(config-if)# pos flag c2 22
Router(config-if)# no shutdown
Router(config)# end
Router# copy running-config startup-config
```

ML-Series POS: Case 1

This example shows a POS configuration between two ML-Series cards. As shown in Figure 11-43, Router A is the routing engine located in the ML-Series card, physically present in ONS 15454 A, and Router B is the routing engine located in the ML-Series card, physically present in ONS 15454 B. The two ONS 15454 nodes are connected via a SONET/SDH ring.

Figure 11-43 *ML-Series to ML-Series Configuration*

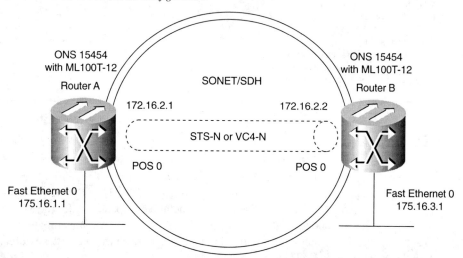

Example 11-4 shows the ML-Series IOS configuration for a POS interface on Router A.

Example 11-4 *Router A POS Configuration (ML-Series to ML-Series)*

```
hostname Router_A
!
interface FastEthernet0
ip address 172.16.1.1 255.255.255.0
!
interface POS0
ip address 172.16.2.1 255.255.255.0
crc 32
pos flag c2 1
!
router ospf 1
log-adjacency-changes
network 172.16.1.0 0.0.0.255 area 0
network 172.16.2.0 0.0.0.255 area 0
```

Example 11-5 shows the ML-Series IOS configuration for a POS interface on Router B.

Example 11-5 *Router B POS Configuration (ML-Series to ML-Series)*

```
hostname Router_B
!
interface FastEthernet0
ip address 172.16.3.1 255.255.255.0
!
interface POS0
ip address 172.16.2.2 255.255.255.0
crc 32
pos flag c2 1
!
router ospf 1
log-adjacency-changes
network 172.16.2.0 0.0.0.255 area 0
network 172.16.3.0 0.0.0.255 area 0
!
```

ML-Series POS: Case 2

This example shows a POS configuration between an ML-Series card and a Cisco 12000 GSR. As shown in Figure 11-44, Router A is the routing engine located in the ML-Series card, physically present in ONS 15454 A, and Router B is the GSR. The ONS 15454 node and the GSR are connected via a SONET/SDH ring. The default encapsulation for the ML-Series card is LEX, and the corresponding default MTU is 1500 bytes. When connecting to an external POS device, it is important to ensure that both the ML-Series switch and the external device share the encapsulation, MTU size, and C2 flag.

Figure 11-44 *ML-Series to GSR Configuration*

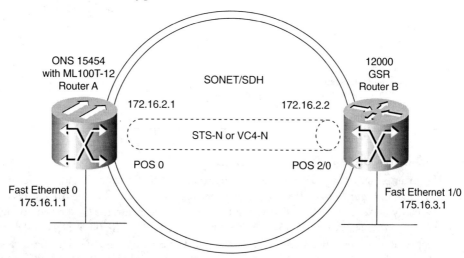

Example 11-6 shows the ML-Series IOS configuration for a POS interface on Router A.

Example 11-6 *Router A Configuration (ML-Series to GSR)*

```
hostname Router_A
!
interface FastEthernet0
ip address 172.16.1.1 255.255.255.0
!
interface POS0
ip address 172.16.2.1 255.255.255.0
encapsulation ppp
pos flag c2 22
crc 32
!
router ospf 1
log-adjacency-changes
network 172.16.1.0 0.0.0.255 area 0
network 172.16.2.0 0.0.0.255 area 0
```

Example 11-7 shows the Cisco 12000 GSR IOS configuration for a POS interface on Router B.

Example 11-7 *Router B (GSR) Configuration (ML-Series to GSR)*

```
hostname GSR
!
interface FastEthernet1/0
ip address 172.16.3.1 255.255.255.0
!
interface POS2/0
ip address 172.16.2.2 255.255.255.0
```

continues

Example 11-7 *Router B (GSR) Configuration (ML-Series to GSR) (Continued)*

```
crc 32
encapsulation PPP
pos scramble-atm
!
router ospf 1
log-adjacency-changes
network 172.16.2.0 0.0.0.255 area 0
network 172.16.3.0 0.0.0.255 area 0
!
```

ML-Series POS Configuration: Case 3

This example shows a POS configuration between an ML-Series card and a G-Series card. As shown in Figure 11-45, Router A is the routing engine located in the ML-Series card, physically present in ONS 15454 A, and Router B is external to the G-Series card located in ONS 15454 B. The two ONS 15454 nodes are connected via a SONET/SDH ring. The default encapsulation for both the ML-Series card and the G-Series card is LEX, and the corresponding default MTU is 1500 bytes. LEX encapsulation uses a C2 value of 0x01 regardless of the scrambling setting.

Figure 11-45 *ML-Series to G-Series Configuration*

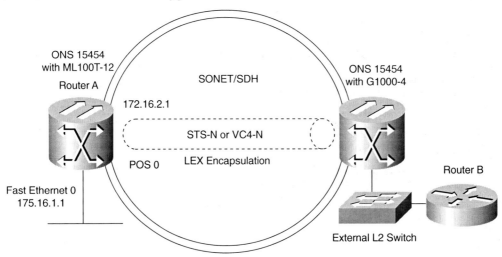

Example 11-8 shows the ML-Series IOS configuration for a POS interface on Router A.

Example 11-8 *Router A Configuration (ML-Series to G-Series)*

```
hostname Router_A
!
interface FastEthernet0
ip address 172.16.1.1 255.255.255.0
!
```

Example 11-8 *Router A Configuration (ML-Series to G-Series) (Continued)*

```
interface POS0
ip address 172.16.2.1 255.255.255.0
crc 32
!
router ospf 1
log-adjacency-changes
network 172.16.1.0 0.0.0.255 area 0
network 172.16.2.0 0.0.0.255 area 0
```

The G-Series card on the adjacent ONS 15454 does not have Layer 3 configuration capability. If you notice on Figure 11-45, the G-Series card is fronted by a Layer 2 Ethernet switch and a router. The Layer 3 configuration will be found on the external router. The Layer 2 switch and router combination can be replaced with a multiple-Layer 2/3 switch if desired.

ML-Series Bridge Configuration

This section describes how to configure bridging for the ML-Series card. IOS supports transparent bridging for Fast Ethernet, Gigabit Ethernet, and POS interfaces. IOS Software functionality combines the advantages of a spanning-tree bridge and a router. This combination provides the speed and protocol transparency of a spanning-tree bridge, along with the functionality, reliability, and security of a router.

The ML-Series card can be configured to include interfaces in a bridge group that are part of the same spanning-tree instance. This bridges all nonroutable traffic among the network interfaces comprising the bridge group. Interfaces not participating in a bridge group cannot forward bridged traffic. If the destination address of the packet is known in the bridge table, the packet is forwarded on a single interface in the bridge group. If the packet's destination is unknown in the bridge table, the packet is flooded on all forwarding interfaces in the bridge group. The bridge places source addresses in the bridge table as it learns them during the bridging process.

A separate spanning-tree process runs for each configured bridge group, and each bridge group participates in a separate spanning tree. A bridge group establishes a spanning tree based on the bridge protocol data units (BPDUs) it receives on only its member interfaces. Follow the steps in Table 11-10 to configure the bridging on the ML-Series card beginning in global configuration mode.

Table 11-10 *ML-Series Bridge Configuration Steps*

Step	Command	Description
Step 1	Router(config)# **no ip routing**	Enables bridging of IP packets.
Step 2	Router(config)# **bridge** *bridge-group-number* **protocol** {**rstp**	**ieee**}	Assigns a bridge group number and defines the appropriate spanning-tree type: either IEEE 802.1D Spanning Tree Protocol or IEEE 802.1W Rapid Spanning Tree Protocol.

continues

Table 11-10 *ML-Series Bridge Configuration Steps (Continued)*

Step	Command	Description
Step 3	Router(config)# **bridge** *bridge-group-number* **priority** *number*	(Optional) Assigns a specific priority to the bridge to assist in the spanning-tree root definition. The lower the priority, the more likely the bridge is selected as the root.
Step 4	Router(config)# **interface** *interface-type interface-number*	Enters interface configuration mode to configure the interface of the ML-Series card.
Step 5	Router(config-if)# **bridge-group** *bridge-group-number*	Assigns a network interface to a bridge group.
Step 6	Router(config-if)# **no shutdown**	Changes the shutdown state to up and enables the interface.
Step 7	Router(config-if)# **end**	Returns to privileged EXEC mode.
Step 8	Router# **copy running-config startup-config**	(Optional) Saves your entries in the configuration file.

ML-Series Bridging Example

This example shows a bridge configuration between two ML-Series cards. As shown in Figure 11-46, Router A is the bridge located in the ML-Series card, physically present in ONS 15454 A, and Router B is the bridge located in the ML-Series card, physically present in ONS 15454 B. The two ONS 15454 nodes are connected via a SONET/SDH ring.

Figure 11-46 *ML-Series Bridging Configuration*

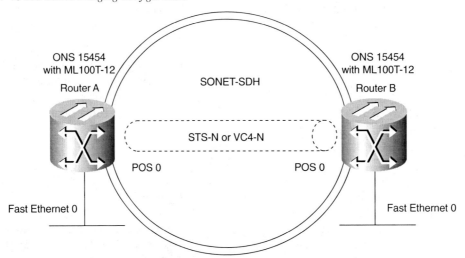

Example 11-9 shows the ML-Series IOS configuration for a bridge group on Router A.

Example 11-9 *Router A Bridge Group Configuration*

```
!
bridge 1 protocol ieee
!
interface FastEthernet0
no ip address
bridge-group 1
!
interface POS0
no ip address
crc 32
bridge-group 1
pos flag c2 1
```

Example 11-10 shows the ML-Series IOS configuration for a bridge-group on Router B.

Example 11-10 *Router B Bridge Group Configuration*

```
!
bridge 1 protocol ieee
!
interface FastEthernet0
no ip address
bridge-group 1
!
interface POS0
no ip address
crc 32
bridge-group 1
pos flag c2 1
```

ML-Series STP and RSTP Configuration

The ML-Series cards implement the IEEE 802.1D Spanning Tree Protocol (STP) as well as the IEEE 802.1W Rapid Spanning Tree Protocol (RSTP).

The STP is a Layer 2 link-management protocol that provides path redundancy while preventing loops in a Layer 2 network. Spanning tree defines a tree with a root switch and a loop-free path from the root to all switches in the Layer 2 network. Spanning tree forces redundant data paths into a standby (blocked) state. If a network segment in the spanning tree fails and a redundant path exists, the spanning-tree algorithm recalculates the spanning-tree topology and activates the standby path.

When two interfaces on a switch are part of a loop, the spanning-tree port priority and path cost settings determine which interface is put in the forwarding state and which is put in the blocking state. The port priority value represents the location of an interface in the network topology and how well it is located to pass traffic. The path cost value represents media speed. The ML-Series supports the per-VLAN spanning tree (PVST+) and a maximum of 255 spanning-tree instances.

RSTP provides rapid convergence of the spanning tree. It improves the fault tolerance of the network because a failure in one instance (forwarding path) does not affect other instances (forwarding paths). The most common initial deployment of RSTP is in the backbone and distribution layers of a Layer 2 switched network; this deployment provides the highly available network required in a service-provider environment.

RSTP improves the operation of the spanning tree while maintaining backward compatibility with equipment that is based on the (original) 802.1D spanning tree. The RSTP provides rapid convergence of the spanning tree by assigning port roles and by determining the active topology. RSTP takes advantage of point-to-point links and provides rapid convergence of the spanning tree. Reconfiguration of the spanning tree can occur in less than 2 seconds (in contrast to 50 seconds with the default settings in the 802.1D spanning tree), which is critical for networks carrying delay-sensitive traffic, such as voice and video. The ML-Series supports per-VLAN Rapid Spanning Tree (PVRST) and a maximum of 255 RSTP instances.

A switch running RSTP supports a built-in protocol-migration mechanism that enables it to interoperate with legacy 802.1D switches. If this switch receives a legacy 802.1D configuration BPDU (a BPDU with the protocol version set to 0), it sends only 802.1D BPDUs on that port. However, the switch does not automatically revert to the RSTP mode if it no longer receives 802.1D BPDUs because it cannot determine whether the legacy switch has been removed from the link unless the legacy switch is the designated switch.

Per-VLAN STP and RSTP are enabled by default on the ML-Series card. STP is enabled by default on VLAN 1 and on all newly created VLANs up to the specified spanning-tree limit of 255. Table 11-11 shows the default STP and RSTP configuration for the ML-Series cards.

Table 11-11 *Default STP and RSTP Configuration*

Feature	Default Setting
Enable state	Up to 255 spanning-tree instances can be enabled
Switch priority	32768 + bridge ID
Spanning-tree port priority (configurable on a per-interface basis—used on interfaces configured as Layer 2 access ports)	128
Spanning-tree port cost (configurable on a per-interface basis)	1000 Mbps: 4 100 Mbps: 19 10 Mbps: 100 STS-1: 34 STS-3c: 14 STS-6c: 9 STS-9c: 7 STS-12c: 6 STS-24c: 3

Table 11-11 *Default STP and RSTP Configuration (Continued)*

Feature	Default Setting
Hello time	2 seconds
Forward-delay time	15 seconds
Maximum-aging time	20 seconds

Disabling and Enabling STP-RSTP

Follow the steps in Table 11-12 to disable or enable STP-RSTP on a per-VLAN basis on the ML-Series card beginning in global configuration mode.

Table 11-12 *Steps to Disable STP-RSTP*

Step	Command	Description
Step 1	Router(config)# **interface** *interface-id*	Enters the interface configuration mode.
Step 2	Router(config-if)# **bridge-group** *bridge-group-number* **spanning disabled**	Disables STP or RSTP on a per-interface basis.
Step 3	Router(config-if)# **no bridge-group** *bridge-group-number* **spanning disabled**	(Optional) Re-enables STP or RSTP on a per-interface basis.
Step 4	Router(config-if)# **end**	Returns to privileged EXEC mode.

Configuring the Port Priority of an Interface

If a loop occurs, spanning tree uses the port priority when selecting an interface to put into the forwarding state. You can assign higher-priority values (lower numeric values) to interfaces that you want selected first, and lower-priority values (higher numeric values) that you want selected last. If all interfaces have the same priority value, spanning tree puts the interface with the lowest interface number in the forwarding state and blocks the other interfaces. Beginning in privileged EXEC mode, follow the steps in Table 11-13 to configure the port priority of an interface.

Table 11-13 *Steps to Configure Interface Port Priority*

Step	Command	Description
Step 1	Router# **configure terminal**	Enters the global configuration mode.
Step 2	Router(config)# **interface** *interface-id*	Enters the interface configuration mode, and specifies an interface to configure. Valid interfaces include physical interfaces and port-channel logical interfaces (**port-channel** *port-channel-number*).

continues

Table 11-13 *Steps to Configure Interface Port Priority (Continued)*

Step	Command	Description
Step 3	Router(config-if)# **bridge-group** *bridge-group-number* **priority** *value*	Configures the port priority for an interface that is an access port. For the priority *value*, the range is 0 to 255; the default is 128 in increments of 16. The lower the number, the higher the priority.
Step 4	Router(config-if)# **no bridge-group** *bridge-group-number* **priority** *value*	(Optional) Returns the interface to its default setting.
Step 5	Router(config-if)# **end**	Returns to privileged EXEC mode.

Configuring the Path Cost of an Interface

The spanning-tree path cost default value is derived from the media speed of an interface. If a loop occurs, spanning tree uses cost when selecting an interface to put in the forwarding state. You can assign lower-cost values to interfaces that you want selected first and higher-cost values to interfaces that you want selected last. If all interfaces have the same cost value, spanning tree puts the interface with the lowest interface number in the forwarding state and blocks the other interfaces. Beginning in privileged EXEC mode, follow the steps in Table 11-14 to configure the cost of an interface.

Table 11-14 *Steps to Configure Interface Path Cost*

Step	Command	Description
Step 1	Router# **configure terminal**	Enters the global configuration mode.
Step 2	Router(config)# **interface** *interface-id*	Enters the interface configuration mode and specifies an interface to configure. Valid interfaces include physical interfaces and port-channel logical interfaces (**port-channel** *port-channel-number*).
Step 3	Router(config-if)# **bridge-group** *bridge-group-number* **path-cost** *cost*	Configures the cost for an interface that is an access port. If a loop occurs, spanning tree uses the path cost when selecting an interface to place into the forwarding state. A lower path cost represents higher-speed transmission. For *cost*, the range is 0 to 65535; the default value is derived from the media speed of the interface.

Table 11-14 *Steps to Configure Interface Path Cost (Continued)*

Step	Command	Description
Step 4	Router(config-if)# **no bridge-group** *bridge-group-number* **path-cost** *cost*	(Optional) Returns the interface to its default setting.
Step 5	Router(config-if)# **end**	Returns to the privileged EXEC mode.

Configuring the Switch Priority of a Bridge Group

You can configure the switch priority and make it more likely that the switch will be chosen as the root switch. Beginning in privileged EXEC mode, follow the steps in Table 11-15 to configure the switch priority of a bridge group.

Table 11-15 *Steps to Configure Bridge Group Priority*

Step	Command	Description
Step 1	Router# **configure terminal**	Enters the global configuration mode.
Step 2	Router(config)# **bridge** *bridge-group-number* **priority** *priority*	Configures the switch priority of a bridge group. For *priority*, the range is 0 to 61440 in increments of 4096; the default is 32768. The lower the number, the more likely the switch will be chosen as the root switch. The value entered is rounded to the lower multiple of 4096. The actual number is computed by adding this number to the bridge group number.
Step 3	Router(config)# **no bridge** *bridge-group-number* **priority** *priority*	(Optional) Returns the switch to its default setting.
Step 4	Router(config)# **end**	Returns to the privileged EXEC mode.

Configuring the Hello Time

You can configure the interval between the generation of configuration messages by the root switch by changing the hello time. Beginning in privileged EXEC mode, follow the steps in Table 11-16 to configure the hello time of a bridge group.

Table 11-16 *Steps to Configure the Bridge Hello-Time Interval*

Step	Command	Description
Step 1	Router# **configure terminal**	Enters global configuration mode.
Step 2	Router(config)# **bridge** *bridge-group-number* **hello-time** *seconds*	Configures the hello time of a bridge group. The hello time is the interval between the generation of configuration messages by the root switch. These messages mean that the switch is alive. For *seconds*, the range is 1 to 10; the default is 2.
Step 3	Router(config)# **no bridge** *bridge-group-number* **hello-time** *seconds*	(Optional) Returns the switch to its default setting.
Step 4	Router(config)# **end**	Returns to privileged EXEC mode.

Configuring the Forwarding-Delay Time for a Bridge Group

You can configure the forwarding-delay time for a bridge group. The forwarding delay is the number of seconds a port waits before changing from its spanning-tree learning and listening states to the forwarding state. Beginning in privileged EXEC mode, follow the steps in Table 11-17 to configure the forwarding-delay time for a bridge group.

Table 11-17 *Steps to Configure the Bridge Group Forwarding Delay*

Step	Command	Description
Step 1	Router# **configure terminal**	Enters global configuration mode.
Step 2	Router(config)# **bridge** *bridge-group-number* **forward-time** *seconds*	Configures the forward time of a VLAN. The forwarding delay is the number of seconds a port waits before changing from its spanning-tree learning and listening states to the forwarding state. For *seconds*, the range is 4 to 200; the default is 15.
Step 3	Router(config)# **no bridge** *bridge-group-number* **forward-time** *seconds*	(Optional) Returns the switch to its default setting.
Step 4	Router(config)# **end**	Returns to privileged EXEC mode.

Configuring the Maximum-Aging Time for a Bridge Group

You can configure the maximum-aging time for a bridge group. The maximum-aging time is the number of seconds a switch waits without receiving spanning-tree configuration messages before attempting a reconfiguration. Beginning in privileged EXEC mode, follow the steps in Table 11-18 to configure the maximum-aging time for a bridge group.

Table 11-18 *Steps to Configure the Maximum-Aging Time for a Bridge Group*

Step	Command	Description
Step 1	Router# **configure terminal**	Enters global configuration mode.
Step 2	Router(config)# **bridge** *bridge-group-number* **max-age** *seconds*	Configures the maximum-aging time of a bridge group. The maximum-aging time is the number of seconds a switch waits without receiving spanning-tree configuration messages before attempting a reconfiguration. For *seconds*, the range is 6 to 200; the default is 20.
Step 3	Router(config)# **no bridge** *bridge-group-number* **max-age** *seconds*	(Optional) Returns the switch to its default setting.
Step 4	Router(config)# **end**	Returns to privileged EXEC mode.

Monitoring STP and RSTP Status

The commands in Table 11-19 enable you to view STP and RSTP information and statistics.

Table 11-19 *STP and RSTP Status Monitoring Commands*

Command	Description
Router# **show spanning-tree active**	Displays STP or RSTP information on active interfaces only.
Router# **show spanning-tree detail**	Displays a detailed summary of interface information.
Router# **show spanning-tree interface** *interface-id*	Displays STP or RSTP information for the specified interface.
Router# **show spanning-tree summary** [*totals*]	Displays a summary of port states or displays the total lines of the STP or RSTP state section.

ML-Series VLAN Configuration

From a service provider perspective, VLANs or bridge groups enable customers to be grouped logically rather than by physical location. This forms a basis of VPN creation and logical separation of customer traffic by VLAN. ML-Series software supports port-based VLANs and VLAN trunk ports. Trunk ports carry the traffic of multiple VLANs. Each frame transmitted on a trunk link is tagged as belonging to only one VLAN. ML-Series software supports VLAN frame encapsulation through the IEEE 802.1Q standard on both the ML100T-12 and the ML1000-2. The Cisco ISL VLAN frame encapsulation is not supported. ISL frames will be broadcast at Layer 2, or dropped at Layer 3.

The ML-Series switching supports up to 900 VLAN subinterfaces per card. For example, 150 VLANs on four interfaces use 600 VLAN subinterfaces. A maximum of 255 logical VLANs can be bridged per card (limited by the number of bridge groups). Each VLAN subinterface can be configured for any VLAN ID in the full 1 to 4095 range. Figure 11-47 shows a network topology in which two VLANs span two ONS 15454s with ML-Series cards.

Figure 11-47 *ML-Series VLANs*

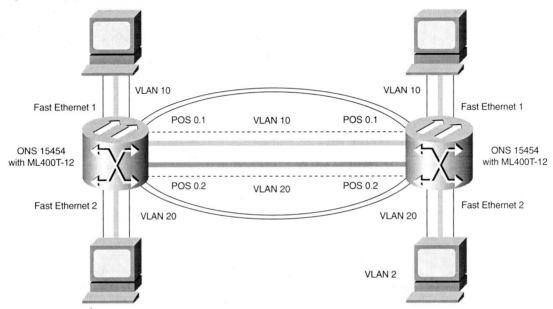

On an IEEE 802.1Q trunk port, all transmitted and received frames are tagged except for those on the VLAN configured as the native VLAN for the port. Frames on the native VLAN are always transmitted untagged and are normally received untagged. On ML-Series cards, the native VLAN is always VLAN ID 1. You can configure VLAN encapsulation on both the ML100T-12 and the ML1000-2. Tagging of transmitted native VLAN frames can be forced by the global configuration command **vlan dot1q tag native**. VLAN encapsulation is supported on both the ML100T-12 and the ML1000-2. VLAN encapsulation is supported for routing and bridging, and is supported on Ethernet interfaces and on POS interfaces with PPP and LEX encapsulation.

To configure VLANs using IEEE 802.1Q VLAN encapsulation, perform the procedure in Table 11-20, beginning in global configuration mode.

Table 11-20 *ML-Series VLAN Configuration Steps*

Step	Command	Description
Step 1	Router(config)# **bridge** *bridge-group-number* **protocol** *type*	Assigns a bridge group (VLAN) number and defines the appropriate spanning-tree type.
Step 2	Router(config)# **interface** *type number*	Enters interface configuration mode to configure the interface.
Step 3	Router(config-if)#**no ip address**	Disables IP processing.

Table 11-20 *ML-Series VLAN Configuration Steps (Continued)*

Step	Command	Description
Step 4	Router(config)# **interface** *type number.subinterface-number*	Enters subinterface configuration mode to configure the subinterface.
Step 5	Router(config-subif)# **encapsulation dot1q** *bridge-group-number*	Sets the encapsulation format on the bridge group to IEEE 802.1Q.
Step 6	Router(config-subif)# **bridge-group** *bridge-group-number*	Assigns a network interface to a bridge group.
Step 7	Router(config-subif)# **end**	Returns to privileged EXEC mode.
Step 8	Router# **copy running-config startup-config**	(Optional) Saves your configuration changes to NVRAM.

ML-Series VLAN Example

This example shows a VLAN configuration between two ML-Series cards. As shown in Figure 11-48, Router A is the multilayer switch located in the ML-Series card, physically present in ONS 15454 A, and Router B is the multilayer switch located in the ML-Series card, physically present in ONS 15454 B. The two ONS 15454 nodes are connected via a SONET/SDH ring. This example configures FE subinterface 0.1 in 802.1Q native VLAN 1, FE subinterface 0.2 in 802.1Q VLAN 200, FE subinterface 0.3 in 802.1Q VLAN 300, and FE subinterface 0.4 in 802.1Q VLAN 400.

Figure 11-48 *ML-Series VLAN Configuration*

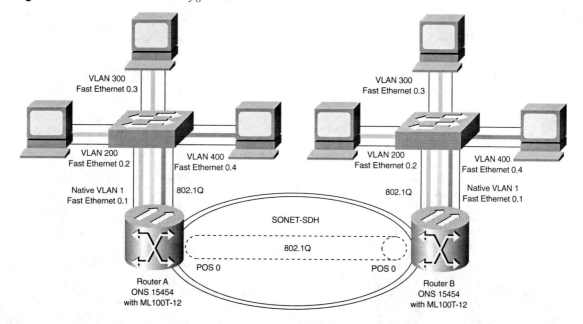

Example 11-11 shows the ML-Series IOS configuration for configuring VLANs on Router A with 802.1Q encapsulation.

Example 11-11 *Router a VLAN Configuration*

```
!
bridge 1 protocol ieee
bridge 2 protocol ieee
bridge 3 protocol ieee
bridge 4 protocol ieee
!
interface FastEthernet0
no ip address
!
interface FastEthernet0.1
encapsulation dot1Q 1 native
bridge-group 1
!
interface FastEthernet0.2
encapsulation dot1Q 2
bridge-group 2
!
interface FastEthernet0.3
encapsulation dot1Q 3
bridge-group 3
!
interface FastEthernet0.4
encapsulation dot1Q 4
bridge-group 4
!
interface POS0
no ip address
crc 32
pos flag c2 1
!
interface POS0.1
encapsulation dot1Q 1 native
bridge-group 1
!
interface POS0.2
encapsulation dot1Q 2
bridge-group 2
!
interface POS0.3
encapsulation dot1Q 3
bridge-group 3
!
interface POS0.4
encapsulation dot1Q 4
bridge-group 4
!
```

Example 11-12 shows the ML-Series IOS configuration for configuring VLANs on Router B with 802.1Q encapsulation.

Example 11-12 *Router B VLAN Configuration*

```
!
bridge 1 protocol ieee
bridge 2 protocol ieee
bridge 3 protocol ieee
bridge 4 protocol ieee
!
interface FastEthernet0
no ip address
!
interface FastEthernet0.1
encapsulation dot1Q 1 native
bridge-group 1
!
interface FastEthernet0.2
encapsulation dot1Q 2
bridge-group 2
!
interface FastEthernet0.3
encapsulation dot1Q 3
bridge-group 3
!
interface FastEthernet0.4
encapsulation dot1Q 4
bridge-group 4
!
interface POS0
no ip address
crc 32
pos flag c2 1
!
interface POS0.1
encapsulation dot1Q 1 native
bridge-group 1
!
interface POS0.2
encapsulation dot1Q 2
bridge-group 2
!
interface POS0.3
encapsulation dot1Q 3
bridge-group 3
!
interface POS0.4
encapsulation dot1Q 4
bridge-group 4
!
```

ML-Series IEEE 802.1Q and Layer 2 Tunneling

The ML-Series card permits large-scale Layer 2 VPN deployment by tunneling customer VLANs. Tunneling is a feature designed for service providers who carry traffic of multiple customers across their networks and are required to maintain the VLAN and Layer 2 protocol configurations of each customer without impacting the traffic of other customers. The ML-Series cards support IEEE 802.1Q tunneling and Layer 2 protocol tunneling. Customers often have specific requirements for VLAN IDs and the number of VLANs to be supported. The VLAN ranges required by different customers in the same carrier network might overlap, and traffic of customers through the infrastructure might be mixed. Assigning a unique range of VLAN IDs to each customer would restrict customer configurations and could easily exceed the VLAN limit of 4096 of the IEEE 802.1Q specification. Using the IEEE 802.1Q tunneling feature, service providers can use a single VLAN to support customers who have multiple VLANs.

Customer VLAN IDs are preserved, and traffic from different customers is segregated within the service provider infrastructure even when they appear to be on the same VLAN. The IEEE 802.1Q tunneling expands VLAN space by using a double-tagging hierarchy and tagging the tagged packets. A port configured to support IEEE 802.1Q tunneling is called a *tunnel port*. When you configure tunneling, you assign a tunnel port to a VLAN that is dedicated to tunneling. Each customer requires a separate VLAN, but that VLAN supports all of the customer's VLANs. Customer traffic tagged in the normal way with appropriate VLAN IDs comes from an IEEE 802.1Q trunk port on the customer device and into a tunnel port on the ML-Series card. The link between the customer device and the ML-Series card is an asymmetric link because one end is configured as an IEEE 802.1Q trunk port and the other end is configured as a tunnel port. You assign the tunnel port interface to an access VLAN ID unique to each customer, as shown in Figure 11-49.

Packets coming from the customer trunk port into the tunnel port on the ML-Series card are normally IEEE 802.1Q tagged with an appropriate VLAN ID. The tagged packets remain intact inside the ML-Series card and, when they exit the trunk port into the service provider network, are encapsulated with another layer of an IEEE 802.1Q tag (called the *metro tag*) that contains the VLAN ID unique to the customer. The original IEEE 802.1Q tag from the customer is preserved in the encapsulated packet. Therefore, packets entering the service provider infrastructure are double-tagged, with the outer tag containing the customer's access VLAN ID, and the inner VLAN ID being the VLAN of the incoming traffic.

When the double-tagged packet enters another trunk port in a service provider ML-Series card, the outer tag is stripped as the packet is processed inside the switch. When the packet exits another trunk port on the same core switch, the same metro tag is again added to the packet. Figure 11-50 shows the structure of the double-tagged packet. When the packet enters the trunk port of the service provider egress switch, the outer tag is again stripped as the packet is processed internally on the switch. However, the metro tag is not added when it is sent out the tunnel port on the edge switch into the customer network, and the packet is sent as a normal

IEEE 802.1Q-tagged frame to preserve the original VLAN numbers in the customer network. The IEEE 802.1Q class of service (CoS) priority field on the added metro tag is set to 0 by default, but can be modified by input or output policy maps.

Figure 11-49 *802.1Q Tunnels in a Service Provider Network*

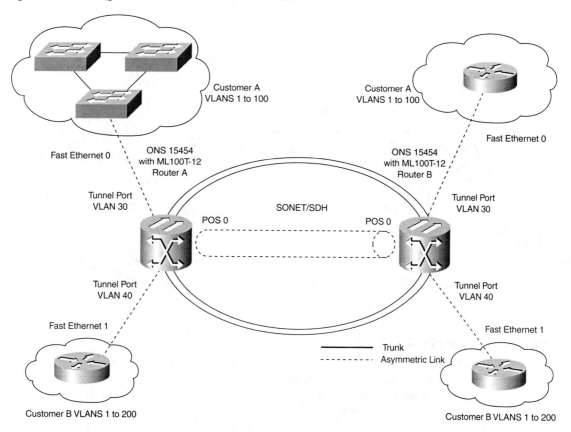

If you plan on using the native VLAN (VLAN ID 1) in the service provider network as a metro tag, it is important that this tag always be added to the customer traffic, even though the native VLAN ID is not normally added to transmitted frames. If the VLAN 1 metro tag were not added on frames entering the service provider network, the customer VLAN tag would appear to be the metro tag. This would result in VPN leaks. The global configuration command **vlan dot1q tag native** must be used to prevent this by forcing a tag to be added to VLAN 1. It is recommended to use VLAN 1 as a private operations support system (OSS) management VLAN in the service provider network and to avoid its usage for transport of customer traffic.

Figure 11-50 *802.1Q Tunneled Ethernet Packet Formats*

DA = Destination Address
SA = Source Address
Len/Etype = Length/Ether Type
FCS = Frame Check Sequence

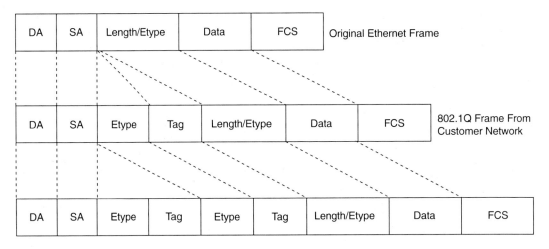

IEEE 802.1Q Tunneling Design Constraints

The designer must keep a few constraints in mind before creating a complete ML-Series-based VPN service blueprint. These constraints could affect the scalability or limit the functionality of service offerings. It is prudent to examine all VPN options and your own requirements before deciding on a technology that's best for you. Listed here are a few design constraints to keep in mind:

- A tunnel port cannot be a routed port.
- Tunnel ports do not support IP ACLs.
- Layer 3 quality of service (QoS) ACLs and other QoS features related to Layer 3 information are not supported on tunnel ports. MAC-based QoS is supported on tunnel ports.
- EtherChannel port groups are compatible with tunnel ports as long as the IEEE 802.1Q configuration is consistent within an EtherChannel port group.
- Port Aggregation Protocol (PAgP) and Unidirectional Link Detection (UDLD) Protocol are not supported on IEEE 802.1Q tunnel ports.

- Dynamic Trunking Protocol (DTP) is not compatible with IEEE 802.1Q tunneling because you must manually configure asymmetric links with tunnel ports and trunk ports.
- Loopback detection is supported on IEEE 802.1Q tunnel ports.
- When a port is configured as an IEEE 802.1Q tunnel port, spanning-tree BPDU filtering is automatically disabled on the interface.

IEEE 802.1Q Tunneling Port Configuration

Beginning in privileged EXEC mode, follow the steps in Table 11-21 to configure a port as an IEEE 802.1Q tunnel port.

Table 11-21   Steps to Configure 802.1Q Tunneling

Step	Command	Description
Step 1	Router# **configure terminal**	Enters global configuration mode.
Step 2	Router(config)# **bridge** *bridge-number* **protocol** *bridge-protocol*	Creates a bridge number and specifies a protocol.
Step 3	Router(config)# **interface fastethernet** *number*	Enters the interface configuration mode and the interface to be configured as a tunnel port. This should be the edge port in the service provider network that connects to the customer switch. Valid interfaces include physical interfaces and port-channel logical interfaces (port channels 1 to 64).
Step 4	Router(config-if)# **bridge-group** *number*	Assigns the tunnel port to a bridge group. All traffic from the port (tagged and untagged) will be switched based on this bridge group. Other members of the bridge group should be VLAN subinterfaces on a provider trunk interface.
Step 5	Router(config-if)# **mode dot1q-tunnel**	Sets the interface as an IEEE 802.1Q tunnel port.
Step 6	Router(config-if)# **no mode dot1q-tunnel**	(Optional) Removes the IEEE 802.1Q tunnel port from the interface.
Step 7	Router(config-if)# **exit**	Returns to global configuration mode.
Step 8	Router(config)# **vlan dot1q tag native**	(Optional) Forces tagging of transmitted native VLAN frames. This command is required if you plan to use VLAN ID 1.
Step 9	Router(config)# **end**	Returns to privileged EXEC mode.
Step 10	Router# **show dot1q-tunnel**	Displays the tunnel ports on the switch.
Step 11	Router# **copy running-config startup-config**	(Optional) Saves your entries in the configuration file.

ML-Series 802.1Q Tunneling Example

This example shows 802.1Q tunneling with multiple customers. As shown in Figure 11-49, Router A is the multilayer switch located in the ML-Series card, physically present in ONS 15454 A, and Router B is the multilayer switch located in the ML-Series card, physically present in ONS 15454 B. The two ONS 15454 nodes are connected via a SONET/SDH ring. Customer A was assigned VLAN 30, and Customer B was assigned VLAN 40. Packets entering the ML-Series card tunnel ports with IEEE 802.1Q tags are double-tagged when they enter the service provider network, with the outer tag containing VLAN ID 30 or 40, appropriately, and the inner tag containing the original VLAN number (for example, VLAN 100). Even if both Customers A and B have VLAN 100 in their networks, the traffic remains segregated within the service provider network because the outer tag is different. With IEEE 802.1Q tunneling, each customer controls its own VLAN numbering space, which is independent of the VLAN numbering space used by other customers and the VLAN numbering space used by the service provider network. At the outbound tunnel port, the original VLAN numbers on the customer's network are recovered. If the traffic coming from a customer network is not tagged (native VLAN frames), these packets are bridged or routed as if they were normal packets, and the metro tag is added (as a single-level tag) when they exit toward the service provider network.

Example 11-13 shows the ML-Series IOS configuration for configuring 802.1Q tunneling on Router A.

Example 11-13 *Router A Tunneling Configuration*

```
!
bridge 30 protocol ieee
bridge 40 protocol ieee
interface FastEthernet0
no ip routing
no ip address
mode dot1q-tunnel
bridge-group 30
!
interface FastEthernet1
no ip address
mode dot1q-tunnel
bridge-group 40
!
interface POS0
no ip address
crc 32
pos flag c2 1
!
interface POS0.1
encapsulation dot1Q 30
bridge-group 30
!
interface POS0.2
encapsulation dot1Q 40
bridge-group 40
```

Example 11-14 shows the ML-Series IOS configuration for configuring 802.1Q tunneling on Router B.

Example 11-14 *Router B Tunneling Configuration*

```
!
bridge 30 protocol ieee
bridge 40 protocol ieee
interface FastEthernet0
no ip routing
no ip address
mode dot1q-tunnel
bridge-group 30
!
interface FastEthernet1
no ip address
mode dot1q-tunnel
bridge-group 40
!
interface POS0
no ip address
crc 32
pos flag c2 1
!
interface POS0.1
encapsulation dot1Q 30
bridge-group 30
!
interface POS0.2
encapsulation dot1Q 40
bridge-group 40
```

Layer 2 Protocol Tunneling

Various Layer 2 protocols, such as STP, CDP, and Virtual Terminal Protocol (VTP), are used by customer networks. These protocols need to be tunneled across the service provider network so that customers can build their own spanning tress, perform device discovery, and ensure a consistent VLAN configuration throughout their network. When protocol tunneling is enabled, edge switches on the inbound side of the service provider infrastructure encapsulate Layer 2 protocol packets with a special MAC address and send them across the service provider network. Core switches in the network do not process these packets, but forward them as normal packets.

Layer 2 protocol data units (PDUs) for CDP, STP, or VTP cross the service provider infrastructure and are delivered to customer switches on the outbound side of the service provider network. This ensures that identical packets are received by all customer ports on the same VLANs with consistent results from a customer perspective. Layer 2 protocol tunneling can be used independently or to enhance IEEE 802.1Q tunneling. If protocol tunneling is not enabled on IEEE 802.1Q tunneling ports, remote switches at the receiving end of the service provider network do not receive the PDUs and cannot properly run STP, CDP, and VTP.

When protocol tunneling is enabled, Layer 2 protocols within each customer's network are totally separate from those running within the service provider network. Customer switches on different sites that send traffic through the service provider network with IEEE 802.1Q tunneling achieve complete knowledge of the customer's VLAN. If IEEE 802.1Q tunneling is not used, you can still enable Layer 2 protocol tunneling by connecting to the customer switch through access ports and enabling tunneling on the service provider access port.

Layer 2 Protocol Tunneling Design Constraints

The designer must keep a few constraints in mind before creating Layer 2 protocol tunnels. These constraints could affect the scalability or limit the functionality of service offerings. Listed here are a few design constraints to keep in mind:

- The switch supports tunneling of CDP, STP (including Multiple STP [MSTP]), and VTP protocols. Protocol tunneling is disabled by default but can be enabled for the individual protocols on IEEE 802.1Q tunnel ports.
- Tunneling is not supported on trunk ports. If you enter the **l2protocol-tunnel interface** configuration command on a trunk port, the command is accepted, but Layer 2 tunneling does not take affect unless you change the port to a tunnel port.
- EtherChannel port groups are compatible with tunnel ports as long as the IEEE 802.1Q configuration is consistent within an EtherChannel port group.
- If an encapsulated PDU (with the proprietary destination MAC address) is received from a tunnel port or access port with Layer 2 tunneling enabled, the tunnel port is shut down to prevent loops.
- Only decapsulated PDUs are forwarded to the customer network. The spanning-tree instance running on the service provider network does not forward BPDUs to tunnel ports. No CDP packets are forwarded from tunnel ports.
- Because tunneled PDUs (especially STP BPDUs) must be delivered to all remote sites for the customer virtual network to operate properly, you can give PDUs higher priority within the service provider network than data packets received from the same tunnel port. By default, the PDUs use the same CoS value as data packets.
- Protocol tunneling has to be configured symmetrically at both the ingress and egress point. If you configure the entry point to tunnel STP, CDP, and VTP, for example, you must configure the egress point in the same way.

ML-Series Layer 2 Protocol Tunneling Configuration

Beginning in privileged EXEC mode, follow the steps in Table 11-22 to configure a port as a Layer 2 tunnel port.

Table 11-22 *Steps to Configure Layer 2 Protocol Tunneling*

Step	Command	Description
Step 1	Router# **conf t**	Enters global configuration mode.
Step 2	Router(config)# **bridge** *bridge-number* **protocol** *bridge-protocol*	Creates a bridge number and specifies a protocol.
Step 3	Router(config)# **l2protocol-tunnel cos** *cos-value*	Assigns a CoS value or values to associate with the Layer 2 tunneling port. The *cos-value* is a number from the 0 to 7 range.
Step 4	Router(config)# **interface fastethernet** *number*	Enters the interface configuration mode and the interface to be configured as a tunnel port. This should be the edge port in the service provider network that connects to the customer switch. Valid interfaces include physical interfaces and port-channel logical interfaces (port channels 1 to 64).
Step 5	Router(config-if)# **bridge-group** *number*	Specifies the default VLAN, which is used if the interface stops trunking. The VLAN ID is specific to the particular customer.
Step 6	Router(config-if)# **mode dot1q-tunnel**	Sets the interface as an IEEE 802.1Q tunnel port.
Step 7	Router(config-if)# **l2 protocol-tunnel [cdp][stp][vtp]**	Sets the interface as a Layer 2 protocol tunnel port and enables CDP, STP, and VTP, which are off by default.
Step 8	Router(config-if)# **exit**	Returns to global configuration mode.
Step 9	Router(config-if)# **end**	Returns to privileged EXEC mode.
Step 10	Router# **show dot1q-tunnel**	Displays the tunnel ports on the switch.
Step 11	Router# **copy running-config startup-config**	(Optional) Saves your entries in the configuration file.

Link Aggregation on the ML-Series Cards

Link aggregation provides logical aggregation of bandwidth along with load balancing and fault tolerance. Link aggregation is implemented as EtherChannel and POS channel on the ML-Series cards.

The EtherChannel interface, consisting of multiple Fast Ethernet, Gigabit Ethernet, or POS interfaces, is treated as a single interface, which is called a *port channel*. You must perform all EtherChannel configurations on the EtherChannel interface (port channel) rather than on the individual member Ethernet interfaces. You can create the EtherChannel interface by entering the **interface port-channel** interface configuration command. Each ML100T-12 supports up to seven Fast EtherChannel (FEC) interfaces or port channels (six Fast Ethernet and one POS).

Each ML1000-2 supports up to two Gigabit EtherChannel (GEC) interfaces or port channels (one Gigabit Ethernet and one POS). EtherChannel connections are fully compatible with IEEE 802.1Q trunking and routing technologies. 802.1Q trunking can carry multiple VLANs across an EtherChannel.

FEC technology builds upon standards-based 802.3 full-duplex Fast Ethernet to provide a reliable high-speed solution for the campus network backbone. FEC provides bandwidth scalability within the campus by providing up to 400-Mbps full-duplex Fast Ethernet on the ML100-12. GEC technology provides bandwidth scalability by providing 2-Gbps full-duplex aggregate capacity on the ML1000-2. POS channel technology provides bandwidth scalability by providing up to 48 STSs or VC4-16c of aggregate capacity on either the ML100-12 or the ML1000-2. From a design-constraint perspective, the reader must note that link aggregation across multiple ML-Series cards is not supported. It is also not possible to traffic shape or police on port-channel interfaces.

NOTE The configuration tasks for Fast EtherChannel and Gigabit EtherChannel are similar to the configuration of Cisco Catalyst multilayer switches. Refer to the appropriate *Cisco IOS Configuration Fundamentals Configuration Guide* for details on the configuration of FEC or GEC.

ML-Series POS Channel Configuration

You can configure a POS channel by creating a POS channel interface (port channel) and optionally assigning an IP address. All POS interfaces that are members of a POS channel should have the same port properties and be on the same ML-Series card. POS channel is supported only with G-Series card compatible (LEX) encapsulation. To create a POS channel interface, perform the procedure in Table 11-23, beginning in global configuration mode.

Table 11-23 *Steps to Create the ML-Series POS Channel Interface*

Step	Command	Description
Step 1	Router(config)# **interface port-channel** *channel-number*	Creates the POS channel interface. You can configure one POS channel on the ML-Series card.
Step 2	Router(config-if)# **ip address** *ip-address subnet-mask*	Assigns an IP address and subnet mask to the POS channel interface (required only for the Layer 3 POS channel).
Step 3	Router(config-if)# **end**	Exits to privileged EXEC mode.
Step 4	Router# **copy running-config startup-config**	(Optional) Saves configuration changes to NVRAM.

NOTE The POS channel interface is the routed interface. Do not enable Layer 3 addresses on any physical interfaces. Do not assign bridge groups on any physical interfaces because doing so creates loops.

To assign POS interfaces to the POS channel, perform the procedure in Table 11-24, beginning in global configuration mode.

Table 11-24 *Steps to Assign POS Interfaces to the POS Channel*

Step	Command	Description
Step 1	Router(config)# **interface pos** *number*	Enters the interface configuration mode to configure the POS interface that you want to assign to the POS channel.
Step 2	Router(config-if)# **channel-group** *channel-number*	Assigns the POS interface to the POS channel. The channel number must be the same channel number that you assigned to the POS channel interface.
Step 3	Router(config-if)# **end**	Exits to privileged EXEC mode.
Step 4	Router# **copy running-config startup-config**	(Optional) Saves the configuration changes to NVRAM.

ML-Series POS Channel Example

This example shows an example of a POS channel configuration between two ML-Series cards. As shown in Figure 11-51, Switch A is the multilayer switch located in the ML-Series card, physically present in ONS 15454 A, and Switch B is the multilayer switch located in the ML-Series card, physically present in ONS 15454 B. The two ONS 15454 nodes are connected via a SONET/SDH ring.

Figure 11-51 *ML-Series POS Channel*

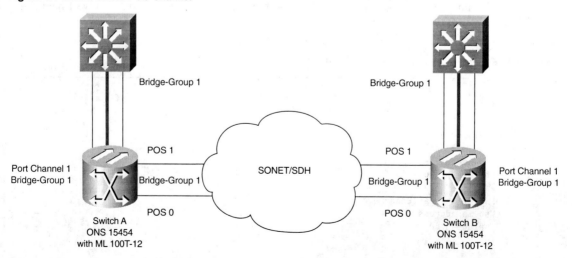

Example 11-15 shows the ML-Series IOS configuration for configuring the POS Channel on Switch A.

Example 11-15 *Switch A POS Channel Configuration*

```
!
bridge irb
bridge 1 protocol ieee
!
interface Port-channel1
no ip address
no keepalive
bridge-group 1
!
interface FastEthernet0
no ip address
bridge-group 1
!
interface POS0
no ip address
channel-group 1
crc 32
pos flag c2 1
!
interface POS1
no ip address
channel-group 1
crc 32
pos flag c2 1
!
```

Example 11-16 shows the ML-Series IOS configuration for configuring the POS Channel on Switch B.

Example 11-16 *Switch B POS Channel Configuration*

```
!
bridge irb
bridge 1 protocol ieee
!
interface Port-channel1
no ip address
no keepalive
bridge-group 1
!
interface FastEthernet0
no ip address
bridge-group 1
```

Example 11-16 *Switch B POS Channel Configuration (Continued)*

```
!
interface POS0
no ip address
channel-group 1
crc 32
pos flag c2 1
!
interface POS1
no ip address
channel-group 1
crc 32
pos flag c2 1
!
```

NOTE When configuring encapsulation over FEC, GEC, or POS, be sure to configure 802.1Q on the port-channel interface, not its member ports. However, certain attributes of port channel, such as duplex mode, need to be configured at the member port levels. Also make sure that you do not apply protocol-level configuration (such as an IP address or a bridge-group assignment) to the member interfaces. All protocol-level configuration should be on the port channel or on its subinterface. You must configure 802.1Q encapsulation on the partner system of the EtherChannel as well.

Configuring VRF Lite on the ML-Series

VRF provides multiple routing instances for MPLS Layer 2.5 VPNs. It provides a separate IP routing and forwarding table to each customer VPN. VRF is used with MP-iBGP (Multiprotocol Internal BGP) between provider equipment (PE) MPLS routers to provide Layer 3 MPLS-VPN. However, the ML-Series VRFs are implemented without MP-iBGP. In an MPLS network, customer equipment (CE) routers normally connect to PE routers or switches.

NOTE For detailed information on the implementation of ATM or router-based MPLS VPNs, refer to the Cisco Press publication *Advanced MPLS Design and implementation*.

With VRF lite, the ML-Series is considered as either a PE extension or a CE extension. It is considered a PE extension because it supports VRFs, whereas it is considered a CE extension because it can serve many customers as a CE box.

A normal MPLS CE can serve only one customer, whereas an ML-Series CE running VRF lite can have multiple interfaces/subinterfaces with a PE for different customers. It holds VRFs (routing information) locally and does not distribute the VRFs to its connected PE(s). It uses VRF information to direct traffic to the correct interfaces/subinterfaces when it receives traffic from customer CPE routers or from service provider PE router(s). Follow the steps in Table 11-25 to configure VRF lite on the ML-Series cards beginning in global configuration mode.

Table 11-25 *VRF Lite Configuration Steps*

Step	Command	Description
Step 1	Router(config)# **ip vrf** *vrf-name*	Enters VRF configuration mode and assigns a VRF name.
Step 2	Router(config-vrf)# **rd** *route-distinguisher*	Creates a VPN route distinguisher.
Step 3	Router(config-vrf)# **route-target** {**import**	**export**	**both**} *route-distinguisher*	Creates a list of import and/or export route target communities for the specified VRF.
Step 4	Router(config-vrf)# **import map** *route-map*	(Optional) Associates the specified route map with the VRF.
Step 5	Router(config-vrf)# **exit**	Exits the current configuration mode and enters global configuration mode.
Step 6	Router(config)# **interface** *type number*	Specifies an interface and enters interface configuration mode.
Step 7	Router(config-vrf)# **ip vrf forwarding** *vrf-name*	Associates a VRF with an interface or subinterface.
Step 8	Router(config-if)# **end**	Exits to privileged EXEC mode.
Step 9	Router# **copy running-config startup-config**	(Optional) Saves configuration changes to NVRAM.

ML-Series VRF Lite Example

This example shows VRF lite configuration on the ML-Series cards. As shown in Figure 11-52, Router A is the multilayer switch located in the ML-Series card, physically present in ONS 15454 A, and Router B is the multilayer switch located in the ML-Series card, physically present in ONS 15454 B. The two ONS 15454 nodes are connected via a SONET/SDH ring. After configuration, Router A and B will end up with three routing tables each; the global routing table, VRF customer_a routing table, and VRF customer_b routing table. Use the **show ip route** command to view the global routing table. The **show ip route vrf** *vrf-name* command displays the VRF routing tables.

Figure 11-52 *ML-Series VRF Lite*

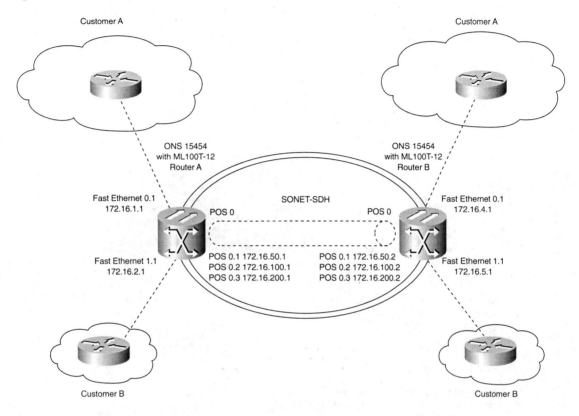

Example 11-17 shows the ML-Series IOS-based VRF lite configuration for Router A.

Example 11-17 *Router A VRF Lite Configuration*

```
!
hostname Router_A
!
ip vrf customer_a
rd 1:1
route-target export 1:1
route-target import 1:1
!
ip vrf customer_b
rd 2:2
route-target export 2:2
route-target import 2:2
!
```

continues

Example 11-17 *Router A VRF Lite Configuration (Continued)*

```
bridge 1 protocol ieee
bridge 2 protocol ieee
bridge 3 protocol ieee
!
interface FastEthernet0
no ip address
!
interface FastEthernet0.1
encapsulation dot1Q 2
ip vrf forwarding customer_a
ip address 172.16.1.1 255.255.255.0
bridge-group 2
!
interface FastEthernet1
no ip address
!
interface FastEthernet1.1
encapsulation dot1Q 3
ip vrf forwarding customer_b
ip address 172.16.2.1 255.255.255.0
bridge-group 3
!
interface POS0
no ip address
crc 32
no cdp enable
pos flag c2 1
!
interface POS0.1
encapsulation dot1Q 1 native
ip address 172.16.50.1 255.255.255.0
bridge-group 1
!
interface POS0.2
encapsulation dot1Q 2
ip vrf forwarding customer_a
ip address 172.16.100.1 255.255.255.0
bridge-group 2
!
interface POS0.3
encapsulation dot1Q 3
ip vrf forwarding customer_b
ip address 172.16.200.1 255.255.255.0
bridge-group 3
!
router ospf 1
log-adjacency-changes
network 172.16.50.0 0.0.0.255 area 0
!
router ospf 2 vrf customer_a
log-adjacency-changes
network 172.16.1.0 0.0.0.255 area 0
network 172.16.100.0 0.0.0.255 area 0
```

Example 11-17 *Router A VRF Lite Configuration (Continued)*

```
!
router ospf 3 vrf customer_b
 log-adjacency-changes
 network 172.16.2.0 0.0.0.255 area 0
 network 172.16.200.0 0.0.0.255 area 0
!
```

Example 11-18 shows the ML-Series IOS-based VRF lite configuration for Router B.

Example 11-18 *Router B VRF Lite Configuration*

```
!
hostname Router_B
!
ip vrf customer_a
 rd 1:1
 route-target export 1:1
 route-target import 1:1
!
ip vrf customer_b
 rd 2:2
 route-target export 2:2
 route-target import 2:2
!
bridge 1 protocol ieee
bridge 2 protocol ieee
bridge 3 protocol ieee
!
!
interface FastEthernet0
 no ip address
!
interface FastEthernet0.1
 encapsulation dot1Q 2
 ip vrf forwarding customer_a
 ip address 172.16.4.1 255.255.255.0
 bridge-group 2
!
interface FastEthernet1
 no ip address
!
interface FastEthernet1.1
 encapsulation dot1Q 3
 ip vrf forwarding customer_b
 ip address 172.16.5.1 255.255.255.0
 bridge-group 3
!
interface POS0
 no ip address
 crc 32
```

continues

Example 11-18 *Router B VRF Lite Configuration (Continued)*

```
 no cdp enable
 pos flag c2 1
!
interface POS0.1
 encapsulation dot1Q 1 native
 ip address 172.16.50.2 255.255.255.0
 bridge-group 1
!
interface POS0.2
 encapsulation dot1Q 2
 ip vrf forwarding customer_a
 ip address 172.16.100.2 255.255.255.0
 bridge-group 2
!
interface POS0.3
 encapsulation dot1Q 3
 ip vrf forwarding customer_b
 ip address 172.16.200.2 255.255.255.0
 bridge-group 3
!
router ospf 1
 log-adjacency-changes
 network 172.16.50.0 0.0.0.255 area 0
!
router ospf 2 vrf customer_a
 log-adjacency-changes
 network 172.16.4.0 0.0.0.255 area 0
 network 172.16.100.0 0.0.0.255 area 0
!
router ospf 3 vrf customer_b
 log-adjacency-changes
 network 172.16.5.0 0.0.0.255 area 0
 network 172.16.200.0 0.0.0.255 area 0
!
```

Configuring IP Protocols for the ML-Series Card

The IOS subsystem on the ML-Series card supports the following routing protocols. For more information on configuring IP routing protocols, refer to the Cisco IOS configuration and command reference publications. Routing protocol support is key to the integration of the ML-Series card in a service provider backbone environment. Interoperability with other Cisco and third-party routers and switches in a core network is possible using standardized Interior Gateway Protocol (IGP) and Exterior Gateway Protocol (EGP) routing protocols. The ML-Series card uses the standard IOS CLI, which enables you to leverage your IOS knowledge to configure routing and switching functions. The ML-Series card also supports standard and extended access lists that enable you to configure control-plane and data-plane access. Table 11-26 indicates the routing protocols supported by the ML-Series card.

Table 11-26 *ML-Series Routing Protocol Support*

Routing Type	Protocol
Unicast IP routing	Static
	RIP v1
	RIP v2
	EIGRP
	OSPF
	BGP
	IS-IS
Multicast routing	PIM

NOTE Certain IOS commands are unique to the ML-Series cards. These commands are covered in Appendix A.

NOTE The ML-Series card also supports integrated routing and bridging (IRB), which enables you to route a given protocol between routed interfaces and bridge groups within a single ML-Series card.

ML-Series Quality of Service

The ML-Series card incorporates QoS and traffic-engineering features that provide control over access to network bandwidth resources. This control enables providers to implement priorities specified in SLAs. The ML-Series QoS provides the capability to classify each packet in the network based on its interface of arrival, bridge group, CoS, IP precedence, and IP differentiated services code points (DSCPs). When packets are classified, further QoS functions can be applied to each packet as it traverses the network.

Policing is provided by the ML-Series card to ensure that no attached equipment submits more than a predefined amount of bandwidth into the network. This feature limits the bandwidth available to a customer and provides a mechanism to support traffic engineering. Priority marking allows Ethernet IEEE 802.1P CoS bits to be marked as they exit the ML-Series card. This feature operates on the outer IEEE 802.1P tag together with 802.1Q double tagging.

Per-class flow queuing is provided to enable fair access to excess network bandwidth, and Low Latency Queuing is supported for voice traffic. It allows allocation of bandwidth to support SLAs and ensure applications with high network resource requirements are adequately served.

Buffers are allocated to queues dynamically from a shared resource pool. The allocation process incorporates the instantaneous system load as well as the allocated bandwidth to each queue to optimize buffer allocation to each queue.

The ML-Series card uses an advanced Weighted Deficit Round-Robin (WDRR) scheduling process to provide fair access to excess bandwidth as well as guaranteed throughput to each class flow. Admission control is a process that is invoked each time that service is configured on the ML-Series card to ensure that the card's available QoS resources are not overcommitted. In particular, admission control ensures that no configurations are accepted where a sum of the committed bandwidths on an interface exceed the total bandwidth of that interface. The QoS bandwidth allocation of multicast and broadcast traffic is handled separately and differently than unicast traffic. Aggregate multicast and broadcast traffic are given a fixed bandwidth commit of 10 percent on each interface, and are treated as best effort for traffic exceeding 10 percent. Multicast and broadcast are supported at line rate.

ML-Series Resilient Packet Ring

Resilient packet ring (RPR) is a standards-based (IEEE 802.17) network architecture designed for metro fiber ring networks. RPR is a MAC protocol designed to overcome the limitations of IEEE 802.1D STP, IEEE 802.1W RSTP, and SONET/SDH in packet-based networks. RPR operates at the Layer 2 level and is compatible with Ethernet and SONET/SDH.

ML-Series RPR Operation

The RPR subsystem of the ML-Series card relies on the QoS features of the ML-Series for implementing classes of service. The ML-Series card QoS mechanisms apply to all SONET/SDH traffic on the ML-Series card, whether passed through, bridged, or stripped. When an ML-Series card is configured with RPR and made part of an SPR, the ML-Series card assumes it is part of a ring. If a packet is not destined for devices attached to the specific ML-Series, the ML-Series card just continues to forward this transit traffic along the SONET/SDH circuit relying on the circular path of the ring architecture to guarantee the packet will eventually arrive at the destination. This eliminates the need to queue and forward the packet flowing through the nondestination ML-Series card. From a Layer 2 or Layer 3 perspective, the entire RPR looks like one shared network segment.

RPR supports operation over protected and unprotected SONET/SDH circuits. On unprotected SONET/SDH circuits, RPR provides SONET/SDH-like protection without the redundant SONET/SDH protection path. Eliminating the need for a redundant SONET/SDH path frees bandwidth for additional traffic. RPR also incorporates spatial reuse of bandwidth through a hash algorithm for east/west packet transmission. RPR uses the entire ring bandwidth and does not need to block ring segments like STP or RSTP. The RPR protocol, using the transmitted packet's header information, allows the interfaces to quickly determine the operation that needs to be applied to the packet. As illustrated in Figure 11-53, an ML-Series card configured with RPR has three basic packet-handling operations:

- Bridge

- Pass through
- Strip

Figure 11-53 *RPR Packet-Handling Operations*

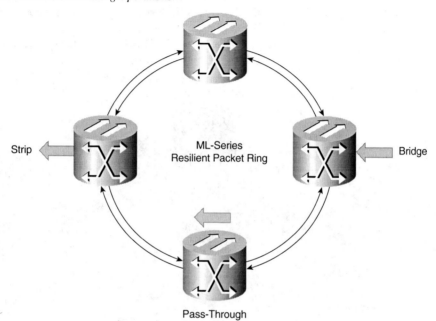

Bridging connects and switches packets between the Ethernet ports on the ML-Series and the POS circuit around the ring. Pass through lets the packets continue through the ML-Series card and along the ring and stripping strips the packet off the ring and discards it. Because STP or RSTP is not in effect between nodes when RPR is configured, the transmitting RPR port recognizes its won packets and strips them after they return from circling around the ring. A hash algorithm is used to determine the direction of the packet around the RPR.

RPR Ring Wrapping

RPR initiates ring wraps in the event of a fiber cut, node failure, node restoration, node insertion, or other traffic issues. This protection mechanism redirects traffic to the original destination by sending it in the opposite direction around the ring after a link-state change or after receiving SONET/SDH path-level alarms. Ring wrapping on the ML-Series card allows sub-50-ms convergence times. RPR convergence times are comparable to SONET/SDH and much faster than STP or RSTP. RPR on the ML-Series card survives both unidirectional and bidirectional transmission failures within the ring. Unlike STP or RSTP, RPR restoration is scalable. Increasing the number of ML-Series cards in a ring does not increase the convergence time. RPR will initiate ring wraps immediately or delay the wrap with a configured carrier delay time. When configured to wrap traffic after the carrier delay, a POS trigger delay time should be added to the carrier delay time to estimate approximate convergence times. The default and minimum POS trigger delay time for the ML-Series Card is 200 ms. A carrier delay time of 200 ms (default) and a POS trigger delay time of 200 ms (default and minimum) combine for a total convergence time of approximately 400 ms. If the carrier delay is set to 0, the convergence time is approximately 200 ms. If the carrier delay time is changed from the default, the new carrier delay time must be configured on all the ML-Series card interfaces, including the SPR, POS, and Gigabit Ethernet or Fast Ethernet interfaces. Figure 11-54 shows RPR ring wrapping on the two nodes adjacent to the fiber cut.

Figure 11-54 *RPR Ring Wrap*

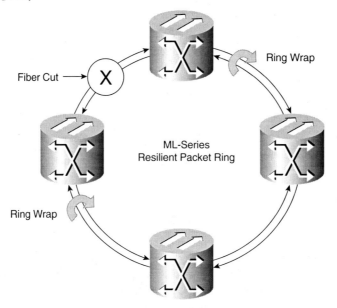

RPR VLAN Support

RPR improves MAC address support because an ML-Series card does not need to learn the MAC address of pass-through packets. The ML-Series card's MAC address table holds only the MAC IDs of packets that have been bridged or stripped by that card. This allows the collective tables of the ML-Series cards in the ring to hold a greater number of MAC addresses. RPR also enhances VLAN support relative to STP and RSTP. In an STP and RSTP, a new VLAN must be configured on all POS interfaces on the ring. In RPR, the VLAN must only be added to the configuration of those interfaces that bridge or strip packets for that VLAN. The ML-Series card still has a 255 architectural maximum limit of VLAN/bridge group per ML-Series card. Because the ML-Series card needs to hold only the VLANs incorporating that card, however, the collective number of VLANs held by all the ML-Series cards in the ring can be much greater.

ML-Series RPR Configuration

RPR configuration on the ML-Series cards is a two-step process. The first step involves configuring point-to-point STS circuits between the ML-Series cards using CTC. The second step involves configuring RPR using the IOS CLI. The bridged ML-Series cards are connected to each other through point-to-point STS circuits, which use one of the first ML-Series card's POS ports as a source and one of the second ML-Series card's POS ports as a destination. All ML-Series cards in an SPR must be connected directly or indirectly by point-to-point SONT/SDH circuits. Keep the following points in mind when configuring a point-to-point circuit on the ML-Series using CTC:

- Leave all CTC Circuit Creation Wizard options at default, except Fully Protected Path on the Circuit Routing Preferences dialog box, which provides SONET/SDH protection. This option should be unchecked. RPR provides Layer 2 protection for SPR circuits.

- Check the option Using Required Nodes and Spans to route automatically on the Circuit Routing Preferences dialog box. If the source and destination nodes are adjacent on the ring, exclude all nodes except the source and destination on the Circuit Routing Preferences dialog box. This forces the circuit to be routed directly between source and destination and preserves STS circuits, which would be consumed if the circuit routed through other nodes in the ring. If there is a node or nodes that do not contain an ML-Series card between the two nodes containing ML-Series card, include this node or nodes in the included nodes in the Circuit Routing Preference dialog box, along with the source and destination nodes.

- Keep in mind that ML-Series card STS circuits do not support unrelated circuit-creation options, such as unidirectional traffic, creating cross-connects only (TL1-like), interdomain (UCP), protected drops, or UPSR path selectors.

When the CTC circuit process is complete, begin an IOS session to configure RPR/SPR on the ML-Series card and interfaces. RPR is configured on the ML-Series cards by creating an SPR interface from the Cisco IOS CLI. The SPR is a virtual interface, similar to an EtherChannel interface. The POS interfaces are the physical interfaces associated with the RPR SPR interface. An ML-Series card supports a single SPR interface. The SPR interface has a single MAC address and provides all the normal attributes of a Cisco IOS interface, such as support for default routes. An SPR interface is considered a trunk port, and like all trunk ports, subinterfaces must be configured for the SPR interface to become part of a bridge group. An SPR interface is configured similarly to a EtherChannel (port-channel) interface. The members of the SPR interface must be POS interfaces. Instead of using the **channel-group** command to define the members, you use the **spr-intf-ID** command. And like **port-channel**, you configure the SPR interfaces rather than the POS interface. ML-Series RPR is supported with LEX encapsulation, which is the default encapsulation for the ML-Series cards.

NOTE In configuring an SPR, if one ML-Series card is not configured with an SPR interface, but valid STS/STM circuits connect this ML-Series card to the other ML-Series cards in the SPR, traffic will not flow between the properly configured ML-Series cards in the SPR, and no alarms will indicate this condition. It is recommended to configure all ML-Series cards in an SPR before sending traffic.

Follow the steps in Table 11-27 to configure RPR on the ML-Series cards beginning in global configuration mode.

Table 11-27 *ML-Series RPR Configuration Steps*

Step	Command	Description
Step 1	Router(config)#**bridge irb**	Enables the Cisco IOS Software to both route and bridge a given protocol on separate interfaces within a single ML-Series card.
Step 2	Router(config)#**interface spr 1**	Creates the SPR interface on the ML-Series card or enters the SPR interface configuration mode. The only valid SPR number is 1.
Step 3	Router(config-if)#**spr station-id** *station-ID-number*	Configures a station ID. The user must configure a different number for each SPR interface that attaches to the RPR. Valid station ID numbers range from 1 to 254.
Step 4	Router(config-if)#**spr wrap {immediate	delayed}**	(Optional) Sets the RPR ring wrap mode to either wrap traffic the instant it detects a link-state change or to wrap traffic after the carrier delay, which gives the SONET/SDH protection time to register the defect and declare the link down. Immediate should be used if RPR is running over unprotected SONET/SDH circuits. Delayed should be run for BLSR or UPSR protected circuits. The default setting is immediate.
Step 5	Router(config-if)#**bridge-group** *bridge-group-number*	(Optional) Assigns the SPR interface to a bridge group. This bridge group bridges the SPR and desired Fast Ethernet or Gigabit Ethernet interfaces.
Step 6	Router(config-if)#**carrier delay msec** *milliseconds*	(Optional) Sets the carrier delay time. The default setting is 200 ms, which is optimum for SONET/SDH protected circuits. The default unit of time for setting the carrier delay is seconds. The **msec** command resets the time unit to milliseconds.
Step 7	Router(config)**interface pos 0**	(Optional) Enters interface configuration mode for POS port 0 to set carrier delay time.
Step 8	Router(config-if)#**carrier delay msec** *milliseconds*	(Optional) Sets the carrier delay time. The default setting is 200 ms, which is optimum for SONET/SDH protected circuits. The default unit of time for setting the carrier delay is seconds. The **msec** command resets the time unit to milliseconds.
Step 9	Router(config)#**interface pos 1**	(Optional) Enters interface configuration mode for POS port 1 to set optional commands.

Table 11-27 *ML-Series RPR Configuration Steps (Continued)*

Step	Command	Description
Step 10	Router(config-if)#**carrier delay msec** *milliseconds*	(Optional) Sets the carrier delay time. The default setting is 200 ms, which is optimum for SONET/SDH protected circuits. The default unit of time for setting the carrier delay is seconds. The **msec** command resets the time unit to milliseconds.
Step 11	Router(config-if)#**end**	Exits to privileged EXEC mode.
Step 12	Router# **copy running-config startup-config**	(Optional) Saves configuration changes to NVRAM.

NOTE The SPR interface is the routed interface. Do not enable Layer 3 addresses or assign bridge groups on the POS interfaces assigned to the SPR interface. When traffic coming in on an SPR interface needs to be policed, the same input service policy needs to be applied to both the POS ports that are part of the SPR interface.

Each of the ML-Series card's two POS ports must be assigned to the SPR interface. The ML-Series STS circuit must have first been provisioned through CTC for the virtual POS ports to appear in Cisco IOS. To assign a POS interface on the ML-Series to the SPR, follow the steps in Table 11-28, beginning in global configuration mode.

Table 11-28 *Steps to Assign POS Interfaces*

Step	Command	Purpose
Step 1	Router(config)#**interface pos** *number*	Enters the interface configuration mode to configure the first POS interface that you want to assign to the SPR.
Step 2	Router(config-if)# **spr-intf-ID** *shared-packet-ring-number*	Assigns the POS interface to the SPR interface. The SPR number must be the same SPR number that you assigned to the SPR interface.
Step 3	Router(config)#**interface pos** *number*	Enters the interface configuration mode to configure the second POS interface that you want to assign to the SPR.
Step 4	Router(config-if)# **spr-intf-ID** *shared-packet-ring-number*	Assigns the POS interface to the SPR interface. The SPR number must be the same SPR number that you assigned to the SPR interface.
Step 5	Router(config-if)# **end**	Exits to privileged EXEC mode.
Step 6	Router# **copy running-config startup-config**	(Optional) Saves the configuration changes to NVRAM.

ML-Series RPR Example

This example shows IOS-based RPR configuration for the ML-Series cards. As shown in Figure 11-55, SPR Station ID 1 is configured on the ML-Series card in node A, SPR Station ID 2 is configured on the ML-Series card in node B, and SPR Station ID 3 is configured on the ML-Series card in node C.

Figure 11-55 *ML-Series RPR Configuration*

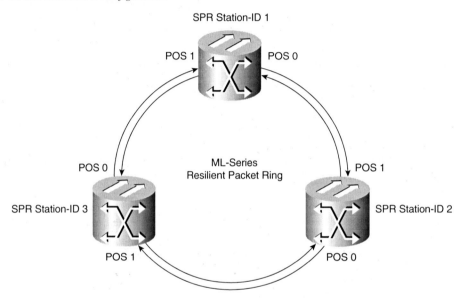

The three ONS 15454 nodes are connected via a SONET/SDH ring. The configuration assumes that ML-Series card POS ports are already linked by point-to-point SONET/SDH circuits configured through CTC.

Example 11-19 shows the ML-Series IOS-based RPR configuration for Node A.

Example 11-19 *Node A RPR Configuration*

```
!
hostname Node A
!
bridge irb
!
interface SPR1
no ip address
no keepalive
spr station-ID 1
hold-queue 150 in
bridge-group 1
!
```

Example 11-19 *Node A RPR Configuration (Continued)*

```
interface POS0
no ip address
spr-intf-ID 1
!
interface POS1
no ip address
spr-intf-ID 1
!
interface GigabitEthernet0
 no ip address
 no ip route-cache
bridge-group 1
!
interface GigabitEthernet1
 no ip address
 no ip route-cache
bridge-group 1
```

Example 11-20 shows the ML-Series IOS-based RPR configuration for Node B.

Example 11-20 *Node B RPR Configuration*

```
!
hostname Node B
!
bridge irb
!
interface SPR1
no ip address
no keepalive
spr station-ID 2
hold-queue 150 in
bridge-group 1
!
interface POS0
no ip address
spr-intf-ID 1
!
interface POS1
no ip address
spr-intf-ID 1
!
interface GigabitEthernet0
 no ip address
 no ip route-cache
bridge-group 1
!
interface GigabitEthernet1
 no ip address
 no ip route-cache
bridge-group 1
```

Example 11-21 shows the ML-Series IOS-based RPR configuration for Node C.

Example 11-21 *Node C RPR Configuration*

```
!
hostname Node C
!
bridge irb
!
interface SPR1
no ip address
no keepalive
spr station-ID 3
hold-queue 150 in
bridge-group 1
!
interface POS0
no ip address
spr-intf-ID 1
!
interface POS1
no ip address
spr-intf-ID 1
!
interface GigabitEthernet0
 no ip address
 no ip route-cache
bridge-group 1
!
interface GigabitEthernet1
 no ip address
 no ip route-cache
bridge-group 1
```

After RPR is configured, you can monitor its status using the **show interface spr** or **show run interface spr** command.

Dual RPR Interconnect

ML-Series RPR includes a mechanism to interconnect rings for protection from matching node failure. The bridge-group protocol, Dual RPR Interconnect (DRPRI), provides two parallel connections of the rings linked by a special instance of RSTP. One connection is the active node, and the other is the standby node. During a failure of the active node, link, or card, a proprietary algorithm detects the failure and causes a switchover to the standby node.

The paired ML1000-2 cards share the same station ID and are viewed by other members of the RPR as a single card. In Figure 11-56, paired cards A and B have the same SPR station ID on RPR-1, and paired cards C and D have the same station ID on RPR-2. The interconnected nodes do not need to be adjacent on the RPR. Bridging, IP routing, IP multicast, policing, and bandwidth allocations can still be provisioned on DRPRI ML1000-2 cards. ML-Series cards A and C are physically housed in node 1, and ML-Series cards B and D are physically housed in node 2.

Figure 11-56 *Dual RPR Interconnection*

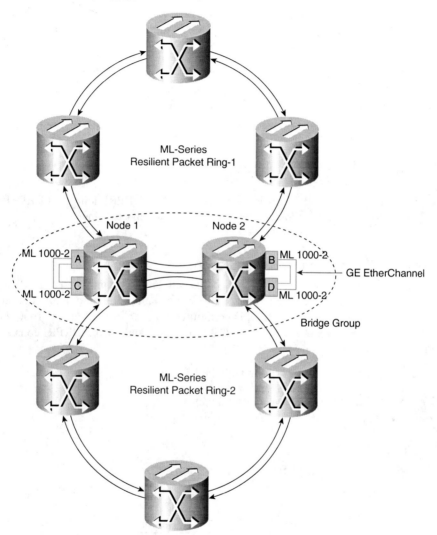

DRPRI has the following characteristics:

- Four ML1000-2 cards are required.
- All four ML1000-2 cards must be part of the same bridge group (VLAN).
- Each paired set of ML1000-2 cards must have the same SPR station ID.
- The bridge group must be configured on SPR subinterfaces.
- The DRPRI bridge group is limited to one protocol, so a bridge group with DRPRI implemented cannot simultaneously implement RSTP or STP.

- On each of the four ML1000-2 cards, both Gigabit Ethernet ports must be joined in GEC and the GEC interface included in the DRPRI bridge group, or one Gigabit Ethernet port must be shut down and the other one included in the DRPRI bridge group. The GEC method is recommended.
- The DRPRI bridge group cannot be used to carry data traffic.
- A DRPRI node can be used only for interconnecting two RPRs. The front ports of the cards should not be used to carry other traffic.
- Non-DRPRI bridge groups carrying traffic between rings should not have STP or RSTP configured.
- Non-DRPRI bridge groups carrying traffic between rings must be configured on each of the four ML-Series cards.
- Q-in-Q and protocol tunnels cannot be started on DRPRI nodes, but DRPRI nodes can bridge Q-in-Q and protocol tunnels across the connected rings.

Dual RPR Interconnect Configuration

DRPRI requires two pairs of ML-Series cards with one pair configured as RPR and belonging to the first RPR and the second pair configured as RPR and belonging to the second RPR, as shown in Figure 11-57. DRPRI is configured on each of the four ML1000-2 cards that connect the two adjacent RPRs. The process of configuring DRPRI consists of the following tasks:

Step 1 Configure a bridge group with the DRPRI protocol.

Figure 11-57 *ML-Series DRPRI Configuration*

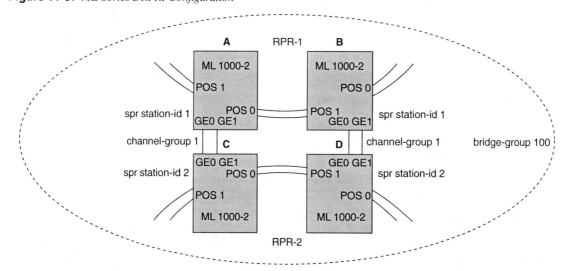

Step 2 Configure the SPR interface:

   (a) Assign a station ID number.

   (b) Assign a DRPRI ID of 0 or 1.

Step 3 Create an SPR subinterface and assign the bridge group to the subinterface.

Step 4 Create a GEC interface.

Step 5 Create a GEC subinterface and assign the bridge group to the subinterface.

To configure DRPRI on the ML-Series cards of the matching nodes, follow the steps in Table 11-29, beginning in global configuration mode.

Table 11-29 *Steps to Configure ML-Series DRPRI*

Step	Command	Purpose
Step 1	Router(config)#**bridge irb**	Enables the Cisco IOS Software to both route and bridge a given protocol on separate interfaces within a single ML-Series card.
Step 2	Router(config)#**bridge** *bridge-group-number* **protocol drpri-rstp**	Creates the bridge-group number shared by the 4 ML1000-2 cards and assigns the protocol for DRPRI to the bridge group. The same command using the same bridge-group number must be given on each of the four cards.
Step 3	Router(config)#**interface spr 1**	Creates the SPR interface for RPR or enters the SPR interface configuration mode on a previously created SPR interface. The only valid SPR number is 1.
Step 4	Router(config-if)#**spr station-ID** *station-ID-number*	Configures a station identification number. The user must configure the same station ID on both the paired cards. Valid station ID numbers range from 1 to 254.
Step 5	Router(config-if)#**spr drpri-ID {0	1}**	Creates a DRPRI identification number of 0 or 1 to differentiate between the ML1000-2 cards paired for DRPRI.
Step 6	Router(config-if)#**interface spr** *shared-packet-ring-sub-interface-number*	Creates the SPR subinterface.
Step 7	Router(config-subif)#**encapsulation dot1q** *vlan-ID*	Sets the SPR subinterface encapsulation to IEEE 802.1Q.
Step 8	Router(config-subif)#**bridge-group** *bridge-group-number*	Assigns the SPR subinterface to a bridge group.

continues

Table 11-29 *Steps to Configure ML-Series DRPRI (Continued)*

Step	Command	Purpose
Step 9	Router(config)# **interface port-channel** *channel-number*	Creates the GEC interface or channel group.
Step 10	Router(config-if)# **interface gigabitethernet** *number*	Enters interface configuration mode for the first Gigabit Ethernet interface that you want to assign to the GEC subinterface.
Step 11	Router(config-if)# **channel-group** *channel-number*	Assigns the Gigabit Ethernet interfaces to the GEC. The channel number must be the same channel number you assigned to the EtherChannel interface.
Step 12	Router(config-if)# **interface gigabitethernet** *number*	Enters interface configuration mode for the second Gigabit Ethernet interface that you want to assign to the GEC subinterface.
Step 13	Router(config-if)# **channel-group** *channel-number*	Assigns the Gigabit Ethernet interfaces to the GEC. The channel number must be the same channel number you assigned to the EtherChannel interface.
Step 14	Router(config-subif)# **interface port-channel** *channel-sub-interface-number*	Creates the GEC subinterface.
Step 15	Router(config-subif)#**encapsulation dot1q** *vlan-ID*	Sets subinterface encapsulation to IEEE 802.1Q. The VLAN ID used should be the same VLAN ID used in step 7.
Step 16	Router(config-subif)#**bridge-group** *bridge-group-number*	Assigns the GEC subinterface to the bridge group.
Step 17	Router(config-if)# **end**	Exits to privileged EXEC mode.
Step 18	Router# **copy running-config startup-config**	(Optional) Saves configuration changes to NVRAM.

ML-Series Dual RPR Interconnect Example

This example shows IOS-based DRPRI configuration for the ML-Series cards. As shown in Figure 11-57, SPR Station ID 1 is configured on the ML-Series cards A and B, and SPR Station ID 2 is configured on the ML-Series cards C and D. All four ML-Series cards are members of bridge group 100. The GE ports are associated with channel group 1. The GE ports of ML-Series card A are connected to the GE ports of ML-Series card C, and the GE ports of ML-Series card B are connected to the GE ports of ML-Series card D. ML-Series cards A and B are connected via a SONET/SDH ring along with other nodes on RPR-1, and ML-Series cards C and D are connected via a SONET/SDH ring along with other nodes on RPR-2. The configuration assumes that the ML-Series card POS ports are already linked by point-to-point SONET/SDH circuits configured through CTC.

Example 11-22 shows the ML-Series IOS-based DRPRI configuration for ML-Series card A.

Example 11-22 *ML-Series Card A DRPRI Configuration*

```
!
hostname ML-Series-A
!
bridge irb
bridge 100 protocol drpri-rstp
!
interface Port-channel1
no ip address
no ip route-cache
 hold-queue 300 in
!
interface Port-channel1.1
 encapsulation dot1Q 10
 no ip route-cache
 bridge-group 100
!
interface SPR1
 no ip address
 no keepalive
 spr station-ID 1
hold-queue 150 in
!
interface SPR1.1
 encapsulation dot1Q 10
 bridge-group 100
!
interface GigabitEthernet0
 no ip address
 no ip route-cache
channel-group 1
!
interface GigabitEthernet1
 no ip address
 no ip route-cache
channel-group 1
!
interface POS0
 no ip address
 spr-intf-ID 1
 crc 32
!
interface POS1
 no ip address
 spr-intf-ID 1
 crc 32
!
ip classless
no ip http server
```

Example 11-23 shows the ML-Series IOS-based DRPRI configuration for ML-Series card B.

Example 11-23 *ML-Series Card B DRPRI Configuration*

```
!
hostname ML-Series-B
!
bridge irb
bridge 100 protocol drpri-rstp
!
interface Port-channel1
 no ip address
 no ip route-cache
 hold-queue 300 in
!
interface Port-channel1.1
 encapsulation dot1Q 10
 no ip route-cache
 bridge-group 100
!
interface SPR1
 no ip address
 no keepalive
 spr station-ID 1
 spr drpr-ID 1
 hold-queue 150 in
!
interface SPR1.1
 encapsulation dot1Q 10
 bridge-group 100
!
interface GigabitEthernet0
 no ip address
 no ip route-cache
 channel-group 1
!
interface GigabitEthernet1
 no ip address
 no ip route-cache
 channel-group 1
!
interface POS0
 no ip address
 spr-intf-ID 1
 crc 32
!
interface POS1
 no ip address
 spr-intf-ID 1
 crc 32
!
ip classless
no ip http server
```

Example 11-24 shows the ML-Series IOS-based DRPRI configuration for ML-Series card C.

Example 11-24 *ML-Series Card C DRPRI Configuration*

```
!
hostname ML-Series-C
!
bridge irb
bridge 100 protocol drpri-rstp
!
interface Port-channel1
no ip address
no ip route-cache
 hold-queue 300 in
!
interface Port-channel1.1
 encapsulation dot1Q 10
 no ip route-cache
 bridge-group 100
!
interface SPR1
 no ip address
 no keepalive
 spr station-ID 2
hold-queue 150 in
!
interface SPR1.1
 encapsulation dot1Q 10
 bridge-group 100
!
interface GigabitEthernet0
 no ip address
 no ip route-cache
channel-group 1
!
interface GigabitEthernet1
 no ip address
 no ip route-cache
channel-group 1
!
interface POS0
 no ip address
 spr-intf-ID 1
 crc 32
!
interface POS1
 no ip address
 spr-intf-ID 1
 crc 32
!
ip classless
no ip http server
!
```

Example 11-25 shows the ML-Series IOS-based DRPRI configuration for ML-Series card D.

Example 11-25 *ML-Series Card D DRPRI Configuration*

```
!
hostname ML-Series-D
!
bridge irb
bridge 100 protocol drpri-rstp
!
interface Port-channel1
 no ip address
 no ip route-cache
 hold-queue 300 in
!
interface Port-channel1.1
 encapsulation dot1Q 10
 no ip route-cache
 bridge-group 100
!
interface SPR1
 no ip address
 no keepalive
 spr station-ID 2
 spr drpr-ID 1
 hold-queue 150 in
!
interface SPR1.1
 encapsulation dot1Q 10
 bridge-group 100
!
interface GigabitEthernet0
 no ip address
 no ip route-cache
 channel-group 1
!
interface GigabitEthernet1
 no ip address
 no ip route-cache
 channel-group 1
!
interface POS0
 no ip address
 spr-intf-ID 1
 crc 32
!
interface POS1
 no ip address
 spr-intf-ID 1
 crc 32
!
ip classless
no ip http server
```

After DRPRI is configured, you can monitor its status using the **show bridge** *bridge-group-number* **verbose** command.

ML-Series Virtual Concatenation

ML-Series virtual concatenation (VACT) enables you to group several noncontiguous STS-1s or VC4s into a single larger virtual STS-N or STM-N. This virtual STS-N or STM-N is referred to as a *virtual concatenated group (VCG)* because it is made up of a group of smaller STS, or VC levels. VACT is used in conjunction with another scheme, link-capacity adjustment scheme (LCAS), which allows members of a VCG to be dynamically added or subtracted to provide additional bandwidth as required. These schemes have now been finalized by the ITU. (G.707 defines VACT and G.7042 defines LCAS.)

ML-Series Switching Database Manager

High-speed forwarding is implemented in the ML-Series cards using the forwarding engine and ternary content-addressable memory (TCAM). The high-speed forwarding information is maintained in TCAM. The SDM is the software subsystem that manages the switching information maintained in TCAM. SDM organizes the switching information in TCAM into application-specific regions and configures the size of these application regions. SDM enables exact-match and longest-match address searches, which result in high-speed forwarding. SDM manages TCAM space by partitioning application-specific switching information into multiple regions.

TCAM identifies a location index associated with each packet forwarded and conveys it to the forwarding engine. The forwarding engine uses this location index to derive information associated with each forwarded packet. The key benefits of SDM in switching are its capability to organize the switching information in TCAM into application-specific regions and its capability to configure the size of these application regions. SDM enables exact-match and longest-match address searches, which result in high-speed forwarding.

SDM partitions TCAM space into multiple application-specific regions and interacts with the individual application control layers to store switching information. SDM consists of the following types of regions:

- **Exact-match region**—The exact-match region consists of entries for multiple application regions, such as IP adjacencies.

- **Longest-match region**—Each longest-match region consists of multiple buckets or groups of Layer 3 address entries organized in decreasing order by mask length. All entries within a bucket share the same mask value and key size. The buckets can change their size dynamically by borrowing address entries from neighboring buckets. Although the size of the whole application region is fixed, you can reconfigure it.

- **Weighted exact-match region**—The weighted exact-match region consists of exact-match entries with an assigned weight or priority. For example, with QoS, multiple exact match entries might exist, but some have priority over others. The weight is used to choose one entry when multiple entries match.

The SDM is responsible for managing TCAM space. TCAM space consists of 65,536 entries, each entry being 64 bits wide. SDM partitions the entire TCAM space for each application region based on user configuration. Although the maximum size of all application regions is fixed, you can reconfigure the maximum size of each application region. The default partitioning for each application region in TCAM is listed in Table 11-30.

Table 11-30 *Default Partitioning by Application Region in TCAM*

Application Region	Lookup Type	Key Size	Default Size	Number of TCAM Entries
IP adjacency	Exact match	64 bits	65,536 (shared)	65,536 (shared)
IP prefix	Longest match	64 bits	65,536 (shared)	65,536 (shared)
QoS classifiers	Weighted exact match	64 bits	65,536 (shared)	65,536 (shared)
IP VRF prefix	Longest prefix match	64 bits	65,536 (shared)	65,536 (shared)
IP multicast	Longest prefix match	64 bits	65,536 (shared)	65,536 (shared)
MAC address	Longest prefix match	64 bits	65,536 (shared)	65,536 (shared)
Access list	Weighted exact match	64 bits	65,536 (shared)	65,536 (shared)

The SDM maximum size for each application region can be configured by performing the procedure in Table 11-31, beginning in global configuration mode.

Table 11-31 *Steps to Configure SDM*

Step	Command	Description
Step 1	Router(config)# **sdm size** *region-name* [**k-entries**] *num-of-entries*	Sets the name of the application region whose size you want to configure. You can enter the size as multiples of 1K (that is, 1024) entries or in absolute number of entries.
Step 2	Router(config)# **end**	Exits to privileged EXEC mode.

Summary

Ethernet services can be transported over SONET and SDH by integrating Ethernet modules in the ONS 15454, 15600, and 15327 chassis. The ONS 15454 supports various Ethernet cards known as the G-Series, E-Series, and ML-Series. The G-Series card supports Gigabit Ethernet and is mainly used for transparent LAN services. The G-Series card does not support an Ethernet switch fabric and normally maps an external Ethernet source to the SONET or SDH STS-N or VC4-N. The E-Series card, on the other hand, supports Ethernet switching, VLANs, IEEE 802.1Q, Spanning Tree Protocol, and IEEE 802.1D. The E-Series card is mainly used to implement Layer 2 customer VPNs using 802.1Q VLANs over the SONET/SDH infrastructure. Scalability issues associated with the use of STP and 802.1Q tagging have been overcome by the

ML-Series cards. The ML-Series cards support full multilayer switching and VRF lite, which enables service providers to create Layer 3 VPNs without having to implement MPLS or IPSec.

The G-Series card in production is known as the G1K-4 card. This card is a four-port Gigabit Ethernet card. The G1K-4 card is not a Gigabit Ethernet switch. It maps up to four Gigabit Ethernet interfaces onto a SONET transport network with bandwidth at the signal levels up to STS-48c per card. The card provides line-rate forwarding for unicast, multicast, and broadcast Ethernet frames. G1K-4 card applications include providing carrier-grade TLS, 100-Mbps Ethernet private line services, and high-availability transport for applications, such as storage over MANs. In a SONET infrastructure, the four ports on the G1K-4 can be independently mapped to any combination of STS-1, STS-3c, STS-6c, STS-9c, STS-12c, STS-24c, and STS-48c circuit sizes, provided the sum of the circuit sizes that terminate on a card do not exceed STS-48c. In an SDH infrastructure, the four ports on the G1K-4 can be independently mapped to any combination of VC4, VC4-2c, VC4-3c, VC4-8c, and VC4-16c circuit sizes, provided the sum of the circuit sizes that terminate on a card do not exceed VC4-16c. The G1K-4 cards use the LEX protocol, which is a derivative of PPP over HDLC for SONET.

The E-Series cards incorporate Layer 2 switching within the card. E-Series cards support VLAN, IEEE 802.1Q, Spanning Tree Protocol, and IEEE 802.1D. Multicard EtherSwitch provisions two or more Ethernet cards to act as a single Layer 2 switch. It supports one STS-6c SPR, two STS-3c SPRs, or six STS-1 SPRs. The bandwidth of the single switch formed by the Ethernet cards matches the bandwidth of the provisioned Ethernet circuit up to STS-6c worth of bandwidth. Single-card EtherSwitch allows each Ethernet card to remain a single-switching entity within the ONS 15454 shelf. This option allows a full STS-12c worth of bandwidth between two Ethernet circuit points. For SDH, the E-Series cards support one VC4-2c or two VC4 circuits. The bandwidth of the single switch formed by the Ethernet cards matches the bandwidth of the provisioned Ethernet circuit up to VC4-2c worth of bandwidth.

The ML-Series cards are independent multilayer switches that can be integrated into the ONS 15454 SONET and SDH chassis. Addition of these cards makes the ONS 15454 Layer 2 and 3 aware, and endows the 15454 with full IP functionality. Two versions of the ML-Series card are in production: the 2-port Gigabit Ethernet card (ML1000-2) and 12-port Fast Ethernet card (ML100T-12). The ML-Series cards can be installed in either a North American ONS 15454 SONET chassis or in an International ONS 15454 SDH chassis. When installed in an ONS 15454 SONET chassis, the card features two virtual ports with a combined STS-48 maximum. The STS or STM circuits are provisioned through the ONS 15454 GUI (CTC) in the same manner as standard OC-N card STS or STM circuits. CTC also provides provisioning, inventory, SONET/SDH alarm reporting, and other standard ONS 15454 card functions for the ML-Series. The ML-Series cards ship with IOS Software that controls the data functions of the card. The Cisco IOS CLI is the primary user interface for the ML-Series card. Most Ethernet and Layer 3 configuration for the card can be done only via the Cisco IOS CLI. SONET/SDH circuits cannot be provisioned through IOS, but must be configured through CTC.

This chapter includes the following sections:

- **Network Design Strategies**—A single multiservice broadband network that can carry time-division multiplexed (TDM) voice, Ethernet, IP, and video with the reliability and restoration times of Synchronous Optical Network/Synchronous Digital Hierarchy (SONET/SDH) along with scalability in the order of terabits per second provides service providers a platform to deliver reliable carrier-class broadband services. The ONS 15000 MSxPs deliver highly available broadband services that include support for voice, TDM, SONET/SDH, Gigabit Ethernet, and all IP applications including voice over IP (VoIP), H.323 video, D1 video, and high-definition television (HDTV). The ONS 15000 MSxP series presents a variety of platforms that can be integrated end to end along with interoperability and compatibility with third-party vendor equipment. A detailed analysis is required to achieve an architecture, design, and solution for an end-to-end optical solution. Various design parameters must be considered prior to deploying such networks. These parameters include customer demographics, capacity planning, scalability, customer service requirements (such as voice, video, and data), oversubscription, customer service level requirements, network availability, protection mechanisms, delay, the fiber plant, physical diverse path routing, link loss budget, technology selection, virtual private networking (VPN) capability, and vendor selection.

- **Case Study: Multiservice Metro Optical SONET/SDH Network**—This section discusses the case study of a service provider involved in the analysis, design, and implementation of a large optical network that provides service to an excess of 350,000 users. The case study presents various design requirements, such as reliability figures, latency, delay, and application requirements. The scenario presented considers the downtown metro area of a large U.S. city and commences with an analysis of service requirements. This discussion is followed up with a capacity planning exercise, fiber plant analysis, delay analysis, technology analysis and comparison, logical design, and finally, it discusses the complete physical design of the network. Scalability considerations are presented followed up with the use of dense wavelength-division multiplexing (DWDM) for scaling bandwidth. DWDM provides a far more cost-effective alternative to multiply the bandwidth versus the cost of laying new fiber. This section presents various scalability options including the use of the ONS 15454 Multiservice Transport Platform (MSTP) for its built-in DWDM capability. The ONS 15454 MSTP provides added value by providing DWDM transport of storage-area network (SAN) applications, such as ESCON, FICON, and Fibre Channel.

- **Case Study: SAN Services**—This is a SAN services case study with a solution discussion.

CHAPTER 12

Optical Network Case Studies

Network Design Strategies

Carriers and service providers have traditionally used various strategies in designing networks. Some of the key business parameters that drive optical network design include the following:

- Low capital expenditure (CAPEX)
- Low operating expenditure (OPEX)
- Low total cost of ownership (TCO)
- Rapid return on investment (ROI)

Networks can be engineered to guarantee a low TCO in the way they are built and operated. TCO can be defined as a sum of CAPEX and OPEX. Some operators prefer the minimum CAPEX model versus the minimum TCO model. In the minimum CAPEX model, the service provider goes for the lowest initial bid proposal from vendors as long as it meets requirements, in an effort to initiate and expand its infrastructure. Minimal consideration is given to operational expenses, network features, and element management. In the minimum TCO model, the service provider goes for minimized OPEX and CAPEX for the network architecture. This requires some short- and medium-term planning. The recurring OPEX over a 5-year period for the minimum TCO model is lower than that for the minimum TCO model. Even though the CAPEX for the minimum CAPEX model is lower initially, the relative CAPEX over a 5-year period is lower than that for the minimum TCO model.

Service operators implementing minimum TCO networks seek to minimize their medium- and long-term expenses. This objective takes into consideration operational costs and network fill, because underused assets turn out to be expensive in terms of cost per bit per mile. They select the lowest-cost bid, provided the vendors satisfy the network operator requirements in terms of features and interoperability. Minimum TCO networks let service providers achieve a faster break-even point than minimum CAPEX networks.

Many carriers have experience in planning TDM voice networks; however, they have difficulty in planning data network infrastructures. They normally end up running parallel voice and data networks. This is normally the case if the carrier has an existing SONET/SDH infrastructure. The OPEX can be truly reduced if the network operator runs a single unified multiservice network. Many pure IP service providers end up running TDM over IP to provide TDM service to their customers. This is an example of highly inefficient bandwidth usage.

A *Single* multiservice broadband network that can carry TDM voice, Ethernet, IP, and video with the reliability and restoration times of SONET/SDH along with scalability in the order of terabits per second is the solution and provides service providers with the best of all worlds. The ONS 15000 MSxPs deliver highly available broadband service including support for voice, TDM, SONET/SDH, Gigabit Ethernet, and all IP application including VoIP, H.323 video, D1 video, and HDTV. The ONS 15000 MSxP series presents a variety of platforms that can be integrated end to end along with interoperability and compatibility with third-party vendor equipment.

A detailed analysis is required to achieve an architecture, design, and solution for an end-to-end optical solution. Let us look at the various parameters and constraints that affect the solution and end result.

Optical network design must take into account the following parameters:

- **Customer demographics**
    — Capacity planning
    — Scalability
- **Customer service requirements**
    — Voice, video, and data
    — Oversubscription
- **Customer service levels**
    — Network availability
    — Mean time between failure (MTBF) and mean time to repair (MTTR)
    — Protection mechanisms
    — Delay
- **Fiber plant**
    — Physical diverse path routing
    — Link loss budget
- **Technology selection**
    — ATM
    — Frame Relay
    — Gigabit Ethernet
    — Multiservice SONET/SDH
    — DWDM
    — VPNs
    — Scalability

- **Vendor selection**
  - CAPEX and OPEX
  - Platform scalability
  - Bandwidth scalability
  - Port densities
  - Protection mechanisms
  - VPN support

Customer Demographics

These requirements are based on the demographic and population model(s) of the area of coverage. It was indicated that coverage was to be provided in the downtown area. This means that the typical customer profile would be that of a medium to large business, with a mix of small business or residential customer(s). The number and classification of customers would drive bandwidth requirements and the capacity model. A unidirectional and bidirectional traffic matrix must be created based on customer applications and count. The customer count also drives the port density and number of service interfaces that would be required at each point of presence (POP) location. A quasi-accurate figure can be determined only after a proper user survey has been performed. These numbers must be extrapolated over a 5-, 10-, and 15-year period.

Customer Service Requirements

A survey of customer service consumption and behavior would influence the technology selected to build the network. For example, most business customers need voice and data services. Voice services would need hunting groups or direct inward dialing (DID) blocks in the case of medium- to large-sized businesses. Data services can be bundled as a package or tariffed separately. These days a service provider can safely assume that the customer would accept Ethernet or IP services from a data perspective. Some customers may have diverse requirements, such as the transport of ATM or ISDN voice. In terms of multitenant units (MTUs), they would need the placement of a Pop access device, co-located with equipment from other carriers that have a presence in the MTU.

Customer Service Levels

Because the proposed network is to be built to service business customers, service level agreements (SLAs) need to be honored. The SLAs translate into network availability in terms of a percentage of uptime (for example, 99.999 percent) and end-to-end latency (for example, 25 ms). Network availability in turn translates into availability of the network elements (NEs) and redundant fiber paths. Network availability takes into consideration not just the MTBF of an NE and its subsystems, but also MTTR of the NE as well as protection mechanisms in terms of 1+1 or 1:N card protection and fiber protection mechanisms, such as automatic protection switching (APS), two-fiber unidirectional path-switched ring (UPSR)/subnetwork connection protection (SNCP), two-fiber bidirectional line-switched ring (BLSR)/multiplex section-shared protection ring (MS-SPRing), or four-fiber BLSR/MS-SPRing.

MTTR is also influenced by the hot-swap capability of cards or modular subsystems within the NE. End-to-end delay also plays a part in service levels that have to be maintained at a consistent level. This delay is introduced in the network as a result of serialization, propagation delay, and NE processing delay. Provider-owned customer premises equipment (CPE), such as routers, also introduces serialization and element processing delay. This also means that critical system components, such as power supplies, common cards, controllers, and trunk links, must be redundant or protected. There should be no single point(s) of failure in the network. There should be less than a 50-ms network failover time in the event of a fiber cut or link failure. Also, the equipment selected must be Network Equipment Business Standards 3 (NEBS 3) compliant.

Fiber Plant

The fiber infrastructure is extremely important in optical network design. The type and number of fibers (strands) available on a particular span or ring determines the type (speed) and number of ports required for the NEs on that ring to achieve the required bandwidth. The physical routing of the fibers determines the kind of ring protection mechanism that can be implemented. If there is physically diverse fiber routing into a building, for example, two-fiber UPSR/SNCP or two-fiber BLSR/MS-SPRing could provide adequate redundancy. If not, chances are that the POP would be lost if ever a backhoe were to plough through the bundled fiber cables at the single ingress/egress point into the building. Physical fiber routing also contributes to the overall SLA and availability figures that can be presented to the customer.

Access to fiber conduits from an inside plant and outside plant perspective is very important for proper fiber maintenance and upgrades including adding fiber if necessary. The number of splice points and fiber quality influences the end-to-end fiber loss that must be within the power budget. Fiber exhaustion in certain spans might also drive requirements for a wavelength-division multiplexing (WDM) solution.

Technology Selection

The service provider has many technology options, such as multiservice SONET/SDH, ATM, Gigabit Ethernet, and pure IP services. The selection of technology must be influenced by the core business model of the service provider and customer service requirements. For example, does a metro service provider intend to be an ISP? In this case, the logical choice is a multilayer (Layer 2/3) Gigabit Ethernet metropolitan-area network (MAN). If the service provider contemplates multiservice and intends to compete in the lucrative voice space, a multiservice SONET/SDH or ATM solution comes to mind. Multiservice platforms offer the option of carrying voice directly over TDM or over IP/Ethernet.

Taking it a step further, the CAPEX and OPEX for an ATM infrastructure would be considerably higher than an MSxP solution, thereby shrinking its advantage in terms of a ROI. Service providers need to serve multiple customers. It is essential that customer traffic remains secure and isolated from each other. Various technologies dictate the nature and creation of VPNs. For example, ATM and Frame Relay VPNs can be created only by defining permanent virtual circuits (PVCs), whereas Ethernet VPNs are created using virtual local-area networks (VLANs). A technology, such as Multiprotocol Label Switching (MPLS) or VRF, enables Layer 2.5 VPNs. Technology scalability is another factor that affects technology selection. If a service provider were to build an Ethernet MAN, for example, single-tagging VLANs would limit the number of customers to 4096. Double-tagged VLANs can theoretically support up to 16,777,216 VLANs. The technology selected must also support oversubscription of services.

Vendor Selection

After the requirements have been firmed up and a technology has been selected, it is time to select a vendor that manufactures equipment that can meet the specifications of the requirements and the technology. One of the first things that comes to mind with vendor selection is the scalability of their equipment. Ideally, upon initial provisioning, the equipment should have adequate port density and bandwidth to handle the number of customers forecast for the first calendar year of operation. Bandwidth scalability mandates that the equipment should be able to scale to customer bandwidth requirements over its lifetime.

The port scalability should be modular, in the sense that addition of subsystems, such as cards, should enable the provisioning of additional ports. It is also desirable that the addition of cards does not involve a downtime. This mandates the use of hot-swappable cards that use independent processors to boot up. 1+1 or 1:N card protection must be considered to provide subsystem-level redundancy. The vendor must support protection technologies, such as automatic protection switching (APS), two-fiber USPR/SNCP, two-fiber BLSR/MS-SPRing, and four-fiber BLSR/MS-SPRing at Layer 1.5. In addition, Layer 2 protection, such as Rapid Spanning Tree Protocol (RSTP) (802.1w) per VLAN, must be supported. The equipment must also support a high MTBF and low MTTR. In terms of VPN support, the vendor must support standards-based, double-tagged VLANs, PVCs, switched virtual circuits (SVCs), tunneling, TDM, MPLS, or VRF. VPN support is normally dictated by the technology selected.

Case Study: Multiservice Metro Optical SONET/SDH Network

A service provider has just been granted right of way to and the irrevocable right of use (IRU) of 30 miles (48 km) of two-strand fiber in the downtown metropolitan area of Washington, D.C. It is the intent of this new competitive local exchange carrier to offer data, voice, and broadband video services in this marketplace. The network operator (service provider) has access to adequate technical, financial, and human resources.

The service provider has performed a user survey and has determined that there is a potential market for up to 10,000 business customers in the downtown area. Of these customers, 8000 can be classified as small businesses with up to 10 employees, and 1800 of these customers can be classified as medium-sized businesses with up to 100 employees. There is also potential for up to 200 large customers having up to 500 employees on average. Customer applications include voice, VPN, and Internet access. There are 300 customers (the 200 large businesses and at least 100 medium-sized businesses) that need connectivity with their out-of-state branch offices and headquarters. There are also about 5000 potential residential and small office/home office (SOHO) customers that need voice and broadband Internet access. There are 250 buildings that need to be lit up by the service provider to provide service to all potential customers. The service provider intends to provide *five nines* or 99.999-percent availability with highly competitive pricing to capture market share.

The service provider has performed a site survey of the downtown area and has determined fiber routing options with diverse ingress and egress from various buildings. Figure 12-1 shows the fiber routing plan for the first seven buildings. The service provider intends to light up these seven buildings in a pilot run. Each of these seven buildings is an MTU, housing up to 35 small businesses and 5 medium-sized businesses in each building. The pilot run will provide voice and Internet access services to the seven buildings.

Figure 12-1 *Fiber Routing Plan*

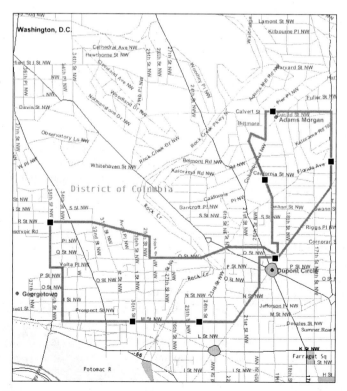

Case Study Solution

This section covers the solution for this case study. The solution covers a requirements analysis, capacity planning, fiber plant analysis, delay analysis, technology analysis, and logical and physical design. It also discusses provisioning of UPSR/SNCP, SONET/SDH, and Ethernet circuits.

Step 1   Requirements Analysis

The customer requirements can be tabulated as shown in Table 12-1. A requirements document must be created that details the customer requirements along with associated specifications and standards.

Table 12-1   *Network Requirements*

Unit	Number	Factor	User Count
Residential	5000	1	5000
Small business	8000	10	80,000
Medium business	1800	100	180,000
Large business	200	500	100,000
Requirement	**Specification**		
Buildings	250		
Network availability	99.999%		
End-to-end latency	25 ms		
Quality of service (QoS)	4 classes of service		
Services	Voice, VPN, and IP Internet access		
CPE service interfaces	DS1/E1 and Ethernet		
Fiber plant	Pilot routing illustrated in Figure 12-1		

Step 2   Capacity Planning

Calculate the estimated bandwidth consumption, using a unidirectional and bidirectional traffic matrix based on the information that the service provider has gathered from the user surveys. The outcome of this capacity plan computation will drive technology and equipment specifications in terms of bandwidth requirements. This study assumes 64 kbps of non-normalized voice bandwidth per user and 128 kbps of non-normalized data bandwidth.

The Dupont Circle location has been designated as the service provider datacenter and network operations center (NOC) because this location co-locates and peers with other ISP autonomous systems, long-distance inter-exchange carriers (IXCs), and the incumbent local exchange carrier (ILEC). The fiber routing indicates that the Dupont Circle node is a good candidate to act as a matching node for a core ring covering the Adams Morgan, California Street, and Florida Avenue buildings. The access ring covers the 34th Street, 30th Street, and the 35th Street nodes.

The total normalized bandwidth is obtained using a 1:10 oversubscription and a similar ratio for GR.303 port-to-service ratio. This case study has considered a GR.303 ratio of 1:10. In reality, a lower figure of 1:4 to 1:7 provides adequate results for GR.303 voice channel concentration. As shown in Table 12-2, the calculations indicate that 114.2 Mbps of normalized bandwidth will be required. The normalized (1:10) bandwidth per building is approximately 16.3 Mbps. However, this traffic is destined for Dupont Circle, because all voice and Internet access traffic will be terminated at the Dupont Circle datacenter. This fact influences the unidirectional traffic matrix shown in Table 12-3.

Table 12-2 *Normalized Bandwidth Calculation*

Location	Small Business Users	Medium Business Users	Small Business Voice Bandwidth	Small Business Data Bandwidth	Medium Business Voice Bandwidth	Medium Business Data Bandwidth	Bandwidth Requirement
35th St	350	500	22.4 Mbps	44.8 Mbps	32 Mbps	64 Mbps	163.2 Mbps
30th St	350	500	22.4 Mbps	44.8 Mbps	32 Mbps	64 Mbps	163.2 Mbps
24th St	350	500	22.4 Mbps	44.8 Mbps	32 Mbps	64 Mbps	163.2 Mbps
Dupont Circle	350	500	22.4 Mbps	44.8 Mbps	32 Mbps	64 Mbps	163.2 Mbps
California Street	350	500	22.4 Mbps	44.8 Mbps	32 Mbps	64 Mbps	163.2 Mbps
Florida Avenue	350	500	22.4 Mbps	44.8 Mbps	32 Mbps	64 Mbps	163.2 Mbps
Adams Morgan	350	500	22.4 Mbps	44.8 Mbps	32 Mbps	64 Mbps	163.2 Mbps
Total Bandwidth							**1142 Mbps**
Total Normalized Bandwidth (1:10) Oversubscription							**114.2 Mbps**

Table 12-3 *Unidirectional Traffic Matrix*

Destination	Source						
	35th Street	30th Street	24th Street	Dupont Circle	California Street	Florida Avenue	Adams Morgan
35th Street		32.6 Mbps					
30th Street			16.3 Mbps				
24th Street							
Dupont Circle	48.9 Mbps					48.9 Mbps	

Table 12-3 *Unidirectional Traffic Matrix (Continued)*

Destination	Source						
	35th Street	30th Street	24th Street	Dupont Circle	California Street	Florida Avenue	Adams Morgan
California Street							
Florida Avenue							32.6 Mbps
Adams Morgan					16.3 Mbps		

Assuming a clockwise traffic direction on the core ring illustrated in Figure 12-1, you will notice that the California Street node adds 16.3 Mbps of traffic to the fiber. However, at Adams Morgan, another 16.3 Mbps is added, giving rise to 32.6 Mbps of bandwidth consumption on the fiber span between Adams Morgan and Florida Avenue. At Florida Avenue, another 16.3 Mbps of traffic is added, giving rise to 48.9 Mbps of bandwidth consumption on the fiber span between Florida Avenue and Dupont Circle.

For the access ring, notice that the 24th Street node adds 16.3 Mbps of traffic to the fiber. However, at 30th Street, another 16.3 Mbps is added, giving rise to 32.6 Mbps of bandwidth consumption on the fiber span between 30th Street and 35th Street. At 35th Street, another 16.3 Mbps of traffic is added, giving rise to 48.9 Mbps of bandwidth consumption on the fiber span between 35th Street and Dupont Circle.

To compute the bidirectional traffic matrix, the reader must appreciate the fact that communications is never symmetrical, in the sense that traffic is always heavier from a server to a client or from a mainframe to a terminal or from the Internet to a PC or workstation. However, some customers have requirements for symmetrical links. For example, large file uploads to the Internet or heavy peer-to-peer traffic works best over symmetrical links. As a service provider, it is better to plan for average-to-worst case and take symmetrical links into account. In that case, the bidirectional plan must be mapped out as shown in Table 12-4.

Table 12-4 *Bidirectional Traffic Matrix*

Destination	Source						
	35th Street	30th Street	24th Street	Dupont Circle	California Street	Florida Avenue	Adams Morgan
35th Street		48.9 Mbps					
30th Street			48.9 Mbps				
24th Street				48.9 Mbps			

continues

Table 12-4 *Bidirectional Traffic Matrix (Continued)*

Destination	Source						
	35th Street	30th Street	24th Street	Dupont Circle	California Street	Florida Avenue	Adams Morgan
Dupont Circle	48.9 Mbps					48.9 Mbps	
California Street				48.9 Mbps			
Florida Avenue							48.9 Mbps
Adams Morgan					48.9 Mbps		

Again assuming a clockwise traffic flow on the core ring, traffic being sourced at Dupont Circle can be dropped off at California Street, Adams Morgan, and Florida Avenue. The span between Dupont Circle and California Street would carry 16.3 * 3 (48.9 Mbps) of traffic, of which 16.3 Mbps would be dropped off at California Street. This leaves 32.6 Mbps of traffic on the span between California Street and Adams Morgan, of which 16.3 Mbps would be dropped off at Adams Morgan. Finally, there would be 16.3 Mbps of traffic on the span between Adams Morgan and Florida Avenue, all of which (16.3 Mbps) would be dropped off at Florida Avenue.

For the collector ring, traffic being sourced at Dupont Circle can be dropped off at 24th Street, 30th Street, and 35th Street. The span between Dupont Circle and 24th Street would carry 16.3 * 3 (48.9 Mbps) of traffic, of which 16.3 Mbps would be dropped off at 24th Street. This leaves 32.6 Mbps of traffic on the span between 24th Street and 30th Street, of which 16.3 Mbps would be dropped off at 30th Street. Finally, there would be 16.3 Mbps of traffic on the span between 30th Street and 35th Street, all of which (16.3 Mbps) would be dropped off at 35th Street.

The bidirectional traffic matrix has just taken the values obtained from the previous unidirectional computation and inserted them into the matrix. However, 48.9 Mbps of traffic on the span between Dupont Circle and California Street has been added to the matrix. The sum total of traffic on the span between California Street and Adams Morgan is now 16.3 Mbps + 32.6 Mbps (48.9 Mbps). Also, the sum total of bidirectional traffic on the span between Adams Morgan and Florida Avenue is 32.6 Mbps + 16.3 Mbps (48.9 Mbps).

For the collector ring, the sum total of traffic on the span between 24th Street and 30th Street is now 16.3 Mbps + 32.6 Mbps (48.9 Mbps). Also, the sum total of bidirectional traffic on the span between 30th Street and 35th Street is 32.6 Mbps + 16.3 Mbps (48.9 Mbps).

The bidirectional traffic matrix indicates a 1:1 TX-RX and RX-TX ratio, which just means that a user would have the same bandwidth available for upload and for download.

The unidirectional and bidirectional traffic matrices indicate that the largest amount of normalized bandwidth consumed on any span is 48.9 Mbps. This indicates that an OC-3/STM-1 should suffice for the pilot deployment. However, this calculation does not take into consideration the remaining 243 buildings. Table 12-5 shows the calculation of normalized bandwidth for all customers, and Table 12-6 shows the total normalized bandwidth that must be considered while designing the network. The computations shown in Tables 12-5 and 12-6 do not detail the flows between the various buildings. This is an exercise that must be performed when information is available for all 250 buildings.

Table 12-5 *Normalized Bandwidth Calculation*

Location	User Count	Voice Bandwidth Per User	Voice Bandwidth	Normalized Voice Bandwidth (1:10)
Residential	5000	64 kbps	320 Mbps	32 Mbps
Small business	80,000	64 kbps	5.12 Gbps	512 Mbps
Medium business	180,000	64 kbps	11.52 Gbps	1.152 Gbps
Large business	100,000	64 kbps	6.4 Gbps	640 Mbps
Location	User Count	Internet Bandwidth Per User	Internet Bandwidth	Normalized Internet Bandwidth (1:10)
Residential	5000	128 kbps	640 Mbps	64 Mbps
Small business	80,000	128 kbps	10.24 Gbps	1.024 Gbps
Medium business	180,000	128 kbps	23.04 Gbps	2.304 Gbps
Large business	100,000	128 kbps	12.8 Gbps	1.28 Gbps

Table 12-6 *Total Normalized Bandwidth for All Customers*

Location	Normalized Voice Bandwidth (1:10)	Normalized Internet Bandwidth (1:10)
Residential	32 Mbps	64 Mbps
Small business	512 Mbps	1.024 Gbps
Medium business	1.152 Gbps	2.304 Gbps
Large business	640 Mbps	1.28 Gbps
Total	2.336 Gbps	4.672 Gbps

continues

Table 12-6 *Total Normalized Bandwidth for All Customers (Continued)*

Location	Normalized Voice Bandwidth (1:10)	Normalized Internet Bandwidth (1:10)
Total Normalized BW (1:10) Year 0	7.008 Gbps	
Compounded growth	15%	25%
Year 1	8.059 Gbps	8.760 Gbps
Year 2	9.268 Gbps	10.950 Gbps
Year 3	10.658 Gbps	13.687 Gbps
Year 4	12.257 Gbps	17.10 Gbps
Year 5	14.095 Gbps	21.386 Gbps
Year 10	28.351 Gbps	65.267 Gbps

The total normalized bandwidth is 2.336 Gbps + 4.672 Gbps (7.008 Gbps). This indicates that an OC-192/STM-64 (9.953 Gbps) core should provide enough bandwidth to cover currently projected customer requirements. A 5-year and 10-year compounded growth projection at 15 percent and 25 percent is also shown in Table 12-6.

From Table 12-6, it is evident that the OC-192/STM-64 bandwidth should meet demand for up to 2 to 3 years. After that, the network would need to be scaled upwardly in terms of bandwidth. Multiple OC-192/STM-64 wavelengths would be required to cover the Year-5 and Year-10 requirements. This translates into a requirement for DWDM in the future.

Step 3 Fiber Plant Analysis

The fiber plant has two components: the inside plant and the outside fiber plant. The inside premise fiber cabling has on average two fusion splice points that form the connection with the ingress or egress conduit to and from the building. The fiber that ingresses or egresses to or from the building terminates at a fiber patch panel. The fiber patch panel connects to the active optical-electronics-optical (OEO) equipment. There are two mechanical connections at either end of the fiber patch cable. Figure 12-2 shows a schematic of the inside plant.

The outside premise cabling extends from the building egress conduit to the east or west run conduit. There are two splice points on average that connect the building fiber to the east and west runs. From an end-to-end perspective between two active OEO NEs, there are eight splice points, two within the inside plant and two splice points on the outside plant at each side. We must not forget the mechanical connections at the connector ends of the single-mode fiber (SMF) patch cables. Figure 12-3 shows a schematic of the outside plant.

Case Study: Multiservice Metro Optical SONET/SDH Network 707

Figure 12-2 *Inside Plant Schematic*

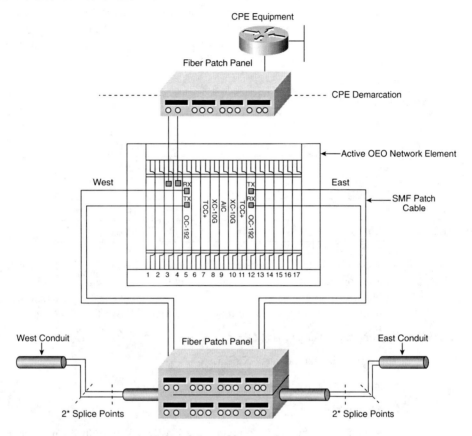

Figure 12-3 *Outside Plant Schematic*

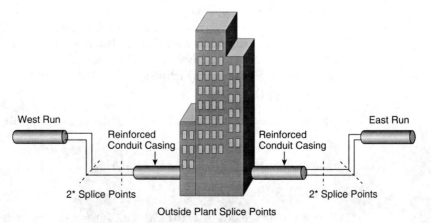

Consider the fiber-optic system in Figures 12-2 and 12-3. We have to determine the optical transmitter launch power and receiver sensitivity at 1550 nanometers (nm). The example assumes inclusion of two patch panels in the path and eight fusion splices, with the system operating over 5 km of SMF cable between buildings. This leads to the fiber loss-budget calculation shown in Table 12-7.

Table 12-7 *Fiber Loss-Budget Calculation*

Component	dB Loss
SMF 1550-nm cable (5 km * 0.25 dB/km)	1.25 dB
FC connectors (4 * 0.5 dB/connector)	2.0 dB
Fusion splices (8 * 0.1 dB/splice)	0.8 dB
Patch panels (2 * 2 dB/panel)	4 dB
Optical safety margin	3 dB
Total Span Loss	**11.05 dB**

As shown in Table 12-7, notice that the total span loss is 11.05 dB. The safety margin compensates for additional splicing that might occur as a result of fiber rerouting or maintenance. The selection of equipment must ensure that the total span loss is within the maximum allowable loss (Z) over the span:

Maximum allowable loss $(Z) = \lambda(X,Y)$ dB

Transmitter launch power = X dBm

Receiver sensitivity = Y dBm

Step 4 Delay Analysis

End-to-end delay has three components. Delay can be introduced in the network as serialization delay, propagation delay, or NE processing delay. Serialization delay has the most significant impact on end-to-end delay.

- **Serialization delay**—The frame sizes or maximum transmission unit (MTU) used by customers plays an important part in the serialization delay. A Layer 2 frame cannot be processed by a receiver until all bits have been received by the receiver and a cyclic redundancy check (CRC) has been performed. Let us examine the serialization delay introduced by SONET/SDH NEs.

  For SONET STS-1:

  (810 bytes/frame * 8 bits/byte)/(51.84 * 10^6 bits/sec) = 125 microseconds

  For SONET STS-3/SDH STM-1:

  (810 bytes/frame * 8 bits/byte)/(155.52 * 10^6 bits/sec) = 41.66 microseconds

  For SONET/SDH STS-12/STM-4:

  (810 bytes/frame * 8 bits/byte)/(622.08 * 10^6 bits/sec) = 10.41 microseconds

For SONET/SDH STS-48/STM-16:

(810 bytes/frame * 8 bits/byte)/(2488.32 * 10^6 bits/sec) = 2.60 microseconds

For SONET/SDH STS-192/STM-64:

(810 bytes/frame * 8 bits/byte)/(9953.28 * 10^6 bits/sec) = 0.65 microseconds

The preceding calculations reveal that the largest delay of 125μs is introduced at the STS-1 rate. However, this is negligible compared to the delay introduced by the CPE device at a lower-bandwidth service interface. If a customer decides to use a 9000-byte frame size over a DS1 interface, for example, the serialization delay can be calculated as follows:

(9000 bytes/frame * 8 bits/byte)/1,536,000 bps = 0.0468 seconds or 46.875 ms

Using such a large MTU has just rendered the SLA requirement of 25 ms impossible to achieve. It is precisely for this reason that SLA agreements must guarantee delay time(s) at certain fixed MTU size(s).

For MPLS-compatible Ethernet frame sizes of 1548 bytes MTU operating at 10-Mbps link speed, the serialization delay can be calculated as follows:

(1548 bytes/frame * 8 bits/byte)/10,000,000 bps = 0.001238 seconds or 1.238 ms

For 1500-byte MTU serial High-level Data Link Control (HDLC) or PPP encapsulation operating over DS1 links, the serialization delay can be calculated as follows:

(1500 bytes/frame * 8 bits/byte)/1,536,000 bps = 0.007812 seconds or 7.8 ms

For 1500-byte MTU serial HDLC or PPP encapsulation operating over E1 links, the serialization delay can be calculated as follows:

(1500 bytes/frame * 8 bits/byte)/2,048,000 bps = 0.00585 seconds or 5.8 ms

- **Propagation delay**—Propagation delay is introduced due to the finite speed of light and the laws of physics. Propagation delay introduces asymmetrical delay in large UPSR/SNCP rings and must be taken into consideration during optical network design. The nominal velocity of propagation (NVP) is defined as the ratio of the speed of light in a vacuum (c) to the refractive index of the material. The value of c is a constant fixed at 300,000 km/second. The refractive index for SMF is 1.5.

NVP = (c)/(Refractive index of fiber)

The NVP for SMF can be calculated as follows:

NVP = (300,000 km/second)/(1.5) = 200,000 km/sec

The inverse of this equation results in the propagation delay of SMF at 0.005 ms/km. This means the propagation delay for SMF in milliseconds can be calculated as follows:

Propagation delay = Fiber length (km) * 0.005 ms/km

- **NE processing delay**—This is the delay measured in subunits of seconds that results from a signal being processed by OEO NEs, such as add/drop multiplexers (ADMs) and 3R regenerators, on the way to its destination. According to ITU-T specifications, the signal must ingress and egress an optical NE within 450 microseconds (0.45 ms). This means the end-to-end NE processing delay in milliseconds can be calculated as follows:

NE processing delay = Total number of NEs on the path * 0.45 ms

The calculation for total delay is as follows:

Total delay = Serialization delay + Propagation delay + NE processing delay

It is our intent to limit the number of end-to-end rings to three. This means that traffic sourced from a collector ring node could traverse over a core ring and then terminate at another collector node. We can also consider a design maximum of 16 NEs (nodes) per ring. This means a signal would traverse over 8 * 3 (24) nodes at worst case. We also have the design premise that the geographic circumference (perimeter) of Washington, D.C. does not exceed 30 km. This means that the signal would not traverse over 30 km/2 (15 km) over a single ring. The maximum distance a signal would traverse is (source collector circumference/2) + (core ring circumference/2) + (destination collector circumference/2) or 15 km + 15 km + 15 km (45 km). The slowest-speed CPE interface the service provider intends to provide customers is the DS1 interface with an MTU of 1500 bytes.

From a design perspective, the equation that needs to be solved is shown here:

Total delay < Serialization delay + Propagation delay + NE processing delay

Total delay < [(MTU * 8 bits/byte)/(Bandwidth of lowest-speed service interface)] + [Fiber length (km) * 0.005 km/s] + [Total number of NEs on the path * 0.45 ms]

Substituting values into the design equation, we get the following:

Total delay < [(1500 * 8)/(1,536,000)] + [45 * 0.005] + [24 * 0.45]

Total delay < [7.8 ms] + [0.225 ms] + [10.8 ms]

Total delay < 18.825 ms

This satisfies our SLA requirement of 25 ms as the end-to-end delay that will not be exceeded.

Step 5 Technology Analysis

The technology selection is based on the requirements defined in Step 1. It is not feasible to provide residential customers that don't have broadband Internet service with IP phones. The first step would involve providing residential subscribers with broadband Internet access via DSL, cable modem, ISDN, satellite dish, or any other broadband technology. These customers can be outfitted with IP phones or Analog Telephone Adapters (ATA) equipment that would attach an analog phone to an Ethernet port. Other factors that must be considered include cost; preservation of existing Class 5 voice/ISDN investment until ROI is realized; inoperability of IP phones in the event of a power outage; absence of reliable Communications Assistance to Law Enforcement Act (CALEA) call-tracing and wire-tap mechanisms; absence of reliable data cabling in many MTU buildings, residential buildings, and houses; and the reluctance of customers to embrace a new technology. This forces us to consider a base technology that will support current TDM/POTS as well as VoIP.

SLA guarantees of 99.999 percent require redundant NEs, subsystems, and fiber paths. It also requires a very high NE MTBF, low NE MTTR, and support for protection mechanisms in terms of 1+1 or 1:N card protection and fiber protection mechanisms, such as APS, two-fiber USPR/SNCP, two-fiber BLSR/MS-SPRing, or four-fiber BLSR/MS-SPRing. MTTR is also influenced by the hot-swap capability of cards or modular subsystems within the NE. This means that all modules of a selected device would need to be hot swappable.

End-to-end delay guarantees of 25 ms means that the technology should support variable-sized MTUs, have optimized fiber runs, and support low-latency processing within the NE. We have also determined the 10.8-dB loss per fiber span (on average) that must be compensated for. Table 12-8 presents a technology analysis and comparison. From the analysis matrix, it is clear that multiservice SONET/SDH is the technology of choice to meet or exceed the requirements.

Table 12-8 *Technology Analysis and Comparison*

Requirement	SONET/SDH	Multiservice SONET/SDH	ATM	Gigabit Ethernet
Scalability	This option provides high levels of scalability. SONET/SDH was originally designed for carrier-wide implementations. Bandwidth options up to OC-192/STM-64 (10 Gbps).	This option provides high levels of scalability. In addition to SONET/SDH, massive upward scalability using DWDM technology (multiples of OC-48/STM-16 or OC-192/STM-64) is possible. VLANs up to 16,777,216 with 802.1Q-in-Q.	Limited scalability, no standard above OC-48. Limited availability of high speed OC-192 interfaces from vendors.	Limited scalability. 10 Gbps not standardized, and WDM is needed to scale above a limited number of 1-Gbps trunks. Will require switch replacement to use 10 Gbps. VLANs limited to 4096.

continues

Table 12-8 *Technology Analysis and Comparison (Continued)*

Requirement	SONET/SDH	Multiservice SONET/SDH	ATM	Gigabit Ethernet
Network Element Processing Delay	<0.45 ms.	<0.45 ms.	<0.45 ms.	<0.45 ms.
VPN Capability	STS-N, VT/VC circuits, and external Layer 2/3 VPN overlay.	STS-N, VT/VC circuits, RPR, 802.1Q VLANs.	ATM PVCs, SVCs, and soft PVCs.	802.1Q VLANs.
QoS and CoS Capabilities	TDM.	TDM, 802.1P, and DSCP.	ATM CoS.	802.1P.
Network Management and Provisioning	CLI, element management, and TL1.	CLI, element management, SNMP, and TL1.	CLI, element management, and SNMP.	CLI and SNMP.
Redundancy and Reliability	APS, two-fiber UPSR/SNCP, two-fiber BLSR/MS-SPRing, four-fiber BLSR/MS-SPRing. 50-ms link failover. NE node and card redundancy (1:1 and 1:N), optical and electrical card redundancy.	APS, two-fiber UPSR/SNCP, two-fiber BLSR/MS-SPRing, four-fiber BLSR/MS-SPRing, and PPMN/ESNCP. 50 ms link failover. NE node and card redundancy (1:1 and 1:N), optical and electrical card redundancy. STP and RSTP (802.1W) per VLAN support.	APS 50-ms link failover. SONET/SDH methods available for ATM over SONET/SDH. ATM NE node and card redundancy (1:1 and 1:N).	Physical redundant links required. 7 to 35 seconds (STP) link failure (recovery + convergence). RSTP (802.1W) support covers this requirement.
Native Ethernet Support	No.	Yes.	No.	Yes.
Native IP Support	No.	Yes.	No.	Yes.
Native TDM Support	Yes.	Yes.	No.	No.
VoIP Support	No.	Yes. VoIP over Ethernet.	Yes. VoIP over ATM.	Yes. VoIP over Ethernet.

Step 6 Logical Design

Based on the design requirements and technology determined in the preceding four steps, we can proceed with an MSxP design. The ONS 15454/15454 SDH MSPPs are the perfect fit. The logical design must account for service transport (voice, VPN, and Internet access), meeting SLA requirements (99.999-percent availability and 25-ms end-to-end latency) with up to four classes of service. The logical design presented in this section addresses only the pilot requirements. Completion of the remaining buildings is an extrapolation of the design exercise.

The selection of platform influences the CAPEX and OPEX and must be made very carefully, based on the requirements we have seen thus far and keeping growth in mind. We can select the 15327 for smaller buildings, the 15454 for medium and large buildings, and the 15454 MSTP for the peering points of this network. The ML100T-12 12-port 1/100BASE-T ML-Series card has been selected to offer Layer 2/3 services on the 15454 MSPP. This card will be present in all nodes. For the 15327, we would need to use a G-Series card with an external Layer 2/3 switch.

Based on the pilot fiber routing illustrated in Figure 12-1, the pilot network lends itself to a two-ring topology with the node at Adams Morgan acting as the matching node. Figure 12-4 shows the logical topology. Based on the capacity plan of Step 2, we have determined that an OC-192/STM-64 core ring should be able to handle the bandwidth requirements. The tributary rings could exist as a combination of OC-12/STM-4 and OC-3/STM-1 rings.

Figure 12-4 *Case Study—Logical Design*

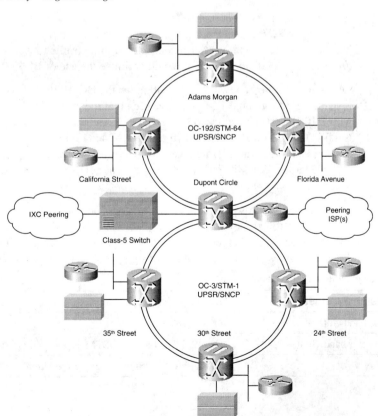

From an optical-link budget perspective, the selection of the four-port OC-3/STM1 IR cards for the collector ring provides a minimum TX power of −15 dBm and minimum RX level (sensitivity) of −28 dBm. This provides a maximum allowable loss of 13 dB. The 11.05-dB loss (with the built-in 3-dB safety margin) obtained from Step 3 is well within this limit. In addition, the selection of the OC-192/STM-64 IR/SH 1550-nm card provides minimum TX power of −1 dBm and a minimum RX level (sensitivity) of −14 dBm. This provides a maximum allowable loss of 13 dB. The 11.05-dB loss (with the built-in 3-dB safety margin) obtained from Step 3 is also well within this limit.

We have two design choices to protect the rings: UPSR/SNCP or BLSR/MS-SPRing. Given the diverse fiber routing options and the nature of the traffic (all traffic gets routed or switched at Adams Morgan), UPSR/SNCP seems to be a logical choice. The design specifications for UPSR/SNCP limit us to 16 UPSR/SNCP rings per ONS 15454 node with the Timing and Control 2 (TCC2) card. In terms of two-fiber BLSR/MS-SPRing, we can terminate two two-fiber BLSR/MS-SPRings or one four-fiber BLSR/MS-SPRing per node. The TCC2 can accommodate up to 32 data communications channels (DCCs) with Release 4.x. The TCC2 card is hardware ready to terminate up to 84 DCCs. Each ring requires two DCCs.

Also, we must remember that BLSR/MS-SPRing protection would consume 96 STS-1 or VC3 channels for protection on the core ring. This would halve the bandwidth, unless there was sufficient interbuilding traffic to warrant *adds* and *drops* on intermediate spans that would optimize the use of BLSR/MS-SPRing.

Let us analyze the bandwidth distribution for the OC-192/STM-64 core UPSR/SNCP ring. From Table 12-6, we have seen that the normalized voice bandwidth with 1:10 GR.303 concentration is 2.336 Gbps, and the normalized data bandwidth is 4.672 Gbps. The ML-Series card uses STS-24c (1.244 Gbps) of bandwidth eastward and westward. To achieve 4.672 Gbps for data, we would need 3.75 (or in tangible terms, four ML-Series cards per core node) to achieve our data bandwidth objectives.

Four ML-Series cards present in each core node indicates that 24c * 4 (96) STS-1s have been assigned for data, out of a possible 192 STS-1s in the OC-192/STM-64 UPSR/SNCP ring. This leaves us with 96 STS-1s (4.976 Gbps) for voice. Table 12-9 shows the various logical design parameters under consideration.

Table 12-9 *Logical Design Specification*

Design Parameter	Description
Core bandwidth	OC-192/STM-64.
Collector bandwidth	OC-12/STM-4 or OC-3/STM-1.
Delay	18.825 ms <= (25-ms design specification).

Table 12-9 *Logical Design Specification (Continued)*

Design Parameter	Description
Optical budget	10.8 dB <= (13-dB design specification).
	OC-3/STM-1 min TX power −15 dBm, min RX level −28 dBm. Optical power budget = 13 dB.
	OC-192/STM-64 min TX power −1 dBm, min RX level −14 dBm. Optical power budget = 13 dB.
Unamplified distance	OC-3/STM-1 15 km => (15-km design specification).
	OC-192/STM-63 40 km => (15-km design specification).
Protection	UPSR/SNCP.
	1:1 electrical card protection.
	1+1 optical card protection (optional).
Platform(s)	15327, 15454 MSPP, and 15454 MSTP. NEBS 3 and OSMINE compliant.
	ML-Series card(s).
	ML-Series ML100T-12 * 4 (STS-24c * 4) [core].
VPN provisioning	802.1Q-in-Q VLANs and VRF.
CPE data interface	100BASE-T.
Classes of service	802.1p and DSCP (=> 8 classes).
CPE voice interface	DS1/E1.

> **NOTE** This design exercise assumes that the core OC-192/STM-64 UPSR/SNCP ring has multiple nodes, and each node is a matching node that interfaces with collector rings with an even bandwidth distribution. It is possible to have nodes on the core ring that do not connect to tributary rings. It is also possible to have collector rings of varying bandwidths.

The two design choices for VPN creation are VLANs or Layer 3 VRF Lite. Some of the ML-Series design specifications are listed here:

- Supported MAC entry per card: 64K
- Number of bridge groups per card: 255
- Number of 802.1W RSTP instances per card: 255
- Supported VLAN IDs: 4K (0–4095)
- Number of active VLANs per card: 255

- Number of 802.Q-in-Q per card: 255
- Number of SPR/RPR instances per card: 1
- Number of cards per SPR/RPR: up to 10

Each customer can maintain his own VLAN numbering. The ML-Series card can accept an 802.1Q trunk from a customer CPE device and add a service provider 802.1Q tag to the frame. Within the service provider network, the service provider tag keeps the traffic distinct. The service provider tag can be stripped at egress.

From a Layer 2 perspective, the ML-Series card can support 64,000 MAC addresses per card. It is good design and an operational practice to interface with CPE routers or multiplayer switches. This limits the number of customer MAC addresses. The ML-Series card can also support 802.1P class of service (CoS) on the inner or outer tag.

From a VLAN-VPN perspective, the ML-Series cards can support up to 4096 customers. It is good design practice to reserve some of those VLANs for management and operation, administration, maintenance, and provisioning (OAM&P)/operations support system (OSS) use. The number of customers can be scaled easily by using multiple ML-Series shared packet ring (SPR)/resilient packet ring (RPR) Packet over SONET (POS) rings. The ML-Series card can support up to STS-24c/VC4-8c of POS bandwidth in either direction. Each POS ring (SPR/RPR) can support 4096 customers. This means that by adding another POS SPR/RPR, an additional 4096 customers can be supported.

Step 7 Physical Design

The logical design derived from the preceding step enables us to build a physical design quite easily. The physical design is the basis for obtaining a detailed bill of materials, creating rack elevations and cross-connect lists for staging purposes. Table 12-10 shows the various cards that have been selected for core ring nodes and their descriptions, and Figure 12-5 shows a fully loaded core node.

Table 12-10 *Physical Design Specification (Core Node)*

Core ONS 15454 MSPP		
Card	Description	Quantity
TCC2	TCC2 card	2
XC10G	XC10G cross-connect cards	2
AIC-I	Alarm interface controller (AIC-I)	1
DS1-14	DS1-14 14-Port DS1 card	2
DS1N-14	DS1N-14 14-port DS1 protect card	2
OC192 IR	OC192 IR/STM64 SH 1550 1-port OC-192/STM-64	2
ML100T-12	ML100T-12 card 12-port Ethernet card	2

Figure 12-5 *Case Study—Core Node*

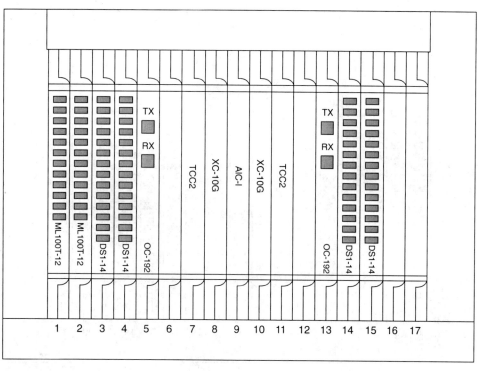

Table 12-11 shows the various cards that have been selected for the matching node and their descriptions, and Figure 12-6 shows a fully loaded matching node.

Table 12-11 *Physical Design Specification (Matching Node)*

Core ONS 15454 MSPP		
Card	Description	Quantity
TCC2	TCC2 card	2
XC10G	XC10G cross-connect cards	2
AIC-I	Alarm interface controller (AIC-I)	1
DS1-14	DS1-14 14-port DS1 card	2
DS1N-14	DS1N-14 14-port DS1 protect card	2
OC192 IR	OC192 IR/STM64 SH 1550 1-port OC-192/STM-64	2
OC8 IR	OC3 IR/STM1 SH 1310-8 8-port OC-3/STM-1 card	2
ML100T-12	ML100T-12 card 12-port Ethernet card	2

Figure 12-6 *Case Study—Matching Node*

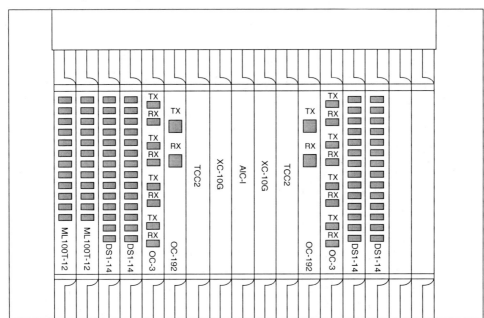

Table 12-12 shows the various cards that have been selected for collector ring nodes and their descriptions, and Figure 12-7 shows a fully loaded collector node.

Table 12-12 *Physical Design Specification (Collector Node)*

Collector ONS 15454 MSPP		
Card	**Description**	**Quantity**
TCC2	TCC2 card	2
XC10G	XC10G cross-connect cards	2
AIC-I	Alarm interface controller (AIC-I)	1
DS1-14	DS1-14 14-port DS1 card	2
DS1N-14	DS1N-14 14-port DS1 protect card	2
OC3 IR	OC3 IR/STM1 SH 1310-8 8-port OC-3/STM-1 card	2
ML100T-12	ML100T-12 card 12-port Ethernet card	2

Figure 12-7 *Case Study—Collector Node*

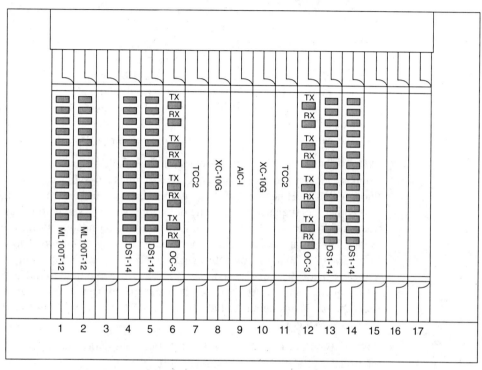

The ONS 15454 SONET/SDH MSPP is a NEBS-compliant shelf assembly that contains 17 card (module) slots, a backplane interface, a fan-tray assembly, a front panel with an LCD, and alarm indicators. ONS 15454 has 17 card slots numbered 1 to 17. All slots are card-ready, meaning that when you plug in a card it automatically boots up and becomes ready for service. The ONS 15454 houses five types of cards including common control, alarm interface, electrical, optical, and Ethernet. The common control cards include the TCC2, cross-connect cards (XCVT or XC10G), and the alarm interface controller card (AIC). The ONS 15454 electrical cards require electrical interface assemblies (EIAs) installed in the rear of the chassis to provide the cable connection points for the shelf assembly, whereas optical and Ethernet cards have frontal faceplate connections rather than backplane connections. The building integrated timing supply (BITS) timing connector, LAN, alarms, modem, and craft port use backplane wire-wrap pins. The various cards used in the core node are as follows:

- **TCC2**—The TCC2s control the main processing functions of the ONS 15454. TCC2s also have an RS-232 craft port for Cisco Transport Controller (CTC). The TCC2 combines timing, control, and switching functions which include system initialization, provisioning, alarm reporting, maintenance, diagnostics, IP address detection and resolution, timing, SONET/SDH DCC termination, and system fault detection.

- **XC10G**—The cross-connect card is the central switching element in the ONS 15454. The ONS 15454 offers two cross-connect cards: the XCVT and XC10G. Circuit cross-connect information is provisioned using CTC or Transaction Language 1 (TL1). The TCC2 then establishes the proper internal cross-connect information and relays the setup information to the cross-connect card. The XC10G card supports STS-192 signal rates. The switch matrix on the XC10G consists of 576 STS-1 bidirectional ports, and its VT matrix can manage up to 336 bidirectional VT1.5s. The XC10G is required to operate the OC-192/STM-64 card, OC-48/STM-16 AS card, G1K-4 Gigabit Ethernet card, ML-Series Ethernet cards, and the four-port OC-12/STM-4 card.

- **AIC-I**—The AIC-I card provides user-provisionable alarm capability and supports local and express orderwire. *Orderwire* is a 64-kbps channel in the SONET/SDH header that installation technicians use to communicate with other technicians in-band. The orderwire ports are standard RJ-11 receptacles. The AIC-I card provides input/output alarm contacts for user-defined alarms and controls, also known as *environmental alarms*. The AIC-I provides up to 16 input contacts, or 12 input contacts and 4 input/output contacts. The AIC-I cards use the backplane wire-wrap field to make the physical connections. An optional alarm expansion panel (AEP) can be installed on the backplane of the AIC-I. The AEP provides 32 input contacts and 16 output contacts.

- **DS1 14-port card (DS1-14)**—The DS1-14 card provides 14 Telcordia-compliant, GR-499 DS1 ports. Each port operates at 1.544 Mbps over a 100-ohm twisted-pair copper cable. The DS1-14 card can function as a working or protect card in 1:1 protection schemes and as a working card in 1:N protection schemes. The DS1N-14 card is identical to the DS1-14 and operates as a protect card in a 1:N protection group. The traffic from an entire DS1-14 card can be grouped and mapped to a single STS-1. Individual DS1 ports can be mapped to a VT1.5.

- **OC-192/STM64 1-Port Card (OC192 IR/STM64)**—The OC192 IR/STM64 SH 1550 card provides one intermediate-reach OC-192/STM-64 port in the 1550-nm wavelength range, compliant with ITU-T G.707, ITU-T G.957, and Telcordia GR-253-CORE. The port operates at 9.95328 Gbps over unamplified distances up to 25 miles (40 km) with SMF-28 fiber limited by loss and/or dispersion. The card supports VT and nonconcatenated or concatenated payloads.

- **OC3 IR/STM1 SH 1310-8 card (OC3 IR/STM1)**—The OC3 IR 4/STM1 SH 1310 card provides eight intermediate or short-range 1310-nm OC-3/STM-1 ports compliant with ITU-T G.707, ITU-T G.957, and Telcordia GR-253-CORE. The port operates at 155.52 Mbps over an SMF span. Each card supports VT and nonconcatenated or concatenated payloads at the STS-1 or STS-3c signal levels. This card provides port-to-port protection.

- **ML-Series 12-port card (ML100T-12)**—The ML100T-12 card provides 12 ports of IEEE 802.3-compliant 10/100 interfaces. Each interface supports full-duplex operation for a maximum bandwidth of 200 Mbps per port and 2.488 Gbps per card. Each port autoconfigures itself for speed, duplex, and flow control. The card provides high-throughput, low-latency packet switching of Ethernet-encapsulated traffic (IP and other Layer 3 protocols) across a SONET/SDH network while utilizing the inherent self-healing capabilities of SONET/SDH protection. Each ML100T-12 card supports wire-speed, Layer 2 Ethernet switching between its Ethernet ports. The IEEE 802.1Q tag and port-based VLANs logically isolate customer traffic. Priority queuing is also supported to provide multiple classes of service. An ONS 15454 with an XC10G card can host the card in any traffic card slot, but an ONS 15454 with an XC or XCVT can host the card only in slots 5, 6, 12, and 13.

Figure 12-8 shows the completed core with an OC-192/STM-64 UPSR/SNCP core ring. The core ring has been provisioned with full OC-192/STM-64 capacity to facilitate the addition of other collector rings as the network gets deployed.

Figure 12-8 *Case Study—Core Ring*

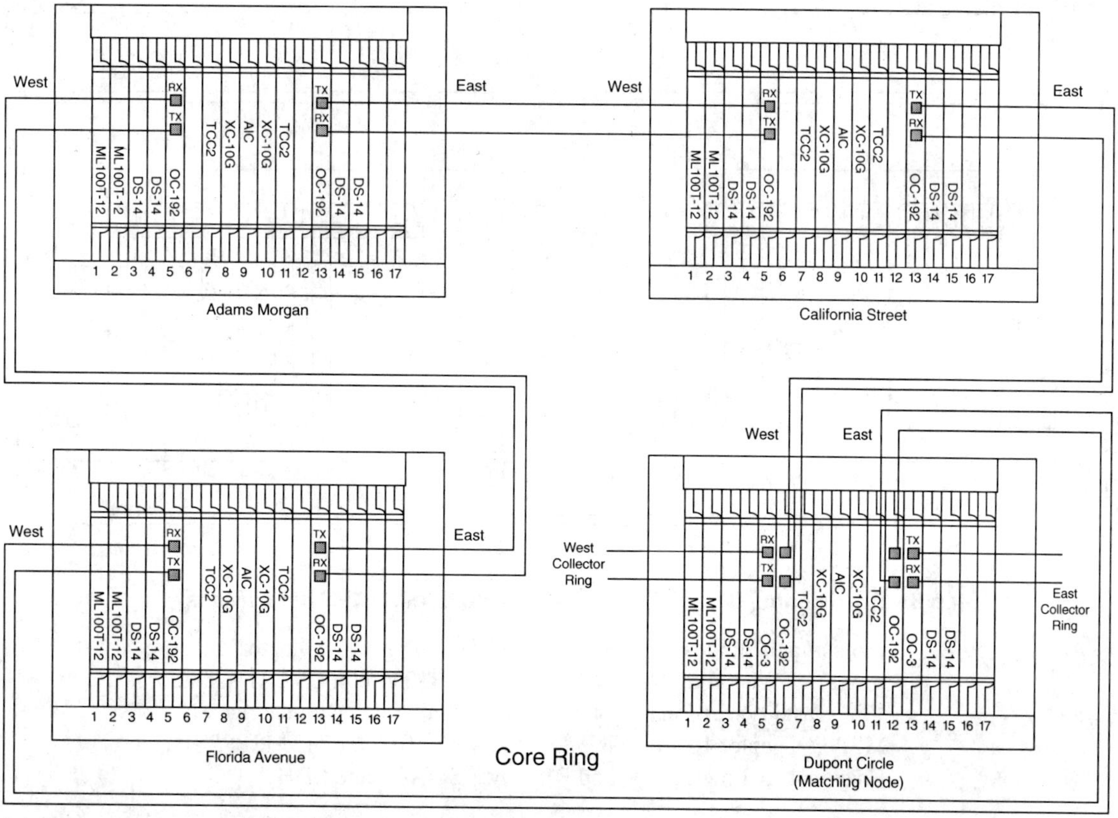

Figure 12-9 illustrates the completed collector with an OC-3/STM-1 UPSR/SNCP collector ring. The collector ring has been provisioned with OC-3/STM-1 capacity that can handle collector bandwidth for the buildings on the collector ring. If additional bandwidth is needed, the OC-3/STM-1 cards can be redeployed onto another ring, and new OC-12/STM-4 cards can be provisioned in the collector ring nodes.

Figure 12-9 *Case Study—Collector Ring*

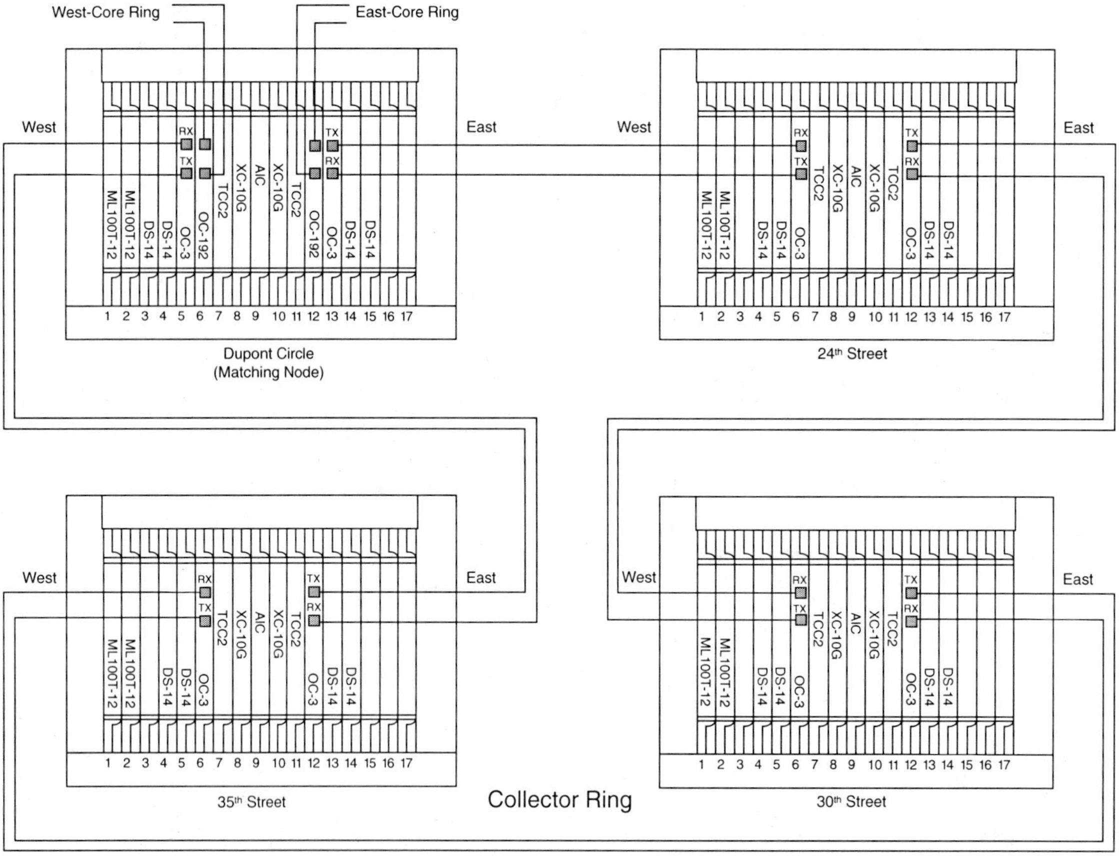

Step 8 Provisioning UPSR/SNCP, SONET/SDH, and Ethernet Circuits

Provision the nodes, timing, and UPSR/SNCP on the collector and core ring(s). Provision SONET/SDH DS1/VC4-LO circuits, VLANs, and/or VRFs, as described in Chapter 9, "Provisioning the Multiservice SONET MSPP," Chapter 10, "Provisioning the Multiservice SDH MSPP," and Chapter 11, "Ethernet, IP, and RPR over SONET and SDH."

Implementing BLSR/MS-SPRing on the Network

The network has been sized to OC-192/STM-64 core bandwidth using UPSR/SNCP protection. If BLSR/MS-SPRing is implemented, however, we must not forget that BLSR/MS-SPRing protection would consume 96 STS-1 or 32 VC-4 channels for protection on the core ring. This would halve the bandwidth, unless there was sufficient interbuilding traffic to warrant adds and drops on intermediate spans that would optimize the use of BLSR/MS-SPRing. In tangible terms, two-fiber BLSR/MS-SPRing would provide (OC-N/2) * (Number of spans) worth of bandwidth, if fully optimized to add and drop traffic at adjacent nodes. For our core ring model, this would result in 4.97 Gbps * 4 spans (19.9 Gbps) of net bandwidth. This number is in theory only.

In actual practice, it is highly improbable that all nodes would drop their entire traffic payload at a neighboring node. To design for two-fiber BLSR/MS-SPRing, the bidirectional traffic matrix must be properly developed to indicate the traffic adds and drops. The design process is similar with the exception of Step 7. Another design rule to keep in mind is that the ONS 15454s can terminate only two two-fiber BLSR/MS-SPRings per node or one four-fiber BLSR/MS-SPRing per node.

Provision the nodes, timing, and BLSR/MS-SPRing on the collector and core ring(s). Provision SONET/SDH DS1/VC4-LO circuits, VLANs, and/or VRFs, as described in Chapters 9, 10, and 11 of this book.

Scaling Up the Network

The compounded growth matrix shown in Table 12-13 indicates that the core bandwidth exceeds 10 Gbps (OC-192/STM-64) in Year 3. The service provider is faced with a few design choices that can alleviate the situation. One choice is to double the fiber count and add additional OC-192/STM-64 cards. However, this approach is not feasible from a financial perspective because far more economical options are available for consideration.

Table 12-13 *Compounded Growth Matrix*

Year	Normalized Bandwidth (1:10)	
	15% Compounded Growth	25% Compounded Growth
Year 0	7.008 Gbps	7.008 Gbps
Year 1	8.059 Gbps	8.760 Gbps
Year 2	9.268 Gbps	10.950 Gbps
Year 3	10.658 Gbps	13.687 Gbps
Year 4	12.257 Gbps	17.10 Gbps
Year 5	14.095 Gbps	21.386 Gbps
Year 10	28.351 Gbps	65.267 Gbps

Scalability Using the 15216 DWDM Filter Mux/Demux

To support the five-year growth plan of 20 Gbps, the core 15454s can be outfitted with 15454-OC192L-1-xx.x cards. The OC-192/STM-64 cards support 32 wavelengths, based on the International Telecommunications Union (ITU) frequency recommendations. The xx.x notation indicates the wavelength transmitted by the card. For example, 15454-OC192L-1-30.3 generates the 1530.33-nm wavelength. Table 12-14 shows the wavelengths supported by the OC-192/STM-64 DWDM card.

Table 12-14 *OC-192/STM-64 ITU Wavelengths (100-GHz Spacing)*

Channel	Wavelength (λ)	Channel	Wavelength (λ)	Channel	Wavelength (λ)	Channel	Wavelength (λ)
59	1530.33	49	1538.19	39	1546.12	29	1554.13
58	1531.12	48	1538.98	38	1546.92	28	1554.94
57	1531.90	47	1539.77	37	1547.72	27	1555.75
56	1532.68	46	1540.56	36	1548.51	26	1556.55
54	1534.25	44	1542.14	34	1550.12	24	1558.17
53	1535.04	43	1542.94	33	1550.92	23	1558.98
52	1535.82	42	1543.73	32	1551.72	22	1559.79
51	1536.61	41	1544.53	31	1552.52	21	1560.61

The wavelengths generated by the OC-192/STM-64 DWDM cards can be multiplexed together using an external optical filter multiplexer/demultiplexer (mux/demux) unit, such as the ONS 15216, which allows bandwidth scaling up to 320 Gbps per fiber.

The ONS 15216 is a metro DWDM system that can accept wavelengths from MSPPs, such as the ONS 15454, and multiplex them onto a single fiber span. Without DWDM, the ONS 15454 would need separate fiber pairs for each and every TX-RX OC-N port. Combining the ONS 15454 with the ONS 15216 reduces the fiber requirement to a single pair. The ONS 15216 supports ITU-T-compliant wavelengths with 100-GHz and 200-GHz spacing.

The ONS 15216 has two components that provide the user with a 16- to 32-wavelength, scalable DWDM solution. The base filter assembly (ONS 15216 Red filter) is a passive unit comprised of a multiplexer and demultiplexer. The multiplexer is a (16 * 1)-wavelength (100-GHz spacing) unidirectional multiplexer, whereas the demultiplexer is a (1 * 16)-wavelength (100-GHz spacing) unidirectional demultiplexer. The multiplexer and demultiplexer operate in the 1530- to 1561-nm frequency band. The (16 * 1) designation indicates that the Red filter can accept 16 input signals and multiplex them onto a single channel or fiber. The (1 * 16) designation indicates that the Red filter can demultiplex a single channel into 16 separate signals.

The multiplexer and demultiplexer systems connect to the two independent fibers in the fiber plant through the common ports on the chassis faceplate. The base Red filter unit integrates two expansion ports, providing the ability to upgrade, in-service, from 16 to 32 wavelengths (100-GHz spacing). The base unit also has two monitor ports that allow the user unobtrusive access to the multiplexed and demultiplexed signals for monitoring or analysis.

The upgrade filter assembly (ONS 15216 Blue filter) is a passive unit comprising a 100-GHz (16 * 1)-wavelength unidirectional multiplexer. It also consists of a 100-GHz (1 * 16)-wavelength demultiplexer. The upgrade filter connects via its front-panel common connector ports to the base unit's upgrade ports via user-supplied optical patch cords.

Table 12-15 indicates the Red and Blue filter ITU-T grid wavelengths supported by the ONS 15216 for the 16-lambda filters.

Table 12-15 *16-Wavelength ITU-T Red and Blue Filter Wavelengths with 100-GHz Spacing*

Base Red Filter Wavelengths	Upgrade Blue Filter Wavelengths
1546.12 nm	1530.33 nm
1546.92 nm	1531.12 nm
1547.72 nm	1531.90 nm
1548.51 nm	1532.58 nm
1550.12 nm	1534.25 nm
1550.92 nm	1535.04 nm
1551.72 nm	1535.82 nm
1552.52 nm	1536.61 nm
1554.13 nm	1538.19 nm
1554.94 nm	1538.98 nm
1555.75 nm	1539.77 nm
1556.55 nm	1540.56 nm
1558.17 nm	1542.14 nm
1558.98 nm	1542.94 nm
1559.79 nm	1543.73 nm
1560.61 nm	1544.53 nm

Optical network designers must keep in mind the insertion loss while integrating the ONS 15216 into the optical network. The transmitting lasers must provide adequate launch power to compensate for the insertion loss. Table 12-16 indicates the insertion loss for the filters and monitor ports of the ONS 15216.

NOTE Even though the 9.0-dB insertion loss for the base Red filter is a significant loss, one must not forget that the OC-192/STM-64 DWDM cards have a minimum output power of +3 dBm and a minimum receiver sensitivity of −20 to −8dBm at +/−1000ps/nm with OSNR of 19 dB at 0.5 RBW (Resolution Bandwidth). This means that the optical power budget per span is 23 dB as opposed to 13 dB for the standard OC-192/STM-64 card. The enhanced power budget easily compensates for the ONS 15216 insertion loss.

Table 12-16 *Insertion Loss for the ONS 15216*

Device or Port	Insertion Loss
Base Red + Upgrade Blue filter	9.0 dB (max, end to end)
Base Red filter	9.0 dB (max, end to end)
Upgrade Blue filter	8.5 dB (max, end to end)
Demultiplexer monitor port	14.6 dB (max)
Multiplexer monitor port	26 dB (max)

The TX ports of the OC-192/STM-64 DWDM cards interface with the mux ports of the 15216. The TX ports and the mux ports must operate at the same wavelength. Similarly, the RX ports of the OC-192/STM-64 DWDM cards interface with the demux ports of the 15216. The RX ports and the demux ports must also operate at the same wavelength. The common in and out ports interface with the existing two-fiber system and carry the multiplexed wavelengths.

NOTE The designer must select 15454-OC192L-1-xx.x cards with wavelengths that match the wavelengths of the ONS 15216 base Red filter. This provides flexibility to add the upgrade Blue filter in-service.

Figure 12-10 shows Year-5 scalability of the system using external 15216 DWDM filters.

Figure 12-10 *Year-5 Scalability Using DWDM*

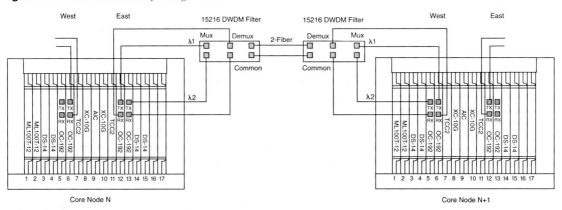

There is a design limit of four 15454-OC192L-1-xx.x cards per 15454 chassis. This means that a single ONS 15454 chassis can support up to 20 Gbps of DWDM bandwidth. This meets the five-year requirements.

Figure 12-11 shows a logical schematic of the Year-5 network. The core ring is OC-192/STM-64 * 2 (20 Gbps) and the collector rings are a mix of OC-48 and OC-192 bandwidths. The northwest ring has been upgraded to OC-192 due to the traffic density present in the NW, whereas the north, north-east, south-west, and south-east rings are OC-48/STM-16 each. If you have noticed, the Dupont Circle node is terminating four UPSR/SNCP rings. Each OC-192/STM-64 wavelength is considered a physical ring. This means that the two DWDM channels are part of the count, even though they ride a single physical fiber pair. The design specifications for UPSR/SNCP limit us to 16 UPSR/SNCP rings per ONS 15454 node with the TCC2 card. In terms of two-fiber BLSR/MS-SPRing, we can terminate two two-fiber BLSR/MS-SPRings or one four-fiber BLSR/MS-SPRing per node. The ONS 15216 DWDM filters are depicted on the core ring of the schematic.

Figure 12-11 *Year-5 Logical Schematic*

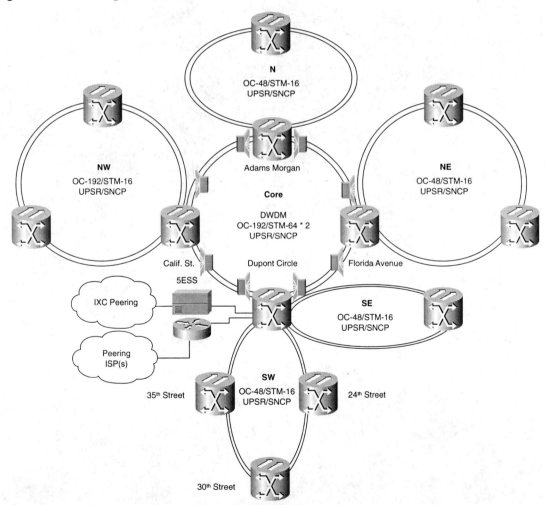

728 Chapter 12: Optical Network Case Studies

To meet 10-year requirements, three ONS 15454 shelves would need to be mounted in a rack, with each shelf providing 20 Gbps of bandwidth, providing a total of 60 Gbps. In such a scenario, it is prudent to redeploy the old fixed-wavelength OC-192-STM-64 in collector rings that need the bandwidth. Higher-bandwidth cards could replace the old OC-48/STM-16 cards in some rings, rendering the older cards obsolete with time. A logical schematic of the Year-10 network is depicted in Figure 12-12. The core ring is OC-192/STM-64 * 6 (60 Gbps) and the collector rings are a mix of OC-192/STM-64 and OC-192/STM-64 * 2 bandwidth. The north-west ring has been upgraded to OC-192/STM-64 * 2 with DWDM due to the traffic density present in the NW, whereas the north, north-east, south-west, and south-east rings are OC-192/STM-64 each. The ONS 15216 DWDM filters are depicted on the core and NW ring(s) of the schematic.

Figure 12-12 *Year-10 Logical Schematic*

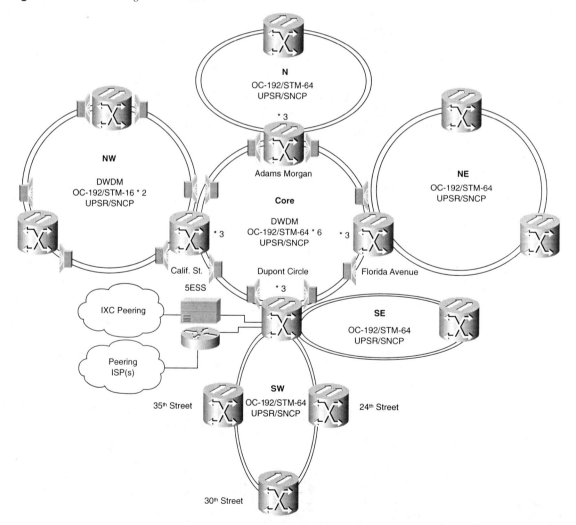

Scalability Using the 15454 MSTP

Another option to provide built-in, Year-10 scalability is to use the ONS 15454 MSTP. The ONS 15454 MSTP is a multiservice ADM that combines SONET/SDH transport and DWDM with various broadband multiservice interfaces such as the following:

- Fibre Channel
- FICON
- ESCON
- Gigabit Ethernet
- 10 Gigabit Ethernet
- OC-192/STM-64
- OC-48/STM-16
- OC-12/STM-4
- OC-3/STM-1
- D1 video
- HDTV

The ONS 15454 MSTP integrates various DWDM transmission elements and also includes the following capabilities:

- DWDM of up to 32 ITU-T wavelengths with 100-GHz spacing
- DWDM of up to 64 ITU-T wavelengths with 50-GHz spacing
- 1- and 4-band flexible OADM functionality
- 1-, 2-, and 4-wavelength OADM
- Optical erbium-doped fiber amplifier (EDFA) pre-amplifier
- Optical EDFA booster amplifier
- Passive OADM functionality
- Amplified OADM functionality
- Optical service channel modules (SCMs)
- Optical SCMs with combiner-separator
- Variable optical attenuation
- Optical power monitoring and equalization
- 100-Mbps to 2.5-Gbps multirate transponder
- 10-Gbps multirate transponder
- 4 * OC-48 10-Gbps transponder

- 2.5-Gbps ESCON/GE/FC muxponder
- GE transponder
- FE user data channel

The Cisco ONS 15454 MSTP integrates full DWDM functionality within the box and reduces dependency on external devices, such as the ONS 15216 filters, DWDM multiplexers, demultiplexers, OADMs, and SCMs. The wavelengths launched by the ONS 15454 MSTP DWDM modules fall in the 1530-nm to 1561-nm range.

An advantage of using the ONS 15454 MSTP is its capability to provide DWDM links for SAN applications, such as ESCON, FICON, and Fibre Channel. Even though these requirements were not defined in the case study presented, service providers can benefit from the 15454 MSTP's capability to provide such services.

Table 12-17 shows the wavelengths launched by the ONS 15454 MSTP DWDM modules. These wavelengths fall in the 1530-nm to 1561-nm range.

Table 12-17 *ONS 15454 MSTP 32-Channel Wavelength Plan (100-GHz Spacing)*

λ (nm)	λ (nm)	λ (nm)	λ (nm)
1530.33	1538.19	1546.12	1554.13
1531.12	1538.98	1546.92	1554.94
1531.90	1539.77	1547.72	1555.75
1532.68	1540.56	1548.51	1556.55
1534.25	1542.14	1550.12	1558.17
1535.04	1542.94	1550.92	1558.98
1535.82	1543.73	1551.72	1559.79
1536.61	1544.53	1552.52	1560.61

The Cisco ONS 15454 MSTP supports the CTC GUI craft interface, which allows point-and-click, A-to-Z wavelength provisioning with nodal control. Integration with EMSs and OSSs is accomplished through the Cisco Transport Manager element management system. In addition, the ONS 15454 MSTP MetroPlanner DWDM optical design tool simplifies network engineering and deployment. It enables designers to confirm designs, automate bill-of-material creation, and it provides node configurations and fiber connections with exportable optical setup files, thereby speeding node turn-ups.

NOTE Another Year-10 option would be to upgrade the racked core 15454 nodes with an ONS 15600. Multiple 15454s are typically interconnected with an OC-192/STM-64 ring. This limits the bandwidth to an OC-192/STM-64 between the racked nodes. Upgrading to a 15600 after Year 5 will alleviate this issue, because the 15600 has a very high port density with support for 4-port OC-192 cards and 16-port OC-48 cards. The 15600 can support up to 64 UPSR rings.

Case Study: SAN Services

A government agency has a requirement to back up large amounts of information at a primary backup disaster recovery (DR) site in Washington, D.C. Security concerns have mandated that a secondary DR site be located 2000 km away from the agency's operational facility. The agency uses a SAN product that has Fibre Channel interfaces and a mainframe that has ESCON connectivity to its front-end processor (FEP). The service provider has been invited to bid on a solution.

Case Study Solution

To provide SAN transport services over large distances, one must consider the Fibre Channel distance limitations of 200 km. An option is to consider running Fibre Channel over IP. However, the drawback to this approach is the nondeterministic nature of IP without hard classes of service in place. The solution is to deploy the ONS 15454/15454 SDH SL-Series card. The SL-Series card can provide any-to-any SAN services over the existing network. In such a scenario, however, we must assume that our service provider has a nationwide network footprint. Figure 12-13 shows the solution for the local DR requirement.

Figure 12-13 *Local DR Solution*

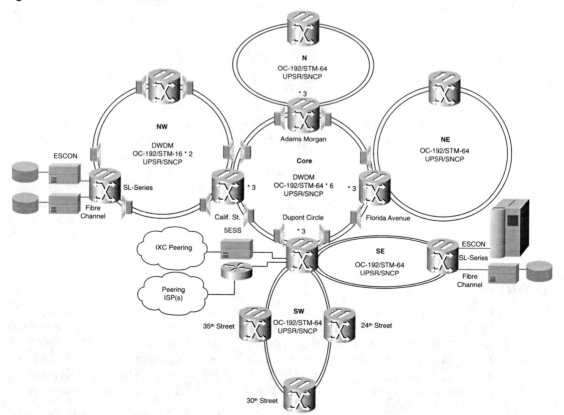

The SL-Series card offers our service provider the opportunity to deliver Fibre Channel and FICON over its resilient SONET/SDH infrastructure with line-rate support for 1-Gbps and 2-Gbps Fibre Channel as well as ESCON interfaces. The SL-Series card offers four client ports, each supporting 1.0625- or 2.125-Gbps Fibre Channel or FICON. The SL-Series card uses Generic Framing Procedure-Transparent (GFP-T) framing to deliver low-latency SAN transport. The card also uses Virtual Concatenation (VCAT) mechanisms to optimize bandwidth consumption over SONET/SDH infrastructures and offer Fibre Channel services at increments of 50 Mbps. Figure 12-14 shows the solution for the remote DR requirement.

Figure 12-14 *Remote DR Solution*

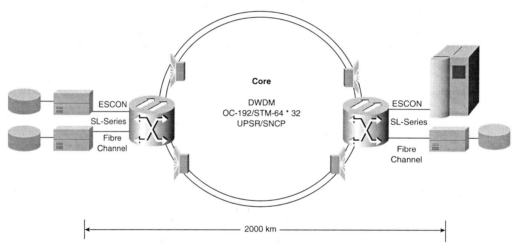

The SL-Series card can be deployed in any of the 4 high-speed (STS-48) slots in the 2.5-Gbps Cisco ONS 15454 systems and any of the 12 interface slots within the 10-Gbps ONS 15454 SONET SDH platform. The card uses buffer-to-buffer credits supported by the connected Fibre Channel switching devices to overcome distance limitations in 1- and 2-Gbps line-rate SAN extension applications. Furthermore, distance extension functions via R_RDY spoofing enables the SL-Series to serve as an integrated Fibre Channel extension device, obviating the need for external SAN extension devices. The SL-Series SAN transport solution supports distances up to 2800 km.

Summary

Carriers and service providers have traditionally used various network-design strategies. A single multiservice broadband network that can carry TDM voice, Ethernet, IP, and video with the reliability and restoration times of SONET/SDH along with scalability in the order of terabits per second provides service providers a platform to deliver reliable carrier-class broadband services. The ONS 15000 MSxPs deliver highly available broadband services including support for voice, TDM, SONET/SDH, Gigabit Ethernet, and all IP applications including VoIP, H.323 video, D1 video, and HDTV. The ONS 15000 MSxP series presents a variety of platforms that can be integrated end to end along with interoperability and compatibility with third-party vendor equipment. A detailed analysis is required to achieve an architecture, design, and solution for an end-to-end optical solution. Various design parameters must be considered prior to deploying such networks. These parameters include customer demographics, capacity planning, scalability, customer service requirements (such as voice, video, and data), oversubscription, customer service level requirements, network availability, protection mechanisms, delay, the fiber plant, physical diverse path routing, link loss budget, technology selection, VPN capability, and vendor selection.

This chapter presented the case study of a service provider involved in the analysis, design, and implementation of a large optical network providing service to an excess of 350,000 users. The case study presented various design requirements, such as reliability figures, latency, delay, and application requirements. The scenario presented considered the downtown metro area of a large U.S. city and commenced with an analysis of service requirements. This discussion was followed up with a capacity planning exercise, fiber plant analysis, delay analysis, technology analysis and comparison, logical design, and finally a discussion of the complete physical design of the network. Scalability considerations were presented followed up with the use of DWDM for scaling bandwidth. DWDM provides a far more cost-effective alternative to multiply the bandwidth versus the cost of laying new fiber. This section presented various scalability options including the use of the ONS 15454 MSTP for its built-in DWDM capability. The ONS 15454 MSTP provides added value by providing DWDM transport of OC-N, Ethernet, and SAN applications, such as ESCON, FICON, and Fibre Channel.

APPENDIX A

ML-Series Command Reference

ML-Series Commands

This section provides a command reference for those Cisco IOS commands or aspects of the commands that are unique to ML-Series cards. For information about the standard Cisco IOS commands, refer to the Cisco IOS configuration and command reference.

[no] bridge *bridge-group-number* protocol {drpri-rstp | ieee | rstp}

To define the protocol used by a bridge group, use the **bridge protocol** global configuration command. If no protocol will be used by the bridge group, this command is not needed. To remove a protocol from the bridge group, use the **no** form of this command with the appropriate keywords and arguments.

Syntax Description:

Parameter	Description
drpri-rstp	The protocol that enables the dual RPR interconnect (DRPRI) feature of the ML-Series card.
ieee	IEEE 802.1D Spanning Tree Protocol
rstp	IEEE 802.1W Rapid Spanning Tree Protocol
bridge-group-number	The identifying number of the bridge group that is assigned a protocol.

Default:

N/A

Command Mode:

Global configuration

Usage Guidelines:

The protocol DRPRI-RSTP is used only when configuring ML-Series cards as part of a DRPRI. A bridge group with DRPRI is limited to one protocol, which means the bridge group cannot implement RSTP or STP at the same time.

Example:

The following example assigns the DRPRI protocol to the bridge group with the bridge group number of 100.

```
Router(config)#bridge 100 protocol drpri-rstp
```

[no] clock auto

Use the **clock auto** command to determine whether the system clock parameters are configured automatically from the TCC+/TCC2. When enabled, both summertime and time zone are automatically configured, and the system clock is periodically synchronized to the TCC+/TCC2. Use the **no** form of the command to disable this feature.

Syntax Description:

This command has no arguments or keywords.

Default:

The default setting is **clock auto**.

Command Mode:

Global configuration

Usage Guidelines:

The **no** form of the command is required before any manual configuration of summertime, time zone, or clock. The **no** form of the command is required if Network Time Protocol (NTP) is configured in Cisco IOS. The ONS 15454 is also configured through CTC to use an NTP or Simple Network Time Protocol (SNTP) server to set the date and time of the node.

Example:

```
Router(config)#no clock auto
```

Related Commands:

- clock summertime
- clock timezone
- clock set

interface spr 1

Use this command to create a shared packet ring (SPR) interface on an ML-Series card for RPR. If the interface has already been created, this command enters SPR interface configuration mode. The only valid SPR interface number is 1.

Default:

N/A

Command Mode:

Global configuration

Usage Guidelines:

The command enables the user to create a virtual interface for the RPR/SPR. Commands, such as **spr wrap** or **spr station-id**, can then be applied to the RPR through SPR configuration command mode.

Example:

The following example creates the shared packet ring interface:

```
Router(config)#interface spr 1
```

Related Commands:

- **spr drpri-id**
- **spr-intf-id**
- **spr station-id**
- **spr wrap**

[no] pos flag c2 *value*

Use this command to specify the C2 byte value for transmitted and received frames. Use the **no** form of the command to return the C2 byte to its default value.

Syntax Description:

Parameter	Description
value	Value of the SONET C2 byte in hex

Default:

When changing the encapsulation on a POS port between LEX and PPP/HDLC, the scrambling and C2 settings will be automatically changed to their default values as follows:

Encapsulation	Scrambling	C2
LEX	pos scramble-spe	pos flag c2 0x01
PPP/HDLC	no pos scramble-spe	pos flag c2 0xCF

In PPP/HDLC encapsulation, changing the scrambling automatically changes the **pos flag c2** to its default as follows. (In LEX encapsulation, changing the scrambling does not affect C2.)

Encapsulation	Scrambling	C2
PPP/HDLC	**pos scramble-spe**	**pos flag c2 0xCF**
PPP/HDLC	**no pos scramble-spe**	**pos flag c2 0x16**

Command Mode:

Interface configuration mode (POS only)

Usage Guidelines:

The C2 byte is normally configured to match the setting on the peer path-terminating equipment (PTE). Using the correct order of operations will avoid having the nondefault settings overridden by the encapsulation change. The recommended order follows:

1 Set encapsulation to PPP/HDLC.

2 Set scrambling (if a nondefault setting is required).

3 Set C2 (if a nondefault setting is required).

Also note that the CRC setting varies among different types of PTE. The default CRC on the ML-Series card is 32 bits, regardless of any other settings. In most circumstances, the default settings should be correct, but users need to verify this with the user documentation for the PTE.

Example:

```
Router(config)#int pos0
Router(config-if)#pos flag c2 0x16
```

Related Commands:

- **pos trigger defects**
- **pos report**

[no] pos report *alarm*

Use this command to specify which alarms/signals are logged to the console. This command has no effect on whether alarms are reported to the TCC+/TCC2/CTC. Use the **no** form of the command to disable reporting of a specific alarm/signal.

Syntax Description:

Parameter	Description
alarm	**all**: All alarms/signals
	pais: Path alarm indication signal
	plm: Path label mismatch
	plop: Path loss of pointer
	ppdi: Path payload defect indication
	prdi: Path remote defect indication
	ptim: Path trace identifier mismatch
	puneq: Path label equivalent to zero
	sd-ber-b3: PBIP BER in excess of SD threshold
	sf-ber-b3: PBIP BER in excess of SF threshold

Default:

The default is to report all alarms.

Command Mode:

Interface configuration mode (POS only)

Usage Guidelines:

The *alarm* value is normally configured to match the setting on the peer path-terminating equipment (PTE).

Example:

```
Router(config)#int pos0
Router(config-if)#pos report all
Router(config-if)#pos flag c2 1
03:16:51: %SONET-4-ALARM: POS0: PPLM
Router(config-if)#pos flag c2 0x16
03:17:34: %SONET-4-ALARM: POS0: PPLM cleared
```

Related Command:

- pos trigger defects

[no] pos trigger defects *condition*

Use this command to specify which conditions cause the associated POS link state to change. These conditions are soaked/cleared using the delay specified in the **pos trigger delay** command. Use the **no** form of the command to disable triggering on a specific condition.

Syntax Description:

Parameter	Description
condition	**all**: All link down alarm failures
	ber_sd_b3: PBIP BER in excess of SD threshold failure
	ber_sf_b3: PBIP BER in excess of SD threshold failure
	pais: Path alarm indication signal failure
	plop: Path loss of pointer failure
	ppdi: Path payload defect indication failure
	prdi: Path remote defect indication failure
	puneq: Path label equivalent to zero failure

Default:

The default conditions for encapsulation of PPP are **ber_sf_b3**, **pais**, and **plop**. For encapsulation LEX, **ppdi** is an additional default (that is, **ber_sf_b3**, **pais**, **plop**, and **ppdi**).

Command Mode:

Interface configuration mode (POS only)

Usage Guidelines:

The *condition* value is normally configured to match the setting on the peer path-terminating equipment (PTE).

Examples:

```
Router(config)#int pos0
Router(config-if)#pos trigger defects all
```

Related Command:

- **pos trigger delay**

[no] pos trigger delay *time*

Use this command to specify which conditions cause the associated POS link state to change. The conditions specified in the **pos trigger defects** command are soaked/cleared using this delay. Use the **no** form of the command to use the default value.

Syntax Description:

Parameter	Description
time	Delay time in milliseconds, 200 to 2000

Default:

The default value is 200 milliseconds.

Command Mode:

Interface configuration mode (POS only)

Usage Guidelines:

The *time* value is normally configured to match the setting on the peer path-terminating equipment (PTE). The time granularity for this command is 50 milliseconds.

Examples:

```
Router(config)#int pos0
Router(config-if)#pos trigger delay 500
```

Related Command:

- pos trigger defects

[no] pos scramble-spe

Use this command to enable scrambling.

Syntax Description:

This command has no arguments or keywords.

Defaults:

The default value depends on the following encapsulation:

Encapsulation	Scrambling
LEX	pos scramble-spe
PPP/HDLC	no pos scramble-spe

Command Mode:

Interface configuration mode (POS only)

Usage Guidelines:

Scrambling is normally configured to match the setting on the peer path-terminating equipment (PTE). This command can change the POS flag C2 configuration.

Examples:

```
Router(config)#int pos0
Router(config-if)#pos scramble-spe
```

Related Command:

- pos flag c2

show controllers pos *interface-number* [details]

Use this command to display the status of the POS controller. Use the **details** argument to obtain certain additional information as described here.

Syntax Description:

Parameter	Description
interface-number	Number of the POS interface (0–1)
details	

Default:

N/A

Command Mode:

Privileged EXEC

Usage Guidelines:

This command can be used to help diagnose and isolate POS or SONET problems.

Example:

```
Router#show controllers pos0 details
```

Related Commands:

- show interface pos
- clear counters

show interface pos *interface-number*

Use this command to display the status of the POS.

Syntax Description:

Parameter	Description
interface-number	Number of the POS interface (0–1)

Default:

N/A

Command Mode:

Privileged EXEC

Usage Guidelines:

This command can be used to help diagnose and isolate POS or SONET/SDH problems.

Example:

```
Router#show interfaces pos0
POS0 is up, line protocol is up
  Hardware is Packet/Ethernet over Sonet
  Description: foo bar
  MTU 4470 bytes, BW 155520 Kbit, DLY 100 usec,
  reliability 255/255, txload 1/255, rxload 1/255
  Encapsulation HDLC, crc 32, loopback not set
  Keepalive set (10 sec)
  Scramble enabled
  Last input 00:00:09, output never, output hang never
  Last clearing of "show interface" counters 05:17:30
  Input queue: 0/75/0/0 (size/max/drops/flushes); Total output drops: 0
  Queueing strategy: fifo
  Output queue :0/40 (size/max)
  5 minute input rate 0 bits/sec, 0 packets/sec
  5 minute output rate 0 bits/sec, 0 packets/sec

     2215 total input packets, 223743 post-HDLC bytes
     0 input short packets, 223951 pre-HDLC bytes
     0 input long packets , 0 input runt packets
     0 input CRCerror packets , 0 input drop packets
     0 input abort packets
     0 input packets dropped by ucode

     0 packets input, 0 bytes
     Received 0 broadcasts, 0 runts, 0 giants, 0 throttles, 0 parity
     0 input errors, 0 CRC, 0 frame, 0 overrun, 0 ignored, 0 abort
     2216 total output packets, 223807 output pre-HDLC bytes
     224003 output post-HDLC bytes

     0 packets output, 0 bytes, 0 underruns
     0 output errors, 0 applique, 8 interface resets
     0 output buffer failures, 0 output buffers swapped out
     0 carrier transitions
```

Related Commands:

- **show controller pos**
- **clear counters**

show ons alarm

Use this command to display all the active alarms on the card.

Syntax Description:

This command has no arguments or keywords.

Default:

N/A

Command Mode:

Privileged EXEC

Usage Guidelines:

This command can be used to help diagnose and isolate card problems.

Example:

```
Router#show ons alarm
Equipment
Active Alarms: None

Port Alarms
  POS0 Active: TPTFAIL
  POS1 Active: TPTFAIL
  GigabitEthernet0 Active: None
  GigabitEthernet1 Active: None

POS0
Active Alarms : None
Demoted Alarms: None

POS1
Interface not provisioned
```

Related Commands:

- show controller pos
- show ons alarm defects
- show ons alarm failures

show ons alarm defect eqpt

This command displays the equipment-layer defects.

Syntax Description:

This command has no arguments or keywords.

Default:

N/A

Command Mode:

Privileged EXEC

Usage Guidelines:

This command displays the set of active defects for the equipment layer and the possible defects that can be set.

Example:

```
Router#show ons alarm defect eqpt
Equipment Defects
Active: CONTBUS-IO-B
Reportable to TCC/CLI: CONTBUS-IO-A CONTBUS-IO-B CTNEQPT-PBWORK
    CTNEQPT-PBPROT EQPT RUNCFG-SAVENEED ERROR-CONFIG
```

Related Command:

- show ons alarm failures

show ons alarm defect port

This command displays the port-layer defects.

Syntax Description:

This command has no arguments or keywords.

Default:

N/A

Command Mode:

Privileged EXEC

Usage Guidelines:

This command displays a set of active defects for the link layer and the possible defects that can be set. Note that the TPTFAIL defect can occur only on the POS ports and CARLOSS can occur only on the Ethernet ports.

Example:

```
Router#show ons alarm defect port
Port Defects
  POS0
  Active: TPTFAIL
  Reportable to TCC: CARLOSS TPTFAIL
  POS1
  Active: TPTFAIL
  Reportable to TCC: CARLOSS TPTFAIL
  GigabitEthernet0
```

```
Active: None
Reportable to TCC: CARLOSS TPTFAIL
GigabitEthernet1
Active: None
Reportable to TCC: CARLOSS TPTFAIL
```

Related Commands:

- show interface
- show ons alarm failures

show ons alarm defect pos *interface-number*

This command displays the link-layer defects.

Syntax Description:

Parameter	Description
interface-number	Number of the interface (0–1)

Default:

N/A

Command Mode:

Privileged EXEC

Usage Guidelines:

This command displays a set of active defects for the POS layer and the possible defects that can be set.

Example:

```
Router#show ons alarm defect pos 0
POS0
Active Defects: None
Alarms reportable to TCC/CLI: PAIS PRDI PLOP PUNEQ PPLM PTIM PPDI
    BER_SF_B3 BER_SD_B3
```

Related Commands:

- show controller pos
- show ons alarm failures

show ons alarm failure eqpt

This command displays the equipment-layer failures.

Syntax Description:

This command has no arguments or keywords.

Default:

N/A

Command Mode:

Privileged EXEC

Usage Guidelines:

This command displays a set of active failures for the equipment layer. If an EQPT alarm is present, the Board Fail defect that was the source of the alarm will display.

Example:

```
Router# show ons alarm failure eqpt
Equipment
Active Alarms: None
```

Related Command:

- show ons alarm defect

show ons alarm failure port

This command displays the port-layer failures.

Syntax Description:

This command has no arguments or keywords.

Default:

N/A

Command Mode:

Privileged EXEC

Usage Guidelines:

This command displays a set of active failures for the link layer.

Example:

```
Router#show ons alarm failure port
Port Alarms
  POS0 Active: TPTFAIL
  POS1 Active: TPTFAIL
  GigabitEthernet0 Active: None
  GigabitEthernet1 Active: None
```

Related Commands:

- show interface
- show ons alarm defect

show ons alarm failure pos *interface-number*

This command displays the link-layer failures.

Syntax Description:

Parameter	Description
interface-number	Number of the interface (0–1)

Default:

N/A

Command Mode:

Privileged EXEC

Usage Guidelines:

This command displays a set (active) of failures for a specific interface at the POS layer. The display also specifies whether an alarm has been demoted, as defined in Telcordia GR-253.

Example:

```
Router#show ons alarm failure pos0
POS0
Active Alarms  : None
Demoted Alarms: None
```

Related Commands:

- show controller pos
- show ons alarm defect

spr drpri-id {0 | 1}

Creates a DRPRI identification number of 0 or 1 to differentiate between the ML-Series cards paired for the dual RPR interconnect (DRPRI) protection feature.

Default:

N/A

Command Mode:

SPR interface configuration

ML-Series Commands

Usage Guidelines:

DRPRI paired sets share the same SPR station ID, so the DRPRI identification number helps identify a particular card in a DRPRI pair.

Example:

The following example assigns a DRPRI identification number of zero to the SPR interface on an ML-Series card:

```
Router(config)#interface spr 1
Router(config-if)#spr drpri-id 0
```

Related Commands:

- **interface spr 1**
- **spr-intf-id**
- **spr station-id**
- **spr wrap**

spr-intf-id *shared-packet-ring-number*

Assigns the POS interface to the SPR interface.

Syntax Description:

Parameter	Description
shared-packet-ring-number	The only valid shared packet ring number (SPR number) is 1.

Default:

N/A

Command Mode:

POS interface configuration

Usage Guidelines:

- The SPR number must be 1, which is the same SPR number assigned to the SPR interface.
- The members of the SPR interface must be POS interfaces.
- An SPR interface is configured similarly to a EtherChannel (port-channel) interface. Instead of using the **channel-group** command to define the members, you use the **spr-intf-id** command. And like **port-channel**, you then configure the SPR interfaces rather than the POS interface.

Example:

The following example assigns an ML-Series card POS interface to an SPR interface with the SPR number of 1:

```
Router(config)#interface pos 0
Router(config-if)#spr-intf-id 1
```

Related Commands:

- interface spr 1
- spr drpri-id
- spr station-id
- spr wrap

spr station-id *station-id-number*

Configures a station ID.

Syntax Description:

Parameter	Description
station-id-number	The user must configure a different number for each SPR interface that attaches to the RPR. Valid station ID numbers range from 1 to 254.

Default:

N/A

Command Mode:

SPR interface configuration

Usage Guidelines:

The station ID differentiates among the SPR interfaces from the different ML-Series cards attached to the RPR.

Example:

The following example sets an ML-Series card SPR station ID to 200:

```
Router(config)#interface spr 1
Router(config-if)#spr station-id 200
```

Related Commands:

- interface spr 1
- spr drpri-id
- spr-intf-id
- spr wrap

spr wrap [immediate | delayed]

Sets the RPR wrap mode to either wrap traffic the instant it detects a link state change or to wrap traffic after the carrier delay, which gives the SONET protection time to register the defect and declare the link down.

Syntax Description:

Parameter	Description
immediate	Wraps RPR traffic the instant it detects a link state change.
delayed	Wraps RPR traffic after the carrier delay time expires.

Default:

The default setting is immediate.

Command Modes:

SPR interface configuration

Usage Guidelines:

The **immediate** parameter should be used if RPR is run over unprotected SONET circuits. Run **delayed** for SONET-protected circuits (BLSR or UPSR).

Example:

The following example sets an ML-Series card to wrap delayed:

```
Router(config)#interface spr 1
Router(config-if)#spr wrap delayed
```

Related Commands:

- interface spr 1
- spr drpri-id
- spr-intf-id
- spr station-id

Cisco IOS Commands Not Supported on the ML-Series Cards

The following Cisco IOS commands are not supported by the 12.1(x) software image on the ML-Series card. This is because they are not tested, or due to hardware limitations of the ML-Series platform. These unsupported commands display when you enter the question mark (**?**) at the CLI prompt.

The following are unsupported privileged EXEC commands:

- **clear ip accounting**
- **show ip accounting**
- **show ip cache**
- **show ip tcp header-compression**
- **show ip mcache**
- **show ip mpacket**

The following are unsupported global configuration commands:

- **access-list aaa {1110-1199}**
- **access-list aaa {700-799}**
- **access-list aaa {200-299}**
- **async-bootp**
- **boot**
- **bridge** *number* **acquire**
- **bridge** *number* **address**
- **bridge cmf**
- **bridge** *number* **bitswap-layer3-addresses**
- **bridge** *number* **circuit-group**
- **bridge** *number* **domain**
- **bridge** *number* **lat-service-filtering**
- **bridge** *number* **protocol dec**
- **bridge** *number* **protocol ibm**
- **bridge** *number* **protocol vlan-bridge**
- **chat-script**
- **class-map match access-group**
- **class-map match class-map**

- class-map match destination-address
- class-map match mpls
- class-map match protocol
- class-map match qos-group
- class-map match source-address
- clns
- define
- dialer
- dialer-list
- downward-compatible-config
- file
- ip access-list log-update
- ip access-list logging
- ip address-pool
- ip alias
- ip bootp
- ip gdp
- ip local
- ip reflexive-list
- ip security
- ip source-route
- ip tcp
- ipc
- map-class
- map-list
- multilink
- netbios
- partition
- policy-map class queue-limit
- priority-list
- queue-list
- router iso-igrp

- **router mobile**
- **service compress-config**
- **service disable-ip-fast-frag**
- **service exec-callback**
- **service nagle**
- **service old-slip-prompts**
- **service pad**
- **service slave-log**
- **subscriber-policy**

The following are unsupported POS interface configuration commands:

- **access-expression**
- **autodetect**
- **bridge-group x circuit-group**
- **bridge-group x input**
- **bridge-group x lat-compression**
- **bridge-group x output**
- **bridge-group x subscriber-loop-control**
- **clock**
- **clns**
- **custom-queue-list**
- **down-when-looped**
- **fair-queue**
- **flowcontrol**
- **full-duplex**
- **half-duplex**
- **hold-queue**
- **ip accounting**
- **ip broadcast-address**
- **ip load-sharing per-packet**
- **ip route-cache**
- **ip security**
- **ip tcp**

- ip verify
- iso-igrp
- loopback
- multilink-group
- netbios
- priority-group
- pulse-time
- random-detect
- serial
- service-policy history
- source
- timeout
- transmit-interface
- tx-ring-limit

The following are unsupported Fast Ethernet or Gigabit Ethernet interface configuration commands:

- access-expression
- clns
- custom-queue-list
- fair-queue
- hold-queue
- ip accounting
- ip broadcast-address
- ip load-sharing per-packet
- ip route-cache
- ip security
- ip tcp
- ip verify
- iso-igrp
- keepalive
- loopback
- max-reserved-bandwidth
- multilink-group

- netbios
- priority-group
- random-detect
- service-policy history
- timeout
- transmit-interface
- tx-ring-limit

The following are unsupported port-channel interface configuration commands:

- access-expression
- carrier-delay
- cdp
- clns
- custom-queue-list
- duplex
- down-when-looped
- encapsulation
- fair-queue
- flowcontrol
- full-duplex
- half-duplex
- hold-queue
- iso-igrp
- keepalive
- max-reserved-bandwidth
- multilink-group
- negotiation
- netbios
- ppp
- priority-group
- random-detect
- timeout
- tx-ring-limit

The following are unsupported BVI interface configuration commands:

- **access-expression**
- **carrier-delay**
- **cdp**
- **clns**
- **flowcontrol**
- **hold-queue**
- **iso-igrp**
- **keepalive**
- **l2protocol-tunnel**
- **load-interval**
- **max-reserved-bandwidth**
- **mode**
- **multilink-group**
- **netbios**
- **ntp**
- **mtu**
- **timeout**
- **transmit-interface**
- **tx-ring-limit**

APPENDIX B

References

Agrawal, G. P. *Non-Linear Fiber Optics, Second Edition*. Academic Press, 1995.

Bertsekas, D. and R. Gallagher. *Data Communication Networks*. McGraw-Hill, 1996.

Bononi, A. *Optical Networking*. Springer Publishers, 1999.

Chlamtac, I., A. Ganz, and G. Karmi. "Lightpath Communications: An Approach to High Bandwidth Optical WANs," *IEEE Transactions on Communications, Vol. 40(7)*, July 1992.

Chow, Ming-Chwan. *Understanding SONET/SDH Standards and Applications*. Andan Publishers, 1996.

Cisco Systems, Inc. website. www.cisco.com.

Goodman, J. W. *Physics of Optoelectronic Devices*. John Wiley and Sons, 1995.

Goralski, Walter. *SONET*. McGraw Hill, 2000.

Green, P. *Fiber Optic Networks*. Prentice Hall, 1993.

Gumaste, Ashwin and Antony Tony. *DWDM Network Designs and Engineering Solutions*. Cisco Press, 2002.

IEEE Communications Magazine—Special Issue on Optical Networks, April 2002.

Kartalopoulos, Stamatios V. *Introduction to DWDM Technologies*. John Wiley and Sons, 1998.

Kartalopoulos, Stamatios V. *Understanding SONET/SDH and ATM*. Wiley IEEE Press, May 1999.

Keiser, G. *Optical Fiber Communications*. McGraw-Hill, 1996.

Liu, M. *Principles and Applications of Optical Communications*. Kluwer Publications, 1996.

Lowery, A. J. et al. "Multiple Signal Representation Simulation of Photonic Devices, Systems, and Networks," *IEEE Journal on Selected Areas in Quantum Electronics, Vol. 6*, April 2000.

Mukherjee, B. *Optical Communications Networks*. McGraw-Hill, 1997.

Mukherjee, B. et al. "Some Principles for Designing a Wide Area WDM Optical Network," *IEEE/ACM Transactions on Networking, Vol. 4(5)*, October 1996.

Narula, A., P. Lin, and E. Modiano. "Efficient Routing and Wavelength Assignment for Re-configurable WDM Networks," *Journal on Selected Areas in Communication*, January 2002.

Ramamurthy, B. *Design of Optical WDM Networks-LAN, MAN and WAN Architectures*. Kluwer Publications, 2001.

Ramaswami, R. and K. Sivarajan. *Optical Networks: A Practical Perspective, Second Edition*. Morgan Kauffman, 2001.

Saleh, B. and M. Teich. *Fundamentals of Photonics*. John Wiley and Sons, 1991.

Shimada, S. and H. Ishio. *Optical Amplifiers and Their Applications*. John Wiley and Sons, 2000.

Siller, Curtis A. Jr. and Mansoor Shafi. *A Sourcebook of Synchronous Networking, SONET/SDH*. IEEE Press, 1996.

Stern, Thomas E. and Krishna Bala. *Multiwavelength Optical Networks: A Layered Approach*. Prentice Hall PTR, May 1999.

Su, Chao, Lian-Kuan Chen, and Kwok-Wai Cheung. "Theory of Burst-Mode Receiver and Its Applications in Optical Multiaccess Networks," *Journal of Lightwave Technology, Vol. 15(4)*, April 1997.

Taub, M. and S. Shilling. *Principles of Communication Systems*. McGraw-Hill, 1992.

Warren, Dave and Dennis Hartmann. *BCMON*. Cisco Press, 2004.

Yoo, M., C. Qiao, and S. Dixit. "Optical Burst Switching for Service Differentiation in the Next-Generation Optical Internet," *IEEE Communications Magazine*, February 2001.

Glossary

A

add/drop. The process in which a part of the information carried in a channel is demodulated (dropped) at an intermediate point and different information is modulated (added) on to the channel for subsequent transmission.

ADM (add/drop multiplexer). Digital multiplexing equipment that adds and drops traffic in an optical network. The ADM also acts as a digital cross-connect switch.

AIS (alarm indicating signal). A code sent downstream indicating that an upstream failure has occurred. SONET defines the following four categories of AIS: Line AIS, STS path AIS, VT path AIS, and DS-N AIS.

AMI (alternate mark inversion). The line-coding format in transmission systems where successive 1s (marks) are alternatively inverted and transmitted with polarity opposite, to that of the preceding mark.

ANSI (American National Standards Institute). A membership organization that develops U.S. industry standards and coordinates U.S. participation in the International Standards Organization (ISO).

APS (automatic protection switching). A 1+1 APS protection-switched architecture is one in which the headend signal is permanently bridged (at the electrical level) to service and protection equipment, to enable the same payload to be transmitted identically to the tailend service and protection equipment. At the tailend, each service and protection optical signal is monitored independently and identically for failures. The receiving equipment selects either the service or protection channel based upon the switching criteria. A 1:n APS protection-switched architecture is defined as an architecture in which any one of n service channels can be bridged to a single optical protection channel. Headend-to-tailend communications are accomplished by using the SONET APS channel, bytes K1 and K2.

ASE (amplified spontaneous emission). Buildup of spontaneous emission in an optical fiber amplifier. ASE typically adds noise to the amplifier system.

asynchronous. A network where transmission system payloads are not network synchronized and each network node runs on its own clock.

asynchronous mapping. These mappings are defined for clear channel transport of digital signals that meet the standard DSX cross-connect requirements, typically DSX-1 and DSX-3 in most practical applications. At the asynchronous mapping interface, frame acquisition and generation are not required.

ATM (Asynchronous Transfer Mode). A switching technique in which information is organized into fixed-length cells with each cell consisting of 53 bytes. ATM cells are carried over logical virtual circuits defined by VPI/VCI values.

attenuation. Reduction of signal magnitude or signal loss, usually expressed in decibels (dB).

availability. The reliability criterion is an end-to-end, two-way availability objective of 99.98 percent for interoffice applications (0.02 percent unavailability or 105 minutes/year down-time). The objective for a loop transport between the central office and the customer premises is 99.99 percent.

AWG (arrayed waveguide grating). Integrated optical circuits formed by a series of curved silica waveguides that can combine (multiplex) or separate (demultiplex) different wavelength signals in an optical network.

B

bandwidth. Information-carrying capacity of a communication channel. Analog bandwidth is the range of signal frequencies that can be transmitted by a communication channel or network.

bend radius. The allowable limit a fiber cable can be bent, measured in subunits of meters.

BER (bit error rate). (1) Percentage of bits in a transmittal received in error. (2) The number of coding violations detected in a unit of time, usually 1 second. (3) Specifies expected frequency of errors.

bidirectional. A transmission and reception system that operates in both directions.

BIP (bit-interleaved parity). A parity check that groups all the bits in a block into units (such as byte), and then performs a parity check for each bit position in the group.

BIP-8 (bit-interleaved parity-8). A method of error checking in SONET that allows a full set of performance statistics to be generated. A BIP-8 creates 8-bit (1-byte) groups and performs a parity check for each group.

BISDN (Broadband Integrated Services Digital Network). A single ISDN that can handle voice, data, and video services.

bit. One binary digit or a pulse of data.

bit stuffing. In asynchronous systems, a technique used to synchronize asynchronous signals to a common rate before multiplexing.

bit synchronous. A way of mapping payload into VTs that synchronizes all inputs into the VTs, but does not capture framing information or allow access to subrate channels carried in each input.

BITS (Building Integrated Timing Supply). BITS generally supplies DS-1 and DS-0 level timing throughout an office. The BITS concept minimizes the number of synchronization links entering an office, because only the BITS will receive timing from outside the office.

BLER (block error rate). One of the underlying concepts of error performance is the notion of errored blocks—that is, blocks in which one or more bits are in error. A block is a set of consecutive bits associated with the path or section monitored by means of an error detection code (EDC), such as bit-interleaved parity (BIP).

bps (bits per second). The number of bits passing a point every second that constitutes the unit of measurement of bandwidth or speed for digital systems.

Brillouin scattering. Stimulated Brillouin scattering is an interaction between the optical signal and the acoustic waves in the fiber that causes the optical power to be scattered backward toward the transmitter. It is a narrowband process that affects each channel in a DWDM system individually. It is noticeable in systems that have channel powers in excess of 5 to 6 dBm. In most cases, SBS can be suppressed by modulating the laser transmitter to broaden the line width.

broadband. Services requiring in excess of 50 Mbps transport capacity. Many service providers loosely use the term broadband for DS-1 and subrate links that exceed 128 kbps in capacity.

byte interleaved. Interleaving of bytes from each STS-1, which are placed in sequence in a multiplexed or concatenated STS-N signal.

byte synchronous. A way of mapping payload into VTs that synchronizes all inputs into the VTs, captures framing information, and allows access to subrate channels carried in each input.

C

Category I. Terminal options that perform an asynchronous multiplex function. Examples of Category I transport NE interfaces are low-speed interfaces to ADMs, digital radio terminals, fiber-optic terminals (excluding terminals that function solely as digital repeaters or regenerators), and low-speed interfaces to DACSs.

Category II. Equipment interfaces whose behavior with respect to timing jitter is governed exclusively by input timing recovery circuitry. Examples of Category II transport NE interfaces are digital terminals at a DLC system, repeaters for metallic cables, regenerators for fiber-optic cables, and high-speed interfaces to DACSs. STS-N and OC-N interfaces to a SONET NE are also considered Category II.

CCITT (Consultative Committee for International Telegraph and Telephone). A technical organ of the United Nations, this specialized agency for telecommunications is currently known as the International Telecommunications Union - Telecommunications.

CEPT (European Conference of Postal and Telecommunications Administrations). The CEPT format defines the 2.048-Mbps European E1 signal, which is made up of 32 voice-frequency channels.

channel. The smallest subdivision of a circuit that provides a type of communication service.

circuit. A communications path or network; usually a pair of channels providing bidirectional communication.

circuit switching. Basic switching process whereby a circuit between two users is opened on demand and maintained for their exclusive use for the duration of the transmission.

cladding. The material surrounding the core of an optical fiber. The cladding must have a lower index of refraction to steer the light in the core.

clock-free-run mode. An operating condition of a clock in which its local oscillator is not locked to an external synchronization reference, and does not use storage techniques to sustain its accuracy.

clock holdover mode. An operating condition of a clock in which its local oscillator is not locked to an external synchronization reference, but does use storage techniques to maintain its accuracy with respect to the last known frequency comparison with a synchronization reference.

CMI (coded mark inversion). This is the STS-3 line code. This is a two-level nonreturn to zero code. A binary 1 is coded by either of the amplitude levels, +A or −A, for one full unit time interval (T) in such a way that the level alternates for successive binary 1s. For a binary 0, there is always a positive transition (−A to +A) at the midpoint of the binary unit interval (T/2).

CMIP (Common Management Information Protocol). The network management protocol defined by OSI. It is used to convey CMIS-defined operations over an OSI network.

CMIS (Common Management Information Services). The portion of the OSI network management specification that defines the management services available to a network management system.

concatenate. The linking together of various elements or data structures.

concatenated STS-Nc. A signal in which the STS envelope capacities from N STS-1s have been combined to carry an STS-Nc SPE. They are multiples of STS-1s concatenated to allow signals at that specific rate to be transported across SONET transmission systems.

concatenated VT. A composite VT that is composed of N [ts] VTs combined; its payload is transported as a single entity rather than separate signals.

CRC (cyclic redundancy check). A technique for using overhead bits to detect transmission errors.

CV (coding violation). A transmission error detected by the difference between the transmitted and the locally calculated bit-interleaved parity.

CWDM (coarse wavelength-division multiplexing). A version of WDM that uses multiple, widely spaced wavelengths, lying anywhere in the transmitting region of an optical fiber (1260 to 1620 nanometers).

D

dark fiber. Dark fiber refers to unused fiber-optic cable that has not been lit.

DCC (data communications channel). OAM&P channels in SONET that enable communications between individual network elements. In the section layer, 3 bytes (D1, D2, and D3) are allocated in STS-1 number 1 of an STS-N signal for section data communications. These 3 bytes are treated as one 192-kbps data channel for the transmission of alarms, maintenance, control, administration, as well as other network element communication needs. In the line layer, 9 bytes (D4–D12) are used as a 576-kbps data channel for similar purposes.

DCS (digital cross-connect system). An electronic cross-connect system that has access to lower-rate channels in higher-rate multiplexed signals as well as the capability to electronically rearrange (cross-connect) those channels.

defect. A limited interruption in the capability of an item to perform a required function.

demultiplexing. A process applied to a multiplexed signal for recovering signals combined within it and for restoring the distinct individual channels of the signals.

DFB laser (distributed feedback laser). Lasers giving out a very sharply defined color of light, similar to Fabry-Perot design, but with the addition of a corrugated structure above the active layer. The DFB feeds back one specific wavelength into the cavity that is then amplified and emitted.

diffraction. The deflecting of a light wave when it travels through an object, such as a grating.

digital signal. An electrical or optical signal that varies in discrete steps. Electrical signals are coded as voltages; optical signals are coded as pulses of light.

dispersion. Different wavelengths of light travel at slightly different speeds in optical fiber, which causes optical pulses to spread out as they travel through a system.

distribution frame. A physical piece of hardware where cross-connects are made.

DLC (digital loop carrier). A device similar to a channel bank that multiplexes a number of local voice lines into a smaller number of lines for transmission to a central office.

drop and broadcast. A cross-connect typically used to enable a broadcast transmission. A signal in the high-speed time slot is used to provide simultaneous drops at more than one node. A distance-learning application would use drop and continue to feed multiple classrooms.

DS-4NA. The DS4-NA (where NA stands for North America) is specified for a 139.264-Mbps interface (not 274 Mbps, as referenced in some literature with regard to DS-4 systems). This specification is compatible with ITU-T Recommendation G.755 for multiplexing 45-Mbps signals into 139-Mbps signals, but does not specify the multiplexing of other signals into the 139-Mbps signal.

DSF (dispersion-shifted fiber). Optical fiber that has a point of minimum dispersion moved toward its point of lowest attenuation (around 1550 nanometers).

DSL (digital subscriber line). A method of providing high-speed data services over the twisted-pair copper wires traditionally used to provide plain old telephone service

(POTS). Types of DSL include ADSL (asymmetric digital subscriber line), HDSL (high-data-rate digital subscriber line), SDSL (single-line digital subscriber line), and VDSL (very-high-data-rate digital subscriber line).

DSX-1. Can refer to either a cross-connect for DS-1 rate signals or the signals cross-connected at a DSX-1 level. DSX-1 is a specific level used to allow customers to check the mask or quality of the DS-1 signal. It also enables users to design their infrastructure to allow maintenance and a specific service level agreement (SLA).

DSX-3. Can refer to either a cross-connect for DS-3 rate signals or the signals cross-connected at a DSX-3 level. DSX-3 is a specific level used to allow customers to check the mask or quality of the DS-3 signal. It also enables users to design their infrastructure to allow maintenance and a specific SLA.

dual-ring internetworking. A topology in which two rings are connected at two different nodes, thereby providing traffic an alternate path from one ring to another.

DWDM (dense wavelength-division multiplexing). A variety of WDM that uses multiple wavelengths (or channels) in the 1550-nanometer region of the infrared spectrum. The wavelengths are closely spaced, usually evenly, on a grid.

E

ECSA (Exchange Carrier Standards Association). An organization that specifies telecommunications standards for ANSI.

EDFA (erbium-doped fiber amplifier). Optical amplifiers made of short lengths of optical fiber doped with the rare Earth element erbium. A pump laser excites erbium ions in the fiber, which pass their energy to optical signals passing through.

edge grooming. The aggregation and segregation of traffic just outside the avenues of a carrier network. This type of grooming is typically accomplished at the lowest possible levels of granularity (currently STS-1) for ease of service provisioning and support of legacy phone traffic.

EMS (element management system). A platform supporting single or multiple optical spans to provide network management services.

envelope capacity. The number of bytes the payload envelope of a single frame can carry; the SONET STS payload envelope is the 783 bytes of the STS-1 frame available to carry a signal.

ES (errored second). A performance-monitoring parameter. ES "type A" is a second with exactly one error, and ES "type B" is a second with more than one and less than the number of errors in a severely errored second (SES) for the given signal. ES by itself means the sum of the type A and B ESs.

ESD (electrostatic discharge). The discharge of a high-voltage transient caused by static charging.

eye diagram-eye pattern. An oscilloscope display in which the optical signal is displayed and compared in real time to a reference eye mask pattern. The eye pattern is a visual description of optical signal quality.

F

Fabry-Perot laser. The most basic design of laser, consisting of two specially designed slabs of semiconductor material on top of each other, with another material between them forming what is known as the active layer or laser cavity. Electric current flows through the device from the top slab to the bottom, and the emission of light occurs in the active layer.

failure. A termination of the capability of an item to perform a required function; a failure is caused by the persistence of a defect.

FBG (fiber Bragg grating). Small sections of optical fiber that act like selective mirrors, only reflecting back specific wavelengths. This reflection is caused by a periodic change of refractive index in the fiber core, which reflects mainly at the Bragg wavelength.

FEBE (far-end block error). A message sent back upstream from a downstream receiving network element that detects errors, usually a coding violation.

FEC (forward error correction). Additional bits added to a signal that can be used to detect and correct errors that occur during transmission through a system.

FERF (far-end receive failure). A signal to indicate to the transmit site that a failure has occurred at the receive site.

fiber plant. Building or underground fiber-optic cable infrastructure.

Fibre Channel. Serial data transfer architecture developed by a consortium of computer and mass storage device manufacturers that has been standardized by ANSI. The most prominent Fibre Channel standard is Fibre Channel Arbitrated Loop (FC-AL).

fixed stuff. A bit or byte whose function is reserved; fixed-stuff locations, sometimes called reserved locations, do not carry overhead or payload.

floating mode. A VT mode that allows the VT synchronous payload envelope to begin anywhere in the VT; pointers identify the starting location of the VT SPE.

framing. Method of distinguishing digital channels that have been multiplexed together.

frequency. The number of cycles of periodic activity that occur in a discrete amount of time, usually a second.

FWM (four-wave mixing). Basically an intermodulation and cross-talk phenomenon that occurs in WDM systems due to the nonlinear nature of the fiber-optic cable. The effect occurs in areas of zero dispersion, because the signals need to travel at the same velocity in the fiber for the effect to occur. FWM does not occur in the 1550-nanometers window unless the fiber's dispersion shifted. It also refers to a nonlinear Kerr effect in which two or more signal wavelengths can interact to create a new wavelength.

G

gain. The amplification factor of an optical amplifier computed as the ratio of output optical power to input power, usually measured in decibels (dB). It usually represents an increase in the amplitude level of an optical signal.

gain flattening. The art of getting equal amounts of amplification over a range of wavelengths in an optical amplifier.

grating. A device designed to allow specific wavelengths to be reflected, while others pass through it.

grooming. Consolidating or segregating traffic for efficiency.

H - I

hot swapping. The process of replacing a module without bringing down the equipment. This process occurs by sliding an active module into a fully powered-up unit.

in-fiber Bragg grating. An optical fiber grating is an optical fiber component consisting of a length of optical fiber wherein the refractive index of the core has been permanently modified in a periodic fashion, generally by exposure to an optical interference pattern as generated by an ultraviolet laser.

insertion loss. The loss introduced into an optical system by the insertion of optical devices and/or splices measured in decibels (dB).

integrated optical circuits. Similar in principle to electrical integrated circuits, but with the combining of many tiny versions of current optical components onto single silicon wafers.

interleave. The capability of SONET to mix together and transport different types of input signals in an efficient manner, thus allowing higher transmission rates.

IR (intermediate reach). IR optical interfaces refer to optical sections with system loss budgets from 0 dB to 12 dB. Typically low-power devices, such as SLM or MLM lasers, are used.

isochronous. All devices in the network derive their timing signal directly or indirectly from the same primary reference clock.

ITU-T (The International Telecommunications Union-Telecommunications). An international body associated with telecommunications standardization.

J

jitter. Timing jitter is the short-term variation of a digital signal's significant instant from its ideal position in time, where short-term implies phase oscillations of frequency greater than or equal to 10 Hz. Significant instants include, for instance, optimum sampling instants. Long-term variations, where the variations are of frequency less than 10 Hz, are called wander.

jitter generation. The process whereby jitter appears at the output port of an individual piece of digital equipment in the absence of applied jitter at the input. When looped back at the high-speed rate, whether or not a standard interface exists at the higher rate, Category I equipment must produce less than 0.3 unit intervals (UIs) of route mean square (RMS) jitter and less than 1.0 UI of peak-to-peak timing jitter at the output of the terminal receiver. This is as specified in TR-499. In TR-253 for SONET, a DS-3 interface will generate jitter less than 0.4 UI peak to peak.

jitter tolerance. For STS-N electrical interfaces, input jitter tolerance is the maximum amplitude of sinusoidal jitter at a given jitter frequency, which when modulating the signal at an equipment input port results in no more than 2 errored seconds cumulative, where these errored seconds are integrated over successive, 30-second measurement intervals. For the OC-N optical interface, it is defined as the amplitude of the peak-to-peak sinusoidal jitter applied at the input of an OC-N interface that causes a 1-dB power penalty.

jitter transfer. This is the relationship between jitter applied at the input port and the jitter appearing at the output port.

K - L

Kerr effect. The optical Kerr effect occurs when the index of refraction of a fiber optic varies with the intensity of the transmitted light. This is a nonlinear process that occurs when the product of the laser power and the effective system length becomes a significant fraction of the nonlinearity coefficient y.

line. One or more SONET sections, including network elements at each end, capable of accessing, generating, and processing line overhead.

line amplifier. A line amplifier is a multiwave device that amplifies signals. It does so to maintain signal strength over long distances. The line amplifier is also referred to as an optical line amplifier (OLA).

locked mode. A VT mode that fixes the starting location of the VT SPE; locked mode has less pointer processing than floating mode.

LOH (line overhead). Eighteen bytes of overhead accessed, generated, and processed by line-terminating equipment; this overhead supports functions such as locating the SPE in the frame, multiplexing, or concatenating of signals.

long reach (LR). LR optical interfaces refer to optical sections with system loss budgets from 10 dB up to 28 dB at OC-3, to 24 dB at OC-12, and to 20 dB at OC-48. Typical of long-haul telecommunication systems, LR interfaces are based on high-power (for instance, 500 uW or −3 dBm), multilongitudinal mode (MLM) or single-longitudinal mode (SLM) lasers.

LTE (line-terminating equipment). Network elements, such as add/drop multiplexers or digital cross-connect systems, that can access, generate, and process line overhead.

M

map-demap. A term for multiplexing, implying more visibility inside the resultant multiplexed bit stream than available with conventional asynchronous techniques.

mapping. The process of associating each bit transmitted by a service into the SONET payload structure that carries the service; for example, mapping a DS-1 service into a SONET VT1.5.

MEMS mirrors (micro-electro-mechanical systems mirrors). Mirrors that are no larger in diameter than a human hair and that can be arranged on special pivots so that they can be moved in three dimensions. Several hundred such mirrors can be placed together on mirror arrays no more than a few centimeters square in size to form an optical cross-connect.

mesochronous. A network whereby all nodes are timed to a single clock source; thus, all timing is exactly the same (truly synchronous).

misalignment loss. The loss of power resulting from angular misalignment, lateral displacement, and end separation.

MPLS (Multiprotocol Label Switching). A method used to direct data traffic in networks in which IP over ATM is used. In MPLS, IP routers at the edge of the network label packets in a way that greatly facilitates their handling by ATM switches at the network core. MPLS is used in core networks to facilitate VPNs, traffic engineering, and QoS. MPLS can also be implemented in IP-only networks.

MS-SPRing (multiplexed section protection ring). MS-SPRing performs ring switching or span switching between nodes. Working and protection traffic is transmitted bidirectionally over spans. The protection traffic can be flexibly used for transmitting extra traffic. Two main redundant topologies are in place: 2-Fiber MS-SPRing and 4Fiber MS-SPRing.

multiplexer. A device for combining several channels to be carried by one line or fiber.

mux-demux (multiplex-demultiplex). Multiplexing allows the transmission of two or more signals over a single channel; demultiplexing is the process of separating previously combined signals and restoring the distinct individual channel.

N - O

narrowband. Services requiring up to 1.5-Mbps transport capacity.

NE (network element). Any device that is part of a SONET transmission path and serves one or more of the section, line, and path-terminating functions.

NZDSF (nonzero dispersion-shifted fiber). This type of fiber was designed to introduce a small amount of dispersion without the zero-point crossing being in the WDM passband. With this type of fiber, you can eliminate, or at least greatly reduce, the degradation due to four-wave mixing, a distortion mechanism that requires the spectral components to be phase-matched along the fiber.

OAM&P (operations, administration, management, and provisioning). Provides the applications, facilities, and personnel required to manage a network. Sometimes called OA&M.

orderwire. A channel used by installers to expedite the provisioning of lines.

OSI reference model (Open System Interconnection reference model). A seven-layer model that provides a standard architecture for data communications.

OSS (operations support system). A network management system used for a single specific purpose, such as billing or alarm monitoring.

OTDM (optical time-division multiplexing). The interleaving of optical signals from different sources to create a higher composite bit rate.

overhead. Extra bits in a digital stream used to carry information besides traffic signals; orderwire, for example, would be considered overhead information.

OXC (optical cross-connect). Device that can move optical signals between different optical fibers, without the need for conversion to electrical signals.

P - Q

packet switching. An efficient method for breaking down and handling high-volume traffic in a network; a transmission technique that segments and routes information into discrete units.

parity check. An error-checking scheme that examines the number of transmitted bits in a block that hold the value 1; for even parity, an overhead parity bit is set to either 1 or 0 to make the total number.

path. A logical connection between the point where an STS or VT is multiplexed up to the point where it is demultiplexed.

payload. The portion of the SONET or SDH signal available to carry service signals, such as DS-1 and DS-3; the contents of an STS SPE or VT SPE.

payload pointer. Indicates the beginning of the synchronous payload envelope (SPE).

photon. The basic unit of light transmission used to define the lowest (physical) layer in the OSI seven-layer model.

plesiochronous. A network with nodes timed by separate clock sources with almost the same timing.

PM (performance monitoring). Measures the quality of service and identifies degrading or marginally operating systems (before an alarm would be generated).

PMD (polarization mode dispersion). Light transmitted down a single mode fiber can be decomposed into two perpendicular polarization components. Distortion results because of each polarization propagating at a different velocity. PMD causes pulse spreading as the polarizations arrive at different times. The longer the span, the worse the PMD. PMD is also frequency dependent.

POH (path overhead). Overhead accessed, generated, and processed by path-terminating equipment; POH includes 9 bytes of STS POH and, when the frame is VT-structured, 5 bytes of VT POH.

pointer. A part of the SONET overhead that locates a floating payload structure; STS pointers locate the SPE; VT pointers locate floating mode VTs; all SONET frames use STS pointers; only floating mode VTs use VT pointers.

poll. An individual control message from a central controller to an individual station on a multipoint network.

PON (passive optical network). Enables a single fiber access line to support a cluster of buildings through the use of a passive splitter close to the cluster. The splitter could also be active. In a PON, the light from a single fiber can be split into multiple signals of similar or different wavelengths and steered to individual buildings via short lengths of fiber.

POP (point of presence). A point in the network where inter-exchange carrier (IXC) facilities, such as DS-3 or OC-N, meet with access facilities managed by local exchange companies (LECs) or other service providers. POP could also indicate customer access points into an LEC or IXC network.

power budget/loss budget. The amount of optical power launched into a system that will be lost through various mechanisms (for instance, insertion losses and fiber attenuation), usually measured in dB.

PTE (path-terminating equipment). Network elements, such as fiber-optic terminating systems, that can access, generate, and process POH.

pulse density. At all digital interfaces, digital bit streams must contain sufficient energy for self-extraction of a timing signal. The level of energy is controlled by ensuring that the signal has a sufficient number of pulses as specified by a pulse density. In general, as the bit rate increases, the desired level of pulse density also increases, resulting in different requirements being applied to different levels in the digital hierarchy.

pump laser. A laser used to excite ions in a material, usually used in optical amplifiers.

QoS (quality of service). Classes of services used by carriers to service providers to guarantee delivery of traffic as per a service level agreement.

R

RAI (remote alarm indication). A code sent upstream in a DS-N network as a notification that a failure condition has been declared downstream; RAI signals were previously referred to as yellow signals.

Raman amplification. Optical amplification process throughout the actual transmission fiber in an optical network, caused by a carefully selected pump-laser wavelength scattering from atoms in the fiber and changing its wavelength to that of the optical signal.

Rayleigh scattering. Light scattering in an optical fiber due to slight changes in the core's refractive index.

RDI (remote defect indication). A signal returned to the transmitting-terminating equipment upon detecting a loss of signal, loss of frame, or AIS defect; RDI was previously known as FERF.

refraction. The change in direction of light due to its passing between two different materials.

refractive index. The property of a material that determines how fast light travels through it.

regeneration (3R). The process of amplifying (correcting loss), reshaping (correcting noise and dispersion), retiming (synchronizing with the network clock), and retransmitting an optical signal.

regenerator. A device that restores a degraded digital signal for continued transmission. It is also referred to as a repeater.

REI (remote error indication). An indication returned to a transmitting node (source) that an errored block has been detected at the receiving node (sink); this indication was formerly known as far-end block error (FEBE).

repeater. A repeater is a device that boosts the power of an optical signal. An optical amplifier does this without any conversion of the light into an electrical signal.

restoration. Action taken to repair it and return its services to an impaired (degraded) or unserviceable telecommunications service or facility.

RFI (remote failure indication). A failure is a defect that persists beyond the maximum time allocated to the transmission system protection mechanisms; when this situation occurs, an RFI is sent to the far end.

RPR (resilient packet ring). RPR is fiber-based ring network architecture in which data is carried in packets rather than over TDM circuits. RPR uses a modification of the Dijkstra algorithm to achieve a loop-free topology across the SONET rings with SONET-like rapid reconvergence on ring break.

S

SDH (Synchronous Digital Hierarchy). The ITU-T-defined world standard of transmission whose base transmission level is 52 Mbps (STM-0) and is equivalent to SONET's STS-1 or OC-1 transmission rate.

section. The span between two SONET network elements capable of accessing, generating, and processing only SONET section overhead; this is the lowest layer of the SONET protocol stack with overhead.

section overhead. Nine bytes of overhead accessed, generated, and processed by section-terminating equipment; this overhead supports functions such as signal framing and performance monitoring.

SES (severely errored second). A second in which a signal failure occurs, or more than a preset amount of coding violations (dependent on the type of signal) occurs.

single-mode fiber. A mode is one of the various light waves that can be transmitted in an optical fiber. Each optical signal generates many different modes, but in single-mode fiber the aim is to only have one of them transmitted. This is achieved through a core that is very small in diameter (usually around 10 micrometers).

slip. An overflow (deletion) or underflow (repetition) of one frame of a signal in a receiving buffer.

SNCP (subnetwork connection protection). SNCP performs path protection switching, wherein working traffic is transmitted in one direction and protection traffic in an opposite direction around the ring. Traffic is selected at each end of the path. SNCP supports multiple rings and is typically used for access and metro networks where a hub node terminates most traffic flows.

SNR (signal-to-noise ratio). Ratio of the amplitude of the optical signal to the amplitude of the noise.

SOA (semiconductor optical amplifier). SOA is similar to regular lasers, but with nonreflecting ends and broad wavelength emission. An incoming optical signal stimulates emission of light at its own wavelength, thereby amplifying it.

soliton pulse. A specially designed optical pulse that takes advantage of nonlinear effects to reverse the effects of dispersion, which enables the pulse to travel through a system while maintaining its shape and integrity.

SONET (Synchronous Optical Network). A standard for optical transport that defines optical carrier levels and their electrically equivalent Synchronous Transport Signals.

SPE (synchronous payload envelope). A major portion of the SONET frame format used to transport payload and STS POH; a SONET structure that carries the payload (service) in a SONET frame or VT; the STS SPE can begin anywhere in the SONET frame.

SPR (shared packet ring). A combination of Ethernet circuits over a SONET infrastructure that forms a loop-free Layer 2 switched network.

SR (short reach). SR optical interfaces refer to optical sections having system loss budgets up to 7 dB. Depending on the SONET/SDH hierarchical level, SR transmitters can be either LEDs or low-power MLM lasers.

SRP (Spatial Reuse Protocol). The SRP protocol derives its name from the spatial reuse capability, in which bandwidth is only consumed on traversed segments of the RPR ring. Unicast packets travel along ring spans between the source and destination nodes only.

SRS (stimulated Raman scattering). Stimulated Raman scattering results from the interaction between the optical signal and silica molecules in the fiber. This process is broadband and applies to the overall optical spectrum being transmitted. SRS manifests itself as a transfer of power from the shorter wavelengths to the longer wavelengths.

STE (section-terminating equipment). Equipment that terminates the SONET section layer; STE interprets and modifies or creates the section overhead.

STM (Synchronous Transport Module). An element of the SDH transmission hierarchy; STM-1 is SDH's base-level transmission rate equal to 155 Mbps; higher rates of STM-4, STM-16, and STM-48 are also defined.

stratum. Level of clock source used to categorize accuracy.

STS-1 (Synchronous Transport Signal level 1). The basic SONET building block signal transmitted at a 51.84-Mbps data rate.

STS POH (STS path overhead). Nine evenly distributed POH bytes per 125 microseconds starting at the first byte of the STS SPE; STS POH provides for communication between the point of creation of an STS SPE and its point of disassembly.

STS PTE (STS path-terminating equipment). Equipment that terminates the SONET STS path layer; STS PTE interprets and modifies or creates the STS POH; an NE that contains STS PTE will also contain LTE and STE.

superframe. Any frame structure made up of multiple frames; SONET recognizes superframes at the DS-1 level (D4 and extended superframe) and at the VT (500 μs STS superframes).

synchronous. A network where transmission system payloads are synchronized to a master (network) clock and traced to a reference clock.

T

T1X1. A subcommittee within ANSI that specifies SONET optical interface rates and formats.

TARP (TID Address Resolution Protocol). This is used on an NE-NE interface when there is a need to translate the TID of TL-1 messages to the CLNP address (network service access point [NSAP]) of an NE. The protocol would typically be used by a GNE in a TL-1/X.25 network that needs to map TIDS to NSAPs in a subtending network.

TDM (time-division multiplexing). A method for transmitting multiple calls over a single line; each call is assigned a recurring time slot on the line, and a small portion of that call gets transmitted over the line each time its assigned time slot is available.

timing recovery-clock recovery. The extraction of clock information from an optical signal.

transmission delay. To control echo and to minimize the effect on digital throughput, the maximum (one-way absolute delay for steady-state operation) of a 100-mile transport system with no intermediate terminals is 1 ms. This applies for all interface options provided. The required maximum delay for shorter systems is to be decreased in direct proportion to the route mileage.

transmission loss. Total loss encountered in transmission through a system.

tunable lasers. Lasers that can be adjusted to emit one of several different wavelengths, usually on the ITU grid.

U - W

UPSR (unidirectional path-switched ring). A method of providing redundancy for fiber-optic lines on a SONET ring. The SONET ring consists of two fiber-optic lines, each carrying the same traffic, but transmitting it in opposite directions around the ring. If one line fails, the backup line is already carrying the same traffic.

VCI (virtual channel identifier). The VCI is a 16-bit field in the header of an ATM cell. Its value defines the logical path to the next destination of the ATM cell. ATM switches use the value of the VPI/VCI fields to identify the next ATM node.

VCSEL (vertical cavity surface emitting laser). Lasers with a vertical cavity that emit light from their surface, in contrast to regular edge emitters.

VPI (virtual path identifier). The VPI is an 8-bit field in the header of an ATM cell. Its value defines the logical path to the next destination of the ATM cell. ATM switches use the value of the VPI/VCI fields to identify the next ATM node.

VPN (virtual private network). A closed user group that uses MPLS, encryption, or tunneling to provide a subscriber with a secure private network that runs over a public network infrastructure.

VT (virtual tributary). A signal designed for transport and switching of sub-STS-1 payloads.

VT group. A 9-row-by-12-column structure (108 bytes) that carries one or more VTs of the same size; seven VT groups can be fitted into one STS-1 payload.

VT POH (VT path overhead). Four evenly distributed POH bytes per VT SPE starting at the first byte of the VT SPE; VT POH provides for communication between the point of creation of a VT SPE and its point of disassembly.

VT PTE (VT path-terminating equipment). Equipment that terminates the SONET VT path layer; VT PTE interprets and modifies or creates the VT POH; an NE that contains VT PTE will also contain STS PTE, LTE, and STE POH.

wander. Long-term variations in a waveform.

wavelength. The length of one complete wave of an alternating or vibrating phenomenon, generally measured from crest to crest, or from trough to trough of successive waves.

WDM (wavelength-division multiplexing). (1) A technique in fiber-optic transmission for using multiple light wavelengths (colors) to send data over the same medium. (2) Two or more colors of light on one fiber. (3) Simultaneous transmission of several signals in an optical waveguide at differing wavelengths.

wideband. Services requiring 1.5- to 50-Mbps transport capacity.

INDEX

Numerics

1+1 protection architecture (SDH), 254
4MD-xx.x cards, 336
10M-L1-xx.x/10M-xx.x service interface cards, 335
10T-L1-xx.x/10T-xx.x service interface cards, 335
32-Channel Wavelength Plan (100-GHz Spacing), ONS 15454 MSTP, 333
32DMX-O cards, 336
32MUX-O cards, 336
1550-nm loss-minimized fiber (ITU-T G.654), 72

A

A1 bytes (SDH), 225
A2 bytes (SDH), 225
access rings, 5
AD-1B-xx.x cards, 336
AD-1C-xx.x cards, 336
AD-2C-xx.x cards, 336
AD-4B-xx.x cards, 336
AD-4C-xx.x cards, 336
ADMs (add/drop multiplexers)
    linear ADMs, 419
        configuration, 356, 418–427, 508-517
        creating, 419–420
    SDH (Synchronous Digital Hierarchy), 242–243
    SONET, 176–177
Advanced MPLS Design and Implementation, 275
AIC-I cards, 720
AIS (alarm indication signal), 164
    E1 signal format, 36
    SDH, 237
    superframes, 31
Alarm Browser, CTM (Cisco Transport Manager), 348–349
alarms
    AIS (alarm indication signal), 164
        E1 signal format, 36
        SDH, 237
        superframes, 31
    E1 signal format, 36–39
    ESF (extended superframe), 31–32
    SDH (Synchronous Digital Hierarchy), 235-236
        AIS (alarm indication signal), 237
        B1 errors, 238
        B2 errors, 238
        B3 errors, 238
        BIP-2 errors, 238
        LOF (loss-of-frame) alignment, 237
        LOP (loss-of-pointer) alignment, 237
        LOS (loss-of-signal) alarm, 237
        LSS (loss of sequence synchronization), 238
        OOF (out-of-frame) alignment, 237
        RDI (remote defect indication), 238
        REI (remote error indication), 237
        RFI (remote failure indication), 238
    SDH (Synchronous Digital Hierarchy), 208
    SF (superframe), 31–32
    SONET, 140, 163
        AIS (alarm indication signal), 164
        B1 errors, 164
        B2 errors, 165
        B3 errors, 165
        BIP-2 errors, 165
        LOF (loss-of-frame) alignment, 163
        LOP (loss-of-pointer) state, 163
        LOS (loss-of-signal) alarm, 163
        LSS (loss of sequence synchronization), 165
        OOF (out-of-frame) alignment, 163
        RDI (remote defect indication), 164
        REI (remote error indication), 164
        RFI (remote failure indication), 164
A-law encoding, analog-to-digital conversion, 24
algorithms, fairness algorithm, RPRs (resilient packet rings), 292
amplifiers, WDM (wavelength-division multiplexing), 121–125
    distributed amplifiers, 124
    EDFAs (Erbium-Doped fiber amplifiers), 123
    hybrid amplifiers, 124
    RFAs (Raman fiber amplifiers), 124
analog signals
    digital signals, converting to, 21–24
    electromagnetic interference (EMI), 21
    generating, 21

processing, 18-24
radio frequency interference (RFI), 21
reception, 21
AOTFs (acoustic optical tunable filters), optical multiplexers, 117–118
applications, fiber-optics, 50
APS (automatic protection switching), 140, 157
SDH (Synchronous Digital Hierarchy), 253–255
SONET, 187–189
armored outside-plant cables, 58
arrayed waveguides, optical multiplexers, 115–116
asymmetrical delays, 198, 264
ATM, 275
ATM/SMDS, 3
A-to-Z provisioning, 429
attenuation
extrinsic attenuation, 66
fiber-optics, 64–67
intrinsic attenuation, 64
wavelengths, compared, 65
AU pointers, SDH (Synchronous Digital Overhead), 226
automatic protection switching (APS). See APS (automatic protection switching)
automatically routed circuits, creating, 429–433
avalanche photodiodes, receivers, 127

B

B1 bytes (SDH), 225
B1 errors
SDH, 238
SONET, 164
B2 errors
SDH, 238
SONET, 165
B3 errors
SDH, 238
SONET, 165
background block error (BBE), E1 signal format, 36
Baird, John Logie, 49
band-separation method, WDM (wavelength division multiplexing), 98
bandwidth
compounded growth matrix, 723
efficiency, improving, 10-11
management, RPRs (resilient packet rings), spatial reuse, 292-293
normalized bandwidth calculations, 705
basic rate interface (BRI), ISDN, 39–40
BBE (background block error), E1 signal format, 36
BDCS (broadband digital cross-connect)
SDH (Synchronous Digital Hierarchy), 243–244
SONET, 178
Bell, Alexander Graham, 49
bend radius, fiber-optic cabling, 76–77
BERs (bit error rates), 127
bidirectional line-switched ring (BLSR). See BLSRs (bidirectional line-switched ring)
bidirectional rings
SONET, 190–191
unidirectional rings, compared, 256–257
bidirectional traffic matrix, 703
bidirectional WDM, 96–98
band-separation method, 98
circulator method, 99
interleaving-filter method, 98–99
BIP, 157
BIP-2 errors
SDH, 238
SONET, 165
birefringence, 67
bit, 154
bit error rates (BERs), 127
bit errors, E1 signal format, 37
bit slips, E1 signal format, 37
bit-interleaved parity (BIP-24) bytes, 226
BITS Out references, 372
BLANK FMECs (front-mounted electrical connections), 325
BLSR Ring Map Change dialog box, 399
BLSR Ring Map dialog box, 399
BLSR/MS-SPRing, implementing, 723
BLSRs (bidirectional line-switched rings), 6, 198-205
configuration, 356, 392–412
DCC terminations, creating, 395–396
K3 bytes, remapping, 397
linear ADMs, converting from, 423–427
nodes
adding, 404–406
failures, 201
removing, 407–409

ports, enabling, 396–397
provisioning, 397–400
subtending from, 417–418
trunk cards
    installing, 394–395
    moving, 409–412
upgrading, 400–402
UPSRs
    subtending from, 415–416
    subtending to, 414–415
Bragg resonance wavelengths, 107
BRI (basic rate interface), ISDN, 39–40
bridge group forwarding, configuration, 650–651
bridges
    interface configuration, 643–645
    SPRs, 279
byte stuffing, 224

C

C2 bytes (SDH), 230
cables (fiber-optic), 48
    attenuation, 64–67
    bend radius, 76–77
    chromatic dispersion, 66–67
    construction, 54–55
    cross-phase modulation (XPM), 69
    four-wave mixing (FWM), 69–70
    glass fiber-optic cables, 55
    interference, 64
    jackets, 56
    loose buffer cable plants, 76
    minimum bend radius, 76
    multifiber fiber-optic cables, 56–58
    optical signal-to-noise ratio (OSNR), 68
    PCS (plastic-clad silica) fiber-optic cables, 56
    physics, 51–53
    plastic fiber-optic cables, 55
    polarization mode dispersion (PMD), 67–68
    propagation modes, 58–63
    self-phase modulation (SPM), 69
    splicing, 75–76
    stimulated Brillouin scattering (SBS), 70
    stimulated Raman scattering (SRS), 70
    submarine cable systems, 77
    tensile loading, 76–77
    termination, 48, 73–75
    tight buffer cable plants, 76
calculations, normalized bandwidth calculations, 705
Call Reference Value (CRV) (1 or 2 octets) field (Q.931 header), 45
call setup, ISDN, 45
campus-to-central office (CO) traffic, 5
capacity planning, network design, 701–706
card view, CTC (Cisco Transport Controller), 342–343
CAS (channel-associated signaling), 26
case studies
    multiservice metro optical SONET/SDH networks, 699–722
        BLSR/MS-SPRing implementation, 723
        capacity planning, 701–706
        circuit provisioning, 722
        delay analysis, 708–710
        fiber plant analysis, 706–709
        logical design, 712–716
        physical design, 716–718
        requirements analysis, 701
        technology analysis, 710–712
        upscaling, 723–730
    SAN services, 731–732
C-band (conventional) optical frequency bands, WDM systems, 96
CCITT (Consultative Committee for International Telegraph and Telephone), 4
CCS (common channel signaling), 27
CDMA (code-division multiple access), 19
cell-switched data networks, 3
Change User dialog box, 366
channel banks, 26
    TDM (time-division multiplexing), 47
channel spacing, WDM (wavelength division multiplexing), 99
channel-associated signaling (CAS), 26
Chappe, Claude, 49
chirp, WDM transmitters, 111
chromatic dispersion compensation
    fiber-optics, 66–67
    WDM (wavelength-division multiplexing), 134–136
Circuit Attributes dialog box, 433

Circuit Creation dialog box, 427, 433
Circuit Destination dialog box, 431
Circuit Name Search dialog box, 438
Circuit Source dialog box, 431
circuits, 428
    automatically routed circuits, creating, 429–433
    manually routed circuits, creating, 433–436
    monitor circuits, creating, 437–438
    ONS 15454 circuits, searching for, 438–439
    path traces, creating, 440–442
    provisioning, 356, 428–444, 518–553
    SNCP circuits, removing, 485
    SONET DCC tunnels, provisioning, 442–444
    unidirectional circuits, multiple drop creation, 436–437
    UPSR circuits, editing, 439–440
Circuits on Span dialog box, 389
circuit-switched networks, 18, 25–26
    TDM (time-division multiplexing) signaling, 26–27
circulator method, WDM (wavelength division multiplexing), 99
circulators, optical multiplexers, 119
Cisco Transport Controller (CTC). *See* CTC (Cisco Transport Controller)
Cisco Transport Manager (CTM). *See* CTM (Cisco Transport Manager)
CLECs (competitive local-exchange carriers), 307
client software, CTM (Cisco Transport Manager), 345–346
    Alarm Browser, 348–349
    Domain Explorer, 346–347
    Network Map, 350–351
    Node View, 348–350
    Subnetwork Explorer, 347–348
clock slips, E1 signal format, 37
coarse wavelength-division multiplexing (CWDM). *See* CWDM (coarse wavelength-division multiplexing)
COBRA (Common Object Request Broker Architecture), 206
code errors, E1 signal format, 37
code-division multiple access (CDMA), 19
collector rings, access rings, aggregation, 5
COMET (Complete Optical Multiservice Edge and Transport), 306-307

common channel signaling (CCS), 27
Common Object Request Broker Architecture (CORBA), 206
communications systems, fiber-optics, 78
    receivers, 81–83
    transmitters, 78–80
compensation, WDM (wavelength division multiplexing), 92, 133–134
competitive local-exchange carriers (CLECs), 307
Complete Optical Multiservice Edge and Transport (COMET), 306-307
compounded growth matrix, 723
configuration
    BLSRs (bidirectional line-switched rings), 356, 392-412
    linear ADMs
        SDH, 446, 508–517
        SONET, 356, 418–427
    MS-SPRing
        SDH, 485–502
        SONET, 446
    OSPF, 380–385, 474–478
    SNCP, SDH, 478–485
    subtending rings, 356, 413–418
    UPSR (unidirection path-switched ring), 356–392
        DCC terminations, 387
console (CTC), 337
Consultative Committee for International Telegraph and Telephone (CCITT), 4
Control Panel, CTM (Cisco Transport Manager), 353
controlled slip (frame), 212
conversions, analog-to-digital conversions, 21–24
CoS (classes of service), RPRs (resilient packet rings), 291
couplers, optical multiplexers, 119
CRC (cyclic redundancy checking), E1 signal format, 36–38
Create Area Range dialog box, 384
Create BLSR dialog box, 398
Create DCC Tunnel Connection dialog box, 443
Create Protection Group dialog box, 368
Create SDCC Terminations dialog box, 387, 395, 415
Create Static Route dialog box, 379

Create User dialog box, 365
Create Virtual Link dialog box, 384
cross-connect systems, 428
    TDM (time-division multiplexing), 47
cross-phase modulation (XPM), fiber-optics, 69
CSU/DSU (channel service units/digital service units), TDM (time-division multiplexing), 46
CTC (Cisco Transport Controller), 306-307, 311, 336-338
    A-to-Z provisioning, 429
    card view, 342–343
    CTC console, 337
    default gateway, 377–378, 470
    initial provisioning tasks
        SDH (Synchronous Digital Hierarchy), 447–449
        SONET, 357-359
    multiple CTCs, static routes, 380–381
    network view, 340–342
    node view, 338–340
    ONS nodes
        same IP subnet, 375–376
        searching for, 438–439
        separate IP subnet, 376
    static routes, 473
CTM (Cisco Transport Manager), 306-307, 343–345
    client software, 345–346
        Alarm Browser, 348–349
        Domain Explorer, 346–347
        Network Map, 350–351
        Node View, 348–350
        Subnetwork Explorer, 347–348
    Control Panel, 353
    NE Explorer, 351–353
customer demographics, design strategies, 697
customer rings, 5
customer service levels, design strategies, 697–698
customer service requirements, design strategies, 697
cutoff wavelengths, 61
CWDM (coarse wavelength-division multiplexing), 92, 100
    ITU grid, 92, 102–104
cyclic redundancy check (CRC) errors, E1 signal format, 38

D

D4 SF (superframe), 30
D5 ESF (extended superframe), 30
DACS (digital access and cross-connect system), 3, 26
    TDM (time-division multiplexing), 47
data communication channel/generic communication channel (DCC/GCC), AIC-I cards, 324
DBR (distributed Bragg reflector) lasers, WDM (wavelength-division multiplexing), 107–108
DCC/GCC (data communications channel/generic communication channel), AIC-I cards, 324
DCCs, 155, 226, 769
    LDCCs (line DCCs), 442
    SDCC (section DCC), 442
    SONET DCC tunnels, provisioning, 442–444
    terminations
        BLSR DCC terminations, creating, 395–396
        MS-SPRing, creating, 488–489
        UPSRs, configuration, 387
DCU (dispersion compensation module), 90
default gateway, CTC (Cisco Trasnport Console), 377–378
Define New Drop dialog box, 436
degraded minutes, E1 signal format, 38
delay analysis, network design, 708–710
Delete Circuit dialog box, 411
Delete User dialog box, 366
dense wavelength-division multiplexing (DWDM) infrasturctures. *See* DWDM (dense wavelength-division multiplexing) infrastructures
depressed-clad fiber, 61
descrambling techniques, 155
design strategies, 696
    customer demographics, 697
    customer service levels, 697–698
    customer service requirements, 697
    fiber infrasturcture, 698
    parameters, 696
    SANs, case study, 731-732
    technology selection, 698–699
    vendor selection, 699

DFB (distributed feedback) lasers, WDM (wavelength-division multiplexing), 106–107
DGD (differential group delay), 67
dialog boxes
　BLSR Ring Map, 399
　BLSR Ring Map Change, 399
　Change User, 366
　Circuit Attributes, 433
　Circuit Creation, 427, 433
　Circuit Destination, 431
　Circuit Name Search, 438
　Circuit Source, 431
　Circuits on Span, 389
　Create Area Range, 384
　Create BLSR, 398
　Create DCC Tunnel Connection, 443
　Create Protection Group, 368
　Create SDCC Terminations, 387, 395, 415
　Create Static Route, 379
　Create User, 365
　Create Virtual Link, 384
　Define New Drop, 436
　Delete Circuit, 411
　Delete User, 366
　Edit Circuit, 437
　Edit Circuits, 436
　Upgrade BLSR, 401
dielectric jackets, fiber-optic cables, 57
differential group delay (DGD), 67
digital access and cross-connect system (DACS), 3, 26
digital loop carriers, SDH (Synchronous Digital Hierarchy), 245
digital signals, analog signals, converting from, 21–24
direct modulation, 111–112
dispersion, WDM (wavelength division multiplexing), 92, 133-134
　chromatic dispersion, 134, 136
　polarization mode dispersion, 136–138
dispersion compensation module (DCU), 90
dispersion-shifter fiber (ITU-T G.653), 71–72
distributed amplifiers, 124
distributed feedback (DFB), 107
diversity, SONET, 184–186
DLP (digital loop carriers), SONET, 179–180

Domain Explorer, CTM (Cisco Transport Manager), 346–347
double tagging, SPRs, 279
doubly clad fiber, 61
DPT (Synamic Packet Transport), 275
　RPRs (resilient packet rings), 288–290
DRI (dual-ring interconnect), 194-196, 260-261
drops, unidirectional circuits, creating for, 436–437
DS framing
　multiframing formats, 29–32
　T-carrier systems, 29
DS1 14-port cards, 720
DS1 signals, SDH (Synchronous Digital Hierarchy) multiplexing, 218
DS1/E1 FMECs (front-mounted electrical connections), 325
DS1-14 electrical cards, 314
DS1N-14 electrical cards, 314
DS2 signals, SDH (Synchronous Digital Hierarchy) multiplexing, 219
DS3 signals, SDH (Synchronous Digital Hierarchy) multiplexing, 220
DS3-12 electrical cards, 314
DS3-12E electrical cards, 315
DS3i-N-12 electrical cards, 327
DS3N-12 electrical cards, 314
DS3N-12E electrical cards, 315
DS3XM-6 electrical cards, 315
dual-ring interconnect (DRI) architecture, 260–261
DWDM (dense wavelength-division multiplexing), 2, 13-16, 92, 101, 276
　dynamic range, span analysis, 84
　infrastructures, 306
　ITU grid, 92, 102–104
　lasers, 113
　tunable lasers, 108–109

E

E1 bytes (SDH), 226
E1 FMECs (front-mounted electrical connections), 325
E1 signals, SDH (Synchronous Digital Hierarchy) multiplexing, 218–219
E-100 ports, provisioning, 566–567
E1000-2-G Ethernet cards, 318, 330

E100T-G Ethernet cards, 318, 330
E1-120NP FMECs (front-mounted electrical connections), 325
E1-120PROA FMECs (front-mounted electrical connections), 325
E1-120PROB FMECs (front-mounted electrical connections), 325
E1-42 electrical cards, 327
E1-N-14 electrical cards, 327
E2 orderwire bytes (SDH), 229
E3 signals, SDH (Synchronous Digital Hierarchy) multiplexing, 219
E3/DS3 FMECs (front-mounted electrical connections), 325
E3-12 electrical cards, 327
E4 signals SDH (Synchronous Digital Hierarchy) multiplexing, 220
E-band (extended) optical frequency bands, WDM systems, 96
E-bit indication, E1 signal format, 38
EC1-12 electrical cards, 315
E-carrier system, 18, 32–34
    alarms, 36, 38–39
    CRC error checking, 36–37
    errors, 36, 38–39
    FAS (frame alignment signal), 34–35
    MFAS (muliframe alignment signal), 35–36
ECC (embedded communications channel), 155
echo cancellation, analog-to-digital conversions, 24
ECSA (Exchange Carriers Standards Association), 4
EDFAs (Erbium-Doped fiber amplifiers), 123
Edit Circuits dialog box, 436
editing
    protection groups, 369
    UPSR circuits, 439–440
    users, 366
electrical card protection, 320, 331
electrical signals
    SDH, 213–214
    SONET, 140, 144–145
electromagnetic interference (EMI), analog signals, 21
embedded communications channel (ECC), 155
embedded operations channel (EOC), 155
encapsulation, Ethernet over SONET/SDH, 277
encoding analog-to-digital conversions, 23–24
Enterprise Systems Connection (ESCON), 3

environmental alarms, 720
EOC (embedded operations channel), 155
Erbium-Doped fiber amplifiers (EDFAs), 123
errored blocks, E1 signal format, 38
errored seconds (ES), E1 signal format, 38
ES (errored seconds), E1 signal format, 38
ESCNP mesh networks, SDH, 517–518
ESCON (Enterprise Systems Connection), 3
E-Series card ports, SPRs, 282
E-Series provisioning
    Ethernet over SDH, 556, 595–623
        EtherSwitch point-to-point circuits, 597–603
        EtherSwitch ports, 589–592
        G1K-4 Ethernet ports, 587–589
        hub-and-spoke Ethernet circuits, 607–611
        MAC table information retrieval, 621
        multicard EtherSwitch manual cross-connect, 615–618
        ports, 596–597
        RMON alarm thresholds creation, 622–623
        single-card EtherSwitch manual cross-connect, 611–615
        SPR (Shared Packet Ring), 603–606
        STP (Spanning Tree Protocol) activation, 620–621
        VLAN membership, 618–620
    Ethernet over SONET, 556, 565–584
        E-100 ports, 566–567
        EtherSwitch point-to-point circuits, 567–571
        EtherSwitch SPR (shared packet ring), 571–574
        hub-and-spoke Ethernet circuits, 574–577
        MAC table information retrieval, 584
        multicard EtherSwitch manual cross-connect, 579–581
        RMON alarm thresholds, 584
        single-card EtherSwitch manual cross-connect, 577–579
        Spanning Tree Protocol activation, 582–584
        VLAN membership, 581–582

ESF (extended superframe), 30–31, 171
    alarms, 31–32
ESNCP (Extended SNCP) mesh networks, 446
Ethernet, 274–276
    encapsulation, SONET/SDH, 11–12
    GE (Gigabit Ethernet), 274–275
    ML-series cards
        accessing, 628–632
        bridge group forwarding, 650–651
        bridge interface configuration, 643–645
        Fast Ethernet interface configuration, 634
        IEE 802.1Q tunneling, 656–661
        IOS command modes, 632–633
        IP protocols, 672–673
        Layer 2 tunneling, 656–658, 661–663
        link aggregation, 663–664
        POS channels, 664–667
        POS interface configuration, 636–643
        provisioning, 624–653, 655–669,
        QoS, 673–674
        RPR (resilient packet ring), 674–691
        STP configuration, 645–650
        switching database manager, 691–692
        VACT (virtual concatenation), 691
        VLAN configuration, 651–655
        VRF lite, 667–669, 671
    MSPPs (multiservice provisioning platforms),
        274
    RPRs (resilient packet rings), 274, 287–290
        bandwidth management, 292–293
        CoS (classes of service), 291
        DPT (Dynamic Packet Transport),
            288–290
        fairness algorithm, 292
        layer management, 303
        MAC (Media Access Control), 294–303
        OAM functions, 303
        rerouting, 293–294
        topology discovery, 290–291
        traffic protection, 293–294
    SONET/SDH, 556
    SPRs (shared packet rings), 274, 277–281
        design constraints, 281–282
            E-Series card ports, 282
            matching nodes, 286
            ML-Series card ports, 283–285

Ethernet over SDH, 557
    E-Series provisioning, 556, 595–623
        EtherSwitch point-to-point circuits,
            597–603
        hub-and-spoke Ethernet circuits,
            607–611
        MAC table information retrieval, 621
        multicard EtherSwitch manual cross-
            connect, 615–618
        ports, 596–597
        single-card EtherSwitch manual cross-
            connect, 611–615
        SPR (Shared Packet Ring), 603–606
        STP (Spanning Tree Protocol) activation,
            620–621
        VLAN membership, 618–620
    G-Series provisioning, 556, 585–595
        EtherSwitch circuits, 589–592
        G1K-4 Ethernet ports, 587–589
        manual cross-connect, 592–595
        RMON alarm thresholds, creating, 622–623
Ethernet over SONET, 557
    E-Series provisioning, 556, 565–584
        E-100 ports, 566–567
        EtherSwitch point-to-point circuits,
            567–571
        EtherSwitch SPR (switched packet ring),
            571–574
        hub-and-spoke Ethernet circuits,
            574–577
        MAC table information retrieval, 584
        multicard EtherSwitch manual
            cross-connect, 579–581
        RMON alarm thresholds, 584
        single-card EtherSwitch manual
            cross-connect, 577–579
        Spanning Tree Protocol activation,
            582–584
        VLAN membership, 581–582
    G-Series provisioning, 556–565
        G1K-4 manual cross-connect
            provisioning, 563–565
        G1K-4 point-to-point circuit provisioning,
            561–563
        G1K-4 port provisioning, 560–561
        point-to-point circuits, 559

EtherSwitch, 568
    point-to-point circuits, provisioning, 567–571, 597–603
    single-card EtherSwitch manual cross-connect, provisioning, 577–579
    SPR (shared packet ring), provisioning, 571–574
Exchange Carriers Standards Association (ECSA), 4
extended superframe (ESF), 171
external cavity tunable lasers, 109
external modulation, 112–113
external timing, SDH (Synchronous Digital Hierarchy), 212
external tunable lasers, 109
extra traffic, 194, 254
extrinsic attenuation, 66

F

F1 bytes (SDH), 226
F2 bytes (SDH), 231
F3 bytes (SDH), 232
Fabry Perot cavity filters, optical multiplexers, 117
fairness algorithm, RPRs (resilient packet rings), 292
FALM (frame alarm), E1 signal format, 38
far-end block error (FEBE), 159, 237
FAS (frame alignment signal), E1 signal format, 34–35, 38
Fast Ethernet, interface configuration, ML-Series cards, 634
FC connectors, fiber-optic cable termination, 73
FDM (frequency-division multiplexing), 19
FEBE (far-end block error), 159
FEC (forward error correction), WDM (wavelength-division multiplexing), 128–130
fiber Bragg grating, optical multiplexers, 115
fiber diversity, SDH (Synchronous Digital Hierarchy), 250–252
fiber infrastructure, design strategies, 698
fiber plant analysis, network design, 706–709
fiber routing
    SDH, 250–252
    SONET, 184–186

fiber-optics
    applications, 50
    attenuation, 64–67
    cables, 48
        bend radius, 76–77
        benefits of, 4
        construction, 54–55
        glass fiber-optic cables, 55
        jackets, 56
        minimum bend radius, 76
        multifiber fiber-optic cables, 56–58
        PCS (plastic-clad silica) fiber-optic cables, 56
        plastic fiber-optic cables, 55
        propagation modes, 58–63
        splicing, 75–76
        submarine cable systems, 77
        tensile loading, 76–77
        termination, 48, 73–75
    chromatic dispersion, 66–67
    communications systems, 78
        receivers, 81–83
        transmitters, 78–80
    cross-phase modulation (XPM), 69
    fiber span analysis, 83
        dynamic range, 84
        margin calculations, 84–86
        MMF span analysis, 86–88
        power budget, 84–86
        receiver sensitivity, 84
        SMF span analysis, 88–89
        transmitter launch power, 83–84
    fiber-span analysis, 48
    four-wave mixing (FWM), 69–70
    history of, 49–50
    interference, 64
    ITU-T G.651 (multimode fiber with 50-micron fiber), 71
    ITU-T G.652 (nondispersion-shifted fiber), 71
    ITU-T G.653 (dispersion-shifter fiber), 71–72
    ITU-T G.654 (1550-nm loss-minimized fiber), 72
    ITU-T G.655 (nonzero dispersion shifted fiber), 72
    loose buffer cable plants, 76
    optical signal-to-noise ratio (OSNR), 68

performance considerations, 53
physics, 51–53
polarization mode dispersion (PMD), 67–68
propagation, 48
self-phase modulation (SPM), 69
splicing, 48
stimulated Brillouin scattering (SBS), 70
stimulated Raman scattering (SRS), 70
tight buffer cable plants, 76
types, 48
fibers, nodes, connecting to, 495
fiber-span analysis, 48
Fibre Channel, 3
FICON (Fibre Connectivity), 3
filtering analog-to-digital conversions, 22
Force command, 389
four-fiber BLSRs, 202–204
    upgrading to, 400–402
four-fiber MS-SPRing, upgrading to, 497–499
four-fiber rings
    SONET, 191–193
    two-fiber rings, compared, 257–259
four-way mixing (FWM), fiber-optics, 69–70
frame alarm (FALM), E1 signal format, 38
frame alignment signal (FAS). *See* FAS (frame alignment signal)
Frame Relay, SONET/SDH, 3
frame-switched data networks, SONET/SDH, 3
framing
    SDH (Synchronous Digital Hierarchy), 208, 220–221
        higher-level framing, 208, 239–241
        pointers, 223
        STM-1 frames, 221–222
        tributaries, 222–223
    SONET, 140, 148–149
        STS-N framing, 151–153
frequencies, E1 signal format, 38
frequency slicers, optical multiplexers, 120
frequency-division multiplexing (FDM). *See* FDM (frequency-division multiplexing)

G

G1 bytes (SDH), 231
G1K-4 Ethernet cards, 319, 330, 557
    Ethernet over SDH, 587–589
        EtherSwitch circuit provisioning, 589–592
        manual cross-connect, 592–595
        manual cross-connect provisioning, 592
    GBIC (gigabit interface converter), 558
    manual cross-connect provisioning, 563–565
    point-to-point circuit provisioning, 561–563
    port provisioning, 560–561
gateways
    CTC (Cisco Transport Console), 377–378, 470
    ONS 15454 gateways, enabling, 377
    SDH gateways, enabling, 468–470
GE (Gigabit Ethernet), 3, 274–275
    interface configuration, ML-Series cards, 635
gigabit interface converter (GBIC), G1K-4 cards, 558
glass fiber-optic cables, 55
glass-clad fibers, invention of, 49
group velocity dispersion (GVD). *See* chromatic dispersion
groups
    protection groups
        creating, 367–369
        deleting, 369
        editing, 369
        provisioning, 366–369, 457–461
    VTs (virtual tributaries), 166–171
G-Series provisioning
    Ethernet over SDH, 556, 585–595
    Ethernet over SONET, 556–565
        G1K-4 manual cross-connect provisioning, 563–565
        G1K-4 point-to-point circuit provisioning, 561–563
        G1K-4 port provisioning, 560–561
        point-to-point circuits, 559
GVD (group velocity dispersion), 134

H

H4 bytes (SDH), 231
Hansell, Clarence W., 49
hardware-ready cards, 413, 503
high priority (Class A) class of service, RPRs (resilient packet rings), 291

high-density polyethylene (HDPE) jackets, fiber-optic cables, 57
higher-level framing, SDH (Synchronous Digital Hierarchy), 208, 239–241
Hopkins, Harold H., 49
hub topologies
    SDH, 247–248
    SONET, 182
hub-and-spoke Ethernet circuit provisioning, 574–577
    Ethernet over SDH, 607–611
hybrid amplifiers, 124

I

IDTs (integrated digital terminals), 179
IEEE 802.1Q tunneling, ML-Series cards, 656–661
ILECs (incumbent local-exchange carriers), 307
in-band FEC, WDM (wavelength-division multiplexing), 128
incumbent local-exchange carriers (ILECs), 307
inital provisioning tasks
    SDH, 447
        basic node information, 449–450
        CTC (Cisco Transport Controller), 447–449
        network information setup, 450–452
        security, 452–457
        users, 452–457
    SONET, 357
        basic node information, 359–360
        CTC Cisco Transport Controller), 357–359
        IP network setup, 361–362
        SDH, 446
        security, 362–366
        users, 362–366
insertion loss, ONS 15216, 726
inside-plant cables, 57
installation
    BLSR trunk cards, 394–395
    MS-SPRing trunk cards, 486–488
    trunk cards, 479–481
    UPSR trunk cards, 386
integrated digital loop carrier (IDLC), 245
integrated digital terminals (IDTs), 179, 245
integrated platforms, benefits of, 4
Integrated Services Digital Network (ISDN). *See* ISDN (Integrated Services Digital Network)
integration, TDM signals, SONET, 141–143
inter-exchange carriers (IXCs), interoperability standards, 4
interference, fiber-optics, 64
interleavers, optical multiplexers, 120
interleaving-filter method, WDM (wavelength division multiplexing), 98–99
internal timing
    SDH (Synchronous Digital Hierarchy), 212
        setting up, 464–466
    SONET, 373–374
International Telecommunication Union Telecommunication Standardization Sector (ITU-T), 4
Internet Protocol Security (IPSec), 275
intersymbol interference (ISI), 134
intial provisioning tasks, 356
intrinsic attenuation, 64
    raylight scattering, 65
inventories, nodes, 356
    SDH, 466–467
inventories, ONS 15454 nodes, 374–375
IOS command modes, ML-Series cards, 632–633
IP (Internet Protocol), 556
    configuration, SDH (Synchronous Digital Hierarchy), 450–452
    initial provisioning tasks, 361–362
    ML-series cards
        accessing, 628–632
        bridge group forwarding, 650–651
        bridge interface configuration, 643–645
        configuration, 672–673
        Fast Ethernet interface configuration, 634
        Gigabit Ethernet interface configuration, 635
        IEEE 802.1Q tunneling, 656–661
        IOS command modes, 632–633
        Layer 2 tunneling, 656–658, 661–663
        link aggregation, 663–664
        POS channels, 664–667
        POS interface configuration, 636–643
        provisioning, 624-682
        QoS, 673–674

RPR (resilient packet ring), 674–691
STP configuration, 645–650
switching database manager, 691–692
VACT (virtual concatenation), 691
VLAN configuration, 651–655
VRF lite, 667–669, 671
OAM&P, SDH nodes, 446, 467–478
ONS nodes for OAM&P, 356, 375–385
IP subnets, CTC (Cisco Transport Console) ONS nodes, 375–376
IPSec (Internet Protocol Security), 275
ISDN (Integrated Services Digital Network), 18, 39
 BRI (basic rate interface), 39–40
 call setup, 45
 layer 1, 41–42
 layer 2, 42–44
 layer 3, 44–45
 link layer establishment, 44
 PRI (primary rate interface), 40–41
ISI (intersymbol interference), 134
isolators, optical multiplexers, 119
ITU grid, 92
 WDM (wavelength-division multiplexing) systems, 102–104
ITU-T (Internation Telecommunication Union Telecommunication Standardization Sector), 4
ITU-T G.651 (multimode fiber with 50-micron fiber), 71
ITU-T G.652 (nondispersion-shifted fiber), 71
ITU-T G.652.C (low water peak nondispersion-shifted fiber), 71
ITU-T G.653 (dispersion-shifter fiber), 71–72
ITU-T G.654 (1550-nm loss-minimized fiber), 72
ITU-T G.655 (nonzero dispersion shifted fiber), 72
IXCs (inter-exchange carriers), interoperability standards, 4

J-K

J0 bytes (SDH), 154, 225
J1 path traces, creating, 440–442
J1 user-programmable bytes (SDH), 230
J2 bytes (SDH), 233
jackets, fiber-optic cables, 56
jitter, 212

K1 bytes (SDH), 227
K2 bytes (SDH), 227
K3 bytes, remapping
 BLSRs, 397
 SDH, 490
Kao, Charles K., 49
Keck, Donald, 50

L

lasing thresholds, 106
Layer 1 (ISDN), 41–42
Layer 2 (ISDN), 42–44
Layer 2 tunneling, ML-Series cards, 656–663
Layer 3 (ISDN), 44–45
Layer 3 (OSI model), 44
layers
 RPRs, managing, 303
 SDH, 208
  multiplex section, 215
  path layers, 215
  photonic layer, 216
  regenerator section layer, 216
 SONET, 140, 146
  line layer, 147
  path layer, 146–147
  photonic layer, 148
  section layer, 147
L-band (long wavelength) optical frequency bands, WDM systems, 96
LC connectors, fiber-optic cable termination, 74
LDCCs (line data communications channels), 205, 442
leased lines
 carrier perspective, 26
 customer perspective, 25
LEDs, 79–80
legacy SONET/SDH, 6–7
Length (1 octet) field (Q.931 header), 44
line data communications channel (LDCC), 205
line DCCs (LDCCs), 442
line layer (SONET), 140, 147
line overhead (LOH), SONET, 155–159
line switching, SONET, 193–194
line timing, SDH, 212
linear add/drop architectures, 181

ML-Series cards **795**

linear ADMs, 419
   BLSRs, converting to, 423–427
   configuration
      SDH, 446, 508–517
      SONET, 356, 418–427
   creating, 419–420, 509–510
   MS-SPRings, converting to, 514–517
   SNCP rings, converting to, 510–513
   UPSRs, converting to, 420–423
linear characteristics, fiber-optics, 64–68
line-terminating equipment (LTE), 147
link aggregation, ML-Series cards, 663–664
link budgets, span loss, 83
link layer establishment, ISDN, 44
Lockout command, 389
LOF (loss-of-frame) alignment
   SDH, 237
   SONET, 163
LOFS (loss of frame seconds), E1 signal format, 38
logical design, networks, 712–716
LOH (line overhead), SONET, 155–159
loop timing, 145, 212
loose buffer cables, cable plants, compared, 76
LOP (loss-of-pointer) alignment (SDH), 237
LOP (loss-of-pointer) state (SONET), 163
LOS (loss of signal) alarm, superframes, 32, 163, 237
LOSS (loss of signal seconds), E1 signal format, 38
loss of frame seconds (LOFS), E1 signal format, 38
low priority (Class C) class of service, RPRs (resilient packet rings), 291
low water peak nondispersion-shifted fiber (ITU-T G.652.C), 71
LSS (loss of sequence synchronization)
   SDH, 238
   SONET, 165
LTE (line-terminating equipment), 147

M

M1 bytes (SDH), 228
MAC (Media Access Control)
   information
      retrieval, 621
      retrieving, 584
   RPRs (resilient packet rings), 294–303

MacChesney, John, 50
Mach-Zehnder interferometers, 118–119
Mandatory and Optional Information Elements (variable length) field (Q.931 header), 45
MANs (metropolitan area networks), building, technology options, 275
manual cross-connect provisioning, 563–565
manually routed circuits, creating, 433–436
margin calculations, span analysis, 84–86
master clocks, 145
matching nodes, 177
   SPRs, 286
material absorption, intrinsic attenuation, 64
Maurer, Robert, 50
mechanical tunable lasers, 108–109
medium priority (Class B) class of service, RPRs (resilient packet rings), 291
medium-density polyethylene (MDPE) jackets, fiber-optic cables, 57
meshed topologies
   SDH, 249–250
   SONET, 184
Message Type (1 octet) field (Q.931 header), 45
MetroPlanner DWDM optical design software, 334
MFAL (multiframe alarm), E1 signal format, 38
MFAS (multiframe alignment signal), E1 signal format, 35–38
MIC-C/T/P FMECs (front-mounted electrical connections), 325
minimum bend radius, fiber-optic cables, 76
ML1000-2 Ethernet cards, 319
ML1000-2 Gigabit Ethernet cards, 330
ML100T-12 Ethernet cards, 319, 330
ML-Series cards
   12-port cards, 721
   accessing, 628–632
   bridge group forwarding, 650–651
   bridge interface configuration, 643–645
   Fast Ethernet interface configuration, 634
   Gigabit Ethernet interface configuration, 635
   IEEE 802.1Q tunneling, 656–661
   IOS command modes, 632–633
   IP protocols, 672–673
   Layer 2 tunneling, 656–663
   link aggregation, 663–664
   POS channels, 664–667
   POS interface configuration, 636–643
   provisioning, 624–680, 682

QoS, 673–674
RPR (resilient packet ring), 674–691
SPRs, 283–285
STP configuration, 645–650
switching database manager, 691–692
VACT (virtual concatenation), 691
VLAN configuration, 651–655
VRF lite, 667–671
MMF span alaysis, fiber-optics, 86–88
modulators, WDM transmitters, 111–113
monitor circuits, creating, 437–438
MR-L1-xx.x/MR-1-xx.x service interface cards, 335
MRP-L1-xx.x/MRP-1-xx.x service interface cards, 335
MSOH (Multiplex Section Overhead), SDH, 226–229
MSP (multiplex section protection), 6
MSPP (multiservice provisioning platforms), 4, 8-12, 274-275
    bandwidth efficiency, 10–11
    Ethernet encapsulation, 11–12
    provisioning, 12
    QoS (quality of service), 11
    signaling, 12
    SDH, 245-246, 446
    SONET, 180, 356-357
MS-SPRing (multiplex section-shared protection rings), 6, 264–271
    configuration
        SDH, 485–502
        SONET, 446
    DCC terminations, creating, 488–489
    linear ADMs, converting from, 514–517
    nodes
        adding, 492–496
        removing, 496–497
    ports, activating, 488–489
    provisioning, SDH, 490–492
    traffic, switching, 493
    trunk cards
        installing, 486–488
        moving, 500–502
    upgrading, 497–499
MSTP (Multiservice Transport Platform), 694
MTP/MPO connectors, fiber-optic cable termination, 75
MT-RJ connectors, fiber-optic cable termination, 74
MTSE (Miltiplex Section-Terminating Equipment), 215

mu-law encoding, analog-to-digital conversions, 24
multicard EtherSwitch manual cross-connect, provisioning, 579–581, 615–618
multifiber fiber-optic cables, 56–58
multiframe alarm (MFAL), E1 signal format, 38
multiframe alignment signal (MFAS) distant alarm, E1 signal format, 38
multiframing formats, DS framing, 29–32
multimode fiber with 50-micron core (ITU-T G.651), 71
multimode graded index, fiber-optic cables, 62–63
multimode step index, fiber-optic cables, 59–60
multiple drops, unidirection circuits, creating for, 436–437
multiplex section (SDH), 215
multiplex section protection (MSP), 6
multiplex section switching, path switching, compared, 259–260
multiplex section-shared protection rings (MS-SPRing), 264–271
Multiplex Section-Terminating Equipment (MSTE). See MSTE (Multiplex Section-Terminating Equipment)
multiplexed section protection rings (MS-SPRing). See MS-SPRing (multiplexed section protection rings)
multiplexing
    ADM (add/drop multiplexers), 94, 176–177
    characteristics, 92
    CWDM (coarse wavelength-division multiplexing), 100
    DWDM (dense wavelength-division multiplexing), 2, 13-16, 101, 306
    FDM (frequency-division multiplexing), 19
    linear ADMs
        configuration, 418–427, 446
        configurations, 508–517
    MSP (multiplex section protection), 6
    MS-SPRing (multiplexed section protection ring), 6
    MSTE (multiplex section-terminating equipment), 215
    optical multiplexers, 113–121
        acousto optical tunable filters, 117–118
        arrayed waveguides, 115–116
        circulators, 119
        couplers, 119
        Fabry Perot cavity filters, 117

fiber Bragg grating, 115
frequency slicers, 120
interleavers, 120
isolators, 119
Mach-Zehnder interferometers, 118–119
periodic filters, 120
TFFs (thin film filters), 114–115
SDH (Synchronous Digital Hierarchy), 216–220
SONET, 140, 174–175
TDM (time-division multiplexing), 2, 18-20, 694
    analog signal processing, 20–24
    circuit-switched networks, 26–27
    E-carrier system, 32–39
    network elements, 46–47
    schematics, 19
    SDH integration, 209–214
    SONET/SDH, 3
    statistical TDM, 20
    synchronous TDM, 20
    T-carrier system, 27–32
TM (terminal multiplexers)
    SONET, 176
WDM (wavelength-division multiplexing), 92-96, 104–105
    amplifiers, 121–125
    bidirectional WDM, 96–99
    channel spacing, 99
    characteristics, 127–133
    compensation, 133–134
    dispersion, 133–138
    ITU grid, 102–104
    multiplexers, 113–121
    need for, 93
    optical frequency bands, 96
    optical multiplexers, 113–121
    optical-fiber media, 125
    receivers, 125–127
    schematic, 94
    systems, 94
    transmission impairments, 92, 127–133
    transmitters, 106–113
    unidirectional WDM, 96–97
multiservice metro optical SONET/SDH networks, case study, 699–722
    BLSR/MS-SPRing implementation, 723
    capacity planning, 701–706
    circuit provisioning, 722
    delay analysis, 708–710
    fiber plant analysis, 706–709
    logical design, 712–716
    physical design, 716–718
    requirements analysis, 701
    technology analysis, 710–712
    upscaling, 723–730
multiservice provisioning platforms (MSPPs). *See* MSPPs (multiservice provisioning platforms)
Multiservice Transport Platform (MSTP). *See* MSTP (Multiservice Transport Platform)

N

N1 bytes (SDH), 232
N2 bytes (SDH), 233
NE Explorer, CTM (Cisco Transport Manager), 351–353
negative stuffing, 224
negative timing justification, 156
NEs (network elements), 140
    SDH, 208
        add/drop multiplexers, 242–243
        BDCS (broadband digital cross-connect), 243–244
        digital loop carriers, 245
        MSPPs (multiservice provisioning platforms), 245–246
        regenerators, 241
        WDCS (wideband digital cross-connect), 244–245
    SONET, 175
        ADM (add/drop multiplexer), 176–177
        BDCS (broadband digital cross-connect), 178
        digital loop carriers, 179–180
        MSPPs (multiservice provisioning platforms), 180
        regenerator, 175
        terminal multiplexers, 241–242
        TM (terminal multiplexer), 176
        WDCS (wideband digital cross-connect), 178–179
    TDM (time-division multiplexing), 18, 46
        channel banks, 47
        cross-connect systems, 47

CSU/DSU, 46
DACS, 47
repeaters, 46
network management
    SDH, 208, 271–272
    SONET, 205–206
Network Map, CTM (Cisco Transport Manager), 350–351
network view, CTC (Cisco Transport Controller), 340–342
networks
    design
        capacity planning, 701–706
        circuit provisioning, 722
        delay analysis, 708–710
        fiber plant analysis, 706–709
        logical design, 712–716
        physical design, 716–718
        requirements analysis, 701
        technology analysis, 710–712
    design strategies, 696
        customer demographics, 697
        customer service levels, 697–698
        customer service requirements, 697
        fiber infrastructure, 698
        prarmeters, 696
        technology selections, 698–699
        vendor selections, 699
new data flags, 224
node view, CTC (Cisco Transport Controller), 338–340, 348-350
nodes
    BLSR
        adding, 404–406
        removing, 407–409
    failures, BLSRs, 201
    fibers, connecting to, 495
    initial provisioning tasks, 359–360
    inventories, 356, 374–375
    IP networking
        for OAM&P, 356
        OAM&P, 375–385
    MS-SPRing
        adding, 492–496
        removing, 496–497
    multiple subtending rings, 413

SDH
    basic information setup, 449–450
    inventories, 466–467
    IP networking for OAM&P, 446, 467–478
SNCP
    adding, 483–484
    removing, 484–485
static routes, 378–380, 470–473
UPSR, adding/removing, 388–392
nondispersion-shifted fiber (ITU-T G.652), 71
nonlinear characteristics, fiber-optics
    cross-phase modulation (XPM), 69
    four-wave mixing (FWM), 69–70
    self-phase modulation (SPM), 69
    stimualted Brillouin scattering (SBS), 70
    stimualted Raman scattering (SRS), 70
nonzero dispersion-shifted fiber (ITU-T G.655), 72
normalized bandwidth, compounded growth matrix, 723
normalized bandwidth calculations, 705
northbound traffic, 308

O

O'Brien, Brian, 49
OAM functions, 303
OAM&P (operation, administration, maintenance, and provisioning), 375
    IP networking, SDH nodes, 446, 467–478
    ONS nodes, IP networking, 356, 375–385
O-band (original) optical frquency band, WDM systems, 96
OC12 IR/STM4 SH 1310 optical cards, 316
OC12 IR/STM4 SH 1310-4 optical cards, 316
OC12 IR/STM4-4 1310 optical cards, 316
OC12 LR/STM4 LH 1310 optical cards, 316
OC12 LR/STM4 LH 1550 optical cards, 316
OC192 IR/STM64 SH 1550 optical cards, 317
OC192 LR/STM64 LH 1550 optical cards, 317
OC192 LR/STM64 LH ITU DWDM optical cards, 318
OC192 SR/STM64 IO 1310 optical cards, 317
OC-192/STM64 1-Port cards, 720
OC-192/STM-64 ITU wavelengths, 724
OC3 IR/STM1 SH 1310-4 optical cards, 315
OC3 IR/STM1 SH 1310-8 cards, 720

OC3 IR/STM1 SH 1310-8 optical cards, 315
OC48 ELR/STM16 EH 100 GHz DWDM optical cards, 317
OC48 IR 1310 optical cards, 316
OC48 IR/STM16 SH AS 1310 optical cards, 316
OC48 LR 1550 optical cards, 317
OC48 LR/STM16 LH AS 1550 optical cards, 317
OC-N line rates, 149
ONS
    craft interface, initial provisioning tasks, 357–359
    nodes, initial provisioning tasks, 359–360
ONS 15100 series products, 308
ONS 15200 series products, 308
ONS 15300 series products, 309
ONS 15400 series products, 311-336
ONS 15454
    alarm interface controller (AIC-I), 314
    card protection, 320–321, 331-332
    circuits, searching for, 438–439
    cross-connect cards, 313, 323-324
    electrical cards, 314–315
    Ethernet cards, 318–319, 329-330
    FMECs (front-mounted electrical connections) cards, 325–326
    gateways, enabling, 377
    MSPP, 311–332
    MSTP, 332–336
        optical transmission elements, 335–336
        scalability, 729
        service interface cards, 335
    nodes
        CTC (Cisco Transport Console), 375–376
        inventories, 356, 374–375
        IP networking for OAM&P, 356, 375–385
        multiple subtending rings, 413
        static routes, 378–380
    optical cards, 315–317, 327–329
    OSPF configuration, 380–385, 474–478
    protection groups, provisioning, 366–369
    SAN (storage area networks) cards, 319–320, 330–331
    SDH security levels, 453
    SDH user idle times, 455
    timing, 356, 370–374, 461–466
    Timing and Control (TCC2) cards, 313, 322–323
ONS 15500 series products, 309
ONS 15600 series products, 310

ONS 15800 series products, 310
OOF (out-of-frame) alarm, superframes, 31
OOF (out-of-frame) alignment
    SDH, 237
    SONET, 163
operation, administration, maintenance, and provisioning (OAM&P). See OAM&P (operation, administration, maintenance, and provisioning)
operations support system (OSS), 308
OPT-BST cards, 335
optical card protection, SDH, 331
optical frequency bands, WDM (wavelength division multiplexing), 96
optical interface layers, SONET, 140, 146-148
optical multiplexers, 113–121
    acousto optical tunable filters, 117–118
    arrayed waveguides, 115–116
    couplers, 119
    Fabry Perot cavity filters, 117
    fiber Bragg grating, 115
    frequency slicers, 120
    interleavers, 120
    isolators, 119
    Mach-Zehnder interferometers, 118–119
    periodic filters, 120
    TFFs (thin film filters), 114–115
optical protection, ONS 15454 cards, 320
optical signals
    SDH, 213–214
    SONET, 140, 144–145
optical signal-to-noise ratio, WDM (wavelength-division multiplexing), 130–133
optical signal-to-noise ratio (OSNR), fiber-optics, 68
optical transmission elements, ONS 15454 MSTP optical transmission elements, 335–336
optical-fiber media, WDM (wavelength-division multiplexing), 125
optical-power measurements, fiber-optics, 53
OPT-PRE cards, 335
OSC-CSM cards, 335
OSCM cards, 335
OSPF configuration, 380–385, 474–478
OSS (operations supprt system), 308
out-of-band FEC, WDM (wavelength-division multiplexing), 129

P

PAM (Pulse Amplitude Modulation), 23
parameters, network design strategies, 696
path layers (SONET), 140, 146–147
path layers (SDH), 215
path overhead (POH), SONET, 159–162
Path Protected Mesh Networking (PPMN), 250
path switching
    multiplex section switching, compared, 259–260
    SONET, 193–194
path traces, creating, 440–442
path-terminating equipment (PTE). *See* PTE (path-terminating equipment)
payload pointers (SDH), 223
PCA (protection channel access), 194, 332
PCM (pulse code modulation), 19, 23
PCS (plastic-clad silica) fiber-optic cables, 56
PDH (plesiochronous digital hierarchy), 143, 210–211
periodic filters, optical multiplexers, 120
permanent virtual circuits (PVCs), 7
phase variation, 145, 212
photo-elastic effect, acoustic waves, 117
photonic layer (SDH), 216
photonic layer (SONET), 140, 148
photon-phonon interaction, acoustic waves, 117
Photophone, invention of, 49
physical design, network design, 716–718, 722
physics, fiber-optics, 51–53
PINs (positive intrinsic negatives), 82, 126
plant ribbon-cable systems, 56
plastic fiber-optic cables, 55-56
plesiochronous digital hierarchy (PDH). *See* PDH (plesiochronous digital hierarchy)
POH (path overhead)
    SDH, 229–233
    SONET, 159–162
    VTs (virtual tributaries), 173–174
point of presence (POP) nodes, 3–5
pointers
    SDH, 223
    SONET, 156
point-to-multipoint topologies
    SDH, 247
    SONET, 181–182
point-to-point circuits, 561–563
    Ethernet over SDH, 597-603
    EtherSwitch point-to-point circuits, provisioning, 567–571
    G-Series provisioning, 559
point-to-point topologies
    SDH, 246–247
    SONET, 181
polarization dependent loss (PDL), 68
polarization mode dispersion, WDM (wavelength-division multiplexing), 136–138
polarization mode dispersion (PMD), fiber-optics, 67–68
polarization mode dispersion compensation, WDM (wavelength-division multiplexing), 138
POP (point of presence) nodes, 3–5
population inversion, 106
ports
    BLSR, enabling, 396–397
    channels, 663
    enabling, 369
    E-Series, provisioning, 596–597
    MS-SPRing, activating, 488–489
    SNCP ports, enabling, 481–482
    UPSRs, enabling, 388
    VLAN membership, provisioning, 581–582
PoS (Packet over SONET), 275
    configuration, ML-Series cards, 664–667
    interface configuration, 636–643
positive intrinsic negatives (PINs), 82
positive stuffing, 224
power budget, span analysis, 84–86
PPMNs (Pat-Protected Mesh Networks), 427-428
PRI (primary rate interface), ISDN, 40–41
primary reference source (PRS), 145
propagation, fiber-optics, 48
propagation modes, fiber-optic cables, 58
    multimode graded index, 62–63
    multimode step index, 59–60
    single-mode step index, 60–62
protection architectures
    SDH, 208
    SONET, 140, 187–189
protection channel access (PCA), 194
protection groups
    creating, 367–369
    deleting, 369
    editing, 369

provisioning, 356, 366–369
    SDH, 457–461
protectional architectures, SDH, 253–255
Protocol Discriminator (1 octet) field (Q.931 header), 44
provisioning
    A-to-Z provisioning, 429
    BLSRs, 397–400
    circuits, 356, 428–444
    E-Series provisioning
        Ethernet over SDH, 556, 595–623
        Ethernet over SONET, 556, 565–584
    G-Series provisioning
        Ethernet over SDH, 556, 585–595
        Ethernet over SONET, 556–565
    initial provisioning tasks, 357
        basic node information setup, 359–360
        CTC (Cisco Transport Controller), 357–359
        IP network setup, 361–362
        security, 362–366
        users, 362–366
    initianl provisioning tasks, 356
    ML-Series provisioning, 624–682
    MS-SPRing, 490–492
    OAM&P (operation, administration, maintenance, and provisioning), 375
        IP networking of ONS nodes, 375–385
    ONS 15454 timing
        SDH, 462–466
        SONET, 37-372
    PPMNs (Pat-Protected Mesh Networks), 427
    protection groups, 356
        SDH, 457–461
        SONET, 366-369
    SDH
        circuits, 518–553
            inital provisioning tasks, 447–457
            MSPP, 446–447
    SONET
        DCC tunnels, 442–444
        MSPP, 357
Proxy ARP, ONS 15454 gateways, enabling, 377
PRS (primary reference source), 145
PSTN (Public Switched Telephone Network), 4
PTE (path-terminating equipment)
    SDH, 215
    SONET, 147

Public Switched Telephone Network (PSTN), 4
pulling tensions, fiber-optic cables, 77
Pulse Amplitude Modulation (PAM), 23
pulse code modulation (PCM), 19, 23
PVCs (permanent virtual circuits), 7

Q-R

Q.931 headers (ISDN Layer 3), fields, 44
QoS (Quality of Service), 11
    ML-Series cards, 673–674
quantization, analog-to-digital conversions, 22–23

radio frequency interference (RFI), analog signals, 21
Raman fiber amplifiers (RFAs), 124
raylight scattering, intrinsic attenuation, 65
RBOCs (Regional Bell Operating Companies), interoperability standards, 4
RDI (remote defect indication)
    SDH, 238
    SONET, 164
RDTs (remote digital terminals), 179
receivers
    fiber-optic communications systems, 81–83
    sensitivity, span analysis, 84
    WDM (wavelength-division multiplexing), 125–127
        avalanche photodiodes, 127
        PIN photodiodes, 126
reception, analog signals, 21
Red CFA (carrier failure alarm), superframes, 32
regenerator section layer (SDH), 216
regenerators
    SDH (Synchronous Digital Hierarchy), 241
    SONET, 175
Regional Bell Operating Companies (RBOCs), interoperability standards, 4
REI (remote error indication)
    SDH, 237
    SONET, 164
remapping K3 bytes, 397, 490
remote digital terminals (RDTs), 179
remote fiber terminals (RFTs), 180, 245
repeaters, 175
    TDM (time-division multiplexing), 46

requirements analysis, network design, 701
rerouting RPRs (resilient packet rings), 293–294
resilient packet ring (RPR). *See* RPR (resilient packet ring)
RFAs (Raman fiber amplifiers), 124
RFI (remote failure indication)
    SDH, 238
    SONET, 164
RFTs (remote fiber terminals), 180
rings
    SDH, 208, 248–271
    SONET, 140, 189–190
        asymmetrical delay, 198
        bidirectional rings, 190–191
        BLSR (bidirectional line-switched rings), 198–205
        DRI (dual-ring interconnect), 194–196
        four-fiber rings, 191–193
        line switching, 193–194
        path switching, 193–194
        two-fiber rings, 191–193
        unidirectional rings, 190–191
        UPSR (unidirectional path switched rings), 196–197
rings/architaectures. *See* ring architectures
BLSRs, configuration, 392–412
MS-SPRing
    adding, 495–496
    DCC terminations, 488–489
    nodes, 492–497
    port activation, 488–489
    provisioning, 490–492
    trunk cards, 500–502
    upgrading, 497–499
SNCP trunk cards, installing, 479–481
subtending rings, 446, 502-508
    configuration, 356, 413–418
UPSRs, configuration, 385–392
RMON alarm thresholds, creating, 584, 622–623
RPR over SONET/SDH, ML-series cards
    accessing, 628–632
    bridge group forwarding, 650–651
    bridge interface configuration, 643–645
    configuration, 674–691
    Fast Ethernet interface configuration, 634
    Gigabit Ethernet interface configuration, 635
    IEEE 802.1Q tunneling, 656–661
    IOS command modes, 632–633
    IP protocols, 672–673
    Layer 2 tunneling, 656–663
    link aggregation, 663–664
    POS channels, 664–667
    POS interface configuration, 636–643
    provisioning, 624–682
    QoS, 673–674
    STP configuration, 645–650
    switching database manager, 691–692
    VACT (virtual concatenation), 691
    VLAN configuration, 651–655
    VRF lite, 667–671
RPRs (resilient packet rings), 274287–288, 290
    bandwidth management, 292–293
    CoS (classes of service), 291
    DPT (Dynamic Packet Transport), 288–290
    fairness algorithm, 292
    MAC (Media Access Control), 294–303
    OAM functions, 303
    rerouting, 293–294
    topology dicovery, 290–291
    traffic protection, 293–294
RSOH (Regenerator Section Overhead), SDH, 225–226
RSTE (Regenerator Section-Terminating Equipment), 216

S

S1 bytes (SDH), 228
sampling oscilloscopes, 127
SANs, case study, 731–732
S-band (short wavelength) optical frequency bands, WDM systems, 96
SC connectors, fiber-optic cable termination, 74
scaling up, optical networks, 723–730
Schultz, Peter, 50
scrambling techniques, 155
SDCC (section data communications channel), 205, 422
SDH (Synchronous Digital Hierarchy), 2-3, 208, 307
    alarms, 208, 235–236
        AIS (alarm indication signal), 237
        B1 errors, 238
        B2 errors, 238

SDH (Synchronous Digital Hierarchy) 803

B3 errors, 238
BIP-2 errors, 238
LOF (loss-of-frame) alignment, 237
LOP (loss-of-pointer) alignment, 237
LOS (loss-of-signal) alarm, 237
LSS (loss of sequence synchronization), 238
OOF (out-of-frame) alignment, 237
RDI (remote defect indication), 238
REI (remote error indication), 237
RFI (remote failure indication), 238
circuits, provisioning, 518–553
electrical signals, 213–214
ESCNP (extended SNCP) mesh networks, 442, 517–518
Ethernet, 274, 556
Ethernet over SDH, 557
    E-Series provisioning, 556, 595–623
    G-Series provisioning, 556, 585–595
framing, 208, 220–221
    higher-level framing, 208, 239–241
    pointers, 223
    STM-1 frames, 221–222
    tributaries, 222–223
gateways, enabling, 468–470
initial provisioning tasks, 447
    basic node information, 449–450
    CTC (Cisco Transport Controller), 447–449
    network information setup, 450–452
    security, 452–457
    users, 452–457
IP, 556
K3 bytes, remapping, 490
layers, 208
    multiplex section, 215
    path layers, 215
    photonic layers, 216
    regenerator section layer, 216
linear ADMs, configurations, 446, 508–517
MSPP (multiservice provisioning platform), 446
MS-SPRing
    configuration, 446, 485–502
    nodes, 492–497
    provisioning, 490–492
    trunk cards, 500–502
    upgrading, 497–499

multiplexing, 216–218
    DS1 signals, 218
    DS2 signals, 219
    DS3 signals, 220
    E1 signals, 218–219
    E3 signals, 219
    E4 signals, 220
multiservice metro optical SONET/SDH case study, 699–722
network elements, 208, 241
    add/drop multiplexers, 242–243
    BDCS (broadband digital cross-connect), 243–244
    digital loop carriers, 245
    MSPPs (multiservice provisioning platforms), 245–246
    regenerators, 241
    terminal multiplexers, 241–242
    WDCS (wideband digital cross-connect), 244–245
network management, 208, 271–272
nodes
    inventories, 466–467
    IP networking for OAM&P, 446–478
    static routes, 470–473
optical signals, 213–214
PDH (plesiochronous digital hierarchy), 210–211
protection architectures, 208, 253–255
protection groups, provisioning, 457–461
PTE (path-terminating equipment), 215
ring architectures, 208, 255–271
rings, subtending, 446, 502–508
SDH MSPP, provisiooning, 447
SNCP (subnetwork connection protection), configuration, 478–485
standards, 4
STM-N (Synchronous Transport Module-N), 213–214
synchronization, 212–213
TDM signals, integration, 208–214
topologies, 208, 246
    fiber diversity, 250–252
    fiber routing, 250–252
    hub topology, 247–248
    meshed topology, 249–250
    point-to-multipoint topology, 247

point-to-point topology, 246–247
ring topology, 248–249
transport overhead, 224
AU pointers, 226
MSOH (Multiplex Section Overhead), 226–229
POH (path overhead), 229–233
RSOH (Regenerator Section Overhead), 225–226
SONET interworking, 233–235
secondary circuits, traffic monitors, creating, 437–438
section data communications channel (SDCC), 205, 442
section layer (SONET), 140, 147
section overhead (SOH), SONET, 154–155
section-terminating equipment (STE), 147
security, initial provisioning tasks
SDH, 452-457
SONET, 362-366
self-phase modulation (SPM), fiber-optics, 69
sensitivity curves, receivers, 82
service affecting, 409
service interface cards, ONS 15454 MSTP service interface cards, 335
service profile IDs (SPIDs), 44
SES (severely errored second), E1 signal format, 38
SF (superframe), 30–32
SH 1310-4 electrical cards, 328
shared packet ring (SPR). *See* SPR (shared paket ring)
shared protection, 201
signals, 12
analog signal processing, 18
analog-to-digital conversions, 21–24
CAS (channel-associated signaling), 26
CCS (common channel signaling), 27
electrical signals, 140, 144–145, 213–214
optical signals, 140, 144–145, 213–214
PAM (Pulse Amplitude Modulation), 23
PCM (Pulse Code Modulation), 23
TDM signals
SDH integration, 209–214
SONET, 140–143
SONET integration, 140
signal-to-noise ratio, WDM (wavelength-division multiplexing), 130–133

single points of failure, eliminating, 4
single-card EtherSwitch manual cross-connect, provisioning, 577–579
Ethernet over SDH, 611–615
single-mode dual-step index, fiber-optic cables, 60–62
SMF span alaysis, fiber-optics, 88–89
SNCP (subnetwork connection protection), 6, 262, 446
circuits, removing, 485
configuration, SDH, 478–485
ESNCP (Extended SNCP) mesh networks, 446
linear ADMs. converting from, 510–513
nodes
adding, 483–484
removing, 484–485
ports, enabling, 481–482
ring traffic, switching, 482–483
subtending rings, 504
trunk cards, installing, 479–481
Snitzer, Elias, 49
SOH (section overhead), SONET, 154–155
SONET (Synchronous Optical Network), 2–3, 307
alarms, 140, 163
AIS (alarm indication signal), 164
B1 errors, 164
B2 errors, 165
B3 errors, 165
BIP-2 errors, 165
LOF (loss-of-frame) alignment, 163
LOP (loss-of-pointer) state, 163
LOS (loss-of-signal) alarm, 163
LSS (loss of sequence synchronization), 165
OOF (out-of-frame) alignment, 163
RDI (remote defect indication), 164
REI (remote error indication), 164
RFI (remote failure indication), 164
C2 byte values, 161
DCC tunnels, provisioning, 442–444
electrical signals, 140, 144–145
Ethernet, 556
Ethernet over SONET, 274, 557
E-Series provisioning, 556, 565–584
G-Series provisioning, 556–565
framing, 140, 148–149
STS-N framing, 151–153

initial provisioning taskes, 356
IP (Internet Protocol), 556
LTE (line-terminating equipment), 147
MSPP (multiservice provisioning platform), 356-357
multiplexing, 140, 174–175
multiservice metro optical SONET, case study, 699–722
NEs (network elements), 140, 175
    ADM (add/drop multiplexer), 176–177
    BDCS (broadband digital cross-connect), 178
    digital loop carriers, 179–180
    MSPPs (multiservice provisioning platforms), 180
    regenerator, 175
    TM (terminal multiplexer), 176
    WDCS (wideband digital cross-connect), 178–179
network management, 205–206
OC-N line rates, 149
optical interface layers, 140, 146
    line layer, 147
    path layer, 146–147
    photonic layer, 148
    section layer, 147
optical signals, 140, 144–145
protection architectures, 140, 187
    APS (automatic protection switching), 187–189
PTE (path-terminating equipment), 147
ring architecture, 189–190
    asymmetrical delay, 198
    bidirectional rings, 190–191
    BLSR (bidirectional line-switched rings), 198–205
    DRI (dual-ring interconnect), 194–196
    four-fiber rings, 191–193
    line switching, 193–194
    path switching, 193–194
    two-fiber rings, 191–193
    unidirectional rings, 190–191
    UPSR (unidirectional path switched rings), 196–197
ring architectures, 140
SDH (Synchronous Digital Overhead), 233–235
SPE, 150–151
standards, 4
STE (section-terminating equipment), 147
synchronization, 144–145
TDM signals
    integration, 141–143
    integration of, 140
TNM (Telecommunications Network Management) model, 140
topologies, 140, 180
    diversity, 184–186
    fiber routing, 184–186
    hub topologies, 182
    meshed topologies, 184
    point-to-multipoint topologies, 181–182
    point-to-point topologies, 181
    ring topologies, 183
transport overhead, 153
    LOH (line overhead), 155–159
    POH (path overhead), 159–162
    SOH (section overhead), 154–155
transpot overhead, 140
VTs (virtual tributaries), 165–166
    groups, 166–171
    POH (path overhead), 173–174
    superframes, 171–173
SONET/SDH (Synchronous Optical Network/Synchronous Digital Hierarchy), 2-7
    DACS (digital access cross-connect systems), 3
    DWDM (dense wavelength-division multiplexing), 13–16
    legacy SONET/SDH, 6–7
    MSPPs (multiservice provisioning platforms), 4, 8–12
        bandwidth efficiency, 10–11
        Ethernet encapsulation, 11–12
        provisioning, 12
        QoS (quality of service), 11
        signaling, 12
    Multiservice SONET/SDH, 3
southbound traffic, 308
span analysis, fiber-optics, 83
    dynamic range, 84
    margin calculations, 84–86
    MMF span analysis, 86–88
    power budget, 84–86
    receiver sensitivity, 84

SMF span analysis, 88–89
transmitter launch power, 83–84
Spanning Tree Protocol (STP). *See* STP (Spanning Tree Protocol)
spatial reuse, RPRs (resilient packet rings), 292–293
SPE, 150–151
SPIDs (service profile IDs), 44
splicing fiber-optic cabling, 75–76
SPRs (shared packet rings), 274, 277
    bridges, 279
    design constraints, 281–282
        E-Series card ports, 282
        matching nodes, 286
        ML-Series card ports, 283–285
    double tagging, 279
    Ethernet over SDH, 603–606
    EtherSwitch provisioning, 571–574
    STP model, 279
    VLAN identifiers, 279
ss bits (SDH), 234
SSM (Synchronization Status Messaging), 370
ST connectors, fiber-optic cable termination, 74
static routes
    CTCs, 473
        multiple CTCs, 380–381
        nodes, 470–473
static routes, ONS nodes, 378–380
statistical TDM, 20
STE (section-terminating equipment), 147
stimulated Brillouin scattering (SBS), fiber-optics, 70
stimulated emission, 106
stimulated Raman scattering (SRS), fiber-optics, 70
STM-1 frames (SDH), 221–222
STM1 SH 1310 optical cards, 328
STM16 EH 100-GHz DWDM electrical cards, 328
STM16 LH AS 1550 electrical cards, 328
STM16 SH AS 1310 electrical cards, 328
STM1E 1, FMECs (front-mounted electrical connections), 326
STM1E-12 electrical cards, 327
STM1SH 1310-8 electrical cards, 328
STM4 LH 1310 electrical cards, 328
STM4 LH 1550 electrical cards, 328
STM4 SH 1310 electrical cards, 328
STM64 LH 1550 electrical cards, 329
STM64 LH ITU DWDM electrical cards, 329

STM-N (Synchronous Transport Module-N), 213–214
Stoke's low-wavelength waves, 70
STP (Spanning Tree Protocol)
    configuration, 645–650
    Ethernet ports, enabling, 582-584, 620–621
Stratum 3 (ST3) clocks, 370
stratum clock hierarchy, 146
STS-N framing, SONET, 151–153
submarine cable systems (fiber-optics), 77
subnetwork connection protection (SNCP). *See* SCP (subnetwork connection protection)
Subnetwork Explorer, CTM (Cisco Transport Manager), 347–348
subtending rings, 5, 446, 502–508
    configuration, 356, 413–418
superframes. *See* SF (superframe) and ESF (extended superframe)
SVCs (switched virtual circuits), 7
switching traffic
    MS-SPRing, 493
    SNCP rings, 482–483
switching database manager, ML-Series cards, 691–692
switching protection, ONS 15454 cards, 320, 331
synchronization
    SDH, 212–213
    SONET, 144–145
Synchronization Status Messaging (SSM), 370
Synchronous Digital Hierarchy (SDH). *See* SDH (Synchronous Digital Hierarchy)
Synchronous Optical Network (SONET). *See* SONET (Synchronous Optical Network)
synchronous TDM, 20
Synchronous Transport Module-N (STM-N). *See* STM-N (Synchronous Transport Module-N)

T

T-carrier system, 18, 27–29
    DS framing, 29
    DS multiframing formats, 29–32
TCC2 cards, 413, 503, 719
TDM (time-division multiplexing), 2, 18-20, 694
    analog signal processing, 20–24
    circuit-switched networks, 26–27

E-carrier system, 32–34
    alarms, 36, 38–39
    CRC error checking, 36–37
    errors, 36, 38–39
    FAS (frame alignment signal), 34–35
    MFAS (multiframe alignment signal), 35–36
network elements, 18, 46
    channel banks, 47
    cross-connect systems, 47
    CSU/DSU, 46
    DACS, 47
    repeaters, 46
schematics, 19
SONET/SDH, 3
statistical TDM, 20
synchronous TDM, 20
T-carrier system, 27–29
    DS framing, 29
    DS multiframing formats, 29–32
TDM signals
    SDH integration, 208–214
    SONET integration, 140–143
technology analysis, network design, 710–712
Telecommunications Network Management (TNM). See TNM (Telecommunications Network Management)
tensile loading, fiber-optic cabling, 76–77
termination, fiber-optic cables, 48, 73–75
TFFs (thin film filters), 114–115
through timing, SDH, 212
tight buffer cable plants, loose buffer cable plants, compared, 76
time slot 16 AIS alarm, E1 signal format, 38
time-division multiplexing (TDM). See TDM (time-division multiplexing)
timing, ONS 15454 timing, 370-374
    SDH, 461–466
    SONET, 356
timing references, 372
TM (terminal multiplexers)
    SDH, 241–242
    SONET, 176
TNM (Telecommunications Network Management), 140
topologies, 2-7
    discoveries, RPRs (resilient packet rings), 290–291

Multiservice SONET/SDH, 3
SDH (Synchronous Digital Hierarchy), 2, 208, 246
    fiber diversity, 250–252
    fiber routing, 250–252
    hub topology, 247–248
    meshed topology, 249–250
    point-to-multipoint topology, 247
    point-to-point topology, 246–247
    ring topology, 248–249
SONET (Synchronous Optical Network), 2, 140, 180
    diversity, 184–186
    fiber routing, 184–186
    hub topologies, 182
    meshed topologies, 184
    point-to-multipoint topologies, 181–182
    point-to-point topologies, 181
    ring topologies, 183
traffic
    bidirectional traffic matrix, 703
    monitoring, secondary circuits, 437–438
    MS-SPRing, switching, 493
    northbound traffic, 308
    SNCP, switching, 482–483
    southbound traffic, 308
    unidirectional traffic matrix, 702
    UPSR traffic, switching, 389
traffic protection, RPRs (resilient packet rings), 293–294
tranmitter launch power, span analysis, 83–84
transceivers, 100
transmission impairments, WDM (wavelength-division multiplexing), 92, 127–133
    FEC (forward error correction), 128–130
    optical signal-to-noise ratio, 130–133
transmitters
    fiber-optic communications systems, 78–80
    WDM (wavelength-division multiplexing), 106
        chirp, 111
        distributed Bragg reflector lasers, 107–108
        distributed feedback lasers, 106–107
        modulators, 111–113
        tunable lasers, 108–109
        vertical cavity surface emitting lasers, 110–111

transport overhead
    SDH, 224
        AU pointers, 226
        MSOH (Multiplex Section Overhead), 226–229
        POH (path overhead), 229–233
        RSOH (Regenerator Section Overhead), 225–226
        SONET interworking, 233–235
    SONET, 140, 153
        LOH (line overhead), 155–159
        POH (path overhead), 159–162
        SOH (section overhead), 154–155
tributaries, SDH, 222–223
trunk cards
    BLSR trunk cards
        installing, 394–395
        moving, 409–412
    installing, 479–481
    MS-SPRing trunk cards
        installing, 486–488
        moving, 500–502
    UPSR trunk cards, installing, 386
tunable lasers, WDM (wavelength-division multiplexing), 108–109
two-fiber BLSRs, upgrading, 400–402
two-fiber MS-SPRing, upgrading, 497–499
two-fiber rings
    four-fiber rings, compared, 257–259
    SONET, 191–193

U

UAS (unavailable seconds), E1 signal format, 38
U-band (ultra-long wavelength) optical frequency bands, WDM systems, 96
UDCs (user data channels), AIC-I cards, 324
unavailable seconds (UAS), E1 signal format, 38
unidirectional circuits, multiple drops, creating, 436–437
unidirectional path-switched ring (UPSR). *See* UPSR (unidirectional path-switched ring)
unidirectional rings
    bidirectional rings, compared, 256–257
    SONET, 190–191
unidirectional traffic matrix, 702

unidirectional WDM, 96–97
Upgrade BLSR dialog box, 401
upgrading
    BLSRs, 400–402
    MS-SPRing, 497–499
    optical networks, 723–730
UPSRs (unidirectional path-switched rings), 6, 356
    BLSRs
        subtending from, 414–415
        subtending to, 415–416
    configuration, 356
    circuits, editing, 439–440
    configuration, 385–392
    linear ADMs, converting from, 420–423
    nodes, adding/removing, 388–392
    ports, enabling, 388
    SONET, 196–197
    traffic, switching, 389
    trunk cards, installing, 386
user data channels (UDCs), AIC-I cards, 324
users
    creating
        SDH, 456
        SONET, 365
    deleting
        SDH, 457
        SONET, 366
    editing
        SDH, 457
        SONET, 366
    initial provisioning tasks, 362–366

V

V5 bytes (SDH), 232
VACT (virtual concatenation), ML-Series cards, 691
Van Heel, Abraham, 48–49
VC payload pointers (SDH), 224
VCSELs (vertical cavity surface emitting lasers), WDM (wavelength-division multiplexing), 110–111
vendors, selecting, 699
virtual tributaries (VTs). *See* VTs (virtual tributaries)
VLANs
    configuration, ML-Series cards, 651–655

Ethernet ports, provisioning for membership, 581–582
identifiers, SPRs, 279
membership, provisioning, 618–620
VRF lite, configuration, ML-Series cards, 667–671
VTs (virtual tributaries), 140
    SONET, 165–166
        groups, 166–171
        POH (path overhead), 173–174
        superframes, 171–173

W-Z

wander, E1 signal format, 38
WANs (wide area networks), building, technology options, 275
wavelength-division multiplexing (WDM). *See* WDM (wavelength-division multiplexing)
wavelengths
    attenuation, compared, 65
    cutoff wavelengths, 61
    OC-192/STM-64 ITU wavelengths, 724
WDCS (wideband digital cross-connect)
    SDH, 244–245
    SONET, 178–179
WDM (wavelength-division multiplexing), 92-96, 104-105
    amplifiers, 121–125
        distributed amplifiers, 124
        EDFAs (Erbium-Doped fiber amplifiers), 123
        hybrid amplifiers, 124
        RFAs (Raman fiber amplifiers), 124
    bidirectional WDM, 96–98
        band-separation method, 98
        circulator method, 99
        interleaving-filter method, 98–99
    channel spacing, 99
    characteristics, 92, 127–133
    compensation, 92, 133–134
    CWDM (coarse wavelength-division multiplexing), 92, 100
    dispersion, 92, 133–134
        chromatic dispersion, 134–136
        polarization mode dispersion, 136–138
    DWDM (dense wavelength-division multiplexing), 92, 101
    ITU grid, 102–104
    multiplexers, 113–121
    need for, 93
    optical frequency bands, 96
    optical multiplexers, 113–121
        AOTFs (acousto optical tunable filters), 117–118
        arrayed waveguides, 115–116
        circulators, 119
        couplers, 119
        Fabry Perot cavity filters, 117
        fiber Bragg grating, 115
        frequency slicers, 120
        interleavers, 120
        isolators, 119
        Mach-Zehnder interferometers, 118–119
        periodic filters, 120
        TFFs (thin film filters), 114–115
    optical-fiber media, 125–127
    receivers, 125-127
    schematic, 94
    systems, 94
    transmission impairments, 92, 127–133
        FEC (forward error correction), 128–130
        optical signal-to-noise ratio, 130–133
    transmitters, 106
        chirp, 111
        distributed Bragg reflector lasers, 107–108
        distributed feedback lasers, 106–107
        modulators, 111–113
        tunable lasers, 108–109
        vertical cavity surface emitting lasers, 110–111
    unidirectional WDM, 96–97

XC10G cross-connect (XC) cards, 314, 323, 720
XCVT cross-connect (XC) cards, 313
XC-VXL-10G cross-connect (XC) cards, 323
XC-VXL-2.5G cross-connect (XC) cards, 324

Yellow CFA (carrier failure alarm), superframes, 32

Cisco Press

SAVE UP TO 20% OFF

Become a Preferred Member and save at ciscopress.com!

Complete a **User Profile** at ciscopress.com today and take advantage of our Preferred Member program. Benefit from discounts of up to **20% on every purchase** at ciscopress.com. You can also sign up to get your first **30 days FREE on InformIT Safari Bookshelf** and **preview Cisco Press content**. With Safari Bookshelf, you can access Cisco Press books online and build your own customized, searchable IT library.

All new members who complete a profile will be automatically registered to win a **free Cisco Press book** of their choice. Drawings will be held monthly.

Register at **www.ciscopress.com/register** and start saving today!

The profile information we collect is used in aggregate to provide us with better insight into your technology interests and to create a better user experience for you. You must be logged into ciscopress.com to receive your discount. Discount is on Cisco Press products only; shipping and handling are not included.

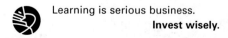